AN INTRODUCTION TO THE DESIGN AND BEHAVIOR OF BOLTED JOINTS

MECHANICAL ENGINEERING

A Series of Textbooks and Reference Books

Editor

L. L. Faulkner

*Columbus Division, Battelle Memorial Institute
and Department of Mechanical Engineering
The Ohio State University
Columbus, Ohio*

Additional Volumes in Preparation

Mechanical Engineering Software

AN INTRODUCTION TO THE DESIGN AND BEHAVIOR OF BOLTED JOINTS

THIRD EDITION, REVISED AND EXPANDED

JOHN H. BICKFORD

Consultant, Bidwell Industrial Group, Inc.
Middletown, Connecticut

MARCEL DEKKER, INC.   NEW YORK · BASEL · HONG KONG

Library of Congress Cataloging-in-Publication Data

Bickford, John H.
    An introduction to the design and behavior of bolted joints / John
H. Bickford. — 3rd ed., rev. and expanded.
        p.   cm. — (Mechanical engineering ; 97)
    Includes bibliographical references and index.
    ISBN 0-8247-9297-1 (alk. paper)
    1. Bolted joints.   I. Title.   II. Series: Mechanical engineering
(Marcel Dekker, Inc.) ; 97.
    TA492.B63B5   1995
    621.8'82—dc20   95-21233
                                                                          CIP

The publisher offers discounts on this book when ordered in bulk quantities. For more information, write to Special Sales/Professional Marketing at the address below.

This book is printed on acid-free paper.

MARCEL DEKKER, INC.
270 Madison Avenue, New York, New York 10016

Current printing (last digit):
10 9 8 7 6 5 4 3

PRINTED IN THE UNITED STATES OF AMERICA

To my bright and wonderful family, in the order of their arrival:
Anne, David, Peter, Leila, and now Zöe.

Preface to the Third Edition

Although their exact birth date is unknown, it's certain that threaded fasteners have been around for at least 500 years. They're still the fastener of choice when we want an easy and relatively low-cost way to assemble anything from a frying pan to a satellite. They're virtually our only choice if we want to create a specific clamping force to hold a joint together. The bolt and its cousins are, in fact, marvelously simple mechanisms for creating and maintaining this force. And, of course, there's no better fastener if we also need to disassemble and reassemble something for maintenance or other purposes. Considering its tenacious hold on life and its persistent popularity, it's interesting that we still don't know all we'd like to about the bolt, or about the bolted joint and the way it behaves in service. Until a few years ago, moreover, most engineers knew virtually nothing about these things, and an "introduction" seemed in order. The response to the first and second editions of this text seems to confirm that premise.

The number of engineers who know the basic concepts of joint behavior, however, has grown substantially; if my seminar students are any indication. The time has come, therefore, to deal with the subject in greater depth than in previous editions. As a result, this third edition contains far more specific information concerning design and behavior; with many new equations and numerical examples. For example, three new chapters, 21–23, are devoted specifically to design, covering joints loaded in tension, gasketed joints, and joints loaded in shear. In addition, several other chapters have been expanded to include specific design procedures or recommendations.

A lot of this new material—especially that dealing with joints loaded in tension—is based on a modified version of the design procedure defined some years ago by the German engineering society Verein Deutscher Ingenieure (VDI). I have modified their equations to account for—or at least to make more visible—such phenomena as elastic interactions, gasket creep, and differential thermal expansion, which can have a major impact on joint life and behavior. My debt to VDI, however, is substantial. If I have misinterpreted or misrepresented them, the fault is mine and I apologize.

This edition also contains a detailed discussion of the new gasketed joint design procedure developed by the Pressure Vessel Research Committee (PVRC) and now being incorporated into the ASME Boiler and Pressure Vessel Code. This procedure is based on the new gasket factors developed by the PVRC, and I have included the most recently published list of these factors. I have also shown, in a numerical example, how to use the new factors, and how the results obtained compare with the results of calculations based upon the historical Code procedure. Chapter 22, on the design of gasketed joints, also includes a discussion of alternate design procedures suggested by other people or groups in Europe.

Chapter 19 deals separately with the behavior of gaskets and contains much new material on important gasket properties such as creep and blowout resistance. Gasket test procedures and the new gasket rating factors proposed by the PVRC are also discussed at length for the first time.

Chapter 23, on the design of shear joints, is based on both VDI techniques and the design recommendations of the AISC, with the latter being more useful and informative for this type of joint. As in the other chapters on design, I have included numerical examples, this time for the design of friction-type, bearing-type and eccentrically loaded joints.

Other new material in this edition includes Chapter 3, on threads—configurations, nomenclature, strength—as well as additional material in Chapters 6, 15, 16, and 17, dealing with joint assembly, fatigue, self-loosening, and corrosion.

This edition, like the previous ones, is based almost entirely upon the work of others, as shown by the many references cited at the end of each chapter. Each and every one of those authors deserves my respect and my thanks. Many of those who contributed the most to my education are listed in the following Acknowledgments. My current debt is so broad, however, that I'll let the references serve for this edition.

I do, however, want to add an acknowledgment that I should have included in both the first and the second editions: my debt to my publisher. I owe a great deal to Graham Garratt, Vice President and Publisher, who first suggested that I write such a text—and who later convinced me that a second and now a third edition were desirable. Writing a book is not a trivial task, and I probably would not have attempted it without his gentle urging and continued support.

Revising a text, I was surprised to find, is more challenging than writing one. Correcting, updating and improving a text while adding new material could challenge more nimble minds than mine; and here I have been blessed with the friendly and helpful guidance of Walter Brownfield, who supervised the production of both the second and third editions.

So, my thanks to both of these people at Marcel Dekker, Inc., and to the many engineers and scientists listed in the references.

John H. Bickford

Acknowledgments

As before, most of the information in this book is not original with me. I am merely passing along to you the education that I have received from so many other people and sources, including those mentioned in the Preface to the first edition. For recent years, I am especially indebted to those listed below.

In the Preface to the second edition, I mentioned the seminar students, who are too numerous to name but to whom I owe a great deal. My education has been advanced even more, I think, by my participation in a number of technical societies and groups that are struggling to resolve a variety of bolting issues. I was involved, for example, with the Subcommittee on Bolted Flange Connections of the Pressure Vessel Research Committee; I was a member of their Task Group on Gasket Testing and was the Chairman of the Joint Task group on the Elevated Temperature Behavior of Bolted Flanges. I have learned a great deal from many of the engineers who have attended these meetings and/or who serve as consultants to these groups. I am especially indebted to André Bazergui of l'Ecole Polytechnique in Montreal, Jim Payne of J. Payne Associates, George Leon of the Electric Boat Division of General Dynamics, and J. Ronald Winter of the Tennessee Eastman Company. Dr. Bazergui was kind enough to review Chap. 18 on gaskets in the second edition. His comments have influenced the final version of that chapter; any remaining errors are mine.

I was also the Vice-Chairman of The Research Council on Structural Connections and have benefited significantly from their work and deliberations, which involves special inputs from John Fisher of Lehigh University, Joseph Yura and Karl Frank of the University of Texas at Austin, Geoffrey Kulak of the University of Alberta, Edmonton, Thomas Tarpy of Stanley D. Lindsey and Associates of Nashville, Peter Birkemoe of the University of Toronto, Michael Gilmor of the Canadian Institute of Steel Construction, and Bill Milek, now retired, formerly with the AISC.

Dr. Kulak revised the valuable and influential text *Guide to Design Criteria for Bolted and Riveted Joints* (Wiley, 1987), originally written by John Fisher and John Struik. I cited this text and my debt to it in the Preface to the first edition.

I also participated in meetings of the Atomic Industrial Forum/Metals Properties Council Task Group on Bolting and was Chairman of a Working Group on Bolting organized by the ASME Committee on Operations and Maintenance (Nuclear Codes and Standards). Both of these groups were established to define and resolve bolting issues which concerned the NRC. Key players to whom I am indebted here include Ed Merrick, at that time with TVA and now with APTECH in California, Russell Hansen of GA Technologies and Joe Flynn, Jr., of INPO. I benefited from information on nuclear bolting problems provided by Ed Jordan, Robert Baer, and William Anderson of the Nuclear Regulatory Commission. I am also grateful for more recent input from Richard Johnson of the same organization.

The Electric Power Research Institute of California played a significant role in the work of the AIF/MPC Task Group, funding much of the group's research. They also funded the development at Raymond Engineering* of a *Reference Manual on Good Bolting Practices* and three training cassettes for bolting engineers and mechanics in the nuclear power industry. I was a coauthor of the manual and participated in the preparation of the cassettes. The material in both was developed with the help of, and was reviewed by, maintenance and operating engineers in a number of nuclear plants as well as by members of the ASME Working Group. Many of the tips and suggestions that found their way into the manual and cassettes have also been included in the second edition.

I am also grateful to the Industrial Fastener Institute of Cleveland. Raymond Engineering was an invited member of that group for several years, and I have learned much from my contacts with their members and from technical discussions with Charles Wilson, their Director of Engineering.

Colleagues at Raymond Engineering contributed to my education as

---

* The Bolting Products Division of Raymond Engineering, Inc., is now the Industrial Tool Division of Bidwell Industrial Group, Inc.

well. Special mention should go to Jesse Meisterling and to Michael Looram.

Stan Johnson of Johnson Gage deserves mention, too, for his input on thread strength, thread gaging, and the like.

Bolting products customers of Raymond Engineering have provided much information about bolting problems in nuclear power, petrochemical, aerospace, automobile, and other industries. Unfortunately the number of "teachers" here is so great that individual mention is impossible, but my debt to them is nonetheless considerable.

Last but not least, I would like to acknowledge this latest of many debts to my wife, Anne, who once again lost a husband to a word processor. Ready or not, Anne, I'm now coming back!

Preface to the Second Edition

When I wrote the first edition of this book, most people, including most engineers, were generally unaware of the importance of the bolted joint in our "high-tech" world. The few who were experts were often considered remnants of that previous age when large iron and steel railroads, ships, tractors, and bridges first evolved.

In recent years, however, a series of newsworthy events, many of them tragic, have made us realize that the threaded fastener still plays a major role in our lives. Oil drilling platforms have tipped over, airplane engines have failed, roofs have collapsed, and astronauts have died because of bolted joint failures. The Nuclear Regulatory Commission has declared "bolting" to be an "unresolved generic safety issue with number one priority," even though no bolt-related accidents or equipment failures have occurred in that industry. And, most recently, the realization that substandard or "counterfeit" bolts are flooding the country, with safety implications for our defense, and our nuclear, aerospace, auto, and other industries, has led to congressional hearings and has even been reported on network television.

Even though our general awareness has been raised, the technology of bolted joints is still in its infancy. We know a lot more than we used to (some of that new knowledge is reflected in this new edition), but we still have a long way to go. Like weather forecasters, bolting engineers must still deal with very large numbers of unknowns and variables. As a result, our predictions and attempts to solve or prevent problems must often be based on past experience, trial and error, overdesign, and so forth, as in the past, rather than on the hard-and-fast answers so preferred by engineers.

Each of us, however, can benefit from the prior experience, the success and failure of others. Years ago, I designed a bolted joint seminar based on the material in the first edition. This seminar, which is still being given, has been sponsored by Raymond Engineering, the University of Wisconsin, and most recently by the ASME. Students have been drawn from the automobile, aerospace, power, marine, heavy equipment, and other industries that face bolting problems. The students have included people who design, build, and use bolted equipment. And I think that, over the years, they have contributed as much to my education as I have to theirs, offering tips, suggestions, and examples of things that have worked and have not worked. Their questions and problems have certainly forced me and the other instructors to dig more deeply than we might have into the literature, and elsewhere, to try to shed light on some of the problems that still plague us.

Much of that digging is reflected in this new edition, in which I have attempted to include information that will answer the most commonly asked questions. The first edition, I'm afraid, raised as many questions as it resolved, and, although neither I nor anyone else at the present time has all the answers to the questions that face bolting engineers, I have attempted to include far more concrete tips and suggestions and data than I did in the earlier edition.

The new material in this edition includes:

Specific suggestions for optimizing the results obtained when assembling bolted joints. Tips are given for assembly procedures based on torque control, torque-turn control, turn of nut, stretch control, ultrasonic measurement of bolt stress and the like (Chaps. 5–11).

A variety of suggestions for how to pick preload (or torque) for a given application, starting with simple methods for relatively unimportant joints and proceeding to more sophisticated methods (Chap. 21).

A new chapter devoted to the material properties that affect the strength of the fastener and/or the stability of the preload or clamping force on the joint in service (Chap. 3). Also, more data on such things as nut factors (Chap. 5), gasket stiffness (Chap. 4), the elevated temperature properties of bolting materials (Chap. 3), gasket creep (Chap. 18), and the relative costs of bolting materials (Chap. 3).

A greatly expanded discussion of stress corrosion and other stress cracking phenomena, with data on the stress corrosion resistance of a variety of bolting alloys (Chap. 19).

A tabulation of key bolting equations in calculator (or computer) format (Appendix H).

A discussion of fastener coatings, with their uses, strengths, and weaknesses, including substitutes for cadmium plating (Chap. 19).

An expanded discussion of fatigue failure, with new data (Chap. 17).

A discussion of a phenomenon I call "elastic interactions," which occurs when we tighten groups of bolts and which can have a significant influence on the amount of clamping force developed in a joint (especially a gasketed joint) during assembly. Most people, myself included, were unaware of this phenomenon when I wrote the first edition. Interactions can cause assembly preloads to vary by 4:1 or more, even if tensioners are used to tighten the bolts (Chap. 6).

A simple procedure that will allow you to make a rough estimate of the stiffness of a bolted joint, a procedure based on experimental data generated by several different groups (Chap. 4). Although the procedure is only approximate, it is much cheaper than the experiments or finite element analysis required for a more exact answer—and it will be good enough for many applications.

A nearly complete revision of the discussion of ultrasonic measurement of bolt stress or strain to reflect the significant advances that have occurred in this technology in recent years (Chap. 11).

Major revisions to and extension of the discussion of gaskets, with a description of recent results of research sponsored by the Pressure Vessel Research Committee and a discussion of the new gasket factors now being proposed as replacements for the m and y factors of the ASME Boiler and Pressure Vessel Code (Chap. 18).

A procedure for estimating the effect of a change in temperature on preload or on the clamping force on the joint (Chap. 14), plus a discussion of the other ways in which elevated temperature can affect a gasketed joint (Chap. 18).

A structured procedure for answering bolted joint questions and/or for predicting results when the joint is assembled and put in service (Chaps. 20–21).

I think that you will find that the information listed above, plus that carried over from the first edition, will help you deal with this complicated thing called a bolted joint.

John H. Bickford

Preface to the First Edition

To "get down to the nuts and bolts" of a topic has always meant to get to the heart of it, and rightfully so. After all, the joints are the weakest element in most structures. This is where the product leaks, wears, slips, or tears apart. I have heard that the "improper use of fasteners"—in joints, of course—is the largest single cause of the warranty claims faced by U.S. automobile manufacturers. An Air Force engineer told me that the cost of a modern military airplane is a linear function of the number of fasteners involved. These claims may be apocryphal, but the problems are real.

In spite of their importance, bolted joints are not well understood. Mechanical engineering students may receive a brief introduction to the subject in a design course, but only a small percentage of them—in school or afterward—will ever get involved enough for a real understanding. The specialists who design things which must not fail—airplanes, nuclear reactors, or heavy equipment costing hundreds of thousands or even millions of dollars—are forced to learn all there is to know about the design and behavior of bolted joints. The rest of the engineering fraternity, even designers, are guided by guesswork, experience, or handbooks. And they still have problems.

As a matter of fact, even sophisticated designers have problems at the present state of the art because the behavior of a bolted joint involves a large number of variables difficult or impossible to predict and/or control. There are widely used design theories and equations, many of which we shall study in this book, but these are usually simplifications and approximations. They have been used, successfully, on all sorts of joints in all sorts of products, but they are not sufficient for critical joints. Most of them, furthermore, have been around for years, and they have fallen behind the demands being placed on contemporary designs, e.g., higher operating temperatures and pressures, new materials, the increased clamor for more safety or environmental protection, and better strength-to-weight ratios. Even the thorough, widely used, and often-modified ASME Boiler and Pressure Vessel Code has failed to keep pace with the needs of the designer.

The engineering societies are aware of these problems, of course, and are currently funding extensive experimental and theoretical studies to advance the science (or is it an art, at present?) of bolted joint design. It is believed that this work will make accurate joint design possible—but not until the end of this decade. That forecast, coming from the most knowledgeable people in this business, gives you an idea of the magnitude of the problem. None of us, of course, can wait 10 years for solutions to our current design problems. We have to function at the current state of the art. Even this is a challenge, given the complexity of the subject, but currently available information can help us minimize joint problems even if we can't eliminate them. Hence, this book will serve as an introduction to the design and behavior of bolted joints and a primer for engineers or students who are struggling with the subject in depth for the first time. It will also help plant engineers, maintenance engineers, production engineers, and other nondesigners understand the nature of and reasons for their bolted joint problems, and give them some help in solving or reducing these problems.

The information in this book has come primarily from two sources. My employer, Raymond Engineering Inc., has manufactured for some years unusual tools and equipment for assembling and disassembling large bolted joints. In a desire to increase our knowledge of bolted joint technology, we commissioned, in 1978, a computerized literature search. This search, directed by Stephen Ford of the Battelle Memorial Institute, uncovered thousands of articles: some unique, some repetitive; some "correct," some ridiculous; some well-written but some not. File drawers full of articles, including, by and large, all that was known, or at least all that had been published, about bolted joints at that time.

Since then, we have sponsored a biweekly computerized "update" search of many different engineering files, including EI, DOE, BHRA, NASA, ISMEC, ASM, INSPEC, CPI, CAC, NTIS, USG, and many others. These updates are made for us by the New England Research Center at the University of Connecticut.

These updates have kept our library current—and our readers busy! This present book is, to a large extent, an overview of the state of the art as revealed by this literature search, so, as the author, I am much indebted to Mr. Ford for starting my education, and to UCONN for continuing it.

I'm even more indebted to the many engineers and scientists who wrote the articles: Bob Finkelston, Gerhard Meyer, and Dieter Strelow of SPS; G. H. Junker of Unbrako-SPS; Nabil Motosh of Asslut University in Egypt; John Fisher of Lehigh University; Wayne Milestone of the University of Wisconsin; Ed Rice of Ingersoll-Rand; and Sam Eshghy of Rockwell stand out as key influences, but there are hundreds of others. Any errors in my book, of course, should not be blamed on them, but rather on my inability to understand.

But there's more to it than that. We're not scientists or academics. We're engineers and businessmen, and although we're deeply interested in the theories and explanations, our goal is to understand and solve, or prevent, field problems. It's nice to know that "the equations don't always work because . . ."; but we still have to tighten those joints, right now, in such a way that they stay put for the life of the product, or at least until the next maintenance shutdown. And so we kept looking for equations, information, rules-of-thumb, divine guidance, or *anything* that would get us there. And this led us in two directions that have produced results.

First, our search for something better led us to an impressive new instrument called the ultrasonic extensometer—invented by Donald Erdman of Pasadena and Howard McFaul of Douglas Aircraft Corporation. This instrument is designed specifically to measure the actual strain in a bolt before, during, and after tightening. Here, for the first time, we had a way to measure "tightness" in bolts, with a high degree of accuracy, under any and all field conditions, statically or dynamically, and across the board. Prior methods were only practical for samples—you strain-gauged a few bolts, for example, used load cells under the heads of a few, or made a laboratory experiment. In many cases you modified field conditions simply by taking the measurements. If nothing else, the results you obtained were unnaturally "good," because the person using the wrench was more careful with those bolts than before or after your experiment.

Several types of extensometer are described at length in Chap. 9. I'm indebted to Donald Erdman, incidentally, for reading and correcting this chapter, as well as for making it possible for us to measure bolt stretch ultrasonically.

The extensometer makes it possible to check unmodified bolts assembled by unsuspecting people, and to monitor such elusive things as dynamic loads or long-term relaxation. Engineers have long been able to measure the inputs to the system, e.g., torque applied to the nut, composition of the lubricant, and angle of turn of the nut. Now we can see the immediate effects and results, as a function of job conditions and/or time. We felt just as the electrical engineer must have when someone handed him the first oscilloscope.

We have used this instrument extensively in our laboratory to study bolt problems and to analyze and check some of the information and theories uncovered by the literature search. But more significantly, we have used it in the field. We organized a bolting services group which sent technicians to many parts of the country—and overseas—to help customers tighten or disassemble problem joints. This hands-on experience provides the second major source of information on which this book is based. I'll tell you what really happens when you tighten various kinds of joints, under often difficult conditions, with a variety of tools and procedures—even if no one, at present, can fully explain *why* they behave this way. And I'll describe some of the techniques we and our customers have used to solve or minimize today's problems.

Most of our work has involved very large bolted joints, e.g., pressure vessels, pipe joints, heat exchangers, engine heads, and helicopter transmissions; but we've also been involved with small aerospace assemblies, and have had some exposure to the tools and techniques used by automobile and other mass producers. So, although the case history emphasis in this book will be on large fasteners, the design and behavior information is applicable to most types of bolted joints.

One warning for those involved in the design and construction of buildings: You will not find much information here on structural steel joints. Many of the topics covered would be pertinent to such joints, but I make little or no attempt to relate them to those applications. This is an area in which my company and I have had very little experience—and it's an area that is very well covered by Fisher and Struik's excellent *Guide to Design Criteria for Bolted and Riveted Joints* (Wiley, 1974). That work, on the other hand, doesn't cover liquid joints, or the problems faced by production engineers.

I am sure that some of you will find the subject of bolted joints as interesting as I do—at least by the time you finish the book. Before we start, let me add one more note of appreciation for my secretary, Tressa Battista, who faced too many drafts with too little time, but did it all.

We're ready. Let's learn about bolted joints.

John H. Bickford

Contents

APPENDICES

AN INTRODUCTION TO THE DESIGN AND BEHAVIOR OF BOLTED JOINTS

I

INTRODUCTION TO BOLTED JOINTS

1
Basic Concepts

This book is intended to give you an introduction to the design and behavior of bolted joints and hopefully will help you become better designers, assemblers, or users of such joints, or help you analyze and prevent joint failures. The subject is a complex one, which is why a text of over 700 pages can be considered only an introduction. The material presented here, however, should be all the information that many or most people need. Numerous references at the end of each chapter lead the way to further details for those who need or want to know more.

This first chapter gives an overview of the material to be covered in the rest of the book; it's an introduction to the introduction, if you will. You might find it useful to come back to this chapter and reread it, if you get bogged down in the detail of subsequent chapters and have trouble seeing how that subject or detail fits the overall picture.

I. TWO TYPES OF BOLTED JOINT

Bolted joints come in two flavors, depending on the direction of the external loads or forces acting on the joint. If the line of action of the forces on the joint is more or less parallel to the axes of the bolt, the joint is said to be loaded in tension and is called a tension or tensile joint. If the line of action of the load is more or less perpendicular to the axes of the bolt, the joint is loaded in shear and is called a shear joint. Both types are illustrated in Fig. 1.1. Some joints support combined tensile and shear

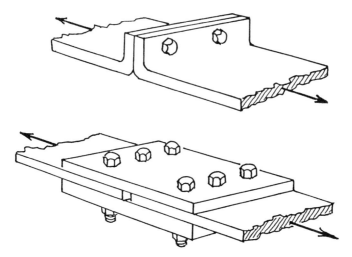

Figure 1.1 Bolted joints are classified by the service loads placed on them. If those loads—forces—are applied in a direction more or less parallel to the axes of the bolts, as in the upper sketch here, the joint is called a tensile or tension joint. If the line of action of the forces is essentially perpendicular to the axes of the bolts, as in the lower sketch, the joint is called a shear joint.

loads and are named after the larger of the loads placed on them, be it tensile or shear.

The distinction between tensile and shear joints is important, because the two types differ in the way they respond to loads, the ways in which they fail, the ways in which they are assembled, etc. In general, the tensile joint is the more complex of the two—as far as behavior and failure are concerned—and it's the more common type of joint. Most of this text, therefore, is devoted to it. Another reason for this bias: Messers Kulak, Fisher, and Struik have written an excellent text, *Guide to Design Criteria for Bolted and Riveted Joints*, second edition (John Wiley & Sons, New York, 1987), which is devoted almost entirely to shear joints.

II. THE BOLT'S JOB

The purpose of a bolt or group of bolts in all tensile and in most shear joints is to create a clamping force between two or more things, which we'll call joint members. In some shear joints the bolts act, instead, primarily as shear pins, but even here some bolt tension and clamping force is useful, if for no other reason than to retain the nuts.

A. Tensile Joints

Specifically, in tensile joints, the bolts should clamp the joint members together with enough force to prevent them from separating or leaking. If the joint is also exposed to some shear loads, the bolts must also prevent the joint members from slipping.

Coincidentally, the tension in the bolt must be great enough to prevent it from self-loosening when exposed to vibration, shock, or thermal cycles. High tension in the bolt can also make it less susceptible to fatigue (but sometimes *more* susceptible to stress cracking). In general, however, we usually want the bolt in a joint loaded in tension to exert as much force on the joint as it and the joint members can stand.

There are two important facts you should keep in mind when dealing with tension joints. First,

The bolt is a mechanism for creating and maintaining a force, the clamping force between joint members.

Second,

The behavior and life of the bolted joint depend very much on the magnitude and stability of that clamping force.

Note that I did not say "magnitude and stability of the preload" or of the tension in the bolt or of the torque applied to the bolt. Those parameters are related to the clamping force, often closely related, but the key issue as far as joint behavior is concerned is the force the two joint members exert on each other (the clamping force), created, of course, by the force the bolts are exerting on them.

The key issue as far as bolt life and integrity are concerned is, however, the tension in it—so we must keep our eye on both interjoint clamping force and bolt tension to be successful.

The clamping force on the joint is initially created when the joint is assembled and the bolts are tightened by turning the nut or the head of the bolt. This act, of course, also creates tension in the bolt; the tension is usually called preload at this stage.

Although there may be some plastic deformation in some of the threads when a bolt is tightened normally, most of the bolt and the joint members respond elastically as the bolt is tightened. The joint members are compressed a slight amount, and the bolt is stretched by a larger amount.

In effect, both joint members are bolts behave like stiff springs, one being compressed and the other stretched as suggested in Fig. 1.2. Like springs, furthermore, they acquire potential (or stored) energy. If we released them after tightening them, they would suddenly snap back to their

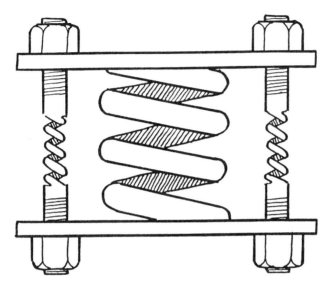

Figure 1.2 Bolts and joint members deform elastically when the bolts are tightened. In effect, they act like stiff springs as suggested by this sketch. This fact, that they act like springs, greatly influences the behavior of the joint.

original dimensions. *It is this stored energy which allows bolts to maintain that all-important clamping force between joint members after we remove the wrench.*

We might even say that the tensile joint, unlike its welded or bonded joint cousin, is "alive," filled with energy and able to do its job only *because* it's filled with energy.

B. Shear Joints

The bolt's main job in a shear joint is to keep the joint from slipping or from tearing apart in the slip direction. If the joint must also support some tensile load, the bolt must resist that too.

In some shear joints, as already mentioned, the bolts resist slip by acting as shear pins, and joint integrity is determined by the shear strength of the bolts and joint members. There are a number of reasons why we will often want to tension these bolts, as we'll see, but the exact amount of tension, or of the energy stored in them, is not a critical factor.

In other shear-loaded joints slip is prevented by friction restraint between joint members. These friction forces are created by the clamping load, which in turn is created by heavily tensioned bolts. Here again, therefore, the bolt is a mechanism for creating and maintaining a force, and the magnitude and life of that force depend on the potential energy stored in the bolts during assembly. Even here, however, we're usually less concerned about creating an exact amount of tension in the bolts during assembly than we are when we're dealing with tensile joints, because service loads don't affect bolt tension and clamping force in shear joints.

III. THE CHALLENGE

The bolted joint presents users and designers with many problems. In part this is because it is "alive"—it keeps changing state in response to service and environmental conditions, as we'll see. A more common source of problems, however, is the fact that the assembly process and the in-service behavior are affected by literally hundreds of variables, many of which are difficult or impossible to control or to predict with accuracy. As a result, when we deal with bolted joints we must inevitably deal with a lot of uncertainty. What follows is a quick review of some of the sources of this uncertainty. We'll take a closer look at most of these things in later chapters.

A. The Assembly Process

Bolts and joint members in both tension and shear joints respond in the same way to the act of tightening the bolts. There are differences in the accuracy with which we must tighten them, but most of the discussion which follows applies to all joints.

As far as all tension and most shear joints are concerned, the goal of the assembly process is to establish an initial clamping force between joint members, to introduce the first energy into bolt and joint springs. And, in tension joints, we're usually interested not just in tensioning the bolts but in tensioning them by a desired amount, because the life and behavior of such joints are so dependent on the "right amount" of clamping force. We want enough clamping force to prevent a variety of failure modes, but we must also make sure that the bolt tension and clamping force do not exceed an upper limit set by the yield strengths of the materials, the anticipated loads to be placed on the joint in service, and other factors. Unfortunately, as already mentioned, hundreds of variables affect the results when we tighten a group of bolts, so predicting or achieving a given clamping force is extremely difficult.

We attempt to control the buildup of clamping force by controlling the buildup of tension or preload in the bolt. In most cases, we do that by controlling the amount of torque applied to the nut or head.

The work we do on the fastener while tightening it is equal to one-half the applied torque times the angle (measured in radians) through which the nut turns. Typically, about 10% of this input work ends up as potential energy stored in the joint and bolt "springs." The rest is lost in a variety of ways.

Most is lost as heat, thanks to friction restraints between the nut and joint surface and between male and female threads. Some energy is used to twist and, often, to bend the bolt a little. Some energy may be lost simply in pulling heavy or misaligned joint members together or dragging a bolt through a misaligned or interference fit hole. More is lost by spreading the bottom of the nut, a process called nut dilation.

A major problem for the designer and assembler is that it is virtually impossible to predict how much of the input work will be lost due to factors such as these. The amount lost can and usually will vary a lot from one bolt to another, even in the same joint.

In spite of these uncertainties and losses, some potential energy is developed in each bolt as it's tightened, and it starts to create some clamping force in the joint. But then the bolt relaxes—loses some energy—for a couple of reasons.

A process called embedment occurs as high spots on thread and joint contact surfaces creep out from under initial contact pressure and the parts settle into each other. More drastically, a previously tightened bolt will relax somewhat when its neighbors in the joint are subsequently tightened. We call this process elastic interaction, and it can eliminate most or even all of the tension and energy created in the first bolts tightened in the joint. We'll examine this phenomenon in detail in a later chapter.

The amount of relaxation a bolt will experience is even more difficult to predict than the amount of initial tension it acquires when first tightened, increasing the challenge of the assembly process.

Anything which reduces the amount of energy stored in a bolt reduces the force it exerts on the joint. Too little torque, too much friction, rough surface finish, twisting, bending, hole interference, relaxation—all can result in less stored energy, less preload, less clamping force.

Anything which increases the energy stored will increase the force. There are a couple of things which can do this during assembly: too much torque or too little friction, thanks perhaps to a better-than-anticipated lubricant.

Again, all of these factors are difficult to predict or control, making it very difficult to achieve a particular amount of preload or clamp force at assembly. Because many factors can give us less preload than desired and only a couple can give us more, we often—perhaps usually—end up with less than expected at assembly.

Bolts in shear joints are subjected to the same assembly problems and variables as are bolts in joints loaded in tension. There's a difference, however.

In tension joints we always care about the amount of preload, tension, clamping force, and potential energy developed during assembly because of the way such joints respond to service loads. We're not so concerned about this when dealing with shear joints. We'll see why when we examine the in-service behavior of such joints.

B. In-Service Behavior

The in-service behavior of tensile joints differs substantially from that of shear joints, and this is reflected in the different ways we design and assemble the two types. Here's a preliminary look at the differences.

Joints Loaded in Tension

We encountered many uncertainties when we assembled a tensile joint. Further uncertainties are introduced when we put such a joint to work—when we load it, expose it to vibration or shock, subject it to change in temperature, anoint it with corrosive fluids, etc. Being alive, it responds to such things, and as it responds the tension in the bolt and the clamping force between joint members change.

First, and most important, *the tensile load on the joint will almost always increase the tension in the bolts and simultaneously decrease the clamping force between joint members.* This is undesirable—and unavoidable. And it is the major reason why we care so much about the exact amount of bolt tension and clamping force developed at assembly.

If the assembly preloads are too high, the bolts may yield or break when they encounter the service loads. On the other hand, if assembly preloads are too small, the clamping force on which the joint depends may all but disappear when service loads decrease it.

Other service factors can also change bolt tension and clamping force and will affect our choice of assembly preload. For example, relaxation processes like embedment—or gasket creep—are increased by loads and by elevated temperatures. Vibration, shock, or thermal cycles can cause the bolt to self-loosen. Differential expansion between bolts and joint members can increase bolt tension and clamping force simultaneously—or can reduce both. In this case heat energy is being used to increase or redistribute the energy stored in the parts.

Chemical energy—exhibited as corrosion—can increase clamping force as corrosion products build up under the face or the nut or head of the bolt.

These factors present an additional challenge to the designer. They increase the difficulty of predicting joint behavior, because the designer can rarely predict the exact service loads or conditions the joints will face. The joint's response, furthermore, will be influenced by such hard-to-pin-down factors as the condition of the parts or the exact dimensions and material properties of the parts. Behavior will also be influenced by the hard-to-predict amount of preload in the bolts, which the designer must somehow specify.

Once again, however, the factors which lead to less clamping force are more common than the ones which can lead to more clamping force. Since this is also true, as we've seen, of the assembly process, we are forced to recognize Bickford's little-known First Law of Bolting: *Most bolted joints in this world are providing less clamping force than we think they are.*

Shear Joints

Shear loads do not affect the tension in the bolts or the clamping force between joint members, at least until such loads become so high that the joint is about to fail. Predicting behavior and avoiding failure are therefore easier when we're dealing with shear joints than when we are dealing with tensile joints. This, in part, explains why people who design airframe, bridge, or building structures rely so heavily on shear joints and avoid using tension joints whenever possible.

I don't mean to imply that shear joints won't respond to service loads and conditions; they will. Bolt tension and clamping force will change if temperatures change. Vibration or shock can loosen the bolts. Parts can rust, and corrosion products can build up and alter bolt and joint stresses. If the loads on the joint are cyclical, the stresses in bolts and joints members will fluctuate. But the in-service uncertainties the designer faces, and their consequences, are usually less than those he'll face when dealing with tensile joints.

IV. FAILURE MODES

The main reason we want to "control" or "predict" the results of the assembly process and the in-service behavior of the joint is to avoid joint failure. This can take several forms.

A joint will obviously have failed if its bolts self-loosen—shake apart—or break. Self-loosening is a complicated process and is described in a separate chapter, along with ways to combat it. In general, however, it's caused by vibratory or other cyclical shear loads which force the joint members to slip back and forth. A major cause of self-loosening is "too little preload"—and hence too little clamping force. Both tensile and shear joints are subject to this common mode of failure.

Bolts in both types of joint can also break because of corrosion, stress cracking, or fatigue—all of which are also covered in later chapters and two of which are encouraged by the wrong preload. Stress cracking occurs when bolts are highly stressed; fatigue is most apt to occur when there's too little tension in the bolts. Even corrosion can be indirectly linked to insufficient preload, if a poorly clamped joint leaks fluids which attack the bolts.

If the bolts fail for the reasons just cited or if they exert too little force on the joint, perhaps because of the assembly or in-service conditions discussed earlier, the shear joint may slip or the tension joint may separate or leak. Each of these things means that the joint has failed.

It's obvious that a leak is a failure, but what's wrong with a little slip or with separation of a joint that doesn't contain fluid?

Slip can misalign the members of a joint supporting shear loads—cramping bearings in a machine, for example. Or it can change the way a structure absorbs load, perhaps overstressing certain members, causing the structure to collapse. Slip can lead to fretting corrosion or to fatigue of joint members. As already mentioned, cyclical slip can lead to self-loosening and perhaps loss of the fasteners. Vibration loosening of bolts and fatigue failure of shear joint members are of particular concern to airframe designers.

Separation of the members of a joint supporting tensile loads can encourage rapid fatigue failure of the bolts. It can also destroy the integrity of a structure or machine. It can allow corrosants to attack bolts and joint surfaces. Separation means the total absence of clamping force, which means, in effect, that the joint is not a joint at all.

A gasketed joint can leak if the initial clamping force between joint members during assembly is not great enough or if the in-service clamping force (which will almost always differ from the assembly clamp) is too low. The joint does not have to separate for leakage to occur.

Note that most joint or bolt failure modes are encouraged by insufficient bolt tension and/or insufficient clamping force. Self-loosening, leakage, slip, separation, fatigue—all imply too little clamp.

A few problems can be caused by too much tension or clamping force, however. Stress corrosion and hydrogen embrittlement cracking of bolts can occur in both shear and tensile joints and are more likely if bolt stresses are high. Joint members and gaskets can be damaged by excessive clamp. Joint members can also be distorted by excessive bolt loads; the "rotation" of raised face, pressure vessel flanges is a common example. Fatigue life can sometimes be shortened by high stress, although more commonly it's caused by insufficient clamping force.

But failures caused by too little clamping force are more common in either tensile or shear joints than are failures caused by too much clamp. And, as we've seen, assembly and service conditions are more apt to give us too little clamp than too much. Welcome to the world of bolting!

V. DESIGN

A. In General

The design of bolted joints, like the design of anything else, involves a detailed consideration of function, shapes, materials, dimensions, working loads, service environment, etc. Every industry has characteristic or "typical" joint configurations and needs, and it would be impossible to detail each in a single text. We can, however, look at some generalities which must apply to most joints, whatever their specific application. And we can review design procedures which have been accepted and used by many.

Specifically we'll focus on a design procedure developed by Verein Deutscher Ingenieure (VDI), a German engineering society, but with some modifications and extensions. We'll also examine in some detail the design rules for flanged, gasketed joints found in the ASME Boiler and Pressure Vessel Code, with emphasis on changes which are currently being introduced to those rules. And we'll take a brief look at the design of structural steel, shear joints.

B. Specific Goals of the Designer

The joint designer, of course, is faced with all the assembly and in-service uncertainties detailed earlier. In spite of these uncertainties, he must do two things when designing a joint which will be loaded all or in part in tension:

1. *He must pick bolt and joint sizes, shapes, and materials which will guarantee enough clamping force to prevent bolt self-loosening or fatigue, and to prevent joint slip, separation, or leakage, when clamping forces are at a minimum (because of the factors we've described) and those hard-to-predict service loads are at a maximum.*

2. *In addition, he wants to select bolts which are able to support a combination of maximum assembly stress plus the maximum increase in stress caused by such service conditions as applied load and differential thermal expansion.*

If his joint is loaded only in shear, and will depend for its strength only on the shear strength of the bolts and joint members, then those strengths will determine the design. Such joints must not be subjected to varying or cyclical loads, or self-loosening and fatigue problems might be encountered. If service conditions permit it, however, such joints are safe—and greatly simplify the design process.

There are other things which the designer must worry about when designing tensile joints and some shear joints. He'll consider the bearing stresses the bolts create on joint surfaces, the amount of change in load the bolts see (which can affect fatigue life), the accessibility of the bolts (which can affect assembly results), and the flexibility or stiffness of bolts and joint members. If he's designing a tension joint he'll be especially interested in the so-called stiffness ratio of the joint, because this affects the way in which a given service load changes bolt tension and clamping force.

In any tension joint and in shear joints where clamping force is important, the designer will want to do everything he can to improve the energy storage capacity of his bolts. He'll find that long thin bolts and thick, metal joint members can store more energy than short stubby bolts or nonmetallic joints; hence our historic problems with sheet metal joints and our emerging problems when we try to bolt composite materials.

As we'll see, the many assembly and service uncertainties the designer has faced have traditionally forced him to overdesign—"oversize" might be better—the bolts and joint members, with resulting penalties in weight and parts cost. We'll quantify this oversizing and suggest ways it can sometimes be reduced.

Although most of the book will deal with subjects like assembly practices, in-service behavior of the joint, and failure modes rather than design specifics, everything relates to and should affect the design of bolted joints.

VI. THE LAYOUT OF THE BOOK

We'll start with some background material on the strength of bolts and of threads, the stiffness of bolts and joint members, and a review of the properties of the materials usually used for these things. We'll focus on properties that affect basic strength and which affect the stability of the parts (several material properties can encourage changes in clamping force).

Next we'll look in considerable detail at the many options we have for controlling the assembly process, looking at torque, torque and turn, strength, direct tension, and ultrasonic control of preload and clamping force.

Then we'll turn our attention to the joint in service: how it responds to service loads and conditions, how it fails, how to improve its response and minimize the chances of failure.

Finally, having learned in detail what we're up against, we'll develop procedures for estimating results and will close with a detailed discussion of a modified version of the VDI design procedure.

Appendices in the back will give you a variety of reference data to aid design activities or analytical calculations.

Enough introduction! Let's begin our serious study of this thing called a bolted joint.

2

Stress and Strength Considerations

We learned in the last chapter that the bolt's job is to clamp the joint together firmly enough to prevent slip, separation, or leakage and that the bolt must be strong enough to support the maximum preload it receives at assembly, plus the maximum additional loads it sees in service as a result of forces applied to the joint, differential thermal expansion, etc.

When designing, evaluating, specifying, or selecting a bolt for a particular job, therefore, one of our first questions will be: "Is this bolt strong enough to clamp this particular joint?" As we're about to see, the question is much simpler than the answer, because there are many aspects to the concept of strength when we're dealing with a bolt.

I. TYPES OF STRENGTH

In general, of course, the strength of any machine part is determined by such things as the size and shape of the part, the material it's made from, the heat treatment of that material, its operating temperature, and its condition (has it been abused, is it corroded, etc.?). Engineers also define strength in a variety of ways. Here are the more important definitions we use when we're dealing with bolts.

A. Tensile Strength

First, we must worry about the capacity of the bolt to generate and sustain a sufficient tensile force, since that force will be one of the main factors which determine the clamping force between joint members. (It's not the only factor, as we'll see in Chap. 6.) In most applications, the room temperature tensile strength of the bolt under static loads will be one of only two strength factors we need to be concerned about.

We'll describe this strength in several ways, however, depending on our needs. We'll use the term ultimate strength, yield strength, or—unique to threaded fasteners—proof strength. Each of these is defined and explained in this chapter. Each term defines the amount of tension we can exert on a bolt before exceeding that definition of its strength. If we apply more than its ultimate strength, the bolt will break.

B. Thread-Stripping Strength

The amount of tension we can create in a bolt depends not only on the strength of its body but also on the shear strength of its threads. If we're designing a nut or deciding how deep to make a tapped hole, we'll want to be sure that the thread engagement length will be great enough to allow us to develop the full ultimate strength of the bolt. A broken bolt is easier to detect than a stripped thread, so we never want the threads to strip.

Thread-stripping strength is the only other strength we'll have to worry about in most static load applications. In most situations we'll "consider" it simply by using a standard nut with the bolt we have selected, more carefully, above. Such nuts have been carefully designed to develop the full strength of the bolts (as long as nuts of the proper material are used with a given bolt, see Table 4.3). For tapped holes or special nuts, however, we'll have to compute thread-stripping strength.

Tensile strength will be considered at length in this chapter, thread-stripping strength in Chap. 3. Before going on, however, you should be aware of other strength considerations which will also be treated in later chapters.

C. Shear Strength

The primary load on most bolts is a tensile load along the axis. In some situations, however, the bolt also sees a load at right angles to the axis, usually called a shear load. This is especially common in structural applications. We'll consider ways to estimate such loads, and their effect on the bolt, in Chap. 14.

D. Brittle Fracture Strength

In the discussion so far we've been talking about the "normal" strength of a "normal" bolt under fairly static loads, at room temperature. Things change, sometimes drastically, when one or more of these qualifications are absent. For example, if the bolt is made of a very hard material it can theoretically support very high tensile loads. If a tiny crack or flaw exists in that bolt's surface, however, it might fail suddenly and unexpectedly under loads well below its theoretical strength. We'll consider brittle fracture in greater detail in Chap. 4.

E. Strengths at High and Low Temperatures

Another deviation from the norm we must often face is an "extreme" temperature. Temperatures that are higher or lower than "normal ambient" (which is usually taken to mean 70°F or 20°C) will alter the tensile and stripping strength of the bolt, because of basic changes in the tensile and shear strengths of the material from which the bolt is made. We'll consider this point in Chap. 4 when we look at some specific bolt materials and their properties.

F. Fatigue Strength

Tensile and stripping strengths are a measure of the resistance of the bolt to static or slowly changing loads. Cyclic loads lead to entirely different types of failure, as we'll see in Chap. 17 when we discuss fatigue. A bolt which promises to be "better" (stronger) from a static tensile strength point of view can actually be "weaker" (more likely to fail) under cyclic loads, a fact which has "burned" many an unsuspecting designer.

G. Stress Corrosion Cracking Strength

Similarly, the resistance of a bolt to stress corrosion cracking (SCC) or hydrogen embrittlement can be inversely proportional to its conventional tensile strength. We'll look at SCC in Chap. 18 when we deal with corrosion failure in general.

So strength is a many-faceted topic which we'll return to again and again in this text. To get started, let's look at what we usually mean when we talk about the strength of a bolt, its room temperature tensile strength.

II. THE BOLT IN TENSION

A. Elastic Curves for Bolts in Tension

If we place a relatively ductile fastener (such as an ASTM A325 or an SAE Grade 5) in a tensile-testing machine and gradually apply a pure tension load between the head of the bolt and the nut, we'll generate a tension versus change-in-length curve such as that shown in Fig. 2.1.

The initial, straight-line portion of the elastic curve for a bolt is called the *elastic region*. Loading and unloading the bolt repeatedly to some point on this portion of the curve will never result in a permanent deformation of the bolt (although it may result in ultimate fatigue failure, as discussed in Chap. 17).

The upper end of the straight line ends at the *proportional limit*, where the line is no longer straight, followed closely by the *elastic limit* (tension loads beyond this point will produce some permanent deformation), followed by the *yield strength point*. Loading the fastener to this last point will create a *particular* amount of permanent deformation—usually chosen as 0.2 or 0.5% of the initial length. A definition of this sort is necessary because the point at which an engineering body can be said to have yielded is not obvious. As we'll see in a minute, some portions of the bolt will have yielded long before the body as a whole has been loaded to its yield strength, and other portions of the bolt are not even close to yield when the bolt has taken a permanent set of 0.5% or the like.

Another point of interest is the *ultimate strength* (often called *tensile strength*) of the bolt. This is the maximum tension which can be created

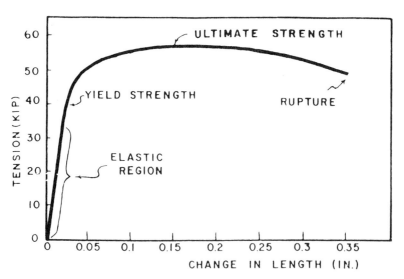

Figure 2.1 Points of interest on the elastic curve of a $\frac{7}{8}$–16, A325 bolt with a 5-in. grip length. The proportional limit and elastic limit are located near each other at the upper end of the elastic region.

by a tensile load on the bolt. It is always greater than the yield strength—sometimes as much as twice yield—but always occurs in the plastic region of the curve, well beyond the point at which the bolt will take a permanent set.

A final point of interest on the elastic curve is the rupture point, where the bolt breaks under the applied load. This and the other points we've discussed are all shown in Fig. 2.1.

B. Elastic Curves Under Repeated Loading

If we load the bolt well into the plastic region of its curve and then remove the load, it will behave as suggested in Fig. 2.2. Note that it returns to the zero-load point along a line that is parallel to the original elastic line, but offset from the original line by an amount determined by the permanent deformation created by the earlier tension load on the bolt [11].

If we now reload the bolt but stay below the maximum tensile load applied earlier, the bolt will follow this new straight line and will again function elastically. In fact, its behavior will be elastic well past the tensile load which caused permanent deformation in the first place. The difference between the original yield strength and the new yield strength is a

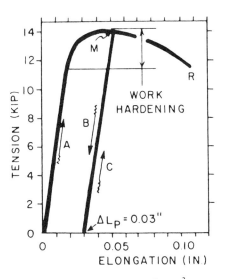

Figure 2.2 Elastic curve for a $\frac{3}{8}$–16 × 4 socket-head cap screw loaded (A) to point M, well past the yield strength, and then unloaded (B) to reveal a permanent deformation L_P = 0.03 in. If reloaded, it will follow path (C).

function of the work hardening which has been done on the bolt by taking it past yield on the first cycle, as noted in Fig. 2.2. Loading it past this new yield point will create additional permanent deformation; but *unless* we take it well past the new yield point we won't damage or break the bolt by yielding it a little—in fact, we'll have made it a little stronger, at least as far as static loads are concerned.

Many bolt materials can be taken past initial, and new, yield points a number of times before they will break. The stronger, more brittle materials, however, can suffer a loss of strength by such treatment, as shown, for example, by Fisher and Struik [1]. Loss of strength in several ASTM A490 bolts, because of repeated cycling past yield (under wind and water loads), has been publicly cited as a contributing factor in the 1979 collapse of the roof of the Kemper Auditorium in Kansas City, for example [2].

C. Stress Distribution Under Tensile Load

Let's place a bolt in a joint and load it in pure tension (this is possible if we use a hydraulic tensioner instead of a wrench to tighten the bolt). If the bolt is perfectly symmetrical, the faces of the head and nut are exactly perpendicular to the axis of the threads, joint surfaces are flat and parallel, etc. (most of which we'll never encounter in practice), loading the bolt and joint this way will produce the stress distribution shown in Fig. 2.3 [3,4].

Two points are worth noting:

Even though the bolt has been loaded in pure tension, it is well supplied with compressive stress, thanks (for example, in the shank) to Poisson's reduction and thanks to the fact that the tension built into the bolt during the tightening process must subsequently be held there by compressive forces in the nut, bolt head, and joint members.

Complex though the picture is, it's a far cry from the "truth" in most applications, where it is complicated by imperfect geometry as well as by the presence of torsional, bending, and shear stresses, as we'll discuss later. Our picture, furthermore, ignores localized, but often significant, stress concentrations in the threads.

D. Stress Concentrations

Figure 2.3 gives us a simplified view of the directions of stress in a loaded bolt and joint. An analysis of stress magnitudes would reveal three danger points, where stress concentrations create stress levels well beyond the average. These points are the fillet, where the head joins the body; the

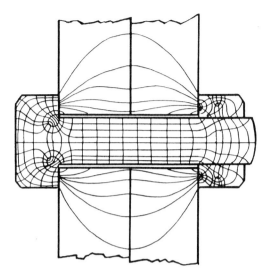

Figure 2.3 Lines of principal tension and compression stress in a bolt loaded in pure tension (and lines of principal compression stress in the joint).

thread run-out point, where the threads meet the body; and the first thread to engage the nut. As we'll see, these are the points at which the fastener will usually fall.

E. Magnitude of Tensile Stress

In much of what follows, we'll find it useful to know something about the magnitude of tensile stress within the fastener. For example, many of our calculations will be based on the assumption that tensile stress is zero at the free ends of the bolt and that it rises uniformly through the head to the level found in the body, as suggested by Fig. 2.4. There's a similar pattern in the threaded end, but the average stress in the threaded section is higher than the average in the body because the cross-sectional area is less in the threads.

A finite element analysis made by General Dynamics–Fort Worth [4], however, suggests a far more complex pattern. Scientists at General Dynamics say that the magnitude of tensile stress along the *axis* of the bolt does approximate that shown in Fig. 2.4, but that the magnitude of tension along other lines parallel to the axis of the bolt looks more like

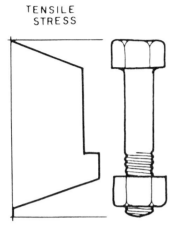

Figure 2.4 The magnitude of tensile stress in a bolt—the simplistic view often assumed in bolt calculations.

that shown in Fig. 2.5, with the peak stresses (at the fillet and nut-bolt engagement points) being two to four times the average stress in the body.

To complicate things still further, the above is true only for long, thin bolts, by which is meant bolts that have a grip length-to-diameter ratio greater than 4:1. In short, stubby bolts, the picture shows a general variation in tension stress from side to side, as well as end to end, as shown

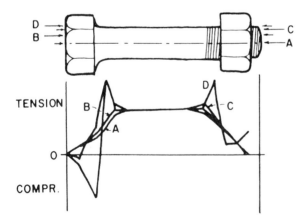

Figure 2.5 More accurate view of the tensile stress along four lines parallel to the axis of the bolt (see also Fig. 2.6).

in Fig. 2.6. There is now no such thing as "uniform stress level," even in the body.

If you stop and think for a minute, you'll realize that 4:1 isn't very short and stubby. A $\frac{1}{4}$–20 bolt having a grip length of 1 in. would be considered short and stubby by this definition, which means that the majority of fasteners we're going to deal with are probably short and stubby.

As a result, in the majority of applications, we're dealing with fasteners in which there is no uniformity of tensile stress. And this has all sorts of implications when we come to compute such things as stress level, preload, spring constants, and elongation, as we'll see in subsequent discussions.

Unfortunately, even the General Dynamics picture is oversimplified, at least as far as stresses in the threaded portion of the bolt are concerned. They do not, for example, take into account the stepped difference between average body stress and average thread stress shown in Fig. 2.4. They also ignore stress concentrations at thread roots and thread run-out points. In fact, they assume a threadless fastener with uniform take-out of load between nut and bolt. This assumption probably doesn't affect their estimates of stress within the body of the fastener as much as it affects their estimates of stress levels at the surface of the fastener, but it's something to keep in mind.

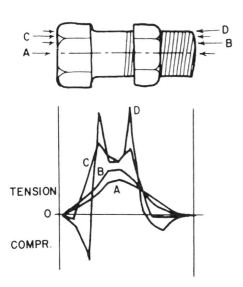

Figure 2.6 Magnitude of tensile stress along four separate paths in a bolt having a length-to-diameter ratio less than 4:1.

Here's a final, and slightly different, look at the tensile stresses in a bolt. The contour lines in Fig. 2.7 are lines of *equal* axial tension; they do not show the direction of tension. They're similar, therefore, to isochromic lines in a photoelastic model.

F. Stress in the Nut

A slightly more accurate plot of the peak stresses in nut or bolt threads is shown in Fig. 2.8 [3, 6]. The fall-off in stress is not linear, as in the previous figures, but curved. Note that adding more threads (a longer nut) doesn't reduce the peak stress by much. The first three threads carry most of the load in any case.

Obviously, this stress picture is not an attractive one. Since most of the load is on the first thread or so, most of the nut isn't doing its share of the work. This situation can be improved in a number of ways—tapering the threads or altering the pitch on either nut or bolt to force more uniformity in load distribution, for example. Perhaps the most popular way is to use a nut that is partially in tension, such as one of those shown in Fig. 2.9 [9]

One study [9] of titanium tension nuts, similar to that shown in Fig. 2.9C, with most threads in tension, resulted in the computed stress distribution shown in Fig. 2.10.

Another analysis, confirmed by experiment [10], shows that if the pitch of the bolt threads is 0.13% longer than the pitch of the nut threads—a difference of only 0.000065 in. per pitch in a $\frac{1}{4}$–20 thread, for

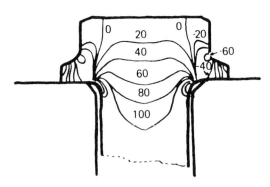

Figure 2.7   Lines of equal axial tension in a $\frac{9}{16}$–18 bolt loaded to 100 ksi tension in the shank [5]. Values given are in ksi.

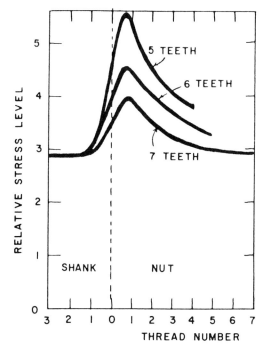

Figure 2.8 Peak stresses in three different nuts, having five, six, and seven teeth, respectively.

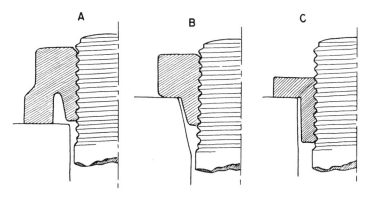

Figure 2.9 Nuts which are partially loaded in tension, such as those shown here, see a more uniform tooth stress distribution than do conventional nuts.

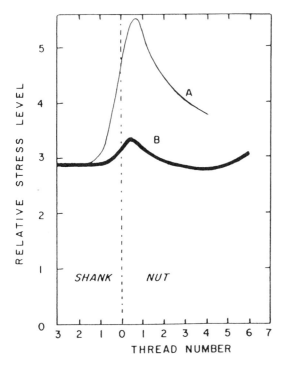

Figure 2.10   Relative stress level in nut and bolt threads for the tension nut shown in Fig. 2.9B (curve B) and for a conventional nut (curve A).

example—then the outermost threads of the nut will be stressed more heavily than the innermost, as shown in Fig. 2.11. This sort of variation must be common in practice, and it helps explain the difficulties of predicting how a given fastener or joint will behave.

Stress distribution similar to that shown in Fig. 2.11 can also occur when the thread engagement is very long. Tapped holes often have more threads than conventional nuts, which, typically, have five threads. I've encountered tapped holes with 18 or 20 threads. Large cylindrical nuts, used for example on large pressure vessel flanges, can have a similar number of threads. Even if both male and female threads are cut or rolled to tight, correct tolerances, the threads will usually have slightly different pitch distances, and this can place the maximum thread contact stress at the far end of the thread engagement, or in the middle, or "somewhere else."

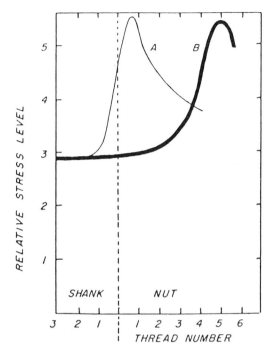

Figure 2.11 Stress distribution in threads with nominal pitch (A) and when there is a +0.13% error in the pitch of the bolt threads (B).

III. THE STRENGTH OF A BOLT

A. Proof Strength

It's useful at this point to look at the procedure used to determine and define such things as the proof strength and yield strength of bolts. We'll need this information when we study the various means by which we control the tightening of bolts. Recognize, however, that this is just an introduction to the important subject of strength (or bolt failure) and that there are many practical factors which can modify the apparent strength predicted by the conventional procedures.

In this chapter we want to estimate the static strength of the body and threads of the fastener, starting with the tensile strength of the threaded portion of the body (the weakest section). Because the stress picture is complicated, computing strength would be difficult, so the engineering societies have devised a way to determine strength experimentally and to base design and manufacturing specifications on the resulting data.

A large number of fasteners, made with well-defined materials and standard thread configurations, were subjected to tensile loads to determine:

The highest tensile force they could withstand without taking any permanent deformation (called the *proof load*). See Table 2.1 for a few examples.
The tensile force which produces 0.2 or 0.5% permanent deformation (used to define yield strength).
The highest tensile force they could support prior to rupture (used to define ultimate tensile strength).

The resulting data were then published. You don't have to worry about stress magnitudes or variations; the tables tell you how many pounds of force a fastener of a given shape and material can safely stand. Fastener manufacturers are required to repeat these tests periodically to make sure that their products meet the original standards.

There was also a need to compute design limits for nonstandard fasteners. Here the path becomes a little more murky but will still give you a useful answer unless your need for accuracy and safety is critical.

The people who tested the fasteners found, again by experiment, that if they divided the experimentally determined load at yield, in pounds, by a cross-sectional area based on the mean of the pitch and minor diameter of the threads, the result was a theoretical tensile *stress* at yield, which agreed closely with the stress at which a cylindrical test coupon of the same material would yield, and with similar results for ultimate strength, proof load, etc. The following data can now be published:

The *stress area* (A_S) of a standard thread, based on the mean of pitch and root diameters

Table 2.1 Proof Loads of a Few Typical Bolts

Bolt	Bolt size (kips)			
	$\frac{1}{4}$–20	1–8	2–8	4–8
ASTM A307 GR A	2.70	20	91.2	364
SAE J429 GR 8	3.82	72.7	NA	NA
ASTM A193 B7	NA	57.3	262	946
H-11	NA	117	537	2146

The *proof* strength of a given thread, in pounds per square inch (psi), computed by dividing the experimentally determined proof strength, in pounds, by the stress area

The *ultimate strength* or *yield strength*, of a given thread, in pounds per square inch, by dividing the experimentally determined ultimate or yield load in pounds by the same stress area

B. Tensile Stress Area

Since the mean of pitch and root diameters works for all thread sizes, it's also possible to write an expression for the stress area (A_S) of a standard 60° thread, as follows [7]:

$$A_S = 0.785 \left(D - \frac{0.9743}{n} \right)^2 \tag{2.1}$$

where D is the nominal diameter and n is the number of threads per inch. Appendix F lists the stress areas of all standard unified and metric threads.

As an example, let's compute the stress area of a $\frac{1}{4}$–20 UNC, Class 2A thread.

$D = 0.25$ in.

$n = 20$

$$A_S = 0.785 \left(0.25 - \frac{0.9743}{20} \right)^2 = 0.0318 \text{ in.}^2$$

This expression for area is used to compute the tensile or shear strength of a fastener. For example, to compute the proof load, in pounds, that your own fasteners should tolerate, it's only necessary to look up (or compute) the stress area of the threads and then multiply this by the published values for the proof strength, in pounds per square inch, of that particular material. And you get the correct answer, for most purposes.

The fallacy, of course, lies in the suggestion that the tensile stress level within the fastener equals the proof stress at the proof load point, or the ultimate stress at the ultimate load point. As we've seen in Figs. 2.5 and 2.6, the actual stress levels vary widely within the fastener, and it's unlikely that much, if any, of the fastener really sees proof stress at proof load. But the procedure is nicely circular. All of the simplifying assumptions are neatly counterbalanced by making the same assumptions in reverse at design time. It works.

A technique similar to that used for bolts is used for specifying the static strength of nuts. The stress area used to compute average nut stress levels is, in fact, the cross-sectional area of the *male* threads, A_S, just discussed (showing again how "artificial" these calculations are). The "strength" of a nut, of course, is simply the stripping strength of its threads. We get to that in Chap. 3.

C. Other Stress Area Equations

Although Eq. (2.1), or its metric equivalent in Eq. (2.4), is the expression most often used for the stress area of a thread, there are times when a more conservative (smaller) area is used for an added factor of safety. For example, before leaving this topic, we should mention that a slightly modified equation for stress area is recommended for fastener materials having an ultimate strength (as determined by coupon tests) in excess of 100,000 psi. This time the stress area is based on the mean of the *minimum* pitch diameter (E_{Smin}) permitted by thread tolerances and the nominal root diameter. This gives a slightly smaller stress area (partially canceling the load benefits you'd derive from an assumption of higher ultimate strength). Anyway, the equation is [25]

$$A_S = \pi \left(\frac{E_{Smin}}{2} - \frac{0.16238}{n} \right)^2 \tag{2.2}$$

To see how this expression for A_S differs from that given in Eq. (2.1), let's use it to repeat our calculation of the stress area of another $\frac{1}{4}$–20 UNC Class 2A thread; this one presumably of a stronger material.

$E_{Smin} = 0.2127$ in.

$n = 20$

$$A_S = 3.14159 \left(\frac{0.2127}{2} - \frac{0.16238}{20} \right)^2$$

$A_S = 0.0301$ in.^{2}

So this A_S is 5% smaller than that we computed earlier.

It's worth noting that an equivalent to Eq. (2.2) was included in the 1974 version of ANSI B1.1 [7] but has been eliminated from the 1984 and 1989 versions of this same standard [19], which, in general, contain a much shortened discussion of thread strength and the related tensile stress area than did the earlier version. Presumably Eq. (2.2) is a more accurate expression for the tensile strength of a high- strength thread than is Eq. (2.1). All such areas are determined—or at least confirmed—by experiment, as already mentioned. The authors of B1.1, however, may have decided not to confuse the issue by suggesting a variety of stress areas.

I leave it for the purists among you to decide which equation to use. Incidentally, Section 2 of Federal Standard H28, a key document when it comes to defining the dimensions, tensile stress areas, and recommended lengths of engagement of threads, currently gives only Eq. (2.1). [24] As a final comment on all this, and probably to complicate matters further, ANSI B1.1-1974 gave the low stress–high stress cutoff point as 180,000 psi, not 100,000 ksi as in *Machinery's Handbook* [18], and this value appeared in other references as well (e.g., British Standard BS 3643). All of this emphasizes the empirical nature of the various equations for A_S. They are attempts to define "equivalent cylinders" for bolts, based on tensile tests of actual bolts.

There's at least one other tensile stress area in common use. The U.S. military uses a stress area based on maximum pitch diameter for UN and UNJ threads when dealing with fastener material strengths between 160 and 260 ksi—and when the threads are rolled after heat treatment. The resulting stress areas are tabulated in military specification NAS 1348. The areas are larger than the A_S values computed by Eq. (2:1): about 17% larger for the smallest screws tabulated (0.600–80 threads) to about 4% larger for the largest (1.500–12). Again, these stress areas must reflect test results which show, in this case, that rolling threads after heat treat leads to a much stronger bolt than does rolling before heat treat or cutting.

As a final note on tensile stress area, Eq. (2.1) gives reasonable results for fasteners of diameter $\frac{1}{4}$ in. and larger. The strength of smaller fasteners will be somewhat less than predicted by Eq. (2.1) [16].

In spite of all these nuances, remember that every time you encounter "tensile stress area" in the literature or a common standard or the like, the authors will be referring to Eq. (2.1) "unless otherwise stated." Most people will never use the alternatives.

There is another, still more conservative stress area that is still widely used, however. It is based on the root diameter of the threads rather than on the mean of pitch and root diameter, as for the "true" stress area. It differs from the areas we've talked about so far in that this one is *not* based on experimental data. It is designed to introduce a factor of safety in thread strength calculations. The designer purposely assumes a stress area smaller than the "real" [Eq. (2.1)] tensile stress area to be sure the bolts aren't overstressed in service.

The use of root diameter area rather than tensile stress area is mandated by the ASME Boiler and Pressure Vessel Code, for example. The expression for this area is

$$A_r = 0.7854 \left(D - \frac{1.3}{n} \right)^2 \tag{2.3}$$

where D and n have the same meaning as in Eq. (2.1).

You'll find tensile stress and thread root areas for UN and UNC threads in Table 2.2 and a complete list of tensile stress areas for all inch series and metric threads in Appendix F. Figure 2.12 can help you visualize the differences between the various stress areas we've been discussing.

All of the stress area equations given so far are for American Standard Unified thread forms with an included angle of 60°. Different equations must be used for 55° Whitworth, Acme, Buttress, or other thread forms.

D. Stress Areas—Metric Threads

The expressions for the stress areas of metric threads differ from those used for inch series threads because in metric threads, the pitch of the

Table 2.2 Thread and Bolt Stress Areas [7, 15] (all areas are in square inches)

Thread	Tensile areas		Stripping areas	
	A_s (stress)	A_r (root)	Class 2A thread	Class 3A thread
$\frac{1}{4}$–20	0.0318	0.0269	0.092	0.096
$\frac{5}{16}$–18	0.0524	0.0454	0.147	0.157
$\frac{3}{8}$–16	0.0775	0.0678	0.216	0.232
$\frac{7}{16}$–14	0.1063	0.0933	0.296	0.321
$\frac{1}{2}$–13	0.142	0.126	0.390	0.427
$\frac{9}{16}$–12	0.182	0.162	0.502	0.548
$\frac{5}{8}$–11	0.226	0.202	0.624	0.681
$\frac{3}{4}$–10	0.334	0.302	0.908	1.01
$\frac{7}{8}$–9	0.462	0.419	1.25	1.38
1–8	0.606	0.551	1.66	1.82
$1\frac{1}{8}$–8	0.790	0.728	2.13	2.329
$1\frac{1}{4}$–8	0.969	0.890	2.65	2.913
$1\frac{3}{8}$–8	1.23	1.16	3.22	3.55
$1\frac{1}{2}$–8	1.49	1.41	3.86	4.26
$1\frac{5}{8}$–8	1.78	1.68	4.55	5.04
$1\frac{3}{4}$–8	2.08	1.98	5.30	5.03
$1\frac{7}{8}$–8	2.41	2.30	6.09	6.81
2–8	2.50	2.30	6.96	7.72
$2\frac{1}{4}$–8	3.56	3.42	8.84	9.83
$2\frac{1}{2}$–8	4.44	4.29	10.95	12.18
$2\frac{3}{4}$–8	5.43	5.26	13.28	14.80
3–8	6.51	6.32	15.84	17.67
$3\frac{1}{4}$–8	7.69	7.49	18.62	20.8
$3\frac{1}{2}$–8	8.96	8.75	21.63	24.15
$3\frac{3}{4}$–8	10.34	10.11	24.79	27.79
4–8	12.18	11.57	28.28	31.64

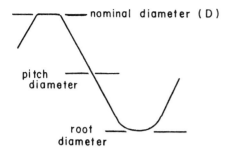

Figure 2.12 Relationship between root diameter, pitch diameter, and outside or nominal diameter in a standard 60° thread. The centerline of the thread is out of the picture, toward the bottom.

threads, rather than the number of threads per inch, is used to define the thread. The following metric equation corresponds to Eq. (2.1), defining, as it does, the common tensile stress area [24]:

$$A_S = 0.7854 \ (D - 0.938p)^2 \tag{2.4}$$

where A_S = tensile stress area (mm)
         p = pitch of the threads (mm)
         D = nominal diameter of the fastener (mm)

The more conservative expression for higher-strength steels (over 686 MPa—99.5 ksi—this time) is [17]

$$A_S = 0.7854(E_{Smin} - 0.268867p)^2 \tag{2.5}$$

And the expression for thread root area is [24]

$$A_r = 0.7854(D - 1.22687p)^2 \tag{2.6}$$

As before, E_{Smin} = minimum pitch diameter (mm); D and p have the same meanings as in Eq. (2.4).

E. Strength of the Bolt Under Static Loads

The bolt or nut will fail under static loads if those loads exceed the strength of the fastener. To understand and avoid static failure, therefore, we must learn how to calculate the static strength of the parts. We'll concentrate on tensile loads; we will compute shear strength of the joint in Chap. 14.

In estimating the static strength of a fastener under tensile loads, we must consider four possibilities:

The body of the bolt will break (usually at one of the three stress concentration points).

The bolt threads will strip (fail in shear).
The nut threads will strip.
Both threads will strip simultaneously.

To determine the static strength, we must explore each of these four possible failure modes. The weakest link, of course, will decide the strength of the chain. We'll consider bolt strength now, and thread strength in the next chapter.

The static strength of the body is usually computed for that section which is threaded, i.e., has a reduced cross-sectional area. The tensile force required to yield or break the bolt (you'll have to decide which of these definitions of failure you prefer) is, simply,

$$F = \sigma A_S \qquad\qquad (2.7)$$

where F = force which will fail the bolt (lb, N)
         σ = ultimate tensile or yield strength of the bolt material (psi, MPa)
         A_S = effective "stress area" of the threads (in.^{2}, mm^2) [see Eq. (2.1) or (2.4)]

This equation may be used to compute any kind of tensile strength—ultimate, proof, yield, etc. As an example, let's compute the room temperature yield strength of an Inconel 600 bolt with a $\frac{1}{4}$–20 UNC Class 2A thread. The yield strength of this (and other) bolt materials is given in Tables 4.1 and 4.4.

   σ = 37 ksi

   A_S = 0.0318 in.^{2} [see the example with Eq. (2.1)]

   F = 37 × 0.0318 = 1177 lb

What would the strength of this same bolt be at 1200°F (649°C)? With reference to Table 4.4,

   σ = 17 ksi

   A_S = 0.0318 in.^{2}

   F = 17 × 0.0318 = 541 lb

Remember that part of the strength of the bolt might be absorbed by torsion stress if we use a wrench to tighten it. This could reduce the apparent tensile yield or ultimate strength (σ) by 10–20%, but only during tightening. This is why many people say, "If it doesn't break when you tighten it, it won't ever break." If loads are static only and are less than the preload developed during tightening, and corrosion or high temperature doesn't enter the picture, the statement is probably true.

The strength of a bolt, of course, determines how much clamping force it can exert on a joint. Table 2.1 gives a few examples of the forces bolts can sustain, to give you a general feeling for the capabilities of individual bolts. The forces tabulated are proof loads or equivalents—about 90% of yield. The forces are given in thousands of pounds (kips). To convert to metric Newtons, multiply the tabulated figures by 4.448×10^3.

The examples in Table 2.1 show the impressive strength of bolts. They suggest that a single $\frac{1}{4}$–20 A307 bolt, properly mounted and supported, could support the weight of a car. Or that one 4–8 H-11 bolt could alone support the weight of nearly 1000 cars! (The strength of these and other materials is discussed at length in Chap. 4)

IV. THE STRENGTH OF THE JOINT

Before looking at combinations of loads on the bolt, let's take a quick look at the joint.

A. Contact Stress Between Fastener and Joint

The contact pressure between the head of the bolt and the joint is not uniform. Neither is the contact pressure between the nut and the joint. For either case, the contact pressure might look something like that shown in Fig. 2.13, as suggested by an experimentally confirmed, finite-element analysis made in Japan [10].

The contact pressure pattern shown in Fig. 2.13, although non-uniform, assumes that the contact surfaces of the bolt head and nut will be perfectly parallel to the surfaces of the joint, at least when loading begins. Since this will rarely be the case—these surfaces are usually not exactly perpendicular to the thread or hole axes, for example—the actual pressure distribution will probably be even more irregular than suggested in the drawing.

The contact pressure between nut or bolt head and the joint can have an important influence on the way in which a loaded fastener retains the potential energy stored in it during assembly (see Chap. 1). Excessive pressure will allow the head and/or nut to embed itself gradually in the joint surfaces, allowing the fastener to relax, to shed some of its stored energy. These contact stresses can be impressive, as suggested by the data in Table 2.3, which shows the stresses generated when the tabulated fasteners are tightened to 75% of their tensile proof load. [20] (For those primarily interested in inch series fasteners, M10, M16, and M22 metric fasteners are approximately $\frac{3}{8}$, $\frac{5}{8}$, and $\frac{7}{8}$ in. in diameter, respectively.) As

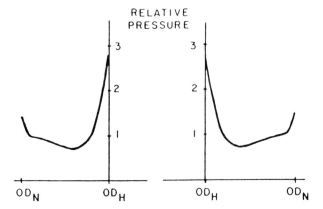

Figure 2.13 Relative contact pressure between head of bolt, or nut, and the joint surface. OD_H is the outer diameter of the hole. OD_N is the outer diameter of the contact surface of the nut (or bolt head).

can be seen from the data, a flanged head or a washer can significantly decrease these contact stresses.

The thickness of the washer can also make a difference. Studies made at Newport News Naval Shipyard confirm the contact stress pattern shown in Fig. 2.14 and show how this nonuniform head- or nut-to-joint stress distribution creates high stress gradients in the joint members.

Table 2.3 Contact Stress Between Bolt Head and Joint When the Bolts Are Tightened to 75% of Proof Load

Fastener grade	Size	Head	Contact stress	
			ksi	N/mm
10–9	M10	Flanged	36.3	250
	M16		39.9	275
10–9	M10	Plain hex	52.2	360
	M16		94.3	650
10–9	M10	Hex with washer	27.6	190
	M16		33.4	230
12–9	M10	Plain hex	63.8	440
	M22		131	900
12–9	M10	Hex with washer	29	200
	M22		50.8	350

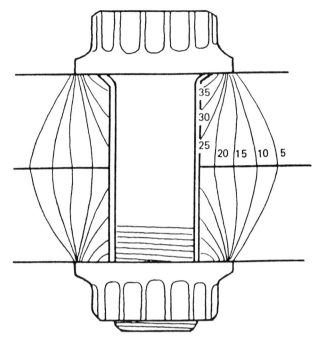

Figure 2.14 Lines of equal compressive stress in joint members when the bolt is loaded to 100 kip. Values given are in ksi.

(We'll look at joint stresses in a minute.) The studies further suggest that standard thickness ANSI washer lack sufficient stiffness to distribute the bolt loads with acceptable uniformity. Thicker washers are more effective; they reduce stresses in washer and joint members and, as a result, reduce joint deformation. The numerical results of these studies have not yet been published. [21]

What's an acceptable stress? This is considered to be a stress slightly higher than the compressive yield strength of the joint material [20, 21]. Table 2.4 lists some experimentally derived data for acceptable contact stresses. The data are taken from the VDI Directive 2230, so the joint materials are described in terms used in Germany. The designations will undoubtedly be meaningful to metallurgists in English-speaking countries, however. [23] Note that VDI suggests that fasteners tightened by hand can be brought to higher contact stresses than those tightened automatically, presumably by pneumatic or electric nut runners or the equivalent (see Chaps. 7 and 8). Why is this? The more sophisticated of such tools can be more accurate than a hand wrench, so the difference can't be explained

Table 2.4   Acceptable Maximum Head-to-Joint Contact Stress for a
Variety of Joint Materials

| | How tightened | | | |
| | Automatically | | By hand | |
Joint material	ksi	N/mm	ksi	N/mm
St37	29	200	43.5	300
St50	47.9	330	72.5	500
C45V	87	600	131	900
GG-25	72.5	500	109	750
GD Mg Al 9	11.6	80	17.4	120
GK Mg Al 9	11.6	80	17.4	120
GK Al Si 6 Cu 4	17.4	120	26.1	180

by the accuracy tolerance. I suspect that it has to do with the speed with
which the fasteners are tightened.

B.   Stresses Within and Between the Joint Members

As a result of all this, the joint is loaded in a nonuniform manner by the
bolt, and these nonuniform stresses are passed down into the joint members. Figure 2.14 shows the resulting, barrel-shaped, lines of equal compressive stress within the joint [5]. Notice that the relative magnitude of
stress varies by 8 to 1 between the most highly stressed region and the
outer rim of the barrel-shaped stress pattern.

The fact that the joint members are stressed nonuniformly means that
the clamping pressure exerted by the bolt through the joint members on
the interface between upper and lower joint members is also nonuniform,
a fact that can cause problems in gasketed joints. A number of finite-
element analyses and experimental studies have been made of the contact
interface pressure. The results of one such study [8] are shown in Fig.
2.15. Note that the contact pressures can be zero only two or three bolt
diameters away from the bolt hole. The solution for this problem, of
course, is to place bolts close together in a multibolt joint so that no
portion of the interface is entirely free of contact pressure. It is never
possible, however, to produce exactly uniform pressure at the joint interface. Too many bolt holes, furthermore, will weaken the flange members
and/or create nut interference and wrenching problems. As a rule of
thumb, therefore, bolt holes are usually placed approximately $1\frac{1}{2}$ diameters
apart.

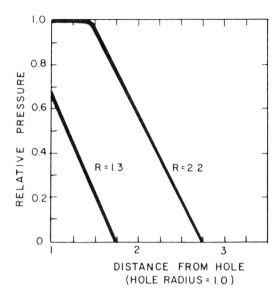

Figure 2.15 Relative interface pressure between joint members as a function of the distance from the edge of the hole. *R* is the ratio between the contact diameter of the head of the bolt (or nut) and the diameter of the hole.

C. Static Failure of the Joint

Static failure of joint members is even less common than static failure of the fasteners, except perhaps in structural joints loaded in shear. Failures here are frequent enough to warrant a brief look. If the designer knows which cross sections might fail, he can usually avoid failure.

The failures shown here are all of what used to be called bearing-type joints (see Chap. 14). Friction joints fail statically, too, but they do so by slipping into bearing and then failing by one of the ways shown below.

Joints in bearing can fail in a number of ways as suggested by Fig. 2.16. The actual failure mode will depend on the relative strength of the bolts versus the strength of the cross section of the joint members at various load points. It will also be affected by the distance between the bolts and the edge of the plates, by the distance between bolts within the group, etc.

Typical failure modes include [14]:

1. Tear-out or marginal failure, where the bolts are located too close to the edge of the plate (Fig. 2.16A)

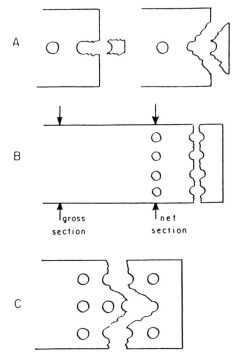

Figure 2.16 Some static failure modes of axial shear joints. (A) Tear-out or marginal failure. (B) Failure through the "net section." (C) Zigzag failure.

2. Failure of the "net section" of the plate because the bolts are spaced too closely, or because the plate is too thin or too soft (Fig. 2.16B)
3. A zigzag failure when there is too short a distance between bolt holes (Fig. 2.16C)

V.   OTHER TYPES OF LOAD ON A BOLT

We have looked, at length, at the stress distribution and strength of a fastener under pure tensile load, because this is by far the most important type of loading. A bolt is always put into severe tension when it is properly tightened. Subsequent external loads won't modify this basic tension load very much if the joint is designed properly, as we'll see in Chap. 12. On occasion, however, we do want to estimate the magnitude of other types of stress which can be imposed on a bolt in use. We'll consider some of these now, one by one, and later suggest a way in which their combined effects might be estimated, or at least visualized.

A. Shear Loads

Shear loads of the sort shown in Fig. 2.17 are often encountered in structural steel joints. Note that there can be either one or several shear planes through the bolt, depending on the nature of the joint and the number of joint members being clamped together. Note, too, that these shear planes could pass either through the body of the bolt or through the threads, or through both.

To compute the shear strength of the bolt we multiply the shear strength of the material (S_U) by the total cross-sectional area of the shear planes, taking all of them into account. For shear planes through threads, we use the equivalent thread stress area (A_S) given earlier in our analysis of tensile loads. For example, for the conditions of Fig. 2.17D,

$$F = S_U \times (2A_B + A_S) \qquad (2.8)$$

where A_B is the cross-sectional area of the body (in.^{2} or mm^2). We'll consider shear loads at greater length in Chap. 14.

B. Torsional Loads

When we tighten a nut, we apply a torsional moment as well as a stretching force to the bolt, thanks to frictional and geometric restraints between

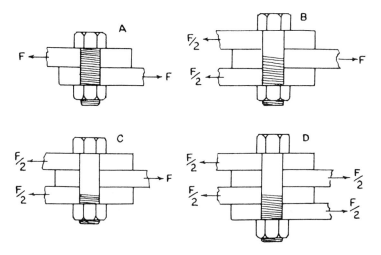

Figure 2.17 Possible shear loading conditions on a bolt in a structural joint.

nut and bolt threads. As we'll see in Chap. 7, the torsional moment (T_{tor}) applied to the bolt is given by the following expression (assuming that there is no prevailing torque):

$$T_{tor} = F_P \left(\frac{P}{2\pi} + \frac{\mu_t r_t}{\cos \beta} \right) \tag{2.9}$$

where P = thread pitch (in., mm)
   μ_t = coefficient of friction between threads
   r_t = effective radius of contact between threads (in., mm)
   β = half-angle of a tooth (30° for UN or ISO threads)
   T_{tor} = torsional moment (lb-in., N-mm)
   F_P = preload (lb, N)

Assuming that we can substitute an equivalent cylinder for the fastener and that the cylinder has a diameter equal to the nominal diameter (D) of the fastener and a length equal to the effective length (L_E) of the fastener (see Chap. 5), we can write the following expressions for the angle of twist (θ_{tw}), the torsional energy (W_{tor}), and the maximum stress (S_{tor}) produced in the outermost fiber of the fastener as follows [12]:

$$\theta_{tw} = \frac{32 L_E T_{tor}}{\pi G D^4} \tag{2.10}$$

$$W_{tor} = \tfrac{1}{2} T_{tor} \times \theta_{tw} = \tfrac{1}{2}(T_{tor})^2 \left(\frac{32 L_E}{\pi G D^4} \right) \tag{2.11}$$

$$S_{tor} = \frac{16 T_{tor}}{\pi D^3} \tag{2.12}$$

where θ_{tw} is in radians
   S_{tor} is in psi or N/mm^2
   G is the shear modulus in psi or N/mm^2
   D and L_E are in in. or mm

C. Bending Loads

If the geometry of the parts forces the bolt to bend as it is tightened, it will see additional tensile stress along its outer flank and less tensile stress along the inner flank, as suggested in Fig. 2.18. Assuming once again that the bolt can be replaced by an equivalent cylinder having an external diameter equal to the nominal diameter of the fastener (D) and a length equal to the fastener's effective length (L_E), we can compute the bending moment (M_b), the maximum stress (S_B) created by pure bending, and the bending energy (W_{bn}) as follows [13]:

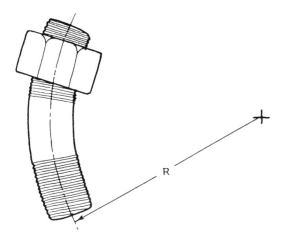

Figure 2.18 Radius of curvature of a bent bolt.

$$M_b = \frac{IE}{R} = \frac{\pi ED^4}{64R} \tag{2.13}$$

$$S_B = \frac{M_b D}{21} = \frac{ED}{2R} \tag{2.14}$$

$$W_{bn} = \frac{1}{2} \int_0^{L_E} \frac{M_b^2}{EI} \, dl = \frac{1}{2} \frac{E\pi D^4 L_E}{64R^2} \tag{2.15}$$

where L_E = effective length of the fastener (in., mm)
E = modulus of elasticity (psi, N/mm^2)
I = moment of inertia (in.^{4}, mm^4)
R = radius of curvature (in., mm)
S_B = maximum bending stress (psi, N/mm^2)
M_b = bending moment (lb-in., N-mm)
D = nominal diameter (in., mm)

VII. COMBINED LOADS ON A BOLT

A. Work-Energy Equation

In most applications a bolt will see a combination of the tension, torsion, shear, and bending stresses discussed earlier. Exactly which combination will depend on the geometry of the given set of parts and on the coefficient of friction. It is never possible in practice to compute the resulting maximum stresses in a given fastener, because we never have enough information about the variables involved. I will not attempt, therefore, to describe how one might combine the possible stresses mathematically into a single maximum.

Because the bolted joint depends on stored energy to do its job (see Chap. 1), it's informative to write an equation describing how the input energy—the work done on the nut by the guy with the wrench—is absorbed by the bolt and joint members. Although this equation still involves all of the variables encountered in practice—and we will therefore never be able to solve it for a particular bolt—it does show us how complex the act of tightening a bolt can be and forces us to keep this fact in mind when we deal with some of the simplified design equations we'll encounter in subsequent chapters.

We'll assume here that there's a linear relationship between input torque and turn of the nut, so that we can define the work into the system (W_{in}) as

$$W_{in} = \tfrac{1}{2}(T_{in} \times \theta_{in}) \tag{2.16}$$

where T_{in} is the maximum torque applied to the nut (in.-lb, N-mm) and θ_{in} is the total turn applied to the nut (rads) *after* snugging. This is only an approximation, because it assumes a linear $T_{in} - \theta_{in}$ relationship. We'll see the real relationship in Chap. 8. But this expression describes most of the input work, and it will make it unnecessary for us to handle each term in the energy equation as an integral. We're after a concept here, not an accurate calculation. In any case, we'll never know enough about a given fastener to achieve accuracy.

Specifically, when we tighten a nut on a bolt and clamp a joint together, we do work on the joint in a number of ways. It's possible to write an expression for each portion of this work and to set their sum equal to the input work. Since we've assumed a linear relationship between T_{in} and θ_{in}, we'll also assume a linear buildup in energy stored in each mode.

We stretch the bolt (tension energy):

$$W_{ten} = \tfrac{1}{2}(\Delta L \times F_P) \tag{2.17}$$

where ΔL = change in length of the bolt (in., mm)
$\quad\quad\quad F_P$ = preload (lb, N)

We twist the bolt (torsional energy):

$$W_{tor} = \tfrac{1}{2}(T_{tor})^2 \frac{32L_E}{\pi GD^4} \tag{2.12}$$

We do work (W_{tf}) against the friction forces between nut and bolt threads (heat energy)—see Chap. 7:

$$W_{tf} = \tfrac{1}{2}T_{tf} \times \theta_R = \frac{1}{2}\frac{\mu_t F_P r_t}{\cos \beta} R \tag{2.18}$$

where θ_R is the *relative* turn between nut and bolt (rad), and the other terms have been defined earlier.

We do work (W_{nf}) against the friction forces between nut and joint members (heat energy)—Chap. 7 again:

$$W_{nf} = \tfrac{1}{2}T_{nf} \times \theta_{in} = \tfrac{1}{2}\mu_n F_P r_n \theta_{in} \tag{2.19}$$

where θ_{in} = total input turn (rad)
r_n = effective contact radius between nut and joint (in., mm)
μ_n = coefficient of friction between nut and joint

We bend the bolt (bending energy) (W_{bn}):

$$W_{bn} = \frac{1}{2}\frac{\pi E D^4 L_E}{64 R^2} \tag{2.15}$$

We compress the joint (compressive energy) (W_{jc}):

$$W_{jc} = \tfrac{1}{2}(\Delta T_j \times F_P) \tag{2.20}$$

where ΔT_j = compression of the joint (in., mm)
F_P = preload (lb, N)

We compress the nut (compressive energy, W_{nc}):

$$W_{nc} = \tfrac{1}{2}(\Delta T_N \times F_P) \tag{2.21}$$

where ΔT_N = compression of the nut (in., mm)
F_P = preload (lb, N)

We can now sum all of these terms together, equate them to the input work, and combine terms. Here's one possible combination. Note that we can eliminate the $\tfrac{1}{2}$ from each term.

$$F_P^2\left[\left(\frac{P}{2\pi} + \frac{\mu_t r_t}{\cos \beta}\right)^2 \left(\frac{32 L_E}{\pi D^4 G}\right)\right]$$
$$+ F_P\left(\Delta L + \Delta T_J + \Delta T_N + \frac{\mu_t \theta_R r_t}{\cos \beta} + \mu_n \theta_{in} r_n\right) \tag{2.22}$$
$$+ \left(\frac{E\pi D^4 L_E}{64 R^2} - T_{in}\theta_{in}\right) = 0$$

As we noted earlier, this equation involves a number of factors which will be difficult or impossible to estimate in a given application—such things as the coefficients of friction, the radius of curvature of the bolt, the relative turn between nut and bolt, and the "deflection" of the nut, for example. The equation also, as we mentioned, assumes a linear buildup of input torque, which makes it only a rough approximation in any event. This equation has never been tested, but it would appear to be a more accurate description of the relationship between applied torque (T_{in}) and achieved preload than is the common, long-form torque-preload equation [Eq. (7.2)], to be discussed at length in Chap. 7. Recently A. R. Srinivas of the Space Application Center in India has shown, in fact, that Eq. (2.22) reduces to Eq. (7.2) when one realizes that the F_P^2 term can be ignored (because the product D^4G is very large); the ΔT_J, bending, and ΔT_N terms are very small; and θ_R is essentially equal to θ_{in}. It's also necessary to realize that $\Delta L/\theta_{in}$ equals $P/2\pi$, where P is the pitch of the threads (the distance between the crests of two thread teeth) [25].

If nothing else, our energy equation shows that it is not a trivial matter to predict how much a bolt is going to be preloaded when we apply torque to the nut. We don't produce "just tension" in the fastener. We create a lot of other stresses. We produce heat. And it's not at all obvious—ever—exactly what results we will achieve.

B. Strength Under Combined Loads

We are interested primarily in the tensile strength of bolts, less often in the shear strength. The tensile strength determines the amount of preload we can safely put into a bolt on tightening it and the amount of tensile working load it can see thereafter. It's important to recognize, therefore, that the tensile strength of a given fastener is reduced if the fastener also sees torsion or shear loads. In an extreme case, for example, let's assume that nut and bolt threads have galled near the end of the tightening operation. This will result in an abnormally high level of torsion in the bolt. If we now apply tension, we'll find that the bolt will break at a tension level well below normal—perhaps even below proof load. The torsion stress has robbed part of the total strength of the bolt. Only if the bolt is loaded, statically, in pure tension can we count on it to support a proof load without deformation, or to yield at a particular level of tension.

Note that torsion stress will rob part of the total strength of the bolt only while the torsion stress is present. As we torque a bolt, for example, it is subjected to some torsion. Such a bolt will yield at a certain level of tensile stress. After we remove the torque wrench from the nut, however, the torsion stress will tend to disappear (thanks to embedment relaxation,

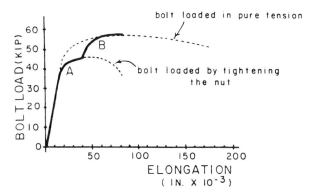

Figure 2.19 A fastener tightened with a torque tool, and therefore exposed to simultaneous torsion and tension, will yield at a slightly lower level of tensile stress (A) than a fastener subjected to pure tension. This same fastener, however, will support a higher tensile stress in service before yielding any further (B), because the torsion stress component will, in general, disappear rather rapidly after initial tightening in most situations. Data are for a $\frac{7}{8}$ A325 bolt with a $4\frac{1}{8}$ grip length [1].

as we will learn in Chap. 6). If we now apply an external tensile load to the fastener, we will discover that it will support a higher level of tension than that which caused it to yield in the first place, as suggested in Fig. 2.19. A number of different fastening tools and strategies take advantage of this fact, as we will see later [1].

REFERENCES

1. Fisher, John W., and J. H. A. Struik, *Guide to Design Criteria for Bolted and Riveted Joints*, Wiley, New York, 1974.
2. Rocking that fatigued bolts felled arena roof, ENR, *McGraw-Hill's Construction Weekly*, August 16, 1979.
3. Wiegand, H., and K. H. Illgner, *Berechnung und Gestaltung von Schraubenuerbindungen*, Springer-Verlag, Berlin, 1962, p. 48.
4. Fastener preload indicator, Contract no. F33615-76-C-5151, Report prepared for Air Force Materials Laboratory, Air Force Systems Command, Wright-Patterson Air Force Base, Ohio, by General Dynamics–Ft. Worth, June 15, 1978.
5. Meyer, G., and D. Strelow, Simple diagrams aid in analyzing forces in bolted joints, *Assembly Eng.*, January 1972.
6. Seika, M., S. Sasaki, and K. Hosono, Measurement of stress concentrations in threaded connections, *Bull. JSME*, vol. 17, no. 111, September 1974.

7. Unified Inch Screw Threads, ANSI Standard B1.1-1974, ASME, New York, 1974.
8. Gould, H. H., and B. B. Mikic, Areas of contact and pressure distribution in bolted joints, *Trans. ASME*, vol. 94, no. 3, August 1972.
9. Motosh, N., Load distribution on threads of titanium tension nuts and steel bolts, *J. Eng. Ind. ASME*, vol. 97, no. 1, pp. 162–166, 1975.
10. Maruyama, K., Stress analysis of a bolt-nut joint by the finite element method and the copper-electro-plating method, *Bull. JSME*, vol. 19, no. 130, April 1976.
11. Finkelston, R. J., Preloading for optimum bolt efficiency, *Assembly Eng.*, August 1974.
12. Roark, Raymond, and Warren C. Young, *Formulas for Stress and Strain*, 5th ed., McGraw-Hill, New York, 1975, p. 287.
13. Eshbach, Ovid W., *Handbook of Engineering Fundamentals*, 2nd ed., Wiley, New York, 1953, pp. 5–28.
14. Fisher, J. W., and J. H. A. Struik, *Guide to Design Criteria for Bolted and Riveted Joints*, Wiley, New York, 1974, p. 84 ff.
15. Blake, Alexander, *What Every Engineer Should Know about Threaded Fasteners*, Marcel Dekker, New York, 1986.
16. Crispell, Cory, A better way to design small fasteners, *Machine Design*, Cleveland, Ohio, Jan. 22, 1987.
17. BS 3643: Part 2: 1966, British Standards Institution, London, 1966.
18. Oberg, Eric, Franklin D. Jones, and Holbrook Horton, *Machinery's Handbook*, 20th ed., Industrial Press, New York, 1975.
19. Unified Inch Screw Threads, ASME B1.1-1989 (Revision of ANSI B1.1-1982), American Society of Mechanical Engineers, New York, 1989.
20. Eccles, William, Bolted Joint Design, *Engineering Designer* (UK), November 1984, p. 11.
21. Barron, Joseph, in report to the Bolting Technology Council, at a meeting in Cleveland, Ohio, April 19, 1993.
22. Junker, G. H., Principle of Calculation of High Duty Bolted Joints; Interpretation of Directive VDI 2230, published by SPS, Jenkintown, PA (no date given)
23. VDI 2230: Systematic Calculation of High Duty Bolted Joints, VDI Society for Design and Development, Committee on Bolted Joints, Issued by the Verein Deutscher Ingenieure (VDI), VDI Richtlinien (Dusseldorf, October 1977, Translated from the German by Language Services, Knoxville, TN, published by Oak Ridge National Laboratories as ORNL-tr-5055, Table 14.
24. Screw Thread Standards for Federal Services, Section 2, Unified Inch Screw Threads—UN and UNR Forms, FED-STD-H28/2B, 20 August 1991, Defense Industrial Supply Center, Philadelphia, PA, p. 57.
25. Srinivas, A. R., Determination of Torque Values for Metric Fasteners—an Approach, Space Payload Group, Space Applications Center, Indian Space Research Organization, Ahmedabad, India, Document 380 053, May 1992.

3
Threads and Their Strength

The threads are obviously an important element of the threaded fastener. They give this sturdy, industrial product its unique ability to be installed, removed, and reinstalled as many times as we wish. They also affect fastener performance in a major way. As we'll see, thread type, thread class, thread configuration, the way in which the threads are produced, and the fit between male and female threads can affect not only thread strength—and, therefore, fastener tensile strength—but also the resistance of the fastener to such things as self-loosening and fatigue. The amount of preload achieved for a given torque can be influenced by thread configuration and by whether the threads have been cut or rolled. All things considered, it's worthwhile to take a close look at threads.

I. THREAD FORMS

A. Thread Forms in General

Literally hundreds of thread forms have been designed, and many are still in common use in a wide variety of applications. Fortunately, we only have to worry about the few that are currently used in threaded fasteners. To clear the decks, however, let's start by taking a quick look at three other forms which are often mentioned and which many beginners to bolting assume that they should know about. These forms are illustrated in Fig. 3.1, along with a currently popular 60° form for comparison. Anyway, the three we *don't* have to worry about are [1, 25]:

The ACME thread: this is used for power transmission, for example, to produce traversing motion on machine tools.

The Buttress thread: used when the thrust on the screw is in one direction only, for example, airplane propellor hubs and columns for large presses [26].

The Whitworth thread: this form, which had a 55° included angle instead of the now universal 60°, was for decades a British standard form but has now been replaced by an ISO inch series. It was the first screw thread form to have rounded roots, I believe.

All of the modern fastener thread forms we need to know about—the metric as well as inch series forms—are based on an arrangement of 60° angles. As we'll see this basic geometry is modified in several different ways, but it's the starting point for all contemporary fastener thread profiles.

B. Inch Series Thread Forms

In the United States our principal inch series thread form standards are ASME B1.1 1989 [2] and Federal Standard FED-STD-H 28/2B [3]. Both

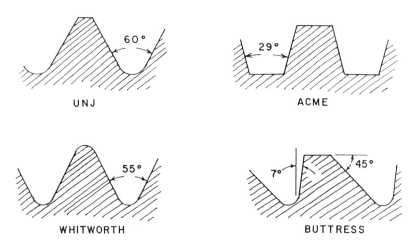

Figure 3.1 Three well-known thread forms which are *not* currently used with threaded fasteners and, for comparison, one which is (the UNJ form). The ACME is a well-known machine tool thread used for traversing screws. The Whitworth is a now-obsolete fastener thread form once used in the U.K. It has now been replaced by a 60-in.-series ISO form.

of these describe the basic Unified Thread Form, identified by the code letters UN/UNR.

A slightly modified version of the UN/UNR thread is defined in Military Specification MIL-S-8879 C and is called the UNJ form [4]. An ANSI/ASME standard for J threads, B1.15, is in preparation.

The differences between these three forms, UN, UNR, and UNJ, are shown in Fig. 3.2. As you can see, the differences are very slight and consist entirely of the way the sharp roots of the teeth are filled in. These differences occur only in the external or male thread form. The same internal thread form is used with each. In any event, the differences are:

The UN form has flat-bottomed or, optionally, slightly rounded roots.
The UNR form must have slightly rounded roots.
The UNJ form has generously rounded roots.

C. Metric Thread Forms

The currently popular metric threads are identified by the code letters M and MJ. The basic geometry of metric and inch series threads is identical,

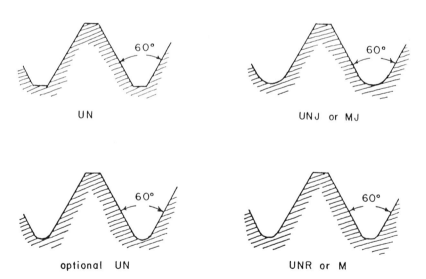

Figure 3.2 These are the thread forms most commonly used in the Western world at the present time. Each is a 60° included angle form. They differ from each other primarily in the way the roots of the external (male) threads are shaped. The UN form has flat, or, optionally, slightly rounded roots. The UNJ and metric MJ forms have generously rounded roots. The UNR and metric M forms have slightly rounded roots.

but the way we define the metric threads differs from the way we define the inch series ones, as we'll see.

U.S. standards for metric threads include ANSI/ASME B1.18M-1982 [22] for standard commercial fasteners and ANSI/ASME B1.13M-1983 (Reaffirmed 1989) for fasteners having nonstandard pitch-diameter combinations and/or special lengths of engagement [5]. Both define M profile threads. Another standard, ANSI B1.21M-1978, defines MJ threads [6]. The M profile does not include an absolutely flat root option (equivalent to the UN form) but only a "radiused" option—flat with rounded fillets blending into the thread flanks—and a "rounded root" option similar to the UNR profile.

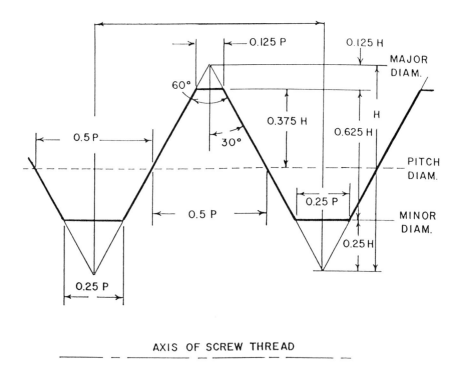

AXIS OF SCREW THREAD

Figure 3.3 The basic profile of the UN, UNR, and metric M thread forms. This is the starting point for the design of those threads. It is also identical to the so-called design profile for all UN and most metric internal threads. The design profiles of external UN or metric threads have slightly altered shapes and dimensions, as explained in the text.

II. THREAD PROFILES

A. Basic Profile

The design of the male and female threads in a given series starts with a "basic profile" which has been called the "permanently established boundary between the provinces of the external and internal threads" [7]. Figure 3.3 shows the basic profile of both external and internal threads for the UN, UNR, and metric M thread forms. Figure 3.4 shows the few differences which convert the standard UN or M basic profile to either a metric J or UNJ profile. Again, only the external J thread's basic profile is modified; the internal thread profile for J as well as the other threads is shown in Fig. 3.3.

B. Design Profile

This profile, also called the "design thread form" [7], defines the maximum material profiles for the internal and external threads of a given

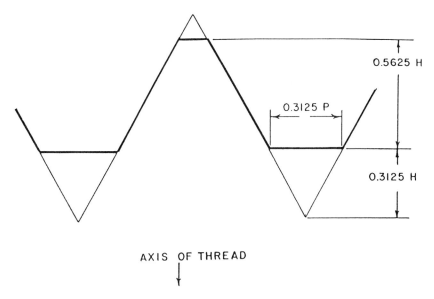

AXIS OF THREAD

Figure 3.4 The basic profile for the UNJ and metric MJ threads differs from the profile shown in Fig. 3.3 only in the dimensions illustrated here. Again, this basic profile for the J threads is identical to the design profiles for the internal threads in those classifications but would be modified slightly for the design profiles of the external threads.

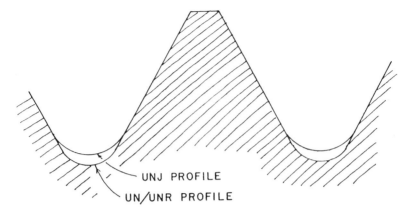

Figure 3.5   The design profiles of the UN/UNR and UNJ (or M and MJ) threads differ primarily in the way the roots of the teeth are shaped, as suggested by Fig. 3.2. This sketch shows a closeup of the difference for Class 2A UN threads. As you can see, the difference appears to be slight, but the UNJ and MJ threads have significant strength and fatigue advantages over the UN/UNR or M threads.

form. Note that if the maximum material profile for both nut and bolt coincided with the basic profile there would be no clearance between male and female threads and they couldn't be assembled. Some clearance must be provided for, and this is generally done by making the design profile for the male thread slightly smaller in diameter than its basic profile. We'll take a closer look at how this is done in Sec. IV of this chapter.

The design profile for most female threads is the same as the basic profile shown in Fig. 3.3 (or Fig. 3.4 for J threads). This is true for all UN threads and for many or most metric threads. The design profile of some internal metric threads can deviate from their basic profile as we'll see.

Design profiles for UN, UNR, and UNJ external threads and their metric equivalents are shown in Fig. 3.2. Figure 3.5 gives us a closer look at the differences between the UNR and UNJ (or M and MJ) profiles.

III.   THREAD SERIES

Design profiles can be applied to threads of any size, and this leads to what are called "thread series." In the Unified Thread system, for example, we have [2]:

1. Threads called just UN/UNR or UNJ: the "constant-pitch" series. Each thread in such a series has the same number of teeth per inch. For example, there's an "8 pitch" series, which means that each thread in the group has eight threads per inch. The angle the helix of the thread makes around the fastener varies with the diameter of the fastener, but the depth of the teeth is constant, regardless of diameter, because of the rigid 60° geometry on which the form is based.

    Altogether there are eight constant-pitch thread series, including those with 4, 6, 8, 12, 16, 20, 28, and 32 threads per inch, and they're available in both the basic UN/UNR and the UNJ forms.
2. In addition to the constant-pitch series, there are several groups classified by "coarseness." This refers not to their quality but to the relative number of threads per inch produced on a common diameter of fastener. For example, the codes UNC, UNCR, and UNCJ identify "coarse-pitch" threads.
3. UNF, UNFR, and UNFJ all designate "fine-pitch" threads.
4. UNEF, UNEFR, and UNEFJ are used for "extra fine" threads.
5. Provision has also been made for a "special" series called UNS, UNRS, or UNJS having pitch-diameter combinations not found in any of the standard series above. Most of us will never have to worry about these.

As I mentioned, the coarseness designations refer to the relative number of threads per inch. For a fastener of a given diameter a UNC thread has fewer threads per inch than a UNF, which in turn has fewer than a UNEF. For example, here is the nomenclature we'd use to define the five standard threads currently specified for a fastener having a nominal (body) diameter of $\frac{1}{2}$ in.

The UNC (coarse) thread for that diameter has 13 threads per inch and is encoded as a UNC $\frac{1}{2} \times 13$ thread. Our other options here would be UNRC $\frac{1}{2} \times 13$ or UNJC $\frac{1}{2} \times 13$.

Next in line we have a UN (constant-pitch) thread which has 16 threads per inch and is called UN $\frac{1}{2} \times 16$ or UNR $\frac{1}{2} \times 16$ etc.

Next there's a UNF (fine) thread with 20 threads per inch, called UN $\frac{1}{2} \times 20$.

Then a UNEF (extra fine) thread with 28 threads per inch, or UNEF $\frac{1}{2} \times 28$.

Finally, there's another constant-pitch thread for $\frac{1}{2}$-in. fasteners, UN $\frac{1}{2} \times 32$. Like all of the others in the group, it can have UNR or UNJ forms too.

These numbers all change as the diameter of the fastener changes. For example, the following group of threads has been codified for 2-in.-

diameter fasteners:

UNC 2 × 4½   UN 2 × 12
UN 2 × 6   UN 2 × 16
UN 2 × 8   UN 2 × 20

All the threads in this group are constant-pitch threads except for the first, coarsest thread. This is true of most "large-diameter" fasteners. In fact, above a 4 in. diameter there are nothing but constant-pitch threads. And once again, each of these threads can have a UNR or UNJ form instead of UN if you wish.

Metric threads generally duplicate the inch series threads, but with threads considered to be only coarse or fine. Instead of using threads per inch to classify them, however, we'll use the pitch distance between two teeth, in millimeters. For example,

M6 × 1

would specify a thread having a nominal diameter of 6 mm and a pitch distance of 1 mm.

IV. THREAD ALLOWANCE, TOLERANCE, AND CLASS

We still haven't finished classifying threads. The next important consideration is the way the male and female threads fit together. Is the fit loose and sloppy, or is it tight? Rough applications require the first, precision ones the second.

The fit between mating threads is determined by the basic clearance between them, a clearance determined by the way the threads are dimensioned and by the tolerances placed on those dimensions. Although the philosophy is the same for both metric and inch series threads, the nomenclature used is different [2, 3, 7], so we'll consider the two types separately.

A. Inch Series Threads

Allowance

The basic clearance is established by a "thread allowance," which determines the minimum clearance between threads of a given class. Another way of saying this is that it determines the minimum distance between male and female threads when both nut and bolt are in their "maximum material" conditions: the fattest bolt and the thickest nut. The lower sketch in Fig. 3.6 shows male and female Class 1A or 2A external threads separated only by the basis allowance.

Tolerance

Manufacturing tolerances are now placed on the allowance. These are always in the direction of less material; they always make the clearance

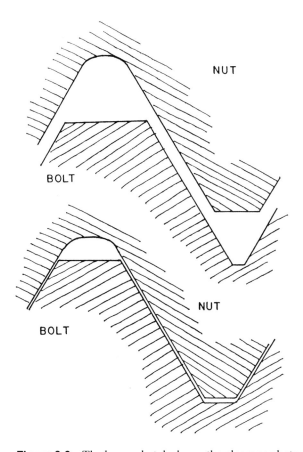

Figure 3.6   The lower sketch shows the clearance between male and female Class 2A UN threads when only the basic allowance separates them. The upper sketch shows how much this clearance increases when the full manufacturing tolerance is added to the allowance. In effect, the lower sketch shows the maximum material condition for bolt and nut; the upper sketch the minimum material condition for both. The bolt thread is reduced in diameter by the allowance and tolerance, and the roots of the teeth are rounded. The diameter of the nut teeth has been increased slightly by the manufacturing tolerance, and the roots of the teeth are rounded.

between nut and bolt threads greater than the clearance determined by the allowance, never less. The upper sketch in Fig. 3.6 shows the clearance between a pair of external 1A or 2A threads when the full tolerance is added to the allowance.

In the Unified Thread System the female threads are always dimensioned to the basic profile; that shown in Figs. 3.3 or 3.4. It's only the male thread which is reduced a little (made smaller in diameter) by the allowance. Tolerances, again always in the direction of less material, have to be placed on both male and female threads, of course.

Class

Three basic fits or "classes" are defined for Unified threads. These are given the codes 1A, 2A and 3A for male threads and 1B, 2B and 3B for female threads. The pair 1A and 1B define the loosest fit; 3A and 3B define the tightest.

Class 1A, 1B threads are used for rough work, for example, where some thread damage can be expected or conditions are very dirty. These are also the easiest threads to assemble.

Class 2A, 2B threads are for "normal" applications. The bolts and nuts you buy in the hardware store are all of this class.

Class 3A and 3B fasteners are used for applications requiring an extra degree of precision.

Class 1A and 2A threads are assigned the same allowance in the Unified system; but more generous tolerances are placed on the 1's than on the 2's, so the average fit is looser.

Class 3 threads are assigned a "zero allowance," so the fit can be line-to-line. A small tolerance on each thread makes assembly possible.

B. Metric Threads

The clearance between male and female metric threads is also determined by a basic allowance and by tolerances in the direction of less material. The number of tolerance and allowance options is greater with metric threads than with inch series threads, and different names are used to describe these things [5–7].

Tolerance Position (the Allowance)

The basic clearance between male and female threads, called the allowance in inch series threads, is called the "tolerance position" for metric threads and is identified by a letter symbol:

G or H for internal threads
e, f, g, or h for external threads

G and e define the loosest fits and the greatest clearance (greatest allowance). H and h define zero allowance, no deviation from the basic profile. Note that although internal metric threads can be assigned zero allowance (H) as are internal UN threads, they can optionally be assigned a tangible allowance (G)—made a little less fat—to accommodate plating or other coatings or perhaps just to provide a looser fit.

Tolerance Grade (the Tolerance)

What we call the tolerance of inch series threads is called the "Tolerance grade" for metric threads and is identified by a number symbol. Seven tolerance grades have been established for external threads and are identified by the numbers 3 through 9 inclusive. Nine defines the loosest fit (the most generous tolerance) and 3 defines the tightest. To complicate our lives further, different groups of options have been established for external and internal threads, and tolerances have been placed on several thread dimensions. The possibilities listed in ANSI/ASME 1.13-1983 are shown in Table 3.1.

Tolerance Class (the Class)

The tolerance grade and tolerance position symbols are now combined into a "tolerance class" code of alphanumeric symbols. Here are some examples:

6g is a general-purpose callout for external threads. It defines the tolerance position and tolerance grade for the major diameter. Used in conjunction with a 6H nut, the fastener would be a reasonable substitute for applications previously using a Class 2A/2B pair.

4g6g is also a general-purpose callout for external threads, this time when a tighter fit is required. It defines the tolerance grade and position for

Table 3.1 Tolerance Grades Assigned to Metric Threads

Dimension controlled	Specified tolerance grades
Minor diam., internal threads	4, 5, 6, 7, 8
Pitch diam., internal threads	4, 5, 6, 7, 8
Major diam., external threads	4, 6, 8
Pitch diam., external threads	3, 4, 5, 6, 7, 8, 9

both the pitch diameter (4g) and the major diameter (6g). Note that in inch series practice these two diameters cannot be toleranced separately as they can here. Class 4g6g, however, is considered to define an external thread which is an approximate equivalent to an inch series Class 3A thread. A nut of tolerance class 6H could be used here.

4H5H defines the tolerance grade and position for a common internal thread which could also be used with the male threads defined above. The 4H classifies the pitch diameter and the 5H the minor diameter. The symbol 5H can also be used alone for applications where a looser fit is acceptable.

C. Inch Series and Metric Thread Classes, Compared

It's useful to be able to equate inch series and metric allowances and tolerances. A couple of examples were given above. Table 3.2 [1] defines some more options. It doesn't define all of the possibilities but only some of the more common "approximate equivalents." Pitch diameter tolerances are not included here for the metric threads, for example. This is not uncommon, but adding them would presumably identify still more approximations.

D. Coating Allowances

If fasteners are to be plated or otherwise coated, some clearance must be provided for the coating. The way this is done is presumably of interest only to fastener manufacturers, who would be guided by the applicable standards. Users (buyers) might find the following summary of interest, however.

Inch series: The allowances specified for Class 2A external threads can accommodate coatings of reasonable thickness. Special provisions must be made for other classes of external thread, all internal threads,

Table 3.2 Approximately Equivalent Classifications: Inch Series and Metric Threads

Inch series		Metric	
Bolts	Nuts	Bolts	Nuts
1A	1B	8g	7H
2A	2B	6g	6H
3A	3B	4h	5H

and threads to be given heavy coatings. Major and pitch diameter limits before and after coating must be specified on engineering drawings, for example [2].

Metric series: External thread tolerance classes 6g and 4g6g provide allowances which can accommodate normal coatings. For heavy coats, or if position h or H tolerance positions are involved, one should consult the standards, e.g., ANSI/ASME B 1.13M, for manufacturing allowances [5].

Some problems have occurred in recent years with mechanically galvanized fasteners. The male thread dimensions had been reduced rather drastically, to accommodate the coating, and that resulted in a significant loss of tooth strength. This is the kind of problem one sometimes encounters when using low-cost suppliers.

E. Tolerances for Abnormal Lengths of Engagement

The allowances and tolerances specified in thread standards assume normal lengths of engagement between male and female threads. The definition of "normal" is spelled out in the specifications, but, typically, it means lengths of engagement ranging from one to one and a half times the nominal diameter of the thread. Lengths for fine-pitch threads are alternately given in number of pitches, with a range of 5 to 15 being considered normal [2].

If the length of engagement is to be abnormally short, then it's wise to reduce the clearance between male and female threads by reducing the allowance or the tolerance. If the engagement is to be unusually long, tolerances must be relaxed or pitch mismatch may make it impossible to assemble the fastener or to run the bolt into a deep, tapped hole.

The standards, again, define the procedures for modifying the allowance or tolerance. ANSI/ASME B1.13M is especially clear on this point. It says that for very short lengths of engagement the tolerance on the pitch diameter of the external thread should be reduced by one number. For example, instead of 4g6g one might specify 3g6g.

For extra long lengths of engagement B1.13M says that the allowance on the pitch diameter should be increased. A normal 4g6g would become 5g6g, for example.

V. INSPECTION LEVELS

Several levels of inspection have been defined for threads, depending upon the nature of the application and the consequences of failure. There is considerable debate at present about which fasteners should be required to pass which tests. We'll take a much longer look at this subject, and the broader subject of thread/fastener strength, later on in the chapter. Since inspection levels are sometimes tacked onto thread designations, it's useful to review them briefly at this point [23].

Level 21 is the least rigorous and is designed to guarantee functional assembly of male onto female threads, plus functional size control of maximum material limits. This level is used with most fasteners at present. The inspection can be performed with fixed GO, NO GO gages.

Level 21A is similar but is used only for metric threads.

Level 22 controls the above and also controls the minimum material size limits over the full length of engagement of the thread.

Level 23 controls all of the above and also controls, within established max-min limits, such things as thread flank angles, lead, taper, and roundness. This kind of inspection can be performed only with indicating gages or optical comparators or other devices which allow the inspector to measure all such parameters.

VI. THREAD NOMENCLATURE

We can now put all of the above together to give the complete alphanumeric code or description of a thread.

A. Inch Series

An example of an inch series external (bolt) thread "code" would be

$\frac{1}{4}$–20 UNC 1A (21)

$\frac{1}{4}$ = nominal diameter in inches

20 = number of threads per inch

UNC shows that this is a UN thread from the Coarse series.

1A shows that this is a loose fitting, external thread (A) with a finite allowance and a maximum tolerance on both pitch and major diameters. (Class 1)

21 shows that the thread is to be inspected with simple GO, NO GO gages.

If that thread were used on a bolt with a 1-in.-long body, the code used to define the fastener would be

$\frac{1}{4}$–20 × 1

Coarseness and fit would not usually be added to the fastener code. The number of threads per inch gives the user coarseness information (a quarter-inch UNF fastener has 28 threads per inch; a quarter-inch UNEF one has 32). A fit of 2A and inspection level 21 would presumably be assumed for such a bolt.

Another example would be

0.2500–32 UNJEF 3A, Safety Critical Thread

Here we see the quarter-inch nominal diameter of the external thread given in decimal form followed by the number of threads per inch, 32; the series, UNJ Extra Fine; and the allowance and tolerance level, 3A. We're also told that this fastener is intended for "safety critical applications." That statement defines the quality control and gaging procedures used with it, namely level 23.

B. Metric Thread

An example of a complete code for an external metric thread would be

MJ6 × 1-4h6h

M shows that this is a metric thread.
J shows that the teeth have rounded roots with larger than standard radii.
6 = nominal diameter in mm.
1 = distance between successive thread crests (i.e., the pitch) in mm.
4h = the tolerance grade (4) and tolerance position (h) for the pitch diameter of the thread. (Position h specifies zero allowance; Grade 4 is used for normal applications.)
6h = the tolerance grade and position for the major diameter. (Again h signifies zero allowance; Grade 6 is also used for normal applications.)

VII. COARSE VS. FINE VS. CONSTANT-PITCH THREADS

Which is best, coarse-pitch, fine-pitch, or constant-pitch threads? It depends on your application. Each has advantages over the other [2–4, 8, 9].

A. Coarse-Pitch Threads

Coarse pitch is generally recommended for routine applications. Such threads will have greater stripping strengths when used with weak nut or joint materials—or when used on larger-diameter fasteners. Some say bolts over 1 in. in diameter should always have coarse threads; others put the crossover point at $1\frac{1}{2}$ in.

It's easier to tap brittle material if coarse-pitch threads are used. Such threads are also easier to use in most cases: easier to start, faster rundown, etc.

B. Fine-Pitch Threads

Fine-pitch threads must be close fitting—made to Class 3 tolerances—to have acceptable stripping strength, but if this is done the bolts these threads are used on can have higher tensile strengths because the thread root and pitch diameters—and therefore the tensile stress area, A_S—are greater than they would be for a coarse-pitch thread on the same nominal diameter. This advantage can be obtained, however, only with a suitably long length of engagement between male and female threads. We'll study this subject later on in this chapter.

Fine-pitch threads are stronger in torsion, which means that they can be loaded to higher preloads before yielding. They also resist self-loosening under vibration or shock, and resist stress corrosion cracking, better than do coarse pitch threads.

C. Constant-Pitch Threads

Constant-pitch threads are designed for applications where there will be repeated assembly and disassembly and/or where it may be necessary to rethread the part in service. They're used for adjusting collars, for thin nuts or threaded sleeves on shafts. They're also used in the design of compact parts [2, 4].

The 8-thread series is used on large-diameter fasteners and was originally intended for bolts used in gasketed joints containing high pressure. It's also widely used as a substitute for coarse series fasteners when the basic fastener diameter exceeds 1 in.

The 12-thread series is used as a continuation of the fine thread series when bolt diameters exceed $1\frac{1}{2}$ inch. It was also originally intended for pressure vessels but has now found wider use.

The 16-thread series is also used on large-diameter fasteners, again for those requiring fine-pitch threads. It's used as well for adjusting collars and as a continuation of the extra fine pitch series for bolt diameters over $1\frac{1}{16}$ inch.

D. Miscellaneous Factors Affecting Choice

We'll see other thread characteristics which may affect our choice of thread as we proceed through the book, but a few miscellaneous comments may be in order here.

A tighter fit, i.e., 3A vs. 2A, gives a 10% increase in thread stripping strength, because there's more root cross section to be sheared. The rounded roots of the J profile will increase the strength still further.

The UNJ or MJ threads also have more resistance to fatigue than do the UN/UNR or M threads [21].

Threads tend to strip before the bolt breaks if the male-female fit is loose [10].

The number of threads in the grip (between the face of the nut and the head of the bolt) affects the ductility and stiffness of the fastener. Since we (usually) want ductility and low stiffness (a more resilient spring for better energy storage) it would seem that we'd usually want fully threaded fasteners. We'll be especially interested in ductility if using yield control to tighten the fasteners. (See Chap. 8.)

Factors like the shear strength of the fastener and its fit with its hole, however, often argue instead for partial threads and an unthreaded body of nominal or reduced diameter.

VIII. THE STRENGTH OF THREADS

There's a surprising amount of disagreement on what parameters determine the strength of a thread and on how best to evaluate the quality—including the strength—of a threaded fastener before use. Let's take a look at some conventional wisdom concerning thread strength and then look at some recent thoughts and concerns about thread strength and quality.

A. Basic Considerations

As we saw in Chap. 1, one of our principal design goals is "a fastener strong enough to support the maximum preload it might receive during assembly, plus the maximum additional loads it might see in service, as a result of forces applied to the joint, differential thermal expansion, etc." The larger the nominal diameter of a fastener, of course, the stronger it will be. As far as static loads are concerned, therefore, we'd like the shank or body of the bolt to be the full, basic, or nominal diameter of the thread, or at least to be greater than the root diameter of the threads [2].

We must then specify a length of thread engagement capable of developing the full strength of that body. This is just another way of saying that we want the body to break before the threads strip, because a broken bolt is easier to detect than a stripped thread.

When the threads strip they do so by shearing in one of three ways. If the nut material is stronger than the material from which the bolt is made, the threads will strip at the roots of the bolt teeth. If the bolt material is stronger, stripping will occur at the roots of the nut threads. If the materials have equal strengths, both nut and bolt threads will strip simultaneously, at their pitch diameters.

Studies made at the National Bureau of Standards many years ago showed that the tensile/shear strength ratio for common fastener materials varied from 1.7 to 2.0. As a result, the stripping areas (A_{TS}) defined in the formulas below (on which the recommended lengths of thread engagement are based) are set at twice the tensile stress area (A_S) of the same thread [5].

If the fastener is to be subjected to fatigue or impact loads, we'd like it to be more resilient than a fastener subjected to static loads. Some recommend a shank (body) diameter about 60% of that used for static loads if the fastener will see impact loads, or a shank diameter of 90% of the static diameter if it will experience fatigue loads (repeated load cycles) [2].

B. The Static Shear Strength of a Thread

We use a very simple equation to estimate the force required to strip (shear) the threads of a bolt or nut:

$$F = S_U A_{TS} \tag{3.1}$$

where S_U = ultimate shear strength of the nut or bolt materials (psi, MPa)

       A_{TS} = cross-sectional area through which the shear occurs (in.^{2}, mm^2) (This is not the cross-sectional area of the body, as we'll see in a minute.)

As mentioned earlier, thread failure will occur in either the nut or bolt threads—or in both simultaneously—depending on the relative strengths of the nut and bolt materials. A different expression must be used to compute the shear stress area for each type of failure [11].

The following equations were taken for the first edition of this book from ANSI B1.1-1974 [11]. Modified versions of the equations for shear areas, A_{TS}, can be found in the current edition of B1.1 [2] but the equations for length of engagement, L_e, have been removed. All of the equations, however, can be found in the current FED-STD-H28/2B [3] but with modified nomenclature.

Simplified formulas for shear areas are also given in H28 and have been included below.

Those who wish to convert these equations to the present nomenclature can use Table 3.3.

Table 3.3 Old vs. New notation Used in
Screw Thread Formulas

Old notation (Used in this book)	New notation
D_S	d
E_S	d_2
K_n	D_1
E_n	D_2
L_e	LE
A_{TS}	AS_S for bolts
	AS_n for nuts

C. Nut Material Stronger than Bolt Material

Failure occurs at the root of bolt threads. The equations for shear area (A_{TS}) and the length of thread engagement (L_e in inches or millimeters) required to develop full strength of the threads are as follows:

$$A_{TS} = \pi n L_e K_{nmax}\left[\frac{1}{2n} + 0.57735(E_{Smin} - K_{nmax})\right] \tag{3.2}$$

$$L_e = \frac{2A_S}{\pi n K_{nmax}[(1/2n) + 0.57735(E_{Smin} - K_{nmax})]} \tag{3.3}$$

where A_{TS} = shear area at root of bolt threads (in.^{2}, mm^2)
n = number of threads per inch
A_S = tensile stress area of bolt (in.^{2}, mm^2)
K_{nmax} = maximum inner diameter (ID) of nut (in., mm)
E_{Smin} = minimum pitch diameter (PD) of bolt (in., mm)

As mentioned above, FED-STD-H28 gives simplified expressions for shear areas. Here's the expression used when the nut material is stronger than the bolt material:

$$A_{TS} = \pi E_S \frac{5L_e}{8} \tag{3.4}$$

where E_S = the basic (or nominal) pitch diameter of the external thread (in., mm) (see comments below).

Let's do an example to see how much difference we get when using the simple vs. the more complex expression and to see where to find the data required to use these equations. Incidentally, most of these equations—and many of the others found in this text—have been repeated in calculator/computer format in Appendix H for your convenience in this electronic age.

We'll take, as our example, a $\frac{3}{4}$–12 UN Class 2A thread and compute A_{TS} for a length of engagement equal to one diameter (the thickness of a heavy hex nut). We get the necessary data from a recent edition of either ASME B1.1 or *Machinery's Handbook* [24]. These tell us that:

$K_{nmax} = 0.678$ in.

$E_{Smin} = 0.6887$ in.

$E_S = 0.6959$ in. (see note below)

Note that the basic or nominal pitch diameter of the bolt is equal to the minimum pitch diameter of the internal (nut) thread, because a zero allowance is assigned to the internal thread which, therefore, conforms to the basic profile at the pitch line and elsewhere.

From Eq. (3.2) in calculator format:

$$A_{TS} = 3.14159 \times 12 \times 0.75 \times 0.678 \times [1/(2 \times 12)$$
$$+ 0.57735(0.6887 - 0.678)] = 0.917 \text{ in.}^2$$

From Eq. (3.4)

$$A_{TS} = 3.14159 \times 0.6959 \times 5 \times 0.75/8 = 1.022 \text{ in.}^2$$

The difference is 12% in this case, with the simplified expression giving a slightly less conservative result (it would take more force to shear 1.022 in.^{2} than 0.917 in.^{2}).

D. Nut Material Weaker than Bolt Material

These are the equations you would use to compute the strength of a tapped hole in a joint material such as aluminum or cast iron. The use of aluminum is increasing in automotive and military applications, for example, as designers struggle to reduce weight. The strengths of various bolt and joint materials will be found in Chap. 4 in Tables 4.1, 4.2, and 4.14.

Failure occurs at the root of nut threads. The equations are

$$A_{TS} = \pi n L_e D_{smin} \left[\frac{1}{2n} + 0.57735(D_{smin} - E_{nmax}) \right] \qquad (3.5)$$

$$L_e = \frac{S_{st}(2A_S)}{S_{nt}\pi n D_{smin}[(1/2n) + 0.57735(D_{smin} - E_{nmax})]} \qquad (3.6)$$

where D_{smin} = minimum OD of bolt threads (in., mm)
 E_{nmax} = maximum PD of nut (in., mm)
 S_{st} = tensile strength of the bolt material (psi, MPa)
 n = threads per inch
 A_S = tensile stress area of bolt (in.^{2}, mm^2)
 S_{nt} = ultimate tensile strength of the nut material
 A_{TS} = shear area of root of nut threads (in.^{2}, mm^2)
 L_e = length of thread engagement required to develop full strength (in., mm)

The simple expression for shear area when the bolt material is stronger than the nut material, again from FED-STD-H28, is

$$A_{TS} = \pi E_n \frac{3L_e}{4} \tag{3.7}$$

where E_n = the basic (minimum) pitch diameter of the nut (in, mm).

E.   Nut and Bolt of Equal-Strength Materials

Failure occurs simultaneously in both parts, at the pitch line. The equations are

$$A_{TS} = \pi E_S \frac{L_e}{2} \tag{3.8}$$

$$L_e = \frac{4A_S}{\pi E_S} \tag{3.9}$$

where E_S = nominal pitch diameter of the bolt (in., mm)
 A_S = tensile stress area of bolt (in.^{2}, mm^2)
 A_{TS} = shear area at pitch line of both threads (in.^{2}, mm^2)
 L_e = length of thread engagement required to develop full strength (in., mm)

F.   Things Which Modify the Static Strength of Threads

We saw earlier that there are a number of factors which can modify the anticipated tensile strength of a bolt—such things as high temperature, corrosion, torsion, or cyclic loading. These things can also modify the strength of threads. So can some other factors which aren't quite as obvious. For example:

Nut dilation [12, 13]. If the walls of the nut are not thick enough, the wedging action of the threads will dilate the nut, partially extracting the nut threads from the bolt threads. This reduces thread engagement and therefore reduces the cross-sectional areas which support the shear load, reducing shear strength. If the ratio between width across flats and nominal diameter is only 1.4:1, for example, strength will be reduced by 25% as shown in Fig. 3.7. (The across flats-to-diameter ratio of standard nuts is 1.5 or 1.6:1.) Note that the reduction applies to both nut or bolt threads, the failure occurring in the weaker of the two.

Relative strength of nut-to-bolt threads [14–16]. As we have seen, the relative strength always determines which members will fail. If there is too big a difference between the two materials, another factor must be considered: The weaker of the two threads will deflect under the relatively stiff action of the other, creating a form of thread disengagement that again reduces the area supporting shear stress. Note that it doesn't matter which thread—nut of bolt—is substantially weaker than the other. The result is shown in Fig. 3.8.

Coefficient of friction. If the coefficient of friction between nut and bolt threads is too low, then both nut dilation and thread bending become

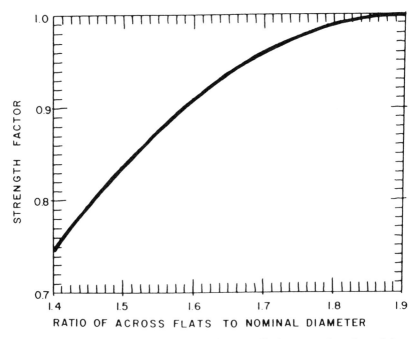

Figure 3.7 Strength reduction factor for nut dilation, as a function of the ratio of the across-the-flats distance to the nominal diameter of the fastener. (Modified from Ref. 13.)

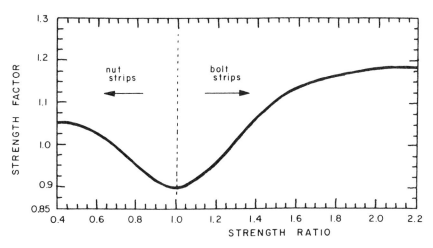

Figure 3.8 Strength reduction factor for thread bending. The horizontal axis gives the ratio of nut strength to bolt strength. (Modified from Ref. 14.)

more likely because the threads can pull apart more readily. A lubricant such as phosphate and oil, for example, is said to reduce resistance to thread stripping by as much as 10% [14].

Rotary motion [14, 17]. Dynamic friction is usually less than static friction. As we have seen above, anything which reduces friction between nut and bolt threads makes it easier for the nut to dilate and/or for the threads to bend. This means that the threads are a little more likely to strip during torquing operations when the nut is moving relative to the bolt than they are under static loads. The reduction in strength is estimated to be approximately 5%.

To compute the modified potential strength of a nut thread, therefore, one multiplies the apparent strength in pounds by the appropriate nut dilation factor from Fig. 3.7 and by the appropriate thread bending factor (for nuts) from Fig. 3.8. If the threads are lubricated, the computed strength should be reduced by an additional 10%; if torque is used to tighten the nuts, a final 5% reduction is required.

Similar calculations are used to estimate the strength of the bolt threads, the only difference being that the thread bending factor used (from Fig. 3.8) will be that for bolts rather than for nuts. As an example, let's compute the strength of the threads for the $\frac{3}{4}$–12 UN, 2A bolt whose thread stripping area we computed a moment ago to be 0.917 in.^{2} (using the long expression for A_{TS}). Let's assume that because of space limitations we're using a nut with a ratio of width across flats to nominal diameter of 1.45:1, a little less than normal.

We used the thread-stripping area formula for "nut material stronger than bolt material." Let's now assume that the nut material is 25% stronger than the bolt material, with a shear strength of 120 ksi. We're going to use it with an ASTM A490 bolt whose shear strength is 96 ksi (Table 4.2).

The bolts are to be lubricated with molydisulfide, an even better thread lube than phosphate and oil (see Table 7.1) and they'll be tightened with a torque wrench.

We use Eq. (3.1) to compute the theoretical force required to strip the threads.

$$F = S_U \times A_{TS}$$

where $S_U = 96$ ksi
$A_{TS} = 0.917$ in.
$F = 88,030$ lb

Now we apply the strength reduction factors as follows:

SR1 = strength reduction factor for nut dilation for a 1.45:1 ratio = 0.8 (from Fig. 3.7).

SR2 = strength reduction factor when the nut material is 25% stronger = 1.1 (from Fig. 3.8)

SR3 = coefficient of friction factor. Let's assume 15% loss of strength (which is probably conservative) because moly is almost 50% more lubricious than is phos-oil (see Table 7.1). So SR3 = 0.85.

SR4 = rotary motion factor, a loss of 5%; so SR4 = 0.95.

The reduced estimate for the strength of our threads is now:

$$F' = F \times SR1 \times SR2 \times SR3 \times SR4$$

$$F' = 88,030 \times 0.8 \times 1.1 \times 0.85 \times 0.95 = 66,550 \text{ lb}$$

(3.10)

A significant difference! Even if we assume a 10% reduction for lubricity instead of 15%, the reestimated strength is only 66,230 lb. However, the ultimate tensile strength of a $\frac{3}{4}$-12 A490 bolt is only 59,670 lb (found by multiplying A_S by the ultimate tensile strength of 170 ksi max from Table 4.1) so the bolt would presumably break before these threads stripped, which is desirable. Not much margin for safety, however.

Is this analysis valid? The reduction factors we've just used come from studies made by E. M. Alexander for the SAE [14] and the results, commonly called the "Alexander model," have been widely accepted and used. Conclusion: the estimate we've just made is valid. If we needed more strength than that implied by the results we should increase the length of engagement between the bolt and our abnormally thin-walled nut.

G. Which is Usually Stronger—Nut or Bolt?

You will find that the proof strength of a standard nut is generally greater than the proof strength of the fastener with which it is supposed to be used. Designers would prefer bolt failure to nut failure because a failure of the bolt is more obvious. For example, the amount of torque we can apply to a bolt with a stripped thread is often greater than that we had applied just before stripping occurred. The increase torque indicates an increase in tension or preload in the bolt, when, in fact, all preload is lost when the thread fails. On the other hand, there's no chance of misreading the situation when the body of a bolt breaks; that's obvious. In an apparent contradiction, nuts are made of weaker (softer) materials than bolts. This encourages plastic yielding in nut threads to bring more threads into play in supporting the load. But the nut as a body will still withstand a higher tensile force than the mating bolt. For the same reason designers will want tapped holes to be deep enough to more than support the full strength of the bolt. Equations (3.3), (3.6), and (3.9) will lead to this result. In fact, most people will use the equations only to find the length of tapped holes. They won't design nuts.

Note that standard nuts come in several configurations. As far as hex nuts are concerned, a regular hex nut has a thread length equal to 0.875 times the nominal diameter of the bolt. Thick and heavy hex nuts have a length equal to the nominal diameter. All three should be able to develop the full strength of the bolt with varying factors of safety, but there can be problems. You'll find a further discussion, and some recommendations on which nut to use, in Chap. 4, Table 4.3.

H. Table of Tensile Stress and Shear Areas

Equations are nice, but it's often handy to have a table of "answers." Table 3.4 gives a summary of the various stress areas we've discussed: tensile stress (A_S), thread root (A_r), and the stripping areas for external threads $[A_{TS}$ from Eq. (3.2)]. The stripping areas for internal threads are 1.3 to 1.5 times those shown in the table, so using the tabulated values gives an added factor of safety. Remember, we want the bolt, not the nut, to fail.

As mentioned earlier, a more complete table, but of tensile stress areas only, will be found in Appendix F.

Table 3.4 Thread and Bolt Stress Areas [11, 18] (all areas are in square inches)

Thread	Tensile areas		Stripping areas	
	A_s (stress)	A_r (root)	Class 2A thread	Class 3A thread
$\frac{1}{4}$–20	0.0318	0.0269	0.092	0.096
$\frac{5}{16}$–18	0.0524	0.0454	0.147	0.157
$\frac{3}{8}$–16	0.0775	0.0678	0.216	0.232
$\frac{7}{16}$–14	0.1063	0.0933	0.296	0.321
$\frac{1}{2}$–13	0.142	0.126	0.390	0.427
$\frac{9}{16}$–12	0.182	0.162	0.502	0.548
$\frac{5}{8}$–11	0.226	0.202	0.624	0.681
$\frac{3}{4}$–10	0.334	0.302	0.908	1.01
$\frac{7}{8}$–9	0.462	0.419	1.25	1.38
1–8	0.606	0.551	1.66	1.82
$1\frac{1}{8}$–8	0.790	0.728	2.13	2.329
$1\frac{1}{4}$–8	0.969	0.890	2.65	2.913
$1\frac{3}{8}$–8	1.23	1.16	3.22	3.55
$1\frac{1}{2}$–8	1.49	1.41	3.86	4.26
$1\frac{5}{8}$–8	1.78	1.68	4.55	5.04
$1\frac{3}{4}$–8	2.08	1.98	5.30	5.03
$1\frac{7}{8}$–8	2.41	2.30	6.09	6.81
2–8	2.50	2.30	6.96	7.72
$2\frac{1}{4}$–8	3.56	3.42	8.84	9.83
$2\frac{1}{2}$–8	4.44	4.29	10.95	12.18
$2\frac{3}{4}$–8	5.43	5.26	13.28	14.80
3–8	6.51	6.32	15.84	17.67
$3\frac{1}{4}$–8	7.69	7.49	18.62	20.8
$3\frac{1}{2}$–8	8.96	8.75	21.63	24.15
$3\frac{3}{4}$–8	10.34	10.11	24.79	27.79
4–8	11.18	11.57	28.28	31.64

In Table 3.4 the thread-stripping areas are given in square inches for a length of thread engagement equal to one nominal diameter of the bolt (the common length for thick or heavy hex nuts).

IX. OTHER FACTORS AFFECTING STRENGTH

So much for the conventional wisdom concerning thread strength. The formulas we've looked at were all based on the assumption that the threads would be manufactured within tolerances specified by ANSI/ASME B1.1

or equivalent. Recent aerospace and other experience has suggested that this may not be enough, that various thread distortions can be produced during manufacture—and may have a significant effect on the thread's strength and performance. The issue is currently being debated, and several research projects to resolve it are under way or planned at this time (mid-1995). In any event, here are some of the factors whose importance is currently being studied.

A. Pitch Diameter

If thread geometry gets too far away from that defined by the ASME B1.1 standard, the equations of this chapter no longer work [19]. And small differences in thread dimensions, angle, etc. may make a significant difference in thread strength [20]. The pitch diameter of a 0.625–18 UNF-3A/3B thread, for example, is a nominal 0.5889 in. ASME B1.1 allows a tolerance of $+0$, -0.0035 in. on that pitch diameter. Assuming that this bolt is used with a nut having a nominal pitch diameter (also 0.5889 in.), a bolt having the minimum pitch diameter allowed by the specification would have 16% less thread strength than a bolt with a nominal pitch diameter.

If the pitch diameter of that bolt is 10 mils less than nominal, it will have less than half the rated strength; at 20 mils, it will have only a quarter of its rate strength; all, still, if used with a nut having nominal PD.

A similar loss of strength occurs if the pitch diameter of the nut is greater than nominal. If both nut and bolt are wrong—the nut being too large and the bolt too small—the loss of strength can be almost complete.

B. Other Thread Parameters

Pitch diameter is not the only geometrical factor we must be concerned about. Anything which reduces the amount of contact between male and female threads will affect their strength. If the bolt or nut is slightly tapered, for example, the threads will be partially disengaged at one end of the engagement length. If the threads on either are slightly out of round, they will not be fully engaged during a portion of each turn. If the pitch of the male threads differs from that of the female threads, they will be in engagement only over a portion of the engagement length. If the helix angle of nut or bolt is irregular, we can get a condition called a "drunken" thread. If the flank (included) angle of the teeth is too great or too small, we will get improper engagement. Some of these problems are illustrated in Figs. 3.9 through 3.11. All of them, again, may cause a significant loss of strength in the threads: but studies are needed to confirm this.

Joint Failure

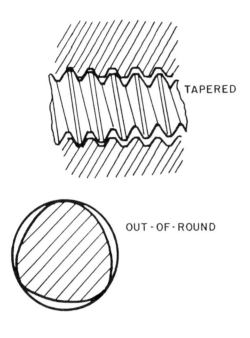

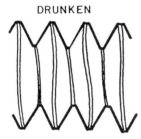

Figure 3.9 Tapered out-of-round or drunken threads all reduce thread-stripping strength.

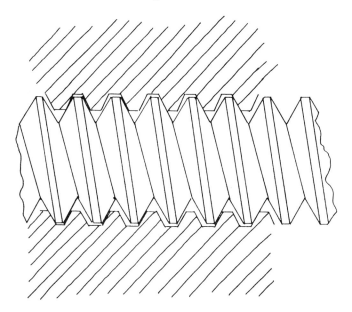

Figure 3.10 If the pitch of the male threads differs significantly from that of the female threads, they may be in contact for only part of the length of engagement.

These problems, incidentally, have been more common in recent years, again thanks to manufacturers of low-cost bolts.

Note that simple GO and NO GO thread gages will not catch such problems as incorrect flank angle or incorrect pitch diameter. They really only check basic root and nominal diameter, so a bolt can pass such gages and still have very little thread strength. "Indicating gages" are available, however, which check all of the necessary geometrical factors.

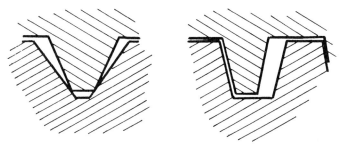

Figure 3.11 Incorrect tooth angles can also result in improper engagement and loss of thread strength.

REFERENCES

1. Sharman, J. M., Threaded fastener selection, *Engineering* (UK), October 1975.
2. Unified Inch Screw Threads UN and UNR Thread Forms, ASME Standard B1.1-1989, ASME, New York, 1989.
3. Screw Thread Standards for Federal Services—Section 2, Unified Inch Screw Threads—UN and UNR Forms, FED-STD-H28/2B, Federal Supply Services, General Services Administration, Philadelphia, August 20, 1991.
4. Screw Threads, Controlled Radius Root with Increased Minor Diameter, General Specification for, MIL-S-8879C, U.S. Government Printing Office, Washington, DC, July 25, 1991.
5. Metric Screw Threads—M Profile, ANSI/ASME B1.13M-1983, ASME, New York, 1983.
6. Metric Screw Threads—MJ Profile, ANSI B1.21M-1978 (Reaffirmed 1991), ASME, New York, 1991.
7. Nomenclature, Definitions, and Letter Symbols for Screw Threads, ANSI/ASME B1.7M-1984, Reaffirmed 1992, ASME, New York, 1992.
8. Thread forms and torque systems boost reliability of bolted joints, *Product Engineering*, December 1977, p. 37ff.
9. Waltermire, W. G., A fresh look at a basic question: coarse or fine threads? *Machine Design*, March 17, 1960.
10. Comments made by Richard T. Barrett of NASA Lewis Research Center at a bolting conference jointly sponsored by the Industrial Fasteners Institute and the Federal Highway Commission at the NASA Lewis Research Center, Cleveland, November 4, 1992.
11. Unified Inch Screw Threads, ANSI Standard B1.1-1974, ASME, New York, 1974.
12. Ellison, H. W., *Effect of Nut Geometry on Nut Strength*, General Motors Corp., Warren, MI, 1970.
13. Formula for calculating the stripping strength of internal threads in steel, Report to ISO/TCI/WG4 by Sweden-Bultfabrike, AB, 1975.
14. Alexander, E. M., Design and strength of screw threads, Trans. of Conf. on Metric Mechanical Fasteners co-sponsored by ANSI, ASME, ASTM, and SAE, Presented at American National Metric Council Conference, Washington, DC, 1975.
15. Gill, P., *The Static Strength of Screw Threads*, G. K. N., UK.
16. Parisen, J. D., *Length of Thread Engagement in Nodular Iron*, General Motors Corp., Warren, MI, 1969.
17. Wiegand, H., and K. H. Illgner, *Boltbarkeit von Schraebenuer-bindungen mit ISO*, Gewindeprofil, Konstruktion, 1967.
18. Blake, Alexander, *What Every Engineer Should Know About Threaded Fasteners*, Marcel Dekker, New York, 1986.
19. Yura, J. A., K. H. Frank, and D. Polyzois, High strength bolts for bridges, PMFSEL Report 87-3, University of Texas, Austin, May 1987.

20. From information received from Stanley P. Johnson of the Johnson Gage Company, Bloomfield, CT, May 1988.
21. McNeill, W., A. Heston, and J. Shuetz, A study of factors entering into the calculation for connecting rod joints according to VDI 2230 guidelines, *VDI Berichte, No. 766*, VDI, Germany, 1989.
22. Metric Screw Threads for Commercial Mechanical Fasteners—Boundary Profile Defined, ANSI B1.18M-1982 (Reaffirmed 1987), ASME, New York, 1987.
23. Screw Thread Gaging Systems for Dimensional Acceptability—Inch and Metric Screw Threads (UN, UNR, UNJ, M, and MJ), ANSI/ASME B1.3M-1986, ASME, New York, 1986.
24. *Machinery's Handbook*, the 14th or any more recent edition, The Industrial Press, New York.
25. ACME Screw Threads, ASME/ANSI B1.5-1988, ASME, New York, 1988.
26. Buttress Inch Screw Threads, ANSI B1.9-1973 (Reaffirmed 1985), ASME, New York, 1985.

4
Materials

In the last two chapters we took an initial look at some of the many ways we could evaluate the strength and integrity of the bolt-nut clamp; and we learned how to calculate the tensile strength of the bolt and the stripping strength of the threads, under static tensile loads. In the discussion it was tacitly assumed that we could always select a bolt material strong enough to fill our needs. Now we're going to look at a number of specific bolting materials to see if that's true.

I. PROPERTIES WHICH AFFECT THE CLAMPING FORCE

What properties are we interested in, when we pick a bolting material? Because the most important purpose of the bolts is to clamp the joint members together, we're interested in any physical or chemical or other properties which affect

the *magnitude* of the clamping force we can create at assembly, and
the *stability* of that clamping force. How will that force be modified by use or age or temperature change or some other mechanism?

From an energy standpoint we want to know how much potential energy we can store in the bolt and how much will be retained by it when it's put to work. The first consideration is related to the magnitude of the force, the second to its stability.

A. Magnitude of the Clamping Force

From a bolting material point of view, the magnitude of the initial clamping force will depend primarily on the basic tensile and shear strength of the material. For a given diameter and thread configuration, a stronger material means a stronger bolt. And the stronger the bolt, the greater the clamping force it can produce. We saw how to estimate the strength of a bolt and its threads in the last two chapters. You'll find tabulations of material strengths in this chapter. I don't mean to imply that we'll always tighten bolts to the limit of their strength, but stronger bolts can and usually are tightened to higher tension than weaker ones. Otherwise, we waste the extra money we spend on better materials.

Also note that there are a *lot* of things other than material which affect the clamping force actually achieved at assembly. We'll look at these factors in depth in later chapters. At present we're considering only the clamping capacity of the bolts, the maximum force they could generate if tightened to their full strength.

B. Stability of the Clamping Force

The stability or reliability of the clamping force is a more complex issue, because a number of material properties can affect it, often without affecting the strength of the parts. For example, the clamping force introduced at assembly can be modified by temperature changes, by corrosion, or by external loads on the joint, depending in part on the way the bolt material responds to these things. We'll look at a number of pertinent properties in this chapter.

Most of the ways in which stability can be affected will be discussed in later chapters. I've put all of the relevant material properties in the present chapter, however, because it's easier to pick a material (once you know the properties you're interested in) if all of the necessary information is in one place. Comparisons and trade-offs are easier.

So, properties are discussed in this chapter, and how to use them in subsequent chapters.

As an introduction to the concept of stability, however, here's a brief summary of some of the ways in which the clamping force can be modified by environmental factors and by our choice of bolt material.

Thermal Expansion or Contraction

A change in temperature will change the length of the bolts and the thickness of joint members. Knowing the thermal coefficients of linear expansion will allow us to estimate how much change each part will experience.

If the parts are made from different materials—or are raised to different temperatures—the clamping force on the joint and the tension in the bolts will be modified by differential expansion or contraction. This can increase or decrease the clamping force. It can also break bolts or totally eliminate the tension in them. We'll see how to estimate these changes in Chap. 13 but will look at some related material data in this chapter (Table 4.5).

Corrosion

The resistance of the bolt material to corrosion will determine how long our clamp will survive in the anticipated service environment. The buildup of corrosion products (e.g., rust) can increase clamping forces; additional corrosion can eat through the bolts. We'll consider corrosion mechanisms and stress corrosion cracking in Chap. 18. Some general guidelines will be found in Table 4.10.

Fatigue Rupture

Many materials have an "endurance limit" which, unfortunately, is only a fraction of their apparent (static tensile) strength. If cyclic stress levels are above this endurance limit, the bolt will eventually break and clamping force will be lost. So, the endurance limit is another property we'll be interested in. Fatigue will be the topic of Chap. 17; we'll look at the related material properties here (Table 4.9).

Loss of Strength with Temperature

As already mentioned, a change in temperature can cause a change in clamp force because of differential expansion between bolts and joint members. Temperature can create problems even if the bolts and joint have identical thermal coefficients and identical temperatures, however. The basic strength of the material can be affected enough by high temperature to put the joint in jeopardy. You'll find the necessary data in Table 4.4.

Loss of Clamping Force with Temperature

Elevated temperature can also lead to stress relaxation (discussed in Chap. 13), which can reduce or eliminate the clamping force without any visible or measurable change in the parts. So resistance to stress relaxation is another material property which can affect the integrity of the clamp. Figure 4.3 and Tables 4.7 and 4.8 provide stress relaxation data on a number of fastener materials.

Stress relaxation can and often does take place over an extended period of time. The lower the temperature, the longer it takes the bolt to shed stress. Clamping force can also, however, be lost very rapidly if common bolt materials are subjected to high temperature—during a fire, for example. Typical results are shown in Fig. 4.4.

Elastic Stiffness of the Parts

The modulus of elasticity is another property we'll often be interested in. Modulus, in part, determines the stiffness of bolts and joint members, and stiffness in turn determines how the clamp force introduced at assembly will change when the joint is put in service. Factors like working loads (pressure, weight, shock, etc.), gasket creep, embedment of thread surfaces, elastic interactions between bolts—all to be considered in later chapters—will work to change those initial clamping forces even at room temperatures. The amount of change will depend on the relative stiffness of bolt and joint members (discussed at length in Chap. 5). You'll find some modulus data in the present chapter, in Table 4.6.

Change in Stiffness with Temperature

The modulus of elasticity is also affected by temperature, so the stiffness of bolts and joint members will change as the temperature changes. As one result, a 10% reduction in modulus means a 10% loss of tension in the bolt because it has become a less stiff spring. Note that a reduction in modulus occurs with an increase in temperature, and this increase may cause differential thermal expansion which partially or wholly offsets the loss in stiffness.

Any change in stiffness may also mean a change in the bolt-to-joint stiffness ratio; and that means a change in the way the system responds to external loads. So there are many ways in which a change in temperature can modify the clamping force, with modulus playing several roles.

Table 4.6 includes extreme as well as room temperature data on the modulus of elasticity of many materials.

Brittle Fracture

Ductility can be another important consideration, especially if the bolts are to be tightened past yield (a common practice in structural steel work, as we'll see in Chap. 8). Very hard materials can be very strong—but brittle. The brittleness often leads to unexpected failure at loads below the theoretical strength of the parts. We'll look at some brittle fracture data in this chapter (Fig. 4.2).

C. Miscellaneous Properties

Although things which determine or threaten the clamping force produced by the bolt will always be our main concern, there are times when other material properties must also be considered. Low-weight fasteners, for example, have always been important in aerospace applications and are of growing importance in automotive design as well (lower weight means lower fuel consumption). Some material weights are given in Table 4.13.

The cost of a fastener often influences our choice. Two brief lists of relative costs are included (Tables 4.14 and 4.15).

The electrical or magnetic properties of the fasteners can also be a consideration, but so rarely that we won't worry about them here.

The strength of joint members is usually not as big an issue as the strength of the bolts, but some data are useful and will be found in Table 4.16.

So, there are a number of material properties which will determine the ability of our bolt to clamp things in service, and other properties which will influence our material decisions in special situations. How can we select an appropriate material? The most obvious source of material information on which to base a decision would be an existing fastener "standard."

II. FASTENER STANDARDS

It is believed that some 500,000 fasteners have been defined by standards of some sort. Certainly hundreds of different specifications, recommendations, etc. are available today. The impact of fastener standards on our economy as a whole must be enormous—when you consider the alternative that "everyone designs and builds his own."

Fastener standards are published by several types of organizations, including the following:

Government organizations—for example, the National Institute of Standards and Technology (NIST), the Army, the Navy, the Air Force.
Engineering societies—important fastener standards are published by the Society of Automotive Engineers (SAE), the American Society for Testing and Materials (ASTM), the American Society of Mechanical Engineers (ASME), etc.
Trade associations—for example, the American Bureau of Shipping and the Association of American Railroads publish well-known fastener standards.

Fastener manufacturers—the principal U.S. source is the Industrial Fastener Institute, an association of fastener manufacturers (who publish, among other things, a complete list of other people's standards).

Standards associations—general-purpose groups that publish standards on all sorts of things. Principal ones at the moment are the American National Standards Institute (ANSI) and the International Standards Organization (ISO).

Trade associations, military services, engineering societies, etc. tend, of course, to publish standards affecting fasteners in which they have a special interest. Groups such as NIST, ANSI, and IFI publish standards on all sorts of fasteners.

In general, standards cover such things as fastener materials, mechanical and physical properties, strengths, configurations, dimensions, usage, definitions, finishes, test procedures, grade markings, and manufacturing procedures. This does not mean that every standard covers each of these things—just that standards exist for all of these topics and for others.

The full names and addresses of the organizations from which you can buy standards are given in Appendix C.

III. SELECTING AN APPROPRIATE STANDARD

The purpose of a fastener standard is to define a group of materials and/or fastener configurations which are appropriate for the "typical" needs of a particular industry or a particular class of applications. The standard then makes it unnecessary for each engineer to be a metallurgist when trying to determine what would be appropriate in his application. This saves a great deal of time and money. Standards also reduce product and inventory costs, control quality, enhance product and system safety, and do other important things. For our purposes, we're interested in them as a source of material information.

If you work for an automobile manufacturer, your first choice of standards is simple—you try those prepared by the SAE. If you're involved with pressure vessels, you'll be guided by the ASME, etc.

What do you do if your own industry has not produced a set of bolting standards? There are two readily available sets that are widely used by "miscellaneous" industries and designers. I'm sure that the most commonly used bolting standard, whether users realize they're using it or not, is SAE's J429, which defines automotive Grades 1 through 8. Bolts of these materials are made in large quantities and are therefore relatively inexpensive as specified or standardized bolts go. They're readily available, and they cover a wide range of strength specifications, with tensile strengths ranging from 60 ksi (414 MPa) for the cheaper materials to 150 ksi (1034 MPa) for the more expensive Grade 8's. They're available in sizes ranging from $\frac{1}{4}$ to $1\frac{1}{2}$ in. in diameter. Therefore, they're widely used for small and medium-sized bolting jobs.

The second most commonly used bolting standards may be those published by the ASTM. (They may be the most common.) Some 40 ASTM standards are available, covering a wide range of threaded fastener materials, sizes, special applications, etc. You can buy individual standards or buy a bound volume containing all of them. These standards will be your first choice if you're working with large equipment or systems—structures, pressure vessels, power plants, and the like.

In most situations your needs will be covered by fasteners defined by either the ASTM or the SAE. In every industry, however, including the automotive, there are special applications where something "better" is required. And in some industries, such as aerospace, the unusual is usual. The tables and descriptions which follow are intended to help you identify available fasteners which are not normally used in your own industry, whether or not these are common in someone else's industry.

I'll start with an apology. I had hoped to be able to list every property of interest for every material listed—tensile, yield, and proof strengths; thermal coefficients of expansion; endurance limits; and all the rest. But this information is just not available. Different industries face different problems and specify only those properties of interest to them. If the shear strength of an A193 B7 bolt isn't of interest to the "normal" user, then no one is going to specify or report it. In spite of this lack of complete data, I think you'll usually find what you need—or an approximation of what you need—in the tabulated material.

Before we get to the data, however, it is useful to discuss what is still a "hot topic" in some parts of the bolting world (in spite of the antiquity of the topic!)—the issue of metric fasteners vs. English (U.S. or Unified or "inch series") fasteners. The tables in this chapter include data on both kinds, where available. But, generically, should you be interested in inch series fasteners, or in metric?

Automotive manufacturers in the United States have "gone metric," as have some other manufacturers who depend a great deal on exports, but most manufacturers here are still using English or inch series fasteners. For those facing a decision, the choice will probably be based on economic, political, or marketing considerations. From a purely technical point of view, inch and metric fasteners are available in the same general strengths, with the same general properties, etc. Currently specified metric fasteners have, however, been designed and/or specified more recently than most of their English counterparts. Past experience has revealed some problems with English materials and geometry; the latest metric standards have attempted, at least, to overcome these shortcomings. So metric fasteners may be slightly "better" technically; but the differences are slight.

Note that the decision to choose metric involves two things: fastener material and fastener configuration (or at least basic dimensions). We're primarily interested in materials in this chapter (and in this book). We'll discuss metric fasteners in Sec. VII of this chapter.

IV. MATERIAL PROPERTY NOMENCLATURE

The "code letters" used to identify different material properties vary a bit from one bolting standard or trade organization to another. Since this can confuse the beginner, a brief discussion is in order.

An uppercase S is often used to signify stress. In the ASME Boiler and Pressure Vessel Code, for example, Sa is used for the allowable bolt stress at room temperature. But other letters are used elsewhere. For example, each of the following is used to designate the ultimate tensile strength (in terms of stress in ksi or MPa) of bolts.

UTS is used, commonly, in bolting literature.
Ftu is used in MIL-Handbook 5.
St is used in the thread standards ASME B1.1 and ANSI/ASME B1.13M.

The use of uppercase F for stress is apparently common, or becoming common, with the U.S. government; I've encountered it several times while working on the third edition of this text. Here are some of the other F's I've run across [63] [64].

Fsu = ultimate shear stress
Fbu = ultimate bearing stress
Fby = yield bearing stress
Fty = tensile yield stress
Fcy = compressive yield stress

V. BOLTING MATERIALS

Soon we'll look at some specific properties of materials. But first, a word of caution: *The data which follows is for general reference only.* The properties of an individual batch of bolts can differ from the norm. The specific heat treatment used by one manufacturer can differ from that used by another, and this will affect properties. In the tables which follow you'll sometimes find different values for a given property of a given material. The ultimate strength, for example, may differ from table to table. This reflects the fact that the data in the two (or more) tables came from different sources, and these reported different values.

Differences of this sort are especially common for the more exotic materials used in aerospace or extreme-temperature applications. The strength of such materials, as a matter of fact, is often tailored for a specific application.

The diameter of the fastener can also affect its strength and other properties, because some bolting materials cannot be through-hardened. Large diameters don't get fully hardened in the center; smaller diameters do; so large diameters have lower average tensile strengths (in ksi or MPa). Material near the outer diameter of the fastener will support more load than material near the center, so the average strength of the larger fastener is less.

The shape of a fastener can also affect its strength. We're going to assume "normal" ANSI or equivalent shapes in this text, and so will ignore the ramifications of special configurations. But you'll usually recognize an unusual configuration on sight and should be wary of assigning normal properties to it.

Another factor causing variation in the data is the slow but fairly continuous evolution of most bolting standards and specifications. The ASTM and other committees who write these documents meet periodically and alter them whenever new experience, new conditions, previous misunderstandings, etc. suggest that a change would be desirable. Some of the data in the following tables comes from standards published 10 or more years ago. I've reviewed more recent versions of most of these documents and, since I have found only minor changes, I've left the earlier numbers alone. Once again, however, *if your designs will affect life or safety, or if failure will have severe economic consequences, you should take your design data from the current editions of the pertinent documents rather than from a text of this sort.* Use the tables which follow only as a "shopping guide."

The data in the tables were taken from the ASTM or other standards or specifications cited unless otherwise noted by reference numbers. The latter refer to the References at the end of the chapter.

VI. MATERIAL SETS

To start with, we'll consider several groups of fasteners, each group characteristically used by a different industry or in a particular set of applications. These tables will help the beginner fish in the correct pool. However, they are not intended to define *every* fastener material used in those industries or applications, merely representative examples. All of the materials listed informally here will be described more completely in one or more of the data tables which follow.

A. General-Purpose/Automotive Group

The most commonly used fasteners which are described by a specification (unspecified ones are even more common) are probably those defined in Standard J429 published by the Society of Automotive Engineers (SAE). Typical examples include:

| | | Yield strength | |
Spec.	Grade	ksi	MPa
SAE	2	57	393
J429	5	92	634
	8	130	896

B. Structural Engineering Group

These fastener materials are used in buildings, bridges, and other structures. Two of them at least, A325 and A490, are commonly tightened past yield, on purpose. (See Chaps. 8 and 13 for further information.) The specifications listed below are all published by the ASTM.

| | Yield strength | |
Spec.	ksi	MPa
A307	est. 36[a]	est. 248
A325	81	558
A354	99–109	683–752
A449	58–92	400–635
A490	130	896

[a] Estimated as 60% of the specified ultimate strength of 60 ksi.

C. Petrochemical/Power Group

The following materials are commonly used in petrochemical and power plants, as well as in marine, mining, manufacturing, and other industries using heavy equipment. The structural steel group of materials is also commonly found in such applications. Again, the specifications listed below are published by the ASTM. The wide range of yield strengths listed here, for a given material, results from the fact that these fasteners are often made in large sizes—up to several inches in diameter—and the materials aren't always through-hardened.

| | | Yield strength | |
Spec.	Grade	ksi	MPa
A193	B7	75–105	515–720
	B16	85–105	585–720
A540	B21	105–150	720–1034
	B24	105–150	720–1034

D. Extreme-Temperature Group

I consider many of the fasteners in this group "exotic"—they're uncommon and often expensive (see Table 4.4). Note that I've included ASTM A193 materials in this group as well as in the petrochemical group. An aircraft engine designer might not consider the A193 temperature limit of 800°F "high temperature"—but the petrochemical engineer does. See Tables 4.9 and 4.10 for additional materials, some of which can be used at temperatures as high as 3000°F (1648°C) and/or can be used at cryogenic temperatures.

| | | Temperature | | Yield strength[a] | | |
Spec.	Grade	°F	°C	ksi	MPa	Ref.
ASTM A193	B7	752	400	76	524	
	B8	800	427	17	117	1
	B16	800	427	76	524	
BS4882	B80A	1400	760	73	503	3
A286[b]		1200	649	88	607	3
MP35N[b]		−320	−196	345 UTS	2379 UTS	2
		1000	538	225 UTS	1551 UTS	2
H-11		1000	538	141	972	3
INCONEL, X-750[b]		1500	816	44	303	3
WASPALOY		1600	871	75	517	3
RENE 41		1600	871	80	552	3

[a] Values given are yield strengths for the cited temperature except for those values marked UTS, where strength is given as ultimate tensile strength instead of yield.
[b] Also recommended for cryogenic applications.

E. Corrosion-Resistant Group

As discussed in Chap. 18, no material will resist all types of corrosion. But ASTM A193, BS4882, A286, and MP35N, listed above, are all considered "corrosion resistant." Other materials with this reputation include those listed below. See Table 4.12 for a more complete list.

Spec.	Grade	Yield strength		Ref.
		ksi	MPa	
Carpenter Gall-Tough	430	40	276	6
NITRONIC 50		70	483	7
INCONEL	600	37	255	3
AISI	316	45	310	
	416-H	95	655	
TITANIUM	6A1-4V	128	883	3
Stainless St.	17-4PH	128–185	883–1276	9

F. Metric Group

There are probably as many different specifications for metric fasteners as there are specifications for English ones. But, as discussed in Section VI below, most metric standards have adopted a common series of strength designations. The data below, for example, can be found in either ASTM A568 or SAE J1199, among other places.

Class	Yield strength	
	ksi	MPa
4.6	35	240
8.8	96	662
10.9	136	938
12.9	160	1103

So much for fastener groups. Again, these tables are far from complete. They're merely intended to help you find a standard or group of standards which might contain the material best suited to your needs. These and many more materials will be defined in detail in Tables 4.1–4.16. Before getting to those, however, let's take a closer look at metric fastener specifications.

VII. METRIC FASTENERS

The current effort to get countries now using "English" units to adopt the more widespread metric system includes, as it must, a new, international metric standard for fasteners. Note that, heretofore, there hasn't been just one metric standard. Engineers could and did write as many different metric standards as we have English ones. But since most countries have to accept some change in their own standards to comply with a new international standard, the current effort is seen as a new—and maybe last—chance to reduce the vast number of fastener types, sizes, materials, etc., now currently available. Having standards of any sort was a start—there was a time when we didn't even have that. But now we have a chance to simplify things still further. In the long run, such simplification could more than partially offset the cost of changing drawings, tools, inventories, manuals, procedures, etc., as we adopt the new fasteners.

The numbers used to define the metric grades (called *classes*) have useful meaning. The first number is equal to the minimum tensile strength of the material, in megapascals (MPa) divided by 100. The second number represents the approximate ratio between minimum yield and minimum ultimate strengths for the material. Hence, Class 5.8 has a minimum ultimate strength of approximately 500 MPa, and its minimum yield strength is approximately 80% of its minimum ultimate strength.

All this is far more useful than calling out "Grade 5" or "B7" and letting the uninitiated struggle to find out what that means—as we have always done in the past.

When we're dealing with metric fasteners, we'll want to use metric units for such things as torque, stress, force, etc. You'll find conversion factors (English to metric and vice versa) in Appendix E.

VII. EQUIVALENT MATERIALS

A review of Section V will show that the sets or groups of materials favored by different industries usually cover approximately the same range of yield strengths, with the strength of common sets ranging from something like 30 ksi to 105 ksi or so. As a result, you'll often find that a material, or at least a room temperature strength rating, in one group is matched by a similar or equivalent material or strength in one or more other groups. Substitutions are sometimes possible. In critical applications, however, you should look at the original specifications for both materials. In many applications "equivalent strength" isn't the only criterion for selection. Response to a change in temperature or to corrosion or other factors may differ.

But equivalent materials are available. As one example, all of the following define basically material of the same strength.

AISI 4140	ASTM A193 B7
ASTM A194—GR 7	ASTM A320—GR L7
Metric 9.9	SAE J429 GR 5+
BS 970-En 19A	BS 1506-621 A
DIN 267—9.9	ASTM SA193—B7

IX. MATERIAL PROPERTIES

Now we'll look at the specific properties of a number of bolting materials. To get started, here's an index to the tables and graphs which follow. As mentioned earlier, most of the data in these tables came from the ASTM, SAE, or other specifications and standards listed in the tables. If other references were used, or also used, they are cited in the table. Reference numbers refer to the References at the end of the chapter.

A. Properties Affecting Basic Room Temperature Strength of Fastener Materials (Initial Clamping Force)

Properties	Table or figure
Proof, yield, and ultimate strengths	Table 4.1
Hardness	Table 4.1
Yield strength vs. hardness	Fig. 4.1
Shear strengths	Table 4.2
Brittle fracture strength	Fig. 4.2
Which nut to use	Table 4.3

B. Properties Affecting Life or Stability of the Clamping Force

Properties	Table or figure
Yield strengths at elevated temperatures	Table 4.4
Thermal coefficients of expansion	Table 4.5
Modulus of elasticity vs. temperature	Table 4.6
Stress relaxation vs. temperature	Fig. 4.3 and Tables 4.7 and 4.8
Service temperature limits	Table 4.9
Cryogenic bolting materials	Table 4.10
Fatigue endurance limits	Table 4.11
Corrosion resistance	Table 4.12

C. Miscellaneous Properties of Fastener Materials

Properties	Table or figure
Weight of fastener materials	Table 4.13
Relative cost of fastener materials	Tables 4.14 and 4.15

D. Properties of Typical Joint Materials

Properties	Table or figure
Yield, ultimate, and shear strengths	Table 4.16

Strength of Bolting Materials

Now let's look at some data. Table 4.1 gives the basic room temperature strengths of male fastener materials (bolts, studs, screws) under static, tensile loads. This strength is expressed in as many as three different ways—proof, yield, and ultimate tensile—depending on the available data. The information was taken from the specifications cited in the table and/or from the references listed in the final column. (See the References at the end of the chapter for the complete reference.)

Other aspects of the room temperature strength of bolts are covered later: namely, shear and brittle fracture strengths. A chart giving the relationship between yield strength and material hardness is also included. This can help you identify (or estimate the strength) of unmarked or suspect materials.

Note that only male fastener materials are included here. Nut materials are covered in Table 4.3.

The strengths in Table 4.1 are given in ksi, even for the metric materials (to aid comparison). To convert to metric MPa units, multiply the number of ksi by 6.895. For example, a yield strength of 80 ksi = 80 × 6.895 = 552 MPa.

It is sometimes necessary to determine what material an unmarked bolt is made from, or to estimate its strength. A metallurgical analysis is required for a completely accurate answer; but it has been shown that there's a rough correlation between the hardness of a bolt and its yield strength (within reason—see the discussion of brittle fracture strength below). Studies of low-alloy quenched and tempered steels (LAQT steels) for example, sponsored by the Atomic Industrial Forum/Metals Properties Council Joint Task Group on Bolting, resulted in the curve shown in Fig. 4.1. Using this curve, plus the hardness data found in Table 4.1, plus, hopefully, other clues, you may be able to identify an unmarked bolt or stud. Examples of an LAQT steel would be AISI 4140, ASTM A193 B7, ASTM A490, SAE J429 GR. 5, etc.

Since most bolts are used as clamps, not as shear pins, most bolting specifications and standards list only one or more forms of tensile strength (proof, yield, or ultimate) and not shear strengths. Table 4.2, "Shear Strengths," lists those few materials for which I've found published or reported shear strengths.

If the material you're interested in is not included in Table 4.2, it might help to know that most of the common steels we'll use, with hardness to 40 HRC or so, will have shear strengths in the neighborhood of 60% of ultimate tensile strength, or, in MIL-Handbook 5 terms, Fsu = 0.6 Ftu [63]. The stainless steels are an exception to this rule of thumb; they have shear strengths which are about 55% of their ultimate strengths, or Fsu = 0.55 Ftu [63].

Note that these are only rules of thumb and that other sources give other rules. The Unified Inch-Series Thread Standard ASME B1.1-1989, for example, says that the shear strength of threads is half the tensile strength of the material from which the external threaded part is made. This reflects the fact, I suppose, that nut materials are supposed to be slightly weaker than the bolt materials they're used with, or there's a small safety factor introduced here to reduce the chances that the threads will strip.

A still different rule comes from "the" book on structural steel joints [49], which suggests that the shear strength of most common (structural steel) joint materials is 70% of the ultimate strength of those materials.

Another material property we often need when dealing with the shear strength of the bolts is the bearing yield strength of the joint material. We want the bearing yield strength of the joint members in a shear-loaded joint to be less than the shear strength of the bolts. If this is the case, the walls of the bolt holes in the joint plates will yield before the bolts shear. This will usually bring more bolts into bearing, reducing the shear loads on the first bolts to contact the walls of their holes.

As another rule of thumb, the bearing yield strength of common joint materials is about 1.5 times the material's tensile yield strength, or, in MIL-Handbook 5 nomenclature, Fby = 1.5 Fty [63].

The strengths in Table 4.2 are given in ksi. To convert to metric MPa, multiply the value given by 6.895. For example, a shear strength of 132 ksi would be the equivalent of 6.895 × 132 = 910 MPa.

Simplistically, a harder steel is always a stronger steel. A small fastener can do a big fastener's job—saving weight, reducing joint size, etc.—if the small one is hard enough. It's appealing!

Table 4.1 Room Temperature Strength of Fastener Materials (ksi)

Spec.	Grade	Size	Hardness	Strength			Ref.
				Proof	Yield	Ultimate	
ASTM A193	B5	—	—	—	80	100	
	B6	—	—	—	85	110	
	B7	$<2\frac{1}{2}$	—	—	105	125	4
		$2\frac{1}{2}$–4	—	—	95	115	
		>4–7	—	—	75	100	
	B7M	$\leq 2\frac{1}{2}$	94–99 HRB	—	80	100	
	B8–Cl 1		90 HRB–28 HRC	—	30–55	75–100	
	B8–Cl 2		35 HRC	—	50–100	80–125	
	B16	$\leq 2\frac{1}{2}$	25–34 HRC	—	105	125	
		$>2\frac{1}{2}$–4		—	95	110	
		>4–7	—	—	85	100	
ASTM A307	GR A	$\frac{1}{4}$–4	69–100 HRB	—	—	60	4
	GR B	$\frac{1}{4}$–4	69–95 HRB	—	—	60–100	
ASTM A320	L7	$\leq 2\frac{1}{2}$	99 HRB	—	105	125	
	L43	≤ 4	99 HRB	—	105	125	
	L7M	$\leq 2\frac{1}{2}$	99 HRB	—	80	100	
	L1	≤ 1	99 HRB	—	105	125	
	B8 CL 1		96 HRB	—	30	75	
	B8 CL 1A		90 HRB	—	30	75	
	B8 CL 2	$\leq \frac{3}{4}$	35 HRC	—	100	125	
		$>\frac{3}{4}$–1	35 HRC	—	80	100	
		>1–$1\frac{1}{4}$	35 HRC	—	65	105	
		$>1\frac{1}{4}$–$1\frac{1}{2}$	35 HRC	—	50	100	

Specification	Grade	Size (in.)	Hardness				Note
	B8M CL 2	$\leq\frac{3}{4}$	35 HRC	—	95	110	4
		$>\frac{3}{4}-1$	35 HRC	—	80	100	
		$>1-1\frac{1}{4}$	35 HRC	—	65	95	
		$>1\frac{1}{4}-1\frac{1}{2}$	35 HRC	—	50	90	
ASTM A325	Type 1,2,3	$\frac{1}{2}-1$	24–35 HRC	85	92	120	
		$>1-1\frac{1}{2}$	19–31 HRC	74	81	105	
ASTM A354	BC	$\frac{1}{2}-2\frac{1}{2}$	26–36 HRC	105	109	125	
		$2\frac{3}{4}-4$	22–33 HRC	95	99	115	
	BD	$\frac{1}{4}-2\frac{1}{2}$	33–39 HRC	120	130	150	
		$2\frac{3}{4}-4$	31–39 HRC	105	115	140	
ASTM A437 and A437M	B4B	—	HB331	—	105	145	
	B4C	—	HB277	—	85	115	
	B4D	$\leq2\frac{1}{2}$	HB302	—	105	125	4
		$2\frac{1}{2}-4$		—	95	110	
		$4-7$	—	—	85	100	
ASTM A449		$\frac{1}{4}-1$	25–34 HRC	85	92	120	
		$1\frac{1}{8}-1\frac{1}{2}$	19–30 HRC	74	81	105	
		$1\frac{3}{4}-3$	—	55	58	90	
ASTM A453	651	—	93 HRB–29 HRC	—	50–70	95–100	4
	660	—	95 HRB–37 HRC	—	85	130	
	662	—	99 HRB–35 HRC	—	80–85	125–130	
	665	—	32–41 HRC	—	120	155–170	
ASTM A490	1	$\frac{1}{2}-1\frac{1}{2}$	33–38 HRC	117–120	130	150–170	
	2	$\frac{1}{2}-1$	33–38 HRC	117–120	130	150–170	
	3	$\frac{1}{2}-1\frac{1}{2}$	33–38 HRC	117–120	130	150–170	

Table 4.1 Continued

Spec.	Grade	Size	Hardness	Strength			Ref.
				Proof	Yield	Ultimate	
ASTM	B21–B24						4
A540	Cl 1	To 8	321–444 HB	—	150	165	
	Cl 2	To $9\frac{1}{2}$	311–415 HB	—	140	155	
	Cl 3	To $9\frac{1}{2}$	293–380 HB	—	130	145	
	Cl 4	To $9\frac{1}{2}$	269–363 HB	—	120	135	
	Cl 5	To $9\frac{1}{2}$	241–321 HB	—	100	115–120	
	B 24V	To 11	293–444 HB	—	130–150	145–165	
ASTM							
A574 and	—	$\leq \frac{1}{2}$	—	140	—	180	
A574M	—	$\geq \frac{5}{8}$	—	135	—	170	
ASTM							
A687	—	—	—	—	105	150	
ASTM	GR30	—	—	—	30	60	
F432	GR55	—	—	—	55	85	
	GR75	—	—	—	75	100	
ASTM	Copper	—	F30–B100	—	10–55	30–130	
F468 and	Nickel	—	B60–C37	—	30–90	70–180	
F468M	Aluminum	—	B40–B90	—	31–50	37–76	
	Titanium	—	HV140–C36	—	30–125	40–165	

ASTM F593 St. st.	1,2,3	$\frac{1}{4}-1\frac{1}{2}$	B65–C36	—	30–90	75–160	
	4	—	B65–B95	—	35	70–100	
	5	Same	C20–C45	—	90–120	110–190	
	6	Same	C25–C48	—	100–140	125–220	
	7	Same	C28–C38	—	105	135–170	
ASTM F835 and F835M		—	C36–C44	—	—	137–150	
ASTM F837 and F837M		—	B70–C43	—	26–120	70–160	
ASTM F879	AF	—	B65	—	—	86	
	CW	—	B96–C33	—	—	90	
SAE J429	GR 1	$\frac{1}{4}-1\frac{1}{2}$	70–100 HRB	33	36	60	
	GR 2	$\frac{1}{4}-\frac{3}{4}$	80–100 HRB	55	57	74	
		$>\frac{3}{4}-1\frac{1}{2}$	70–100 HRB	33	36	60	4, 10
	GR 4	$\frac{1}{4}-1\frac{1}{2}$	22–32 HRC	65	100	115	
	GR 5	$\frac{1}{4}-1$	25–34 HRC	85	92	120	
		$>1-1\frac{1}{2}$	19–30 HRC	74	81	105	
	GR 7	$\frac{1}{4}-1\frac{1}{2}$	28–34 HRC	105	115	133	
	GR 8	$\frac{1}{4}-1\frac{1}{2}$	33–39 HRC	120	130	150	
	GR 8.1	$\frac{1}{4}-1\frac{1}{2}$	—	120	130	150	
	GR 8.2	$\frac{1}{4}-1$	—	120	130	150	
Stainless steels	303	—	—	—	40–80	80–125	
	410	—	—	—	110	170	11
	431	—	—	—	140	180	

Table 4.1 Continued

Spec.	Grade	Size	Hardness	Proof	Yield	Ultimate	Ref.
				Strength			
	17-4PH	—	—	—	180	200	
	17-7PH	—	—	—	185	200	
	AM 350	—	—	—	162	200	
	PH15-7MO	—	—	—	155	200	
High-performance materials	A286	—	—	—	95–119	214–260	
	Hasteloy-X	—	—	—	52	—	3, 5, 12, 13
	H-11	—	—	—	215	262	
	Inconel 600	—	—	—	37	90–300	
	Inconel 718	—	—	—	260	226	
	Inconel X750	—	—	—	92	162	
	Maraging 300	—	—	—	260	—	
	MP35N	—	—	—	230	286	58
	Nimonic 80A	—	—	—	90	145	
	Rene 41	—	—	—	157	206	
	MP159	—	—	—	250	260	58
	Astroloy	—	—	—	—	190	
	Rene 95	—	—	—	—	230	
	K-Monel 500	—	—	—	—	155	61
Titanium	Ti-6A1-4V	—	—	—	128	200	
	Waspaloy	—	—	—	115	185	

Category	Grade	Size	Hardness				
Copper alloys	Al bronze	—	—	—	—	93	63
	Si bronze	—	—	—	—	93	57
	Copper	—	—	—	—	93	
Fastener grade aluminum	2024-T4	—	—	—	—	68	
	6061-T6	—	—	—	—	45	
	6262-T9	—	—	—	—	58	
	7050-T73	—	—	—	—	75	
	7075-T6	—	—	—	—	83	
	7075-T73	—	—	—	—	73	
Metric materials							
ISO 898/1	4.6	M5-M100	67-95 HRB	33	35	58	
OR	4.8	M1.6-M16	71-95 HRB	45	49	61	
ASTM	5.8	M5-M24	82-95 HRB	55	61	75	
F568	8.8	M16-M72	23-34 HRC	87	96	120	
OR	9.8	M1.6-M16	27-36 HRC	94	104	131	
SAE	10.9	M5-M100	33-39 HRC	120	136	151	
J1199	12.9	M1.6-M100	38-44 HRC	141	160	177	
ASTM A325M	1,2,3	M16-M36	23-34 HRC	87	96	120	
ASTM A490M	1,2,3	M12-M36	33-39 HRC	120	136	151	
ASTM F738M	—	—	B81-C48	—	30-138	73-174	
ASTM	Al-50	—	B88	—	—	84	
F879M	Al-70	M3-M14	B96-C33	—	—	76	
St. St.	Al-70	M16-M20	B83-C30	—	—	64	

Sources: ASTM specifications A193, A307, A320, A325, A354, A449, A453, A325M, A490M, F568, A490, A540; SAE J429; plus the references cited at the end of the chapter; plus ISO 898/1 and SAE J1199.

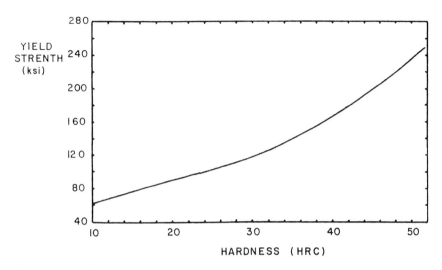

Figure 4.1   Yield strength vs. hardness for low-alloy, quenched, and tempered steels, such as ASTM A193 B7.

Harder fasteners tend to be more brittle, however, and are therefore more susceptible to unexpected brittle fracture, the failure originating from an undetected crack or flaw in the surface. The harder the fastener, the greater the chances of such failure. Attempts have been made, therefore, to specify the safe limits of hardness and strength which we can count on [14, 15]. Figure 4.2 shows the results for LAQT (low-alloy quenched and tempered) steels such as AISI 4140, A193 B7, and similar materials.

Curve A shows the yield strength of LAQT steels as a function of increasing hardness, a repeat of Fig. 4.1. An increase in hardness produces an apparently limitless increase in strength [14].

These results are supported, in the HRC 40–50 region, by curve B, which shows the results of a series of tensile tests made on alloy steel bolts heat-treated to different hardnesses and tested in a tensile machine [15]. This time the information plotted is ultimate strength, not yield strength, so curve B is slightly above curve A in the region of overlap.

As before, curve B shows a steady increase in strength as hardness is increased—but only up to the point the author of Ref. 15 describes as the "critical hardness." Increasing the hardness progressively beyond this point (or region: the data showed quite a bit of scatter) resulted in brittle failure at progressively lower and lower tensile loads. Presumably stress concentrations were created by cracks or flaws, and the material became less and less able to shed or reduce these concentrations by local-

Table 4.2   Shear Strength

Material	Ultimate strength (ksi)	Shear strength (ksi)	Notes	Ref.
ASTM A325	132	79		16
ASTM A490	165	96		16
Stainless steel				
A286	232	119		17
431	230	128		17
Custom 455	269	151		17
PH12-9Mo	280	148		17
PH13-8Mo	243	142		33
Cryogenic materials				
A286	214	116	AT 70°F	18
A286	291	282	AT 423°F	18
Unitemp 212	214	132	AT 70°F	18
Unitemp 212	278	166	AT 423°F	18
Inconel 718	226	138	AT 70°F	18
Inconel 718	291	168	AT 423°F	18
Miscellaneous materials				
H-11	238	135		33
H-11	269	159		33
Marage-300	277	149		33
MP35N	283	162		33
Ti 6A1-4V	200	180		18
Beryllium	75	40		18
Steel	200	240		18
MP159		132		58
Fastener grade aluminum				
2024-T4	68	41		57
6061-T6	45	30		
6262-T9	58	35		
7050-T73	75	45		
7075-T6	83	48		
7075-T73	73	44		

Sources: ASTM A325, A490, A286, plus the references cited at the end of the chapter.

ized plastic yielding in the vicinity of the crack. Eventually the bolts would break. They were truly supporting high stress levels, as an increase in hardness says they should, but only in a local area. The average stress across the full bolt, when it broke, was less than the average stress a softer, more ductile bolt would support.

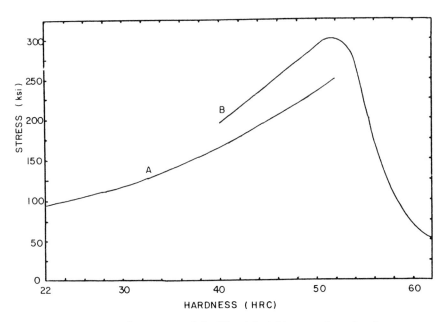

Figure 4.2 Curve A is a repeat of the Fig. 4.1 yield strength vs. hardness curve for low-alloy, quenched, and tempered steels. Curve B is a plot of ultimate strength vs. hardness for similar materials. The dramatic fall in curve B beyond hardness HRC 52 reflects the tendency of very hard steels to fail "early" by brittle fracture.

Nut Selection

A threaded fastener consists of a bolt and nut (or tapped hole). Knowing the strength of the bolt alone is not sufficient, since it's never used alone. We also want a nut or tapped hole which will develop the full strength of the bolt.

Specifications such as ASTM A194 and A563, SAE J995, etc., tabulate the proof, yield, and/or ultimate strengths of nuts the way ASTM A193, A449, and SAE J429 tabulate bolt strengths. Listing all of that information would nearly double the number of tables in this chapter. But I don't think that's necessary.

Most designers and users focus on the strength and other characteristics of the bolt or stud. Having selected a bolt, they then want to choose an appropriate nut, or design a suitable tapped hole. As far as tapped holes are concerned, you can use the thread strength equations of Chap. 3 to select such things as length of thread engagement. When it comes to nuts, most designers will use those recommended in the pertinent bolt specification they're using—or in the nut standard to which the bolt standard refers.

Table 4.3, "Which Nut to Use," summarizes these bolt-nut recommendations. If you want to confirm that the recommendations are correct for your application, you can refer to the standards cited in the table. If you're bothered by the fact that several grades of nuts are sometimes listed for a single bolt material—that's the way the nut or bolt standard does it too. Any of the several nuts listed are acceptable for that bolt. You can base your choice on availability, cost, your desire to standardize your own inventories, or on a special requirement for your application.

In some cases you'll be using a bolt which is not included in Table 4.3. Nut recommendations are not made, usually, for the more exotic bolt materials, for example. In such cases, you'll find the following general guidelines of use.

1. In general, we want the nut or tapped hole to support *more* load than the bolt, because bolt failures are easier to detect than stripped nuts. One rule of thumb: the nut's proof load should approximately equal the ultimate tensile strength of the bolt [21].
2. In an apparent contradiction of the above, nuts are usually (but not always) made of slightly softer (less strong) material than the mating bolts, so that the nut threads will yield locally and better conform to the bolt threads when loaded. This better distributes the stresses in both parts.
    Note that the strength of a part depends on dimension, shape, etc., as well as choice of material. This is why it is possible to make a "stronger" nut from a "weaker" material.
3. If you need a nut for a bolt made of a special material such as titanium, you'd presumably make the nut of the same material. The nut might be heat-treated differently, however, to satisfy point 2 above. Again, this is not a universal custom.
4. The strength of a nut depends, in part, on the length of thread engagement. To increase safety or reduce failure rates, you can consider increasing the length of engagement. Doubling the length of the nut, however, won't double its strength; the relationship is more complex than that, as discussed in Chap. 3.
5. A nut's strength also depends on its width "across flats." Nuts with thinner walls "dilate" more—partially disengage themselves from the bolt under load—and so have less strength. This is one of the reasons why standard nuts are available in several configurations—hex nuts, thick hex nuts, and heavy hex nuts. These differ in height (thread length) too. Your choice will be based on economics and on the fact

Table 4.3 Which Nut to Use

Bolt spec.	Grade	Nut spec.	Grade	Ref.
ASTM A193	B5	ASTM A194	Any	
	B6		Any	
	B7		2H,4,7, or 8	
	B7M		2HM,4M,7M, 8M	19
	B8 CL 1		8	
	B8 CL 2		8	
	B16		4,7, or 8	
ASTM A307	A or B	ASTM A563	A,B,C,D,DH, DH3	
ASTM A320	L7	ASTM A194	4 or 7	
	L43		4 or 7	
	L7M		7M	
	L1		4 or 7	
	B8 CL 1		8	
	B8 CL 1A		8	
	B8 CL 2		8	
	B8M CL 2		8	
ASTM A325		A194 or A563	2 or 2H	
A354			C3,D,DH,DH3	
A449		A563	C3,D,DH,DH3	
A453		A563	C3,D,DH,DH3	
A490		A453		
		A563 or A194	DH,DH3	
			2H	
A540		A540		
SAE J429	1	SAE J995	2	
	2		2	
	4		2,5	
	5		5,8	
	7		5,8	
	8		8	
Metric	4.6		4	
	4.8		4	
	5.8		5	28
	8.8		8	
	9.8		9	
	10.9		12	
	12.9		12	

Table 4.3 Continued

Bolt spec.	Grade	Nut spec.	Grade	Ref.
Special combinations to resist galling				
Stainless steel	316	Stainless steel	400 series	
	Nitronic 60		Nitronic 50	
	Cold drawn		Cold drawn	
	316		316	
Low-alloy steel A193	B7,B16	Stainless steel	300 series	

Sources: A194, A563, A453, A540; SAE J995; plus the references cited at the end of the chapter.

that most bolts (and nuts) aren't loaded anywhere near the limit of their strengths. As a result, a regular hex nut—or equivalent in a tapped hole—is sufficient for most applications. Thicked, heavier nuts would be preferred if the bolts are to be loaded to target preloads beyond 60–70% of yield perhaps, or if the consequences of failure suggest that an extra degree of safety would be prudent.

Note a curious fact that is beginning to bother bolting engineers. There are no standards which define the strength or behavior or properties of a bolt-and-nut system. Instead, we have separate standards for male and for female fasteners. And in some situations the dimensional tolerances on threads are such that the resulting thread engagement can be poor, sometimes resulting in a bolt-nut system which will be unable to develop the full load either could support if tested alone. Mechanically galvanized inch series ASTM A325 bolts and nuts, for example, can cause problems. See Ref. 16 for a lengthy discussion of the nut-bolt specification problem, as related to A325 and A490 fasteners and their metric counterparts.

One frequently asked question is "What nut should I use to minimize galling between nut and bolt?" Table 4.3 ends with a few combinations I have heard recommended by petrochemical or nuclear power engineers. They may be worth trying in your application.

Effects of Temperature on Material Properties

Table 4.4 is the first table of data concerning the "stability" of the clamping force first exerted by the bolt on the joint at the time of assembly. If the operating temperature of the system is higher than room temperature, then the strength of the bolt will be less than it was at assembly. As a result, too high a temperature can cause a bolt—especially a heavily

Table 4.4 Yield Strength (ksi) vs. Temperature

Spec.	Grade	Temp. °F (°C)									Ref.
		70 (20)	400 (204)	600 (316)	800 (427)	1000 (538)	1200 (649)	1400 (760)	1600 (871)	1800 (982)	
ASTM A193	B6	85	76	72							1, 8
	B7	75–105	65–92	60–85	53–74						
	B8–Cl 1	30	21	18	17						
	B16	85–105	79–98	75–93	67–83						
ASTM A307	GR B	36	31	27							9
ASTM A320	L7,L43,L7A	105	92	84	73						1
ASTM A325	Type 1	81	70								
ASTM A354	BC	94–109	87–96	81–89							1
	BD	125	110	102							
ASTM A453	651	50–70	13–56	36–51	33–46						8
	660	85	82	81	80						
ASTM A540	B21–B24										
	Cl 1	150	134	124							
	Cl 2	140	125	116							

Cl 3	130	116	108						1
Cl 4	120	108	99						
Cl 5	100–105	89–94	83–87						
Stainless steels									
310	40–80					>22			3, 9
420	80	71	66						
431	140								
17-4PH	128–185	119–168	115–156						
17-7PH	185								
AM 350	162								
PH15-7MO	155			104	105				
High-performance materials									
A286	95–119					88–90			21
Hasteloy-X	52								3
H-11	215				141				
Inconel 600	37						17		
Inconel 718	260						138		
Inconel X750	92						>44		
Maraging 300	260								
Nimonic 80A	90						73		
Rene 41	157							80	
Titanium									
Ti-6Al-4V	128		95					75	
Waspaloy	115								

Sources: The references cited at the end of the chapter.

preloaded one—to fail. This is certainly a type of "instability" we want to avoid.

The data in the table are in ksi units. To convert to metric units (MPa) multiply the number tabulated by 6.895. For example, a yield strength of 85 ksi would be the equivalent of 85 × 6.895 = 586 MPa.

Table 4.5 provides data on one of the most common types of clamping force instability—thermal expansion. Since few joints operate at a perfectly constant temperature, thermal expansion or contraction of bolts and joint members is virtually universal. This is only a problem, however, when we're dealing with dissimilar materials having different coefficients, or are dealing with a system in which bolts and joint members reach different temperatures in operation. Generally speaking, the change in temperature must be "significant," too, before we run into problems. Normal changes in ambient temperature don't, in my experience at least, cause problems.

The coefficients given below are the average coefficients involved in a temperature change from 70°F to the higher temperature cited. They are not the coefficients at that higher temperature. You'll find the average to be more useful than the specific coefficient, as we'll see in Chap. 13.

The data in the table are in inches/inch/°F. To convert to metric units of mm/mm/°C, multiply the tabulated value by 5/9. For example, a coefficient of 6.5 in./in./°F would be the equivalent of 6.5 × 5/9 = 3.61 mm/mm/°C.

Table 4.6 provides data for another common form of clamping force instability. The modulus of elasticity, in part, determines the stiffness of bolts and joint members. Stiffness and stiffness ratios affect the degree to which the clamping forces introduced at assembly will change when the joint is subsequently subjected to external loads, temperature changes, and the like, as we'll see in Chaps. 6, 12, and 13, for example.

The modulus itself also changes as temperature changes; so the response of the bolt and joint to loads, etc., is different at different temperatures.

The temperature-induced change in modulus can alone change the tension in a bolt, all other effects aside, as we'll see in Chap. 13. So, modulus with and without temperature change affects the stability of the clamping force in several ways.

The data in Table 4.6 are expressed in × 10⁶ psi. A modulus of 30, therefore, is 30,000,000 psi. To convert these data to metric units in GPa, multiply the numbers tabulated by 6.895. A modulus of 30, therefore, equates to 30 × 6.895 = 207 GPa.

Stress relaxation is a cousin to the more familiar phenomenon called creep. Creep involves the slow change in dimension (the strain) of a part subjected to a heavy load (stress). If we threaded one end of a stud into a ceiling, hung a heavy weight from it, and turned up the temperature in the room, the stud would slowly stretch and eventually break.

Stress relaxation, on the other hand, involves the slow shredding of load (stress) by a part under constant deflection (strain). A bolt which has been tightened into a joint, for example, is held in a constant, "stretched" condition by the joint members. The initial tension in the bolt will gradually disappear if stress relaxation occurs. Again, high temperature encourages the process. So stress relaxation is another possible source of "instability" of the clamping force created on the joint when we first tightened those bolts.

We'll look at stress relaxation in Chap. 13. As we'll see, it's only a problem at elevated temperatures.

Figure 4.3 and Tables 4.7 and 4.8 give data on stress relaxation for a number of fastener materials. The data in Table 4.7 are from an ASTM complilation of stress relaxation data [30]. Data in Table 4.8 are from BS 4882 on petrochemical bolting materials [20].

Figure 4.3 shows the residual stress in a bolt after 1000 hr of exposure to the temperatures shown. As an example, a carbon steel bolt will lose approximately 30% (retain 70%) of its initial preload if exposed to 300°C for 1000 hr. The material references in Fig. 4.3, B7, B8, B8M, and B16, are from BS 4882:1973, and correspond to equivalent materials in ASTM A193. B17 is an AISI 660 austenitic alloy. B80A is more commonly known by its trade name, NIMONIC 80A. Neither of these materials is found in A193.

Stress values in Tables 4.7 and 4.8 are given in ksi. To convert to metric MPa units, multiply the value shown by 6.895. For example, a stress of 30 ksi would be the equivalent of 30 × 6.895 = 207 MPa.

A final word about Table 4.8. This table shows the maximum service temperatures which BS 4882 and equivalent materials can experience continuously without losing most of their preload or tension [20]. Comparison of some of these data with Fig. 4.3 will show that these "upper service limits" define the final knee in the residual stress-temperature curves for the materials. Some stress relaxation will occur below these temperatures, but it won't be as extreme as the relaxation which will occur above them. See the discussion of Fig. 4.3 for identification of B7, B8, B17, etc.

In spite of stress relaxation there are many exotic aerospace bolting materials which can be used at very high temperatures and still retain enough strength and energy storage capacity to be useful. A number of these are listed in Table 4.9.

We're usually concerned about stress relaxation if our bolted joints will be exposed, on purpose, to high in-service temperatures for long periods of time. We use high-temperature materials for such applications and take their stress relaxation into account when estimating in-service

Table 4.5 Coefficient of Linear Expansion ($\times 10^{-6}$ in./in./°F)

Spec.	Grade	Temp. °F (°C)									Ref.
		70 (20)	400 (204)	600 (316)	800 (427)	1000 (538)	1200 (649)	1400 (760)	1600 (871)	1800 (982)	
ASTM A193	B5	6.5	7.0	7.2	7.3						22, 23
	B6	5.9	6.4	6.5	6.7						
	B7	5.6	6.7	7.3	7.7						
	B8	8.5	9.2	9.5	9.8						
	B16	5.4	6.6	7.2	7.6						
ASTM A307		6.4	7.1	7.4	7.8						22, 23
ASTM A320	L7	5.6	6.7	7.3	7.7						22, 23
	L43	6.2	7.0	7.3	7.6						
	L7M	6.2	7.0	7.3	7.6						
	B8 CL 1	8.5	9.2	9.5	9.8						
ASTM A325		6.2	7.0	7.3	7.6						22, 23
ASTM A354		6.2	7.0	7.3	7.6						22, 23
ASTM A449		6.2	7.0	7.3	7.6						22, 23
ASTM A453	651	9.1	9.7	10.0	10.2						22, 23

Category	Material											Ref.
ASTM	A490	6.2	7.0	7.3	7.6							22, 23
ASTM A540	B21	5.4	6.6	7.2	7.6							22, 23
	B22	5.6	6.7	7.3	7.7							
	B23	6.2	7.0	7.3	7.6							
	B24	6.2	7.0	7.3	7.6							
Stainless steels	310	8.4	8.8	9.0	9.3	9.5						24
	410	5.5		5.6		6.4	6.5					
	17-4PH	6.0	6.1	6.3	6.5							
	17-7PH	5.7	6.6	6.8	6.9							
	AM 350	6.3		6.8		7.2						
	PH15-7MO	5.0	5.4	5.6	5.9	6.1						
High-performance materials	A286	9.17	9.35	9.47	9.64	9.78	9.88	10.32				13, 24
	Hasteloy-X	7.7	7.82	7.9	8.15	8.39		8.56	8.81	9.02	9.20	
	H-11	6.1	6.4		6.68	6.9		7.11				
	Inconel 600	7.4	7.7	7.9	8.1	8.4		8.6	8.9	9.1	9.3	
	Inconel 718	7.1	7.5	7.7	7.9	8.0		8.4	8.9			
	Inconel X750	6.96	7.14	7.46	7.76	8.10		8.41	8.84	9.33	9.75	
	MP35N		7.6			8.7						
	Nimonic 80A	7.0	7.2	7.4	7.6	7.7		7.9	8.2	8.6		
	Rene 41	6.7	6.8	7.0	7.2	7.5		7.8	8.2	8.7	9.3	
Titanium	Ti-6A1-4V	4.8	5.0	5.1	5.2	5.3		6.1				
	Waspaloy	6.8	7.1	7.3	7.6	7.8		8.0	8.5	8.9	9.8	
	Inconel 725			7.3								62
	K-Monel 500			8.4								62

Table 4.5 Continued

Spec.	Grade	70 (20)	400 (204)	600 (316)	800	1000	1200	1400 (760)	1600 (871)	1800 (982)	Ref.
General-joint materials	Aluminum	12.8		16							25, 26
	Copper	9.3									
	Gray cast iron	5.9									
	Mild steel	6.7	8.3	8.8							
	Bronze	10			15	16					
	Concrete	5.6–7.8									
	Hard rubber	44.4									25, 26
	Brick	3.4–5.3									
	18-8 stainless steel	9.9									
	Wood (with grain)	1.4–5.3									
	Wood (across	18–34									
	Plain carbon steel	6.41	7.07	7.42	7.76						22
	2024 aluminum	12.3	13.3								22
	Ferritic ductile iron		6.6–7.0								54
	Pearlitic ductile iron		6.5–6.6								54
	Austenitic ductile iron		2.2–10.5								54
	Carbon-silicon steels	5.6	6.7	7.3	7.7						22
	High-alloy steels (TYP)	8.5	9.2	9.5	9.8						22

Sources: The references cited at the end of this chapter.

Table 4.6 Modulus of Elasticity at Various Temperatures ($\times 10^6$ psi)

Spec.	Grade	Temp. °F (°C)											Ref.
		−325 (−198)	−200 (−129)	70 (20)	400 (204)	600 (316)	800 (427)	1000 (538)	1200 (649)	1400 (760)	1600 (871)	1800 (982)	
ASTM A193	B5	32.9	32.3	30.9	29.0	28.0	26.1						23, 27
	B6	31.2	30.7	29.2	27.3	26.1	24.7						
	B7	31.6	31.0	29.7	27.9	26.9	25.5						
	B8–CL 1	30.3	29.7	28.3	26.5	25.3	24.1						
	B16	31.6	31.0	29.7	27.9	26.9	25.5						
ASTM A307		31.4	30.8	29.5	27.7	26.7	24.2						23, 27
ASTM A320	L7	31.6	31.0	29.7	27.9	26.9	25.5						23, 27
	L43	31.6	31.0	29.7	27.9	26.9	25.5						
	B8	30.3	29.7	28.3	26.5	25.3	24.1						
ASTM A325	Type 1,2,3	31.4	30.8	29.5	27.7	26.7	24.2						23, 27
ASTM A354		31.2	30.6	29.3	27.5	26.5	24.0						23, 27
ASTM A449		31.2	30.6	29.3	27.5	26.5	24.0						23, 27
ASTM A453		30.3	29.7	28.3	26.5	25.3	24.1						23, 27
ASTM A490		31.2	30.6	29.3	27.5	26.5	24.0						23,27

Table 4.6 Continued

Spec.	Grade	Temp. °F (°C)											Ref.
		−325 (−198)	−200 (−129)	70 (20)	400 (204)	600 (316)	800 (427)	1000 (538)	1200 (649)	1400 (760)	1600 (871)	1800 (982)	
ASTM A540	B21,B22	31.6	31.0	29.7	27.9	26.9	25.5						23, 27
	B23,B24	29.6	29.1	27.8	26.1	25.2	23.0						
SAE J429	GR 1,2,4	31.4	30.8	29.5	27.7	26.7	24.2						27
	GR 5,7,8	31.2	30.6	29.3	27.5	26.5	24.0						
High-performance materials	A286			29.1				23.5	21.1	18.7	18.9		13,24
	Hasteloy-X			28.6								18.5	13,24
	H-11			30.6				23.0	16.0				13,24
	Inconel 600			31.4							23.1		13,24
	Inconel 718			29.0									13,24
	Inconel X750			31.0								20.0	13,24
	MP35N			33.6		30.8							13,24
	Nimonic 80A			31.2							22.7		13,24
	Ti 6Al-4V			16.5									13,24
	Rene 41								25.9				13,24
	Waspaloy			30.6				26.7			23.6		13,24
	Inconel 725			30							22.7		62
	K-Monel 500			28									62

								Ref.
Metric	4.6/4.8/5.8	31.4	30.8	29.5	27.7	26.7	24.2	27
SAE J1199	8.8/9.8/10.9	31.2	30.6	29.3	27.5	26.5	24.0	
Joint materials								
Steels	Low-carbon	31.4	30.8	29.5	27.7	26.7	24.2	20.4	53
	Med-carbon	31.2	30.6	29.3	27.5	26.5	24.0	20.2	53
	Carbon-moly.	31.1	30.5	29.2	27.4	26.4	23.9	20.1	53
	Chrome-moly.	31.6	31.0	29.7	27.9	26.9	25.5	23.9	53
	Austenitic	30.3	29.7	28.3	26.5	25.3	24.1	22.8	53
	2024 aluminum	11.7	11.4	10.6	9.2				53
SAE J158	Malleable iron castings			25–26[a]				
SAE J434	Ductile iron castings			22				
	Cast aluminum			8–10					52
	Brass, cold rolled			13.1					52
	Cast iron			12–14					52
	Wrought iron			26–29					52
	Magnesium			6.1					52
	Concrete			2.0					52
	Wood (with grain)			1.2					56

[a] Modulus of iron can vary very substantially at room temperature. Remains fairly constant through 800°F. (Ref. 54.)
Sources: The references cited at the end of this chapter.

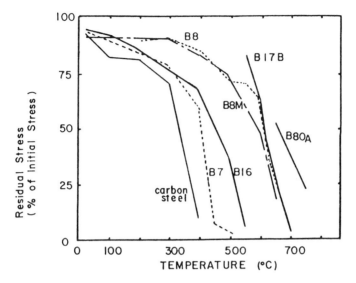

Figure 4.3 Stress relaxation of petrochemical bolting materials as a function of service temperature. Exposure in each case was for 1000 hr at the temperatures shown [19].

clamping forces. We face different problems if our products are subjected to high temperature accidentally, during a fire, for example.

Figure 4.4 shows the response of structural steel bolts when exposed to high temperatures for only 1 hr. These tests were conducted in Japan, but I assume that their structural steel bolts would be comparable to our A325 and A490 materials.

The people who conducted these tests conclude from the data summarized in Fig. 4.4 that a slip critical structural steel joint (see Chap. 14 for definition of a slip critical joint) could safely be exposed to a fire temperature of 350°C (662°F) and still retain structural integrity [59]. Bolts tightened to a "standard preload" would lose perhaps 10% of this preload in 1 hr. Bolts tightened to only 90% of standard preload would lose as much as 20%, however.

By further experiments they found that there's a cumulative time and temperature effect here. A clamping force loss of 20% was encountered in bolts exposed to any of the following combinations of time and temperature [60].

250 hr at 250°C (482°F)

20 hr at 300°C (572°F)
0.7 hr at 350°C (662°F)

Other bolted joints showed a 30% loss of clamping force after:

4100 hr at 250°C (482°F)
880 hr at 300°C (572°F)
3.2 hr at 350°C (662°F)

As Fig. 4.4 shows, higher temperatures resulted in greater loss; 1 hr at 500°C (932°F) reduced tension in bolts to only 10% of their initial preload, for example. Rapid cooling—e.g., in water—increased the loss, incidentally. I still think it would be a good idea to let the firemen squirt water on your burning buildings!

So much for "elevated" temperatures. We're also sometimes concerned with extreme temperatures in the other direction. Table 4.10 and Fig. 4.5 provide some data about the strength of certain exotic materials to very low, cryogenic temperatures [58]. Unfortunately, my source gave only the room temperature strengths for most of these materials, as you'll see from Table 4.10. Figure 4.5 suggests that their cryogenic temperature strengths could be higher than their room temperature strengths, but that there will be exceptions such as the particular titanium alloy shown in the chart. You'll have to contact aerospace bolt manufacturers for further information.

Fatigue Properties

Cyclic loads can cause a bolt to fatigue and break, destroying the clamping force on the joint. The number of load cycles a bolt will tolerate before failure is called its fatigue life.

This fatigue life depends on many factors, as we'll see in Chap. 17. Table 4.11, "Fatigue Data," is included only to give you a rough idea of the extent to which cyclic loads can degrade the apparent tensile strength of a threaded fastener. The data would apply directly to your application only if the environmental and load conditions for your joint were identical to those seen by the bolt when tested. The bolts would have to be identical as far as geometry, heat treatment, surface finish, condition, etc., were concerned—and that will never be the case. Even if conditions were identical, furthermore, fatigue test data always exhibit a considerable amount of "scatter." So don't take the data in Table 4.11 as gospel, but the table is informative.

The data are given in two different ways; as an "endurance limit" or as a "fatigue strength." The endurance limit is the maximum tensile stress

Table 4.7 Stress Relaxation

Bolt information	Test temp. °F (°C)	Initial stress (ksi)	Residual stress (ksi) 100 hr	1000 hr	10,000 hr	Final Stress (ksi)	Final Time (hr)
ASTM A193 B7 1-in. diam.	850 (454)	30.0		13.2	(10.0)		1920
ASTM A453 GR 651 1¼-in. diam. as rec'd plus 1200°F (650°C) for 48 hr	1200 (649)	32		5	(2.85)		1012
ASTM A453 GR 655: 2-in. diam.	1100 (593)	19.9	14.3	15.9		16.0	3769
Type 431 st. steel (AN6C)	700 (371)	95.8	68			67.8	150
	900 (482)	95.8	23			20	150
	900 (482)	63.25	20			18.8	150
Austenitic st. steel DIN 17006: X5CrNiMo 1810 M12 bolt. Nut: DIN 931 or 934: rolled threads	842 (450)	50.6	46.5	44.7			
	1022 (550)	50.6	34.1	22.9			
A286 bolt: 5/16-in. diam. -MS 9035-19 configuration -HiR thread form per AMS 7478	1200 (649)	70	52.9			100	100
		70	34			100	100

Material	T (°F/°C)						
Nimonic 80A: 3.5-in. diam.	1202 (650)	39.5	30.4	24.2	17.8	14.2	21,167
	1292 (700)	37.1	28.5	22.1		15.7	2,900
	1499 (815)	29.6	9.8			5.7	290
Nimonic 80A: $1\frac{1}{8}$-in. diam.	1157 (625)	33.1	28	24.2	18.6	15.7	27,200
Bolt: AISI 8740 steel (M12) Nut: German aircraft spec. LN 1.4944	662 (350)	45	39	31.5			
		85.2	67.5	52.5			
Vascojet 1000 (modified H-11 steel) $\frac{1}{2}$-20 studs	900 (482)	117	78				
MP35N: $\frac{1}{4}$-28 bolt Nut of same alloy	700 (371)	132				113	150
Rene 41 bolt: 10–32 threads Waspaloy nut	1400 (760)	99.7				25.5	50
Same: $\frac{1}{4}$-threads	Same	73				32.9	50
Same: $\frac{1}{2}$-20 threads	Same	73				35.1	50
Same: $\frac{1}{2}$-20 threads	Same	46.4				28.3	50
Udimet 500 bolt $\frac{5}{16}$-18 threads	1100 (593)	50				36.5	12
Waspaloy nut and bolt MIL-S-8879 thread form: $\frac{1}{4}$-28 threads	1400 (760)	100				7.7	50

Source: Ref. 30.

Table 4.8 Stress Relaxation High-
Temperature Service Limit

Material (from BS 4882)	Temp. limit	
	°F	°C
Carbon steel	572	300
Mild steel	572	300
B7	752	400
B6	932	500
B16	968	520
B8, B8T, B8C	1067	575
B8M	1112	600
B17	1202	650
B80A	1382	750

Source: British Standard BS 4882:1973.

the bolt can stand for an infinite number of cycles, infinite life. Fatigue strength is the maximum stress the bolt will stand for an average life of the number of load cycles cited in Table 4.11. As an example, an AISI 4340 bolt, heat-treated to a room temperature ultimate tensile strength of 165 ksi, can be expected, on the average, to tolerate 1000 cycles of load if that load never exceeds 10 ksi. Your bolts might stand more or less load than this, for 1000 cycles; but your results would probably be similar to those reported here. You'd be unlikely to get 10 times this much life, or one-tenth of it.

The data in Table 4.11 are given in ksi. To convert to metric MPa units, multiply the numbers tabulated by 6.895. For example, a fatigue strength of 10 ksi would equate to $10 \times 6.895 = 69$ MPa.

Terms used in the table include:

TRBHT = threads rolled before heat treatment
TRAHT = threads rolled after heat treatment
SMIN, SMAX, SMEAN = minimum, maximum, and mean stress seen by the bolts during the test (ksi)

Corrosion

Predicting the useful life of a fastener in a corrosive environment is perhaps even more difficult than predicting its life under cyclic, fatigue loads. Much depends on the nature of the electrolyte, which can take an almost infinite number of forms, concentrations, etc. Temperature can play a

Materials

Table 4.9 Service Temperature Limits

Material type	Name	Room temp. UTS (ksi)	Temperature limit °F	°C	Ref.
Iron-base alloys	4340[a]	180	450	238	58
	PH 13-8 M	220	650	344	
	Custom 455	220	650	344	
	Marage 300[a]	260	900	483	
	H-11[a]	260	900	483	
	A-286	200	1200	649	
	TD-Ni-Cr	18[b]			
Nickle-base alloys	MP35N	260	700	372	58
	Inco 718	180	1200	649	
	Rene 95	230	1200	649	
	Rene 41	150	1400	760	
	Waspaloy	150	1400	760	
	Astroloy	190	1600	871	
Titanium-base alloys	Ti 1-8-5	200	300	148	58
	Ti 6-6-2	180	500	260	
	Ti 6-4	160	500	260	
				1093	
Haynes alloys	HA 188	45[b]	1800	982	58
	HA 214		2000	1093	
	HA 230		2000	1093	
Columbium base	Cb 752	35[b]	2500	1371	58
	C 1294	35[b]	2500	1371	
Tantalum base	T-222	20[b]	3000	1648	58
	Ta-10W	18[b]	3000	1648	

[a] Protective coating required.
[b] At maximum service temperature, not at room temperature.

large role. So can geometry of the parts, stress levels, stress concentrations, the properties of mating parts, surface flaws, crevices between parts, fluid velocity, and many other factors—as we'll see in Chap. 18.

When selecting a fastener material, however, it can be useful to have a checklist of available materials which are considered "corrosion resistant" in general. Table 4.12 includes such a list. A list of the relative costs of corrosion-resistant fastener materials is given in Table 4.15 as well.

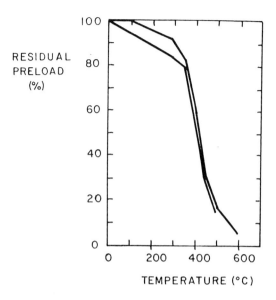

Figure 4.4 The residual preload in a group of structural steel bolts, as a percentage of the initial, room temperature preload, after a 1-hr soak at the various temperatures shown. From this data the investigators concluded that fire temperatures above 350°C (662°F) and/or "soak" times greater than 1 hr could seriously damage the integrity of a structure. Note that the diameter of the bolt influences the rate at which it loses preload. The upper curve is for an M30 bolt, 30 mm (about 1.2 in.) in diameter. The lower curve is for a smaller M16 bolt [59].

Table 4.10 Cryogenic Bolting Materials Used to −423°F (−253°C)

Type	Name	Room temp. UTS (ksi)	Ref.
Iron-base alloys	A-286	140	58
	C.R. A-286	200	
	PH 13-8 Mo	220	
	Custom 455	220	
Nickle-base alloys	Rene	150	58
	Waspaloy	150	
	Inco 718	180	
	Astroloy	190	
	C.R. Inco 718	220	
	MP35N	260	
Titanium-base alloys	Ti 5Al-2.5Sn	110	58

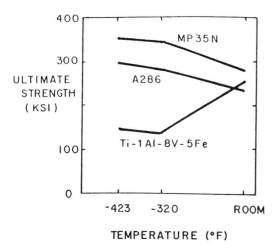

Figure 4.5 The ultimate strength of three exotic bolting materials at cryogenic temperatures. As suggested here, the strength of some materials is increased by a very low temperature, but it would be unwise to assume that this is true for all materials. The strength of another titanium alloy, Ti-5Al-5Sn-5Zr, increases between room temperature and −320°C but then decreases a bit if further cooled to −423°C. Its strength at that temperature is about 210 ksi, still above its room temperature ultimate of about 160 ksi [58].

Neither list is "complete," but they can help you start a search for a suitable material.

The ultimate tensile strength (UTS) data given in Table 4.12 are in ksi. To convert to metric MPa units, multiply the values shown by 6.895. For example, a UTS of 70 ksi would equal 70 × 6.895 = 483 MPa.

Miscellaneous Considerations

The data in Tables 4.13, 4.14, and 4.15 usually influence the selection of a fastener material and may dominate the selection in some cases. Weight is especially important in aerospace applications and is coming to mean more and more in automotive work. Cost is always of concern to the responsible designer.

As far as cost is concerned, however, remember that it is influenced by many factors—the quantities purchased, the popularity of a particular size or configuration of fastener, economic conditions at the time of purchase, the competition for a particular market, trade laws and tariffs, etc. The data in Tables 4.14 and 4.15, therefore, should be used with considerable caution.

Table 4.11   Fatigue Data

Bolt	Ultimate tensile strength (ksi)	Endurance limit (ksi)	Fatigue strength (ksi)	10^3 cycles	S_{min}/S_{max} ratio	Test conditions	Ref.
Metric M16, CL 8.8	126	10.2			0.1	TRBHT	35
Metric M14 × 1.5	142–170	7.1–11.4				Rolled threads	44
SAE J429, GR 8 $\frac{3}{8}$ unf. thread	150		9.2	5.96	0	Preload = 1420 lb	39
Same	150		9.2				39
SAE J429, GR 8	150	$4500 \times D$ (D = nom. dia.)		4654.0	0	Preload = 8420 lb	40
SAE J429, GR 8	150	18					37
Metric CL 10.9 M12 × 1.25	150	8				S_{mean} = 100 ksi; TRBHT	34
Same	150	25.7				S_{mean} = 100 ksi; TRAHT	34
AISI 4340 stud 1½-in. diam.	128		63	35.6	0.59		38
Same	156		50	300	0.08		
Same	157		67	22	0.33		
Same	159		50	127	0.1		
AISI 4340	165		10	1			4

Material						Comments	Ref.
M10, grade 12.9	180	6.9			0.1	Preload = 5.24 lb	45
M10, grade 12.9	180	10.7			0.1	Preload = 8.75 lb	45
Steel, $\frac{1}{2}$-in. diam.	180–200	11			−1.0	Coarse threads, TRBHT	36
Steel, $\frac{1}{2}$-in. diam.	180–200	8			−1.0	Fine threads, TRBHT	36
Steel, $\frac{1}{2}$-in. diam.	180–200	70			−1.0	TRAHT	36
Steel, 3-in. diam.	180–200	30			−1.0	TRAHT	36
MP35N	190		82	65 min.		Tested as a shear bolt	2
MP35N	270		135	65 min.		Tested as a tension bolt	2
MP35N	260		117	5000	0.1		33
MP35N	260		135	500		Tested at room temp.	2
MP35N	345		135	2300		Tested at 320°F (−196°C)	2, 43
Custom 455	220		99	500	0.1		33
PH 13-8 Mo	220		99	5000	0.1		33
PH 12-9 Mo	220		99	2000	0.1		33
Inconel 718	220		99	2000	0.1		33
H-11	220		99	5000	0.1		33
H-11	260		117	300	0.1		33
Marage 300	260		117	156	0.1		33

Source: The references cited at the end of this chapter.

Table 4.12   Corrosion-Resistant Materials

Material	Notes
Steel, coated	UTS 80–125 ksi
	Low-carbon, medium-carbon, and low-alloy steels can be made more or less resistant to atmospheric corrosion, or equivalent, by coating them with paint or by plating them. (See Chap. 19.) Examples of such materials: A193 materials except the B8 series; A325, A490, SAE J429 materials, metric materials 4.6–12.9, etc.
Austenitic st. steel	UTS 75–120 ksi
	Most common of the stainless steels and more corrosion resistant than the three listed below. Nonmagnetic. Can't be heat-treated but can be cold-worked. Good high- and low-temperature properties. 321 can be used up to 1500°F (816°C), for example. Examples: A193 B8 series, A320 B8 series, any of the 300 or 18-8 series materials, such as 303, 304, 316, 321, 347, etc. [21].
Ferritic st. steel	UTS 70 ksi
	Can't be heat-treated or cold-worked. Magnetic. Examples include 430 and 430F [21].
Martensitic st. steel	UTS 70–180 ksi
	Heat treatable, magnetic. Can experience stress corrosion if not properly treated. Examples: 410, 416, 431 [21].
Precipitation hardening st. steel	Typical UTS, 135 ksi
	Heat treatable. More ductile than martensitic stainless steels. Examples: 630, 17-4PH, Custom 455, PH 13-8 Mo, ASTM A453-B17B, AISI 660 [21,46].
Nickel-based alloys:	
Nickel-copper	UTS 70–80 ksi
	Can be cold-worked, but not heat-treated. Example: Monel [21].
Nickel-copper-aluminum	UTS 130 ksi
	Can be both cold-worked and heat-treated. Good low-temperature material. Example: K-Monel [21].

Table 4.12 Corrosion-Resistant Materials

Material	Notes
Titanium	UTS 135–200 ksi
	Good corrosion resistance. Low coefficient of expansion. Has a tendency to gall more readily than some other corrosion-resistant materials. Expensive. Example: Ti6A1-4V [21].
Superalloys	UTS 145–286 ksi
	High-strength materials with excellent properties at high and/or low temperatures. Primarily used in aerospace applications. Expensive. Some, such as MP35N, are virtually immune to marine environments and to stress corrosion cracking. Examples: H-11, Inconel, MP35N, A286, Nimonic 80A, etc. MP35N, Inconel 718, and A286 are especially recommended for cryogenic applications [33,46].
Nonferrous materials	There are many nonferrous fastener materials which can provide outstanding corrosion resistance in applications which would rapidly destroy more common bolt materials. The main drawback to these materials is a general lack of strength; but lack of strength can sometimes be made up by using fasteners of larger diameter or by using more fasteners. Here are a few of the many materials available:
Silicon bronze	UTS 70–80 ksi;
Aluminum	UTS 13–55 ksi;
Nylon	UTS 11 ksi.

Source: The references cited at the end of this chapter.

Note that this is confirmed by the tables themselves. They come from two different sources at two different times. Table 4.15 contains data compiled by a large "mass producer" in 1980 [42]. The data in Table 4.14 were assembled more recently [28]. The contradictions emphasize the difficulty of publishing such information. But the tables should serve to remind you that costs can vary substantially and that a "better" fastener may not be worth the expense.

Table 4.13 Density of Fastener
Materials (lb/in.^{3})

Fastener material	Relative weight
Nylon	0.05
Berylium	0.066
Aluminum (2024-T4)	0.100
Titanium 6A1-4V	0.16
Steel	0.28
H-11	0.282
Waspaloy	0.286
A286	0.286
Rene 41	0.298
Inconel X-750	0.298
Inconel 600	0.301
Naval brass	0.304
K-Monel	0.305
Phosphor bronze	0.320

Source: Ref. 11, 18, 24, 25.

On the other side of the argument, however, remember that the material cost of a fastener is only part of the job cost. If you use a more expensive (stronger) material, you may be able to use fewer and/or smaller fasteners. Joint members can also be scaled down, in many situations. Assembly costs may be reduced because you don't have to tighten as many bolts and/or can open preload tolerances. Warranty and liability

Table 4.14 Approximate Relative Costs of Fastener Materials

Generic material	Typical examples	Relative cost
Medium-carbon steel	ASTM A449, metric 5.8, SAE GR 5, ASTM A325	1
Stainless steels	304, 316 ASTM A193 B8	2.5
Austenitic iron-based alloys	A-286	12
Nickle-copper alloys	Monel	20
Titanium	6A1-4V	75
Austenitic nickle-based alloys	Nimonic 80A, Rene 41, Waspaloy, Inconel X-750	100

Source: Ref. 42.

Table 4.15 Relative Costs of Corrosion-Resistant Fastener Materials

Material	Ultimate tensile strength (ksi)	Relative cost
Nylon	11	1
6061 aluminum	45	1
2024-T4 aluminum	60	1
410 stainless steel	70–150	1
416 stainless steel	85–150	1
18-8 stainless steel	80	2
Copper	45	2
Silicon bronze	85	2
Brass	60	2
Naval bronze	70	2
316 stainless steel	80	3
Carpenter 20	80	4
309 stainless steel	80	4
Monel 400	75	4
Monel K500	105–150	5
Inconel 600	85	6
Hastelloy C	125	6
Titanium, commercial	65	6
Titanium 6A1-4V	130–160	6

Source: Ref. 28.

claims may be reduced by fewer fastener failures. So "true cost" is a complex issue. But Tables 4.14 and 4.15 give you some data to start with.

Joint Materials

We've already considered some joint material properties—the coefficient of linear expansion and the modulus of elasticity, for example (Tables 4.5 and 4.6)—because these things affect the stability of the clamping force. Now we're going to look at the room temperature strength of some typical joint materials. I call these "typical" because the choice of joint material is virtually unlimited. But data in Table 4.16, "Room Temperature Strength of Typical Joint Materials," can be found in ASME, SAE, or other standards.

    The strength of joint materials is usually not a major factor in bolted joint analysis or design. If the bolt holes are placed too close to the edge of joint members, however, they can tear out under shear loads, as we saw in Chap. 2. More commonly, we'll manage to strip the threads from a tapped hole, another failure in shear of the joint material.

Table 4.16 Room Temperature Strength of Typical Joint Materials
(ksi)

Joint material	Strength			
	Yield	Tensile	Shear	Ref.
Structural steels				
Low-carbon steels (A36, Fe37)	33–36	58–80	41–56	48
High-strength steel (A588)	42–50	63–70	44–49	48
High-strength, low-alloy steel	40–65	60–80	42–56	48
(A242, A441, A572, Fe52)				
Quenched and tempered carbon steel (A537)	50–60	70–100	49–70	48
Quenched and tempered alloy steel	90–100	100–130	70–91	48
(A514, A517)				
Automotive materials				
Steels				
SAE J414				
1010 hot rolled	26	47	35	
1010 cold drawn	44	53	37	
1020 hot rolled	30	55	41.2	
1020 cold drawn	51	61	43	
1035 hot rolled	39.5	72	54	
1035 cold drawn	67	80	56	
Aluminum die castings	14–24	46	19–29	
SAE J453				
Grade 303, 306, 308, 309				
Gray iron castings				
SAE J859				
G1800	—	18	—	
G2500	—	25	31	55
G3000	—	30	38	
etc.				
Malleable iron castings				
SAE J158				
M3210	32	50		
M4504	45	65		
M5003	50	75		
M5503	55	75		
M7002	70	90		
M8501	85	105		
Ductile iron castings				
SAE J434				
D4018	40	60		
D4512	45	65		
D5506	55	80		
D7003	70	100		

Table 4.16 Continued

Joint material	Strength			
	Yield	Tensile	Shear	Ref.
Steel castings				
SAE J435				
ASTM A27	30	60–65		
ASTM A148 low alloys	50–60	80–90		
ASTM A148 high-strength alloys	85–145	105–175		
Pressure vessel and piping flanges				
Ferritic steels for high-temperature service				
ASTM A182, GR. F1, F2, F5, F7, F11, F12, etc.	40	70	49	50
ASTM A182, GR. F6a Cl 2	55	85	60	50
GR. F6a Cl 3	85	110	77	50
GR. F6a Cl 4	110	130	91	50
GR. F6b, F6NM	90	110–135	77–95	50
Austenitic stainless steels for high-temperature service				
ASTM A182, GR. F304, F310, F316, F347, F321, etc.	30	75		50
Grades F304N, F316N	35	80		50
Grade FXM-19	55	100		50
Gray iron castings				
ASTM A278 Cl 20	—	20		50
Cl 30	—	30		50
Cl 40	—	40		50
Cl 50	—	50		50
etc.				
Ductile iron castings				
ASTM A395	40	60		50
Alloy steel forgings for low-temperature service				
ASTM A522	75	100		50
Austenitic ductile iron castings for low-temperature service				
ASTM A571-71	30	65		50

Sources: ASTM specifications A36, Fe37, A588, A242, A441, A572, Fe52, A537, A514, A517; SAE specifications J414, J453, J859, J158, J434, J435; plus Refs. 48, 50, and 55.

Although such failures are reasonably common, you'll find that most bolted joint failures can be traced to failure of the bolts, or, even more commonly, to poor assembly practices. Nevertheless, some joint strength data are useful.

Incidentally, although many of the joint material failures we'll experience are in shear, information on shear strength is especially difficult to find. Apparently, shear tests are difficult to make and give scattered results. In Table 4.16 therefore, I've often resorted to Fisher's comment that the shear strength of most common steels is usually about 70% of the tensile strength [49]. He's dealing with structural steels, so I wouldn't extend that rule of thumb to castings or aerospace materials or other things which may be common in other industries. But I must confess I haven't found much data.

The information in Table 4.16 comes from the standards cited and/or from the references given in the final column.

The values in the table are given in ksi. To convert to metric MPa units, multiply the number shown by 6.895. For example, a yield strength of 33 ksi would be equivalent to $33 \times 6.895 = 228$ MPa.

REFERENCES

Note: In addition to the references listed below, a great deal of information was taken from the ASTM, SAE, ASME, and other standards cited in the various tables.

1. *ASME Boiler and Pressure Vessel Code*, Section III-Div 1, 1983 ed., Table I–13.3, ASME, New York, 1983.
2. Hood, A. Craig, MP35N—a multiphase alloy for high strength fasteners, *Metal Progress*, May 1968.
3. *1984 SAE Handbook*, Vol 1, *Materials*, SAE, Warrendale, PA, 1984, pp. 10.140–10.142.
4. Chung, Y., Threshold preload levels for avoiding stress corrosion cracking in high strength bolts, Report No. 0284-03EV Rev 1, Bechtel Group, Inc., San Francisco, April 1984.
5. McIlree, A. R., Degradation of high strength austenitic alloys X-750, 718 and A-286 in nuclear power systems, International Symposium on Environmental Degradation of Materials in Nuclear Power Systems—Water Reactors, Electric Power Research Institute, Palo Alto, CA, August 22–24, 1983.
6. Carpenter Gall-Tough Stainless, Brochure of Carpenter Technology Corp., Reading, PA, 1987.
7. Fighting corrosion with new specialty steels, *Product Engineering*, November 1979.

8. *ASME Boiler and Pressure Vessel Code*, Section III, Div. 1, 1983 ed, Table I–1.3, ASME, New York, 1983.

9. Winter, J. R., exerpts from Bolt document, privately published, 1974.

10. Grade and material markings, Part VIII, *Fastener Technology*, pp. 50–58, December 1986.

11. Hood, A. C., Corrosion in threaded fasteners—causes and cures, *Machine Design*, December 1981.

12. Hood, A. C., and R. L. Sproat, Ultrahigh strength steel fasteners, Special Technical Publication no. 370, ASTM, Philadelphia, 1965.

13. Brochure No. 3275-681-2M, MP35N, SPS Technologies, Jenkintown, PA, 1987.

14. From data presented at a meeting of the Atomic Industrial Forum, Metal Properties Council Joint Task Group on Bolting, Knoxville, TN, 1983.

15. Gnuchev, V. S., Investigation of the strength of bolts, *Problemy Prochnosti*, no. 4, pp. 113–115, April 1977.

16. Yura, J. A., K. H. Frank, and D. Polyzois, High strength bolts for bridges, PMFSEL Report no. 87-3, University of Texas, Austin, TX, May 1987.

17. Zelus, P., and Peter M. Rush, Survey of fastener technology, *Metals Engineering Quarterly*, vol. 10, pp. 1–5, August 1970.

18. Sproat, Robert L., The future of fasteners, *Machine Design*, vol. 38, no. 30, pp. 107–122, December 22, 1966.

19. Standard BS 4882:1973, British Standards Institution, London, 1973.

20. Mountford corrosion resistant fasteners, GKN Fasteners Corrosion Lab Reprint 32, Fred Mountford, Ltd., Warley, UK.

21. Blake, Alexander, *What Every Engineer Should Know About Threaded Fasteners*, Marcel Dekker, New York, 1986.

22. *ASME Boiler and Pressure Vessel Code*, Section III-Div. 1, 1983 ed., Table 1–5.0, ASME, New York, 1983.

23. Ibid., Table 1–7.3.

24. *1984 SAE Handbook*, vol. 1, *Materials*, SAE Warrendale, PA, pp. 10.137–10.139, 1984.

25. Ibid., p. 10.22.

26. *Handbook of Chemistry and Physics*, Ed. Charles D. Hodgman, Chemical Rubber Publishing Co., Cleveland, OH, 1943.

27. *ASME Boiler and Pressure Vessel Code*, Section III-Div. 1, 1983 ed., Table 1–6.0, ASME, New York, 1983.

28. *Machinery's Handbook*, 20th ed., Industrial Press, New York, 1975, p. 1181.

29. Bickford, J. H., and Michael Looram, Good bolting Practices: a reference manual for nuclear power plant personnel: vol 1., large bolt manual, Document no. NP-5067, Electric Power Research Institute, Palo Alto, CA, 1987, p. 44.

30. Compilation of stress relaxation data for engineering alloys, ASTM, Data Series, Publication DS 60, Publ. Code #05-060000-30, ASTM, Philadelphia, 1982.

31. Private conversation with Paul Bonenberger, General Motors Corp., Detroit, MI, 1981.
32. Private conversation with Robert Finkelston, SPS Laboratories, Jenkintown, PA, 1980.
33. Patel, Sumant, and Edward Taylor, New high strength fastener materials resist corrosion, *Metal Progress*, September 1971.
34. Finkelston, R. J., and P. W. Wallace, Advances in high performance mechanical fastening, Society of Automotive Engineers, no. 800451, Warrendale, PA, 1980.
35. Yoshimoto, I., K. Maruyama, K. Hongo, and T. Sasaki, On improvement of fatigue strength of high strength bolts, Bulletin P.M.E., no. 43, Tokyo Institute of Technology, Tokyo, Japan, March 1979.
36. Crispell, C., New data on fastener fatigue, *Machine Design*, Penton Publications, Cleveland, OH, p. 74, April 22, 1982.
37. Thread forms and torque systems boost reliability of bolted joints, *Production Engineering*, p. 37, December 1977.
38. Fritz, R. J., Cyclic stresses for bolts and studs, General Electric Co., Schenectady, NY.
39. Tensioning of fasteners (studs and bolts), Boiler and Machinery Engineering Bulletin no. 4.3, American Insurance Association, New York, 1981.
40. Aaronson, Stephen F., Analyzing critical joints, *Machine Design*, pp. 95–101, January 21, 1982.
41. Crispell, Corey, New data on fastener fatigue, *Machine Design*, pp. 71–74, April 22, 1982.
42. From a seminar on threaded fasteners, presented by a FORTUNE 500 manufacturing concern, September 1980.
43. List and description of various aerospace fastening systems used by the Boeing Commercial Airplane Co., Seattle, WA.
44. Erker, A., Design of screw fastenings subject to repeated stresses, Proc. of International Conference on Fatigue of Metals, September 10–14, 1956, Institution of Mechanical Engineers, London, 1956.
45. Chapman, Ian, John Newnham, and Paul Wallace, The tightening of bolts to yield and their performance under load. *Journal of Vibration, Acoustics, Stress and Reliability in Design*, Transactions of the ASME, vol. 108, p. 213, April 1986.
46. Hood, A. Craig, Corrosion in threaded fasteners—causes and cures, *Machine Design*, pp. 153–156, December 7, 1961.
47. Snow, A. L., and B. F. Langer, Low cycle fatigue of large diameter bolts, Transactions of the ASME, *Journal of Engineering for Industry*, pp. 53–60, February 1967.
48. Kulak, Geoffrey L., John W. Fisher, and John H. A. Struik, *Guide to Design Criteria for Bolted and Riveted Joints*, 2nd ed., Wiley, New York, 1987, pp. 9–11.
49. Ibid., p. 114.

50. *ASME Boiler and Pressure Vessel Code*, Section II, Part A, ASME, New York, 1980.
51. *Manual of Steel Construction*, 8th ed., AISC, Chicago, 1980, pp. 6–7.
52. *Handbook of Chemistry and Physics*, 27th ed., Chemical Rubber and Publishing Co., Cleveland, OH, 1943, p. 1631.
53. *ASME Boiler and Pressure Vessel Code*, 1983, Section VIII, Part UF, Table UF-27, ASME, New York, 1983.
54. Elevated temperature properties of cast iron, SAE standard J125, SAE, Warrendale, PA, 1969.
55. Juvinall, Robert C., *Engineering Considerations of Stress, Strain and Strength*, McGraw-Hill, New York, 1967, p. 559.
56. Blake, Alexander, *Practical Stress Analysis in Design*, Marcel Dekker, New York, 1982, p. 12.
57. Design factors for threaded aluminum fasteners, *Machine Design*, June 26, 1980, p. 109.
58. From notes for a bolting seminar prepared by Ray Toosky of McDonnell Douglas, 1991.
59. Wakiyama, Kozo, and Akio Tatsumi, Residual force in high strength bolts subjected to heat, *Transactions of the A.I.J*, No. 313, March 1982, p. 19 ff.
60. Wakiyama, Kozo, and Akio Tatsumi, Residual force in high strength bolts subjected to cyclic heat, reprinted from *Technology Reports of the Osaka University*, vol. 29, no. 1515, 1979, p. 533 ff.
61. Raymond, Louis, Titanium alloys, *American Fastener Journal*, May/June 1992, p. 8 ff.
62. Faller, Kurt, and William A. Edmonds, The incentives for titanium alloy fasteners in the Navy/Marine environment, *American Fastener Journal*, September/October 1992, p. 44 ff.
63. Barrett, Richard T., Design criteria, Part eight of a ten part series . . . fastener basics, *American Fastener Journal*, May/June 1992, p. 56.
64. Conversations with Ray Toosky of McDonnell Douglas, August 1993.

5
Stiffness and Strain Considerations

As we learned in Chap. 1, both bolt and joint members are, in effect, stiff springs. They deflect under load. They relax when the load is removed. They store potential energy and can create an appropriate clamping force—and therefore function as an effective joint—only as long as they retain "enough" of that energy.

Because of all this, one of the most important properties of the bolt and joint members is their stiffness. Less stiff—I'll often call them "softer"—springs can often store energy more effectively than very stiff ones, so we'll be interested in how stiff they are. In addition, the "joint stiffness ratio," which I'll define near the end of this chapter, is an extremely important design parameter which affects they way the bolts and joint members absorb external loads, respond to changes in load, respond to changes in temperature, etc.

Let's look in detail, then, at this concept of stiffness and at the related deflection of or strain in the bolt and joint members. We'll start with the bolt, then examine the joint, and then take a brief look at the stiffness ratio.

I. BOLT DEFLECTION AND STIFFNESS

A. Basic Concepts

Let's apply equal and opposite forces to the ends of a rod of nonuniform diameter, as shown in Fig. 5.1. If the tension stress created in the rod is

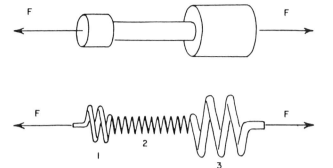

Figure 5.1 Rod of nonuniform diameter, loaded in tension, and equivalent spring model.

below the proportional limit, we can use Hooke's law and the relationship between springs in series to compute the change in length of the rod.

The combined change in length of the rod will be equal to the sum of the changes in each section:

$$\Delta L_C = \Delta L_1 + \Delta L_2 + \Delta L_3 \qquad (5.1)$$

Hooke's law tells that the change in one section will be

$$\Delta L = \frac{FL}{EA} \qquad (5.2)$$

where ΔL = change in length (in., mm)
   A = cross-sectional area (in.^{2}, mm^2)
   L = length of the section (in., mm)
   E = modulus of elasticity (psi, GPa)
   F = applied tensile force (lb, N)

Since the various sections are connected in series, they each see the same force, so we can combine the two equations above and write

$$\Delta L_C = F\left(\frac{L_1}{EA_1} + \frac{L_2}{EA_2} + \frac{L_3}{EA_3}\right) \qquad (5.3)$$

Now, the spring constant of a body is defined as

$$K = \frac{F}{\Delta L} \qquad (5.4)$$

where K = spring constant or stiffness (lb/in., N/mm)
         ΔL = change in length of the body under load (in., mm)
         F = applied load (lb, N)

The spring constant of a group of bodies, connected in series, is

$$\frac{1}{K_T} = \frac{1}{K_1} + \frac{1}{K_2} + \frac{1}{K_3} \qquad (5.5)$$

where K_T = combined spring constant of the group (lb/in., N/mm)
         $K_1, K_2,$
               etc. = spring constants of individual members of the group (lb/in., N/mm)

Now, the equation for the spring constant of a body can be rewritten as

$$\Delta L = \frac{F}{K} \quad \text{or} \quad \Delta L = F\left(\frac{1}{K}\right) \qquad (5.6)$$

Comparing our equation for the spring constant for a group of bodies to the equation for the stretch or change in length of a group of bodies, we see that

$$\frac{1}{K_T} = \frac{L_1}{EA_1} + \frac{L_2}{EA_2} + \frac{L_3}{EA_3} \qquad (5.7)$$

Note that the stiffness of either a plain or complex body is very much a function of the ratio between length and cross-sectional area—it's a function, in other words, of the *shape* of body just as much as it is a function of the material from which the body is made. If we take one piece of alloy steel and make two bolts from it, one a short, stubby bolt and the other long and thin, and we then place the bolts in tension and plot elastic curves for them, we will end up with two curves such as shown in Fig. 5.2.

The equations used for a rod having several different diameters are basically the equations we would use for computing change in length and stiffness of bolts. If we can compute or predict the lengths, cross-sectional areas, and modulus of the material, we should be able to compute the deflection under load. There is some ambiguity, however, about each of these factors when we're dealing with bolts. Let's take a closer look.

B. Change in Length of the Bolt

Effective Length

Tensile loads are not applied to bolts "from end to end"; they're applied between the inner face of the nut and the undersurface of the head. The

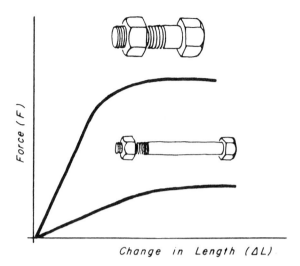

Figure 5.2 Elastic curves for a short, stubby bolt and a long, thin bolt cut from the same material.

entire bolt is not loaded, therefore, the way the test rods are. There is zero tensile stress in the free ends, for example.

There is, however, some stress in portions of both the head and the threads (see Fig. 2.5). We cannot assume that the bolt is merely a cylinder equal in length to the grip length. Instead, we have to make some assumption concerning the stress levels which will allow us to estimate an "effective length" for the bolt which is somewhere between the true overall length and the grip length.

We know from Chap. 2 that tensile stress in a bolt is maximum near the inner faces of the head and nut, and that tensile stress is zero at the outboard faces of nut and head. Assuming that there is a uniform decrease in stress from inboard to outboard faces of the head, as suggested by Fig. 2.4, we can make the assumption that the average stress level in the head of the bolt is one-half the body stress; or we can make a mathematically equivalent assumption and say that one-half of the head is uniformly loaded at the body stress level and that the rest of the head sees zero stress. Similarly, we can say that one-half of the threads engaged by the nut are loaded at the "exposed thread" stress level. We are now in a position to say that the effective length (L_E) of the fastener is equal to the length of the body (L_B) plus one-half the thickness of the head (T_H) added to the length of the exposed threads (L_T) plus one-half the thickness of the nut (T_N), as suggested by Fig. 5.3.

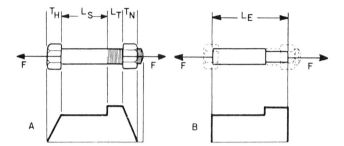

Figure 5.3 Illustration of actual bolt configuration and average tensile stress levels (A) and the equivalent configuration and stress distribution assumed for calculation purposes (B).

Compare the "actual" stress levels sketched in Fig. 5.3 with those shown in Figs. 2.5, 2.6, and 2.7. We have taken the simplest case for estimating the effective length of our bolt. There's really no simple way we could deal with the *true* stress distribution—that would involve a finite-element analysis or the like and would require more information about the geometry of a particular bolt and joint than we'll ever have in practice. We'll find, however, that the assumptions we have made give us reasonable predictions in many applications, because the bulk of the fastener is stressed at, or near, the levels we have assumed. It is only the surfaces of the fastener that exhibit the maximum deviations from these averages.

At least that's true as far as long, thin bolts are concerned. As the length-to-diameter ratio of the bolt decreases, and the bolt becomes more and more short and stubby, our assumption of effective length becomes more and more suspect. More about this in Chaps. 9 and 11.

I have suggested that we use one-half the thickness of the head and one-half the thickness of the nut to compute the effective length of the equivalent fastener. I should mention in passing that other sources recommend slightly different correction factors, such as 0.4 × nominal diameter for the head, and another 0.4 × D for the nut [1]; or 0.3 × D for each [2]; or nothing for the head and 0.6 × D for the nut [3]. The thickness of a standard heavy hex nut, incidentally, is equal to the nominal diameter, so the correction I have suggested is equal to 0.5 × D for a heavy hex nut and a little more for a light nut.

Cross-Sectional Areas of the Bolt

We also have to make some assumptions concerning cross-sectional areas of the bolt when computing change in length. The body area is no mystery;

it's merely equal to $\pi \times D^2 \div 4$, where D is the nominal diameter of the fastener.

For the cross-sectional area of the threads, however, we must use the effective or "stress area" discussed in Chap. 2.

Computing Change in Length of the Bolt

We can now compute the approximate change in length of the bolt under load:

$$\Delta L_C = F_P \left(\frac{L_{be}}{EA_B} + \frac{L_{se}}{EA_S} \right) \tag{5.8}$$

where L_{be} = the effective length of the body (true body length plus one-half the thickness of the head of the bolt) (in., mm) (see Appendix G)

L_{se} = the effective length of the threads (length of exposed threads plus one-half the thickness of the nut) (in., mm) (see Appendix G)

ΔL_C = combined change in length of all portions (in., mm)

A_S = the effective stress area of the threads (see Chap. 2) (in.^{2}, mm^2) (see Appendix F)

A_B = the cross-sectional area of the body (in.^{2}, mm^2)

E = the modulus of elasticity (psi, N/mm^2) (Table 4.6)

F_P = tension in bolt (lb, N)

If the fastener has a more complex shape, as shown in Fig. 5.4, then additional sections must be computed, but there is otherwise no change in the procedure. The change in length for the fastener shown in Fig. 5.4, for example, would be

$$\Delta L_C = F_P \left(\frac{L_1}{EA_1} + \frac{L_2}{EA_2} + \frac{L_3}{EA_3} + \frac{L_4}{EA_4} + \frac{L_5}{EA_5} + \frac{L_6}{EA_6} \right) \tag{5.9}$$

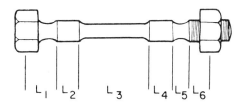

Figure 5.4 Each cross section of a complex fastener must be computed separately.

C. Bolt Stiffness Calculations

Once we know how to compute the change in length of the fastener, we can also estimate the spring constant or stiffness, using the relationship

$$K_B = \frac{F_P}{\Delta L_C} \qquad (5.10)$$

Before using this equation let me mention that Eq. (5.12) gives us an alternative—and perhaps more convenient—way to estimate bolt stiffness. But for now let's continue with Eq. (5.10).

D. Example

Let's compute the stiffness and change in length of a $\frac{3}{8}$–16 $\times$ $1\frac{1}{2}$ SAE, Grade 8 hex bolt (shown in Fig. 5.5) in a joint having a 1-in. grip length (L_G). We can get the nominal dimensions we'll need by measuring a sample bolt, or, more safely, by referring to the pertinent specifications, or to the data in Appendixes F and G. These tell us that:

Specification	Dimension
SAE J 104 (nut)	Height of nut (T_N) = 0.3285 in.
SAE J 105 (bolt)	Height of head (T_H) = 0.2345 in.
	Thread length (L_T) = 1.000 in.
ANSI B1.1-1974 (thread)	Tensile stress area of threads (A_S) = 0.0775 in.[2]

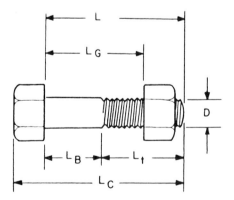

Figure 5.5 Bolt whose stiffness and elongation under load are computed in the text.

From the description of the bolt we already know, of course, that

Nominal body diameter (D) = 0.375 in.
Nominal shank length (L) = 1.500 in.

We must now compute the nominal cross-sectional area (A_B),

$$A_B = \frac{\pi D^2}{4} = \frac{\pi(0.375)^2}{4} = 0.1104 \text{ in.}^2$$

the nominal length of the body (L_B),

$$L_B = L - L_t = 1.5 - 1.0 = 0.5 \text{ in.}$$

and the effective lengths of body and threaded sections, remembering that we're interested only in the threads that are actively engaged in carrying load:

$$L_{be} = L_B + \frac{T_H}{2} = 0.5 + 0.1173 = 0.6173 \text{ in.}$$

$$L_{se} = L_G - L_B + \frac{T_N}{2} = 1.0 - 0.5 + 0.1643 = 0.6643 \text{ in.}$$

We're now ready to compute the reciprocal of the spring constant of this bolt in this joint:

$$\frac{1}{K_B} = \frac{L_{be}}{EA_B} + \frac{L_{se}}{EA_S} = \frac{0.6173}{30 \times 10^6 \times 0.1104}$$

$$+ \frac{0.6643}{30 \times 10^6 \times 0.0775} = 0.4708 \times 10^{-6} \text{ in./lb}$$

Once we have this, we can compute the change in length for a given force. What force should we use? A typical tightening specification would be "60% of yield strength," which, for an SAE Grade 8 bolt, would be 0.6 × 130,000, or 78,000 psi. (You'll find yield strengths in Tables 4.1 and 4.4)

To convert this desired stress level to axial force, we multiply the stress value by the tensile stress area of the threads:

$$F_P = \sigma \times A_S = 78,000 \times 0.0775 = 6045 \text{ lb}$$

We can now compute the change in length which this tensile force would create in this bolt in this joint,

$$\Delta L_C = F_P\left(\frac{1}{K_B}\right) = 6045 \times 0.4708 \times 10^{-6} = 0.00285 \text{ in.}$$

and, for later reference, we note that

$$K_B = \frac{1}{0.4708 \times 10^{-6}} = 2.124 \times 10^6 \text{ lb/in.}$$

E. Actual vs. Computed Stretch and Stiffness

The equations we have given are widely used, but the stretch and stiffness they predict can be quite inaccurate. The exact dimensions of a given bolt won't be the nominal dimensions in most cases, because of manufacturing variations. The modulus of elasticity will vary a little. Stress concentrations of the sort discussed in Chap. 2 can make a large difference in the actual relationship between applied force and change in length in a given bolt. Bending can also distort the relationship. These and other factors will be discussed at length in Chaps. 9 and 11, where we discuss stretch measurement as a way to control preload.

F. Stiffness of Bolt-Nut-Washer System

So far we have considered only the stiffness of the bolt itself. The joint is never clamped by "a bolt," however; it's clamped by a bolt-and-nut "system"—or by a bolt-nut-washer system. The stiffness of this combination of parts is found by

$$\frac{1}{K_T} = \frac{1}{K_B} + \frac{1}{K_N} + \frac{1}{K_W} \qquad (5.11)$$

where K_T = total stiffness of the system (lb/in., N/mm)
K_B = stiffness of the bolt (lb/in., N/mm)
K_N = stiffness of the nut (lb/in., N/mm)
K_W = stiffness of washer (lb/in., N/mm)

As we'll see in Chap. 13 when we examine the nonlinear behavior of a joint, the fact that the joint is clamped by a bolt-nut-washer system, instead of by the bolt alone, makes a big difference. The stiffness of the system, for example, is only about half the stiffness of the bolt alone [6]. Interactions between parts also make the behavior—including the apparent stiffness—drastically nonlinear, especially at low load levels.

None of this is taken into consideration in classical joint design, however, which assumes linear, elastic behavior and which assumes that the stiffness of the clamping element will be that of the bolt alone. We'll know better by the time we study Chap. 13.

Before leaving the subject of washers we should note that the washer can have a significant impact on the stiffness of the joint. A large-diameter, heavy washer will allow a bolt to apply clamping force to more joint material than will a light washer—or no washer. The more joint material involved, the more force it takes to compress the joint by a given amount; i.e., the joint becomes a stiffer spring. As we'll see in later chapters, this has many implications for the design and behavior of the bolted joint.

G. Alternative Expression for Bolt Stiffness

Although the procedure we've used to compute bolt stiffness is correct and emphasizes the all-important deflection or change in length of the bolt, it's cumbersome. If we're interested in the stiffness of a conventional bolt we can often use a simpler expression as follows.

We start, again, with Eq. (5.10).

$$K_B = \frac{F_p}{\Delta L}$$

But now we use Hooke's Law to eliminate the F term.

Hooke's law $E = \dfrac{F_p/A_s}{\Delta L/L_e}$

Rewriting it,

$$F_p = \frac{EA_s\Delta L}{L}$$

Substituting this expression for F_p in Eq. (5.10) gives us

$$K_B = \frac{EA_s}{L_e} \tag{5.12}$$

where E = modulus of elasticity of the bolt (psi, GPa)
   A_s = tensile stress area of the bolt (in.^{2}, mm^2)
   L_e = effective length of the bolt (in., mm)

This equation assumes that the stiffness of the body of the bolt is the same as the stiffness of the threaded section—or that the bolt is fully threaded. If neither is true, Eq. (5.12) is not as accurate as Eq. (5.10). Nevertheless, it's widely used and gives us a convenient way to approximate K_B. When we computed the stiffness of the bolt shown in Fig. 5.5 we used the following equation:

$$\frac{1}{K_B} = \frac{L_{be}}{EA_B} + \frac{L_{se}}{EA_s}$$

As you can see, this is an extended version of Eq. (5.12). Like Eq. (5.12), furthermore, it doesn't require us to compute the change in length of the bolt. Although a little more complicated than Eq. (5.12), it's still easy to use and gives us a far more accurate estimate of K_B than does the more common Eq. (5.12).

H. The Energy Stored in the Bolt

Lest we forget!—we're interested in the bolt as an energy storage device. It can create that all-important clamping force on the joint only as long as it retains potential energy. Although we'll always be interested in whether or not our bolts are "good" or "bad" energy storage devices, we won't often have to compute the exact amount of that energy. If we do want to compute it, we use the following expression:

$$PE_B = \Delta L \times \frac{F_B}{2} \tag{5.13}$$

where F_B = tension in the bolt (lb, N)
ΔL = deflection in the bolt (in., mm)

And PE_B is the potential energy stored in the bolt (in-lb, mm-N). Note that if we were to plot the tension in the bolt vs. its deflection as we tightened the nut, we'd generate a curve like that shown in Fig. 5.6: a

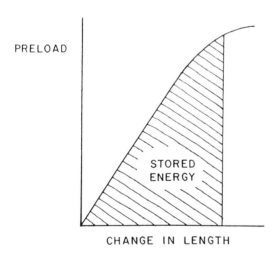

Figure 5.6 A plot of the tension or preload in a bolt vs. its deflection. The curve will be a straight line as long as the bolt deforms elastically. When it yields, the line becomes curved, as shown. The energy stored in the bolt is equal to the area under the curve.

straight line as long as the bolt deforms elastically, curving over at the top if we tighten it so much that it yields (deforms plastically). Equation (5.13) is good only for the straight-line portion of the curve; it defines the area under that curve. We'd have to use graphical techniques, calculus, or a computer to estimate the energy stored in a bolt which had deformed plastically—but this would still be equal to the area under the F_B-ΔL curve.

II. THE JOINT

A. Basic Concepts

We can also treat joint members as springs in series when we compute joint stiffness and deflection. The loads, of course, are compressive rather than tensile, but the basic equations are the same. For example, if we apply equal and opposite forces to a pair of blocks as shown in Fig. 5.7, the change in thickness (ΔT_J) of the system of blocks and the spring constant (K_J) will be

$$\Delta T_J = \Delta T_1 + \Delta T_2 = F\left(\frac{T_1}{EA_1} + \frac{T_2}{EA_2}\right) \tag{5.14}$$

and

$$K_J = \frac{F}{\Delta T_J} \tag{5.15}$$

where

$$\frac{1}{K} = \frac{1}{K_1} + \frac{1}{K_2} = \frac{T_2}{EA_1} + \frac{T_2}{EA_2} \tag{5.16}$$

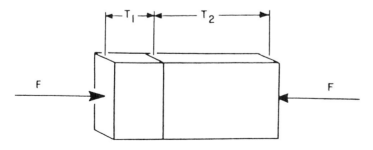

Figure 5.7 Two blocks in compression.

Theoretically, the relationship between applied compressive force and deflection for our pair of blocks should be linear as long as the force stays within the elastic limit of the material. In practice, however, we will often find that the stiffness of a joint is *not* linear and may not be fully elastic. Some report the preload-compression relationship shown in Fig. 5.8 [1]. Others report a variety of nonlinear effects. We'll look at some of these in Chap. 13. Before we consider these complexities, however, it is useful to review the "classical" theories which have been used to evaluate joint behavior in the past. Although simplified, they are often used as a basis for more complex theories. They're also "good enough" for many applications. So let's take a look at them now.

B. Computing Joint Stiffness

We assumed a simplified, equivalent body shape for a bolt, to make routine calculations of stiffness and deflection less complicated. We have to do the same sort of thing for the joint.

That portion of the joint which is put in compressive stress by the bolt can be described as a barrel with a hole through the middle, as suggested in Fig. 2.3. Some workers, therefore, have substituted an "equivalent barrel" for the joint [4], but more common substitutions are hollow cylinders [1] or a pair of frustum cones [5] as in Fig. 5.9.

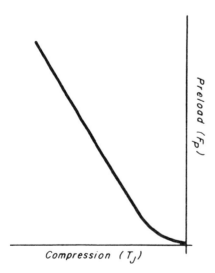

Figure 5.8 The deflection (T_J) of joint members can be nonlinear at low load levels (low F_P).

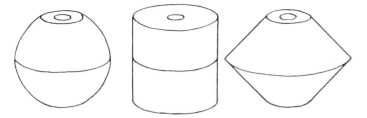

Figure 5.9 Equivalent shapes substituted for joint members in calculating joint stiffness and deformation.

Stiffness of Concentric Joints

A discussion of eight proposed ways to estimate the stiffness of a hard (nongasketed) joint is given by Motosh [4]. The equivalent cylinder approach is described at length by Meyer and Strelow [1]. Unfortunately, each of these techniques assumes that:

1. Joint behavior will be linear and fully elastic.
2. There is only one bolt in the joint and it passes through the center of the members being clamped together (this is called a *concentric* joint).
3. The external load applied to the joint is a tension load and it is applied along a line that's concentric with the bolt axis.

Your own experience, I'm sure, will tell you that limitations 2 and 3 are substantial ones and mean that these equations and recommendations may not apply—at least not very accurately—to many of the joints with which we will be dealing. They're our only choice at the present state of the art, however, except as noted below. At least they're our main *theoretical* choice. If the approximations they give us aren't good enough, we have to determine joint stiffness experimentally.

We will use the equivalent cylinder approach, in this book, to estimate stiffness. This involves the general equation

$$K_J = \frac{EA_C}{T} \tag{5.17}$$

where K_J = stiffness of joint (lb/in., N/mm)
   E = modulus of elasticity (psi, MPa)
   A_C = cross-sectional area of the equivalent cylinder used to represent the joint in stiffness calculations (in.^{2}, mm^2)
   T = total thickness of joint or grip length (in., mm)

Note the similarity of this equation to Eq. (5.12). The big difficulty here is A_C, the cross-sectional area of the equivalent cylinder. The equations we'll use for A_C are summarized in Fig. 5.10. Note that there are three different equations, depending on the diameter of contact (D_B) between the bolt head (or washer) and the joint, and its relationship to the outside diameter of the joint (D_J) [1,7]. If the joint has a square or rectangular cross section, its "diameter" is the length of one side (or of the shortest side of the rectangle). D_H is the diameter of the hole.

Stiffness of Eccentric Joints

Most bolts don't run through the centerline of the joint and/or external tension loads don't align themselves with bolt axes. If bolt and/or load lie

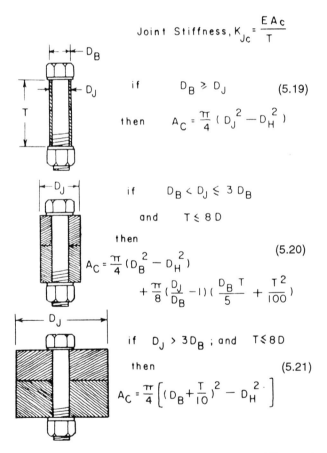

$$\text{Joint Stiffness}, K_{Jc} = \frac{E A_C}{T}$$

if $\qquad D_B \gtrless D_J \qquad$ (5.19)

then $\qquad A_C = \frac{\pi}{4}(D_J^2 - D_H^2)$

if $\qquad D_B < D_J \leqslant 3\,D_B$

and $\qquad T \leqslant 8\,D$

then $\qquad\qquad\qquad\qquad$ (5.20)

$$A_C = \frac{\pi}{4}(D_B^2 - D_H^2)$$
$$+ \frac{\pi}{8}\left(\frac{D_J}{D_B} - 1\right)\left(\frac{D_B\,T}{5} + \frac{T^2}{100}\right)$$

if $\quad D_J > 3D_B$; and $\quad T \leqslant 8D$

then $\qquad\qquad\qquad\qquad$ (5.21)

$$A_C = \frac{\pi}{4}\left[\left(D_B + \frac{T}{10}\right)^2 - D_H^2\right]$$

Figure 5.10 Equations used to compute the stiffness of concentric joints using the equivalent cylinder method. We'll call this stiffness K_{jc}.

away from the joint centerline, the joint is called *eccentric* and our choice of stiffness equations is diminished still further. The German engineering society, Verein Deutscher Ingenieure (VDI), however, has published equations that can be used to estimate the stiffness of eccentric joints as long as the cross-sectional area of that portion of a joint which is loaded by one bolt is not much larger than the contact area between bolt (or nut or washer) and joint [7]. With reference to Fig. 5.11, the area we assume to be loaded by the bolt is A_J. The stiffness equations which follow assume that

$$A_J = b \times W \quad \text{if} \quad W \leq (D_B + T_{min}) \tag{5.18a}$$

$$A_J = b \times (D_B + T_{min}) \quad \text{if} \quad W > (D_B + T_{min}) \tag{5.18b}$$

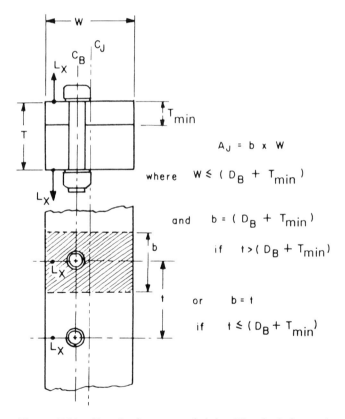

Figure 5.11   Sketch of an eccentric joint. The shaded area, $b \times W$, can be considered that portion of the joint interface which is loaded by a single bolt. See text for the equations used to estimate this area, A_J.

In each case:

$$b = t \quad \text{if} \quad t \leq (D_B + T_{\min})$$

$$b = (D_B + T_{\min}) \quad \text{if} \quad t > (D_B + T_{\min})$$

where W, t, b, and $T_{\min}$ are illustrated in Fig. 5.11 (all in in., mm)
   D_B = diameter of contact between bolt head (or washer) and the joint (in., mm)
   D_H = diameter of the bolt hole (in., mm)

If joint dimensions exceed the limits suggested above (for W), the equations given in Fig. 5.12 don't apply. If the joint satisfies the limitations, then its stiffness may be estimated from the equations given in Fig. 5.12,

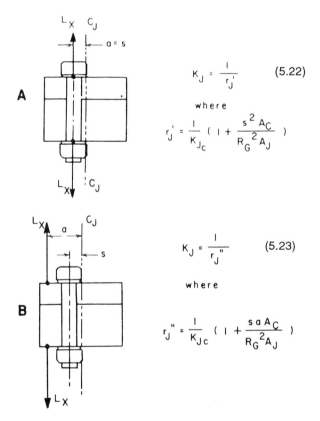

$$K_J = \frac{1}{r_J'} \qquad (5.22)$$

where

$$r_J' = \frac{1}{K_{J_C}} \left(1 + \frac{s^2 A_C}{R_G^2 A_J} \right)$$

$$K_J = \frac{1}{r_J''} \qquad (5.23)$$

where

$$r_J'' = \frac{1}{K_{J_C}} \left(1 + \frac{s a A_C}{R_G^2 A_J} \right)$$

Figure 5.12 Equations used to compute the stiffness of eccentric joints when the line of action of the external load (L_X) coincides with the bolt axis (A) and when it does not (B).

where

C_J = centerline of joint

L_X = external load (lb, N)

a = distance between external load and joint centerline (in., mm)

s = distance between bolt axis and joint centerline (in., mm)

A_C = cross-sectional area of equivalent concentric cylinder (see Fig. 5.10) (in.^{2}, mm^2)

k_{jc} = stiffness of equivalent concentric cylinder (see Fig. 5.10) (lb/in., N/mm)

K_J = stiffness of eccentric joint (lb/in., N/mm)

r'_J = resilience of eccentric joint when load and bolt are coaxial (in./lb, mm/N)

r''_J = resilience of eccentric joint when load and bolt fall along different axes (in./lb, mm/N)

R_G = radius of gyration of joint area A_B (in., mm)

A_J = cross-sectional area of joint (see Fig. 5.11) (in.^{2}, mm^2)

For reference, the radius of gyration for a square cross section is [8]

$$R_G = 0.289d \tag{5.24}$$

where d = length of one side

For a rectangular cross section it is

$$R_G = 0.209d \tag{5.25}$$

where d = length of longer side

For a circular cross section it is

$$R_G = \frac{d}{2} \tag{5.26}$$

where d = diameter of circle

Stiffness in Practice

Experience shows that the stiffness of a "typical" joint (whatever that may be) is about five times the stiffness of the bolt which would be used in such a joint. Very thin joints—sheet metal and the like—will be substantially stiffer, although the stiffness of the bolt will also increase rapidly as it gets shorter, as suggested in Fig. 5.13. In this figure, incidentally,

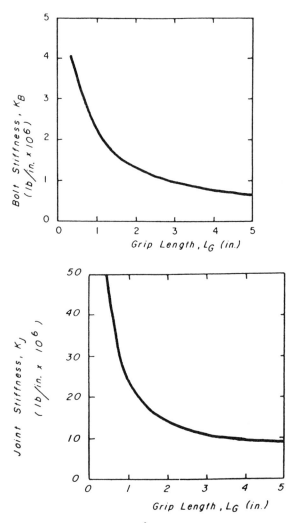

Figure 5.13   Stiffness of $\frac{3}{8}$–16 bolts (K_B) and of the joints (K_J) they must be used in. Bolt hole to joint edge distance has been assumed constant at 1.25 in. Note 10:1 difference in vertical scale.

we have used the equivalent cylinder approach to estimate the possible stiffness of a concentric, hard joint.

A Quick Way to Estimate the Stiffness of Nongasketed Joints

Here's another way to use Motosh and VDI data to estimate the stiffness of nongasketed joint. Both sources have published charts on which are plotted the joint-to-bolt stiffness ratio (K_J/K_B) as a function of the bolt's "slenderness ratio" L/D, where L = the effective length of the bolt and D = nominal diameter.

Figure 5.14 shows a combined version of the published data for slenderness ratios varying from 1:1 to 16:1. The straight line represents the Motosh data; the curved line is from VDI. As you can see, they're in good agreement above a slenderness ratio of about 4:1.

Figure 5.15 shows a similar plot for thinner joints, with L/D ratios of 1.2:1 or less [9]. Projections of the lower end of the VDI and Motosh curves are also shown in Fig. 5.15, showing that the agreement in the data

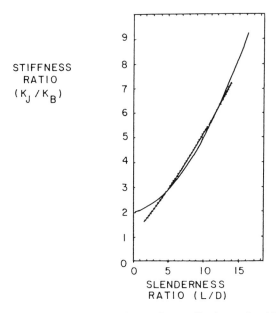

Figure 5.14 Plots of experimentally determined joint-to-bolt stiffness ratio (K_J/K_B) vs. the length-to-diameter slenderness ratio of the bolt (L/D) for L/D ratios up to 16:1. The straighter line is plotted from data published by Motosh, the curved line from VDI data.

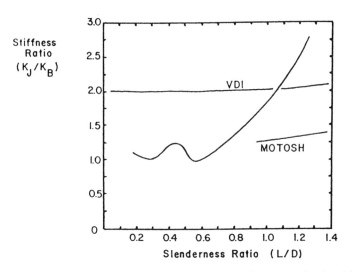

Figure 5.15 Plot of stiffness ratio vs. slenderness ratio for thin joints. These data, like those shown in Fig. 5.14, can be used to estimate the stiffness of nongasketed joints, as explained in the text. The lower ends of the VDI and Motosh curves of Fig. 5.14 are repeated here for comparison.

for thin joints is less than perfect. Nevertheless, for any slenderness ratio, this is the best information I'm aware of.

These curves can be used to estimate joint stiffness as follows:

1. Use Eq. (5.10) or (5.12) or a version thereof to compute the stiffness of your bolts (K_B).
2. Compute the L/D ratio of your bolt, using the effective length (L_{be} + L_{se}) for L.
3. Enter Fig. 5.14 or 5.15 with your L/D ratio and find the corresponding K_J/K_B stiffness ratio.
4. Multiply the K_B computed in step 1 by the K_J/K_B ratio to estimate K_J.

Note that the data in Figs. 5.14 and 5.15 is good only for steel bolts used in nongasketed steel joints. If your joint is made of something else, complete the above steps and then modify the estimate of stiffness as follows.

$$K_J' = K_J \frac{E_m}{30 \times 10^6} \tag{5.27}$$

where K_J = stiffness of a steel joint as estimated from the procedure above (lb/in., N/mm)

K'_j = stiffness of the same joint, but made from an alternate material (lb/in., N/mm)

E_m = modulus of elasticity of the alternate material (psi, MPa)

III. GASKETED JOINTS

A. In General

We've analyzed both the bolt and the joint as groups of springs in series. In such an arrangement, if the stiffness of one spring is substantially less than the stiffness of the others, the "soft" one will dominate the behavior of the group. Minor changes in the stiffness of springs A and C in Fig. 5.16, for example, won't have much influence on the overall deflection of the train of springs under applied load F. By the same token, changes in the applied force will create a much larger change in the deflection of spring B than it will in the deflection of A or C.

Gaskets are relatively soft bodies compared to other joint members; they have to be in order to do their job of plugging leak paths. As a result, the stiffness of a gasketed joint is essentially equal to the stiffness of the gasket. This creates fatigue and other problems, as we'll see in later chapters. And this is true, incidentally, even if the deformation of the gasket is basically plastic, rather than elastic, as is often the case.

As we'll soon see, the force-deflection behavior of a gasket is strongly nonlinear and irreversible (the gasket exhibits hysteresis and creep). Its "stiffness," therefore, varies as it is loaded and then unloaded. In estimating joint behavior we'll be interested in the gasket's behavior as it is unloaded (by pressurizing the vessel, for example, and partially relieving the joint, as explained in Chaps. 12 and 19). Fortunately, its stiffness is similar if it is reloaded, by turning off the pressure and/or by pressure cycles.

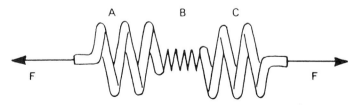

Figure 5.16 The overall stiffness and deflection of a group of springs in series is dominated by one spring (B) if that spring is much less stiff than the others.

Although the earlier statement is usually correct—that the stiffness of a gasketed joint tends to be dominated by the stiffness of the gasket—there are times when we wish to confirm this (and times when it's not true!). The stiffness of a gasketed joint can be estimated as follows.

1. Use the procedure described earlier to estimate the stiffness of a nongasketed joint (K_J of K_J') having the same dimensions as your gasketed one.
2. Now use the following equation, similar to Eq. (5.11), to estimate the stiffness of the gasketed joint.

$$\frac{1}{K_{TG}} = \frac{1}{K_J} + \frac{1}{K_G}$$ (5.28)

where K_{TG} = stiffness of gasketed joint (lb/in., N/mm)
   K_J = stiffness of nongasketed joint (lb/in., N/mm)
   K_G = stiffness of the gasket as it is unloaded or reloaded (lb/in., N/mm)

Note that this discussion, and Eq. (5.28), assumes that the gasket is located between two concentric cylinder joint members which would otherwise be in metal-to-metal contact, a joint such as that is shown in Fig. 5.17A. A joint of the same grip length, using bolts of the same L/D ratio, but having a raised face configuration, such as that shown in Fig. 5.17B, would be less stiff. Fortunately, we now have some data about the

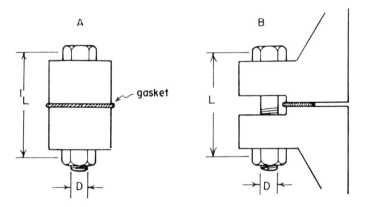

Figure 5.17 The text describes a procedure for estimating the stiffness of a gasketed joint in continuous contact, as in A above. A raised face flange having the same grip length (B) would presumably be less stiff.

bolt-to-joint stiffness ratio for raised face flanges. We'll look at that data in a minute.

To use Eq. (5.28) you need to know the stiffness of the gasket as it is unloaded and/or reloaded in use by pressure cycles, thermal effects, external loads, etc. Such data are not readily available and, if an accurate answer is essential, may require a test. Unfortunately, measuring the deflection of a gasket under load is not an easy thing to do; so even "a test" can be a problem. Perhaps the gasket manufacturer can supply the information.

One of the main reasons that values for gasket stiffness are not published is that the stress-deflection behavior of gaskets is nonlinear, as already mentioned. To define stiffness you must know which part of the behavior curve you're interested in. Figure 5.18 illustrates this problem. It includes stress-deflection curves for several gaskets, as they are initially loaded (part A of each curve) and subsequently unloaded and reloaded in use (part B). Typically the B portion of the curve has a very steep slope

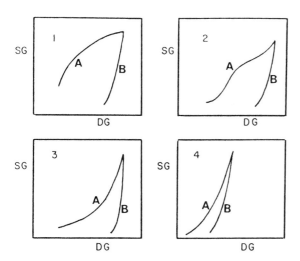

Figure 5.18 Curves of gasket stress (SG) vs. gasket deflection (DG) for four types of gasket: (1) spiral wound, flexible, graphite filled; (2) spiral wound, asbestos filled; (3) stainless steel, double jacketed, with flexible graphite envelope; and (4) compressed asbestos. In each plot the A curve shows the behavior of the gasket as it is first loaded. Curve B shows the behavior as it is unloaded and reloaded. As can be seen, the behavior is nonlinear, with gasket stiffness varying from point to point.

at the upper end (the gasket stiffness is high), but the gasket becomes progressively less stiff as the stress is reduced. It's not uncommon for the stiffness to vary by two or three to one, or more, along the part B curve. Incidentally, the behavior of many compressed fiber, asbestos substitute gaskets is similar to the behavior of the compressed asbestos gasket shown in curve 4 of Fig. 5.18.

Like everything else in bolting, furthermore, the behavior of gaskets is "scattered." The stress-deflection characteristics of one will differ from those of another, even if they're the same type of gasket and were made by the same manufacturer at the same time. The same type, but made by different manufacturers, will vary even more. In addition, the stiffness will usually increase under repeated load cycles and in some cases will change as gasket temperature changes. More about this in Chap. 13.

All of this suggests that an author would be unwise or worse to publish a table of gasket stiffnesses—but here goes!

The data in Table 5.1 has been derived from a long and careful series of room temperature tests sponsored by the Task Group on Gasket Testing of the Pressure Vessel Research Committee (PVRC). The gaskets tested were all designed for pressure vessel use; I don't know of any comparable, publicly available data for automotive or other types of gasket. But the materials tested by the PVRC include compressed asbestos, asbestos substitutes, solid metals, and a number of composites, most of which are also found in the gaskets used by other industries. I believe, therefore, that the data tabulated will be of use to most of you who attempt to estimate the stiffness of a gasketed joint. After all, you'll need a gasket stiffness estimate of some sort, and no matter how approximate these data may be for your application, I doubt that you'll be able to find a better number without making a test of your own.

Note that the stiffnesses in Table 5.1 are expressed in stress (psi) per inch of deflection rather than in force (lb) per inch of deflection. A gasket with a large surface area obviously will require more force to compress than will a smaller gasket. For our purposes, computing the stiffness of that portion of the gasket which will be compressed by a single bolt, we must decide how to define an appropriate, single-bolt, joint-interface area. I suggest the shaded area A_J of Fig. 5.11 would be appropriate, since it will also be used in estimating the stiffness of the metallic joint members. An alternate procedure, however, would be to use one of the assumptions of Fig. 5.9 to define the area of gasket loaded by each bolt.

Once you have an estimate of the area loaded by a bolt, you multiply that area by a stiffness value from Table 5.1 to get the per bolt stiffness of the gasket in lb/in. Equation (5.28) is then used to get the combined,

Table 5.1 Gasket Stiffness

Gasket type	Width, in. (mm)	Thickness, in. (mm)	Stiffness, psi/in. $\times 10^5$ (MPa/mm)	Ref.
Compressed asbestos	0.50 (12.7)	0.125 (3)	12–35 (326–952)	10
Compressed asbestos	0.9 aver. (23)	0.062 (1.6)	22–120 (598–3264)	11, 15
Double-jacketed st. steel, flex. graphite envelope	0.50 (12.7)	0.180 (4.57)	17–20 (454–549)	10
Double-jacketed st. steel, asbestos filled	0.50 (12.7)	0.125 (3)	33–65 (898–1768)	16
Double-jacketed low carbon st., mica filled	0.50 (12.7)	0.125 (3)	32–44 (870–1197)	14
Spiral wound asbestos filled	0.689 (17.5)	0.180 (4.6)	10–66 (272–1795)	17
Spiral wound, flex graphite filled	0.563 (14.3)	0.180 (4.6)	7–48 (190–1306)	17
Spiral wound, chlorite graphite asbestos subst.	0.563 (14.3)	0.180 (4.6)	8–69 (218–1877)	17
Sheet PTFE	0.938 (23.9)	0.100 (2.54)	6–26 (163–707)	12, 17
Metal O-ring (hollow), silver plated, 321 st. st. (wall thickness: 0.3 in.)	0.17 ring diam. (4.3)		4–43 (109–1170)	12, 17
Nonasbestos auto head gasket, laminated on metal core	—	0.045 (1.13)	42–97 (1142–2638)	(a)
Beater added, compressed aramid fiber (asbestos subst.), premium quality	0.50 (12.7)	0.064 (1.6)	7–19 (190–517)	13
Compressed aramid fiber (asbestos subst.), service quality	0.50 (12.7)	0.064 (1.6)	8–18 (218–490)	13

Table 5.1   Continued

Gasket type	Width, in. (mm)	Thickness, in. (mm)	Stiffness, psi/in. × 10^5 (MPa/mm)	Ref.
Flexible graphite sheet	0.75 (19.1)	0.061 (1.55)	12.3–18.6 (335–506)	12
Flexible graphite sheet	0.938 (24)	0.10 (2.5)	8–22 (218–598)	17
Low carbon steel, flat	0.50 (12.7)	0.125 (3.18)	40–47 (1088–1278)	10, 14
Stainless steel, flat	0.50 (12.7)	0.125 (3.18)	42–100 (1142–2720)	14
Soft copper, flat flat flanges	0.375 (9.5)	0.063 (1.60)	64–81 (1741–2203)	10
Soft copper, flat 0.064 nubbin, one flange	0.375	0.063	44–69	10
Soft copper, corrugated	0.50 (12.7)	0.020 (0.51)	43–53 (1170–1442)	10
Low carbon steel, corrugated	0.50 (12.7)	0.20 (0.51)	43–54 (1170–1469)	10
18Cr-8Ni st. steel, corrugated	0.50 (12.7)	0.020 (0.51)	33–41 (898–1115)	10

[a] From private correspondence with a gasket manufacturer.
Source: The references cited at the end of this chapter.

per bolt stiffness of the gasket and joint members. Equation (5.10) or (5.12) has already been used to estimate the stiffness of a bolt. You're now ready to compute the various bolt-to-joint stiffness ratios you'll need in Chaps. 12, 13, etc.

B.   Stiffness of a Raised Face Flanged Joint

Recently, Dr. Bernard Nau, one of the members of the PVRC's Committee on Bolted Flanged Connections, performed some calculations to determine the stiffness of a raised face flange [20]. The calculations were made for joints containing spiral wound, asbestos-filled and compressed asbestos gaskets for British Standard BS 1560 flanges of Class 150 to Class 2500 and for nominal pipe diameters ranging from 1 in. (25.4 mm) to 25 in. (635 mm). Other calculations were made for taper hub API 605 flanges

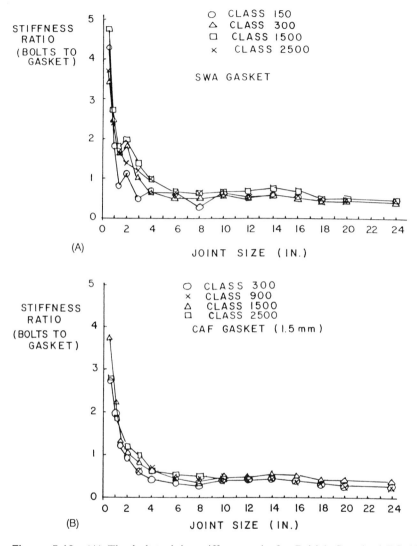

Figure 5.19 (A) The bolt-to-joint stiffness ratio for British Standard BS 1560 raised face flanges containing spiral wound, asbestos-filled gaskets. The stiffness of the joint is assumed to be equal to (dominated by) the stiffness of the gasket. (B) The bolt-to-joint stiffness ratio for BS 1560 raised face flanges containing compressed asbestos gaskets 1.5 mm thick. Again, the gasket stiffness determines the joint stiffness.

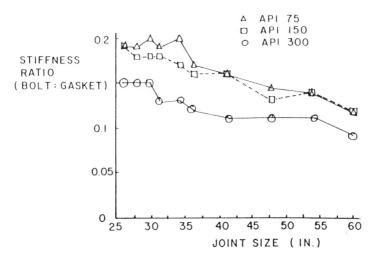

Figure 5.20 The bolt-to-joint stiffness ratio of API 605 raised face flanges containing compressed asbestos gaskets 1.5 mm thick. As before, the gasket stiffness determines the joint stiffness.

of 25 to 60 in. (635 to 1524 mm) in diameter. Some of the results are shown in Figs. 5.19 and 5.20.

Dr. Nau based his calculations on the assumption that the flanges did not rotate (see Chap. 19) and that the stiffness of the joint was determined by the stiffness of the gasket. The charts, in other words, show the ratio of bolt stiffness to gasket stiffness, but we can safely follow his lead and use this as the bolt-to-joint stiffness ratio. Note that for most of the conditions analyzed the BS 1560 joints (gaskets) turn out to be about twice as stiff as the bolts. The bolt stiffness dominates only for the smallest joints. With the API flanges the gasket stiffness is 5–10 times that of the bolts.

In his referenced paper Dr. Nau also roughed out a considerably more complex procedure for estimating the stiffness of a raised face flange if it does rotate. He proposes four simultaneous equations which contain several "difficult" (hard to estimate) terms. I've heard that others have used finite-element analysis to make these estimates.

IV. AN ALTERNATE WAY TO COMPUTE JOINT STIFFNESS

Shoberg and Nassar [19] have recently shown that the stiffness of the joint and the stiffness ratio can be determined in an experiment which measures the torque applied to the nut and the angle through which the nut turns as it is being tightened through the straight line portion of the torque-turn curve shown in Fig. 5.21, 6.3, or 8.1. The equation they have derived is:

$$K_J = \frac{\Delta T / \Delta\theta}{\left(\dfrac{KDP}{360} K_B\right) - \dfrac{\Delta T}{\Delta\theta}} \tag{5.29}$$

where P = pitch of the threads (in., mm)
  ΔT = increase in the torque applied to the nut (lb-in., N-m)
  $\Delta\theta$ = resulting increase in turn of the nut (degrees)
  K = nut factor defining the torque to preload relationship (see Chap. 7)
  D = nominal diameter of the fastener (in., mm)
  K_B = stiffness of the bolt (lb/in., N/mm)
  K_J = stiffness of the joint (lb/in., N/mm)

I'll give the derivation of this equation in Chap. 8, where we'll take a close look at the torque-turn behavior of the joint during assembly.

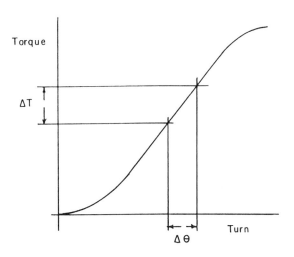

Figure 5.21 This curve shows the typical relationship between applied torque and the turn of the nut when a bolt is tightened (see Chap. 8). The middle portion of the curve is usually an approximately straight line. A change in torque (ΔT) along this straight line, and the corresponding change in angle ($\Delta\theta$), can be used to estimate the stiffness of the joint, as explained in the text.

V. THE JOINT STIFFNESS RATIO OR LOAD FACTOR

Now that we know how to compute or estimate the stiffness of the bolts and of the joint members we're ready to use this data to compute an important design factor called the "joint stiffness ratio" of the bolted joint. Note carefully that this is not simply the bolt-to-joint stiffness ratio; it's more complicated than that. In this book, and in much of the bolting literature, this ratio is expressed in terms of the stiffness of the joint elements, or

$$\phi_K = \frac{K_B}{K_B + K_J} \tag{5.30}$$

where ϕ_K = joint stiffness ratio or load factor (a dimensionless constant)
K_B = stiffness of the bolt (lb/in., N/mm)
K_J = stiffness of the joint material around a bolt (lb/in., N/mm)

In other places, for example, in the German VDI Directive 2230 [18], the stiffness ratio, called a "load factor," is expressed in terms of the resilience of the parts. Resilience is the reciprocal of stiffness (i.e., $r = 1/K$), so

$$\phi_K = \frac{r_j}{r_s + r_j} \tag{5.31}$$

where ϕ_K = joint stiffness ratio or load factor (a dimensionless constant)
r_s = resilience of the bolt (in./lb, mm/N)
r_j = resilience of the joint (in./lb, mm/N)

We'll see how to use this ratio or load factor in Chap. 12 and in the various chapters devoted to the design of the joint.

VI. STIFFNESS—SOME DESIGN GOALS

A. The Energy Stored in the Joint Members

Once again we could plot a curve showing the relationship between the deflection of the joint and the clamping force on it, as in Fig. 5.22; and once again the energy stored in this "spring" is equal to the area under the curve. Theoretically that curve will consist of a straight line with a curve at the upper end if the joint members—or gasket—start to deform

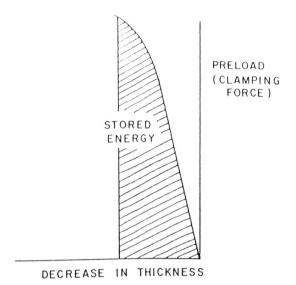

PRELOAD
(CLAMPING
FORCE)

STORED
ENERGY

DECREASE IN THICKNESS

Figure 5.22 A plot of the deflection of the joint members as a function of the clamping force exerted on the joint by the bolts. The plot shown here assumes that joint deflection is linear as clamping force starts to build up, then becomes nonlinear when the joint yields and starts to deform plastically. This assumption is often used for design purposes but doesn't reflect the fact that the true force-deformation curve of the joint is nonlinear throughout, especially if the joint contains a gasket.

plastically. As suggested in Fig. 5.8 and as we'll see in more detail in Chaps. 13 and 18, this picture is a very much simplified illustration of the actual behavior of the joint spring, especially if this is a gasketed joint; but Fig. 5.22 can be taken as a first approximation and is used in most cases to analyze the behavior of a bolted joint. We'll use it, for example, in Chaps. 12 and 22 when we first look at the response of a bolted joint to tensile loads, and in Chap. 22 when we learn how to design a nongasketed joint.

So now—in Figs. 5.6 and 5.22—we have pictures of the energy stored in both bolts and joint members. How do we use this information? The answer: we use it only to remind ourself that stiffness is an important design consideration—because it affects the amount of energy stored in the bolt-joint system. The more energy we can store there, the more abuse the joint will be able to withstand before it fails. And stiffness is the key to the "amount stored." We won't have to measure or compute the energy itself; we'll spend our time worrying about, computing, and manipulating stiffness. Let's take a look.

B. The Relationship Between Stiffness and Stored Energy

Take another look at Fig. 5.6. Let's replace the bolt theoretically illustrated there with a bolt having a larger diameter. Let's tighten the new bolt to the same preload as we tightened the old one. Because the larger bolt is stiffer than the old one, it will deflect less than the old did. Its preload-deflection curve will be steeper—"more vertical" if you will. This means that the area under the curve, estimated by Eq. (5.13), will be less. And *this* means that the same loss of deflection will mean a greater loss of preload in the new, fatter bolt than in the original one. The same amount of thermal change or vibration loosening or relaxation will cause a greater loss of preload in the bolt which has stored less energy.

Alternately, of course, a bolt of smaller diameter taken to the same preload would be less sensitive to the changes mentioned above. If we couldn't take a thinner bolt to this preload we might achieve the same effect by using two thin bolts in place of the original fat one, or by using a longer bolt of the original diameter. More length means less stiffness as suggested by Eq. (5.12).

As an example, consider two $\frac{3}{8}$–20 SAE Grade 5 bolts tightened to proof load (85 ksi). Same applied torque; same preload (7106 lb in these bolts having a tensile stress area of 0.0836 in.^{2}). But one bolt is $\frac{1}{2}$ inch long, the other $\frac{3}{4}$ inch—and they're used in joints of those thicknesses. As we'll see in Table 9.2, these bolts will have been stretched approximately 0.003 in. per in. of grip length, when taken to proof load (just under yield). So the short bolt has stretched 0.0015 in., the longer bolt 0.0023 in. If any of the various relaxation effects reduces the deflection of the shorter bolt by 0.001 in. its preload would drop to 33% of proof load, a residual clamping force of only 2345 lb. The same 0.001-in. reduction in the stretch of the longer bolt would drop its preload to 56% of proof; it retains 3979 lb of clamping force.

The bottom line is this: we almost always want "less stiff" bolts if we have a choice. By this means we hope to avoid the chronic problems often associated with short, stubby fasteners—such as sheet metal screws—which loosen so readily in service.

What about the joint? Do we want it to be as resilient as possible? Probably not. First of all, the bolt, being less stiff, will almost always store much more energy than the joint. You might think of the bolt as the active element in the system, with the joint as the passive or resistive element. Second, and more important, we also want a "good" stiffness ratio in our joints.

C. The Stiffness Ratio

We can't go into it at this point, but as we'll see in Chap. 12 and in the chapters devoted to joint design, we usually want a low stiffness ratio. The lower this ratio the less the clamping force will be affected by external loads, by thermal change, by vibration, by fatigue, etc. So we'll often try to minimize the stiffness ratio. Fortunately, it helps to minimize the bolt stiffness at the same time; for once we're not trying to achieve conflicting design goals.

REFERENCES

1. Meyer, G., and D. Strelow, Simple diagrams aid in analyzing forces in bolted joints, *Assembly Eng.*, January 1972.
2. *Fastener Preload Concepts*, Contract no. F33615-76-C-5251, for Air Force Materials Laboratory, Air Force Systems Command, WPAFB, Dayton, Ohio, by General Dynamics, Fort Worth, June 15, 1978.
3. Sawa, T., and K. Maruyama, On the deformation of the bolt head and nut in a bolted joint, *Bull. JSME*, vol. 19, no. 128, February 1976.
4. Motosh, N., Determination of joint stiffness in bolted connections, *Trans. ASME*, August 1976.
5. Osgood, C. C., How elasticity influences bolted joint design, *Machine Design*, pp. 92ff, February 24, 1972.
6. Pindera, J. T., and Y. Sze, Influence of the bolt system on the response of the face-to-face flanged connections, Reprints of the 2nd International Conference on Structural Mechanics in Reactor Technology, Berlin, Germany, September 10–14, 1973, vol. III, part G–H.
7. Junker, G. H., Principle of the calculation of high duty bolted joints; interpretation of directive VDI 2230, Unbrako technical thesis, published by SPS, Jenkintown, PA, p. 7.
8. Oberg, E., F. D. Jones, and H. L. Horton, *Machinery's Handbook*, 20th ed., Industrial Press, New York, 1975, p. 378.
9. Osgood, Carl C., *Fatigue Design*, Wiley-Interscience, New York, 1970, pp. 196–197.
10. Derenne, Michel, and Andre Bazergui, Results of production tests on different gasket styles, Preliminary report to the Task Group on Gasket Testing of the Pressure Vessel Research Committee, October 1985.
11. Bazergui, Andre, and Luc Marchand, PVRC milestone gasket tests, first results, Welding Research Council Bulletin no. 292, WRC, New York, February 1984.

12. Bazergui, A., L. Marchand, and H. Raut, Draft of report on the development of a production test procedure for gaskets, subsequently published as Welding Research Council Bulletin no. 309, WRC, New York, September 1984.

13. Raut, H. D., Report to Task Group on Gasket Testing, Pressure Vessel Research Committee, January 26, 1988.

14. Bazergui, Andre, Additional results of production tests of gaskets, Preliminary report to the Task Group on Gasket Testing of the Pressure Vessel Research Committee, New York, May 1986.

15. From an expanded plot of Part B test results, gasket stress vs. gasket deflection, compressed asbestos gasket, provided to the Task Group on Gasket Testing of the Pressure Vessel Research Committee by Andre Bazergui, New York, October 1981.

16. Derenne, M., A. Bazergui, and L. Marchand, Draft report on short term mechanical gasket tests at elevated temperatures, Presented to the Joint Task Group on the Elevated Temperature Behavior of Bolted Flanges of the Pressure Vessel Research Committee, New York, October 1986.

17. From enlarged plots of test data produced during the Milestone and Production Test Series, by A. Bazergui, given to the author in 1985.

18. Junker, G. H., Principle of the calculation of high-duty bolted joints: Interpretation of VDI Directive 2230, SPS, Jenkintown, PA, date unknown.

19. Shoberg, Ralph S., *Engineering Fundamentals of Torque-Turn Tightening*, R. S. Technologies, Ltd., Farmington Hills, MI, 1991 or 1992.

20. Nau, B. S., On the design of gasketed joint, 12th International Conference on Fluid Sealing, Brighton, UK, May 10–12, 1989.

II
TIGHTENING THE JOINT
Establishing the Clamping Force

6
Introduction to Assembly

That all-important clamping force which holds the joint together—and without which there would be no joint—is not created by a good joint designer, nor by high-quality parts. It is created by the mechanic on the assembly line or job site, using the tools, procedures, and working conditions we have provided him with. The force is brought into being as energy in the mechanic or power tool is converted to potential energy stored in the joint and bolt members. The correct amount of force cannot be created if the design is faulty or the parts don't fit together properly or they break; but getting all this right, while necessary, is not enough. The final, essential *creator* of the force is the mechanic, and the time of creation is during assembly. So it's very important for us to understand the assembly process.

Because of this, the next six chapters will be devoted to assembly, starting, in this chapter, with an overview of the process. How do the bolts and the joint members respond as we tighten the bolts? We'll see that the behavior of the parts, during assembly, is complex. We'll take a close look at several unseen, difficult to detect, difficult to quantify factors which can have a significant impact on the results, on the amount of clamping force developed in the joint. This in turn will teach us that it isn't easy to control the build-up of clamping force in the joint and that those mechanics need all the help we can give them, both as product designers and as production engineers.

In the following five chapters, we'll look at the many options we have for control of the bolt-tightening process, starting with relatively simple, crude methods and proceeding on to ever more elaborate and accurate ones. Our knowledge of the behavior of the bolts and joint during assembly will help us evaluate the merits of these options.

In still later chapters we'll learn why accurate control of the clamping force is so necessary, how too much or too little clamp can degrade the behavior and life of the joint in service. As we learned in Chap. 1, if the bolts and joint members don't contain the correct amount of stored energy and therefore create the correct amount of clamping force, we'll have joint problems. In other words, proper assembly is essential.

I. INITIAL VS. RESIDUAL PRELOAD

The clamping force a bolt exerts on the joint is usually called—or equated to—the so-called *preload* in the bolt. This term is used in general in most of the literature on bolting to describe the tension in the bolt at any time, but this, in my opinion, is a mistake. I like to think of the preload created in an individual fastener when it is first tightened as *initial* preload, even though that term may be redundant. As you'll see, the effects we're about to discuss will frequently modify this preload as the fastener relaxes and/ or as we tighten other fasteners in the joint. I call the final preload in the bolts the *residual* preload.

When the joint is put into service, a variety of things can act to modify the preload in individual fasteners still further. This could be called *in-service tension in the bolts*.

Each of these preloads or tensions is directly proportional to the amount of potential energy stored in the bolt: as it is first tightened, or after relaxation occurs, or in service. In most cases these preloads or tensions will also be directly proportional to the clamping force between joint members; but there are exceptions, as we'll see.

But—now that we have these definitions under our belt—let's get on with it. What happens during assembly?

II. STARTING THE ASSEMBLY PROCESS

We're going to assemble a hypothetical joint, using as our example a round, gasketed, pipe flange joint held together by 16, $1\frac{1}{8}$–8, ASTM A193 B7 bolts. (See Fig. 6.1.) The large diameter and the presence of a gasket make this assembly a little more difficult than most, but therefore allow us to look at a more complete range of assembly problems than would a simpler example. Most of the discussion would apply to joints in general.

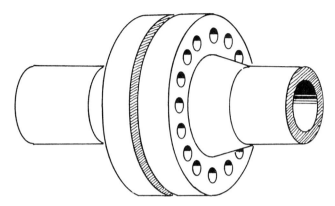

Figure 6.1 This is a sketch of the large-diameter, pressure vessel joint used as an example in this chapter. We see what happens when we install and tighten the bolts.

We're also going to measure the torque we apply to the nuts to control the build-up of initial preload in these bolts. This is the most common, and one of the simplest, types of control. It will be the subject of Chap. 7, so we won't go into a lot of detail here about it's pros, cons, etc. We'll just use it for now.

A. Assembling the Parts

We start by roughly aligning the flanges so that we can insert the bolts by hand. When we finish pushing and pulling on the flanges their mating *1* surfaces are not exactly parallel and the holes aren't aligned perfectly, so *2* we have to tap a few of the bolts with a hammer to get them through their holes, and some of them stick out a little farther, on the nut end, than do *3* others. Now we're going to apply a preliminary "snugging torque" to run the nuts down and pull the flanges together.

B. Tightening the First Bolt

To load the joint (and gasket) evenly we'll apply the snugging torque in a cross or star pattern, as shown in Fig. 6.2. We'd use a similar pattern on a square or rectangular joint if the bolts were all around the edge. In a rectangular, structural joint pattern, with several rows of bolts, we'd start snugging at the center of the bolt pattern and work our way out to the free edges.

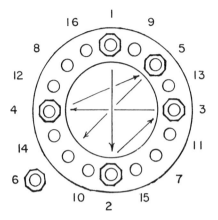

Figure 6.2 We'll tighten the bolts of our example joint in the "star pattern" sequence shown here. We'll use three passes, at one-third, two-thirds, and final torque, following the same sequence on each pass.

$1\frac{1}{8}-8$ B7 Studs

We'll use 225 lb-ft of torque for this first, snugging pass. This is about a third of the final torque we're planning to use, and we'll follow it with a second pass at two-thirds of final torque, and then a third and final pass at full torque. In a structural steel joint we would follow the snugging pass with a second (last) pass at the final torque. Note that in each case we're following basically a two-step procedure: pull the joint together and then tighten it. Because this is a learning experiment we'll use ultrasonic equipment (Chap. 11) to measure the preload in each bolt as we tighten it. We'll also measure the angle through which the nut turns after it contacts the surface of the joint, and we'll measure the amount by which the bolt stretches and the amount by which the joint is compressed.

We now apply the snugging torque to the first bolt and use the resulting preload, torque, and turn data to plot the curves shown in Fig. 6.3. We're doing work on this fastener as we tighten it. The amount of work is equal to the area under the torque turn curve (measured in lb-ft or N-m times radians). Ideally, all of this work would be converted to potential energy in the bolt and in those portions of the joint members which surround it. If that were the case all of the work we do on this fastener would end up contributing to the clamping force. Unfortunately and unavoidably, most of our input work is lost.

Typically about 90% of the work we do on a nut is converted to heat, thanks to the frictional resistance between the face of the nut and the surface of the joint and between male and female threads. About 50% is

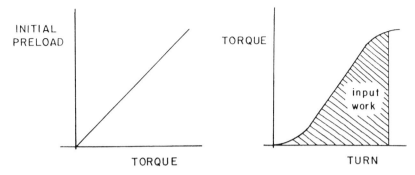

Figure 6.3 As we tighten the first bolt in our example joint we plot the buildup of initial preload vs. applied torque (left-hand diagram) and applied torque vs. the angle through which the nut turns (right diagram). The area under the torque-turn curve is equal to the amount of work we're doing on the nut and to the energy delivered to the fastener-joint system.

lost under the nut, about 40% within the threads, as shown in Fig. 6.4. Only 10% of the input work typically ends up as potential energy in the bolt; so only 10% ends up as bolt preload or as clamping force between joint members.

We'd like to apply a given torque to each bolt and create a given amount of initial preload—the same amount—in each bolt. But the fact that most of the work we do on the nuts is converted to heat makes this virtually impossible, because these frictional losses are extremely difficult to predict or control. Let's assume, for example, that this first nut we're tightening is a little drier than average. As a result, let's assume that 52% of the input work is converted to heat at the nut-joint interface, rather than the "typical" 50%. A 4% increase in friction—from 50% to 52% of the input work—is easy to come by.

This 4% increase in friction loss—that extra 2% of the input work going into heat—means that 2% LESS of the input work will be converted to the thing we're interested in, the tension in the bolt. We started with the assumption that an average of only 10% of the input would be going into preload; now we've lost a fifth of that. This bolt will, therefore, end up with only 80% as much preload as we expected it to. A 4% swing in friction has caused a 20% change in assembly preload, a very bad "leverage" situation. And, as we'll learn in the next chapter, there are a lot of factors which can cause this kind of variation in friction.

Although in our learning experiment we're measuring both torque and bolt tension, we won't attempt to compensate for the frictional differences

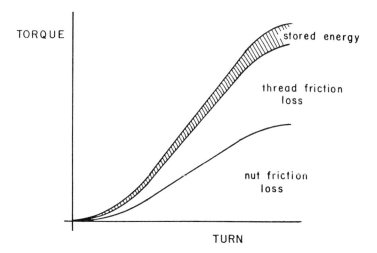

Figure 6.4 This diagram shows the approximate way in which the energy delivered to the fastener-joint system is absorbed by it. About 50% of the input is lost as friction-generated heat between the face of the nut and the surface of the joint. Another 40% is lost as heat between male and female thread surfaces. Only about 10%, on the average, ends up as potential energy stored in the bolt and joint "springs"; and only that 10%, therefore, ends up as preload in the bolt and clamping force on the joint.

between bolts; we'll apply the snugging torque of 225 lb-ft to the first bolt and let the initial preload end up where it may.

We also plot the deflections in the bolt and in the joint material surrounding the bolt vs. the preload we create in the bolts and the presumably equal and opposite clamping force on the joint. See Fig. 6.5. We then combine these force-deflection curves, plotting the preload on a common axis, as also shown in Fig. 6.5. This creates what the bolting world calls a "joint diagram." The pure of heart among you may complain that the preload in the bolt and the clamping force on the joint are equal and opposite, action and reaction forces, and that both should not be shown as positive values, but this joint diagram is a great convenience so we'll draw it as shown.

Since the diagram records the forces developed in bolt and joint and the deflection of each part, it also gives us a visual indication of the stiffness of the bolt and the stiffness of the joint. These are proportional to the slopes of the two straight lines, and, as we saw in Chap. 5, can be computed as follows:

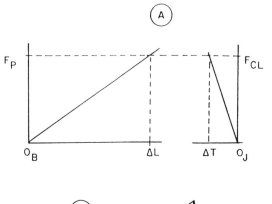

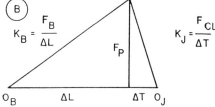

Figure 6.5 As we tighten the first bolt we also plot the buildup of preload (F_P) in the bolt vs. the increase in length (ΔL) of the bolt, and the buildup of clamping force on the joint (F_{CL}) vs. the compression or change in thickness (ΔT) of the joint. At this point we assume that the preload will be equal to the clamping force. These two plots are shown at the top of this illustration. We then combine those two plots, as shown at the bottom of this illustration, to start constructing what we'll call a "joint diagram."

$$K_B = \frac{F_P}{\Delta L} \tag{6.1}$$

$$K_J = \frac{F_P}{\Delta L} \tag{6.2}$$

where F_P = preload in the bolt and joint (lb, N)

ΔL = deflection (stretch) of the bolt (in., mm)

ΔT = deflection (compression) of the joint (in., mm)

K_B = stiffness of the bolt (lb/in., N/mm)

K_J = stiffness of the joint material being loaded by this bolt (lb/in., N/mm)

Note that the areas under the bolt and joint curves also equal the amount of energy stored in these parts, as shown in Figs. 5.6 and 5.22 in the last chapter. So this simple diagram contains a lot of useful information. We'll extend this diagram in Chap. 12 to add the effects of external loads on the joint. We'll also use joint diagrams when we design joints. For now, however, we're merely interested in using the joint diagram to illustrate the preloading of the bolts.

Our boss, who has already read a previous edition of this book, uses Eq. (7.4) in the next chapter to compute the average preload he expected us to get in this first bolt when we applied 225 lb-ft of torque to it. He tells us that we should have created 12,000 lb of tension in the bolt. Because of the slightly higher than average friction loss described earlier, however, this bolt has ended up with only 80% of that preload, or 9600 lb.

Has this really created 9600 lb of clamping force between joint members? Our joint diagram assumes it has, but the correct answer is "probably not"—at least as far as this first bolt is concerned. Remember that we had to tap some of those bolts into their holes? This implies that there was contact between the sides of those bolts and their holes—bolt/hole interference. Furthermore, the flange surfaces were not pulled into full contact when we first assembled the parts. They were slightly misaligned, as we could tell by the fact that some of those bolts stuck out farther on the threaded end than did others. Before we go on to snug tighten the remaining 15 bolts in our example joint, let's talk a look at how hole interference and/or nonparallel flanges might affect the buildup of clamping force during the assembly process.

III. BOLT PRELOAD VS. CLAMPING FORCE ON THE JOINT

The main purpose of the bolts is to clamp the joint members together. A common misconception is that there is always an equal and opposite action-reaction relationship between the tension in the bolts in a joint and the thing we're interested in, the clamping force between the joint members. If there are eight bolts in the joint and an average tension of 10,000 lb in each of the eight bolts, then, simplistically, the joint is clamped together with an interface force equal to eight times 10,000, or 80,000 lb. That's usually true, but there can be some significant exceptions.

A. Effects of Hole Interference

Consider the situation shown in Fig. 6.6. We are tightening this stud by turning the upper nut. Our goal is to clamp the joint members together. The hole in the upper joint member is undersized, so the stud is a press fit in this member.

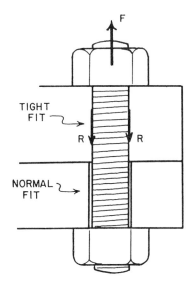

Figure 6.6 If there is interference between the bolt and the hole, the clamping force between joint members may be less than the tension in the bolt, changing the torque–clamping force relationship.

Thanks to frictional and/or embedment constraints between the sides of the bolt and the walls of the hole, it will take some positive force to pull the bolt through the hole—and then to stretch it within the hole. Where does this force come from?

The force is created, obviously, when we turn the nut, creating tension in the bolt. Thanks to hole interference some of this tension will not end up as clamping force between joint members. Part of it will be lost as the bolt fights its way past the walls of the hole.

I'm sure you can envision the extreme case in which the holes are grossly undersized and the torque normally specified for a bolt of this diameter is insufficient even to bring the joint members into contact, much less provide any real clamping force. Note that misalignment between the holes of upper and lower joint members could create a similar hole-bolt interface problem.

This is obviously not a "good" situation, but hole misalignment, undersized holes, press-fit fasteners, etc. are relatively common in the bolting world.

Hole interference is used on purpose by the airframe industry, for example, to reduce the possibility of fatigue failure of the shear-loaded joints used in airframe structures. (Compressive stress built up in the walls of the bolt holes fights the formation or growth of fatigue cracks.)

The holes are purposely drilled smaller than the diameter of the bolts to create this interference. There is, of course, a manufacturing tolerance on the diameters of both holes and bolts, so the amount of interference varies. The greater the interference, the greater the force required to pull the bolt through its hole. It also requires more force to pull a bolt through a thick plate than through a thinner one, for a given amount of interference.

One airframe manufacturer recently measured the amount of force required to pull bolts of a given nominal diameter through holes drilled in plates of varying thickness. Some of the results are shown in Figs. 6.7 and 6.8. The bolt and hole diameters used in the experiment varied through the full range of the manufacturing tolerance. It was found that the force required to pull some of these bolts through their holes exceeded the average preload which would be developed, by the specified torque, in a bolt of that diameter, even if the bolts were used in regular holes with normal clearance. In other words, the specified torque could not be

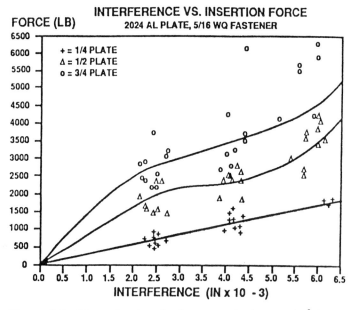

Figure 6.7 Chart showing the force required to push $\frac{5}{16}$-in. fasteners through interference fit holes in aluminum plates varying in thickness from $\frac{1}{4}$ to $\frac{3}{4}$ in. Bolt diameters are larger than the hole diameters by the amounts shown on the horizontal axis. Note the wide scatter in the results, which are summarized in Fig. 6.8.

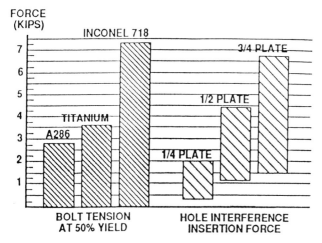

Figure 6.8 The three vertical bars on the right side of this chart show the range of force required to thrust $\frac{5}{16}$-in. fasteners through interference fit holes in aluminum plates of various thickness, summarizing some of the data shown in Fig. 6.7. The three bars on the left side of the chart show the nominal preload which would be generated by $\frac{5}{16}$-in. bolts if tightened to 50% of yield. Three bolt materials are shown. Note that only the Inconel bolts would have enough preload to overcome the worst-case interference forces and still provide some interface clamping force on the joint if $\frac{1}{2}$-in. or $\frac{3}{4}$-in. plates were involved.

counted on to pull all of the bolts through their holes, much less go on to develop any clamping force between joint members. In spite of this the same torque is specified by the airframe manufacturer for all bolts of a given diameter, without regard to the amount of hole interference seen by a given bolt—or the thickness of the plates in which the bolt is used.

As a result, torque is *not* used in this application to pull the bolts through the holes. A bolt puller does that job, and temporary clamps are used to hold the joint members together until the bolts have been installed and tightened. But in some joints the act of tightening the bolts is supposed to create some clamping force between upper and lower joint members, and the amount of this clamp must vary widely.

B. Resistance from Joint Members

Another factor which can rob from the clamping force between joint members is shown in Fig. 6.9. A heavy cover is being lifted up against a flange on a pressure vessel. As shown in the figure, the joint members have not

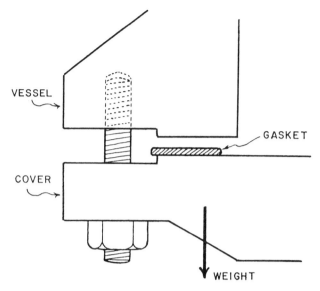

Figure 6.9 Heavy and/or misaligned or warped joint members can also affect the relationship between the torque applied to the fastener, the tension in the fastener, and the clamping force between joint members.

yet been brought into contact, but we are turning the nut on the stud to bring them into contact. At the moment shown in the figure, there is already tension in the stud equal to the weight of the cover. As a result, it will take torque to advance the cover up against the flange, thanks to the normal frictional constraints between male and female threads and between the nut and the cover.

So, at the point shown, there is torque, there is tension, there is friction loss; there probably will be torsion in the stud—but there is zero clamping force between joint members.

Eventually the two joint members *will* be brought into contact. Further torque, at this point, will be required to create a clamping force between joint members to load the gasket. That torque should presumably be added to the torque required to pull the joint members together in the first place—but it rarely is, in practice.

There aren't many applications in which a heavy weight is raised against a mating joint member by tightening the bolts, of course (although we have encountered some such situations). It is, however, common for large joint members to be misaligned or nonparallel as the assembly process starts. Getting a pipe flange, for example, to mate with the flange of a pump or valve often requires a lot of motion in the flange members. The forces required to align such systems will have the same effect that the force created by the weight in Fig. 6.9 has on that joint.

I wish I could tell you how much extra torque to add for misalignment or the like, but I can't. I have never seen anything in the literature on this subject, either. Yet I'm convinced that this factor can seriously degrade the relationship between tension in the bolts and clamping force between joint members, especially in large joints, as one example. Warped or nonflat (e.g., wavy) joint members, incidentally, could create the same sort of problem.

I once met a maintenance supervisor in a large petrochemical plant who took this problem seriously. He insisted that his crews align gasketed flanges within 12 mils before bolting them together. Bolting nonparallel flanges, he said, was a waste of time; "they'll always leak."

The Puget Sound Naval Shipyard has also studied this problem [18]. They made theoretical calculations and conducted experiments to determine the forces, stresses, and moments in pipes and flanges when the flanges are misaligned. They used an hydraulic tensioner, for example (see Chap. 10), to measure the force required to pull misaligned flanges together, as shown in Fig. 6.10. The chart in Fig. 6.11 shows some of the

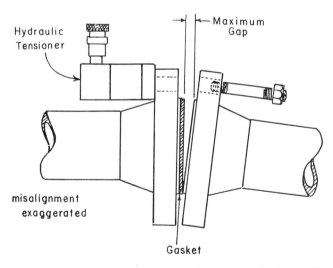

Figure 6.10 Puget Sound Naval Shipyard used a hydraulic tensioner as shown in this sketch to learn how much force would be required to pull misaligned flanges together. The data was recorded as a function of the size of the "gap" shown here.

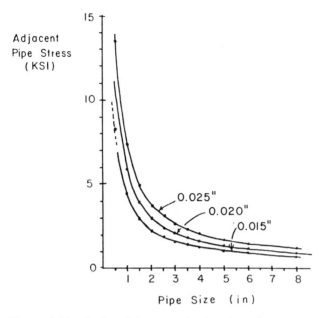

Figure 6.11 A plot of the stress in the pipe adjacent to a misaligned flange vs. the nominal diameter of the pipe, for misalignment gaps of 0.015, 0.020, and 0.025 in.. It is assumed that the nearest rigid support for the pipe is located 100 pipe diameters away, along the pipe, from the misaligned flange. The moment on the flange and the force required to pull the two halves into full contact would be proportional to the pipe stress.

results of their work: the stresses in the pipe adjacent to the flange as a function of nominal pipe diameter and with flange gaps of 0.015, 0.020, and 0.025 in. The forces (and torques) required to pull the flanges together would be proportional to these stresses.

In each case documented in Fig. 6.11 it is assumed that there is a length of pipe equal to 100 times the nominal pipe diameter between the flange and the first rigid pipe support (a rod hanger or intersection with a larger pipe).

As a result of these studies, Puget Sound developed some flange parallelism criteria, designed, as I understand it, to keep pipe stress below 3 ksi near any flange which is connected to turbines or other rotating equipment. Pipe stresses beside flanges located 50 or more pipe diameters away from rotating equipment were allowed to go slightly higher. The resulting criteria are shown in Table 6.1.

Table 6.1   Flange Parallelism Criteria[a]

	Flanges adjacent to rotating equipment			Line flanges
Pipe size	1 in. or less	$1\frac{1}{4}$ through 8 in.		All sizes
Wall thickness	All	SCH 40 or below	Over 40 through SCH 80	All thicknesses
Raised face	0.10 in.	0.025 in.	0.020 in.	0.005 in./in. of contact diameter
Flat face	0.020 in.	0.035 in.	0.030 in.	0.005 in./in. of contact diameter

[a] If the pipe diameter is over 10 in. or wall thickness greater than schedule 80, the specifications say that a special analysis is required.

As a final note, if misalignments and stresses are too high they are sometimes reduced by the use of bellows-type expansion joints at some point in the pipeline.

IV.   CONTINUING THE SNUGGING PASS

We're now going to continue tightening the 16-bolt gasketed, flange joint we're taking as an example in this chapter. We apply the snugging torque of 225 lb-ft to each of those bolts, following the cross-bolting pattern shown in Fig. 6.2. We measure the preload, turn, and deflection in and around each bolt as we tighten it and we record these data. We find that, on average, we've created the anticipated preload in these bolts but that individual bolt results vary from the average by ±30%. That, our boss assures us, is a typical result of torque-tightening a group of unlubricated, as-received, steel bolts and nuts against steel joint surfaces. Everyone's happy, so we now go out and have lunch.

When we come back from lunch we find our boss's boss on the job site, reviewing our data. He wants to see how we managed to measure the tension or preload in these bolts, so we plug in our ultrasonic instrument and remeasure the preload in bolt number one, the first bolt we tightened to a preload of 9600 lb. To our embarrassment we find only 900 lb of tension in that bolt. And we find a wide range of residual preloads in the other 15 bolts. We think that our instrument is misbehaving, but the boss disagrees. He gives us a lecture on bolt relaxation, which some people call "torque loss." Here's what he says.

V. SHORT-TERM RELAXATION OF INDIVIDUAL BOLTS

Whether or not there's a one-to-one relationship between bolt tension and interface clamping force, there will often be some initial loss of tension in individual bolts after they are initially tightened. Let's call this "short-term" relaxation, to distinguish it from other effects, to be discussed in Chap. 13, which will cause further loss of tension over a long period of time.

In general, short-term relaxation occurs in a bolted joint because something has been loaded past its yield point and will creep and flow to get out from under the excessive load. This can be a component, such as a soft bolt or a gasket; more commonly it's only a portion of a component, such as the first threads in a nut. Let's look at some examples.

A. Sources of Short-Term Relaxation

Here are some things which can cause relatively short-term relaxation, starting with the most common of all, embedment.

The surfaces of the threads in the nut, the bolt, and the faying surfaces of the structural members, washers, etc. are never perfectly flat, even if such parts are given a high polish—which is rarely the case with industrial structures and fasteners. Under a microscope they are a series of hills and valleys.

When such parts are first loaded, they contact each other only through high spots on the metal surfaces. Even a small bolt, however, is able to exert extremely high surface pressures on structural members or on its own threads. Thread dimensions have been selected to support these high loads, but only if a significant percentage of the total thread surface shares that load.

Since initial contact areas are relatively small, the metal at the contact points cannot stand the pressures. Plastic deformation occurs until enough of the total thread surface has been brought into play to stabilize the situation and support the load without further deformation.

The same thing happens in the faying surfaces of the structures, though perhaps to a lesser extent because larger surfaces are involved and initial contact areas are larger. Embedment is illustrated in Fig. 6.12.

Many of these surface high spots are smoothed away during the tightening process [1]—at least they will be if the fasteners are torqued. Hydraulic tensioners don't load the active threads, or even the joint surfaces underneath the nut or washer, until they let go of the bolt. As a result, there is often more embedment relaxation after tensioning than there is after torquing.

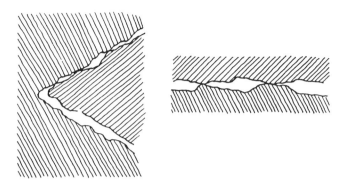

Figure 6.12 High spots on thread and other contact surfaces will yield and creep under initial contact forces. As a result, the surfaces will settle into each other until enough surface area has been brought into contact to stabilize the joint. The process is called embedment.

Embedment is worse on new parts than on reused ones. In critical applications it can be minimized by tightening, loosening, and retightening the fasteners several times. This is done on camera mounts in space satellites, for example [2].

Poor Thread Engagement

If the bolt is undersized, or the nut oversized, thread contact areas will be less than those planned by the designer, and substantial plastic deformation may occur, as shown in Fig. 6.13 [3].

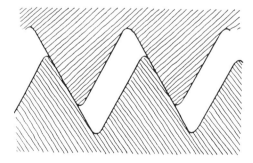

Figure 6.13 Poor thread engagement may be a major source of plastic deformation and therefore joint relaxation.

Thread Engagement Too Short

The length of thread engagement for steel fasteners should be at least 0.8 times the nominal diameter of the fastener. If the engagement length is too short (too few threads support the load), thread contact areas are again smaller than those intended by the fastener manufacturer and excessive relaxation can result. One author claims that if thread engagement length is greater than 1.25 times the nominal diameter, "permanent set is negligibly small" [1].

Soft Parts

If parts are softer than intended by the designer—perhaps because of improper heat treat or incorrect material—they may creep and relax substantially even if the geometry is correct and loads are normal.

Bending

If the fastener is bent as it is tightened, it will see higher stresses along one side than along the opposite side, as we have seen in previous chapters. These higher stresses mean more plastic flow and therefore greater-than-normal embedment or thread relaxation.

Nonperpendicular Nuts or Bolt Heads

The contact faces of nuts and bolt heads are never exactly perpendicular to the axis of the threads or to the axis of the bolt hole, as suggested in Fig. 9.1. This means that only a portion of the contact surface of the nut or bolt head is loaded when we first tighten the fastener. These abnormally loaded surfaces will creep until enough additional contact area has been involved to reduce contact pressures and stabilize the joint.

Fillets or Undersized Holes

If the head-to-body fillet contacts the edge of the bolt hole as shown in Fig. 6.14, the edge of the hole will break down under initial contact pressures. This may result in a complete loss of preload, since such effects are usually large compared to the amount by which the bolt was stretched when it was initially tightened [4].

Oversized Holes

Undersized holes can be a problem; so can oversized holes. Now there is too little contact between nut and joint surface or between bolt head and

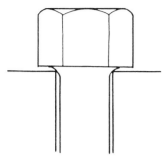

Figure 6.14 Oversized fillets and/or undersized holes may result in total relaxation of a preloaded fastener.

joint surface. Unless a washer or something is used to distribute contact pressures and limit contact stresses, the head and/or nut will embed itself in the joint surfaces, as suggested in Fig. 6.15 [3]. The amount of relaxation will, of course, depend on the strength of the surface supporting the nut or washer. The oversized or slotted holes used to aid the assembly of structural joints, for example, don't increase relaxation appreciably [5].

Conical Makeups

Surface irregularities will exist on conical joint surfaces as well as flat ones. The effect on axial tension in the fastener, however, is magnified if the embedment occurs on conical surfaces. A given amount of relaxation

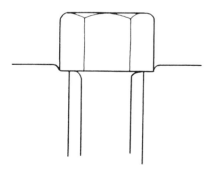

Figure 6.15 Oversized holes may also increase contact stress levels and therefore increase embedment relaxation.

perpendicular to the surface may mean substantially greater relaxation in the axial direction, as suggested in Fig. 6.16 and Eq. (6.3).

$$r = \frac{e}{\sin \theta} \qquad (6.3)$$

where e = embedment relaxation perpendicular to the surface of a conical joint member (in., mm)

r = resulting relaxation parallel to the axis of the fastener (in., mm)

θ = half-angle of the cone (deg)

B. Factors Affecting Short-Term Relaxation

Overstressed parts relax. Overstressing may be created in a number of different ways, as we have seen above. The amount of relaxation which overstressing causes in a given bolt and joint, however, can depend on a number of secondary factors. Here are some of them.

Bolt Length

Long, thin bolts will relax by a smaller *percentage* than short, stubby ones. The total embedment relaxation or the like will be the same for a given initial preload, but that embedment will be a different percentage

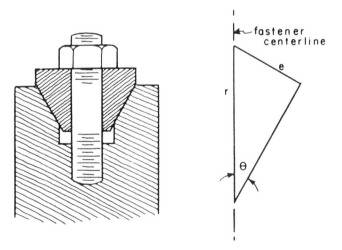

Figure 6.16 Conical or tapered joints usually relax more than flat ones for reasons given in the text.

of the total length of the bolt and therefore will mean a different percentage loss in length. Preload loss will be proportional to the change in length (see p. 61 in Ref. 5).

Many people take advantage of this fact. They add "bushings" above and below flange surfaces for example, as shown in Fig. 6.17. This makes it possible for them to use longer bolts on a given joint.

Belleville Washers

Another common way to reduce the change in clamping force produced by a given amount of embedment is to use Belleville washers, as also shown in Fig. 6.17. These springs have a very flat rate, compared to the stiffness of either bolt or joint members. They will therefore determine (limit) the preload and clamping force in the system (see Fig. 4.15). Because of their flat rate, a small deformation in the bolt or joint won't make an appreciable difference in force levels. For the same reason—and more commonly—Bellevilles are used to compensate for the effects of differential temperature expansion. Spring manufacturers have trouble controlling the stiffness of Bellevilles, however, so this solution to a temperature or relaxation problem can increase the basic scatter in preload.

Number of Joint Members

Increasing the number of surfaces in a joint may increase relaxation effects because there are now more high spots to embed and settle in together.

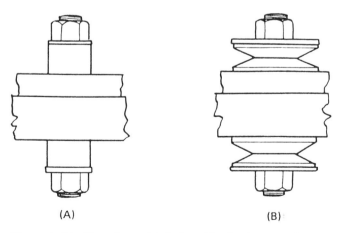

(A)   (B)

Figure 6.17   Since long, thin bolts will relax by a smaller percentage than short, stubby ones, many people use bushings as shown in (A) to reduce percentage relaxation in a given joint. Stacks of Belleville washers (B) are also effective.

Doubling the number of contact surfaces, for example, will almost double relaxation in many cases [6].

Tightening Speed

Creep and flow take time. Fasteners which are tightened very rapidly won't have time to settle in together during the tightening process and will relax more after tightening [7]. This is one of the advantages of hesitation tightening (Fig. 8.8) or torque-recovery tightening (Fig. 8.7). Tighten with high-speed tools, but pause to give the parts time to relax.

Tightening bolts in a series of passes, rather than applying full torque on the first pass, allows time for relaxation. This procedure also pulls the joint together uniformly. For both of these reasons, progressive tightening is a virtual necessity on large gasketed joints.

Simultaneous Tightening of Many Fasteners

Some experiments [8] have suggested that tightening a group of fasteners one at a time results in more relaxation in a given fastener than does tightening several or all of them at once. Presumably a fastener tightened before its fellow sees higher stress concentrations than it does if it is tightened simultaneously with the rest and all share the developing load.

Bent Joint Members

If joint members are soft or warped or bent, tightening one fastener can cause relaxation (or additional stress) in other fasteners. This sort of "cross-talk" between fasteners is very common, although it is not usually seen or recognized. More about this in Section IV.

C. Amount of Relaxation to Expect

The factors which cause and contribute to relaxation are many and hard to predict. Although attempts have been made to write equations for the amount of relaxation to expect [6,8,9], in most cases the amount must be determined experimentally. And, as is our common fate when dealing with bolted joints, it won't be "an" amount, but rather a distribution of values around some anticipated mean.

In general, fasteners relax rapidly following initial tightening, then relax at a slower rate, following the pattern shown in Fig. 6.18. The amount of relaxation varies greatly, depending on the condition of the parts, finishes, initial and local tension levels, fit of parts, and all of the other factors discussed earlier. Here are some of the relaxation amounts and

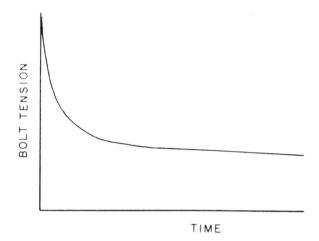

Figure 6.18 Most short-term relaxation occurs in the first few seconds or minutes following initial tightening, but continues at a lesser rate for a long period of time.

times found in the references. Fisher and Struik report that tests of A325 and A354 Grade BD bolts in A7 structural steel showed a loss of 2–11% of preload immediately after tightening, followed by another 3.6% in the next 21 days, followed by another 2% in the next 11.4 years [5, p. 61]. Bethlehem Steel reports that only 5% of the initial tension will be lost in structural bolts set by turn-of-nut techniques—5% over the total life of the structure [10]. Chesson and Munse report a variety of results on a variety of structural bolts, different types of bolt heads, different nuts, with and without washers, etc. [11]. As one example, an A325 bolt with a regular (not heavy) head, a flanged nut, and no washer relaxed 2.6% in the first minute after tightening (most of this in the first 15–20 sec). It had relaxed by 6.5% after 5 days. Hardiman reports that most relaxation occurs in the first few seconds, but that relaxation, usually, never stops [13]. Southwest Research Institute suggests that fasteners lose an average of 5% right after tightening, "because of elastic recovery" [14]. The $2\frac{3}{4}$–8 $\times$ $12\frac{3}{4}$ Nitronic 50 top guide studs in a BWR relaxed an average of 43% after tensioning. Grip length was 4.75 in.; studs were hydraulically tensioned to 160,000 lb; nuts were run down with a measured torque of 500 in.-lb. This is not all embedment, as we'll see in Sec. IV.

Gasketed joints will relax substantially, whether the bolts are torqued or tensioned. This is especially true during preliminary passes, when loss of as much as 80–100% of initial tension is not at all uncommon for reasons to be discussed soon. Gaskets will eventually stabilize, however, and will retain the tension introduced in final tightening operations.

D. Torsional Relaxation

We've been looking at preload or tension relaxation. This is of prime importance to us, because of the general importance of correct preload. We mustn't forget, however, that torsional stress is also built up in a fastener as it is tightened, and that this stress is also subject to varying amounts of relaxation. Many people, in fact, will insist that torsional stress disappears immediately and completely when the wrench is removed from the fastener. Others find that it doesn't disappear until a breakaway torque is applied [15]. Our experience indicates that, like tension relaxation, torsional relaxation depends on many factors; the amount and rate of torsional relaxation will vary substantially from bolt to bolt as well as from application to application.

Figure 6.19 shows the tension and torsion relaxation we measured in an experiment with a $2\frac{1}{4}$–8 × 12 B16 stud which had been lubricated with moly. Torsion relaxed 50% when the wrench was removed; tension actually *increased* 1–2% during this period. We have subsequently seen this phenomenon on many other types and sizes of bolt. Our guess is that, as embedment allows relaxation of both tension and torsion stress to occur, some of the torsional stress is turned into a little more tension stress. The twisted bolt screws itself farther into its own nut. The exchange is encouraged if you lubricate the threads but do not lubricate the face of the nut.

Sizable relaxation of tension, occurring as the torsional stress disappears, can, of course, mask this exchange of torsion for tension. Thus we've observed this exchange only on hard joints. But it's relatively easy to reproduce, and we think its common. Many bolts in large joints mysteriously "grow" a little between passes, for example, even when there is no temperature change or the like to explain the growth, and even when neighboring bolts remain at constant length. Presumably elastic interactions (Section IV) play a role here as well.

A torque wrench appears to respond to torsional stress levels in the bolt as well as to preload levels. The torque required to restart a nut can be less than that required to tighten it in the first place, even if there has been no loss in preload in the meantime—*if* torsional stress has disappeared or been reduced. Repeated "torque recovery" can, as a result, gradually increase the tension in a fastener until it is substantially above initial anticipated levels. In one set of measurements on tank tread end connector bolts, for example, we first tightened the $\frac{5}{8}$–18 × $1\frac{3}{4}$ Grade 8

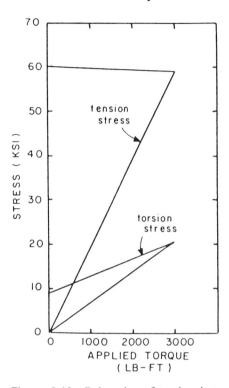

Figure 6.19 Relaxation of torsional stress in a bolt can be accompanied by an actual increase in tension. The bolt screws itself into its nut. Data shown were taken in tests on a $2\frac{1}{4}$–8 × 12 B16 stud.

bolts with a torque of 150 lb-ft. This stretched the bolt 0.0015 in. This was a tapered joint, so the bolt now relaxed to a stretch of only 0.001 in. We reapplied 150 lb-ft of torque, and the stretch returned to 0.0014 in. Incidentally, it took only 100 lb-ft to restart the nut.

The bolt now relaxed again, was tightened again, relaxed some more, etc., as shown in Fig. 6.20. Final preload (stretch) was 33% greater than that achieved in the first pass, even though we never applied more than the initial 150 lb-ft of torque. The final restarting torque was still only 87% of the rated torque. Many mechanics would conclude from this that preload was still 13% below the initial value. This possible interaction between torsional and tension stress further complicates the task of predicting how much a given fastener will relax, of course. It's another complex situation.

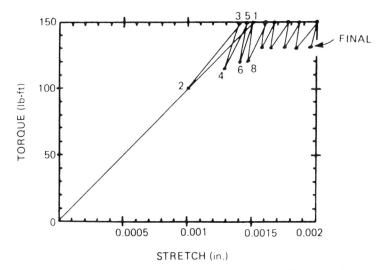

Figure 6.20 Torque-stretch-relaxation history of a $\frac{5}{8}$–18 $\times$ 1$\frac{3}{4}$ Grade 5 bolt. A torque of 150 lb-ft was applied repeatedly to this fastener, with a pause for relaxation between each pass. Final preload was 33% greater than that achieved on the first pass.

VI. ELASTIC INTERACTIONS BETWEEN BOLTS

Even if we can avoid the problems cited above, and can count on achieving a certain amount of preload in the bolts we tighten at assembly one by one, there are going to be times when that preload will be significantly modified as we tighten other bolts in the same joint, thanks to "elastic interactions" between bolts. Let's look at an example.

Let's assume that we're planning to tighten a circular, flanged joint which contains eight bolts. We're going to use ganged hydraulic wrenches or hydraulic tensioners to tighten these bolts two at a time. The bolts we tighten simultaneously, of course, will be opposite each other on the flange, 180° apart.

To explain the process of elastic interactions, we're going to think of the joint as a large spring connected by rigid top and bottom plates to the bolts (smaller springs), which are going to be used to clamp it. This arrangement is suggested in Fig. 6.21, which shows the first two bolts to be tightened.

Now, let us assume that we have tightened bolts 1 and 2 in this joint, and that magically we have achieved *exactly* the initial preload we wanted

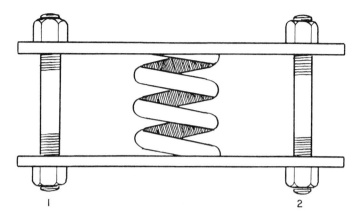

Figure 6.21 A simulated model of a bolted joint, in which the joint members are represented by a large spring, here "loaded" by the first two bolts to be tightened.

in each bolt. Let's say that this preload is 10,000 lb of tension in each bolt.

The 20,000 lb of force which these two bolts are creating on the joint partially compresses the joint. We now go on to tighten bolts 3 and 4 located 90° away from bolts 1 and 2. Again, our tools work magic for us and we create exactly 10,000 lb of initial preload in bolts 3 and 4 when they are tightened (Fig. 6.22).

We now have four small springs (bolts) compressing the joint spring rather than the two small springs we had a moment ago. If bolts 1 and 2

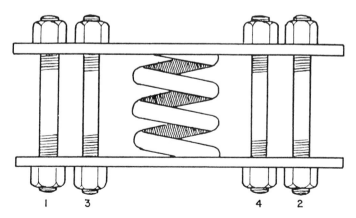

Figure 6.22 The joint model of Fig. 6.21, but now with four bolts to be tightened.

had retained their full preload, we would now have 40,000 lb of force on this joint instead of 20,000. Doubling the compressive force on the joint spring would, of course, double the amount by which it is compressed. But what happens in bolts 1 and 2 when we tighten 3 and 4? Bolts 1 and 2 are allowed to relax a little as the joint is compressed by bolts 3 and 4.

At this point in the process, therefore, bolts 1 and 2 have a slightly lower amount of preload in them than bolts 3 and 4—even though each of the four bolts started with the same initial tension of 10,000 lb. When we now go on and tighten bolts 5 and 6 in a third step, bolts 3 and 4 will relax a little, and bolts 1 and 2 will relax further. Tightening bolts 7 and 8 to complete the assembly will create relaxation in each of the six bolts tightened earlier.

The result is four different levels of *residual* preload in the eight bolts when they are tightened two at a time, even though the *initial* preload in each one was identical to start with. And this is not just a theoretical possibility; it's a very common occurrence. Most people are not aware of this interaction, however, which is visible only if you use ultrasonics or strain gages or something to monitor the tension in the bolts.

Figures 6.23–6.25 show some actual elastic interaction data. A raised face gasketed joint with a spiral-wound, asbestos-filled gasket was tightened in three passes using a cross-bolting pattern and tightening one bolt

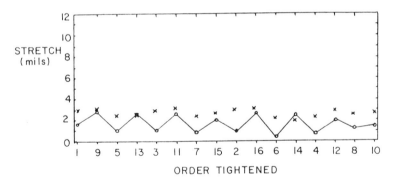

Figure 6.23 The elongation or stretch achieved in the 16 bolts of a gasketed flanged joint as the bolts are initially tightened one by one (x's) and after all have been tightened (solid line). Numbers on the horizontal axis show the location of the bolts, and the order in which they were tightened. The second bolt tightened, #2, is located 180° away from #1. The third and fourth bolts tightened are halfway between bolts 1 and 2, etc. The difference between the x's and solid line shows the loss of initial preload in the bolts as a result of elastic interactions.

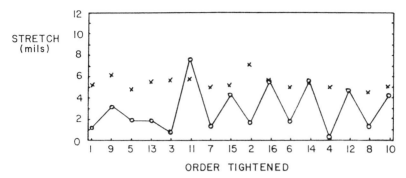

Figure 6.24 Initial and residual preloads in the 16 bolts of the joint shown in Fig. 6.23 after a second tightening pass at a higher torque.

at a time. Figure 6.23 shows the results after the first pass in which 100 ft-lb of torque was applied to each bolt.

The isolated x's show the initial change in length achieved in each bolt as it was tightened individually. This change in length, or stretch, is proportional to the tension in the bolt, as we saw in Chapter 5 (or will see at greater length in Chapter 9). The sawtooth line shows the pattern of *residual* preload (stretch) in all of the bolts in this joint following completion of the first pass. The numbers on the horizontal axis of the figure define the sequence in which the bolts were tightened and their relative position around the joint. For example, bolt 1 was tightened first; bolt 2,

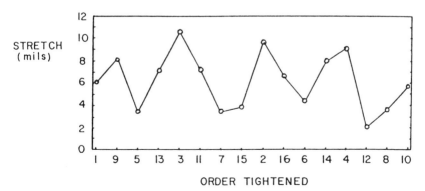

Figure 6.25 Final tension in the 16 bolts of the joint of Figs. 6.23 and 6.24 after a final cross-bolting pass at a final (highest) torque.

180° away from bolt 1, was tightened second; bolt 3, halfway between bolts 1 and 2, was tightened third; etc.

Note that when bolt 1 was originally tightened, approximately 3 mils of stretch was created in that bolt by the 100 ft-lb of torque. After all the bolts in the joint had been tightened, however, the tension in bolt 1 was remeasured, and it was found to have only about 1.5 mils of stretch. Bolt 3 started and ended about the same place. Bolt 6 started with about 2 mils, but lost all but 0.25 mil as the other bolts in the joint were tightened.

In pass 2, 200 ft-lb of torque was now applied to each of the bolts, again one at a time and in the same cross-bolting pattern used for the first pass. Figure 6.24 shows the tension created in individual bolts during this pass (x's) and the final residual tension in each bolt at the end of the pass (line). Note that the sawtooth curve, which was very regular in Fig. 6.23, has now started to break up and become more erratic. Note, too, that at this point in the procedure the scatter between max and min residual tension in the bolts is nearly 20:1, ranging from less than 1 mil of stretch in the bolt tightened fourth to nearly 8 mils of stretch in the eleventh bolt tightened—and this despite the fact that the scatter between max and min tensions created in individual bolts as they were first tightened is fairly normal. In pass 1, for example, the scatter between applied torque and achieved initial preload was roughly ±30%, as anticipated for as-received steel-on-steel bolts. But by the end of pass 2, the scatter in residual tension is 20:1.

Figure 6.25 shows the results after a third and final pass at 275 lb-ft of torque. The solid line shows the final residual preload in all bolts. The max to min range of preload is about 5.5 to 1, a far cry from ±30%.

As mentioned, the joint we have just been examining contained a spiral-wound, asbestos-filled gasket. The fact that a gasket was involved certainly contributed to the amount of elastic interaction observed here. But even hard metal-to-metal joints show this effect to some extent.

The fact that we were using a torque wrench to tighten these bolts contributes somewhat to the initial bolt-to-bolt scatter in tension, of course. But the elastic interaction contributes far more to the final scatter, and has nothing to do with the type of tool used. Figure 10.11, for example, shows similar results obtained when a large joint was tightened with four hydraulic tensioners. The range between max and min residual bolt stretch in this joint was 5 mils to 28 mils, thanks largely to elastic interactions. This was a foundation joint with no gasket.

Although usually invisible, elastic interactions of the sort shown here have always been with us. They are one of the reasons why joints have always had to be grossly overdesigned to function properly. Although worse in gasketed joints than in metal-to-metal joints, they may be less significant in the gasketed joint because this joint has a "memory," as we will see in Chap. 19.

Is there any way to prevent or eliminate elastic interaction?

Attempts have been made to define a torquing procedure which will minimize the effects of elastic interactions [16,17]. The authors suggest that if different amounts of torque were applied to each of the groups of bolts in the example joint illustrated in Fig. 6.22, for example, it would be possible to end up with the same amount of tension in each bolt after a single pass. One might, for example, apply 400 ft-lb of torque to bolts 1 and 2; 300 to bolts 3 and 4; 200 to bolts 5 and 6; and 100 to bolts 7 and 8. Bolts 1 and 2 relax three times; bolts 3 and 4 relax twice; bolts 5 and 6 relax once; and all end up with the same amount of tension finally created in bolts 7 and 8.

Several of us have attempted to do this, but without success. One problem is that the amount of torque applied to the groups tightened earliest can be astronomical—unless a large number of bolts are tightened simultaneously. More important, however, the elastic behavior of the joint and individual bolts is basically unpredictable. Tightening the same bolts in the same joint repeatedly usually produces different final patterns of residual preload.

This is illustrated in Fig. 6.26, which shows the results achieved in another experiment on the joint described in Figs. 6.23–6.25. The bolts were tightened in three passes as before, but the final pattern of preload differs significantly from that shown in Fig. 6.25.

This time, in an experiment to reduce the residual scatter, we made a fourth pass at the final torque of 275 lb-ft, but in reverse order. The bolt tightened last on pass 3 was tightened first on pass 4. The bolt tightened next to last was tightened second, etc. As you can see, this reduced the scatter in residual from about 7:1 to about 3:1, which could be helpful in some situations, but which is probably not worth the effort in others. Note that even 3:1 is a long way from the "±30%" often claimed for torque procedures in general.

We've tried other torquing procedures to reduce final scatter. For example, we've tried making a final pass at 75% of final torque on the odd-numbered bolts only; tried using a large number of passes (15 or 20) at the final torque; tried making several passes at the final torque, using a clockwise pattern, rather than cross-bolting, for these final passes, etc.

Most of the things tried seem to help somewhat, but the main key seems to be "more passes," regardless of what pattern or format is used. Scatter is reduced slowly as the number of passes increases. But scatter cannot be eliminated by any *torquing* procedure I know of. If ultrasonics or datum rods or load cells can be used to monitor results, a different

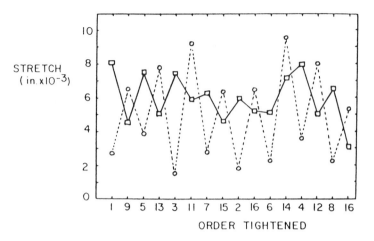

Figure 6.26 The dashed line shows the pattern of final, residual tension in the 16 bolts of the joint described in Figs. 6.23–6.25 after that joint has been loosened and then retightened with the same torques and procedure used earlier. The solid line shows the change in pattern of residual tensions after a fourth and final pass in reverse order (the last bolt was tightened first, etc.).

torque can be applied to each bolt and scatter can then be reduced to ± 10% or less, depending on the time you spend on it and the skill of your mechanics. But this isn't a torque control procedure; it's stretch or strain control. Pure torque control cannot avoid the type of results described.

The only sure way to eliminate elastic interactions entirely is to tighten all the bolts in the joint simultaneously. That is actually done on some reactor pressure vessels and on automotive engine heads, for example. The closer you can come to simultaneous tightening, the less elastic interaction effect you will get. For example, if you can tighten half the bolts in the joint simultaneously, you will see less interaction than if you tighten them individually or tighten one-quarter of them at a time.

What saves most of us is the fact that it's not *necessary* to eliminate or compensate for elastic interactions. The life and behavior of most joints—including most gasketed, pressurized ones—depends more on the average tension in the bolts than on the range of tension or minimum tension. Knowledge of elastic interactions can help you solve a chronic leak or other joint problem, but in most cases, as we'll see in Chap. 19, the common solution is an increase in average preload rather than a decrease in preload scatter, although both are sometimes required.

We can now modify our joint diagram to show the loss in preload created by embedment and by the average elastic interaction effect, as in Fig. 6.27. The dashed lines show the situation—bolt and joint forces and deflections—before relaxation. The solid line shows the residual preload situation at the end of the assembly operation. It is this solid triangle we'll build on in later chapters as we add the effects of external loads etc.

As a final note, Dr. George Bibel of the University of North Dakota has been conducting studies of elastic interactions sponsored by British Petroleum America and by the Pressure Vessel Research Committee [19, 20]. He has found a way to achieve final bolt tensions, with a scatter of only ±2% or so, in large-diameter gasketed or ungasketed joints. (His tests involved flanges with diameters of 16 and 24 in.) Using a cross-bolting pattern he first tightens the bolts to an arbitrary, experimental tension, which he monitors with strain gages or ultrasonics. During this pass he determines the amount of interaction between each bolt and (primarily) its neighbors. He then loosens the bolts and retightens them in a single, cross-bolting pass, determining the amount each is to be tightened by use of a matrix of interaction coefficients derived from the data taken during the experimental pass. Once again, tension not torque is used to

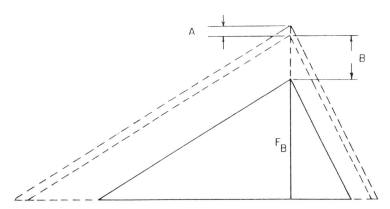

Figure 6.27 The joint diagram of Fig. 6.5 modified to show the effects of embedment and elastic interaction on initial preload. The largest triangle reflects the initial, "just tightened" situation. Embedment reduces the initial preload by an amount shown as A. The average loss in a group of bolts, caused by elastic interactions, is shown as loss B. We use the average loss instead of the individual loss because joint behavior will more likely be determined by average loss than by worse case. The final, average, residual preload in this group of bolts is represented by the solid triangle.

determine the initial preload developed in each bolt during this final pass. He believes that this procedure could, in one possible application, reduce the radiation exposure of maintenance workers in nuclear power plants.

Dr. Bibel reports that elastic interactions during a conventional three-pass bolting procedure can result in an average loss of preload of 25–50%. This is the case, at least, if relatively thick, spiral-wound gaskets are involved. Thinner gaskets reduce the average loss, as does metal-to-metal contact—but some loss occurs in each case.

He also reports that the elastic interactions in a given joint appear to be repeatable and, therefore, predictable, which would suggest that the interaction coefficients would have to be determined only once for a given joint. Finally, he reports that the first bolt tightened in a cross pattern can be overloaded by up to 50% when later bolts are tightened, again, during a normal bolt-up procedure.

VII. THE ASSEMBLY PROCESS REVIEWED

After learning about elastic interactions, we make one more pass around our 16-bolt joint, in the reverse order, to reduce the scatter in residual preload. We remeasure the preloads in each bolt after that pass and find that each has changed. We accept the final results and then draw the block diagram shown in Fig. 6.28 to summarize the things we now know can affect the amount of clamping force created on a multibolt joint when it's tightened. We include a few things learned in earlier chapters or elsewhere; for example, the fact that some bolts will be bent slightly if tightened against nonparallel joint surfaces (as discussed in Chap. 2). We also include some minor factors not discussed, such as the very small heat loss in the tools and drive bars we're using. We also include the possibility of heat loss through "prevailing torque"—when interference of some sort between male and female threads makes it necessary to use a wrench simply to run a nut down, before the nut contacts the joint. We'll learn more about this in Chap. 16.

Because of our interest in stored energy, we also list the many ways the input work we do on the nuts is absorbed by the bolts and joint members. The work:

Heats the parts and tools
Enlarges interference fit holes
Bends and twists the bolts
Dilates the nuts (see Chap. 3)
Pulls the joint members together

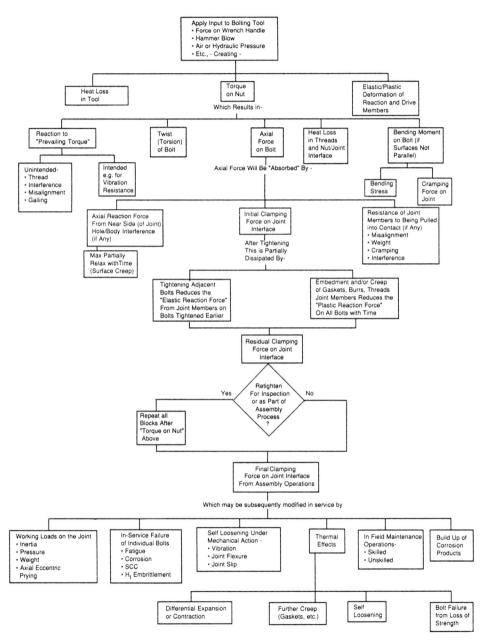

Figure 6.28   Block diagram showing most of the factors which affect the relationship between the input work done by the tool on the bolt or nut, and the subsequent in-service clamping force between joint members.

Overcomes prevailing torque
Deforms the parts plastically (embedment and gasket creep)

and—deforms the bolts and joint members elastically. It stretches the bolts and compresses the joint. Only the work done deforming the parts elastically ends up as stored energy, creating clamping force, but then some of this work is lost as the parts embed and relax.

We now realize why accurate control of the assembly process is so difficult. We would have to predict or control such unpredictable things as the friction forces between parts, the effort required to pull the joint together, the amount of hole interference, and the elastic interactions between bolts to achieve exactly the same amount of residual preload in each bolt. I think that it's safe to say that we'll never be able to afford the effort required to predict or control the many variables involved.

VIII. OPTIMIZING ASSEMBLY RESULTS

This doesn't mean, however, that all is lost. A great deal of work has been and is still being done on assembly tools and techniques. There are ways to get better results than we achieved with our example joint in this chapter. In the next five chapters we're going to look at most of these techniques. This education will allow us to pick the most appropriate assembly methods for our own applications.

Regardless of the type of control we adopt, furthermore, there are a number of relatively simple things we can do to improve our chances of success during assembly operations. In fact, the things listed below can often make more improvement than the use of elaborate or expensive preload control equipment. As you can see, most of the items on the list would tend to make your assembly practices and procedures more consistent. Although you can't predict or control the many variables affecting results, you can reduce their variation by being consistent.

1. Be as consistent as possible in your choice and use of tools, procedures, calibration frequency, etc.
2. Train and supervise the bolting crews. This can help far more than a "better" lubricant or more expensive tool. Let the people know how important good results are, and how difficult they are to obtain. Enlist their help.
3. Make sure the fasteners are in reasonable shape. Wire-brush the threads if they're dirty and rusted. Use stainless steel bristles on alloy steel materials. Chase threads with a tap or die if they're damaged.

4. Use hardened washers between the turned element (nut or bolt head) and joint members.
5. If lubricants are to be used, make sure they're clean and fresh. Apply them consistently, the same amount to the same surfaces by the same procedure. Preload scatter will be minimized if lubes are used on both thread and nut (or head) contact surfaces.
6. Run the nuts down by hand. If you can't, the threads may need to be cleaned or chased.
7. Hold wrenches perpendicular to the axes of the bolts.
8. Apply torque at a smooth and uniform rate. Avoid a "stick-slip" situation as you approach the final torque. If necessary, back the nut off a little so that the specified torque can be reached with the wrench in motion.
9. If hydraulically powered wrenches are used, be sure that adequate reaction points are used, and that the tools aren't twisting or cramping as a result of cocked or yielding reaction surfaces.
10. Snug the joint first, with a modest torque; then tighten it. Try to align the joint members before tightening the bolts.
11. Tighten from the center of bolt patterns toward the free edges if the bolt pattern is rectangular (as in a structural steel joint). Work in a cross-bolting pattern on circular or oval joints.
12. Keep good records of the tools, operators, procedures, torques, and lubricants used. Bolting is an empirical "art" at present. If you record details of your "experiment" you'll have the information you need to make a controlled modification in the procedure the next time, if the first procedure doesn't give the results you want.
13. If possible, develop your own nut factors, experimentally, rather than relying on a table. Perform the experiment under actual job conditions if possible, though a lab test on your joint (or a simulation of your joint) is better than nothing.

REFERENCES

1. Landt, Richard, *Preload Loss and Vibration Loosening*, SPS Technologies, Jenkintown, PA, 1979.
2. Conversation with Carl Osgood, consultant, Cranbury, NJ, 1974.
3. Friesth, E. R., Performance characteristics of bolted joints in design and assembly, SME Technical Paper AD77-715, 1977.
4. High strength bolted joints, *SPS Fastener Facts*, SPS Technologies, Jenkintown, PA, sec. IV-C-4.

5. Fisher, J. W., and J. H. A. Struik, *Guide to Design Criteria for Bolted and Riveted Joints*, Wiley, New York, 1974, p. 179.
6. Meyer, G., and D. Strelow, How to calculate preload loss due to permanent set in bolted joints, *Assembly Eng.*, February and March 1972.
7. Ehrhart, K. F., Preload—the answer to fastener security, *Assembly and Fastener Eng. (London)*, vol. 9, no. 6, pp. 19–23, June 1971.
8. Prodan, V. D., et al., More precise load-relief factor for tightened screw connections, *Vest. Mashin.*, vol. 54, no. 1, pp. 27–28, 1974.
9. Junker, G., *Principles of the Calculation of High Duty Bolted Connections—Interpretation of Guideline VDI 2230*, VDI Berichte, no. 220, 1974, published as an Unbrako technical thesis by SPS, Jenkintown, PA.
10. *High Strength Bolting for Structural Joints*, Booklet 2867. Bethlehem Steel Co., Bethlehem, PA, p. 9.
11. Chesson, E., Jr., and W. H. Munse, *Studies of the Behavior of High Strength Bolts and Bolted Joints*, University of Illinois College of Engineering, Engineering Experiment Station Bulletin 469, 1964.
12. Mayer, K. H., Relaxation tests on high temperature screw fastenings, *Wire*, vol. 23, pp. 1–5, January–February 1973.
13. Hardiman, R., Vibrational loosening—causes and cures. Presented at the Using Threaded Fasteners Seminar at the University of Wisconsin—Extension, Madison, May 8, 1978.
14. Investigation of threaded fastener structural integrity, Report prepared by Southwest Research Institute, San Antonio, Texas, under contract no. NAS9-15140, DRL Tii90, CLIN 3, DRD MA 129 TA, SWRI project no. 15-4665, October 1977.
15. Blake, T. C., and H. J. Kurtz, The uncertainties of measuring fastener preload, *Machine Design*, September 30, 1965.
16. Van Campen, D. H., A systematic bolt-tightening procedure for reactor pressure vessel flanges, ASME First International Conference on Pressure Vessel Technology, Delft, Netherlands, September 29–October 2, 1969, Part 1, pp. 131–141.
17. Rumyantsev, D. V., V. D. Produn, and A. F. Pershin, A tightening procedure for fastening bolts applicable to high-pressure apparatus, *Khimicheskoe i Neftyanoe Mashinostroenie*, no. 2, pp. 3–5, February 1973.
18. Tripp, Owen S., and Earl D. Stevens, Flange parallelism criteria, presented to the PVRC Subcommittee on Bolted Flanged Connections, San Francisco, CA, January 1983.
19. Bibel, G., and R. Ezell, An improved flange bolt-up procedure using experimentally determined elastic interaction coefficients, *Journal of Pressure Vessel Technology*, vol. 114, November, p. 439ff., 1992.
20. Bibel, G., and R. Ezell, Bolted flange assembly: preliminary elastic interaction data and improved bolt-up procedures, draft of paper submitted to the Pressure Vessel Research Committee, May 18, 1993.
21. Bibel, G. D., and D. L. Goddard, Preload variation of torqued fasteners, a comparison of frictional and elastic interaction effects, *Fastener Technology International*, February 17, 1994.

7
Torque Control of Bolt Preload

We have now reviewed the assembly process and have seen why the control of that all-important clamping force on the joint is so difficult. During this review we assumed that we were using the torque applied to the nut to control the tension or preload built up within the bolt. Now we're going to take a much closer look at torque control itself. We'll learn how to estimate the initial preload we'll achieve by applying a given torque to a fastener; we'll take a brief look at torque tools; and we'll look at "restarting" or "breakaway" torque as a means of evaluating the residual preload in a bolt.

In later chapters we'll study other means of controlling assembly preloads. But make no mistake: at this point in time, torque is king. It's by far the best known, most common, and usually the least expensive way to control preload. It doesn't do a perfect job, but it produces results which are good enough for the vast majority of applications requiring control. So—it certainly deserves our full attention.

I. THE IMPORTANCE OF CORRECT PRELOAD

In earlier chapters we learned that the main purpose of most bolts is to clamp the joint members together. The clamping force is created when we tighten the bolts during assembly, when we "preload" the bolts, create tension in them, by turning the nut or bolt while holding the other.

During initial assembly of an individual bolt there's a one-to-one relationship between the tension in a bolt and its preload. As we saw in the last chapter, however, the tensile load in a bolt will change as we tighten other bolts and/or put the joint in service, so the preload is an initial and short-lived affair. Nevertheless, it's extremely important.

As I mentioned in the last chapter, we're interested in the tension in a bolt at three different times.

Initial preload: the tension created in a bolt when it is first tightened
Residual preload: the tension remaining in a bolt at the end of the assembly
    process, after all bolts have been tightened
The tension in a bolt in service

We're most interested in the last one on this list, the tension in the bolts while they're in service, but in this and the next four chapters we're going to focus on the initial preload we get when we first tighten a bolt. This is the preload over which we have the most direct control, and it will usually determine and often dominate the residual and in-service conditions. Unless the joint is very poorly designed, subsequent tightening of other bolts and/or working loads, thermal loads, vibration loads, etc. won't modify the initial preloads enough to cause joint problems. In other words, there'll usually be a direct relationship between this transient, initial preload and the ultimate behavior of the joint. Correct initial preload is essential.

A. Problems Created by Incorrect Preload

We'll take a closer look at these points in later chapters, but here's an introduction to the problems created by incorrect preload.

1. *Static failure of the fastener*: If you apply too much preload, the body of the bolt will break or the threads will strip.
2. *Static failure of joint members*: Excessive preload can also crush or gall or warp or fracture joint members such as castings and flanges.
3. *Vibration loosening of the nut*: No amount of preload can fight extreme transverse vibration, but in most applications, proper preload can eliminate vibration loosening of the nut.
4. *Fatigue failure of the bolt*: Most bolts which fail in use do so in fatigue. Higher preload does increase the mean stress in a fastener, and therefore threatens to shorten fatigue life. But higher preload also reduces the load excursions seen by the bolt. The net effect is that higher preload almost always improves fatigue life.
5. *Stress corrosion cracking*: Stress corrosion cracking (SCC), like fatigue, can cause a bolt to break. Stresses in the bolt, created primarily

by preload, will encourage SCC if they're above a certain threshold level, as we'll see in Chap. 18.

6. *Joint separation*: Proper preload prevents joint separation; this means that it reduces or prevents such things as leaks in a fluid pipeline or blow-by in an engine. The latter, of course, means that proper preload allows the engine to produce more horsepower.

7. *Joint slip*: Many joints are subjected to shear loads at right angles to the axis of the bolt. Many such joints rely for their strength on the friction forces developed between joint members, forces created by the clamping force exerted by the bolt on the joint. Again, therefore, it is preload that determines joint integrity. If preload is inadequate, the joint will slip, which can mean misalignment, cramping, fretting, or bolt shear.

8. *Excessive weight*: If we could always count on correct preload, we could use fewer and/or smaller fasteners, and often smaller joint members. This can have a significant effect on the weight of our products.

9. *Excessive cost*: The cost of many products is proportional to the number of assembly operations. Correct preload means fewer fasteners and lower manufacturing costs—as well as lower warranty and liability costs.

Notice that in all of the above we want "correct" preload, not just preload. Too much will hurt us—so will too little.

In most situations we also want uniform preload, although this will usually be less important than a particular preload. Loading a group of fasteners irregularly can warp or damage joint members or gaskets. Non-uniform preload will also mean that only a few of the bolts carry the external loads. If they don't share the burden planned for them by the designer, the joint may fail.

B. How Much Preload?

We always want the maximum possible preload, but in choosing this, we must consider:

The strength of the bolt and of the joint members under static and dynamic loads

The accuracy with which we expect to tighten the bolts

The importance of the joint, i.e., the factor of safety required

The operating environment the joint will experience in use (temperature, corrosive fluids, seismic shock, etc.)

The operating or working loads which will be placed on the joint in use

Again, we'll consider details of these requirements in later chapters, and will then reconsider the important question, "How much preload do I want?" in Chaps. 20 through 23. In the next few chapters, however, we'll examine the problem of tightening the bolts accurately; we want enough preload, but we can't stand too much. This implies control of some sort.

C. Factors Which Affect the Working Loads in the Bolts

We'll start our detailed study of the control problem soon, but first, to see how control-at-assembly enters the picture, let's quickly summarize the main factors which will determine the *working* loads in our bolts. Important though control at assembly is, we don't want to lose sight of the fact that it's only one of our many concerns.

All of the following will influence our results.

1. *Initial preloads*: The preloads we develop in those bolts when we first tighten them one at a time. Initial preload will be the principal subject of this and the next four chapters.
2. *Sequence/procedure*: The procedure with which a group of bolts are tightened can affect final results substantially. Procedure includes such things as the sequence with which they're tightened, whether they're tightened with a single pass at the final torque or in several passes at steadily increasing torques, etc.
3. *Residual preloads*: The preloads left in the bolts after embedment and elastic interactions.
4. *External loads*: External loads add to or subtract from the tension in the bolts, and therefore from the clamping force on the joint. Such loads must be predicted and accounted for when the joint is designed, and when the "correct" preload is chosen. External loads are created by such things as pressure in the pipeline or engine, snow on the roof, inertia, earthquakes, the weight of other portions of the structure, etc.
5. *Service conditions*: Severe environments can affect operating conditions in the joint and bolts. This is especially true of operating temperatures. These can create differential expansion or contraction which can significantly alter bolt tensions and clamping force. Corrosion can cause change as well. Contained pressure will affect clamping forces.
6. *Long-term relaxation*: There are some long-term relaxation effects which must also be considered: relaxation caused by corrosion, or stress relaxation or creep, or vibration. And again, we want correct bolt loads for the life of the joint, not just for a while.

7. *The quality of parts:* We won't get correct preload, or satisfactory performance from the joint, unless the parts are the right size, are hardened properly, are in good condition, etc. This factor can't be handled separately; it gets in the act by affecting the others. If the bolts are soft, for example, we won't get the expected preload for a given torque, and relaxation will be worse. If joint members are warped or misaligned, it may take an abnormal amount of tension in the bolts (created by an abnormal amount of preload) to create the necessary clamping force between joint members.

That's only a partial list but it covers the main factors. Obviously, we've got our work cut out for us if we expect to overcome all of the possible problems. But we have no choice except to try—or change professions. The more we know about the possibilities and probabilities, the better our chances of success. We're going to start by looking at the most common and popular way to control that all-important initial preload on which all the other factors depend.

II. TORQUE VS. PRELOAD—THE LONG-FORM EQUATION

Whenever we tighten a bolt we perform two acts. We *do* the job—we tighten that bolt; and we *control* the job—we tighten it to a certain point and then make a decision to stop. Since the threaded fastener is designed to be tightened by twisting the nut with respect to the head, or vice versa, we usually find it convenient to *do* the job by applying a torque. This does not mean, however, that we must *control* the job with torque. We have, as you will see, many options when it comes to control. Let's look at some.

When we tighten a bolt:

We apply *torque* to the nut,
The nut *turns*,
The bolt *stretches*,
Creating *preload*.

As we'll see, we can enter this sequence of events at any point to control the thing we're interested in—which is always preload. We can control, in other words, through torque or through turn or through stretch or through a combination of these things. In a few special situations we can control through preload directly.

In general, the closer we enter the chain to preload itself, the more accuracy we can achieve—and the more it will cost us.

We will find it easiest and least expensive to control preload with torque and/or turn, because these are the inputs to the system. Obviously, we'd like to be able to predict results; to predict the amount of preload we get when we tighten a bolt. There are two well-known equations which give us a shot at doing this, though, as we'll see, with less than perfect results. The first, which we'll look at now, is called the long-form torque-preload equation. As Srinivas has shown, this is a simplification of the more complete, but virtually impossible to use work-energy equation [Eq. (2.29)] given in Chap. 2. As a matter of fact, even this long-form simplification presents the user with some problems, but it is, nevertheless, widely used. Let's take a look at it. Experience and theoretical analysis say that there is usually a linear relationship between the torque applied to a fastener and the preload developed in a given fastener, as in Fig. 7.1. In other words,

Torque = (preload) × (a constant)

$$T_{in} = F_P \times C$$

(7.1)

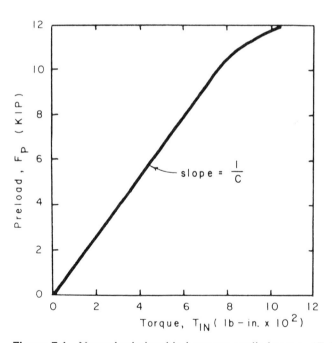

Figure 7.1 Normal relationship between applied torque (T_{in}) and achieved preload (F_P). Note that the straight line becomes a curve at the top when something—the bolt or joint—starts to yield. The bolt is an SAE J429, Grade 8, $\frac{3}{8}$–16 × 2.

But what's the constant? A number of equations have been derived which attempt to define it. Here's one which has been proposed by Motosh [1]:

$$T_{in} = F_P \left(\frac{P}{2\pi} + \frac{\mu_t r_t}{\cos \beta} + \mu_n r_n \right) \tag{7.2}$$

where T_{in} = torque applied to the fastener (in.-lb, mm-N)
F_P = preload created in the fastener (lb, N)
P = the pitch of the threads (in., mm)
μ_t = the coefficient of friction between nut and bolt threads
r_t = the effective contact radius of the threads (in., mm)
β = the half-angle of the threads (30° for UN or ISO threads)
μ_n = the coefficient of friction between the face of the nut and the upper surface of the joint
r_n = the effective radius of contact between the nut and joint surface (in., mm)

This equation involves a simplification, and so the "answer" it gives us is in error by a small amount. But I think it is the most revealing of the so-called long-form equations that have been proposed.

The equation shows that the input torque (left side of the equation) is resisted by three reaction torques (three components on the right-hand side). These are as follows:

$F_P \dfrac{P}{2\pi}$ is produced by the inclined plane action of nut threads on bolt threads. This could be called the *bolt stretch component* of the reaction torque. It also produces the force which compresses the joint and the nut, and it is part of the torque which twists the body of the bolt.

$F_P \dfrac{\mu_t r_t}{\cos \beta}$ is a reaction torque created by frictional restraint between nut and bolt threads. It also provides the rest of the torque which twists the bolt (unless there is some prevailing torque—discussed later—which would add a third component to the twist torque).

$F_P \mu_n r_n$ is a reaction torque created by frictional restraint between the face of the nut and the washer or joint.

If you plug typical fastener dimensions into Eq. (7.2), then assume a coefficient of friction and input torque, you can compute the magnitude of each of the three reaction components separately. If you do this, you will find that the nut friction torque is approximately 50% of the total reaction; thread friction torque is another 40%, and the so-called bolt stretch component is only 10%, as shown in Fig. 7.2.

Note that Fig. 7.2 is just another way of illustrating the point made in Fig. 6.4, where we were primarily concerned with charting the way in which input work was absorbed by the bolt and joint. Since input work is equal to torque times turn of nut, there's a linear relationship between energy and torque. Both Figs. 6.4 and 7.2 give an oversimplified—but very useful—summary of the real situation, however, by suggesting that the only energy (or torque) losses are those due to thread or under-nut friction. In fact, as the energy equation Eq. (2.29) shows, there are other losses. We're about to discuss some of these. As Srinivas has shown, however, if the other losses are negligible (and they usually/often are) the energy equation becomes the long-form torque-preload equation and Figs. 6.4 and 7.2 illustrate the results.

Whether we use the torque equation or the energy equation, however, we're concerned not with how much of the input goes into preload, but by the degree of control we can maintain over this process. And here we run into all sorts of difficulties.

Remember the "leverage" problem we discussed in Chap. 6? Let's assume that the coefficient of friction between nut and joint surface is 10% greater than average. As we'll see, a 10% variation in friction is common; it can be caused by all sorts of things. This 10% increase raises the torque required to overcome nut-joint friction from 50% to 55% of the input torque. Taken alone, that would be no problem; but where does that extra torque some from? It can't come from the operator because he has no way of telling that this set of parts is absorbing more torque between nut and joint. It won't come from the thread friction component, unless

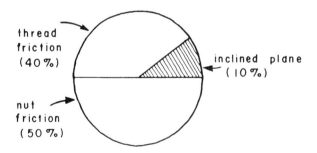

Figure 7.2 Relative magnitudes (typical) if the three reaction torques which oppose the input torque applied to a nut.

the coefficient of friction at that point decreases magically to offset the other increase. So the only place it can come from is the inclined plane or bolt stretch component, the preload component—the thing we're after when we tighten the bolt. But taking an extra 5% of the input torque away from the stretch component reduces that component from 10% of the input to 5%. That's a 50% loss in bolt tension, thanks to a 10% increase in nut friction—a very bad "leverage" situation. And finding things which increase nut or thread friction is not difficult, as we'll see next.

III. THINGS WHICH AFFECT THE TORQUE-PRELOAD RELATIONSHIP

Now let's take a closer look at some of the many factors which affect the amount of initial preload we get when we tighten a fastener. Many of these factors were included in our discussion in Chap. 6—and in the block diagram of Fig. 6.28—but we'll consider them at greater length now, and will add some new ones!

A. Variables Which Affect Friction

The coefficient of friction is very difficult to control and virtually impossible to predict. There are some 30 or 40 variables that affect the friction seen in a threaded fastener [30]. These include such things as

The hardness of all parts
Surface finishes
The type of materials
The thickness, condition, and type of plating, if present
The type, amount, condition, method of application, contamination, and
    temperature of any lubricants involved
The speed with which the nut is tightened
The fit between threads
Hole clearance
Surface pressures
Presence or absence of washer(s)
Cut vs. rolled threads

A word about the last item, cut vs. rolled threads. J. Ron Winter of Tennessee Eastman has presented data showing the results he obtained by tightening separate groups of bolts: one group having cut threads, the other rolled threads. Some of his results are shown in Fig. 7.3. The bolts with rolled threads achieved a higher average preload for a given torque,

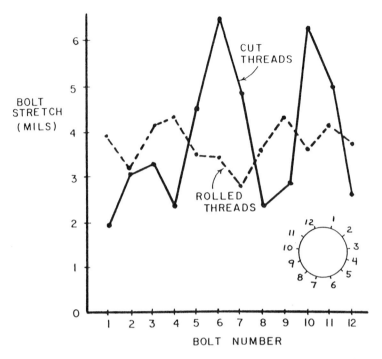

Figure 7.3 This shows the pattern of residual preload on two large, pressure vessel joints after a two-pass bolt-up procedure. Studs with cut threads were used in one joint, studs with rolled threads in the other. The engineer reported that the studs with rolled threads ended up with a higher average preload and with less scatter in preload than did those with cut threads.

and the bolt-to-bolt preloads were less scattered than the bolts with cut threads. Since a lower average and more scatter are characteristic results for any effect which increases thread friction, I assume that this item should be included in the above list.

In any event, you can see that there are many factors which affect friction, and therefore preload. Some of these factors can be controlled to some degree—but complete control is impossible.

Together, all of the factors listed above determine what I call CONTROL ACCURACY, the ability of the variable we've selected for control—in this case, torque—to create the thing we're after, which is always preload. As far as torque is concerned, conventional wisdom—and much experience—suggests that a given torque will create a given preload with a scatter of ±30% in a large group of as-received bolts.

B. Geometric Variables

Friction is often cited as the only villain in the torque-preload equation, but that is not the case. Although we think we know the pitch of the threads, or the half-angle, or the effective contact radii between parts, in practice we have surprising variations in all of these things.

The bolt is *not* a rigid body. It is a highly stressed component with a very complex shape and severe stress concentrations. The basic deformation is usually elastic, but there are always portions of the bolt—for example, in thread roots and the like—which deform plastically, altering geometric factors r_t and pitch.

The face of the nut is seldom exactly perpendicular to the axis of the threads. Holes are seldom drilled exactly perpendicular to the surface of the joint, so contact radius r_n is usually unknown. Some experiments indicate that these factors can introduce even more uncertainty than does friction.

Recently, for example, the Bolting Technology Council sponsored a research study of the torque-preload relationship [29]. The study was done at École Polytechnique in Montreal. The purpose of the study was to find an economically viable way to conduct bolting experiments, the problem being that very large numbers of variables are often going to affect results.

Eleven variables which were believed to have a significant impact on the torque-preload relationship were selected for the École Polytechnique study. Taguchi methods were used to statistically design the experiment. The variables chosen included "perpendicularity." Some bolts were tightened against parallel joint surfaces; others were tightened against joints having a 5° taper. Results of the experiment, which were confirmed by a second round, showed that perpendicularity affected the amount of preload achieved for a given torque more than did any other variable—including whether or not a lubricant had been applied to the threads and nut face.

C. Strain Energy Losses

All of the above variables are at least visible in the long-term equation. There are other sources of error, however, to which this torque-balance equation is blind. When we tighten a nut, we do work on the entire nut-bolt-joint system, as we saw in Chap. 2. Part of the input work ends up as bolt stretch or friction loss, as suggested by the long-form equation; but other portions of the input work end up as bolt twist, a bent shank, nut deformation, and joint deformation. The "true" relationship between input torque and bolt preload, therefore, must take these outputs into account, as we attempted to do in Eq. (2.29). In one extreme case, for example, if the threads gall and seize, input torque produces just torsional strain and no preload at all. The long-form equation would suggest that "all is lost in thread friction torque" for this situation (infinite coefficient of friction in the threads). This isn't true, but thread friction torque is a twist component, so the equation doesn't lie. It's just that the result is strain energy, not heat. In fact, I don't mean to imply that the torque-balance equation is incorrect. It does, indeed, describe the action and reaction torques on the system correctly. But you will get into trouble if, as is common, you then add, "every part of the input energy which is not converted to preload must end up as friction loss because the equation shows only preload or friction terms." The *torques* are only cam action or friction torques, but what this means from an energy distribution standpoint is not revealed by the torque equation.

D. Prevailing Torque

Another factor which is not included in the long-form equation is prevailing torque: the torque required to run down a lock nut which has a plastic insert in the threads, for example. The insert creates interference between nut and bolt threads, and thereby helps the fastener resist vibration. The torque required to overcome this interference doesn't contribute to bolt stretch. It might be considered an addition to the thread friction component of torque, but it is a function of the design of the lock nut, and of the materials used, as well as the geometry, so it's best handled as a separate term, as suggested below.

$$T_{in} = F_P \left(\frac{P}{2\pi} + \frac{\mu_t r_t}{\cos \beta} + \mu_n r_n \right) + T_P \tag{7.3}$$

where T_P = prevailing torque (lb-in., N-mm), and all other terms were defined earlier.

Note that the prevailing torque is not a function of preload, the way all the other terms are. Note too that prevailing torque may not be a constant; it may change as the lock nut is run farther down the bolt, or is reused.

E. Weight Effect

Heavy or misaligned joint members resist being pulled together. This may not affect the trorque-preload relationship, but it will reduce the amount of input torque which ends up as clamping force between joint members.

F. Hole Interference

If the hole is undersize or misaligned it will take some effort merely to pull the bolt through the hole. This, too, can reduce the amount of torque available to create bolt preload.

G. Interference Fit Threads

If threads are damaged, or if they're designed to have zero clearance, it can take some torque to run the nut down on the loose bolt.

H. The Mechanic

Lest we forget, there are people involved here too. The results we get for a specified torque will depend very much on whether or not the person using the wrench has been well trained, knows what he's doing, cares about doing it right, can reach the bolts easily, can see the dial gage on his wrench, etc. The operator can be a more important factor than all of the others combined.

I. Tool Accuracy

We must also remember that tools aren't perfect. They will produce a requested torque with some tolerance or error, depending upon their construction, the accuracy with which the gage reports their output, how recently they have been calibrated, etc.

J. Miscellaneous

There are many other factors which have been found to have some effect on the torque-preload relationship, although generally a smaller effect than the factors listed above. Some years ago, for example, R. Stewart of the Wright-Patterson Air Force Base gave me a list of 75 variables which they had found had a statistically significant impact on the torque-preload relationship. [30] The list included most—but not all—of the factors listed above, plus things like type, thickness, and consistency of plating; the type of bolt head; the treatment of the hole, hole finish, hole concentricity, hole size, countersunk angle; gaps; burrs; type of wrench used to tighten the bolts; whether it was torqued from head or nut ends; number of times the bolt and nut had been used; number, type, and size of the washers used if any; etc. The list ended with item 76: "Any combination of the above."

The École Polytechnique study mentioned above also included, as test variables, such factors as lubricant, grade of bolt, type of wrench used, whether or not the fastener was covered with rust, whether or not it was plated, the number of times it was tightened, full vs. partial thread engagement, and the stiffness of the joint in which the bolts were tightened. All were found to have some effect on the amount of preload achieved for a given torque, but most had a relatively—and sometimes surprisingly—small effect. Corrosion, joint stiffness, and amount of thread engagement were expected to have a fairly large effect, for example, but did not.

Since each of the effects listed above is itself affected by many secondary variables, you can see that literally hundreds of factors can influence the results when we tighten a single bolt. As we saw in the last chapter, additional factors—such as elastic interactions—further complicate our lives when, as usual, we tighten not just one but a group of bolts.

IV.   TORQUE VS. PRELOAD—THE SHORT-FORM EQUATION

There's another equation which I feel is more useful than the long-form equation. It's called the short-form torque-preload equation and it boils everything down to the "fact" that the initial preload created in a bolt is equal to the applied torque divided by a constant. Simple—but only if we know the constant!

A.   The Equation

Remember, we are trying to define the constant in the torque-preload equation. The so-called short-form equation gives us

$$T_{in} = F_P(KD) \tag{7.4}$$

where T_{in} = input torque (lb-in., *not* lb-ft; N-mm)
F_P = achieved preload (lb, N)
D = nominal diameter (in., mm)
K = "nut factor" (dimensionless)

If a prevailing torque fastener is used, the equation must be written

$$T_{in} = F_P(KD) + T_P \tag{7.5}$$

The discussion which follows assumes no prevailing torque, which is usually the case.

Note that the nut factor K is *not* a coefficient of friction. It is, instead, a general-purpose, experimental constant—our old friend the bugger factor—which says that "when we experimentally applied this torque to the fastener and actually measured the achieved preload, we discovered that the ratio between them could be defined by the following expression":

$$\frac{T_{in}}{F_P} = KD \qquad (7.4)$$

The nice thing about K is that it summarizes anything and everything that has affected the relationship between torque and preload in our experiment—including friction, torsion, bending, plastic deformation of threads, and any other factor that we may or may not have anticipated.

The un-nice thing about K, of course, is that it can only be determined experimentally, and experience shows that we really have to redetermine it for each new application. Even then it is not a single number. Experience shows that for accurate prediction we have to make a number of experiments to determine the mean K, standard deviation, etc. Having done this, however, we can indeed predict the minimum and maximum preload we're going to achieve for a given input torque, at a predictable confidence level. We cannot do this with the long-form equation.

Figure 7.4 will give you some idea of the variations you might expect to encounter in the K value for as-received steel fasteners. The histogram in Fig. 7.4 is a compilation of the K's reported in the literature obtained in our monthly computerized search, plus the K's reported by our field service engineers on maintenance and construction sites. The standard deviation for this K data is 0.05, and the mean is 0.199. Plus or minus three

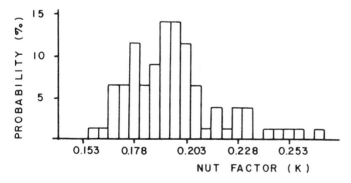

Figure 7.4 Histogram of K values reported for as-received steel fasteners from a large number of sources.

standard deviations, therefore, takes us from a K of 0.15 to a maximum K of 0.25.

The data suggest that we can expect a scatter of $\pm 26\%$ in the preload achieved at a given input torque. That is very close to the $\pm 30\%$ figure you will see quoted by many authors; and a 0.199 mean is equal to the 0.2 value usually cited for steel.

It's possible to combine the long-form and short-form equations mathematically, and so end up with an expression for K in terms of fastener geometry and the coefficient of friction [2]. There's nothing wrong with this approach from a mathematical standpoint, but I think it destroys all that is useful about K. As long as we treat K as an experimental constant it can help us estimate the relationship between torque and preload—if we make the right experiments and if we interpret them correctly. These are big "if's." Finding a mathematical equivalent for K, however, just turns the short-form equation into a quicky long-form equation; and since we can't solve one, we really can't solve the other. Let's free K from any preconceptions about where it comes from. Emerging ultrasonic techniques, to be discussed later, make on-site measurement of K a practical reality for the first time, and so drastically increase the usefulness of K and of the short-form equation.

B. Nut Factors

Note that since the nut factor sums up everything which affects the torque-tension relationship it constitutes an ideal way to report on or analyze the results of torque control procedures. If we had a handy table of the mean K, and scatter in K, associated with various procedures or lubricants or types of tools or combinations of such things, we'd have everything we need to pick tools, procedures, lubricants, etc., for our own applications. As already mentioned, however, an accurate K for a new application must be derived by experiments which measure torque and tension, and therefore derive K, on that application itself, not on similar applications.

Many investigators have found, in fact, that nut factors determined on a sample or prototype joint, in a laboratory, can often differ significantly from nut factors determined on the actual joint in the field or on a production line. This merely reflects the fact that the nut factor does indeed summarize such things as tool accuracy, operator skill, bolting procedure, etc., as well as the more obvious factors such as lubricity and condition of the threads.

If you need further proof that the nut factor is a "soft" number, which must be used with caution, consider the following case histories:

1. Brookhaven National Laboratories measured the coefficient of fric-
    tion between heavily loaded metal surfaces coated with a number of
    thread lubricants. They tested, for example, molydisulfide lubricants
    received from four different sources; and tested each one wet and dry
    to simulate bolted joint conditions. The coefficients of friction ranged
    from 0.026 to 0.273 [3], a variation of over 10:1. Other tests reported
    by the NRC [4] on graphite-based lubricants revealed a 3:1 spread in
    coefficient: copper-graphite and nickle-graphite lubes had a 2:1 range.
    These are coefficient of friction tests, not nut factor tests, but had
    the same lubes been tested in bolts the nut factor would presumably
    have shown similar—or even greater—dispersion.
2. A diesel engine manufacturer reported privately that the torque re-
    quired to achieve a desired preload in engine head bolts increased by
    50% with four reassembly operations using the same parts (and relub-
    ing them each time). Preloads were measured accurately, with ultra-
    sonic equipment which will be described in Chap. 11. The nut factor,
    in this case, increased 50% with reuse of the fasteners. Field studies
    on a large number of 3-in.-diameter fasteners in a nuclear power plant
    revealed the same trend: a steady increase in the torque required to
    achieve a given, ultrasonically measured preload (meaning a progres-
    sively higher nut factor) when the fasteners were reused.
3. An aerospace manufacturer, however, obtained very different results.
    A given $\frac{7}{8}$-in. MP35N bolt was tightened, loosened, and retightened
    repeatedly to a load of 46,000 lb. The torque required during 20 such
    retightenings dropped from 500 lb-ft to less then 200 lb-ft, as shown
    in Fig. 7.5. The nut factor in this case DECREASED with repeated
    retightenings.

A completely trustworthy nut factor, therefore, is like the Holy
Grail—something everyone yearns for, but which no one will ever find.
All of this means that we're back in a "box" which is becoming painfully
familiar. We'd like to pick a material property or a gasket stiffness or now
a nut factor from a handy table and then proceed with our design or
maintenance work; but we find every table preceded by warning labels
saying, "For General Use Only: Make Experiments of Your Own to Determine the True Values for Your Application."

Given the fact that most bolted joints are grossly overdesigned, however, and/or the fact that an occasional failure would be acceptable and/
or the fact that most people can't afford the time or money required to
make experiments, a table is still welcome. You'll find some nut factors,
therefore, in Table 7.1. These data merely report results obtained by oth-

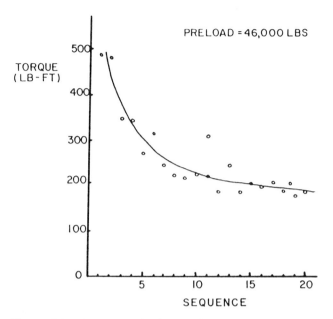

Figure 7.5 Torque required to produce the same 46,000 lb of tension in a single bolt, tightened 20 times in the same hole. Note that the bolt, a $\frac{7}{8}$-in. MP35N fastener, became easier to tighten as it was reused.

ers under conditions which will usually differ from the conditions you face. In most instances not enough information is given on the original "conditions" to allow much comparison anyway. But the data will give you an order-of-magnitude feeling for the way various lubes, fastener materials, rust, etc., affect the torque-tension relationship.

The data in the table are organized by thread "coatings"—lubricants, rust, platings, or "as-received" surfaces. This reflects the fact that lubricity has a major influence on the nut factor. A slippery surface will lead to a lower nut factor than a stickier one. As a result, many people want to assign a different nut factor to each type of lubricant, and "if other things are equal," this is possible. I do it in Table 7.1. But we must remember the earlier discussion of things which affect the initial preloads and/or the working loads in bolts. Many were lubricity factors, but not all. Operator errors and tool calibration were also involved, for example. And we also have the sobering Brookhaven studies, in which they measured the coefficient of friction between heavily loaded (and undoubtedly controlled) metal surfaces—test blocks of some sort, not bolts. They found

Table 7.1 Nut Factors (K)

Fastener materials and coatings[a,f]	Min	Mean	Max
	Reported nut factors		
Pure aluminum coating on AISI 8740 alloy steel [23]	0.42	0.52	0.62
Electroplated aluminum on AISI 8740 alloy steel [23]	—	0.52	—
As received, mild or alloy steel on steel	0.158	0.2	0.267
As received, stainless steel on mild or alloy steel	—	0.3	—
As received, 1-in.-diam. A490 [7]			
as received	—	0.179	—
very rusty[b]	—	0.389	—
with Johnson 140 stick wax	—	0.275	—
Black oxided $\frac{7}{8}$ A325 and A490, slightly rusty[c] [5]	0.15	—	0.22
Black oxide	0.109	0.179	0.279
Cadmium plate (dry)	0.106	0.2	0.328
Vacuum cadmium + chromate [23]	—	0.21	—
Copper-based antiseize	0.08	0.132	0.23
Cadmium plate (waxed)	0.17	0.187	0.198
Cadmium-plated A286 nuts and bolts [33]	0.15	—	0.23
Cadmium plate plus cetyl alcohol on A286 nuts and bolts [33]	0.11	—	0.16
Cadmium-plated nuts used with MP35N bolts [33]	0.18	—	0.29
Dag (graphite + binder)[d] [25]	0.16	—	0.28
Dicronite (tungsten carbide in lamellar form)[d,e]	0.045	—	0.075
Emralon (PTFE + resin)[d] [25]	0.10	—	0.15
Everlube 810 (MoS$_2$/graphite in silicone binder)[d] [24]	0.09	—	0.115
Everlube 811 (MoS$_2$/graphite in silicate binder)[d] [24]	0.09	—	0.115
Everlube 6108 (PTFE in phenolic binder)[d] [24]	0.105	—	0.13
Everlube 6109 (PTFE in epoxy binder)[d] [24]	0.115	—	0.14
Everlube 6122	0.069	0.086	0.103
Fel-Pro C54	0.08	0.132	0.23
Fel-Pro C-670	0.08	0.095	0.15
Fel-Pro N 5000 (paste)	0.13	0.15	0.27
Mechanically galvanized A325 bolts [5]			
as received	0.35	—	0.49
clean and dry	—	0.46	—
slightly rusty[c]	0.36	—	0.39
Mechanically galvanized A325 bolts; lubed with 1 part water, 1 part Jon Cote 639 wax [5]	0.11	—	0.26
Hot-dip galvanized 7/8 A325 [5]			
as received	0.14	—	0.31
slightly rusty[c]	0.09	—	0.17
clean and dry	0.09	—	0.37
Hot-dip galvanized 7/8 A325 lubed with 1 part water, 1 part Jon Cote 639 wax [5]	0.10	—	0.16

Table 7.1 Continued

Fastener materials and coatings[a,f]	Reported nut factors		
	Min	Mean	Max
Gold, on stainless steel or beryllium copper [21]	—	0.4	—
Graphitic coatings	0.09	—	0.28
Lube Lok 1000 (ceramic bonded coating)[d,e]	0.275	—	0.31
Lube Lok 2006 (silicone resin bonded MoS_2 + graphite)[d,e]	0.075	—	0.25
Lube Lok 4253 (high silver and indium content coating)[d,e]	0.21	—	0.24
Lube Lok 4856 (contains powdered tin + lead)[d,e]	0.21	—	0.25
Lube Lok 5306[d,e]	0.04	—	0.18
Machine oil	0.10	0.21	0.225
Microseal 100-1 (graphite plus inorganic binder)[d] [24]	0.09	—	0.115
Molydag (MoS_2 + binder)[d] [25]	0.16	—	0.40
Moly paste or grease	0.10	0.13	0.18
Neolube	0.14	0.18	0.20
Never-Seize (paste)	0.11	0.17	0.21
N5000, threads only, not nut [6]	0.097	0.106	0.117
Pepcoat 6122 on nut, threads, and washer [6]	0.080	0.085	0.089
Pepcoat[e] [26]	0.09	—	0.11
Phos-Oil	0.15	0.19	0.23
PTFE plus binder[d] [25]	0.10	—	0.15
SermaGard (aluminum particles in ceramic binder)[e]			
#902 or 846; basecoat only	—	0.50	—
Basecoat + SermaLube 1000	0.09	0.15	0.40
Basecoat + 751 wax	0.18	0.23	0.90
Misc. Sermagard systems with modified silicone topcoat	—	0.17	—
SermeTel-W (phosphate-bonded, aluminum-rich coating) [23]	—	0.30	—
Solid film PTFE	0.09	0.12	0.16
Zinc plate (waxed	0.071	0.288	0.52
Zinc plate (dry)	0.075	0.295	0.53

[a] All test involve steel bolts in steel joints, unless otherwise noted.
[b] Exposed outdoors for 2 weeks.
[c] Dipped in water, air-dried 24 hr twice.
[d] Use with extra caution. These nut factors were computed from published coefficients of friction (μ) data, using:

$$K = \mu \frac{r_t}{D \cos \beta} + \frac{r_n}{D} + \frac{P}{2D\pi}$$

See Eq. (7.2) for terms and units. Threads ranging from $\frac{1}{4}$–20 to 2–$4\frac{1}{2}$ were assumed in the calculations.
[e] From manufacturer's literature.
[f] See Chap. 18 for list of coating suppliers and their addresses.

a 10:1 variation in lubricity of things called molydisulfide by a variety of manufacturers. So surface coatings are important, and are used to define our table, but we should remember that this is only one of the many factors which determine K.

As a result, the nut factor data in Table 7.1 are "typical" only. They may not represent your own application. *In critical applications you should always develop your own nut factors by an appropriate experiment.*

C. The Coefficient of Friction and the Nut Factor

Lubricant manufacturers and test laboratories are usually interested in friction in general, not just friction in threaded fasteners. Because of this they report test results in terms of coefficient of friction (μ) instead of nut factor. We could use Eq. (7.2) to convert coefficient data to a nut factor, but it's usually less trouble to use the following approximation:

K is approximately equal to μ plus 0.04

For example, if we're told that a new thread lube provides a steel-on-steel coefficient of friction of 0.12, we can estimate that a fastener using that lube would have a nut factor of about 0.16.

Again, this is only an approximation, but many factors in addition to friction affect the nut factor for a given application, so an approximation is useful.

V. TORQUE CONTROL IN PRACTICE

A. What Torque Should I Use?

Perhaps the most common question in bolting is "What torque should I use on this bolt?" The purpose of picking a "good" torque is to end up with a "good" preload, leading to a "good" clamping force on the joint. So, in order to answer the torque question we must first decide what clamping force is required in our application.

We're not ready to do that yet. We must first learn about bolting tools and the way they affect the results. And, more important, we must learn about bolt and joint failure modes, and the correlation between bolt tension, clamping force, and failure. We'll get around to an answer for the "what torque?" question—but not until Chap. 20.

We have, however, acquired one piece of information which will be vital to our final choice of assembly torque. We've learned that we can't count on that torque achieving a single value of tension in all the bolts we tighten. The resulting tensions will be scattered; we'll end up with a range of tensions as broad, perhaps, as one of the ranges of nut factor reported in Table 7.1. To pick a torque to use during assembly we'll

pick an optimum or target preload, selected to avoid or minimize subsequent joint problems. Then we'll
identify the mean nut factor which best defines the conditions of our application. We'll
use that nut factor and target preload in the short-form equation [Eq. (7.4)]
   to define the torque we'll specify for assembly purposes.

We'll also, if the joint is important, use the reported scatter in possible nut factor value to estimate the max-min range of preload that torque might give us. And we'll decide whether those max and min tensions could lead to joint problems. If so, then we'll have to find a better way to control the assembly process (to be discussed in the next few chapters)—or we'll have to redesign the joint to be more tolerant of scatter in clamping force.

The following discussion gives further case histories on nut factor and preload scatter, using actual data. But first: When discussing preload scatter we must be careful to distinguish between the initial preload achieved bolt by bolt as we first tighten them and the residual preloads in that same or a different group of bolts after all in the group have been tightened and embedment and elastic interactions have done their work. Many experimenters report only the first and lull us into a false sense of security. Residual scatter is always greater—and sometimes *much* greater—than the initial scatter.

B. Initial Preload Scatter

Figure 7.6 shows initial preload results for a large number of fasteners tightened one at a time. Note that the distribution of preload achieved for a given torque is skewed right in this case, rather than being Gaussian or normal [15].

If your sample is smaller, there is little you can predict about the distribution or range of results beyond what you know about the possible (but not certain) range in K. Since this can vary by $\pm 26\%$, the preload can vary from -21% to $+35\%$ of the value corresponding to the mean K. But the distribution can be almost anything, as shown in Fig. 7.7, which shows the preloads in a group of bolts tightened to two different torques: 400 ft-lb and 800 ft-lb. But there's a hitch here. This time the preloads plotted are residual preloads measured after all bolts in the joint had been tightened. The studs were loosened between the two tests. These were the same studs, on the same joint, tightened by the same operator, using the same tool, on the same day. Yet the distribution in one case is skewed

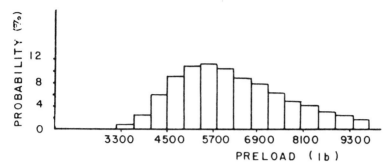

Figure 7.6 Histogram of the initial preload achieved in a large number of $\frac{5}{16}$–24, SAE Grade 8 fasteners with a 2.3-in. grip length tightened to a uniform torque level. (Modified from Ref. 15.)

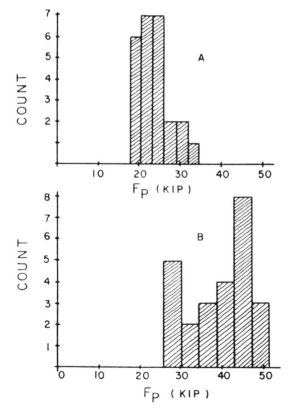

Figure 7.7 Histograms of the residual preload produced in 25, $1\frac{3}{8}$–8 × 10, B7 studs coated with solid-film PTFE when torqued to 400 lb-ft (A) or, in a second test, to 800 lb-ft (B).

right; in the other case, left. Note, too, that the maximum preload obtained at 400 lb-ft overlaps the minimum obtained at twice that torque.

At first glance it looks as if the nut factors for most of these bolts changed between the pass at 400 lb-ft and the pass at 800 lb-ft. If the nut factors had remained the same, and if there were no elastic interactions, the *shape* of the histogram would have remained unchanged; only the mean would have changed. I introduce this case history, however, not to say that nut factors can change (they can!) but to warn you that the short-form torque-tension equation, and the nut factor, give less and less useful results the farther one gets from the tightening of *individual* bolts. The joint just discussed can serve as an example.

The short-form equation is useful when we want to describe the relationship between applied torque and achieved preload *as the fastener is being tightened*, or at the instant the tightening process is completed. A tightening experiment might show the sort of scatter in nut factor suggested by the data in Table 7.1. If we record the final torque applied to a fastener, however, and then measure the residual some time *after* tightening, and then use the short-form equation to compute K, we'll usually conclude that the nut factor varies far more than "usual." The scatter in final tension may be much greater than that shown in Table 7.1.

The reason: postassembly relaxation effects further disperse the already scattered tensions created in bolts as they were tightened individually. Embedment relaxation and elastic interactions between bolts in a group are the main culprits. Both are discussed in the last chapter. As we saw, the resulting scatter in residual (as opposed to initial) tensions in the bolts can be 4:1 or 10:1 or worse. The short-form equation can be applied to such data; you're free to define the terms and conditions for your experimentally derived nut factors any way you like; but the scatter (and therefore the uncertainties) in the results becomes so great as to make your calculations almost useless. It's best to use nut factors only to describe the initial tightening of individual fasteners—groups of them, sure, but one at a time.

C. Low Friction for Best Control

Friction affects the efficiency with which input work on the nut is converted to preload. We're primarily interested, however, in the *accuracy* of this process. Does a lubricant on the threads increase or decrease the preload scatter for a given input torque?

If we use the long-form torque equation to compute the preload achieved in a given fastener for a given input torque at different coefficients of friction, we'll get the results shown in Fig. 7.8. At first glance,

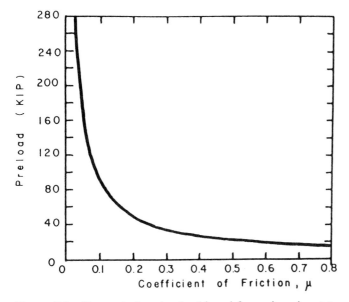

Figure 7.8 Theoretical preload achieved for a given input torque as a function of the coefficient of friction.

Fig. 7.8 suggests that a higher coefficient would provide more accurate control than a lower coefficient. A ±20% uncertainty in a coefficient of 0.4, for example, would mean a 14-kip scatter in preload; while a ±20% uncertainty in a coefficient of 0.1 would mean a 28-kip scatter in preload, although percentage changes in preload would remain about the same.

A similar curve can be drawn for the relationship between K and preload for a given torque. So, does low friction mean poor control?

No. In practice, lubricants usually reduce scatter in preload. Histogram A in Fig. 7.9, for example, shows the scatter in torque required to bring 140, M12 steel bolts to the ultimate tensile point when the bolts were lubricated with machine oil [8]. Note that the torque required for a given preload has a normal distribution, at least in this case.

Histogram B in Fig. 7.9 shows the higher torques—and slightly greater scatter in torque—measured during similar tests on 140, M12 steel bolts which had been cleaned in gasoline and dried. Since all 280 bolts had been taken to ultimate strength, the final preload was approximately the same in each, and "more scatter in torque" means "more variation in μ or K." The difference is not great, however, because machine oil isn't a very good thread lubricant.

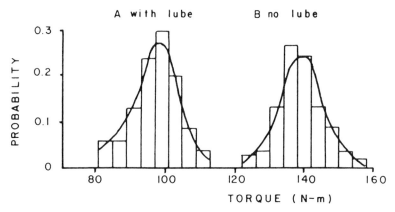

Figure 7.9   Histogram of torques required to tension 140 lubricated bolts (A) and 140 unlubricated bolts (B) to ultimate tensile strength. Bolts were M12, Steel St. 3. (Modified from Ref. 8.)

Figure 7.10 shows more dramatic results obtained in experiments at Raymond Engineering [9] on $2\frac{1}{4}$–8 $\times$ 12, B16, inner liner studs from a large steam turbine. The cluster of curves labeled A were obtained in tests on studs lubricated with molybdenum disulfide grease, a good thread lubricant; cluster B shows results obtained on cleaned and dried studs. There's a substantial difference in the torque-preload scatter. The lubrication obviously helps repeatability here.

Here's some practical advice:

If certainty you need in *clamp*,
Just make sure those bolts are *damp*.

D.   The Lines Aren't Always Straight

The equations predict, and many experiments confirm, that the relationship between applied torque and achieved preload is usually linear until something in the joint yields. There are many times, however, when this is not true. Sometimes thread lubricants migrate or break down, and/or minor galling occurs, so that it requires larger and larger increments of torque to produce the next increment of preload, as shown in Fig. 7.11A [9, 10, 12, 13]. In other cases [10–12], it becomes *easier* to produce the next increment of preload as surfaces smooth, as in Fig. 7.11B [10–12]. There's not much you can do about this, but you should be aware of it. It usually won't make as big a difference in your results as will other uncertainties.

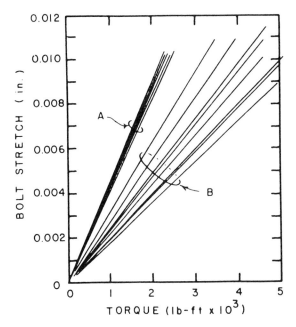

Figure 7.10 Torque-preload tests on lubricated (group A) and unlubricated (group B) $2\frac{1}{4}$–8 × 12, B16 studs.

E. Other Problems

We can summarize the things we've talked about so far—the things which affect the torque-preload relationship—with an acronym, FOGTAR, as follows:

    *F*riction
    *O*perator
    *G*eometry = FOGTAR
    *T*ool *A*ccuracy
    *R*elaxation

Friction includes not only lubricant but surface finish, speed of tightening, type of materials involved, and many, many more variables. Geometry includes not only the manufacturing tolerances on parts but that important perpendicularity between nut face, hole axis, and joint surface. Relaxation includes embedment and elastic interactions, both of which occur as we tighten a group of bolts. Operator and tool accuracy are self-explained.

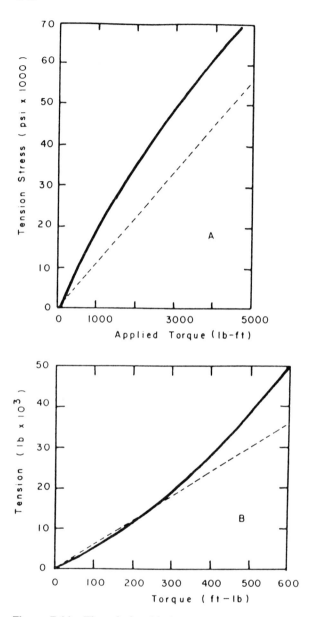

Figure 7.11   The relationship between torque and preload is not always a straight line, as these experimentally derived curves show. Curve A is for a $2\frac{1}{4}$–8 $\times$ 12, B16 stud. Curve B is for a 1-in.-diameter ASTM A325 bolt with a $2\frac{1}{4}$-in. grip length. The dashed line in each case is the theoretical line for nut factors of 0.132 and 0.2, respectively.

FOGTAR! It's probably the Tibetan word for trouble, and it reminds us that it's not just variation in friction which makes torque control less than perfect. It's FOGTAR and all that implies.

Remember this and you can't go wrong.
Your tools are weak and the FOGTAR's strong!

And that's not all. We must also worry about poor design, plus damaged or incorrect parts, if we're the project manager rather than the tool user. Figure 7.12 shows an example of the problems "other things" can cause. A study [14] of joint failures during live missions on the Skylab program showed that only 14% of the failures were caused by what the author of the final report called "incorrect torque" [14]. "Incorrect preload" would probably have been a more accurate term, and it would be the result of most of the factors we've discussed in this chapter.

In any event, 86% of the failures were traced to poor design, bad parts, or operator problems—and this on a program where the quality control activity must have been intensive, to say the least.

Our choice of the long-form or short-form equation, our guess for μ or K, and the accuracy with which we measure torque may have only a small influence on the number of joint failures we must face. To solve the total problem we must control *D*esign, FOGTAR, and *P*arts—DEFOG TARP? Anyway, as we'll see later, some tools and control systems are designed to compensate for some, at least, of the FOGTAR problems—and even for faulty parts. Tool designers have gone well past tool accuracy in contemporary designs.

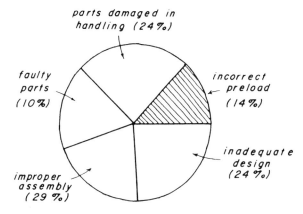

Figure 7.12. A summary of the causes of bolted joint failure on the Skylab program. All fastners have been torqued. (Modified from Ref. 14.)

F. Case History

I think that it's also useful to look at a detailed case history to see how the theories relate to the real world of bolting. Here's an example where torque was used to control the tightening of 120 large bolts in a nuclear power plant. Although torque was used as the control variable, the results were monitored with an ultrasonic extensometer. These devices, to be discussed at length in Chap. 11, measure the change in length of fasteners as they are tightened. As we saw in Chap. 5, change in length measurements can be related to fastener preload, using Hooke's law and/or suitable experiments. Since the extensometer can be used on virtually every fastener in the field, it gives us, for the first time, an opportunity to determine the actual results obtained with a variety of tools, lubricants, procedures, etc., as well as to determine the behavior of joints following tightening and in use. It certainly reveals the limitations as well as the good points of conventional bolting tools, procedures, and theories.

Our first case history illustrates our present topic: the accuracy of torque control in a field situation. Note that this time we're measuring *residual* preload in a group of bolts after all have been tightened.

Nuclear Foundation Bolts, 3–8 × 26, B23, 120 Bolts in Two Concentric Circles

The customer used hydraulic wrenches to tighten these studs. Prior to erection, the original equipment manufacturer had sponsored a careful study to determine the amount of preload which would be produced by the suggested input torque. A few of the studs were strain-gaged for this purpose; tests were performed in an engineering laboratory. A preload of 680 kip ±10% was specified.

As a result of these laboratory tests, an input torque of 13,375 lb-ft was recommended, with a moly lubricant.

In spite of the preliminary tests, the engineers in charge of erection had some concern about the possible torque-preload relationship. As a result, an ultrasonic extensometer was used to measure the actual change in length of each stud as it was tightened, with 0.052 ± 0.005 in. established as the goal. Calculations confirmed by calibration experiments related this change in length to the desired preload of 680 kip.

The actual relationship, measured in the field, between applied torque and achieved stretch is shown in the scatter diagram of Fig. 7.13. Note that none of the 120 studs reached the desired stretch of 0.052 in. at the specified input torque. In fact, the mean torque required was 34,611 lb-ft, with a minimum of 18,000 and a maximum of 42,000 lb-ft. The distribution is shown by the histogram of Fig. 7.14.

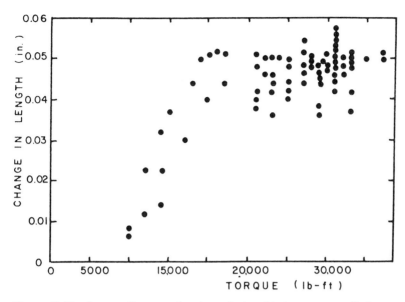

Figure 7.13 Scatter diagram showing relationship between applied torque and achieved preload on 120, 3–8 × 26, ASTM A540, Class 1, Grade 23 nuclear foundation bolts.

It's difficult to explain the large difference between the mean preload achieved in the field and the mean preload achieved by the laboratory tests on the same parts, with the same lubricants, etc. A large part of the *scatter*, however, is relatively easy to understand; it can be laid to operator error.

Four hydraulic wrenches, operated in parallel from one pump, were used to tighten four studs simultaneously. Each wrench was manned by a crew of four workers. Work was performed around the clock, meaning three shifts of workers. Each individual and each crew approached the job a little differently. When it came to lubrication, for example, our technician reported that some of the workers applied the molycote lubricant very carefully. Others applied the lubricant only to the near side of the stud; still others lubricated everything in sight, including wall, ceiling, and fellow workers!

The four crews were synchronized by a whistle, blown by a foreman. Because of the nature of the structure being assembled, at least two crews were out of the foremen's sight at any one time. The whistle signals—which were sometimes neglected—notified the crews that it was

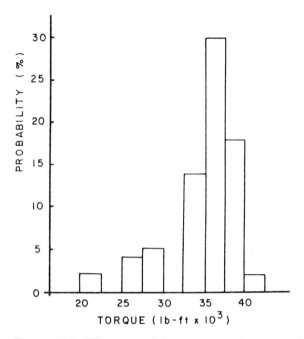

Figure 7.14 Histogram of the torque required to produce the desired 0.052-in. ± 0.005 in. stretch in the foundation bolts described in Fig. 7.13.

time to move the wrench to the next stud or to reset it for another stroke on the same stud, or that pressure was about to be applied to the hydraulic cylinders. (The foreman operated the pump.) Missed strokes were very common.

But missed strokes don't affect the final torque-preload scatter because the extensometer revealed these problems and allowed the teams to compensate for them. Variations in lubricating technique, however, could not be prevented, although they could also be compensated for by applying more torque until the desired stretch had been achieved.

Thanks to the use of the ultrasonic equipment, each stud was finally tightened to the desired preload. Subsequent measurements after a period of 2 days showed, furthermore, that the studs had retained the final stretch.

If pure torque control had been used, as originally planned, the results obtained here would have been drastically different.

VI. TOOLS FOR TORQUE CONTROL

A. In the Beginning—A Search for Accuracy

Manual Torque Wrenches

Man's quest for more reliable bolted joints probably started with the manual torque wrench—an attempt to get improved torque accuracy. Today, after years of development, there are many kinds of torque wrench, ranging in output from a few ounce-inches to 1000 lb-ft. Output torque accuracies range from ±2% to ±20% of full scale.

As soon as man had acquired some semblance of torquing accuracy, he discovered that "accuracy" was not enough. Operator problems—lack of skill, carelessness, etc.—often wiped out the gains made in accuracy. Many of the wrenches on the market, therefore, have been designed to reduce operator problems by providing such things as better gauges, presets, signal lights, audible outputs (clicks), etc. The latest manual wrenches are equipped with electronic measuring systems and digital readout. Figure 7.15 shows some of the many types on the market today.

Shear Loads Created by Torque Wrenches

A manual torque wrench does not just create torque on the nut or bolt head. It also creates a side load which is equal and opposite to the force with which the mechanic pulls the handle of the wrench. This reaction side load is necessary to establish what the professors call "equilibrium." (We better call it that, too!) The input torque—mechanic's pull times the length of the wrench—must equal the reaction torque created by thread and nut friction and by the inclined plane action of the threads, turning torque into stretch of the bolt. At the same time the sum of forces in the x, y, and z planes must each equal zero; action forces must be balanced by reaction forces. Fortunately, we only have to worry about the x plane if we choose our coordinates properly.

Let's look at an example. The mechanic is pulling on the handle of the wrench in Fig. 7.16 with a force of 20 lb. The distance between the center of the bolt and the mechanic's hand is 2 ft. Input torque is clockwise and is

$$T_{in} = 20 \times 2 = 40 \text{ ft-lb}$$

For equilibrium the input moment—which equals the input torque—must equal the reaction moment or

$$\sum M = 0$$

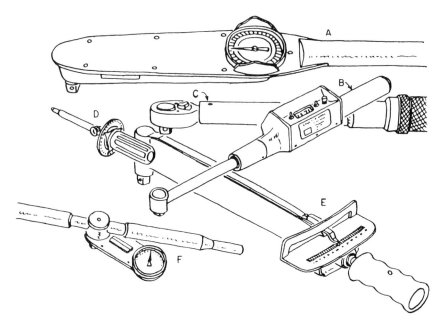

Figure 7.15 A few of the many types of torque wrench being manufactured today. Those shown are made and/or sold by (A, C, F) Snap-On, (B) GSE, and (D, E) Mountz. See Appendix D for the complete names and addresses of these and other manufacturers of bolting tools.

The fastener reacts with a 40 ft-lb torque in the counterclockwise direction.

At the same time the fastener must support the sideways pull of 20 lb. If it didn't, the bolt, mechanic, and wrench would just move sideways. Looking at the free-body diagram of the nut we see that the act of tightening it has created a side or shear load of 20 lb between the nut and the bolt. (This is supported by a 20-lb force between the bolt and its hole.)

Anything which increases the contact pressure between male and female threads increases the frictional drag as the nut is turned around the bolt. The axial forces created here, as the bolt is stretched, will be much greater than this 20-lb side load, but this example illustrates the forces that are created when any unbalanced torque tool is used to tighten a bolt. There are some situations, as we'll see later, where these reaction side loads can become very high.

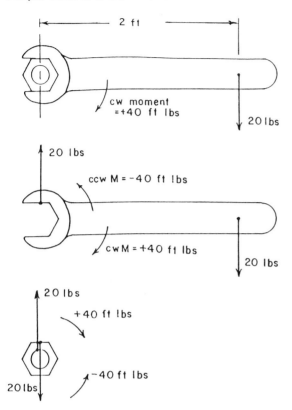

Figure 7.16 A manual torque wrench of any kind produces not only torque on a nut, but also a reaction side load equal to the force exerted by the mechanic on the handle of the wrench. The middle drawing here is a free-body diagram of the wrench. The lower drawing is a free-body diagram of the nut and shows that the reaction side load must be supported by the bolt (which in turn is supported by the joint member). Although the 20-lb side load shown in this example is trivial and would have a negligible effect on the torque-preload relationship, other types of torque tools can create very large side loads of this sort, as described in the text.

B. More Torque for Very Large Fasteners

The higher the torque, of course, the more difficult it becomes to produce it manually. Several tools, therefore, have been designed to help tighten larger fasteners. Those of us who assemble heavy equipment often need one or more of the following tools.

Torque Multipliers

Torque multipliers are gearboxes that multiply the torque produced by a manual torque wrench (see Fig. 7.17). Typical ratios are 4:1 or 10:1, with a few up to 100:1 or so. Thanks to friction losses in the gear trains, multipliers tend to decrease the accuracy of a manual wrench somewhat, but they produce output torques of up to 83,000 lb-ft.

Note that the reaction side loads will tend to increase as the applied torque increases. Note too that something has to hold the gearbox—the multiplier—from rotating or moving sideways as it applies torque to the fastener. Multipliers, therefore, are equipped with a reaction arm which leans against another bolt in the group (as for example the one shown in Fig. 7.17C) or against some part of the structure being assembled (Fig. 7.17 B, D, and E). Larger multipliers are equipped with pins in their base which engage a reaction arm designed for a particular application or engage holes drilled in the joint member itself. (An example is shown in Fig. 7.17A.)

Geared torque wrenches combine the readout of a torque wrench with the gearing of a torque multiplier. This time the readout is of output torque (rather than input to a gear train), and the multiplication ratio is higher—ranging from 125:1 up to 2400:1 on wrenches producing outputs of 600 lb-ft up to 20,000 lb-ft. The high gear ratio means that input torques are very low; the tools can be driven by nut runners. So in fact these are

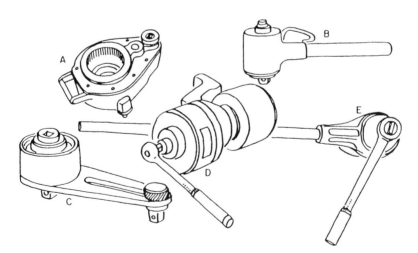

Figure 7.17 Typical torque multipliers, manufactured by (A, B) B. K. Sweeney, (C) Mountz, (D) Plarad, (E) X-4 Corp. See Appendix D for more information.

"nut runner multipliers" rather than "torque wrench multipliers." See Fig. 7.18.

Like multipliers, geared wrenches must be provided with a reaction arm of some sort. Those shown in Fig. 7.18A and C have two pins in their base, equidistant from the output spindle. If provided with a double-sided reaction arm—and if that reacts against two points in the structure which are also equidistant from the drive spindle—the tool will produce a pure torque couple on the fastener. As a result, such tools have been used to tighten the main spindle nuts of high-speed jet engines and other rotating equipment, where the concentricity of the final assembly is very important (to prevent vibration) and it's very useful, therefore, to avoid large side loads on the fastener and fastened parts during assembly.

Hydraulic Wrenches

Another way to get a lot of torque in a small space is to use a hydraulically actuated tool in which a piston drives a short, stubby ratchet wrench through as many cycles as necessary to tighten a bolt. This is probably the most popular type of production tool when torques in the range 1000 to 5000 lb-ft are required, although the tools are available in torques of up to 100,000 lb-ft of output. Output torque accuracies of $\pm 2\%$ to 10%

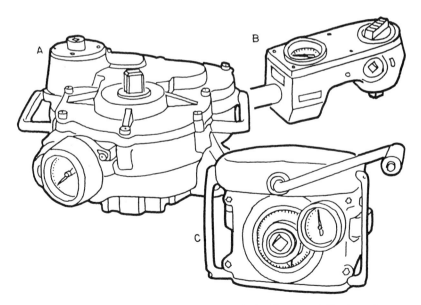

Figure 7.18 Geared torque wrenches made by Raymond Engineering.

of full scale are possible in most cases. This is one of the few power tools available for extreme torques (Fig. 7.19).

Most of these tools have one-sided reaction arrangements and do, therefore, exert fairly large side loads on the fasteners being tightened.

C. Toward Higher Speed

None of the tools described so far are very fast, obviously. For high-speed production applications we must have something else. The most common ''something else'' is an air-powered impact wrench or nut runner.

Impact Wrenches

Tiny hammers within impact wrenches give repeated blows on an output anvil, allowing a relatively small, lightweight, inexpensive tool to produce surprisingly high output torques—tens of thousands of pound-feet in some cases—at least on stiff joints. Impact wrenches are notoriously noisy and

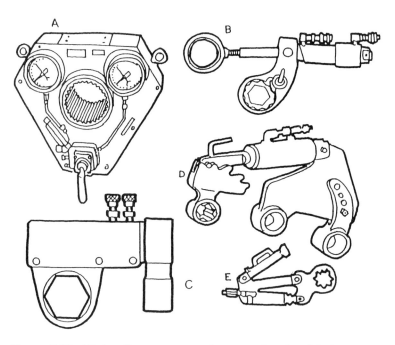

Figure 7.19 Hydraulic torque wrenches come in a bewildering variety of sizes and shapes. Those shown are made and/or sold by (A) Advance Hydraulics, (B, C) Raymond Engineering, (D) B. K. Sweeney, and (E) Plarad.

inaccurate, although some improvements have been made in accuracy in recent years. The amount of torque produced by a given tool on a joint, however, depends very much on the springiness of that particular joint. Even changing the length of the drive bit between wrench and socket can change output substantially—a short bit is a stiffer spring than a long one. Torque accuracies of ±20% are claimed, but in most situations are probably less than this [16,17]. A few of the dozens of styles available are shown in Fig. 7.20.

Nut Runners

Most nut runners contain a small rotary-vane "turbine" whose low torque and high speed are converted to usable output torque and speed by a multistage planetary gear train. Nut runners are available with outputs of up to 650 lb-ft or so. They are quieter than impact wrenches and, even in their simplest form, more accurate. We'll consider them in more detail in a moment.

One way to get higher speed, of course, is to tighten several fasteners simultaneously. Most air-tool manufacturers, therefore, provide multi-

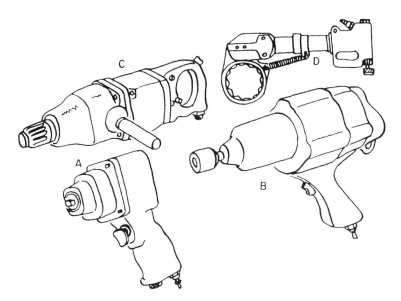

Figure 7.20 All of the impact wrenches shown here are air-powered. A few electrically actuated tools are also available. Those sketched here are provided by (A) ARO, (B, C) Chicago Pneumatic, and (D) Torque Systems, Inc.

spindle as well as single-spindle tools. Both single- and multiple-spindle nut runners are shown in Fig. 7.21.

Multispindle tools react against one nut while turning another. Each nut being turned, in fact, acts as a reaction point for the other nuts being turned. This is convenient and makes it possible to tighten groups of fasteners very rapidly. But the short reaction distances can create substantial side loads. Consider the situation, for example, in which a two-spindle nut runner is applying 600 ft-lb of torque to two 1-in.-diameter bolts which are 3 in. apart.

$$\text{Torque} = 600 \text{ ft-lb}$$
$$\text{Reaction arm} = 3 \text{ in.} = 0.25 \text{ ft}$$
$$\text{Reaction force} = \frac{600}{0.25} = 2400 \text{ lb}$$

If the bolts are only 2 in. apart the reaction force rises to 3600 lb.

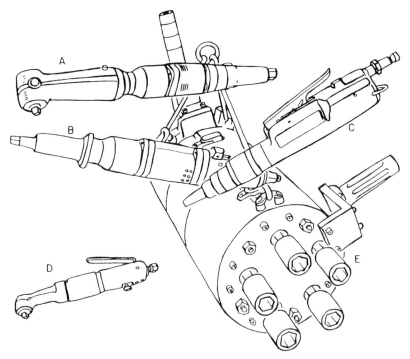

Figure 7.21 Single- and multiple-spindle air-powered nut runners available from (A, B, D) Chicago Pneumatic, (C) Mountz, and (E) Thor. There are many other sources for such tools. See Appendix D for names and addresses.

If the tool is tightening a large number of bolts simultaneously—the tool shown in Fig. 7.21E, for example—my guess is that each bolt will see a different amount of reaction force, that these forces will be statically indeterminate, and that one or a couple of them could be even higher than would be suggested by considering them in pairs.

Air-Powered Torque Wrenches

Nut runners produce only a limited amount of torque. Large impact wrenches produce very high torques, but, as mentioned, with relatively poor accuracy. Several manufacturers, therefore, have combined nut runners with the geared torque wrenches or multipliers to produce high torque at moderate speed. Some of these tools are shown in Fig. 7.22. Output torques of up to 50,000 lb-ft (65,500 N-mm) are available this way.

D. Powered Tools With Improved Control

The early air-powered tools gave people the increase in production speed they were looking for, but with much less torque control than that avail-

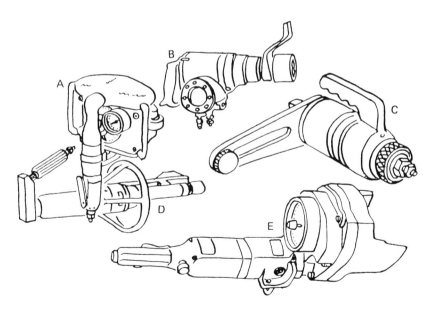

Figure 7.22 Air-powered geared wrenches or multipliers are an attempt to combine the high torques of the geared tools with the high speed of nut runners. Those sketched here can be obtained from (A) Raymond Engineering, (B) Kyokuto Boeki-Kaisha, (C) Norbar, (D) Atlas-Copco, and (E) B. K. Sweeney. See Appendix D.

able from a good torque wrench. Air-tool designers, therefore, were asked to improve the control, a process that is still going on [18].

At the present time a number of different levels of control are available. *Stall-torque nut runners*, for example, are controlled by input air supply. When input energy balances output (plus internal loss), the tools stall. If input air pressure and flow are controlled, the tool will produce output torque with an accuracy of ±10% or so, although ±15% to 20% is probably more common as the tools wear, etc. [16].

Other nut runners rely on one or more slip clutches in the output train to prevent overtorquing. Still others have single- or multiple-toothed clutches which automatically pop open when a preset level of torque has been reached.

Impact wrenches, in the same way, come with a variety of mechanisms designed to give varying degrees of control over output torques. There is so much activity, and so many choices, for both impact wrenches and nut runners that you'd best consult the manufacturers for accuracy specifications, but the range is probably ±5–15%.

Note that each type of control has strengths and weaknesses. A disengaging clutch, for example, may provide maximum accuracy for a gradually tightened joint, but could have problems with a self-tapping screw or prevailing torque lock nut (it would disengage prematurely).

E. Add Torque Calibration or Torque Monitoring

Stall-torque or clutched air tools give some degree of output torque accuracy only as long as they are properly adjusted and within calibration. Air tools tend to wear rapidly, however, and good control on input air supply is not always possible. In some cases, operator skill and carelessness can also make a difference. One of the next steps on the road to better accuracy, therefore, is torque monitoring [16, 19].

Torque calibration is the simplest way to monitor torque. The user periodically measures the output torque being produced by the tool, using a calibration stand of some sort. Several kinds are shown in Fig. 7.23.

The length of time between calibrations depends on such things as the importance of the joint being tightened, the environment in which the tools are being used, the stability of the tools, etc. In some cases the calibration period is specified by a standard-setting body. Structural steel torque wrenches, for example, must be recalibrated daily, per the specification on structural joints issued by the Research Council on Structural Connections of the Engineering Foundation.

The calibrator most frequently used in structural work is shown in Fig. 7.24. The heart of the device is a short hydraulic cylinder with a hole through the middle. A bolt is run through the hole, a nut and/or washers

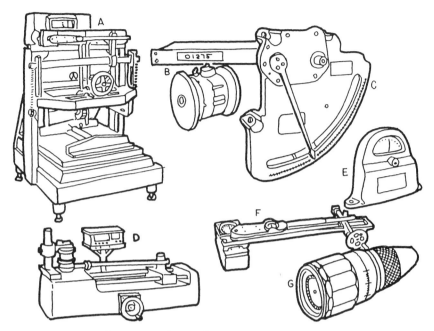

Figure 7.23 Equipment used to calibrate torque tools comes in many different sizes and shapes. Traditional calibrators are mechanical devices, but electronics is rapidly taking over (as in B and D). Equipment shown here is available from (A and F) Snap-On, (B) Himmelstein, (C) B. K. Sweeney, (D) AKO, and (E and G) Mountz.

are added, and the bolt is tightened. This raises the pressure in the cylinder, and a suitably calibrated pressure gauge interprets the increase in pressure in terms of clamping force. It's a simple, effective, and popular tool [20]. It's certainly accurate enough for most purposes, but the torque versus preload relationship is affected somewhat by the stiffness of the joint in which the fastener is used, as we'll see in later chapters. Since the stiffness of the hydraulic cylinder is a lot less than the stiffness of most joints, the accuracy of this calibrator would be a problem in critical applications.

One problem with calibration is the fact that the torque produced by some tools can be influenced by the characteristics (e.g., the stiffness) of the joint. This is especially true of impacting tools, but it can also be true for other tools. Since the stiffness of the calibrator is never the same as that of the joint, frequent calibration is not as reliable a means of control as you might think.

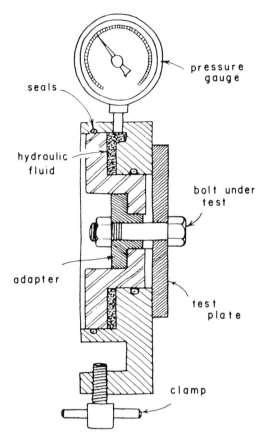

Figure 7.24 Skidmore-Wilhelm calibrator. A fastener is mounted in the calibrator (rather than in the actual joint) and is tightened by applying torque to head or nut. The instrument shows the actual tension achieved in the fastener under these conditions.

A good way to monitor the torque being delivered to the joint is to measure the output of the air tool as it actually tightens the fasteners. Sophisticated torque-monitoring systems are available for this purpose—electrical or electronic systems give a digital or control signal readout as the fastener is being tightened. Some systems are used only to monitor a few bolts once in a while—a sample. Other systems are designed to monitor all of the bolts tightened at a given production station. Monitored systems still use ± 10% air tools, but since they are monitored they really do produce ± 10% [16]. Figure 7.25 shows some of these monitoring systems.

Still other systems are intended to monitor the *tension* in sample joints as well as the torque applied to the fastener. One such, for example, interposes a tension load cell between the head of the fastener (or nut) and the joint. A torque transducer is interposed between the normal production tool and the drive socket. The readout system—which is portable—now watches applied torque *and* achieved tension. The system is not intended to be used on every assembly. It is used by quality control inspectors or engineers to evaluate the results achieved on a sample fastener in a sample joint—perhaps to recalibrate a production tool or to adjust it in the first place. Figure 7.25D shows such a system.

Here's another way to monitor results: Some tools and systems automatically mark the parts after a desired torque has been produced, for a rough visual inspection.

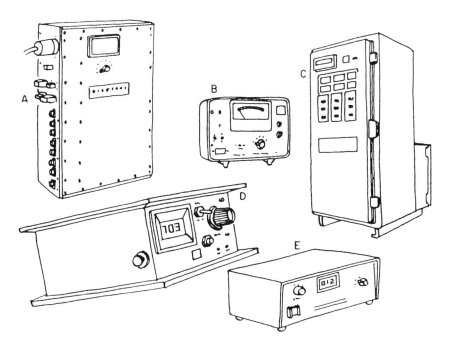

Figure 7.25 Electronic control systems for monitoring and/or controlling the output of air-powered nut runners. These are made by (A) Ingersoll-Rand, (B) GSE, (C) Himmelstein, (D) Thor, and (E) Norbar.

F. Add Torque Feedback for Still Better Control

A good torque monitor will usually show that you're not getting consistent output torque. The next step toward accuracy, therefore, is to provide some sort of feedback control based on torque. The transducer signal used for monitor purposes is now massaged, amplified, and used to actually control the tool—probably through an electrically actuated valve of some sort. Torque accuracies of $\pm 1\%$ to $\pm 5\%$ are now possible [16].

A variety of such tools are available with different principles of operation, response times, control accuracies, etc. Most controls are based on output torque, but some more recent systems base the control on the output speed of the tool. Some versions of the control systems shown in Fig. 7.25 provide air-tool control as well as digital or signal outputs.

VII. FASTENERS WHICH LIMIT THE APPLIED TORQUE

Assembly torque is usually controlled by a torque tool of some sort, but this is not always the case. The structural steel and airframe industries favor, instead, special fasteners which limit the amount of torque which can be applied to them, and, thereby, limit and control the preload developed in them during assembly. The accuracy with which torque—or, more important, preload—can be controlled by such fasteners is less than the accuracy of the best torque tools, but these two industries deal primarily with joints loaded in shear, and the accuracy requirements are less severe than they are for critical joints loaded in tension. Some of these special fasteners are described below. Note that all suffer permanent deformation in some way as the limiting torque is reached. This prevents the mechanic from tightening them further. It also allows an inspector to decide, from visual observation alone, whether or not they have been fully torqued. And it prevents them from being (easily) loosened. Neither industry needs to loosen and reinstall the bolts in the joints they're used in, however, so this is not a problem. This fact does limit their use in other applications, however.

A. The Twist-off Bolt

The twist-off bolt cannot be held or turned from the head. (You'll note in the figure that it has an oval head.) Instead, the bolt is held by the assembly tool from the nut end. An inner spindle on the tool grips a spline section connected to the main portion of the bolt by a turned-down neck. An outer spindle on the tool turns the nut and tightens the fastener, with the tool reacting against the spline section. When the design torque level has been reached, the reaction forces on the spline snap it off, as shown in sketch 3 in Fig. 7.26. The building inspector can determine whether or not a minimum amount of torque was applied to the fastener by looking to see whether or not the spline sections have indeed been snapped loose from the bolts.

If, between calibration and use, the bolts are allowed to become rusty or in any other way suffer a change of lubricity, then the amount of tension actually achieved in field assembly can be quite different from that achieved in the calibration stand, as suggested by some of the data in Table 7.1.

The fact that this fastener can be calibrated in the as-used condition, however, and, even more important, the fact that the inspector has a way to determine whether or not a minimum torque was applied to the fastener make this a popular item.

B. Frangible Nut

The twist-off bolt is used only in the structural steel industry, as far as I know. The airframe people use the same principle, but this time applied to the nut. The outer section of the nut is unthreaded but has a normal hexagon shape which is driven by the assembly tool. It's fastened by a reduced cross-section, breakaway collar to a cylindrical inner section which is threaded. When the limiting torque is reached the hex section

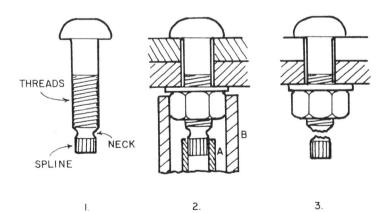

Figure 7.26 A twist-off bolt designed to indicate that a minimum amount of torque has been applied to the bolt. The tool holds and reacts against spline (A), while turning the nut (B).

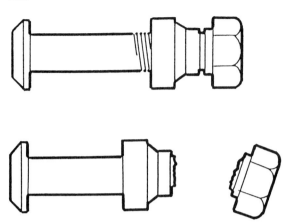

Figure 7.27 This type of twist-off fastener is used extensively on airframes. The cylindrical end of the nut is threaded; the outer, hex section is not. When the torque applied to the fastener reaches a predetermined value the hex end snaps off, as shown in the lower sketch, leaving the cylindrical end fully engaged with the bolt. Only visual inspection is required to prove that a given torque has been applied to the fastener.

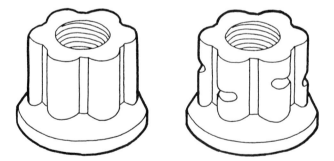

Figure 7.28 This unusual nut must be tightened by a special drive socket which contains hardened steel balls. When a predetermined torque has been reached the balls embed their way through the flanks of the nut, leaving the grooves shown in the sketch to the right. This limits the torque which can be placed on the nut, and leaves marks which allow an inspector to tell that the desired torque has, indeed, been applied.

breaks away from the cylindrical section, which remains to hold the bolt. This device is shown in Fig. 7.27.

C. Ball Drive Socket

Figure 7.28 shows a specially heat-treated nut which is driven during assembly by a drive socket containing hardened balls. When the limiting torque is reached these balls push their way through the corners of the hexagonal nut, eliminating further motion of the nut [35].

VIII. IS TORQUE CONTROL NO GOOD?

Does the fact that there are control problems mean that torque control is no good? No. Torque is still the most versatile and easiest control means. Design engineers, furthermore, are aware of the limitations of torque control and have long since learned to overdesign their products to offset the scatter one will normally get in preload—except in critical joints.

Using torque to control preload is somewhat like driving a car to work. Statistics prove that every time you get behind the wheel there is some possibility that you will not arrive at your destination. This doesn't mean that you should walk instead—or take a commercial airplane (which has a better per-passenger-mile safety record than the automobile). These other options are just not practical or economical in most cases—they don't suit other aspects of your total "get-to-work" problem.

If you know the inherent dangers and limitations of automobile travel, however, and do things to compensate for them, there's a better chance that you will arrive at your destination.

Torque control of preload also has obvious limitations and dangers, but in most cases it will be the only practical or economical choice. Knowing its limitations will improve your chance of success. And success, in this case, is not "preload accuracy"; it's "joints which don't fail." There can be a big difference between those two, and we're doing the person who pays the bills a big disservice if we ignore that fact. We don't want "accuracy" simply because it's technically challenging or interesting. We want it only if we need it.

Besides, there are a lot of things we can do to improve the results we get when we use torque control. Some of these things are listed at the end of the last chapter. I've put them there because most of them can be used to improve assembly results in general, regardless of which type of tool or control system we use.

IX. BREAKAWAY TORQUE

A. The Breakaway Torque Test

Torque is used mainly to control the buildup of initial preload or tension in a bolt during assembly. It is also sometimes used to determine whether or not a fastener has been properly tightened. Clockwise torque is applied to a previously tightened fastener and is slowly increased until the nut "breaks away" or "restarts" in the clockwise direction. Sometimes the operator judges breakaway by noise; the nut will sometimes "snap" during this process. More commonly the operator decides that breakaway occurs when the torque reading on his wrench gage stops increasing and either holds steady or decreases a little.

Note that the inspection torque is always applied in the clockwise or tightening direction. One reason for this: the torque required to loosen a freshly tightened fastener is usually less than the torque required to tighten it, probably because torsional windup in the bolt is now aiding rather than fighting the torque wrench. One reference reports calculations which suggest that the torque required to loosen a fastener with UNC threads will be only 70% of that required to tighten it, the ratio for a fastener with UNF threads being 89%. [31]

This test is only "fair" on recently tightened fasteners. Bolts which have been tightened a long time ago can take a *lot* more torque to loosen—or to restart—than was originally applied to them, especially if they have been exposed to heat and/or corrosive fluids. On the other hand, if they've been exposed to cyclical shear loads from vibration or thermal changes or the like they can take very little torque to restart or loosen, as we'll see in Chap. 16.

Assuming that we use breakaway torque only on recently tightened fasteners, is this a fair test? Does it accurately evaluate the torque used to tighten the bolt or the tension in it? Thanks to Ralph Shoberg of RS Technologies in Farmington Hills, Michigan, we now know the answer to this question [32]. Shoberg used a torque wrench with an electrical readout attached to a strip-chart recorder, plus a test machine of his own design which incorporated an ultrasonic instrument to monitor bolt tension at all points during the test. Here's what he found when he applied a breakaway torque to a well-lubricated bolt.

B. Well-Lubricated Bolt

With reference to Fig. 7.29, a breakaway torque is slowly built up on a previously tightened bolt. When the breakaway torque reaches a level

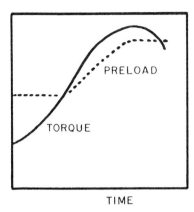
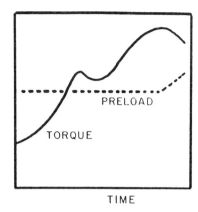

Figure 7.29 Curves showing the buildup of torque on and the change in preload in a previously tightened bolt when it is subjected to a breakaway torque test. The left-hand curve is for a well-lubricated bolt, the right-hand one for a poorly lubed bolt. The response of the bolts to this test is described at length in the text.

that would, during assembly, create the tension currently in the bolt, the friction restraint between the face of the nut and the joint surface is overcome. No noise occurs at this point and the torque being applied to the nut does not level off or fall off; instead it continues to climb. The operator cannot, in fact, detect the nut-joint-friction break point—but that occurs very close to the installation torque level.

The breakaway torque applied to the bolt is increased further. The bolt now starts to twist and the tension in the bolt starts to increase, even though there's no relative motion, yet, between bolt and nut threads.

At some higher, peak torque the friction restraint between male and female threads breaks. The nut starts to move relative to the bolt and the torque reading on the wrench gage drops off. This point can be detected by the operator and he calls this the breakaway torque. It's higher than installation torque, and the tension in the bolt is now higher than the installed preload (and the higher tension will remain when the test wrench is removed). The test is "fair" to the assembler in that it ends up giving him credit for applying more torque than he actually applied.

Unless—and it's a big unless—embedment and elastic interactions have reduced the tension in that bolt to something below the actual, initial preload created when that bolt was first tightened. In this case a blind breakaway torque test—while accurately evaluating the current tension in the bolts—will be a very unfair way to judge the performance of the mechanic.

C. Poorly Lubricated Bolt

Shoberg found that the sequence of events is different if the bolt being tested is poorly lubed, as also shown in Fig. 7.29. The nut-joint friction still breaks first, but this time at a point slightly above installation torque (again assuming no relaxation since installation). The nut will frequently make a sharp noise at this point. And the torque being applied to the bolt may drop off. This is a false peak because the threads haven't broken free from each other yet.

There is no increase in tension this time as higher torque is applied—until the true peak is reached and the male-female thread friction breaks.

Again the test tends to give the mechanic a fair shake if the fastener hasn't relaxed. And it's also a fair judge of the tension in the fastener. So breakaway torque, a popular way to test assembled joints, passes its own test with good marks. This is fortunate because breakaway torque is one of the very few ways we have to inspect or evaluate assembled, bolted joints.

X. THE INFLUENCE OF TORQUE CONTROL ON JOINT DESIGN

Remember that a bolted joint is a mechanism for creating a force, the clamping force designed to hold two or more pieces of a structure or product together. That being the case, anything which affects or helps determine that clamping force must be of interest to a conscientious bolted-joint designer. And the type of control we use at assembly is certainly such a factor.

If we use torque control, with its typical scatter of $\pm 30\%$, some bolts in a large group will acquire 30% more initial preload than the target or average value; other bolts will start their active lives with 30% less; and most will start with something in between. We could estimate this average, of course, by using the short-form torque preload equation [Eq. (7.4)].

The thoughtful designer, therefore, cannot assume that the torque he specifies will achieve average initial preload in every bolt. He must assume that the initial preload seen by some bolts will be the average preload plus 30%. At the same time he must assume that the initial preloads seen by other bolts will be average less 30%.

It's important to understand that, as a result of this, the designer must select bolts and joint members large and strong enough to support the stresses encountered at average initial preload plus 30%. At the same time he must include enough bolts so that there's still sufficient clamping force to hold the structure together—after embedment and elastic interaction relaxation occur—even if some of the bolts have as little as average minus 30%.

Theoretically, from a simple, statistical point of view, the average preload in any group of bolts would be the target preload, with as many of those bolts above average as below average. The clamping force on the joint would just be the average preload times the number of bolts—less embedment and elastic interaction losses, of course. If things were as simple as that, we really wouldn't have to worry about this ±30% business.

The trouble comes from the fact that most joints don't contain enough bolts to guarantee a statistical average preload. And, in a small lot of bolts, obtained from the same source at the same time, or all subjected to the same prior service conditions, all may be more, or all less, lubricious than "average." As a result, and at least in safety-related joints, the safest course of action is to assume that average initial preloads will be something less than the short-form equation would lead us to believe. And then to subtract the further reduction in average preload caused by relaxation and interactions. What's left will be our estimate of the average residual preload; and this is all we can count on to generate the clamping force between joint members as we put that joint to work. All this is illustrated in Fig. 7.30, an extension of Fig. 6.27.

Should we assume that the average initial preload will be a full 30% less than the theoretical average? Only an experiment on the actual product can answer that question with certainty, but 30% would certainly be a defensible number. Consider it a safety factor, if you will, and increase or decrease it depending on the consequences of failure, and on your reading of the care with which you expect this product to be assembled and used. Before considering a more optimistic average, however, you should consider the following.

In Section III of this chapter we looked at some of the many factors which can affect the amount of preload we'll get for a specified applied torque. Some of these scatter factors can give us either more or less than average preload; examples include friction, tool accuracy, and the mechanic. None of the factors can give us only more than average preload. But—and here's the problem—many of the things listed can only give us less than average. These include things like joints that are difficult to pull

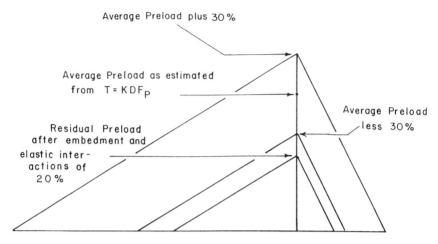

Figure 7.30 This is an extension of the joint diagram shown in Fig. 6.27. This one now includes the effect of scatter or uncertainty in the relationship between applied torque and the initial preload created in the bolt. Conventional wisdom says that the scatter in preload for a given torque, applied to as-received steel-on-steel fasteners, will be ±30%, as shown here. Embedment relaxation and elastic interactions then reduce the initial preloads to some lesser residual value; an average loss of 20% is suggested above. Bolts and joint members must be sized to support the maximum force of average plus 30%; yet we must assume a much lower residual preload when estimating worst-case clamping forces on the joint.

together, nonparallel joint surfaces, nonperpendicular holes, interference fits, and embedment relaxation. When we consider that the other major source of scatter in assembly results—elastic interactions—is also a "less only" factor, we're forced to conclude that it's far safer to assume that we'll end up with a less than average residual clamping force than with a greater than average one.

This forces us to "overdesign" the joint. We must include many, large-diameter bolts and equally large joint members in order to be able to count on a relatively modest residual clamping force. And the accuracy—or lack of it—of the assembly control system we have selected affects the difference between the size of the parts required to support maximum assembly preloads and the amount of clamping force we can count on.

REFERENCES

1. Motosh, N., Development of design charts for bolts preloaded up to the plastic range, *J. Eng. Ind.*, August 1976.
2. Shigley, J. E., *Mechanical Engineering Design*, 3rd ed., McGraw-Hill, New York, 1977.
3. Czajkowski, C. J., Investigations of corrosion and stress corrosion cracking in bolting materials on light water reactors, *Int. J. Pres. Ves Piping*, vol. 26, pp. 87–96, 1986.
4. Anderson, W., and P. Sterner, Degradation of threaded fasteners in reactor coolant pressure boundary of pressurized water reactor plants, U.S. NRC, NUREG-1095, May 1985.
5. Yura, Joseph, Karl Frank, and Dimos Polyzois, High strength bolts for bridges, PMFSEL Report no. 8-3, University of Texas, May 1987.
6. Brenner, Harry S., Results of special evaluation of Pepcoat 6122 and nuclear grade N5000 anti-seize lubricant systems on torque-tension performance of ASTM A325 series hex-head bolts submitted by Power and Engineered Products Company, Inc. (PEPCO), South Plainfield, NJ, Report no. C 16034-1, Almay Research and Testing Corp, Los Angeles, October 25, 1982.
7. Tests on 1 inch A490 conventional and twist-off high strength bolts and direct tension indicators conducted in the as-received condition and after weathering, Report no. 870702, Pittsburgh Testing Laboratory, Pittsburgh, PA, April 10, 1986.
8. Bezborod'ko, M. D., et al., How to secure the reliable tightening of screwed joints, *Vest. Mash.*, vol. 58, no. 1, 1978.
9. Bickford, J., Study of the tension, torsion and bending stresses in $2\frac{1}{4}$–8 × 12, B16 studs during torquing operations, Raymond Engineering, Inc., in-house report, April 1976.
10. Brenner, H. S., Development of technology for installation of mechanical fasteners, Almay Research and Testing Corp., Los Angeles, Report prepared for NASA under Contract no. NA3-3-20779, 1971.
11. Chesson, E., Jr., N. L. Faustino, and W. H. Munse, High strength bolts subjected to tension and shear, *J. Structural Div.*, *Proc. ASME*, October 1965.
12. Breeze, W. D., Tightening torque and clamping force relationship for a range of threaded fastener sizes, Naval Ship Engineering Center, Hyatsville, MD, April 1974.
13. Zmieuskii, V. I., et al., Greases for stainless-steel threaded joints, *Vest. Mash.*, vol. 54, no. 1, 1974.
14. Investigation of threaded fastener structural integrity, Final report, Contract no. NAS9-15140, DRL-T-1190, CLIN 3, DRD MA 129 TA, Southwest Research Institute, Project no. 15-4665, October 1977.
15. Eshghy, S., The LRM fastening system, *Fastener Technol.*, July 8, 1978.
16. Rice, E. E., Reliability in threaded fastener tightening, Society of Manufacturing Engineers, Paper no. AD75-769.

17. Ehrhart, K. F., Preload—the answer to fastener security, *Assembly and Fastener Eng.*, June 1971.
18. Lesner, R. S., Power tool concepts for precision torque control, Society of Manufacturing Engineers, Paper no. AD76-638, 1976.
19. Wakefield, Scott B., Portable electronic joint analysis, Society of Manufacturing Engineers, Paper no. AD75-769, 1975.
20. Wilhelm, Jack, Torque-tension testing equipment, *Compressed Air Mag.*, vol. 75, no. 8, pp. 20–23, August 1970.
21. Hung, N., Clamping force for electrically conductive fasteners, *Machine Design*, October 1984.
22. From information received from the E/M Corporation, West Lafayette, IN, March 1988.
23. Taylor, Edward, SerMeTel W (TM) aluminum coating for aerospace fasteners, SPS Laboratory Report no. 5392, SPS Technologies, Inc., Jenkintown, PA, October 30, 1980.
24. Gresham, Robert M., Bonded solid-film lubricants for fastener coatings, *Fastener Technol. Int.*, April/May 1987.
25. Emrich, Mary, Corrosion protection for fasteners, Part 1, *Assembly Eng.*, October 1982.
26. Pepcoat, a study in protection and performance, G*Chemical Corp., Wayne, NJ, undated manufacturer's literature.
27. Investigation of the torque-tension relationship for Monel, K-Monel and steel threaded fasteners, Report MPR-920, MPR Associates Inc., prepared for Attack Submarine Acquisition Program, NAVSEA PMS393TM6, Washington, DC, June 1986.
28. Report of torque tests to determine the torque-axial load relationship for monel and steel bolts lubricated with Neolube and MIL-L-24479B, Report MPR-1092, MPR Associates, Inc., prepared for Attack Submarine Acquisition Program, NAVSEA PM393TM6, Washington, DC, October 1988.
29. Clement, Bernard, and Andre Bazergui, A study of the preload relationship in bolting technology: experimental design and analysis, Prepared for the Bolting Technology Council, New York, October 1989.
30. Stewart, Richard, Torque/tension variables, list prepared at Wright-Patterson Air force Base, April 16, 1973.
31. Toosky, Raymond, Threaded Fasteners, Torque and Tension Control, notes for a seminar given at McDonnell-Douglas, Long Beach, CA, December 1991.
32. From a presentation made by Ralph Shoberg of R.S. Technologies, Farmington Hills, MI, to the Bolting Technology Council, May 1991.
33. Crispell, Corey, Torque vs induced load of A-286 and MP35N nuts and bolts with cadmium, dry film and cetyl alcohol lubricants, Request No. 8-1-7-EH-07694-AP29-B, NASA Contract No. NAS8-32525, November 14, 1978.
34. Srinivas, A. R., Determination of torque values for metric fasteners—an approach, Spacecraft Payload Group, Space Applications Center, Indian Space Research Organization, Ahmedabad, India, May 1992.
35. From information provided by the manufacturer, VSI.VOI-SHAN, Redondo Beach, CA, 1992.

8
Torque and Turn Control

I. BASIC CONCEPTS OF TURN CONTROL

We have tried to use torque as a means of controlling the preload in the fastener and have found problems. Even perfect input torque can give us a ±25–30% variation in preload.

But when we apply torque, the nut turns. Can we use turn instead of torque to control preload? At first glance this looks very promising. After all, when we turn the nut on a machine-tool lead screw 360°, the screw advances or retracts with a linear displacement equal to exactly one pitch of the threads. Won't a bolt stretch by this amount when we rotate the nut one turn? If so, we could use the "lead screw equation" to relate bolt stretch to turn of the nut, or

$$\Delta L = P \frac{\theta_R}{360} \qquad (8.1)$$

where ΔL is bolt stretch (in., mm), P is the pitch of the threads (in., mm), and θ_R is nut rotation (deg) with respect to the bolt. We could then get bolt preload very easily, assuming that we knew the spring constant or stiffness of the bolt, from

$$F_P = K_B \times \Delta L = K_B P \frac{\theta_R}{360} \qquad (8.2)$$

where K_B is bolt stiffness (lb/in., N/mm) and F_P is preload (lb, N).

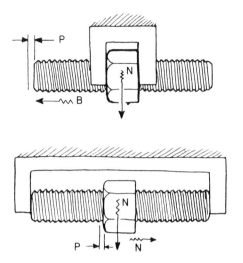

Figure 8.1 In a lead screw, one revolution of the nut will produce a linear displacement of one pitch in the screw *or* in the nut, depending on which is restrained and which is free to move.

But life isn't this simple. As illustrated in Fig. 8.1, the lead screw moves a distance of one pitch when we turn the nut only if the nut is rigidly restrained and the lead screw is perfectly free. If the nut were free and the screw restrained, of course, it is the nut which would move, as it does when we're running a free nut down against a joint. We still have relative motion, between the bolt and nut, equal to one pitch; but turning a loose nut obviously produces no preload whatsoever.

After the nut has bottomed, further turning will indeed stretch the bolt. But during this portion of the tightening process neither nut nor bolt is rigidly restrained. Rotation of the nut produces, in part, displacement of the nut downward into the joint, as joint members and nut compress. It also produces displacement of the bolt upward—elongation of the bolt, in other words. The relative displacement between nut and bolt is still one pitch for one input turn (if we ignore bolt twist), but only a portion of that relative displacement is bolt elongation.

If the spring constant of the body of the bolt equals the combined spring constant of everything else in the joint (including the nut), then half of the displacement (or, more accurately, the deformation) will occur in the bolt and the rest will be distributed in the nut and joint members. Figure 8.2 shows an analog of this situation. The forces on bolt and joint are equal and opposite. If the spring constants are equal, then it follows

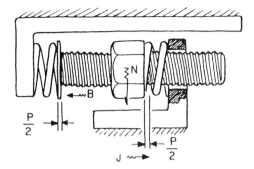

Figure 8.2 Analog of a bolted joint in which the spring constant of the bolt equals the spring constant of joint members. The sketch assumes rigid bodies restrained by equal springs. Displacement of the bolt to the left equals displacement of the nut to the right. The bulkhead restraining the nut spring (joint) is shown partially cut away for clarity.

that deformations produced by equal forces must also be equal. Rotating the nut one turn will only stretch the bolt one-half a pitch.

If the bolt is less stiff than the joint, then it will absorb a larger share of the overall deformation and the joint will see less. This is the situation we will find in a heavy joint with a hard makeup. Forces on bolt and joint are still equal and opposite. Deformations will be in inverse relationship to bolt and joint stiffness. Rotating the nut one turn will stretch the bolt more than one-half a pitch, but less than one pitch.

If the joint is relatively soft compared to the bolt—as it sometimes is if a gasket is involved—then most input turn will be absorbed in deformation of the joint rather than of the bolt, as suggested in Fig. 8.3. Rotating the nut one turn stretches the bolt less than half a pitch—usually much less for large gasketed joints.

Note that we can never generate more force in one spring than we can in any other spring to which it is connected in series. This is why the softest member of a joint—whether it's the bolt or a gasket or something else—will dominate the behavior of the joint, both during tightening and in use. This is true not only when behavior is elastic, such as above, but also if the softer member yields, limiting the force it will support. More of this when we discuss turn-of-nut procedures later in this chapter.

In any event, we now realize that before we can use input turn as a control means, we must be able to predict the relative stiffness of bolt and joint members. As we saw in Chap. 5, computing bolt stiffness is a relatively simple procedure, at least for long thin bolts. Computing joint

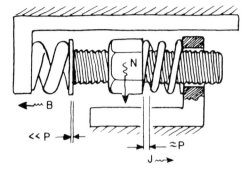

Figure 8.3 In a gasketed joint, the spring constant of the bolt is greater than the combined spring constant of joint members, so turning the nut produces more deformation in the joint than in the bolt.

stiffness is more difficult; it must be determined experimentally in most cases. In fact, there will always be some uncertainty, especially if short bolts, thin joint members, and/or gaskets are involved (see Chap. 9). And any uncertainties in our estimates of the stiffness ratio lead to similar uncertainties in our predictions of the relationship between turn and preload in a given application.

But that's not the crux of the turn control problem. Let's take a closer look at what happens when we apply turn to a nut.

II. TURN VS. PRELOAD

A. Common Turn-Preload Relationship

Let's assume that we're tightening a hard joint. Here's what happens step by step as we rotate the nut. We'll assume that turn with respect to the body of the machine ("ground") equals the turn with respect to the bolt.

The first few turns of the nut produce no preload at all, because the nut has not yet been run down against joint members and they are therefore not yet involved. This situation is shown in Fig. 8.4A.

Finally the nut starts to pull joint members together. There may be frictional restraint between joint members and surrounding structures. Joint members may not be perfectly flat. There may be a bent washer. As a result, although we start to produce some tension in the bolt, most of the input turn is absorbed by the joint and the bolt sees only a small increase in preload, as suggested by Fig. 8.4B. This process is called

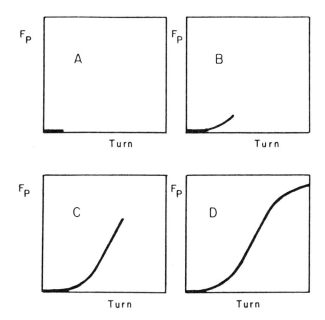

Figure 8.4 Step-by-step buildup of preload in a joint when the turn of the nut relative to the bolt equals turn with respect to "ground." (A) Nut run-down produces no preload. (B) Snugging pulls joint members together, flattens washers, etc., and produces a little preload. (C) Bolt and joint members all deforming elastically. (D) Something in the joint—usually the bolt or gasket—yields to limit further buildup of preload.

snugging the joint, and the amount of turn required varies unpredictably, even between apparently identical bolts or joints.

After the joint has been snugged, all bolts and joint members start to deform simultaneously, with individual deformations in inverse proportion to individual spring constants. Preload now starts to build more rapidly in the bolt, following a straight line whose slope is equal to

$$\text{Slope} = \frac{\Delta F_P}{\Delta \theta_R} = \left(\frac{K_B K_J}{K_B + K_J}\right) \frac{P}{360} \tag{8.3}$$

where K_B and K_J are the spring constants of bolt and joint members (lb/in., N/mm), P is the pitch (in., mm), F_P is preload (lb, N), and θ_R is the input turn in degrees. Note that $(K_B K_J) \div (K_B + K_J)$ is the combined stiffness of the total bolt-joint system, so this equation reduces to a revised version of the F_P-θ equation (8.2).

During this straight-line portion of the process, there is usually a linear relationship between additional input turn and additional preload, as shown in Fig. 8.4C. If we could predict the spring constants involved—and if we could determine where this straight-line portion of the curve starts—measuring turn would give us good control of preload. Unfortunately, however, we will find it very difficult to determine where the straight-line portion of this curve starts. It will vary from bolt to bolt, from assembly to assembly, and from application to application, adding to the uncertainties in spring constant.

If we continue to turn the nut, something in the joint will eventually start to yield. This ends the linear build-up of preload in the joint as suggested in Fig. 8.4D. More important for control purposes, however, it ultimately limits the preload created by turning the nut further. This is really a torque-turn technique, rather than a turn technique, so we'll leave it for later.

Note that the equation given for the straight-line portion of the turn-preload curve could also be applied to the upper and lower, curved portions of this curve if we knew the operating stiffnesses of bolt and joint at every moment. If a bent washer is being snugged, for example, we'd have to know the stiffness of the washer, since its spring constant dominates that of the joint. At the other end, we'd have to know the stiffness of the bolt when it started to deform plastically. None of this, of course, is practical, so the equation isn't of any use except in the straight-line portion. In fact, it even has limitations there, as we'll see in Sec. III.

B. Other Turn-Preload Curves

The S-shaped curve shown in Fig. 8.4 describes the torque-preload behavior of a conventional, moderately stiff joint. There are, however, other possibilities.

Sheet Metal Joint

If joint members are very thin, they are also very stiff. This can mean a nearly vertical rise in the elastic portion of the torque-turn curve, again creating control problems in high-speed, automatic operations. Figure 8.5, for example, shows the typical curve for a bolt tightened against sheet metal [1].

Gasketed Joint

Gasketed joints are often tightened in stages, since irregular loading can distort and damage a gasket, and it is desirable to pull the joint together

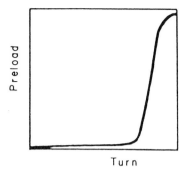

Figure 8.5 Turn-preload curve for a sheet metal screw. (Modified from Ref. 1.)

simultaneously to the extent possible. A typical procedure is to tighten all fasteners to 30% of the final desired preload, then go around again and tighten them all to 60%, and then go around again and tighten them to 100%. Gaskets will usually creep and relax between passes, however, and gasketed joints will experience the elastic interaction relaxation discussed in Chap. 6, so if one were to plot the relationship between the turn applied to a particular fastener and the preload produced in that fastener, he would usually see a pattern such as that shown in Fig. 8.6.

If all of the bolts in a gasketed joint are tightened simultaneously all the way, we would expect the turn-preload curve to be relatively normal—except that the elastic region would have a very low slope, and there would be relaxation following tightening, as shown in Fig. 8.7.

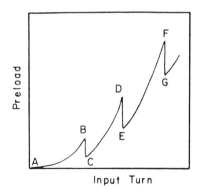

Figure 8.6 The preload in gasketed joints partially relaxes between passes, because of gasket creep and relaxation. The first pass took the bolt from point A to point B, the second pass from C to D, etc.

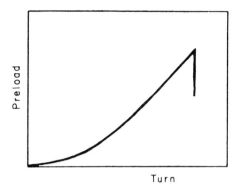

Figure 8.7   If all bolts in a gasketed joint are tightened simultaneously, relaxation occurs only after final tightening—but it still occurs.

III. FRICTION EFFECTS

It is commonly believed that turn control is better than torque control because the relationship between turn and preload is independent of friction. All we have to worry about are those spring constants and "where do we start measuring turn?"

But this is not true. It's a popular misconception based on the fact that most of the turn control equations you'll find in the literature are based on the *relative* turn between nut and bolt (θ_R). If we could indeed measure relative turn, we could ignore friction.

I don't know of any way to do that, however, except in a laboratory. In practice, we always measure turn of nut relative to the frame of the machine, or the floor, or some fixed reference (θ_G) other than the body of the bolt. From our frame of reference, we could give the nut one-half revolution after snugging and produce no preload at all if the threads had seized. On the other hand, one-half revolution could stretch the bolt past yield if someone had used a better-than-normal lubricant on the threads. In-between coefficients of friction in the threads would give us in-between results, as far as the relationship between input turn and preload is concerned.

As a result, we will find that when we measure turn with respect to the same reference point from which we measure input torque, the floor on which we are standing, the turn-preload curve becomes a family of curves like the torque-preload family—a different curve for each coefficient of friction. And the slope of the center portion of each curve is no longer just a function of spring constants; it is also affected by the amount

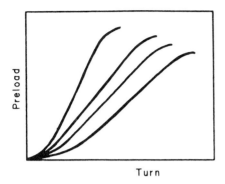

Turn

Figure 8.8 If one measures input turn relative to the machine rather than relative to the bolt, the preload-turn relationship becomes a family of curves, again depending on the coefficient of friction in the threads.

of input work that is going into torsional energy or heat loss in the system (see Fig. 8.8).

As a result of all this, pure turn is no more accurate than pure torque control. But all is not lost. We apply torque and the nut turns. If we use both torque *and* turn to control the process, we do gain some accuracy over using either torque or turn alone, as we'll see.

Note that the slope of the straight-line portion of these curves is no longer given by Eq. (8.3), because the angle of turn used in Eq. (8.3) is the relative angle between nut and bolt, and we're now using turn with respect to ground. The correct equation for the present curve is far more complex as we'll see.

IV. TORQUE AND TURN IN THEORY

A. The Equations

If we tighten a fastener and plot the relationship between applied torque and resulting turn of nut, we will produce an S-shaped curve which looks very much like the turn-preload curve, as you can see by comparing Fig. 6.4 or 8.11 to Fig. 8.4. If we make a number of such experiments, we will find that we face all of the uncertainties which plagued us with either torque or turn control—such things as coefficient of friction, variations in geometry, difficulties of predicting spring constant or snugging point, etc.

From a mathematical point of view, we can confirm this by combining the long-form torque equation with the equation we have used for the straight-line portion of the S-shaped turn-preload curve [Eq. (8.3)], the only portion of that curve for which we can predict combined bolt-joint stiffness with any confidence.

This time, however, we want to express turn of nut in radians, rather than in degrees, so that we can rewrite the turn-preload equation as

$$F_P = \theta_R \left(\frac{K_B K_J}{K_B + K_J} \right) \frac{P}{2\pi} \tag{8.4}$$

We can now replace the preload term of the long-form equation with the above and get the following expression, as you can see by comparison with Eq. (7.3):

$$T_{in} = \theta_R \left(\frac{K_B K_J}{K_B + K_J} \right) \frac{P}{2\pi} \left(\frac{P}{2\pi} + \frac{\mu_t r_t}{\cos \beta} + \mu_n r_n \right) + T_P \tag{8.5}$$

Before we get too carried away with all this, however, let's recognize that we've started with a couple of highly suspect equations and have derived another of the same. The long-form equation describes the theoretical torque balance between rigid, perfect parts; the turn-preload equation describes only a small, theoretically straight-line portion of the complete turn-preload process. At best, the latest equation will describe only the straight-line portion of the torque-turn curve; and it will do that only when we really know the coefficients of friction, the actual geometry for this set of parts, etc. Note, too, that we must know the relative turn between nut and bolt, not the turn with respect to ground, since that's what's used in the turn-preload equation. So the equation isn't of much use in most situations.

Dr. S. Eshghy of Rockwell International, however, has successfully utilized this equation (and a number of other bolt-joint relationships) to produce the interesting logarithmic rate method (LRM) bolt-tightening system we'll look at in a moment [4].

The above equation attempts to describe the relationship between turn and torque. In the process of combining our original torque-preload and turn-preload equations, however, we eliminated, mathematically, the thing we're most interested in: preload. Can we write an expression for all three at once? It would seem that we can if we write the turn preload equation as

$$\frac{P}{2\pi} = \frac{F_P}{\theta_R} \left(\frac{K_B + K_J}{K_B K_J} \right) \tag{8.6}$$

and use the resulting expression to replace the $P/2\pi$ term of the long-term equation, with the following result:

$$T_{in} = F_P \left[\frac{F_P}{\theta_R} \left(\frac{K_B + K_J}{K_B K_J} \right) + \frac{\mu_t r_t}{\cos \beta} + \mu_n r_n \right] + T_P \qquad (8.7)$$

B. Torque, Turn, and Energy

The area under the torque-turn curve defines the amount of energy being delivered to the fastener during the tightening process, as shown in Fig. 6.3. If we have a computer-generated curve we can use standard softwear to compute that area—that energy—by integration. Those without this capability can estimate it by plotting the torque-turn curve, using a planimeter or "counting the squares" between the curve and the turn axis to estimate the area (as we all had to do in the not-so-good old days).

No one estimates the input energy at present, but people who design and build automated, production torque-turn equipment come closer to doing this than they realize. We'll take a look at this equipment in Sec. VII of this chapter.

The fact that torque-turn information defines the input energy is one reason why torque-turn control is significantly better than torque control. I think that more specific use will be made of energy information in the future.

C. Torque-Turn-Preload Cube

Let's pause a moment to recognize the fact that we can never deal with *just* torque versus preload or *just* turn versus preload as we tighten a fastener. We apply torque, the nut turns, and the bolt stretches, creating preload. All of this is going on simultaneously. Preload is being developed in the fastener as a function of both torque and turn simultaneously. We can plot the resulting "space curve" on the three axes of a cube as shown in Fig. 8.9. Each face of this cube gives us a different two-dimensional view of the total preload-versus-torque-and-turn curve, as suggested in Fig. 8.9. Note that the turn-preload and torque-preload views of the space curve agree with Fig. 8.4 and with Fig. 7.1 as well.

Anything which affects one view of this space curve will affect the other views as well. Figure 8.10 shows what a change in friction does to the torque-turn-preload cube. Note that the torque-versus-turn view of this process (looking down from the top) is basically S-shaped—like the turn-preload curve but generally with a different slope, etc.

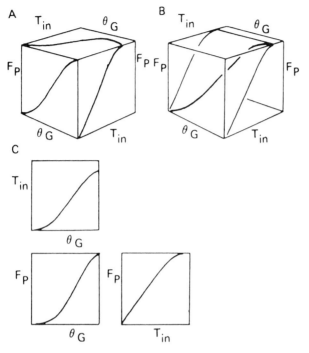

Figure 8.9 Torque and turn cannot be isolated from each other. Together they produce preload. View C is an orthographic projection of the cube. View A shows these projections on the surfaces of the cube, and view B shows the true T-θ-F_P curve snaking around inside the cube.

D. The Broader View

Equation (8.7) suggests that we have just involved more variables and therefore have made things worse by trying to measure torque and turn at the same time, but that's not true in the real world.

There are, in fact, at least three different ways in which measuring torque and turn simultaneously can improve our control over preload:

Torque and turn give us sufficient information to tighten safely until something in the joint yields. The *yielding* limits, and therefore controls, preload. One such technique is called *turn-of-nut* control.

Torque and turn information allows us to spot a large number of serious practical problems. We haven't considered these yet, because they don't enter into the "equations," and neither torque nor turn can reveal them alone. These problems include such things as blind holes,

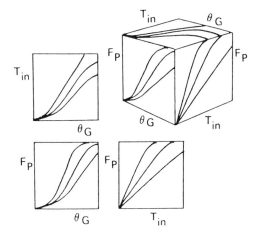

Figure 8.10 Anything—such as a change in friction—which affects one view of the torque-turn-preload process must affect the other views as well.

wrong parts, crossed threads, etc. and are of major concern in automatic production operations. They usually cause less trouble in manual operations, but only if operators are well trained, properly motivated, and careful.

Torque and turn information can be fed to microprocessor or computer-control systems where sophisticated analysis of of the information can be used in a couple of different ways to give us more accurate control of preload than would torque or turn information alone.

Let's look at these possibilities in detail.

V. TURN-OF-NUT CONTROL

A. The Theory

The so-called turn-of-nut method is widely used, especially in structural steel applications. Historically it was the first torque-turn control technique; it is a manual technique, although computer-control equivalents are now available, as we'll see.

In the original turn-of-nut procedure, the nut is first snugged with a torque which is expected to stretch the fastener to a minimum of 75% of its ultimate strength. The nut is then turned "three flats" (half a turn) or the like, which stretches the bolt well past its yield point, as shown in Fig. 8.11 [5].

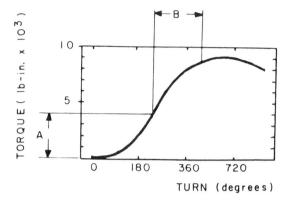

Figure 8.11 In turn-of-nut techniques the nut is first tightened with an approximate torque (A) and then further tightened with a measured turn (B). $\frac{7}{8} \times 5\frac{1}{2}$ ASTM A325 bolt.

The preload produced by the snugging torque, of course, varies because it's affected by all the normal variations in friction, geometry, etc. Subsequently turning the nut past yield, however, always produces about as much tension as the bolt can support. Final variation in preload in a large number of bolts is probably closer to ±5% than to the ±25–30% which would be the case if we used torque or turn control alone.

Torque instead of turn is used as the control means during the snugging process because torque is better able to compensate for start-up variables. If the head of the bolt slips, for example, when we first start to tighten it, we merely keep turning the nut until we have produced enough torque to guarantee that everything is truly snugged and we can start measuring turn.

Turn is used for final control, however, since it is a more accurate way of determining that we have really stretched the bolt past yield. The bolt really does behave like a lead screw—an elastic one—during some portion of its behavior. It *has* to stretch past yield if we turn the nut 180° past almost any point on the linear portion of the torque-turn curve.

Note that the final *torque* required to yield the bolt can vary drastically, making torque a poor means of determining yield. We could do it by looking at the rate of change in torque as a function of turn, but this is an awkward thing to do manually. It is, however, the basis for some computer-controlled techniques, as we'll see later.

Note that this classical turn-of-nut procedure cannot be used on brittle bolts. It can be used safely only on ductile bolts having a long plastic regions, such as the ASTM A325 fasteners used in structural steel work.

Furthermore, it should never be used unless you can predict the working loads to which the bolt will be subjected in use. Anything which loads the bolt above the original tension will create additional plastic deformation in the bolt. If the overloads are high enough, the bolt will break. I don't mean to suggest that A325 bolts tightened this way will be unable to support tensile loads. As shown in Fig. 2.21, the bolts will yield under combined torsional and tensile stress when first tightened. The torsional stresses will disappear soon after tightening, thanks primarily to embedment relaxation. This returns some tensile load capacity to the bolt (perhaps 5–10% of its yield strength). Subsequent tensile loads would have to exceed this value to cause further yielding or rupture.

B. The Practice

Structural Steel

The turn-of-nut procedure for structural bolts was first proposed in the mid-1950s by the Association of American Railroads, influenced, perhaps, by similar techniques being used by the automobile industry [6]. The technique was later modified by Bethlehem Steel Corporation and subsequently adopted in that form by the Research Council on Riveted and Bolted Joints of the Engineering Foundation. In the present form bolts are first tightened with an impact wrench until the tool starts to impact. The bolt has now been snugged. The position of the socket, which is marked at 90° intervals, is now noted and/or marked, and the impact wrench is used to turn the drive socket another 180–270°, depending on bolt length and whether or not the surfaces of the joint are perpendicular to the axis of the bolt threads or are sloped (as is common in structural steel). Beveled washers are often used to compensate for sloped surfaces. Table 8.1 shows the amount of turn recommended by the Research Council [5].

As an alternative, bolts can be snugged by "the full effort of a man using a spud wrench"; they are then turned the amount shown in Table 8.1 by an impact wrench. This procedure is not considered safe on bolts less than $\frac{3}{4}$ in. in diameter, however, as a man can tighten small bolts past the yield point with the spud wrench. Smaller wrenches and/or a different procedure must then be used [7].

The success with which the turn-of-nut procedure controls bolt tension can be seen by reference to Fig. 8.12. Note that this is a tension-elongation curve, not a torque-turn or preload-turn curve, so it is not S-shaped. It is, instead, the elastic curve for this $\frac{7}{8}$-in. A325 bolt, and is based on tests reported by Fisher and Struik [7]. The histogram under the graph shows the elongations produced, in a sampling of bolts, by $\frac{1}{2}$ turn

Table 8.1 Nut Rotation from Snug Tight Condition for Turn-of-Nut Procedure[a,b]

Bolt length (underside of head to end of bolt)	Both faces normal to bolt axis	Disposition of outer face of bolted parts	
		One face normal to bolt axis and other sloped not more than 1:20 (beveled washer not used)	Both faces sloped not more than 1:20 from normal to the bolt axis (beveled washer not used)
Up to and including 4 diameters	$\frac{1}{3}$ turn	$\frac{1}{2}$ turn	$\frac{2}{3}$ turn
Over 4 diameters but not exceeding 8 diam.	$\frac{1}{2}$ turn	$\frac{2}{3}$ turn	$\frac{5}{6}$ turn
Over 8 diameters but not exceeding 12 diam.	$\frac{2}{3}$ turn	$\frac{5}{6}$ turn	1 turn

[a] Nut rotation is relative to bolt regardless of the element (nut or bolt) being turned. For bolts installed by $\frac{1}{2}$ turn and less, the tolerance should be $\pm 30°$; for bolts installed by $\frac{2}{3}$ turn and more, the tolerance should be $\pm 45°$.
[b] Applicable only to connections in which all material within the grip of the bolt is steel.
Source: Ref. 5.

past snug torque. Projecting maximum and minimum elongations up to the tension-elongation curve shows that there was very little variation in the preload achieved in this group of bolts, because A325 bolts are made of a ductile material having a long, flat plastic region.

The Turn-of Nut Procedure in Operations Production

Modifications of the original turn-of-nut procedure are often used in industry; a torque-then-turn procedure is used, but the bolts aren't stretched past their yield point. Though not truly "turn of nut," such procedures can often provide better control than can torque used alone, but because the fasteners aren't taken past yield, they can't match the high degree of preload achieved in structural steel work. (Torque-turn provides manufacturing engineers with other benefits, however, as we'll soon see.)

Figure 8.13 shows the results of an experiment in which a large number of $\frac{5}{16}$-24, SAE Grade 8 bolts, with a grip length of 2.3 in., were tightened with torque-turn air tools against a load washer and a pair of steel blocks

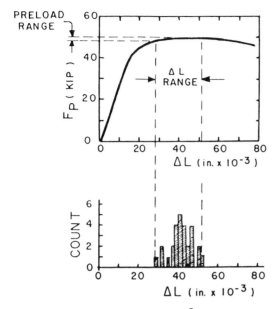

Figure 8.12 Elastic curve for a $\frac{7}{8}$-in. A325 bolt with a 4-in. grip length, and a histogram of the elongations produced in 28 such bolts by a $\frac{1}{2}$-turn-past-snug tightening procedure. Note the small variation in preload (tension) in spite of the almost 2:1 scatter in elongation. (Modified from Ref. 5.)

[4]. Here the scatter in preload is about 1.7:1. This is substantially better than the scatter obtained in a similar group of bolts with torque control, however, as you can see by reference to Fig. 7.6.

The Turn-of-Nut Procedure in Aerospace Assembly

A number of knowledgeable companies have developed manual torque/turn procedures which they call "turn of nut," but which do not involve tightening the fasteners past the yield point. Experience shows that some of those systems provide additional accuracy over systems using torque or turn alone. Here's an example from the aerospace industry.

   An aircraft engine manufacturer applies a seating torque of 3000 lb-in. to a large nut located on the central axis of the engine. This is one of several nuts along the axis; they clamp together the rotating parts of the turbine engine. The nut is now loosened completely and is then retightened to 5000 lb-in.

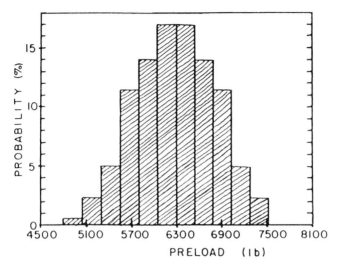

Figure 8.13 Histogram of the preload achieved in a group of $\frac{5}{16}$–24 SAE Grade 8 bolts with a 2.3-in. grip length when they were tightened to less than the yield point by torque-turn procedures. (Modified from Ref. 11.)

The nut is loosened a second time and is now retightened to a snug torque of only 500 lb-in. A turn protractor on the wrench is now set at 0°, and the nut is turned a specified number of degrees and minutes. This final turn can require as much as 120,000 lb-in. of torque.

The final torque is measured and recorded as a cross-check, to make sure that nothing has gone wrong, but turn and turn alone is used for the final control.

There are probably two reasons why preload accuracy is improved here, even though the fastener is not taken anywhere near yield by this particular torque-turn procedure. First, the initial cycling of the fastener will reduce subsequent embedment relaxation. Also, the initial cycling guarantees that all parts of the assembly have been pulled together *before* the official snug torque is applied. This procedure therefore provides a more stable starting place for the final cycle, reducing some of the major uncertainties of the torque-turn procedure. Note, however, that the technique is being used on a high-quality assembly, whose parts are subjected to far more quality control than are the parts of most bolted joints.

VI. CONTROL OF PRODUCTION PROBLEMS

A. Torque-Turn Control

In most of the foregoing we have treated preload as if it depended only on the torque or turn accuracy of the tools, and on those physical properties of a fastener included in the equations—such things as the coefficient of friction, fastener geometry, the elasticity of various parts, etc. In practice, however, the engineer—especially the production engineer—faces many serious problems which have nothing to do with the theoretical behavior of an ideal fastener. These problems include

Blind holes
Holes not tapped deep enough
Wrong size holes
Wrong size bolts
Dirt in holes
Crossed threads
Partially stripped threads
Soft parts
Gross misalignment of parts
Chips under the head or in the hole
Tool malfunctions
Warped mating parts
Burrs

to which must be added the sometimes gross relaxation of parts which have been properly preloaded to start with.

Measuring both torque and turn makes it possible to spot problems of this sort. In one technique, for example, an electronically controlled air tool will produce first a preset torque, then a preset turn on the fastener. The control system will examine the final torque required to produce the final turn. If everything is all right, the torque and turn values will fall somewhere within an acceptable "window" of values, as suggested by Fig. 8.14. If there is a blind hole or the threads are galled, however, then it will take far too much torque to produce the desired turn (and the tool will shut off when it exceeds the maximum acceptable torque). If the bolt is too soft, or the hole is grossly oversized, on the other hand, it will require much less than the rated torque to produce the anticipated turn, and the tool will again cut off. A typical torque-turn control system is shown in Fig. 8.15.

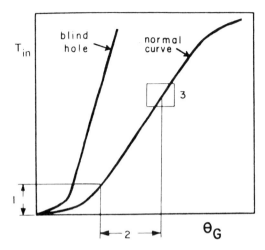

Figure 8.14 A typical torque-turn system will first apply a threshold torque (1), then move the nut a controlled angle of turn (2), and then inspect final (cutoff) torque and turn to see if final results lie within the "window" (3) of desired tolerance. If there's a blind hole, for example, the curve will miss the window, as shown.

Measuring torque and turn simultaneously also gives a little better control over preload, simply because you have more information to work with and "everything can't go wrong at once." But I think the main advantage of torque-turn is that it sees the gross problems one can't detect with torque control alone. This is possible, furthermore, even during high-speed, automatic assembly operations.

Incidentally, torque-turn control can be fooled by certain combinations of the problems listed above. The combination of a soft bolt with a high-friction surface, for example, will often be interpreted by the control system as an acceptable fastener. Bolts which are too hard can't be detected, either. So, the equipment is very useful, but not infallible. We'll look at several case histories in a moment.

B. Torque-Time Control

Another, closely related, approach is to monitor the amount of time required for torque to build up to the desired level. It's easier and cheaper to measure time than turn, and it does approximately the same thing—it allows you to spot gross aberrations caused by crossed threads, soft parts, etc. Unlike torque-turn control, however, torque-time control probably

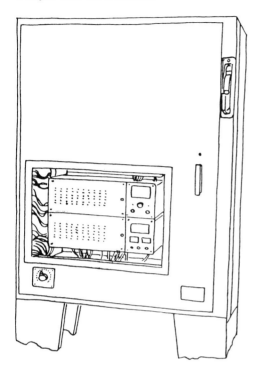

Figure 8.15   Ten-spindle torque-turn control system.

doesn't give you improved control over preload when everything is "normal." A typical torque-time curve is shown in Fig. 8.16.

C. Further Refinements

Hesitation Tightening

Neither turn nor time measurement does much about the relaxation problem. The only cure for this is to "give it time to relax and then retighten it." Air-tool designers have attempted to cope with this problem by hesitation tightening of some sort. The tool tightens a fastener part way, and then hesitates to give the fastener some time to relax before tightening it the rest of the way. Figure 8.17 shows the torque-time curve for a nut runner control system in which torque pulses, separated by brief relaxation times, are applied to the fastener after initial tightening to compensate for relaxation in the fastener and joint.

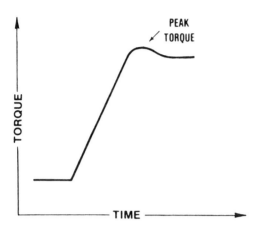

Figure 8.16 Torque-time curve of a standard stall-type air motor. A torque-time control system inspects final results to make sure that they are within the desired "window" of values specified by the engineers. (Courtesy of Ingersoll-Rand.)

Hesitation or pulsed tightening also gives better control at higher speeds. By turning the tool off automatically, whether or not it has reached final torque, you provide your control system with a number of opportunities to say "no more." Control decisions are always made with the tool at rest, and there is much less danger of overrunning the set point because of inertia effects or time delays in the control components.

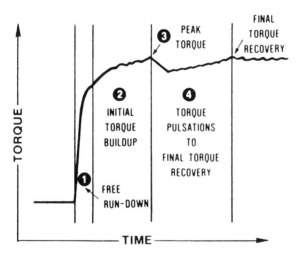

Figure 8.17 Equi-torq motor. Torque pulses are applied to the fastener after initial tightening, to compensate for relaxation. (Courtesy of Ingersoll-Rand.)

Some tools do *all* their tightening in a series of pulses, rather then continuously. The tool shown in Fig. 8.18 also tightens the fasteners to their yield point, a technique we'll consider in more detail in a minute.

Higher Speeds

Torque-turn, torque-time, and hesitation tightening all reduce assembly problems by spotting gross aberrations, partially controlling relaxation in the joint, and/or improving preload control slightly. While all these developments were going on, however, tool manufacturers were still faced with the demand for still higher speeds. In part, they responded to such demands by improving the torque control or response times of basic tools, as well as by providing the more complex systems we will look at in a moment. Faster-acting clutches were developed for nut runners, for example, some of which can now respond to a shut-off signal in as little as 5 msec. Variable-speed tightening also became available. The nut is run down at the highest possible speed; then the tool slows to a more controllable speed for final tightening [8].

The desire to do the job more rapidly also led to multispindle tightening. You do the job faster if you tighten all of the fasteners at once. Multispindles provide other benefits, too. Tightening several fasteners in a given joint simultaneously reduces joint relaxation following tightening—although multispindles are certainly not a complete solution to relaxation.

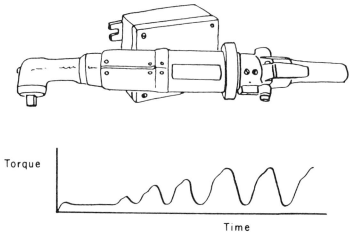

Figure 8.18 A simple air-tool control system in which the bolt is tightened in a series of pulses (Thor).

VII. COMPUTER-CONTROLLED SYSTEMS

A.   The Problem

The techniques described above give some control over gross aberrations and preload, but not enough control of either for really critical joints, especially when very high speed is a must. People soon found, furthermore, that attempts to apply electronically controlled tools to a wide range of production applications were only partially successful. The controls would work fairly well on a "normal" joint. This means one with a relatively ductile bolt, having a conventional shape and a length-to-diameter ratio that is at least 2:1, used in a joint that contains no gaskets and has a "reasonable" volume of joint material. These same systems, however, would not always work when we were dealing with short, stubby bolts, very high-strength bolts, (low-ductility) sheet metal screws, some types of prevailing torque, etc. In situations of this sort the basic S-shaped torque-turn curve becomes distorted, as suggested by Fig. 8.19.

Each of these situations is characterized by a relatively high run-down torque, followed by a very sharp and sudden rise to final torque. In effect, the fastener sees enough starting torque to trigger many shut-off systems, and then slams into a torque wall, putting excessive demands on the response time of control systems and components [12].

The best way to cope with demanding joints of this sort is to provide more sophisticated control systems, to be described next. These systems also provide substantially improved preload accuracy for normal joints as well.

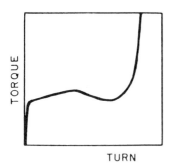

Figure 8.19 Torque-turn curve for a fastener having a prevailing torque lock nut. (Modified from Ref. 12.)

B. Tension or Joint Control—Tighten to Yield

Some of the most sophisticated control systems available today are designed to tighten every fastener to the threshold of yield. Using various strategies, they watch the torque and turn relationships building up in a given fastener, recognize and measure the straight-line portion of the curve, and then shut off when they reach the upper bend in the curve—the point at which something starts to yield. Note that since the torque-preload or torque-turn or turn-preload curves all "flatten out" at the yield point, further input to the fastener does not produce much more preload. This is why tightening to yield gives more accurate control of preload: ± 3–5% is often possible. See Fig. 8.20.

Note that tightening a group of fasteners to the yield point does not mean that each one will be tightened to exactly the same preload. Geomet-

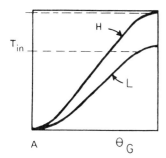

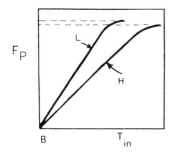

Figure 8.20   (A) A yield-point control system will cut off at point L on the torque-turn curve if friction is low. It will cut off at H if friction is high. (B) Looking at the same cutoff points on the torque-preload view of the torque-turn-preload curve, we can see that yield control does improve preload accuracy substantially. High friction still results in lower preload at yield because the additional torsion stress robs some of the bolt's tension capability.

ric and material variations will introduce some scatter. So will variations in friction, however. As we learned at the end of Chap. 2, torsion stress will rob some of the strength of a bolt. If the friction is high, therefore, torsion stress in the bolt will be high at the end of the tightening operation and the bolt will yield at a lower tensile stress level, i.e., lower preload. The differences probably won't be great in normal operation, but they could be substantial if you inadvertently forgot to lubricate normally lubricated bolts or the like.

The fact that torsion stress absorbs part of the strength of the bolt during tightening is, in fact, useful in this situation. Remember that the fastener will recover its full tensile capacity after the torsion stress has disappeared. And it *will* disappear shortly after the tightening operation, thanks to embedment relaxation, at least in most applications. The tensile stress required to further yield the fastener is higher than that required to yield it in the first place. This means that the fastener which has been "tightened to yield" can support additional static and cyclic working loads without yielding any further—a very important factor in many applications. It can't support an indefinite increase in load, of course, but it can support perhaps 10% more than was required to yield it in the first place.

Figure 8.21 shows a multispindle version of the well-known SPS tighten-to-yield system. The control system shown in Fig. 8.21 monitors

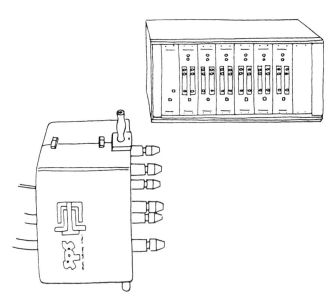

Figure 8.21 Multispindle air-tool control system designed to tighten fasteners to the yield point (SPS).

the rate of change of torque as a function of turn. This derivative peaks as the tool climbs the straight-line portion of the torque-turn curve; then it falls suddenly as the yield point is passed [9]. Variations in run-down torque or snugging angle of turn are ignored, because the computer has been programmed to look for a change only after it has seen a relatively high and relatively constant torque-turn ratio. The control curve can be seen, superimposed on the torque turn curve, in Fig. 8.22.

The system shown in Fig. 8.21 also uses torque-turn windows to spot gross problems such as crossed threads or soft bolts.

One of the advantages usually cited for yield systems is that they tighten each fastener to the maximum safe preload available from that fastener: to the yield point. This means, however, that for a given fastener they would offer the designer no choice of preload. He could not, for example, get ''60% of yield'' to accommodate dynamic loads, temperature changes, or the like. As a result, currently available yield control systems have been designed to operate in other torque-turn modes as well (torque-turn windows, for example).

Note that a ductile fastener tightened to yield still has a substantial useful life. It hasn't been ''damaged'': it has just been work-hardened a little. Thanks to joint relaxation following preload, the designer can often count on purely elastic behavior from a bolt originally tightened to yield, as we have seen. One manufacturer has taken this point one step further

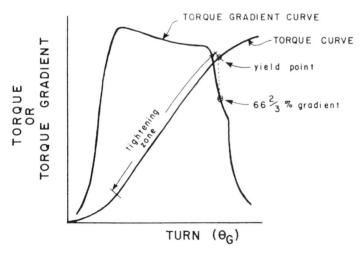

Figure 8.22 Torque-turn curve, and the rate of change of torque as a function of turn. This latter function is used to detect the yield point of the fastener in torque-turn-to-yield control systems. (From information received from SPS.)

and has patented a control system which tightens just past the yield point of the fastener, then deliberately loosens the fastener slightly to bring the average tension back down to some point well within the elastic range, as suggested by Fig. 8.23 [10]. Since the yield point can be found more accurately than the snugging point, this torque-to-yield-and-back-again system improves the accuracy with which the ''60% of yield'' point, or the like, can be found.

C. The Logarithmic Rate Method System

One manufacturer currently produces an even more sophisticated torque-turn control system that allows a fastener to be tightened accurately to any point on the elastic curve from 50 to 100% of yield, not just to the yield point [11]. In effect, this system uses a unique mathematical approach to the torque-turn curve to establish an imaginary threshold torque that is paradoxically more accurate than the thresholds used in conventional torque-turn systems. This imaginary threshold is found by projecting the straight-line portions of the torque-turn curves (a different curve for each coefficient of friction) downward, as shown in Fig. 8.24. The tool finds this point by computing the rate of change of the log of net torque on the fastener versus angle of turn. The ''log rate'' curve peaks at a turn angle that coincides with the threshold torque (which is called *offset torque*).

Like all of the other computer or electronically controlled systems, this one must be calibrated, experimentally, for each new application.

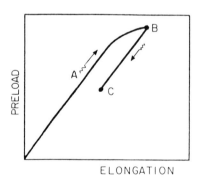

ELONGATION

Figure 8.23 In one recent variation on torque-to-yield control, the fastener is tightened (A) to the yield point (B) and then deliberately loosened slightly (C) to reduce final tension to some percentage of yield. External loads will now work the fastener elastically rather than plastically. Note that we've illustrated this with a preload-elongation curve rather than torque-turn.

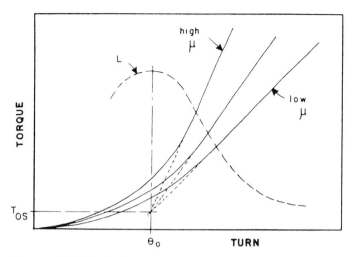

Figure 8.24 In the LRM system the fastener is first tightened to an imaginary threshold torque (T_{os}) which is directly under the peak of the log rate curve (L). The tool then gives the fastener a given turn, starting from θ_0. This mathematically sophisticated approach to torque-turn gives excellent control of preload. (Modified from Ref. 8.)

The behavior of bolts and joints is so complex that it is impossible—not just difficult—to predict exactly how they're going to behave, even if you have a computer helping you. The yield point of the fasteners in a given joint is found experimentally for the SPS or other torque-to-yield systems. The offset torque must be determined before the logartithmic rate method (LRM) system can be used.

Note that none of the truly sophisticated control systems base their decisions on a single parameter or single set of circumstances. The LRM system uses turn from the peak of the log rate curve for the final decision only. The computer examines and evaluates dozens of factors and relationships, working with the torque-turn information. It looks for crossed threads, blind holes, normal rates of change, windows, etc. Only when it's satisfied that this is a normal joint does it turn final control of the tool over to the LRM portion of the program. Systems made by SPS, Ingersoll Rand, and others do the same thing, ending, however, with different strategies for the final decision.

Used on "normal" joints, the LRM system can control preload to within ±2–5%. Figure 8.25 shows one version of the system.

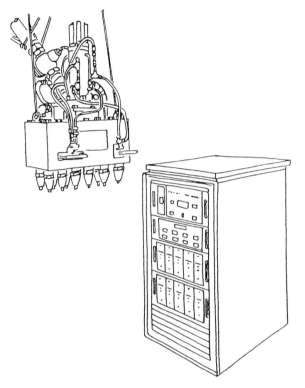

Figure 8.25 LRM console used to control a multispindle air tool of the sort shown (Rockwell).

The LRM strategy is based on mathematical manipulation of the classical long-form torque-preload equation (Chap. 7), combined with the turn of nut with respect to the bolt equation [Eq. (8.2)], as you'll see in the references. The assumption is made that only the coefficient of friction will change, from one fastener to another, in a given application. The system can, therefore, be confused by changes in fastener or joint dimensions, by nonperpendicular contact between parts, and by changes in elasticity of fasteners and/or joint members. In many applications, however, these changes are relatively minor and the system will provide excellent control of preload. These problems emphasize, however, that *all* production control systems of the fastener control systems used in production today are based on our old friends torque and/or turn, because these are the only practical control parameters available to the tool designer.

D. Plus—Permanent Records

Once you have a computer, of course, it becomes possible to manipulate
the torque and/or turn and/or preload data in a number of different ways.
Fastener assembly systems can be tied to larger computers used for production-control purposes, for example. Some systems provide a continuous statistical analysis of the problems encountered on the assembly line.
This helps quality assurance inspectors spot things like a high percentage
of faulty parts, improper procedures at previous stations on the line, etc.
Hard-copy records of torquing operations can also be kept for warranty
or liability protection purposes, as well as for production-control purposes
(Fig. 8.26).

E. Meanwhile, Out in the Field

Many manufacturers are reluctant to switch to a new production method
of fastener control if they cannot provide the same sort of control to field
service and maintenance people. New controls, for one thing, mean new
software: engineering drawings, specifications, product manuals, etc.
These add to the cost of adopting a new system. If production people are

Figure 8.26 Fully computerized tool control system can be used for production
control, statistical analysis, and the like.

working with one set of specifications and drawings, and service people in the field are working with a different set, all sorts of confusion and uncertainty can arise.

Fortunately, the microprocessor makes it possible for companies that manufacture sophisticated production systems to offer the same "brain power" in semiportable or hand-held tools, so that field people and production people can, indeed, use the same tightening strategies (Fig. 8.27). Such equipment is not inexpensive, of course. For critical joints, however, where "accuracy" can mean better product life, safety, strength-to-weight ratios, fewer warranty claims, etc., a switch to the new systems can often be justified.

Manual turn-of-nut techniques are often used in the field to retighten fasteners originally tightened to the yield point, if the mechanic does not have the new microprocessor-controlled tools.

Other types of field equipment are becoming available. SPS has offered an impact gun with yield control; and they also offer several microprocessor-controlled hand wrenches which can be used for torque or torque-angle control (Fig. 8.27).

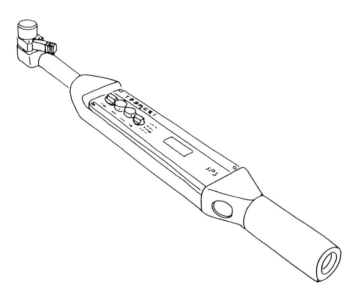

Figure 8.27 The SENSOR I wrench made by SPS. A microcomputer contained in the wrench allows it to be used for torque, or torque-angle, or yield control. The tool can be connected to a recording device such as a strip-chart recorder.

VIII. MONITORING THE RESULTS

The only thing we need to complete the picture is the ability to monitor the results achieved with the production or field systems. We need some means that is more accurate than the assembly systems—something which can be used by quality control or engineering personnel to set up the systems, calibrate them, recalibrate them, etc. Since the systems themselves have now become more accurate than manual torque wrenches, a tension-monitoring device of some sort is almost mandatory.

Tension load cells such as those described in Chap. 10 can sometimes be used, but these tend to alter the characteristics of the joint enough to affect the accuracy of their results (when compared against the inherent accuracy of the best control systems). Nevertheless, this is probably the most popular way to go at the present time.

In the not-too-distant future I expect that ultrasonic devices will be used for this monitor function. Ultrasonic extensometers, which measure the stress or strain in a bolt, are starting to emerge, as discussed in Chap. 11. Such equipment is already being used by a few companies for quality control evaluation and analysis of joints tightened by the LRM and other torque-turn systems. Results have been impressive, and we expect that this technology will play a large part in the future.

The fact that torque-turn systems cannot be monitored on line, in many situations, has led to the development of laboratory techniques for adjusting and supporting them. The equipment is used to assemble samples of the actual joint, in a laboratory where strain gages, ultrasonic equipment, load cells, and other test equipment can be used to inspect results. The torque and turn settings revealed by the laboratory tests are used on the production line. These settings are rechecked in the laboratory at frequent intervals, sometimes with each new batch of bolts. By this process the equipment is indirectly monitored.

It is also useful to subject the equipment to many load cycles, to determine whether or not set points shift, for example. The ideal way to do this would be to use the equipment to tighten many bolts, since that would perfectly duplicate the anticipated load patterns it will encounter in use. This would be very time consuming and expensive, however, so engineers have sought ways to simulate loads which absorb torque and require turn.

Tools can be used to drive electrical generators, or to fight partially engaged clutches or brakes. Even with a small air tool, however, such tests will generate a significant amount of heat energy which must be disposed of. And the load devices tend to be short-lived.

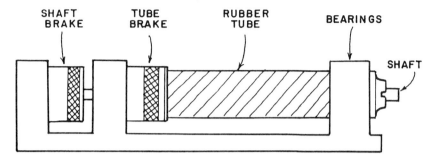

Figure 8.28 Diagram of a device used to load a torque-turn tool repeatedly for life tests. The load is provided by a long shaft and/or by a rubber tube, which are twisted by the tool as described in the text.

One interesting load device which minimizes these problems is diagrammed in Fig. 8.28. One end of a shaft, 4–5 ft in length, is connected to an air brake. A rubber tube, nearly the same length, is held by another brake. The tool to be tested engages the far end of both shaft and tube. Depending on which brake is energized, the tool twists the shaft or tube until the tool stalls or is otherwise turned off. By varying the diameter of the shaft and the wall thickness or durometer of the rubber tube, operators can adjust the amount of torque and/or turn required from the tool. The equipment was developed by the Ford Motor Company, but is now owned and operated by the Lawrence Technological University in Detroit.

IX. TORQUE-TURN CASE HISTORIES

A. Simple Torque-Turn Experiment

Fasteners: 1–8 × $3\frac{1}{2}$, Grade 8, Cadmium-Plated, 2-in. Grip

In an experiment which was repeated many times, four cadmium-plated fasteners were tightened on a test block that was $1\frac{7}{8}$ in. thick. Hard washers, $\frac{1}{8}$ in. thick, were used under the nut; 400 ft-lb of torque was applied to each fastener, after the nuts had been run down "finger tight" with a hand wrench. The turn required to achieve final torque was measured from the finger-tight point. All four fasteners were mounted in the same test block; all fasteners used in the test were obtained at the same time from the same manufacturer, and appeared to be identical.

Test results are plotted on a torque-turn-preload curve in Fig. 8.29. The torque-turn and turn-preload curves are not S-shaped because neither

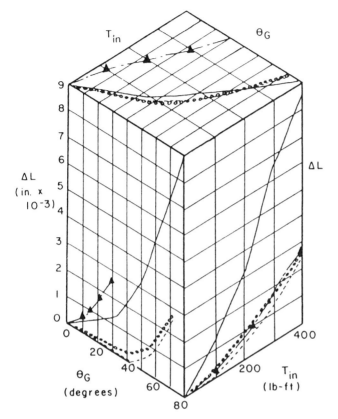

Figure 8.29 Torque-turn preload plot of tests made with four cadmium-plated, 1–8 × 3½, SAE Grade 8 bolts, with a 2-in. grip. Bolt number: (ooooo) 1; (▲ — — — ▲) 2; (--------------) 3; (————) 4.

the fastener nor the test block was loaded past the yield point, but the lower curve of the S is visible in most cases. And the torque-preload plots are approximately straight. In fact, there's nothing particularly noteworthy about any one of these curve sets, taken one at a time.

Taken together, however, they don't encourage torque, or turn or torque-turn control, do they? Note, for example, that the torque-turn behaviors of bolts 1, 3, and 4 are nearly identical. But the torque-elongation plot shows that bolt 4 stretched (was preloaded) to a value three times that of bolts 1 and 3. Or note that, while bolts 1, 2, and 3 behaved alike from a torque-preload point of view, the turn produced by 400 lb-ft on bolt 2 was less than half that produced on the other two. The behavior of this group of four bolts is, in fact, even more erratic than usual. It's a good example of why we often have unexpected problems.

A calculation of the stretch which would be expected in these bolts when 400 lb-ft of torque is applied suggests that each should have been stretched approximately 0.0025–0.0061 in. if used in the as-received cadmium-plated condition. It seems likely, therefore, that bolt 4 had inadvertently been lubricated (although there was no obvious difference in lube condition between fasteners).

Same operator, same tool, same test block, apparently identical fasteners, same torque, yet a three-to-one difference in achieved preload. Let me warn you, again, that there is a great deal of uncertainty associated with the tightening of threaded fasteners, and that you must accept this and take it into account whenever you are designing critical bolted joints; or you must use something better than torque and/or turn for control.

B. Comparison of Torque, Turn, and Torque-Turn Control

Fasteners: 1–5 × 4 SAE, Grade 8

A number of fasteners were tightened by a variety of control means. All were tightened into a Skidmore-Wilhem calibration stand (see Fig. 7.24).

In one series of tests, all of the fasteners were lubricated with molydisulfide paste, and hard washers, also lubricated, were placed between nut and test fixture.

In a second series of tests, some of the fasteners were lubricated with moly and some were not, and a variety of washers and gaskets were placed between nut and test fixture.

The tables on p. 305 show the preload scatter produced by the various control means.

C. Field Studies: Bolts Tightened by Turn-of-Nut Methods

Several years ago I was the chairman of the Committee on Bolt Load Determination of the Research Council on Structural Connections. This Committee was interested in an answer to the question "With what accuracy are bolts tightened by turn-of-nut procedures under actual field conditions?" As mentioned earlier, turn-of-nut is capable of controlling preload within ±5% or so. Careful laboratory tests have demonstrated this, and were used as the basis for the turn-of-nut procedures currently specified in Ref. 5. A number of people, however, have questioned whether or not similar results are obtained under field conditions.

Test Series No. 1: Moly Lube and Washers

Control means	Ratio of maximum to minimum tension measured during tests
Applied 300 lb-ft of torque	1.25:1
Applied 100° of turn past "finger tight"	2.75:1
Applied 200 lb-ft of torque to snug the nut, then 20° of turn	1.5:1
Measured bolt elongation with a micrometer	1.03:1

Test Series No. 2: Variety of Lube Conditions, from Moly to Clean and Dry, and a Variety of Washers and Gaskets

Control means	Ratio of maximum to minimum preload measured during tests
Applied 300 lb-ft of torque	4.6:1
Applied 100° of turn past "finger tight"	7.3:1
Applied 200 lb-ft of torque to snug the nut, then 20° of turn	2.8:1
Measured bolt elongation with a micrometer	1.25:1

In an attempt to answer the question, the Committee sponsored studies in which bolts previously installed by regular crews under normal conditions were selectively loosened and retightened by technicians. The loss of tension in each bolt was measured, ultrasonically, as it was loosened. Tests were made on several types of structural joint.

The Committee also sought and found results of similar tests conducted by other organizations. The results were compiled and compared. The structures involved included several large buildings and a bridge.

Combining these preliminary reports we have data on only 177 bolts, installed by, at most, four or five crews. About a third of the bolts were galvanized, the rest coated with black oxide. Bolt sizes ranged from $\frac{7}{8}$ to $1\frac{1}{4}$ in. in diameter. Our sample, therefore, is too diverse and too limited to be taken as definitive; but I think that the results are interesting.

Only 17 of our 177 bolts were found to have the minimum "snugging" tension specified in Ref. 5. None had more than the minimum. The remaining 160 had less than the minimum, in some cases only half the minimum.

All of these measurements were made on joints which had been assembled anywhere from several weeks to several years prior to the test, so the low tension values may, in part, reflect short- and long-term relaxation effects. In at least one case, however, procedural problems must have contributed. This particular structure was a multistory office building. Steel erection was nearly complete when it was discovered that the wrong washers had been used under the nuts. Hardened $\frac{5}{16}$-in.-thick washers should have been used, since the holes in the joint were oversized. The requirement for heavy washers had only recently been added to the AISC specification, however, and the contractor had inadvertently missed it.

Tests showed that the thin washers were probably responsible for the low levels of tension in these bolts. Nut turn well beyond that called out in the specification was necessary to bring these bolts to minimum tension, because the washers "dished" as the bolts were tightened. It was felt, however, that thin washers alone did not explain the findings, so visits were made to several other job sites to observe bolting crews in action [13]. It was discovered that most of the crews were not following the specified turn-of-nut procedure.

For example, the AISC specification says that bolt-up should start at the most rigid part of the bolting pattern, then proceed toward the free edge of the joint. In fact, the crews started at arbitrary points. The specification also says that the joint should first be snugged, then final-tightened; but the crews tightened each bolt in a single pass. Match marks on the joint were not used, so the crews, in effect, were not controlling nut rotation with any accuracy. Whether or not these practices had been used on the original "thin washer" job—or on the other jobs reviewed in the Committee's study—is unknown; but factors such as these would certainly explain the differences between laboratory studies of the turn-of-nut method and results under field conditions.

As mentioned, our studies found the typical tension in a bolt to be less than the minimum specified. We also found the bolt-to-bolt scatter in preload to exceed ±5% suggested by the laboratory tests. The coefficient of variation found in the field ranged from a low of ±3.5% (a three-sigma scatter of ±10.5%) to 55%, the worst case being the "thin washer" job. The average for the "normal" jobs was ±6.6% (a three-sigma ±19.8%).

Considering the fact that most of the bolts in the sample had probably not been tightened past yield (because so few were found to have the specified minimum tension), a three-sigma scatter of only 19.8% is very good, better than we'd have expected if torque control had been used, for example.

D. Torque-Turn Control in Engine Production

As mentioned earlier, torque-turn control is widely used in automotive and off-road equipment production lines. Several manufacturers now use ultrasonic equipment to set up and calibrate such equipment and to do production audits. Here are some of the things they report.

A manufacturer of diesel engines used torque-turn window control on head bolts. He reported that the amount of turn required to create the desired preload in the bolts varied from about 75° to over 90° and that the angle setting of the equipment had to be adjusted for each new lot of bolts, even if they came from the same manufacturer.

An automotive engine manufacturer reported that year-long tests showed that the torque-turn, window control equipment used on his production line was able to control initial preload within ± 15% if the engine head and block were made of cast iron and gaskets and bolts were obtained from the same suppliers. Scatter could be held within ± 10% on nongasketed joints, e.g., bearing caps mounted on the cast iron block.

If the block and head were made of aluminum, however, and the gaskets and/or bolts came from several sources, initial preloads varied by ± 25%. If the block was cast iron and only the head was aluminum, the scatter was reduced to ± 20%.

E. Torque-Turn Assembly of Automotive Transmission

An automotive laboratory used multispindle, torque-turn–controlled air tools to assemble transmissions. There were about 12 bolts in the assembly, varying in length but each about $1\frac{1}{4}$ in. long. They joined two complex aluminum castings which were separated by a stiff metal and rubber-fiber gasket. Some of the bolts were SAE Grade 2; the rest were Grade 8.

Each spindle of the air tool was adjusted separately, for threshold torque and final turn, to give the same target tension. It was found that the settings required for each spindle differed from the settings required for the others. Threshold torques ranged from 9.7 to 14 lb-ft. Angle settings ranged from 7° to 17°, the larger angles being used in conjunction with the smaller threshold torques. Final torques ranged from 12 to 17 lb-ft.

The laboratory personnel also noticed that a significant amount of relaxation occurred in the tightened assembly. The Grade 8 bolts relaxed as much as 25% in 5 days, "just sitting there." The Grade 2's relaxed even more—40% loss in some cases. Relaxation was not uniform, so it was blamed more on thread and other embedment effects than on gasket creep. (This was, at least, the preliminary hypothesis.)

X. PROBLEMS REDUCED BY TORQUE-TURN CONTROL

The block diagram Fig. 6.28 illustrates the many factors which affect the results when we use a torque wrench to tighten a group of bolts. To what extent does the use of torque-turn control reduce the uncertainties refined there?

Torque-turn doesn't eliminate any of the factors shown in the diagram, but it does help us estimate and cope with several of them. For example, in the third row of boxes, having both torque and turn information will tell us whether or not we're encountering unintended prevailing torque, or excessive friction loss, or, sometimes, an abnormal amount of bending.

In the next row, it should help us detect the fact that there's bolt-hole interference and/or significant resistance to the joint members being pulled together. All this could be very useful, for each of these five factors can be a major source of uncertainty during the assembly process.

Torque-turn control does not, on the other hand, eliminate embedment or elastic interactions (although hesitation tightening will help). It does not change the effects which such things as working loads, vibration, thermal changes, etc. have on the clamping force in the joint. That's not unexpected. It's unfortunate, however, that this better assembly control technique does not provide us with a more accurate way to compensate, after initial tightening, for relaxation or in-service effects. We can't go back to a joint previously tightened by torque-turn and reapply the original torque-followed-by-turn, to compensate for postassembly changes and reestablish the desired clamping force on the joint. In this respect, torque-turn is as "blind" as pure torque. But for initial tightening of individual bolts at least, it can give us some significant advantages over torque control.

XI. HOW TO GET THE MOST OUT OF TORQUE-TURN CONTROL

In Chapter 6 we reviewed some of the things we could do to optimize torque control (see Sec. 6-VII). Many of those steps are also appropriate for torque-turn control. You should, for example, train and supervise the crews, keep good records, make sure that bolts and joint members are in reasonable condition, use properly calibrated tools, etc.

In addition, the following steps can improve the results you get with torque-turn.

1. Determine the torque and turn you should use on your application by making one or more calibration tests on the actual joint, or on a model that simulates the actual joint as closely as possible. Use strain gages or ultrasonics or micrometer measurements to estimate bolt preloads. Repeat the tests periodically; for example, if you're about to use a new lot of bolts (especially if they've come from a new supplier).

2. Be sure that the bolt doesn't rotate while you're applying a measured turn. The turn values specified by the AISC are based on relative rotation between nut and bolt. Bolt twist is ignored (and is usually a small factor). If you define your own turn specifications, they should be based on the same assumptions.

3. Control the lubricity of your fasteners as well as possible. You may not need the degree of control you do for pure torque control, but you want to be sure that your snugging torques develop the desired minimum preloads, yet don't raise bolt tension so high that subsequent turn will break the bolts. For example, protect the bolts used on structural steel jobs so they don't rust before use.

4. Be sure that joint members are properly aligned and in good contact before applying final snugging torques and final turns. Torque-turn controls can sense hole interference or joint resistance problems, but not overcome such problems. Use drift pins to align structural joints, for example, and snug the joint with a few bolts before inserting the final bolts. It you can't insert these easily, give more attention to alignment.

5. Be alert for anything which changes the stiffness of bolts or joint members. Earlier we learned that thin washers, used on oversized holes, had resulted in low tension in the bolts. Warped joint members, or springy joint members which have not been brought into full contact, can do the same thing.

6. In manual torque-turn operations, such as structural steel work, be sure to mark the joint surface and the nut before applying the final turn, to assure that the correct amount of turn has been used.

XII. USING TORQUE-TURN DATA TO ESTIMATE JOINT STIFFNESS

As I mentioned in Sec. IV of Chap. 5, we can use torque-turn data to estimate the stiffness of the joint. The procedure was defined by Ralph

Shoberg and Dr. Sayed Nassar. [14] I gave you their final equation [Eq. (5.30)] in Chap. 5 and promised you the derivation of the equation in this chapter. Now that we have our torque-turn equations in place, we're ready to derive Eq. (5.30).

We start with Eqs. (8.3) and (7.4)

$$\frac{\Delta F_P}{\Delta \theta_R} = \left(\frac{K_B K_J}{K_B + K_J}\right) \frac{P}{360} \tag{8.3}$$

$$T = KDF_P \tag{7.4}$$

Therefore:

$$\Delta T = KD(\Delta F_P)$$

$$\Delta F_P = \frac{\Delta T}{KD}$$

where K = nut factor (dimensionless)
K_B = bolt stiffness (lb/in., N/mm)
K_J = joint stiffness (lb/in., N/mm)
F_P = preload (lb, N)
T = torque (lb-in., N-m)
P = pitch of the threads (in., mm)
θ_R = angle of turn (degrees)
D = nominal diameter of bolt (in., mm)

We combine these to produce:

$$\frac{\Delta T}{\Delta \theta_R} = \left(\frac{K_B K_J}{K_B + K_J}\right) \frac{PKD}{360} \tag{8.8}$$

and rewrite it as

$$K_J = \frac{\Delta T/\Delta \theta_R}{(KDP/360)K_B - (\Delta T/\Delta \theta_R)} \tag{8.9} \quad \text{or} \quad \text{(5.30)}$$

We can further substitute Eq. (5.12) if we haven't already computed bolt stiffness.

$$K_B = \frac{A_S E}{L_{eff}} \tag{5.12}$$

where A_S = tensile stress area of the bolt (in., mm)
E = modulus of elasticity of the bolt (psi, GPa/mm)
L_{eff} = effective length of the bolt (in., mm)

Since torque-turn information is often more readily available than joint stiffness test data, this procedure should find wide use.

REFERENCES

1. Ehrhart, K. F., Preload—the answer to fastener security, *Assembly and Fastener*, June 1971.
2. Goodman, E., T. Hogland, and J. Miller, Establishment of standardization data for Monel and K Monel fasteners, Value Engineering Co., Report no. 8-RD-65, Contract no. NObs-90493, AD 643 667, April 27, 1965.
3. Rice, E. E., Reliability in threaded fastener tensioning, Society of Manufacturing Engineers, Paper no. AD75-795, 1975.
4. Eshghy, S., The LRM fastening system, Society of Manufacturing Engineers, no. AD77-716, 1977.
5. Specification for Structural Joints Using ASTM A325 or A490 Bolts, Research Council on Riveted and Bolted Structural Joints of the Engineering Foundation, November 13, 1985.
6. *High Strength Bolting for Structural Joints*, Bethlehem Steel Corp.
7. Fisher, J. W., and J. H. A. Struik, *Guide to Design Criteria for Bolted and Riveted Joints*, Wiley, New York, 1974, p. 57.
8. High torque nutrunner combines high speed with torque accuracy, *Plant Operating Management*, p. 31, May 1973.
9. Boys, J. T., and G. H. Junker, Modern methods for controlling the tightening of fasteners with power tools, *Design Eng.*, p. 21ff, January 1975.
10. E. E. Rice, Method for Fastener Tightening, U.S. Patent no. 4,016,938, April 12, 1977.
11. Eshghy, S., The LRM fastening system, *Fastener Technol.*, pp. 47ff, July 8, 1978.
12. Ehrhart, K. F., Preload—the answer to fastener security, *Assembly and Fastener Eng.*, June 1976.
13. Notch, J. S., A field problem with preload of large A490 bolts, *The Structural Engineer*, vol. 64A, no. 4, pp. 93–99, April 1986.
14. Shoberg, Ralph S., Engineering fundamentals of torque-turn relationship, R.S. Technologies, Inc., Farmington Heights, MI, 1991 or 1992. (Prepared as a proposed article for *Assembly Engineering* magazine.)

9
Stretch Control

We applied torque, the nut turned, stretching the bolt and creating preload. We have seen that control through torque or turn, or both, did not give us ideal control of preload. We need something better for critical applications. How about control through change in length or stretch of the bolt?

I. THE CONCEPT

With torque and/or turn, we're trying to control the tightening process through the forces applied to, or the motion of, the *nut*. What we're really interested in is the bolt, however, since this is the thing which is being stretched to produce the clamping force on the joint.

As we saw in Chap. 5, we can consider the bolt to be a stiff spring. The relationship between the change in length of the bolt and the preload within it can be described by

$$\Delta L = F_P \left(\frac{1}{K_B} \right) \tag{9.1}$$

where ΔL = the change in length of the bolt (in., mm)
K_B = the stiffness of the bolt (lb/in., N/mm)
F_P = preload in the bolt (lb, N)

This equation says that the change in length of a bolt is equal to the preload within it times a constant. We can use Hooke's law to determine the constant in terms of the bolt properties and dimensions, as we saw in Chap. 5. For a common hex bolt, with a body, for example,

$$\Delta L_C = F_P \left(\frac{L_{be}}{EA_B} + \frac{L_{se}}{EA_S} \right) \tag{9.2}$$

where L_{be} = effective body length (in., mm) (see Chap. 5)
A_B = body cross-sectional area (in.^{2}, mm^2)
E = modulus of elasticity (psi, MPa)
L_{se} = effective thread length (in., mm) (see Chap. 5)
A_S = effective cross-sectional area of threads (in.^{2}, mm^2)
ΔL_C = combined change in length of all sections (in., mm)

Note that we have now apparently eliminated most of the factors which deal with the relationship between two or more parts in the system—factors which gave us major control problems when we were dealing with torque and/or turn. We've eliminated, for example, the friction between nut and bolt threads and that between nut and workpiece. We've eliminated the ratio of the spring constant or stiffness of bolt with respect to the joint. We've eliminated the need to measure the turn of nut with respect to the bolt or the workpiece. When dealing with stretch control we are looking, basically, at the bolt alone, and this provides an enormous simplification—and a subsequent improvement in control accuracy. Equation (9.2) suggests that if we can measure the change in length of the bolt accurately, we can determine the preload with the same degree of accuracy.

One interaction between bolt and joint does remain, however. The effective length of the threads depend on the grip length of the joint—on the combined thickness of joint members. The relationship between bolt stretch and bolt preload will be affected by variations in grip.

Note that all of the variables which remain could be measured and controlled if we needed the ultimate in preload accuracy. Such unmeasurables as the coefficient of friction or relative turn between nut and bolt have all been eliminated.

One attractive feature about stretch control is the fact that we can use it to measure residual preloads long after the fastener has been tightened. To so this we must keep a permanent log of the original length of each fastener, which can be a nuisance. But, when necessary, we can track preload during the life of the bolt merely by comparing its present length to its initial length. By comparison, we can never return to a previously tightened bolt and measure residual preload, accurately, with torque or turn tools.

Another attractive feature of stretch control, for those of us who like to monitor the energy content of bolts and joint members, is the fact that stretch measurements, combined with the related preload estimates, give us our best estimate of the amount of energy stored in the bolt. Take another look at Fig. 5.6 and Eq. (5.13) to see what I mean. The world at large doesn't base preload control or maintenance on energy estimates yet, but those who use various forms of torque-turn control are coming close to doing this. Those who use stretch control come even closer, whether they realize it or not. I suspect that future designers of bolted joints and automated production tooling will pay more attention to energy content and loss than they do at present.

II. THE PROBLEMS

At first glance it looks as if we have now achieved our goal and have found a practical way to measure preload. The bolt is, after all, a relatively simple shape. The modulus of elasticity is a well-known and well-defined quantity. Bolt dimensions are defined and controlled by a variety of different specifications. Our problem is solved.

Or is it? Unfortunately—but not surprisingly—we find that we are still faced with variables and uncertainties. Let's see why.

A. Dimensional Variations

Every dimension on a fastener is, of course, subjected to manufacturing variation or tolerance, even for something as highly standardized as a bolt. Table 9.1 lists typical tolerances on things like diameter, thread length, shank length, head and nut thickness, etc., for a variety of bolts covered by several different specifications. Note the following:

1. Diameters are controlled fairly closely, varying only a percent or so in the worst case and a fraction of a percent at best.
2. Tolerances on overall bolt length are reasonable, varying from a couple of percent for the largest bolts to a maximum of perhaps 10% for the smallest.
3. Tolerances on body and thread lengths, however, are very wide in most cases. The body of a $\frac{3}{8}$–16 × $1\frac{1}{2}$ finished hex bolt can vary by $+0$ to -46% in length, and the threaded portion of the bolt by $+19$ to -0%. These tolerances are perfectly acceptable for normal use, but could be a problem if you are trying to control the preload in a bolt by measuring the change of length, because body and threaded

Table 9.1 Dimensional Tolerances: Hex Bolts

	1/4–20 × 1/2 finished hex bolt	3–8 × 12 finished hex bolt	3/8–16 × 1 1/2 finished hex bolt	1–8 × 3 1/2 finished hex bolt	M10 × 1.5 35-mm-long precision hex bolt
References					
Bolt	ANSI B18.2.1-1972	ANSI B18.2.1-1972	SAE J105	SAE J105	British standard 3692-1967
Nut	B18.2.2-1972	B18.2.2-1972	SAE J104	SAE J104	3692-1967
Threads	B1.1-1974	B1.1-1974	ANSI B1.1-1974	ANSI B1.1-1974	3643-1966 (ISO)
Typical percent tolerance on:					
Body diam.	+0, −2%	+0, −0.4%	+0, −1.6%	+0, −1%	+0, −2.2%
Major diam. of threads	±1.7%	±0.3%	±1.2%	+0, −1.5%	+0, −1.5%
Bolt length	+4%, −6%	±0.7%	+0, −2.6%	+0, −2.9%	±1.4%
Thread length	+4%, −6%	N.A.[a]	+19.0%, −0	+13.9%, −0	+12%, −0
Head height	±4%	±3%	±3.6%	±3.0%	±2.6%
Nut height	±3.2%	±2%	±3.6%	±3.3%	±2.3%
Computed max. variation in:					
Body length	[d]	[e]	+0, −46%[f]	+0, −33.0%[g]	+5.6%, −38.9%[h]
Eff. body length	b	c	+0.7%, −37.9%	+0.6%, −27.1%	+4.7%, −28.7%
Eff. thread length	±6.2%	c	+44.4%, −9.6%	+29.0%, −4.4%	+25.6%, −10.5%

[a] Tolerance not given in specification.
[b] Bolt fully threaded, so there's no body.
[c] Computation not possible because of a.
[d] The calculations were based on the assumption that the grip length varied from 0.230 to 0.270 in.
[e] Assuming a grip length variation of 8.625 to 8.875 in.
[f] Assuming a grip length variation of 0.940 to 1.060 in.
[g] Assuming a grip length variation of 2.440 to 2.560 in.
[h] Assuming a grip length variation of 23.5 to 26.5 mm.

regions stretch by significantly different rates. "More threads and less body" in a bolt means "less preload for a given stretch."
4. We eliminate problem 3, variations in thread and body lengths, in bolts which have no body—which are fully threaded.
5. In general, there is a smaller percentage tolerance on all dimensions in larger bolts than there is in smaller ones, making stretch control of large bolts more accurate than stretch control of small ones (unless the small ones are fully threaded—no body—which can compensate for the difference).
6. Tolerances tend to be a little tighter on the new ISO metric bolts than on those defined by English system specifications such as those published by ANSI or the SAE. They're also tighter, of course, on military or aerospace bolts.

B. Effect of Dimensional Tolerances on Preload Scatter

Let's take a number of $\frac{3}{8}$–16 × $1\frac{1}{2}$ bolts and tighten each one until it has been stretched by exactly the same amount. How much variation would we expect to find in the preload within the bolts for this amount of stretch, if the dimensions of our sample bolts vary as much as permitted by the specifications?

One way to determine the theoretical scatter in preload would be to introduce the maximum and minimum dimensions into the stretch-preload equation (9.2). The results of such a calculation can be quite startling. As an example, the preload in a group of $\frac{3}{8}$–16 × $1\frac{1}{2}$ SAE J105, Grade 8 bolts could vary by $+6.4$ to -15.8% because of permitted dimensional tolerances alone. This assumed a nominal grip length of 1.000 in. which varied by a maximum of ± 0.060 in. (the application involved a gasket). The calculations were made for a "finished" hex bolt. There is no tolerance on thread length on commercial hex or heavy hex bolts, so the variation could be even greater there (unless the bolt is fully threaded).

This maximum and minimum dimension approach, however, is very misleading, and the results are not supported by actual tests. You may find a full tolerance spread in a given dimension if you test enough bolts, from enough suppliers, over a long period of time; but the probability of getting the right combination of maximum and minimum dimensions to give you a maximum or minimum stretch for a given preload is very low.

The actual situation is better analyzed by statistical methods. The procedure is beyond the scope of this book, but involves the partial derivatives of preload as a function of each of the variables in Eq. (9.2), taken one at a time. These partial derivatives are multiplied by the experimentally determined variances for each variable; and then all terms are added to compute the variance for preload as a function of stretch.

Expressed mathematically, this is

$$\sigma_{F_P}^2 = \sum \left(\frac{\partial F_P(Y_1, Y_2, Y_3, \ldots, Y_N)}{\partial Y_K}\right)^2 \sigma_K^2 \tag{9.3}$$

where $\sigma_{F_P}^2$ = the variance of preload (the standard deviation squared)
σ_K^2 = the variance of each parameter taken one at a time (the parameters are effective lengths, the areas, the modulus, etc.)

and the remainder of the expression defines the sum of the series of partial derivatives.

As I said, the procedure is beyond the scope of this text. Full details, however, can be found in the references [1].

We have measured the actual dimensions in a sample of $\frac{3}{8}$–16 × $1\frac{1}{2}$ bolts received from a single source and have thereby determined statistical variances for each dimension separately. We've assumed a three-standard-deviation value in the modulus of elasticity of ±1.5% (more about this variation later) and have used the statistical procedure described above to compute the possible scatter in preload in all such bolts, for a given stretch, because of dimensional tolerances. The calculations indicated that there would be a scatter of only ±2.7%, substantially less than that predicted by the maximum-minimum dimension approach, and more in line with our field experiences with many different kinds of fasteners. A sample received from several sources would presumably show more scatter, but you'll usually be dealing with single sources on a particular joint.

Our experience suggests that dimensional variations, while they could be a problem, are usually not a major source of error when we're using stretch to control preload.

C. The Effect of a Change in Temperature

The dimensional variations discussed so far dealt with room temperature variations, i.e., manufacturing tolerances. A change in temperature of the bolt can create another type of dimensional variation: thermal expansion or contraction. Such changes can be estimated from the following equation:

$$\Delta L = (T_2 - T_1)\rho \tag{9.4}$$

where ΔL = the change in length of the bolt (in., mm)
T_2 = the current temperature of the bolt (°F, °C)
T_1 = the temperature at assembly (°F, °C)

ρ = the linear coefficient of expansion of the bolt material (in./in./°F, mm/mm°C) (see Table 4.5)

If you're going to monitor bolt stretch over a period of time, recheck the stretch after the bolt has been in service. For example, you should keep a log of the bolt temperature when you assemble the joint, and again when you make postassembly measurements of length. Any thermally induced change should be subtracted algebraically from the total change before estimating bolt tension with Eq. (9.2).

If the bolts and joint members are made from dissimilar materials, or are at different temperatures, then the problem of relating bolt stretch to bolt tension becomes more complicated. We'll discuss that situation at greater length in Chap. 13.

D. Variations in Elastic Modulus

Although the modulus of elasticity of bolting steels is always given as 30×10^6 psi in handbooks and the like, in practice it can be as low as 29×10^6. This can introduce $+0$, -3% variation in the preload for a given stretch.

It should be mentioned that the modulus of bolting materials will be reduced substantially by an increase in temperature, as shown in Table 4.6. The increase itself has to be substantial, however, and you won't be tightening the bolts at those temperatures unless you're resorting to "hot bolting" to correct a problem.

E. Gross Plastic Deformation of the Bolt

Our equation assumes elastic behavior of the bolt. If we tighten it past yield, the modulus of elasticity will be only a fraction of 30×10^6, and if we're unaware that the bolt has yielded, our estimate of preload will be grossly inaccurate. This could be the case if the latest batch of bolts have been improperly heat treated, for example, and are soft. We stretch them the specified amount, but will produce only a fraction of the anticipated preload.

F. Variations in Stress Level Within the Fastener

Our equation assumes that the bolt is a pair of cylinders, in series, and that each cylinder is subjected to uniform stress, that in the body cylinder being F_P/A_B, that in the threads being F_P/A_S (see Fig. 5.3, etc.).

In practice, each bolt sees a nearly unique pattern of stress distribution, with wide variations in stress magnitude, because of large and small variations in geometry and material and heat treatment. Each microscopic region of the bolt will yield in proportion to the stress level it sees—obeying Hooke's law if the stress level is below the elastic limit, deforming plastically at higher stress levels. All of these localized strains add together to determine the overall change in length or stretch of the bolt under an applied load. In order to predict stretch perfectly, we would have to know what was happening at every point within the bolt. A finite-element analysis would be better than our "equivalent cylinder" approach, but even this wouldn't be perfect. To predict the behavior of a particular bolt would require more information about it than we'll ever have—we'd have to know such things as the exact radius of each thread root, plus total information about material or heat treat variations from point to point.

Coarser dimensional differences, such as in the fit between nut and bolt threads, can have a still larger effect on stress distribution, as suggested by Fig. 2.11. The length we've assumed for the "equivalent cylinder" for the threaded section (exposed threads plus half the thickness of the nut) is too short under these conditions. Even a small change in thread pitch can have a significant effect on actual results. One of our aerospace customers used a tensile-testing machine to determine the relationship between applied tension and bolt elongation in a large number of Inconel bolts, about $\frac{3}{8}$ in. in diameter and a couple of inches long. He noticed that the slope of the force-elongation curve for a given bolt depended on the number of times that bolt had been loaded. It appeared to be a little less stiff with each successive run. He also found a variation in behavior from one bolt to another, on the first run for each, for example. The total scatter in preload for a given stretch was an alarming 30%.

He traced some of the variation to dimensional tolerances, primarily in bolt diameter. The major factor, however, turned out to be variations in the fit between nut and bolt threads. Each time tension was reapplied to a fastener, the threads yielded a little more, distributing the load over more threads. Whereas only the first one or two threads carry the load during the first test, as many as five would share the burden after several tests, gradually increasing the effective length of the bolt.

He reduced the scatter to an acceptable level (less than 10%) by first tightening each bolt into a test block with a torque wrench, before measuring its stiffness in the tensile machine. The torquing operation "wiped in" the threads, forcing them to conform. This eliminated virtually all of the drift from test to test.

He reduced the effect of the dimensional variations by measuring the stiffness of a sample group of fasteners, selected at random from each shipment received from a supplier. The mean slope of the sample was then used to establish the stretch to be used as the control value for that batch of fasteners. Maximum and minimum deviations from the mean now fell within the tolerance established on preload scatter ($\pm 5\%$).

In another set of experiments, reported in Ref. 2, it was found that the length of the equivalent body and thread cylinders would have to include 0.38 of the thickness of the head, and 0.4 of the thickness of the nut, to make the computed relationship between stretch and preload agree with actual measurements. Using the conventional assumption of one-half the thickness of head and nut resulted in a computed stretch that was 16% greater than actual. The tests were run on an M24 steel bolt, with a grip-to-diameter ratio of 2:1. So it isn't safe to assume that the errors will always be toward longer equivalent cylinders.

Uncertainties in stress distribution can therefore introduce a large and often unacceptable difference between anticipated and actual preload for a given stretch—a control error that can be as great as those we faced with torque or turn control, at least in some cases. There's a way around the problem, however. Before we consider it, let's complete the list of problems.

G. Bending and Nonperpendicular Surfaces

Because joint and nut surfaces are never exactly perpendicular to thread axes and/or because of the influence of gaskets, etc., a bolt almost never stretches uniformly when it is tightened; instead it bends to some degree, as suggested in Fig. 9.1.

What is the change in length of the fastener under these conditions? Is it the change along the centerline? We're really interested in preload, of course, not in change of length per se, so a proper question would be "What is the relationship between the net clamping force on the joint and the change in length along the centerline of the fastener, if the fastener bends as it is tightened?"

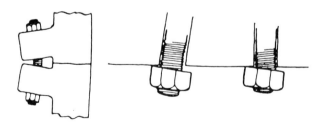

Figure 9.1   Few bolts or studs stretch uniformly when tightened, thanks to the fact that nut, head, and joint surfaces neither start nor end up perpendicular to thread axes.

I don't have a simple answer for you. We've measured stretch ultrasonically for over 10 years, but I'm not aware of any situation in which bending was severe enough to cause a preload problem. It can, however, make stretch measurement more difficult, whether micrometers or ultrasonics or other means are used. Knowing that bending alters the stress patterns within the bolt and knowing that apparently minor differences in geometry can result in substantial differences in stress pattern, and hence in stretch behavior, I expect that we would find that a lot of bending will make a lot of difference, but, as mentioned, we've never had a serious problem.

H. Grip Length

A longer bolt will stretch more than a shorter one, given the same general diameter, shape, and load. As a result, the relationship between computed preload and measured stretch is a function of the grip length—the combined thickness of the joint members. There will be manufacturing tolerance on these dimensions, of course, just as there were on fastener dimensions.

It's unlikely that this will be a major problem in most applications, but it might make enough difference to show in a large gasketed flange joint, if there can be a big difference in the gap between flange members from one side of the joint to the other. A 0.040-in. difference in gap, for example, would mean a 1% difference in grip length on a flange joint that had a combined flange thickness of 4 in. And a 1% difference in grip would mean a nearly 1% difference in the preload achieved for a given bolt stretch. If the bolt or stud has a body, furthermore, the effect might be magnified, since an increase or decrease in gap would change the effective length of the threaded portion of the fastener, not the effective length of the body; and the threaded portion stretches more readily than does the body.

I. Combined Effects

We're talking, still, about the difficulties of *predicting* the relationship between bolt stretch and preload. A number of factors can cause substantial differences between computed and actual results. How do we estimate the total possible error—add the individual error factors?

We could do this, of course, but the results would be unrealistic. It's highly unlikely that everything which could increase the preload for a given stretch will "happen" at the same time. Rather, some factors will tend to increase it, and others will partially or wholly cancel that tendency.

A more realistic approach, therefore, would be to use the statistical approach described briefly in Sec. II.B. You would need to know or estimate the variance for each factor separately, but even guessing these variances would probably give you a more realistic estimate of the possible error than would simple addition of all worst-case possibilities.

J. Computed vs. Actual Results in Practice

Does actual field experience reveal the differences between actual and expected results suggested in the foregoing? Yes.

Thanks to our work with the ultrasonic extensometer, we have determined the difference between computed and actual stretch or stiffness for many different studs and bolts. Some of these results are shown in the scatter diagram of Fig. 9.2. As you can see, computed results can be larger or smaller than actual results; and the scatter gets worse at smaller grip-to-diameter ratios.

Differences of ± 5 or $\pm 10\%$ are common; differences as great as 25% have been seen on studs with small grip-to-diameter ratios (remember that

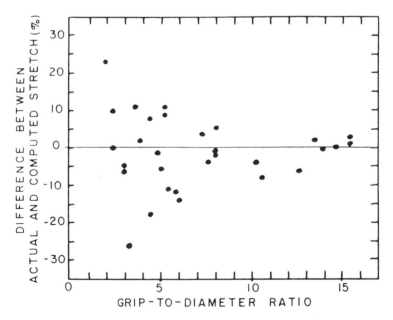

Figure 9.2 A scatter diagram showing the percentage difference between computed and measured spring constants (stiffness) for a variety of ASTM A193 B7 and B16 studs, as a function of grip-to-diameter ratio.

the equivalent cylinder model becomes less and less realistic as the bolt gets shorter). So it looks as if stretch control isn't much better than torque and/or turn control. This time, however, we can reduce errors in practice.

K. Reducing Errors in Practice

The problems we have examined are those of prediction, not control. We can't be certain, ahead of time, how a certain group of fasteners will behave in a given joint (grip length). But we can measure that behavior in one or a sample. Having done this, we'll find that we can predict, with a high degree of confidence, how the rest of this group will probably behave, and can even predict, with somewhat less confidence, how apparently identical fasteners received at a different time, perhaps even from a different source, will behave in the same joint. We can't do this sort of thing with torque or turn control because friction, or snugging turn, is virtually unpredictable. Measuring these on a sample does not help much in defining a larger population. The elastic behavior of bolts, by comparison, is very dependable.

Our field experience indicates, for example, that there is relatively little scatter in the preload achieved for a given stretch in a small group of bolts or studs—those found on a given flange joint, for example. Results of a typical experiment are shown in the histogram of Fig. 9.3. The $\pm 2\sigma$ deviation (95%) is only $\pm 3.8\%$ of the mean.

A similar test was run on 17, $1\frac{3}{8}$–8 × 10 ASTM A193 B7 studs which had been received from two different sources. Applying uniform tension to the studs resulted in a stretch scatter of only $\pm 5.5\%$ (at the two-standard-deviation level). And these results are typical.

Note, again, that the factors which could result in preload scatter—dimensional tolerances and modulus of elasticity—could be measured and controlled, if we needed that additional degree of confidence (and could afford it!). The coefficient of friction could never be measured and controlled, no matter how much we desired it.

In summary, then, although there are several factors which will make accurate *predictions* difficult, we can reduce the in-practice uncertainties by determining the preload-stretch relationship of our fasteners experimentally—by calibration, if you will. Since the behavior is repetitive, this "calibration" procedure will give us a degree of control over preload which is virtually impossible by conventional torque or turn means. (Some of the more sophisticated, computer-controlled torque-turn systems may come close, at least on "easy" joints.)

Calibration doesn't solve all of our problems, however. It tells us what stretch to seek for a given preload. But we now must face certain

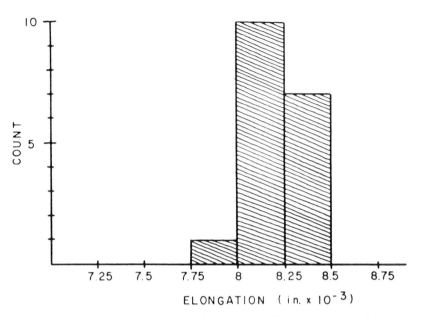

Figure 9.3 Histogram of the scatter in stretch required to achieve a given preload in 18, 1–8 × $3\frac{1}{2}$ SAE Grade 8 studs. A small sample of this sort, presumably all received from the same supplier at the same time, will usually show very little scatter.

practical difficulties in achieving accurate stretch control, just as we had to overcome practical problems (remember FOGTAR?) as well as theoretical problems when dealing with torque control.

L. The Wrong Bolt Problem

One practical problem which can destroy the accuracy of stretch measurement is the "wrong bolt." Remember that we're not interested in measuring stretch alone. Our goal is to estimate the preload in the bolt, and that estimate will be based upon a number of assumptions about the bolt. If the bolt we're measuring is not made from the assumed material, if it's soft (e.g., hasn't been heat treated properly), or if it's the wrong diameter, its stretch-to-preload relationship will be much different than we expect it to be. Stretch measurement is totally blind to differences of this sort.

One way to avoid being misled by wrong bolt problems is to combine stretch measurement with torque or torque-turn measurements as you tighten the bolts. Torque-stretch or torque-turn "windows" (see Fig. 8.14) will make it possible for you catch the bad ones.

That's not our only practical problem, however. We run into other potential difficulties when we start to measure bolt stretch in the field. Let's look next at how we do that and at what can go wrong.

III. STRETCH MEASUREMENT TECHNIQUES AND PROBLEMS

There are several traditional ways to measure bolt elongation in practice. None of them is practical for mass production operations, but they are used quite frequently for critical joints in heavy equipment, process systems, construction work, and the like.

A. Micrometer Measurements

The traditional way to measure bolt elongation, of course, is with a micrometer. If we have access to both ends of the bolt, as suggested in Fig. 9.4, we can use a C-micrometer. There are a number of problems associated with micrometer measurements, however.

Irregular Measurement Surfaces

As we've seen, most bolts bend slightly as they are tightened, and/or if the ends of the bolt are not flat and parallel to each other as received. The bending is always invisible to the operator, but, in effect, it means

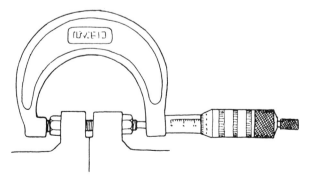

Figure 9.4 Using a C-micrometer to measure the change in length of a flange bolt.

that he must deal with a situation such as that shown (exaggerated) in Fig. 9.5.

Careful measurements of a number of stretched bolts have shown us that the distortions can be even more extreme than suggested in Fig. 9.5. A combination of stress concentration effects, nonperpendicularity between several parts, perhaps nonhomogeneity in the materials, etc., can produce end patterns which look more like those shown in Fig. 9.6. These saddle distortions can be introduced by the tightening process; they are not necessarily present in the unstressed bolt.

Many as-received bolts have concave ends, of course. Some have raised grade markings. All of these things greatly complicate the job of taking accurate measurements with a micrometer.

One way to reduce irregular end uncertainties, at least in experimental situations, is to embed small bearing balls into the ends of the fastener, as shown in Fig. 9.7 [3]. Sometimes the balls are driven into place, sometimes they are cemented into holes drilled in the bolt.

Operator Feel

A certain amount of skill and "feel" is also required for C-micrometer measurements of conventional bolts or studs. On long bolts the operator has considerable difficulty in deciding whether or not the anvils of the micrometer are indeed flat against the ends of the fastener, even if the ends of the fastener are parallel. The result is that different operators will get different results. We ran an experiment recently in which five trained machinists, each well accustomed to making micrometer measurements, were asked, independently, to measure the change in length of a $1\frac{3}{8}$–8 × 10 stud as it was tightened in four stages. The measurements were then tabulated and plotted, as shown in Fig. 9.8. Interestingly enough, a linear regression line run through these micrometer measurements coincides exactly with a regression line run through elongation data taken with the far more accurate ultrasonic extensometer, "proving" that if you take enough

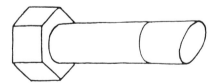

Figure 9.5 Bending, or bolt ends which are not parallel as received, can increase the difficulty of taking accurate micrometer measurements.

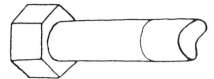

Figure 9.6 Saddlelike distortions are not uncommon in the ends of a stressed bolt.

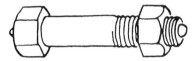

Figure 9.7 Mounting small balls in the ends of the fastener improves the accuracy of micrometer measurements.

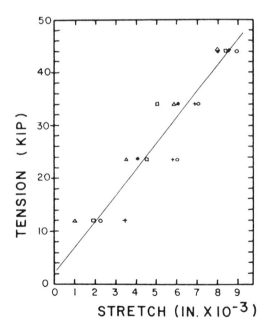

Figure 9.8 Measurements made, independently, by five different machinists on a $1\frac{3}{8}$–8 × 10 B7 stud as it was being tightened. Mechanic: (●) A; (○) B; (□) C; (△) D; (+) E.

micrometer measurements—and use a number of different people to take them—you can indeed measure the change in length of a bolt accurately by this method. It's more common, however, to take single readings, using a single operator; we feel that such measurements are highly suspect.

Measurement Accuracy Required

A fairly high level of accuracy is required, especially if you are measuring small bolts (although the smaller the bolt, the more accuracy you can get with micrometer measurements). Figure 9.9, for example, shows what a 0.001-in. error means as a percent total elongation and as a function of the grip length of the bolt. If you're trying to control preload within ± 10%, at 50% of yield (of a Grade 5 or B7 bolt), then ± 0.001-in. accuracy is

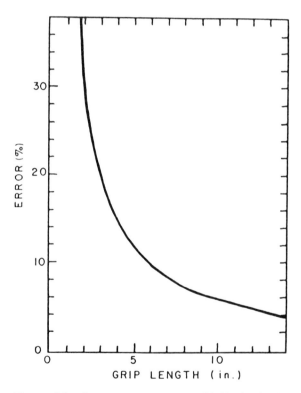

Figure 9.9 A measurement error of 0.001 in. leads to a larger percentage error in predicted preload in a short fastener than it does in a long one. The curve assumes that all fasteners are loaded to 50% of yield (SAE Grade 5 or ASTM B7 fasteners).

acceptable only for bolts more than $5\frac{1}{2}$ in. in effective length. If you want $\pm 2\%$ control of preload in this bolt, then elongation measurements must be made to the nearest 0.0002 in. If the effective length is only 1 in., then $\pm 2\%$ control would require an accuracy of ± 0.00004 in. And in each case we're assuming that the only errors are measurement errors, which is highly unlikely.

Depth Micrometers

It isn't common to use C-micrometers to control preload in bolts. It is common, however, to use depth micrometers to control preload in very large studs, especially those tightened by heater rods or hydraulic tensioners. A loose rod having ground and parallel ends is placed down inside a hole which has been gun-drilled through the center of the stud, as shown in Fig. 9.10. The lower end of the hole is capped by a threaded plug; or the rod has an enlarged end which is threaded, if you can afford or must have one rod per stud.

The micrometer is now used to measure the distance between the end of the stud and the rod in the center. As the stud is stretched, this distance increases because the rod, being loose, is not stretched. The procedure is simple and the accuracy quite good. The distance measured is small, even when a large stud is involved, so problems of "feel," etc., are much less than with a large C-micrometer. The reference anvil of the depth micrometer should always be oriented in the same clock position to minimize errors from nonparallel or nonflat surfaces, however.

It is not necessary for the gage rod to run the entire length of the bolt. If it does not, of course, you will only be measuring stretch in a portion of the bolt and must take this into account. But that is usually easy to do. Under these circumstances, it is necessary to control the depth of the

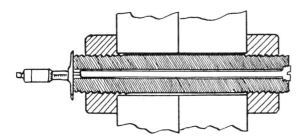

Figure 9.10 Using a depth micrometer and gage rod to measure the change in length of a stud.

gage hole from bolt to bolt so that you will be measuring the same amount of bolt each time and don't have to make a separate calculation for each one. This is especially true if the hole ends in the threaded region of the bolt, since these portions of the bolt stretch more than does the body.

Except in very long bolts, I think it's better to run the gage rod through to the other end, as suggested in Fig. 9.10. It can be threaded into the far end of the bolt or can be retained by a small threaded plug.

Note that the gage rod makes it possible for us to measure the stretch in one end of the bolt with respect to the other end. This is important if the bolt or stud is tightened into a blind hole and we have access to only one end. More important, however, it allows us to measure stretch without introducing the sort of errors we would encounter if we tried to use any other surface as a reference point. The bolt as a whole can move toward the nut as the nut is tightened, flange surfaces deflect, nuts embed themselves into flange surfaces, etc. The measurement accuracy required here demands that the other end of the bolt itself must somehow be used as the reference point.

Note that gage rods provide a built-in "record" of the change in length of the stud. A depth micrometer can be used at any time after assembly to determine the residual stretch in the stud. With other stretch measurement techniques, such as C-micrometers or ultrasonics, it's necessary to keep a log of the original lengths of the studs, to compare against present lengths, for postassembly measurements.

B. Dial Gages

Dial gages can be used instead of depth micrometers on very-large-diameter studs or bolts. They're used to measure the distance between the end of the stud and the end of an internal gage rod, in the same fashion that a depth micrometer was used.

Note that a dial gage cannot be set on the flange surface and used to check the motion of the end of the bolt as it is tightened because, again, of deflections and general motions in nut, bolt, flange surfaces, etc., and change in length only. The gage must be set on the bolt.

C. Screw and Washer Gage

Figure 9.11 shows another version of a gage rod bolt. This time the initial height of the gage rod can be adjusted (it's a screw). The operator sets the underhead of the gage screw at a desired distance above a cruciform washer, this distance being equal to the amount of stretch to be developed in the bolt. The bolt is then tightened until the washer cannot be rotated,

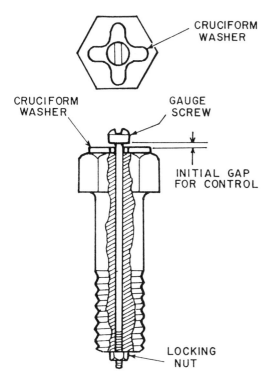

Figure 9.11 Gage screw patented by NASA and available for license. The head of the gage screw is set at a desired gap above a cruciform washer. The bolt is then tightened (stretched) until the washer is tight and cannot be moved by hand.

by hand, under the head of the gage screw. To my knowledge this system is not commercially available. It has been patented by NASA and could be used under license arrangements with them.

D. Commercially Available Gage Bolts

Figures 9.12 and 9.13 show a couple of commercially available gage rod bolts with associated measuring devices. The first (Fig. 9.12), available from SPS, consists of a digital electronic system for measuring the displacement between gage rod and bolt head. SPS manufactures bolts and studs and can provide gage pins in any size or type of bolt the customer is interested in.

Figure 9.13 shows a more elaborate bolt provided by RotaBolt Limited, in England. A rotating "load indicator" is preset at a prescribed air

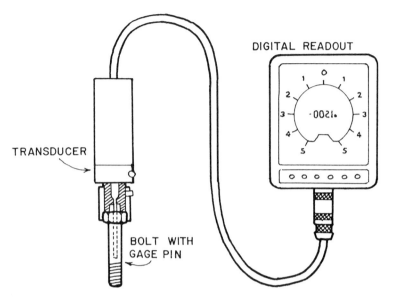

Figure 9.12 A commercially available gage rod system. The portable transducer and readout can be provided with a large variety of bolt types and materials (SPS).

gap above the end of the fastener. The fastener is tightened until a control cap can no longer be turned by hand. The entire measurement system, load indicator, control cap, etc., remain a part of the fastener and can be used to recheck for residual stress at any time after assembly.

E.   Ultrasonic Measurements

Any sort of micrometer measurement is clumsy and time-consuming and leaves something to be desired in accuracy. These facts have minimized the number of applications in which bolt stretch is used to control preload. A new technology, now emerging, involves the ultrasonic measurement of the change in length of bolts, and is expected to make stretch control far more common. Since the technology is new, it should be discussed at length and will, therefore, be left for Chap. 11. It's enough to say at this point that ultrasonics overcome many of the practical problems of microm- eter control, without eliminating any of the prediction uncertainties con- sidered earlier. They also introduce some new problems, of course. But stretch control promises a new level of preload accuracy, as well as the ability to monitor residual and working loads in fasteners, so it's well worth pursuing.

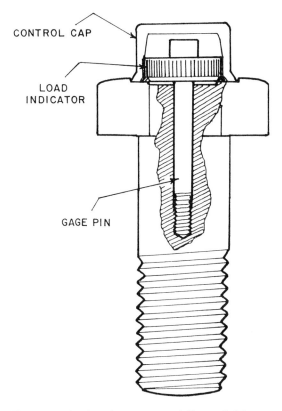

CONTROL CAP

LOAD
INDICATOR

GAGE PIN

Figure 9.13 Another commercially available gage rod bolt. This one retains its indicator for future inspection. (Rotabolt Ltd., England.)

IV. TORQUE-TURN CONSIDERATIONS

Stretch control doesn't introduce any *new* variables. The things which make a difference in the relationship between bolt elongation and achieved preload also affect the relationship between torque and preload or turn and preload. As the torque-turn-preload "cube" of Chap. 8 suggests, anything which affects one of these parameters will have some sort of influence on the others. For example:

Dimensional changes in bolt diameter would affect the torque required to produce a given preload. Smaller diameter, less torque; and the reverse.

Smaller diameters also mean lower bolt stiffness, and this means that more
  turn is required to produce a given preload (the bolt has to be stretched
  farther because it's a weaker spring).
A decrease in the modulus of elasticity would also increase the turn required for a given preload, but probably wouldn't affect torque very
  much.
Variations in stress distribution within the fastener also affect the stiffness
  of the fastener, and hence the turn required.
Bending and nonperpendicular contact (especially between nut and joint
  surface) can affect both torque and turn drastically because of high
  (if localized) contact pressure, localizing plastic deformation, etc.
Grip length affects stiffness and hence turn.

So the things which cause uncertainty in the stretch-preload relationship
also affect torque or turn controls. The things which cause the biggest
problems with torque or turn, however, friction and snugging turn, for

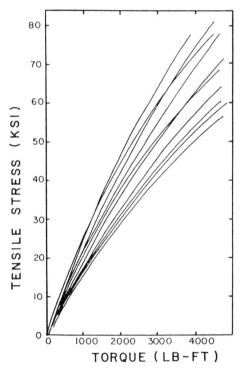

Figure 9.14   Torque vs. tension stress for a number of $2\frac{1}{2}$–8 × 12 B16 studs
which had been lubricated with a colloidal copper antiseize compound.

example, don't influence stretch at all. Even though some of the things which affect the accuracy of stretch control—especially variations in stress pattern—probably affect the stretch-versus-preload relationship more than they affect the accuracy of torque or turn control, the net gain is large.

As an example, Fig. 9.14 shows the results of a large number of tests made on a group of $2\frac{1}{2}$–8 × 12 B16 studs which had been lubricated before each test run with a colloidal copper antiseize compound. Tension stress levels were measured with strain gages. As you can see, there is a "normal" amount of scatter in the torque-preload (stress) behavior of these studs.

The change in length of these studs was also measured, using a depth micrometer (these were heater studs) and ultrasonics. The stretch-vs.-preload behavior of all the studs is plotted in Fig. 9.15. Their stretch behavior isn't perfect—there is some variation between studs and/or be-

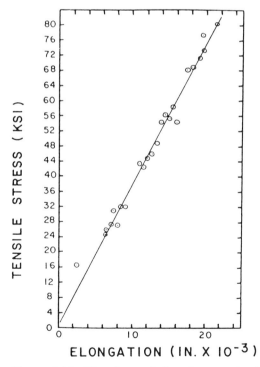

Figure 9.15 The change in length-vs.-preload behavior of the $2\frac{1}{4}$–8 × 12 B16 studs whose torque-preload behavior is plotted in Fig. 9.14.

tween tests. Stretch results are a lot more consistent than the torque-preload results of Fig. 9.14, however.

V. HOW MUCH STRETCH?

The amount of stretch you want will, of course, be determined by the amount of preload you want in the fastener. We won't be able to answer

Table 9.2 Typical Elongation Chart for Common Bolting Materials

Bolting material	20% of yield	40% of yield	60% of yield	80% of yield	100% of yield
Monel 40K psi y.s.	0.3	0.5	0.8	1.1	1.3
SAE GR 2 55K psi y.s.	0.4	0.7	1.1	1.5	1.8
SAE Grade 3; B7 and B16 over 4 in. diam. 80K psi y.s.	0.5	1.1	1.6	2.1	2.7
SAE Grade 5; A325; B7 and B16 up to 4 in. diam. 96K psi y.s.	0.6	1.3	1.9	2.6	3.2
SAE Grade 8; A490 120K psi y.s.	0.8	1.6	2.4	3.2	4.0
Inconel 718 180K psi y.s.	1.2	2.4	3.6	4.9	6.1
4340 steel, RC47 200K psi y.s.	1.3	2.7	4.0	5.3	6.6
Best available high-strength bolt material 240K psi y.s.	1.6	3.2	4.8	6.4	8.0
Titanium (6A14V) 134K psi y.s. $E = 17 \times 10^6$	1.6	3.2	4.8	6.4	8.0

Indicated elongation figures (in thousandths of an inch) are for various percentages of yield strengths (y.s.) of different bolts with a 1-in. grip length. (Modulus of elasticity assumed to be 30×10^6 unless otherwise noted.) To obtain desired elongation for a particular metal, read the elongation figure under the appropriate percentage of yield and multiply by the grip length in inches. For example, to obtain the expected elongation for a SAE Grade 5 bolt stretched to 80% of yield, with a 5-in. grip length, select the appropriate figure, which in this case is 2.6, and multiply by 5. The answer is 0.013 in. Note and warning: Many factors determine the "correct" stretch for a given fastener and application. Use this table with caution.
Source: Courtesy of Raymond Engineering Inc., Middletown, CT.

the question until after we've considered working loads on the bolt, and failure modes. Table 9.2, however, shows the amount of stretch you might see in various bolts, per inch grip length, if they were loaded to 50% of their yield strength. These are "typical" values, the way that the torque or nut factor values given in Chap. 7 are typical. It would be advisable for you to calculate the expected stretch in your application by using Eq. (9.2) or, better still, be determining the preload-stretch relationship experimentally, especially if your bolt has a grip-to-diameter ratio below 4:1.

We'll consider the question "How much stretch (preload)?" at greater length in Chaps. 20 through 23.

VI. PROBLEMS REDUCED BY STRETCH CONTROL

Figure 6.28 lists the factors affecting in-service clamping force. All of these factors were potential problems when we used torque to control the tightening operations. Although some of the factors were made less threatening by torque-turn control, none of them were eliminated. What about stretch control? Does it help reduce the uncertainties?

It does help. A lot. With stretch control we can estimate the tension remaining in the bolt after the assembly tools have been removed. This "feedback" allows us to ignore large groups of variables—or to compensate for their effects. Prevailing torque, friction loss, bolt-hole interference, embedment relaxation, elastic interactions—all will still affect the amount of torque we must apply to the fastener to tighten it, but with stretch control, we merely apply as much torque as is required to create the final, residual tension we want in each bolt. We won't have to measure the applied torque (unless we're concerned about galling or other severe problems). In practice, if we do keep track, we'll probably find that it takes a different amount of torque on each fastener to stretch them by the same amount; but who cares?

If we use gage rods, or keep a log of initial lengths, we can also return to the fastener after some in-service life, and estimate the effects of such things as external loads, vibration, thermal effects, etc. So stretch control is a significant improvement over torque-turn control. At least it is as far as the measurement of residual preload or in-service tension is concerned. It's probably NOT an improvement as far as job cost is concerned, however, especially in production operations. So don't use it unless it's genuinely necessary.

Note that stretch control is still blind to any factor which affects the relationship between tension in the bolt and clamping force between joint members. If the joint members resist attempts to pull them into contact, stretch control will accurately reflect the effort the bolts are making, but won't be able to tell us if the tension in the bolts is merely fighting the joint members or is also, for example, providing pressure on a gasket. In many situations, of course, other observations or measurements (e.g., of the gap between flanges) may warn us that we have a problem, but usually won't tell us how much of a problem.

And before we get too carried away, remember that stretch measurements do not provide perfect estimates of bolt tension. Variations in such things as grip length, bolt geometry, the modulus of elasticity, etc., will still introduce error (scatter) in our preload estimates.

Nevertheless, from an accuracy point of view, stretch is significantly superior to torque or torque-turn in most situations.

There are, however, some things which torque-turn can do better than stretch control can do. A stretch control system can't detect a soft bolt, for example, or a bolt which is the wrong size or made of the wrong material. Torque-turn systems can usually spot these problems. So, each has its strengths and weaknesses. In general, however, when preload accuracy is the main concern, stretch control is usually superior to torque or torque-turn control.

VII. HOW TO GET THE MOST OUT OF STRETCH CONTROL

To maximize stretch control you should do many of the things suggested for torque or torque-turn control: keep good records, train and supervise the crews, be sure that the parts you're using are in good shape, be as consistent as possible in your bolting techniques and procedures, calibrate the tools frequently, etc. And there are some items which should be given special attention.

1. Monitor any and all dimensions which affect the stiffness of the bolt and, therefore, the relationship between bolt tension and elongation. Body length, thread length, and grip length can be especially important.
2. For the same reason, it's useful to monitor the modulus of elasticity of the bolts. Don't assume that the modulus is 30 million just because the fasteners are steel. Knowing and/or controlling the modulus is especially important on long bolts (several feet in length, for example).
3. Be alert to variations in the flatness and/or parallelism on bolt ends. Dished ends, burrs, damaged surfaces, etc., can introduce errors when mechanical or ultrasonic techniques are used to measure stretch.

4. Monitor bolt temperatures if the "before" and "after" measurements are to be made at different times.

REFERENCES

1. Michalec, George W., Statistical backlash error prediction for gear trains, Paper no. 302, Presented at the International Symposium of the Japan Society of Mechanical Engineers, Tokyo, Japan, September 4–8, 1967.
2. Sawa, T., and K. Maruyama, On the deformation of the bolt head and nut in a bolted joint, *Bull. JSME*, vol. 19, no. 128, pp. 203–211, February 1976.
3. Cornford, A. S., Bolt preload—how can you be sure it's right?, *Machine Design*, vol. 47, no. 5, pp. 78–82, March 6, 1975.

10
Preload Control

We applied torque, the nut turned, and the bolt stretched. We tried to control the tightening process through torque, turn, and stretch—and found errors and uncertainties involved with each. Is there any practical way to control stress or preload directly?

In most applications the answer is "No." But a number of techniques are emerging which come close to tension control, and which are usually claimed to be such. For example, there are "tension control" systems which are based on the measurement of torque and turn, or on the measurement of strain. We looked at the torque-turn techniques in Chap. 8 and will now look at some of the strain measurement techniques—many of which are useful in special situations—and then we'll look at some as-yet-unavailable ways in which actual stress could, theoretically, be measured. Finally, we'll look at a useful device that is widely, but erroneously, believed to be a tension-control device: the hydraulic tensioner.

My definition of what constitutes true and direct preload or stress control, and what does not, will seem overrestrictive to some. For the purposes of this book, however, I think it's important for you to be able to tell the difference.

I don't mean to imply that the techniques to be described in this chapter are not useful, or that they are somehow false. All truly control tension with some degree of accuracy—but so does a torque wrench. The fact remains that there is no practical way to measure bolt stress or tension directly, at the present time, in most situations.

I. PRESENT MEASUREMENT TECHNIQUES

A. Strain-Gaged Bolts

One way to determine stress, of course, is to use strain (not stress) gages. The technology here is well advanced, and with the proper procedures and instruments, you can determine stress with far more precision than will usually be required. You'll be measuring the strain at a specific point on the surface of the fastener, however, and so must be careful in locating your strain gages if what you're interested in is average tensile stress (preload). You can also determine separately such things as bending stress or torsion stress by proper positioning of groups of strain gages. Used properly, strain gages are probably the most accurate way to measure bolt tension at the present time. Preload accuracies of ± 1–2% are reported [1, 2].

One problem with strain gage techniques, of course, is that they are expensive and time-consuming. This is perhaps the most accurate way at present to determine the nominal bolt stress in a specific application, or to do experimental work, but it's not considered a practical way to measure stress in each fastener in a production application. One of the things that contributes to the cost is the fact that disconnecting and reconnecting the readout instrument to the strain gages can change the resistance measured by the system substantially (thanks to change in contact resistance in the connectors). For best accuracy, therefore, the readout instruments must be connected to the gages at all times—which is obviously prohibitive. I'm sure there are ways around this problem, but it's the sort of thing that must be faced when using strain gages to make stress measurements. Another practical problem is the delicacy of the wires used to connect the gages to the readout instrument; it's very easy to break these as the bolt is tightened.

Another problem with strain gages. They must be mounted on a portion of the bolt that is stressed. You cannot, in other words, mount them on an easily accessible end of the bolt. You must have access to the shank of the bolt to take a measurement, and this will be a problem in many applications. On large-diameter bolts the gages can sometimes be mounted inside a hollow core. [At least one company (Strainsert, W. Conshohocken, PA) sells bolts and studs in which strain gages have been mounted.]

B. Strain-Gaged Force Washers

Another way to measure preload is to use a force washer—a compressible ring that has been provided with strain gages. These "preload load cells" can be used to measure preload continuously while the fastener is being tightened. An obvious disadvantage is their cost—they have to be left in place after use to be meaningful. As a result, force washers are useful only for experimental measurements and/or for very special applications. Like strain-gaged bolts, however, they are a very accurate way to measure bolt loads; they also have the same disadvantages as strain-gaged bolts.

C. Direct Tension Indicators

The structural steel industry makes wide use of the turn-of-nut bolting procedure. In this procedure, as we saw in Chap. 8, the nut is first snugged with a torque which is intended to create a minimum tension of 75% of the ultimate strength in the bolt. The nut is then rotated through a predetermined angle for final tightening, and is normally tightened past yield.

This bolting procedure is appropriate for structural steel work, because the bolts they use are ductile enough to be taken well past yield, and because, as we'll see in Chaps. 14 and 23, the integrity of such joints often depends on the friction forces developed between joint members. The bolts *must* provide a minimum clamping force to guarantee a minimum friction force. Because the bolts used are ductile, there is less concern about the maximum tension created in the fasteners.

This interest in a guaranteed minimum preload has led the industry to adopt several new types of fastener which improve the chances that the fasteners will be preloaded properly and/or make it easier to inspect previously tightened fasteners for minimum tension. These fasteners are formally classified as either ''alternate design bolts,'' to be discussed soon, or fasteners which allow ''direct tension indicator tightening.''

The most common type of direct tension indicator (DTI) at the present time is a washer with ''bumps'' on its upper surface. In one of several assembly procedures a DTI washer is interposed between the head of the bolt and the surface of the joint. As the nut is tightened, the bumps on the DTI washer yield plastically, reducing the gap between the head of the bolt and the washer. A feeler gage is used to measure this gap. When the gap has been reduced below a preselected maximum value, the tightening process is stopped. No subsequent turn-of-nut is required in this case—this is a substitute procedure for turn-of-nut. In an alternate procedure, shown in Fig. 10.1, the DTI is used under the nut rather than under the head of the bolt.

Several studies have been made to evaluate the accuracy with which the DTI washer controls initial preload in a fastener. In one series of experiments, the accuracy of the device, when used between parallel joint surfaces, ranged from $+4\%$, -6% to $+12\%$, -10%. When used on non-

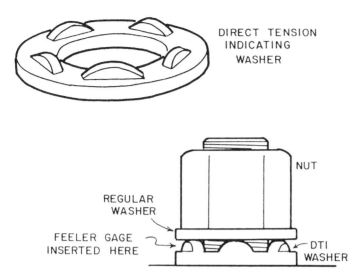

DIRECT TENSION
INDICATING
WASHER

NUT

REGULAR
WASHER

FEELER GAGE
INSERTED HERE

DTI
WASHER

Figure 10.1 Direct tension indicator (DTI). This tension-indicating washer can be used under the head of a bolt or, as shown here, under the nut. If it is used under the nut, it is best to interpose a conventional washer between the DTI and nut. (J. & M. Turner, Inc.)

parallel surfaces (structural steel members are often tapered), the best-case accuracy was $+15\%$, -11% and the worst-case, $+23\%$, -15%. In every case, however, the minimum tension required in structural steel work was achieved [3].

Note that each DTI washer is built to compress the proper distance at a preselected preload. This means that you must use the right DTI to get the right preload. There have been situations in which steel erectors used DTIs built for ASTM A325 bolts in assemblies actually using ASTM A490 bolts. The results were low preloads—at A325 levels—instead of the higher preloads which were intended by the building designer.

Another type of crush washer, this one used by the aerospace industry and called a preload indicating washer (PLI), is shown in Fig. 10.2. The washer consists of four parts: two conventional washers, plus an inner ring and an outer ring with radial clearance in between. The inner ring is slightly thicker than the outer ring. As preload is built up in the bolt, the inner ring compresses. The operator stops applying torque to the bolt when the inner washer has been compressed to the thickness of the outer washer, and the outer washer can no longer be turned around the inner washer when a small pin is inserted in the capstan hole.

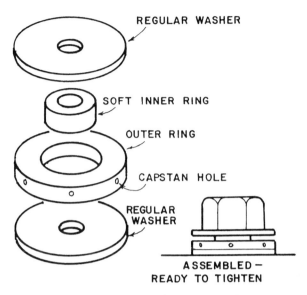

REGULAR WASHER

SOFT INNER RING

OUTER RING

CAPSTAN HOLE

REGULAR
WASHER

ASSEMBLED —
READY TO TIGHTEN

Figure 10.2 Preload-indicating (PLI) tension control system used in aerospace applications.

Again, studies have been made to determine the accuracy with which PLIs control initial preload. One investigator reports that preload varies from 65 to 95% of yield [4]. Others report characteristic accuracies of ±10% of desired preload [5, 6].

Figure 10.3 shows an entirely different type of direct tension-indicating fastener, a type also intended for the structural steel industry, but never, I believe, put into production. The bolt in this case has a flanged head, which is "wavy" when unloaded. The nut is tightened until the waves in the flange have been flattened against the upper surface of the joint, as shown in the figure. The amount of tension required to flatten the waves is known; an inspector can check to see whether or not the waves have been flattened; so the fastener provides some degree of tension control and the possibilities of postassembly inspection. Unlike the DTI washer described earlier, the deformation in this fastener is fully elastic. If the nut loosens or the bolt loses some tension, the waves will rise up again above the joint surface, showing that the fastener should be retightened. The user could detect and compensate for elastic interactions, for example, which could be a significant advantage.

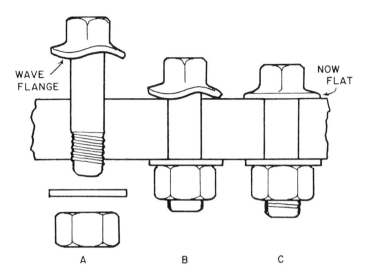

WAVE
FLANGE

NOW
FLAT

A   B   C

Figure 10.3 Direct tension-indicating bolt. (RB&W Corporation.)

D. Alternate Design Bolts

The AISC specification describes "alternate design bolts" as those which incorporate a design feature intended to indirectly indicate tension—or automatically provide it. Figure 10.4 shows what may be the most common form of this type of bolt at the present time—a "twist-off bolt."

The twist-off bolt cannot be held or turned from the head. (You'll note in the figure that it has an oval head.) Instead, the bolt is held by the assembly tool from the nut end. An inner spindle on the tool grips a spline section connected to the main portion of the bolt by a turned-down neck. An outer spindle on the tool turns the nut and tightens the fastener, with the tool reacting against the spline section. When the design torque level has been reached, the reaction forces on the spline snap it off, as shown in sketch 3 in Fig. 10.4. The building inspector can determine whether or not a minimum amount of torque was applied to the fastener by looking to see whether or not the spline sections have indeed been snapped loose from the bolts.

Note that this is really a torque control system, not a direct tension control system. The relationship between torque and tension, therefore, must be calibrated in a Skidmore-Wilhelm (Fig. 7.24) or equivalent. If, between calibration and use, the bolts are allowed to become rusty or in

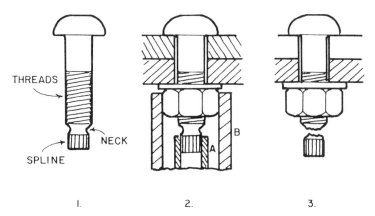

Figure 10.4 A twist-off bolt designed to indicate that a minimum amount of torque has been applied to the bolt. The tool holds and reacts against spline (A), while turning the nut (B).

any other way suffer a change of lubricity, then the amount of tension actually achieved in field assembly can be quite different from that achieved in the calibration stand, as suggested by some of the data in Table 7.1.

The fact that this fastener can be calibrated in the as-used condition, however, and, even more important, the fact that the inspector has a way to determine whether or not a minimum torque was applied to the fastener make this a popular item manufactured by a number of bolt suppliers. I include it here with fasteners which control tension more directly because the structural steel industry, at least, has accepted it as a bolt which "automatically provides" tension.

Another type of alternate design bolt, called a lockbolt, is shown in Fig. 10.5. At first glance it looks similar to a twist-off bolt, but is, in fact, quite different. In one configuration the nut is replaced by a swaging collar. The extended end of the bolt, called a pintail, has annular grooves on it rather than threads. The assembly tool grabs the pintail and pulls on it, creating a modest amount of tension in the bolt (which is called a "pin" in this case). The tool then swages the collar against annular grooves in the end of the bolt, creating substantial further tension. The tool now breaks the pintail section free from the bolt, once again providing an easy way for postassembly inspection. Although considered a "bolt" by many, this fastener could also be defined as a tension-controlled rivet. Another version, however, is more like a bolt.

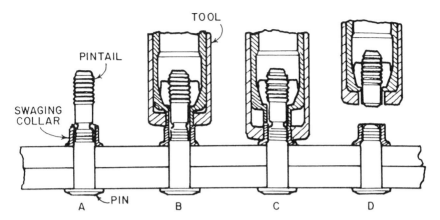

Figure 10.5 A "lockbolt." The tool applies tension by pulling on the pintail. A swaged collar, rather than a nut, is used to develop and retain further tension. The pintail is then broken free. (Huck Manufacturing Co.)

This boltlike lockbolt, which is manufactured by the Huck Manufacturing Company, has threads instead of annular grooves on the pin. Once again, a collar is swaged onto the end of the fastener, but this time swaged into the threads. Also, this time, the swaged collar is hexagonal at its base, and so can be engaged with a conventional wrench if disassembly of the joint is necessary.

E.   Another Load-Indicating Bolt

A still different type of load-indicating fastener, recently developed, is shown in Fig. 10.6 [7]. A gage pin run through the center of the bolt supports a red indicator disk within the head. An envelope of fluid is interposed between the indicator disk and a transparent window in the head of the bolt. Before the bolt is tightened, only a thin film of fluid lies between the window and the disk; so a red spot is visible from the end of the bolt. When the bolt is tightened (stretched), the gage pin moves away from the window and additional liquid is interposed between window and disk. By now the spot visible through the window has changed from red to black. Since the process is reversible, this DTI, like the bolt with the wavy flange, would also reveal relaxation or vibration loosening of the fastener.

In another "calibration gage" version of this fastener, the spot changes from blue (loose) to green (correct tension) to yellow (over-ten-

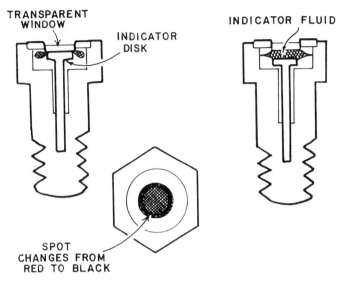

Figure 10.6 Indicator bolt. The spot on the end of the bolt changes from red to black as the bolt is tightened (stretched). (Stress Indicators, Inc.)

sion). The color changes in both types are gradual, not abrupt, but most of the change occurs near the design tension of 90% of proof load. (The change is complete at 100% of proof load.) Preloading accuracies of 2% of minimum tensile strength are claimed by the manufacturer for precision grades of this fastener.

F. Computerized Tension-Control Systems

Several air-tool companies sell computerized systems that are offered as "tension-control" systems. They do control tension—with varying degrees of accuracy—but through sophisticated use of torque and turn information, not through direct knowledge of bolt stress. Several of them have been discussed at length in Chap. 8.

G. Ultrasonic Measurement of Stress in a Bolt

Ultrasonic instruments can now be used to estimate the stress or tension in a bolt. This emerging technology is important enough—and complex enough—to deserve a separate chapter, Chap. 11.

II. THEORETICAL POSSIBILITIES

None of the presently available "tension-measurement" systems really measures stress, although some of them, like strain gages or computer control systems, give excellent control of tension. Are there any theoretical ways to measure stress or tension itself? A number of possibilities have been explored.

A. Magnetic Measurements

Several of the magnetic properties of a material are directly affected by changes in the stress level. For example, if you have access to a stressed portion of the fastener, you can measure the change in magnetic permeability of the bolt material that occurs when the bolt is loaded. If you place magnetic pickups in intimate proximity to principal stress trajectories on the bolt, and connect the pickups to a suitable bridge circuit, you can measure stress at that point.

The technique has been used for years to measure the torque transmitted through the propeller shafts of naval vessels [8]. It has been studied as a possible technique for measuring the tension in mine roof bolts. At the present time, however, no practical equipment is available for making such measurements in an average application; and, as with strain gage measurements, one would still face the necessity of locating the pickups beside a stressed region of the bolt instead of against an accessible end.

It would also be necessary to control the gap between pickup and bolt very precisely; a small difference in gap could mask or magnify a large difference in stress level in the bolt.

Other magnetic properties, such as coercive force, flux density, and hysteresis, are also affected by stress level [5], but, to my knowledge, no successful attempt has been made to use these changes to measure bolt stress.

B. Vibrating-Wire Gauge

In vibrating-wire gauges, one or more pieces of steel wire are fastened, under tension, to the opposite ends of a short cylinder (a "flat washer") placed under the head of the bolt. As the bolt is loaded, the tension in the wires decreases, decreasing their resonant frequency [9].

The technique has been tried experimentally on mine roof bolts. I suppose that a wire run down the center of a hollow bolt might, theoretically, be used as well.

C. Acoustic Emission

Most work on acoustic emission has involved large-scale plasticity, or the detection and analysis of crack initiation and crack propagation. In some work which was done at MIT, however, it was shown that acoustic emission can also be used to determine stress level in a body which has not cracked and which is behaving elastically [9]. The work was not related to bolts.

D. Miscellaneous Possibilities

In 1977 NASA asked the Southwest Research Institute to look for new ways to measure bolt stress or preload [5]. A dozen or more possibilities were explored, including the ultrasonic and magnetic techniques discussed in this chapter and in Chap. 11. Some of the other techniques were

X-ray diffraction
Nuclear magnetic resonance
Exoelectron emission
Eddy currents
Barkhausen noise analysis
Mechanical impact techniques

The experimenters felt, at first, that mechanical impact techniques might work. Most of the other possibilities—with the exception of ultrasonic measurements—were rejected after brief studies and/or theoretical analysis.

The mechanical impact technique consisted of hitting one end of the tightened bolt with a hammer or a steel ball; an attempt was made to correlate preload to the acoustic response of the bolt to the blow. "Voice prints" were recorded and studied.

The experiments were unsuccessful, however, so it was concluded that only ultrasonics (the methods considered in Chap. 11) offers "something better than torque or turn" at the present time.

III. BOLT TENSIONERS

A. The Hardware

In all of the control strategies we have discussed so far, we have assumed that we used torque to tighten the fastener even if we used some other parameter to control the tightening. Because we were not measuring and controlling the buildup of bolt tension itself, we usually had to face a lot of uncertainty in the relationship between our control parameter—torque, for example—and the preload or tension we were after. It would be great if we could apply and control the tension itself. This is the intent of that family of tools called bolt tensioners. As we'll soon see, these offer some significant advantages over the techniques we have talked about so far, but like the others are far from perfect.

Figure 10.7 shows a typical tensioner. Its operation can best be understood by reference to Fig. 10.8. To start the tensioning process a threaded section of the tensioner, called a "thread insert" in Fig. 10.7, is run down by hand over threads on the end of the bolt or stud to be tightened. Note that these stud threads extend beyond the threads engaged by the stud's own nut. This first step is diagrammed in Fig. 10.8A.

Hydraulic fluid under pressure is now pumped into the tensioner, extending its piston and stretching or tensioning the stud, as shown in Fig. 10.8B. At this point in the process the stud contains a very precisely controlled amount of tension. Errors in the control of fluid pressure and/

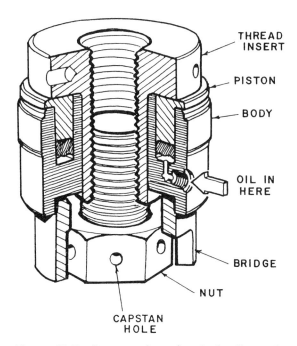

THREAD
INSERT

PISTON

BODY

OIL IN
HERE

BRIDGE

NUT

CAPSTAN
HOLE

Figure 10.7 Cutaway view of an hydraulic tensioner. Oil is introduced under pressure to apply tension to the bolt. The nut is then run down with a capstan bar. (Raymond Engineering.)

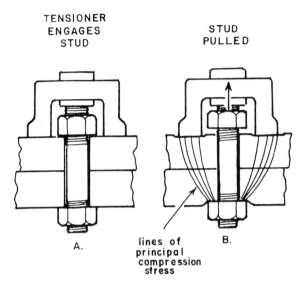

TENSIONER
ENGAGES
STUD

STUD
PULLED

A.

lines of
principal
compression
stress

B.

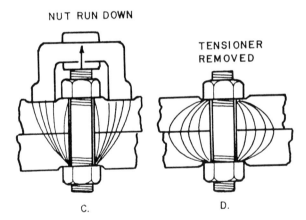

NUT RUN DOWN

TENSIONER
REMOVED

C.

D.

Figure 10.8 Diagram showing the sequence of operations by which a tensioner preloads a stud. See text for explanation.

or friction between the piston and the rest of the tensioner, etc., will introduce a little uncertainty, but it will be minor.

If we could walk away from the system at this point, we would indeed have found a way to preload bolts with almost perfect precision. Instead, however, we now shift control of that tension from the tensioner itself to something else—in this case the stud's nut. We then rely on the nut to retain the tension introduced by the tool.

To accomplish this with the system shown in Fig. 10.7, we insert a steel rod or a capstan bar in a hole in the stud's nut and use the capstan bar to run the nut down against the top surface of the joint, using as much torque as we can generate with the capstan bar (which isn't a great deal). This step is shown in Fig. 10.8C.

We now depressurize the tensioner and remove it from the stud. At this point only the stud's nut is retaining the tension introduced by the tool, as suggested in Fig. 10.8D.

Another type of tensioner is shown in Fig. 10.9. This one is very similar to that shown in Fig. 10.7. Their basic operation is identical. Only the means of running down the nut are different. In the tensioner shown

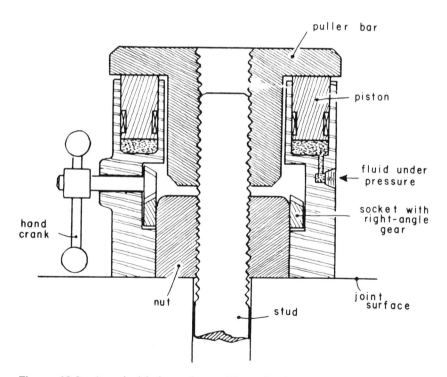

Figure 10.9 A typical bolt tensioner. The puller bar engages exposed threads on the end of the stud or bolt. The hydraulic piston stretches the fastener by forcing the puller bar upward. The fastener's nut is then run down against the joint by the hand crank, working through right-angled gearing. (From information received from Biach Industries, Inc.)

in Fig. 10.7 a capstan bar is used to run the nut down. In Fig. 10.9 a right-angle gear train has been provided for rundown.

B. Tensioner Accuracy

Elastic Recovery

The accuracy with which we can preload bolts with a hydraulic tensioner depends primarily on the accuracy with which we can transfer control of the preload in the stud from the tool to the nut or other device used to retain the tension. This is the case, at least, if we're relying on the tensioner both to *do* the job—to tighten the bolts—and to *control* the final preload. Let's see first what happens if we rely entirely on the tensioner and then look at alternate control means.

A number of factors affect the amount of tension retained in the stud when we remove the tensioner. First, there's an embedment relaxation process tensioner manufacturers call "elastic recovery." This process is illustrated in Fig. 10.8, where I have exaggerated the "embedment" of the tensioner feet and stud nuts in the joint surfaces. In Fig. 10.8A, nothing is embedded in the joint surfaces. In Fig. 10.8B, the tensioner is pulling on the stud. At this point the feet of the tensioner have embedded themselves slightly into the top surface of the joint. The nut on the far end of the stud has also been pulled up into the bottom surface of the joint. The deformations on top and bottom joint surfaces will be primarily elastic.

Figure 10.8B also shows the lines of principal compressive stress which exist in the joint at this step in the process. The joint is compressed between the feet of the tensioner and the far nut. If we were to compute the stiffness of the joint at this point, we would have to include all of the joint material which was included between the feet of the tensioner and far nut.

If we were to compute the stiffness of the bolt or stud at this point, we would have to use an effective length which ran from the midpoint of the far nut (or tapped hole) to the midpoint of the "thread insert" in Fig. 10.7 or "puller bar" in Fig. 10.9 (the things which are pulling on the top ends of the studs).

When we run the nut down against the top surface of the joint, as in Fig. 10.8C, there will be some compressive stress between the upper and lower nuts. At this point, however, most of the work is still being done by the tensioner itself; so compressive stress still exists between the feet of the tensioner and the far nut.

When we remove the tensioner, the upper nut must now pick up the full compressive load previously supported by the feet of the tensioner. The only way it can do this is to embed itself elastically in the top surface of the joint, as shown in Fig. 10.8D. The length of the bolt—and, therefore, the tension in it—is reduced slightly in proportion to the amount which the upper nut has to sink into the top surface of the joint to pick up the load. This is the "elastic recovery" referred to by the tensioner manufacturers.

Note that after the tensioner has been removed, the amount of joint material under compressive stress has been reduced. If we were to compute the joint stiffness at this point, it would be less than it was when the tension was still engaged and pulling on the stud.

At the same time the amount of bolt involved has also been reduced, the effective length now running from the midpoint of the lower nut to the midpoint of the upper nut. The exposed threads above the upper nut previously grabbed by the tensioner are now, of course, unloaded. A shorter effective length for the bolt means that the bolt has become stiffer. So, the bolt stiffness has increased, the joint stiffness has decreased, and the bolt-to-joint stiffness ratio has changed. All of this affects the way in which the bolt and joint distribute the energy put into the system by the tensioner, as we'll see in greater detail in Chap. 12.

The amount by which the stud or bolt shrinks tends to be a constant for a given stud diameter, assuming, of course, that both bolts and joint members are made of steel. In a stud having a short length-to-diameter ratio, this loss of stretch can be a significant portion of the total stretch introduced by the tensioner. The loss is a significant percentage of the original tension. In a long stud the same amount of embedment constitutes only a small fraction of the initial stretch, so the percentage loss of tension is small.

The obvious way to compensate for this elastic recovery process is to introduce additional tension with the tool. Table 10.1 shows the amount

Table 10.1 Overtension Required to Compensate for Elastic Recovery upon Removal of Tensioner

Length-to-diameter ratio of stud	Amount of overtension required (% of desired residual tension)
2	69
4	38
6	26
8	19
10	15
12	12
14	10.5
16	9
18	7.9
20	6.9

of overtension which one reference [10] says is required as a function of the length-to-diameter (L/D) ratio of this stud. If the L/D ratio is 8:1, for example, it is recommended that the tensioner be set to pull on the stud with a tension that is 19% above the residual tension or preload desired.

The tensioner should never be allowed to pull on the stud with a force which would exceed the yield strength of the fastener. If 19% of this initial tension is going to be lost in the example given above, this means that the maximum residual tension which can be expected in this stud is 81% of the yield strength of the stud. If the L/D ratio were only 2:1, then the maximum we could expect would be only 31% of the yield strength of the stud. In general, therefore, the longer the stud, the more effective the tensioning process.

Thread Embedment

In addition to elastic recovery, there will also be some thread embedment when we tighten a stud with a tensioner. Such embedment is probably a little greater than it is when we use a torque wrench, furthermore, since as we torque the stud we load the nut and at least partly smooth out the high spots on the thread surfaces during the tightening process. With a tensioner we suddenly dump the full load on nut and stud surfaces which may have been unloaded up to this point (if they are new parts). As described in Chap. 6, this embedment process will occur over a period of time, with most of it probably occurring in the first few minutes.

The amount of embedment relaxation will depend, to some extent, on the hardness of the bolt or stud and its nut. One observer reports that a five-point difference in Rockwell hardness of these parts can make a significant difference in the final tension retained by the stud after the tensioner has been deenergized.

The Effect of Variations in Nut Rundown Torque

There was a time when tensioner manufacturers instructed their customers to run the stud's nut down "finger-tight" against the upper surface of the joint. Users were, of course, warned to watch out for thread imperfections, etc., which might prevent complete rundown. (Obviously, if the nut is not touching the joint when the tensioner lets go, then little or no tension will be retained in the stud.) But no emphasis was placed on the amount of torque applied to the nut on rundown.

Experiments have shown, however, that in some cases the amount of tension retained can be directly proportional to the amount of torque used for rundown; doubling the rundown torque doubled the amount of tension retained in one experiment, for example (Fig. 10.10). Others report

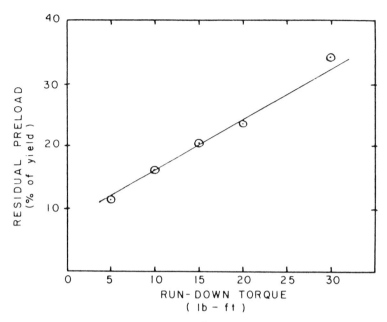

Figure 10.10   Residual tension in a $\frac{3}{4}$–10 × $4\frac{1}{2}$ hex-head cap screw as a function of the rundown torque applied to the nut. The screw was stretched hydraulically to 50% of yield before the nut was run down.

that this is true only up to a certain point, beyond which further rundown torque does not seem to make much difference.

In any event, it's certainly useful to run the nut down as hard as you conveniently can against the joint.

Elastic Interactions

Using a tensioner instead of a torque wrench does not eliminate elastic interactions between bolts in a group as they are tightened one by one or in subgroups. Figure 10.11, for example, shows some actual data of the residual tension in a group of 3–8 × 24, B24 foundation bolts tightened four at a time by hydraulic tensioners. A controlled rundown torque was used on the nuts. The bolts were tightened in a cross-bolting pattern. The max-min range in residual tension in these bolts after one pass is roughly 6:1, about what we'd expect to get with a torque wrench. This scatter can be (and was) reduced by subsequent passes with the tensioner. But the figure illustrates the fact, I think, that tensioners are not immune to

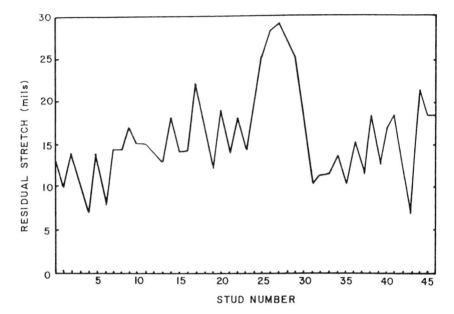

Figure 10.11   Residual tension in 46, 3–8 × 24, B24 foundation bolts when they were tightened, four at a time, with an hydraulic tensioner. The results reveal a significant amount of "elastic interaction."

some of the things which affect the accuracy with which preload can be obtained by other types of tool.

One advantage of tensioners, however, is that they can be coordinated. If you use several of them, you can power them from the same hydraulic pump and, as a result, simultaneously apply the same amount of tension to several bolts in the joint. This can be a big help if you are pulling together a large-diameter and/or gasketed joint, for example. This also reduces elastic interactions, and the residual preload scatter caused by interactions.

Figure 10.12 illustrates this last point. It shows the tension achieved in the bolts of a 4-in., 2500-lb ANSI flange when half the bolts in the joint were tightened simultaneously by ganged tensioners [11]. The first half were originally tensioned to point A on the solid line. The nuts were run down and the bolts relaxed, on the average, to tension B. Two more, identical, passes raised the residual tension to C in this group. The relaxation to this point was caused by elastic recovery and embedment.

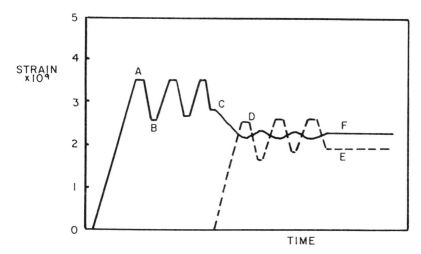

Figure 10.12 Initial and residual tension pattern in the bolts of a 4-in. ANSI flange when 50% of the bolts were tightened, simultaneously, in one step, and the second half were tightened simultaneously in a second step. The results show elastic recovery (e.g., in the tension drop between point A and point B) and elastic interaction (in the drop between C and F).

The tensioners were now moved to the remaining half of the bolts, located between the first half. This time less tension was applied (to point D in Fig. 10.12). Again, there was elastic recovery and embedment to contend with, but these bolts were brought to a final tension, represented by point E. As these studs were being tightened, however, the tension in the first group of studs relaxed to point F—thanks to elastic interaction between joint members and the two groups. The second group helped the first group to compress the joint, allowing the first group to relax.

This is the same process we saw in Chap. 6 when we used torque to tighten a gasketed joint. The difference this time is that, with tensioners, the operators were able to predict and compensate for the elastic interactions (by using less tension on group 2). They used ultrasonics, however, to measure achieved tensions and so to pick the correct initial tensions for the two groups. They couldn't eliminate elastic interactions, but they could adjust their procedure to accommodate them.

If you can afford to mount a separate tensioner on each stud in the joint, in fact, and tighten them all simultaneously, there will be *no* elastic interactions—and *that* can make a *big* difference in the max-min scatter of residual preload left in the joint.

C. Exposed Thread Problems

As described earlier, the studs to be tensioned must be extra long; they must have exposed threads beyond the stud's nut which the tensioner can engage when pulling on the stud. This means that you cannot suddenly decide to switch from a torquing procedure to a tensioning procedure unless you are also prepared to change to longer studs. This is true, at least, of the tensioners discussed so far. It is not true of those shown in Figs. 10.13 and 10.14, which we'll discuss soon.

The exposed threads on the end of the stud can lead to a couple of problems. In many situations they tend to rust, making removal of the stud and nut difficult or impossible. To protect against this, most people lubricate that portion of the stud and then cover the exposed end with a rubber cap, or with extra nuts.

Another problem: The extra threads can make it impossible to engage the stud's nut with a conventional socket. The socket isn't deep enough. As a result, a standard wrench often cannot be used to loosen the nuts during detensioning.

D. Alternate Control Means

We talked about the accuracy with which we can tighten bolts using a tensioner alone. As we've seen, there are still substantial possibilities for error. As a result, many tensioning operations involve some alternate control means, the most popular being stretch measurement using a gage

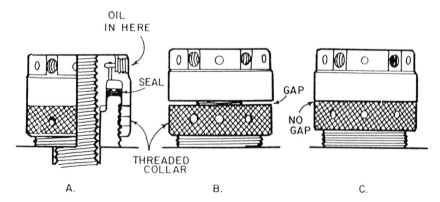

Figure 10.13 Hydraulic nut. This device both tensions a stud and then remains in place to retain that tension. In operation the nut is run onto the stud (A), and then pressurized to tension the stud (B). A threaded collar is now run up to retain that tension (C) and the hydraulic pressure is turned off. (Raymond Engineering.)

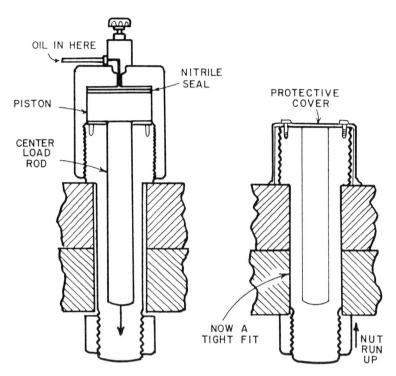

Figure 10.14   Morgrip bolt. A rod is forced down the center of the bolt to tension it. The nut on the far end is then run down to retain tension.

rod of some sort, such as that shown in Fig. 9.10. By using a gage rod and a micrometer, we get the improvement in accuracy and advantages offered by stretch measurement.

Stretch measurement, in fact, is more frequently used with tensioners than with torque wrenches, for some reason, with the tensioners getting credit for the resulting accuracy. That accuracy, however, often comes from the fact that stretch is used to control the tightening process, or that ganged tensioners reduce elastic interactions, rather than from any inherent accuracy of the tensioner itself.

IV.   OTHER TYPES OF TENSIONER

Figure 10.13 shows another type of tensioner. This one not only creates the tension, it remains on the stud to retain it. It's sometimes called an hydraulic nut.

As shown in the illustration, the nut (tensioner) is first run down onto the stud (Fig. 10.13A). Hydraulic pressure now extends the body of the nut, tensioning the stud (Fig. 10.13B). A threaded collar on the lower end of the tensioning nut is now run up under the body to retain the tension (Fig. 10.13C). The hydraulic pressure is now turned off.

As with the tensioners we examined earlier, there will be some loss of tension when control is transferred from the hydraulic piston to the threaded collar. But the same amount of joint and stud is involved during and after tensioning; so elastic recovery should be negligible. Elastic interactions could still occur, but these nuts are generally used on all the studs in a joint, and they can all be energized simultaneously. This should virtually eliminate elastic interactions. These nuts, therefore, should be able to provide higher and more uniform residual preloads than the more conventional tensioners discussed earlier, unless those too are used simultaneously on every bolt in the joint.

Figure 10.14 shows an entirely different approach. This time the tensioner stretches the bolt by pushing a metal rod down into a hole drilled along the axis of the bolt. The bolt's own nut, on the far end, is now run up hard against the joint to retain the stretch put in the bolt by the tensioner.

The tensioner is now removed and a sealing cap is placed over the end of the bolt to protect it.

One advantage of this bolt is that it can be stretched by the tensioner before being inserted into the hole in the joint. Thanks to the Poisson reduction, stretching it will also reduce its diameter. If, when stretched, it is a running fit in its hole, when the tensioner is removed, it will be a press fit in the same hole as it expands partially in response to some loss in stretch. Some additional stretch and reduction in cross section must, of course, be retained if you wish to retain some tension in the stud. If it is to be used merely as a shear pin, then an interference fit is sufficient.

V. BOLT HEATERS

A bolt heater accomplishes the same thing as a bolt tensioner, although the equipment and procedures used are vastly different. A heating rod, such as that shown in Fig. 10.15, is inserted in a hole drilled down the axis of the bolt. The bolt expands (increases in both length and diameter) as it heats up. After it has increased in length by a desired amount, the nut is placed on the bolt and is run down hard against the top surface of the joint. The nut is supposed to retain the change in length of the stud. Preload builds up in the fastener as the bolt cools.

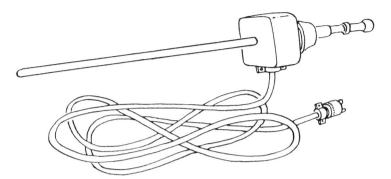

Figure 10.15 The bolt heater is inserted in a hole through the center of the bolt. It raises the temperature of the bolt faster than the temperature of surrounding parts, and so creates a differential change in length.

This is obviously not as well controlled a process as hydraulic tensioning. But people who use the technique report that a skilled crew can usually achieve the desired amount of residual tension in 60% of the studs in a joint with a first pass. Heater rods are then put back in the remaining studs in a second attempt to get the desired tension in them. By the third round, we're told, all studs have been brought to an acceptable tension.

The accuracy of the heater technique depends, of course, on the amount of stretch introduced into the fastener by the heater in the first place; then, on the way in which the nut is run down; then, on the way in which the nut creates and retains tension as the bolt cools. Embedment, elastic interactions, etc., will all still occur unless all bolts are tightened simultaneously and by the same amount, which probably never happens.

One big advantage of the bolt heater is that it is very inexpensive. The larger the stud (in diameter), the greater the cost advantage of the heater over the wrench or tensioner large enough to do the job. A disadvantage is that this is a relatively slow procedure, requiring a fair amount of skill on the part of the operators. Another possible disadvantage is the fact that decarbonization of thread surfaces can sometimes occur when the stud is heated. As we'll see in Chaps. 17 and 18, decarbonization can increase the chances of fatigue failure or stress corrosion cracking.

VI. PROBLEMS REDUCED BY DIRECT TENSION CONTROL

Refer to Fig. 6.28 (p. 209), the block diagram showing factors which affect the in-service clamping force in a bolted joint if we use torque control during assembly. Which of these variables are eliminated or made less troublesome by direct tension control?

The answer depends to some extent, of course, on the type of direct tension control you're talking about. Let's take them in the order in which we have discussed them.

A. Direct Tension Indicators

The direct tension-indicating fasteners, shown in Figs. 10.1 through 10.3, would eliminate most of the problems diagrammed in row 3: such things as bolt twist, heat loss in the threads, and reaction to prevailing torque. They wouldn't be able to compensate for severe bending of the bolt, but this is rarely a real problem.

These indicating fasteners cannot, however, distinguish between tension in the bolt and clamping force on the joint interface, so they do not eliminate problems which would be caused by axial reaction force from the near side of the joint, by resistance of the joint members to being pulled together, etc. Recently, for example, a structural steel erector reported that the DTIs had flattened before the (misaligned) joint members had been pulled together. It was suggested that he continue tightening until the joint was snugged, then remove the bolts one by one, and replace and retighten them with fresh bolts and fresh DTIs.

The bolt with the wavy flange would show whether or not elastic interactions, embedment relaxation, vibration loosening, etc., had occurred because the deformation of its flange is elastic. The DTI washer would not reveal these problems. It could tell us that the bolt had been tightened correctly at one time, but would not show whether or not the bolt had retained that tension. In structural steel situations this is not a significant problem—but it could be a problem in nonstructural applications.

B. Alternate Design Bolts

The fastener shown in Fig. 10.4 is really, of course, a torque control device, not a tension control device. It would, therefore, be subject to all of the uncertainties shown in Fig. 6.28.

The swaged collar lockbolt of Fig. 10.5, however, would get around all of the problems of row 3 of Fig. 6.28. Like the direct tension-indicating fasteners, however, it will be blind to such things as resistance of joint members to being pulled together, elastic interactions, etc.

C. Hydraulic Tensioners

If used alone, without auxiliary control means, hydraulic tensioners again will only get around the problems shown in row 3 of the diagram. If, as is common, they're used in conjunction with a stretch control technique of some sort, they will also be able to compensate for such things as elastic interactions and/or subsequent vibration loosening. With or without stretch control they will be unable to detect the difference between bolt tension and clamping force on the joint interface, however. They will be blind to axial reaction forces from the joint or resistance of the joint members to being pulled together. Note that tensioners used without stretch control cannot detect elastic interactions, but they can eliminate them (if all studs in the joint are tensioned simultaneously) or reduce them (if several studs are tightened simultaneously).

D. Bolt Heaters

Bolt heaters reduce most of the problems reduced by hydraulic tensioners—the factors shown on row 3 of Fig. 6.28. They are also as blind as tensioners to the differences between bolt tension and clamping force in the joint interface. Finally, they can probably reduce elastic interactions if all bolts in the joint are heated simultaneously. Since heaters cannot be coordinated as accurately as tensioners, however, they probably won't fully eliminate these interactions.

VII. GETTING THE MOST OUT OF DIRECT PRELOAD CONTROL

As with the other control techniques we have discussed, there are certain universally useful things which you should do to get the most out of direct preload control: things like keeping accurate records of your tools, procedures, results; training and supervising the operators; making sure that the joint members are properly aligned and pulled together before final tightening; tightening the joint in several passes rather than in a single pass; working from the center or most rigid part of the joint outward toward the free edges; etc. See the end of Chap. 6 for further details.

There are also some special things you can do, depending on the type of direct tension control you are using.

A. Alternate Design or Direct Tension-Indicating Bolts

Each of the fasteners shown in Figs. 10.1 through 10.5 require different procedures and precautions. The tools used in some cases are very special and should, of course, be those provided by the fastener manufacturer. The correct feeler gage should be used with a DTI washer. In general, you should obtain and follow the special instructions provided by the manufacturers of these products.

B. Bolt Tensioners

1. Be sure that the nut turns freely on the stud during rundown, *while* the stud is under tension. Remember that only the male threads are stretched by the tensioner. If close-fitting male and female threads are being used, stretching the male threads may actually create interference which can affect the amount of torque required to run the nut down. It's best, therefore, to use coarse threads and to avoid a class 3 fit.

2. Run the nuts down with as uniform a torque as possible. Measure this torque if you can.

3. Tighten as many fasteners as you can afford simultaneously. The ideal thing would be to tighten them all at once, but this is often impossible. Nevertheless, the more the merrier, because this reduces elastic interaction effects.

4. Verify that the specified hydraulic pressure is applied to each tensioner. Make sure, for example, that all hydraulic connections have been properly made. Sometimes a press-fit hydraulic connection can appear to be completed, but not be, so that a tensioner could get no pressure at all.

5. Use data from the tensioner manufacturer and/or from Table 10.1 to determine how much overtension you should introduce to compensate for elastic recovery.

6. Be sure that the base of the tensioner sits squarely on the joint surfaces, so that the tensioner will pull directly along the axis of the stud. A distorted base can cause interference with the nut, preventing smooth nut rundown.

7. For the same reason, be sure that studs are perpendicular to the joint surface. Tensioning will bend studs if this is not the case, again binding the nut during rundown and probably reducing the amount of residual tension created in the studs. Shimming can be used to compensate for nonperpendicularity.

8. If you are not tightening all studs simultaneously, make a final pass at the final tensioner pressure to compensate for elastic interactions. Apply the final rundown torque once more to each nut while its stud is under tension. If you get a lot of nut motion during this pass, it is probably wise to have still another pass under the same conditions.

9. Use a good thread lubricant on the fasteners to increase the preload created by a given torque.
10. Use thick, large-diameter washers at both ends of the bolt to increase the stiffness of the joint.
11. Elastic interactions, elastic recovery, and other relaxation effects will be reduced if the bolts are less stiff. You can make them less stiff by turning down the bodies, by gun-drilling them, or by using longer bolts of the same diameter (placing collars or Belleville springs under the nuts).

C. Bolt Heaters

1. Use as many heaters as possible simultaneously to minimize elastic interactions.
2. Go for the final stretch (preload) in a single pass to minimize the amount of time the heat is applied (and therefore minimize the possibilities of decarbonizing the studs).
3. Use gage rods, or ultrasonics, or some secondary control means to measure the residual preloads after the bolts have cooled.
4. Reheat and retighten those which aren't right after the first pass, second pass, etc.
5. Run the nuts down with a uniform torque, preferably measured.
6. Use less stiff fasteners and/or heavy washers, as with tensioners.

REFERENCES

1. Donald, Eric, Fatigue-indicating fasteners ("bleeding bolts"): a new dimension in quality control, *Fastener Technol.*, March 4, 1979.
2. Osgood, Carl, How elasticity influences bolted joint design, *Machine Design*, March 1978.
3. Struik, John H. A., and John W. Fisher, Bolt tension control with a direct tension indicator, *AISC Eng. J.*, vol. 10, no. 1, pp. 1–5, first quarter 1973.
4. Brenner, H. S., Development of technology for installation of mechanical fasteners, Report prepared under Contract no. NAS 3-20779 by Almay Research and Testing Corp., Los Angeles, for the Research and Technology Division, Manufacturing Engineering Lab, NAS 8-20779 CR-103179, 1971.
5. Investigation of threaded fastener structural integrity, Report prepared by Southwest Research Institute, San Antonio, Texas, under Contract no. NAS9-15140, DRL T1190, CLIN 3, DRD MA 129 TA, SWRI Project no. 15-4665, October 1977.
6. Rodkey, Edwin, Making fastened joints reliable . . . ways to keep 'em tight, *Assembly Eng.*, pp. 24–27, March 1977.

7. Bolt shows tightness without instruments, *Machine Design*, p. 62, April 7, 1988.
8. Mechanical Technology, Inc., Latham, NY.
9. Williams, James H., Jr., Stress analysis by acoustic emission, Paper 12-77-78, Industrial Liaison Program of the Massachusetts Institute of Technology, Cambridge, MA.
10. Malmstedt, W. D., Tensioning of fasteners (studs and bolts), American Insurance Association, New York, 1981.
11. Torqueing and tensioning: ignorance still a cause for concern, *Nuclear Eng. Int.*, UK, pp. 45–47, November 1987.

11
Ultrasonic Measurement of Bolt Stretch or Tension

In the last several chapters we have discussed a variety of ways in which we can control the tools used to tighten bolts. In each case our goal has been to control the amount of tension produced in the bolts during assembly—or, even more important, to control the amount of clamping force created between joint members during assembly.

Every method we have considered—torque, torque-turn, stretch, and tension control—has had drawbacks and limitations. But each of these methods is good enough for many applications, thanks to the fact that most bolted joints are overdesigned and/or to the fact that we are usually not too concerned about the consequences of an occasional failure. In more and more applications, however, we *are* becoming concerned and would like to find a better way to control bolt tension and/or clamping force. Fortunately, a better way to control bolt tension, at least, is emerging at the present time; namely, the use of ultrasonic techniques. These techniques allow us to get past dozens of the variables which affect the results we achieve with torque and/or turn control. They allow us to see and compensate for the elastic interactions and other factors which limit the accuracy with which we can tighten bolts with hydraulic tensioners. In short, they give us a new, more accurate, and often more convenient way to get the advantages of stretch control or strain gage control.

I. BASIC CONCEPTS

A. Principle of Operation

The basic concepts behind ultrasonic control of bolt stretch or tension are relatively simple. Although, as we'll see in a moment, several different types of ultrasonic instruments have been proposed or built, the most common systems available today are what ultrasonic engineers would call "pulse-echo" or "transit time" instruments. A small acoustic transducer of some sort is placed against one end of the bolt being controlled (see Fig. 11.1). An electronic instrument delivers a voltage pulse to the transducer, which emits a very brief burst of ultrasound (typically two to three cycles). This burst passes down through the bolt, echoes off the far end, and returns to the transducer. The electronic instrument measures very precisely the amount of time required for this burst of sound to make its round trip in the bolt.

As the bolt is tightened, the amount of time required for the ultrasound to make its round trip increases for two reasons:

1. The bolt stretches as it is tightened, so the path length increases.
2. The average velocity of sound within the bolt decreases because the average stress level has increased. Both of these changes are linear functions of the preload in the fastener, so the total change in transit time is also a linear function of preload, as shown in Fig. 11.2.

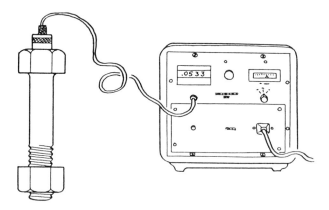

Figure 11.1 An acoustic transducer is held against, or mounted on, one end (either will do) of the fastener to measure the fastener's change in length as it is tightened.

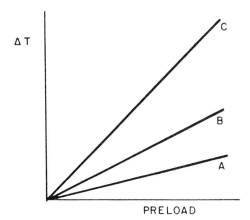

Figure 11.2 A transit time ultrasonic instrument responds to a change in bolt length (curve A) and a change in the velocity of sound through the bolt (curve B). Both changes increase the transit time linearly, as stress (or preload) in the bolt increases. Curve C is the total change in transit time as the fastener is loaded.

The instrument has been designed to measure the change in transit time which occurs during tightening and to interpret and report the results as either a change in length of the fastener, a change in stress level within the threaded region of the fastener, or a change in the tension within the fastener.

The preceding description of "how it works" is obviously simplified. We'll take a far more detailed look when we discuss currently available instruments and the accuracy with which we can measure bolt stress or strain at the present state of the art. In fact, we'll take a much closer look at how ultrasonic instruments work than we did at torque or tensioning equipment because ultrasonic techniques are only "emerging" at the present time, and relatively little information on how they work, and on the factors which affect their accuracy, is available. A detailed knowledge of their operating principles and behavior is often required for best results.

First, however, I want to give you a quick overview on how the instruments are used, and on what kinds of applications.

B. How It's Used

To make this text flow more smoothly, I'm going to call our transit time ultrasonic instrument a bolt gage, a term used by Raymond Engineering Inc. to describe its simpler ultrasonic products. Using such an instrument is simple. A drop of coupling fluid (glycerine, water, oil, grease, resins, etc., can be used) is placed on one end of the fastener to reduce the acoustic impedance between transducer and bolt. The transducer is placed on the puddle of fluid and is held against the bolt, mechanically or magnetically. The instrument is then zeroed for this particular bolt, because each one will have a slightly different acoustic length (even if their physical lengths are the same). If you wish to measure residual preload, or relaxation, or external loads at some later date, you record the length of the fastener (at zero load) at this time. Next, the bolt is tightened. If the transducer can remain in place during tightening, it will show you the buildup of stretch or tension in the bolt. If it must be removed, it is placed on the bolt again, after tightening, to show you the results achieved by the torque, turn, or tension used.

If, at some future time, you wish to measure the present tension in the bolt, you enter the original length of that bolt in the computer in the bolt gage, then place the transducer back on the bolt. The instrument will show you the difference in length or stress between present and zero stress conditions.

It's also possible to use a bolt gage to measure the residual tension in bolts whose preload was controlled by something other than ultrasonics. Let's assume, for example, that the bolts in a bridge were tightened years ago, with turn-of-nut being used to control preload, and the highway department now wants to know how much tension is left in the bolts. This was, in fact, a question which was asked recently—and which was answered with the help of a bolt gage.

In this case the bolt gage is used in a normal fashion, but the sequence of measurements is reversed. A transducer is placed on a previously tightened bolt and its acoustic length is measured. The bolt is now loosened, completely, and its acoustic length remeasured. The loss in length will be proportional to the tension which existed in the bolt before it was loosened. This can be determined by the methods to be described in this chapter, the most accurate of which would involve a calibration test using a sample of the actual bolts.

If you wish to measure the residual tension in several bolts in the same joint you must reinstall and retighten the first one, to its original tension, before measuring and loosening a second bolt. Elastic interactions between the bolts will change the residual preloads in others in the group when you remove the first, as we'll see in a case history near the end of this chapter (see Fig. 11.13, for example) so we must replace each bolt before removing the next.

The instrument whose use has just been described is relatively simple. As we'll see later, more sophisticated equipment is also available today—equipment which automatically records initial lengths, for example, so that these do not have to be reentered manually by the operator. The basic steps taken by the instrument, however, are still those described above.

C. Where It's Used

Measuring or controlling bolt stress or strain ultrasonically is not common at the present time. The equipment and techniques are relatively new, and operator training is required. In some applications an oscilloscope must be used in conjunction with the ultrasonics (at least with the simpler instruments). In any event, operator involvement is required, so the equipment cannot be used "automatically" on a production line.

As a result, ultrasonic equipment is used primarily in applications involving relatively few bolts in critically important joints, or is used for quality control audits, or to set up and calibrate other types of assembly equipment (torque, torque-turn, etc.), or to conduct laboratory or field experiments.

Currently available ultrasonic equipment is ideal for the QC and laboratory roles because it can monitor bolt tension not only as it is being developed in a bolt during assembly, but at any time afterward. It can measure residual preloads after embedment or elastic interactions, it can observe changes in bolt tension as external loads are applied, it can monitor the effects of temperature change, etc. It's a low-cost alternative to strain-gaged bolts in all sorts of studies—"low cost" because such bolts can cost several hundred dollars each.

Currently available instruments can be used on virtually any kind of bolt material, including steel, aluminum, titanium, Inconel, MP35N, and other exotic materials. There are, however, some practical limitations at the present time.

The instruments cannot be used on bolts whose surface temperatures exceed 400°F, because higher-temperature transducers are not available. Both ends of the fastener must have relatively large, flat surfaces on which to mount the transducer and/or against which to generate an echo. Socket head screws can be a problem, therefore, unless they are large enough to mount the transducer down within the socket. The bottom of the socket hole must be machined flat, furthermore.

Although large, flat bolt surfaces are highly desirable, recent developments now make it possible to read some bolts which have dished ends, raised grade markings, and the like. Machining or grinding bolt ends is still common, but no longer universally necessary.

The instrument can be used in conjunction with any type of assembly tool, including torque wrenches, slugging wrenches, impact guns, hydraulic tensioners, right-angle nut runners, etc.

Typical applications at present include such things as the following:

Measuring the residual tension in pressure vessel, reactor, heat exchanger, and other bolts in petrochemical plants.
Controlling the tightening of nuclear steam generator manway studs.
Calibrating and then monitoring the performance of multispindle, torque-turn air tools used to assemble automotive engines (e.g., head bolts, main bearing bolts, connecting rods).
Controlling the tightening of bolted joints in the engines of the Space Shuttle.
Monitoring the tension in the 40-ft-long tie rods in large die-casting machines.
Monitoring the tension in the studs that support the vertical cables on suspension bridges.
Conducting field studies to determine the residual tension in structural steel bolts tightened by calibrated torque or turn-of-nut procedures.
Controlling the tightening of 6-ft-long studs used to brace the legs of off-shore oil platforms. This work was done at depths of up to 200 ft, in the North Sea, in February.
Conducting routine qualification tests on fasteners purchased for automotive production applications (eliminating the need to strain-gage a sample of the fasteners).

The discussion above describes the present situation.

Note, however, that the use of ultrasonics to monitor or control bolt stretch or tension is an emerging technology. New capabilities are being developed at an encouraging pace. New applications are being found all the time. Check with equipment suppliers or other users before concluding that ultrasonics is or is not "practical" for your own applications.

This same "warning" is pertinent for our next topic—currently available equipment. Here again, designs and concepts are still evolving. Even if newer equipment has become available, however, knowledge of earlier products can help you understand the basic techniques, and problems, of this technology.

II. THE INSTRUMENTS

A. Early Transit Time Instruments

Ultrasonic Thickness Gage

In the beginning we had only conventional, ultrasonic "thickness gages" to work with. Even though they're no longer used for these applications, it's instructive to see why these instruments are less than perfect when used with bolts. These instruments operate on either the transit time or resonant frequency principle (to be discussed later) and they can be used on bolts. They are *not* designed, however, to make measurements as the stress level in the body increases, so they respond to any change in transit time as if it were a change in dimension—even if part of the change is caused by the fact that the acoustic velocity of the ultrasound has decreased.

As a result, a thickness gage will give you a change-in-length readout that is about three to four times the actual change. Unfortunately, the exact ratio between readout and actual stretch depends on a number of factors—including alloy content of the bolt, heat treat, grip-to-diameter ratio, and stress concentrations—so you can't safely use these instruments on bolts without first determining the proper ratio for your application.

You can determine the ratio experimentally, however. You must measure the actual stretch by some other means and compare it to the instrument reading to get the correction factor. It's best to test several bolts rather than one, and use statistics to define the average behavior. You now have a correction factor which is used as follows.

You measure the length of the first bolt at zero stress. You note this length. You tighten the bolt and measure its new length. You subtract the first reading from the second and divide the difference by your correction ratio. The result will be the actual change in length. If you've chosen your ratio properly, the results are quite accurate (nearest 0.001 in.).

Since thickness gages are not designed for bolts, you must watch out for things such as limitations on the range of lengths they'll measure and temperature response (many are designed only for indoor use). Some of them get confused between echoes from the far end of the bolt, which we're interested in, and echoes from nearer threads, because they're designed to work on castings or pressure vessels having more breadth than thickness, rather than on long, thin bodies like bolts.

Within their design limits, however, some of these instruments can be used successfully on bolts.

Computerized Thickness Gage

The arithmetic required when you use a thickness gage on bolts is obviously undesirable, and a potential source of errors. One response is to connect the instrument to a computer, and let it do the calculations. A system of this sort has been proposed by General Dynamics [1].

The General Dynamics system is a very sophisticated one, intended for mass production applications. A group of high-speed, air-driven tools will be controlled by a dedicated computer that is fed ultrasonic information. The computer is fed a mathematical program involving lengthy equations which predict the exact relationship between the change in transit time of an ultrasonic wave in a fastener and the average stress level within that fastener. By necessity, these equations involve such things as the second- and third-order elastic constants of the material. Here, for example, is one of the basic equations [1]:

$$\frac{\Delta T}{S \times \text{LE}} = \frac{2}{v_0} \left[\frac{(\mu + \lambda)(10\mu + 4\lambda + 4m)}{6\mu K(2\mu + \lambda)} + \frac{(\lambda + 2l)}{6K(2\mu + \lambda)} + \frac{1}{E} \right]^{-1}$$

(11.1)

where ΔT = change in transit time of the ultrasonic signal as the fasteners is tightened (sec)

S = average stress within the fastener (psi)

E = Young's modulus (psi)

v_0 = velocity of the ultrasonic wave in an unstressed bolt (cm/sec)

LE = effective length of bolt (in., mm)—equals grip plus 0.6 nominal diameter (see Ref. 1)

K = compressibility in dynes/cm^2

μ, λ, l, m = the second- and third-order elastic constants

A dedicated DEC PDP-11 computer was used for the prototype system, but the designers suggested that a microprocessor would be sufficient for the production version. A calibration stand is included in the equipment. It has an interface to the computer, so the user is only required to place a fastener in the calibrator and activate the system to provide the computer with all the information it needs about a given fastener to convert the (incorrect) thickness gage estimates of the bolt stretch into correct estimates. Readout is in terms of average bolt stress rather than in stretch.

Note that the ultrasonic measurements are combined with the torque-turn measurements of an air-tool control system. The torque-turn information is used to spot the gross problems—such as incorrect or soft bolts—that stretch measurement alone cannot detect.

The U.S. Air Force sponsored the development of this system, so the resulting design information is available to any interested company. To date no hardware has been placed on the market.

One interesting feature of the proposed system is that transit time is measured between the first and second echoes of the ultrasonic burst rather than between the main bang and the first echo. This eliminated errors caused by variations in the thickness of the coupling film between bolt and transducer. It also means the system must work with much weaker echo signals, which could be a problem.

It's very instructive to read the progress reports on this development (see the references). The designers chose to follow an "intelligent system" route, in which the controlling computer knows, ahead of time, exactly how a fastener behaves under stress—the physics involved, if you will. After receiving calibration information about the idiosyncrasies of a particular batch of fasteners (to correct for the exact stress pattern of this fastener), the computer uses its prior knowledge to convert the incorrect information received from a conventional ultrasonic thickness gage into an accurate estimate of the average stress in the threaded region of the fastener. In learning how this system works, you'll learn how truly complex the stress-strain relationship is in a bolt, and why it's essential to calibrate any ultrasonic system on your own bolts, before depending on it for accurate control.

Ultrasonic Stress Analyzer

The Harrisonics Division of the Stavely Corporation markets an oscilloscope that can be used to measure bolt stress as well as surface and bulk stress in other types of body. It can determine the principal axes of stress as well. It is basically a transit time instrument. It measures the transit time between the first and tenth echo pulses, for improved accuracy, and is primarily a laboratory instrument [2, 11]. It's shown in Fig. 11.3.

The Erdman-McFaul Extensometer

The Erdman-McFaul Extensometer was the first ultrasonic bolt-control instrument to reach the marketplace. It was developed jointly by Howard McFaul of Douglas Aircraft Corporation and Donald Erdman of Erdman Instruments, Inc., of Pasadena, California, and was designed as a nondestructive testing or quality control instrument rather than as a mass production control device. It was a transit time instrument. The primary readout was in terms of change of length of the fastener, so it was called an "extensometer." One version of this instrument is shown in Fig. 11.3C.

The designers of this instrument took a simple approach, which can be loosely explained as follows. The instrument generated a precision voltage ramp—voltage as a function of time. The slope of this ramp was adjusted, when the instrument was calibrated for a given bolt material, to coincide with the slope of the line representing the change in length of the fastener as a function of transit time. See Fig. 11.4.

A measurement cycle began when the instrument generated a "main bang"—an abrupt voltage pulse that "rang" an ultrasonic transducer that had previously been placed against either end of the bolt. The transducer

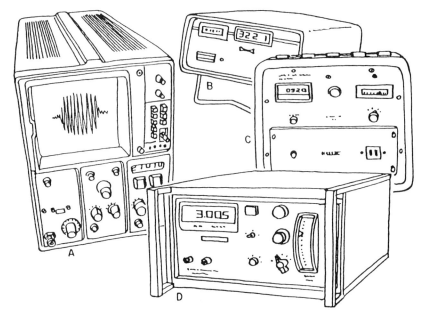

Figure 11.3 A few of the early ultrasonic instruments used to measure bolt stress or strain. Those shown were (or are) available from Harrisonics (A), Torque Systems International (B), and Raymond Engineering (C, D). See Appendix D for manufacturers' addresses.

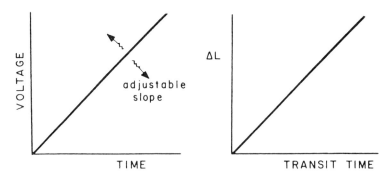

Figure 11.4 In the Erdman-McFaul extensometer, a voltage ramp was adjusted to imitate the transit time-vs.-ΔL curve for the fastener being controlled. The instrument thus accommodated both velocity and dimensional changes in the bolt.

would now oscillate for a few cycles at a frequency that was typically 5 MHz but which could be modified for a particular application.

The burst of ultrasound generated by the ringing of the transducer traveled down through the bolt to the opposite end. It then reflected off the end and returned through the bolt to the transducer.

The voltage ramp, meanwhile, had also been started by the main bang. The returning echo stopped the growth of the ramp. The final ramp voltage was passed through suitable decoders and was displayed by the instrument in terms of a change in length of the fastener. The instrument displayed the change in length of the fastener to the nearest 0.00001 in. [3–5].

The above is a gross simplification of the way the circuits within the instrument actually functioned, but I think it's a reasonable explanation of the general philosophy behind the design of the instrument. It required some considerable sophistication to measure time and to generate an adjustable voltage ramp with the accuracy required for five-digit ΔL readout. But the basic approach was a simple one. Unlike the General Dynamics system, the instrument didn't know anything about the physics of a fastener; it just knew how to give you the "correct answer."

B.   Current Transit Time Instruments

Change-of-Length Measurement

Figure 11.5 shows some of the most recently introduced transit time instruments. These are digital electronic systems designed to measure the transit time of ultrasonic signals with great precision, and then to use those measurements to compute the change in length of the bolt, the stress in the threaded region, and/or the tension in the bolt. To do this the instrument must also be fed certain other information concerning the geometry or stiffness of the bolt, and the properties of the material from which it is made. This information is entered manually, by keyboard, or the instrument is fed a data storage card on which the data have been recorded. The instrument can also be up- or down-loaded from a PC or other computer.

The basic principle of operation is that described earlier: The electronics generates brief bursts of ultrasound, which travel through the fastener, echo off the far end, and return to the sending/receiving transducer (which can be mounted on either end of the bolt). Transit time is measured either from main bang to first echo, or between the first and some subsequent echo (e.g., the fifth). The main bang technique involves larger signals, which can make life easier for the operator, but bolt ends must often be premachined to eliminate dishing, nonparallel surfaces, raised grade markings, and the like. This is still the most common way to measure

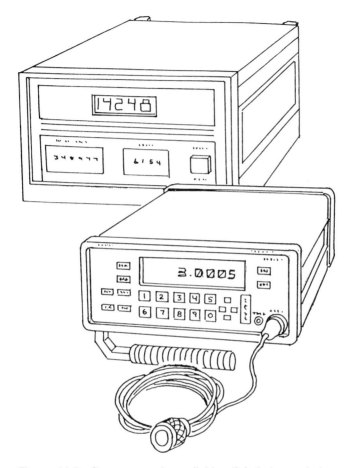

Figure 11.5 Some currently available, digital ultrasonic instruments from Raymond Engineering. The nearer one includes a built-in oscilloscope for monitoring pulse and echo signals, and for displaying digital information. The farther one is designed specifically for use with high-speed-production air tools.

larger fasteners, however (those over $\frac{1}{2}$ in. in diameter and over $\frac{3}{4}$ in. in length).

Measuring transit time from first to a later echo requires the instrument to detect and respond to weaker signals, but allows the equipment to be used on as-received and/or very short fasteners (including those too short to strain-gage). It cannot be done on very long bolts (several feet in length).

In any case, a preliminary measurement of transit time is made before the bolt is tightened. The instrument interprets this as an "initial length" for that bolt. This information is displayed by the instrument, and is also stored in its memory. (The instruments can be programmed for any tightening sequence or strategy of interest to the user.)

The bolt is now tightened, in some cases while the transducer is still mounted on one end (either the far end or inside the drive socket). In other cases the transducer is removed, the bolt is tightened, and then the transducer is replaced. The transit time in the tightened bolt is, typically, 0.3% greater than it was in the untightened bolt.

A microprocessor in the instrument, or a computer connected to it, now compares the transit time in the unloaded bolt to the final transit time. It then uses the following algorithm to compute the change in length of the bolt which occurred as the bolt was tightened:

$$\Delta L_S = \left[\left(\text{MV}\,\frac{T_2}{2} - \Delta t\left(\text{MV}\,\frac{T_2}{2}\right)\text{TF} - \left(\text{MV}\,\frac{T_1}{2}\right)\right]\text{SF} \qquad (11.2)$$

where ΔL_S = the change in length of the bolt which has been created by the tension in the bolt (i.e., ignoring temperature-induced change) (in., mm)

MV = the velocity of sound in the unloaded bolt (in./sec, mm/sec)

T_2 = transit time in the tensioned bolt (sec)

T_1 = transit time in the unloaded bolt (secs)

Δt = the change in temperature of the bolt, if any, between transit time measurements (°F, °C)

TF = a material "temperature factor" (per °F, per °C)

SF = the "stress factor"

The equation involves three calibration constants which must be determined beforehand and fed to the instrument, as previously described. These are MV, SF, and TF. Let's take a closer look at them.

The "material velocity" (MV) is the velocity of sound in the unstressed bolt, at room temperature. It is typically about 250,000 in./sec. This velocity is used to calculate the new length of the bolt when under tension, even though the actual velocity at that point will have decreased by something like 0.2%. This would introduce a slight error if it weren't for the fact that MV is always multiplied by the important "stress factor" (SF), to be discussed soon. SF becomes a catchall calibration constant which automatically takes any change in actual sonic velocity into account.

The "temperature factor" (TF) is a "per °F" (or per °C) multiplier which introduces the two effects which a change in temperature will have on the transit time of the ultrasonic signals. One change involves the linear coefficient of expansion; if the temperature increases, the bolt will get longer. We're only interested in changes in length which have been accompanied or caused by a change in tension within the bolt, so we want to factor out changes caused solely by temperature.

An increase in temperature will also cause a decrease in the velocity of sound within the bolt, as suggested in Fig. 11.6. Like thermal expansion, this phenomenon will cause an increase in transit time which is unrelated to a change in tension in the bolt, so it, too, must be factored out. Both these changes are essentially linear functions of the change in temperature, and so can be combined into one constant which we determine during a calibration exercise (to be described later).

We can, and do, usually use a "typical velocity" for MV; and we will need a temperature factor only if a significant time elapses between measurement of initial and final transit times; so a "typical" or "theoretical" TF is often acceptable. But there's no avoiding the final calibration constant, the "stress factor" (SF).

The stress factor is the ratio between the change in transit time created by the increase in length of the bolt and the total change in transit time

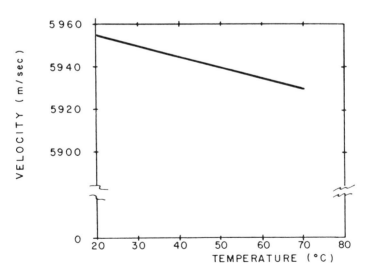

Figure 11.6 Velocity of ultrasound in alloy steel as a function of bolt temperature. (Modified from Ref. 9.)

created by the change in stress and the change in length. As illustrated in Fig. 11.2, therefore, it could be expressed as the slope of line A divided by the slope of line C.

As already mentioned, the transit time in a typical bolt increases by something like 0.3% when the bolt is tightened; and about 0.2% of the change has been contributed by a reduction in ultrasonic velocity which, in turn, was caused by the increase in stress within the bolt. This means that the tension-induced increase in length of the bolt has contributed 0.1%, and the stress factor for this bolt is

$$\text{SF} = \frac{\Delta T_\epsilon}{\Delta T_T} = \frac{0.1}{0.3} = 0.33 \tag{11.3}$$

where   SF = stress factor
ΔT_ϵ = change in transit time created by the strain (i.e., change in length) of the bolt
ΔT_T = total change in transit time (caused by both the velocity and length changes)

The terms ΔT_ϵ and ΔT_T can be expressed as absolute changes in transit time, in seconds, or, as I've done above, as percentages of the original transit time through the unloaded bolt.

Stress factors usually range between 0.25 and 0.35 for most common bolts.

Theoretically the stress factor is a material constant, whose value depends on three material properties as follows:

$$\text{SF} = \frac{\text{MV}}{\text{MV} - K_T E} \tag{11.4}$$

where MV = the velocity of sound in the unstressed bolt (in./sec, mm/sec)
E = the modulus of elasticity of the material (psi, GPa)
K_T = the (negative) slope of the velocity-stress curve expressed as a decimal (in./sec per in./sec per psi, mm/sec per mm/sec per kPa)

The velocity-stress curve is shown in Fig. 11.7. A typical value for the slope of the curve, for a low-alloy bolting material, might be -8×10^{-3} in./sec per in./sec per psi (-29.5×10^{-3} mm/sec per mm/sec per kPa).

I called Eq. (11.4) the "theoretical" expression for SF. In practice, remember, we select SF during a calibration exercise, and we use it, in part, to compensate for the change in sonic velocity which occurs as the

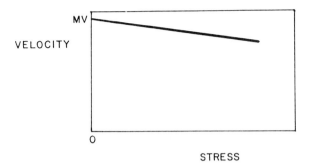

Figure 11.7 Velocity of the ultrasonic wave through a bolt as a function of the stress in the bolt. We are primarily interested in the slope of the curve, which is typically -8×10^{-3} in./sec for a low-alloy bolting material.

bolt is stretched (because there is no independent way for the bolt gage to detect or estimate the velocity shift). As a result, the SF we use in practice could more properly be called a "calibration constant" and, although there's some debate on this, it sometimes appears to be a function of fastener geometry. In any event, and unfortunately, it's clearly affected by residual manufacturing stresses within the fastener, since these affect the no-stress sonic velocity (MV).

We'll look at calibration procedures for determining SF in a later section, and will also take a closer look at the residual stress problem.

Tension or Stress Measurements

We've looked in some detail at the procedure the instruments shown in Fig. 11.5 use to compute the change in length of the bolt as a function of the change in transit time. In most cases, depending on the version of the instrument being used and/or the custom-designed software controlling its operation, the instrument now goes on to estimate the tension in the bolt or the stress in the threaded region. There are several basic ways for it to do this. Let's look at a couple. To simplify things we'll assume, for the moment, that the fastener is fully threaded. We'll look at more complicated situations later.

We can program the instrument to read out a stress (or preload) which approximates the anticipated values. Then, in calibration, we enter typical MV and TF values, apply a known force to a sample bolt, and adjust the SF "calibration constant" until the instrument correctly reports this force. Bolt identification and SF information can now be stored in the instrument for future use. (The appropriate SF will be called up by the instrument when the bolt identification information is reentered.) In effect, the algorithm we're using is simply

$$F_P = F_E \times SF_F \tag{11.5}$$

where F_P = actual tension in bolt (lb, N)
F_E = preestimated tension (lb, N)
SF_F = stress factor used to "correct" the instrument for tension readout

Stress can then be computed by dividing preload by the tensile stress area (A_S) of the fastener (see Chap. 2 for A_S).

There is significant memory capacity in currently available instruments, so a large number of bolts can be programmed this way, with readout in stress or preload.

A second procedure is to base the computation of preload on the previously computed change in length of the fastener, using Hooke's law as follows:

$$\sigma_{AV} = \frac{(\Delta L)E}{L_{IL}} \tag{11.6}$$

where σ_{AV} = average tensile stress in the fastener (psi, N/mm)
ΔL = change in length computed in Eq. (11.2)
L_{IL} = initial, unloaded, effective length of the fastener (in., mm)
E = modulus of elasticity of the material (psi, GPa)

One advantage of this procedure is that no new calibration constants are allowed; MV, TF, and SF are used to compute ΔL and that is then used to compute stress. If we also want preload, we instruct the instrument to multiply stress by the tensile stress area to display tension in pounds (or newtons).

Again, we can program the instrument to store the information required for it to make these computations for many different bolts.

When we get involved with force or stress calibration and readout, however, we'll find that things are a little less certain than they were when we were only dealing with bolt stretch. The relationships between the force on a fastener and its elongation will depend on its elasticity, its behavior as a spring. We studied this relationship in detail in Chap. 5.

The change in transit time in the fastener will depend on the change in length and the change in average stress level. A given change in length will be accompanied by a greater change in stress in a short fastener than in a long one (assuming the same diameter for each). As we see in Eq. (11.6), the "initial, unloaded, *effective*" length of the fastener has entered the picture. In Chap. 5 we learned that effective length is usually taken as the grip length plus half the thickness of the nut and bolt head (or two nuts). This is a reasonable assumption if we're dealing with a fastener having a length-to-diameter ratio of 4:1 or more; but the assumption becomes more important—and has a larger effect on our calculations—as the grip length and L/D ratio are reduced.

Remember that our instrument can only detect the *total* change in transit time in the bolt. Because neither grip nor any other length is involved when it sorts the stress change from the length change, it can accurately estimate stretch. But when we try for preload, we must either tell it, accurately, the effective length of a fully threaded bolt [as in Eq. (11.6)]; or we must calibrate it for the exact effective length it will have to deal with on the job. The procedure described in Eq. (11.5), in other words, works only if SF_F is selected for the specific grip length or effective length to be encountered when the instrument is used.

Since calibrating separately for each grip length can be a real nuisance, the procedure defined by Eq. (11.6) is usually preferred. We use an algorithm which allows the instrument to be calibrated on any convenient grip length, and then can use the instrument on any other grip or effective length. We must, in that case, tell the instrument, before use, what the effective length will be; but at least we've avoided multiple calibrations.

All of this becomes even more complicated if the fastener is not fully threaded. Now we will no longer be interested in the *average* stress within the fastener; we'll want to know the stress in a given region; usually the threaded region. If the fastener has only two cross sections, body and threads, we can have the instrument compute the average stress in the tensile stress cross section (A_S) by multiplying the stress computed in Eq. (11.6) by the expression

$$\frac{L_{IL}}{L_S} \times \frac{L_S/A_S}{L_B/A_S + L_S A_S} \tag{11.7}$$

where L_{IL} = initial unloaded length (in., mm)
L_B = effective length of body (in., mm)
L_S = effective length of threads within grip (in., mm)
A_B = cross-sectional area of body (in.², mm²)
A_S = tensile stress area (in.², mm²)

An instrument using these algorithms can be preprogrammed with all the dimensional and calibration information required for a given job—or can be preprogrammed to request bolt length and diameter (and to obtain A_S, L_B, L_S, etc., from its memory) or preprogrammed to request manual input of all necessary information, depending on the application. In any event, the three original calibration constant, MV, TF, and SF, are universally valid for a given bolt material.

The algorithms described about are typical, but many others can be used. Since most users are more interested in force (preload) readout than in stress, the following is popular; for example:

$$F_P = \Delta L_A \times K_B \times SF$$

where ΔL_A = apparent change in length of the bolt (in., mm)
  K_B = stiffness of bolt (lb/in., N/mm)
  SF = stress factor
  F_P = preload (lb)

Our other two calibration constants (MV and TF) are involved in ΔL_A, of course. (Note that $\Delta L_A \times SF$ equals the true change in length.)

Miscellaneous Features of Current Instruments

In addition to storing calibration information about a variety of bolt materials and/or bolts, presently available instruments can store and print out assembly results, using hard-copy printers, tape, or disk recorders. The instruments can feed information to strip chart recorders for real-time monitoring of air tools. They can feed PCs or larger computers, to accumulate assembly data or keep records for statistical studies, warranty protection, production records, etc. They can prompt operators in assembly procedures, and/or alert them to assembly problems. They can control hydraulic wrenches, hydraulic tensioners, or some air tools. They can also spot and warn the operator of gross cracks in a bolt, although they aren't true "flaw detectors" and can't see the tiny defects a true flaw detector can see. In some applications current instruments can accurately measure the change in length or preload in bolts tightened past yield.

C. Other Types of Instrument

The ROUS System

Dr. Joseph Heyman, at NASA-Langley, has developed the ROUS system. This is not a transit time system; it is a resonating or continuous wave (CW) system. The acronym ROUS stands for reflection oscillator ultrasonic spectrometer. In the original design, one transducer, placed against either end of the bolt, generates a band of ultrasonic frequencies. A second transducer, mounted concentrically with the first (or it could be mounted against the other end of the bolt) looks for that particular frequency which has resonated within the bolt. This will be that frequency which sees the length of the bolt as an exact multiple of its wavelength. Actually, the instrument is designed to measure the change in resonant frequency from the unstressed to the stressed condition of the bolt, just as the instruments we considered earlier were designed to measure a change in transit time rather than transit time itself. In Dr. Heyman's latest design, incidentally, one time-shared transducer serves as both transmitter and receiver.

Although ROUS does not require as substantial a computer as the General Dynamics unit, it is still, basically, an equation-solving system. It works with the following equation [6]:

$$\Delta F_j = F_j \left(\frac{1}{v} \frac{dv}{dS} - \frac{1}{E} \right) \Delta S \qquad (11.8)$$

where F_j = the mechanical resonant frequency in an unstressed bolt (Hz)
ΔF_j = change in resonant frequency as the bolt is stressed (Hz)
v = the velocity of ultrasound within the bolt (in./sec)
S = average stress within the bolt (psi)
ΔS = change in stress as the bolt is tightened (psi)
E = Young's modulus (psi)

The resonant frequency is affected by both the change in stress and the change in length of the fastener, as shown in Fig. 11.8. Both changes are essentially linear functions of the tension in the bolt; and the stress change creates about two-thirds of the total change in transit time. Although the hardware and algorithms are different, therefore, this system has much in common with transit time systems. In fact, one could be called the inverse of the other; the transit time system measures "time per cycle"; the ROUS measures "cycles per time."

It is believed that a resonant system would be about 100 times more accurate than transit time systems. This inherent accuracy may allow simplification and cost reduction of the instrument, but this level of accuracy won't be required in most practical applications.

Note that ROUS interprets the change in resonant frequency as a change in average stress level in the fastener, rather than as stretch or strain.

Thread Strain Analyzer

An even more elaborate system has been built for the Central Electricity Generating Board (U.K.). A motor-driven transducer is mounted on the top of a bolt and is angled to produce echoes from individual threads [8]. Echoes from each thread are received at slightly different times, the time

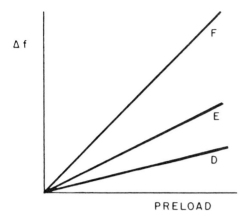

Figure 11.8 Like the transit time instrument, the resonant-frequency ultrasonic instrument responds linearly to changes in bolt length (D) and to the stress levels in the bolt (E). Curve F reflects the total change in resonant frequency as the bolt is loaded.

differences being proportional to thread pitch. As the bolt is loaded, the pitch increases in proportion to local stress levels. This gives the operator a very detailed look at the loads on the threads, as in Fig. 11.9.

The transducer is manipulated by three motors, as shown in Fig. 11.10, so that it can be made to sweep the threads, both parallel to the bolt axis and around it. The space between the transducer and head of the bolt is filled with a liquid couplant such as ethylene glycol.

Direct Ultrasonic Measurement of Stress

There are several different types of ultrasonic waves. The equipment we have discussed so far uses only longitudinal waves, which travel in a direction perpendicular to the surface against which the transducer is placed. By using a different sort of transducer, however, it is also possible to generate what are known as shear waves. And, for reasons too complex to consider here, the difference in velocity between shear and longitudinal waves is directly proportional to the tensile stress level within the body [9]. Although this is called a "direct measurement of stress," we're still looking at measurements, or at least calculations, based on ΔTs.

One of the suggested advantages of this ultrasonic system is that it can be used to estimate stress levels in the plastic region as well as in

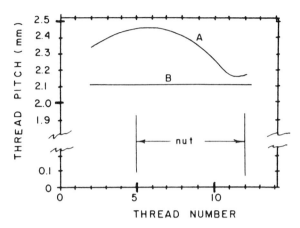

Figure 11.9 Thread pitch measurement on three $\frac{3}{4}$-in.-diameter, BSF mild steel bolts under load (A) and unloaded (B). The increase in pitch under load can be used to estimate the magnitude of the load. The nut is engaged with threads 5 through 12. (Modified from Ref. 8.)

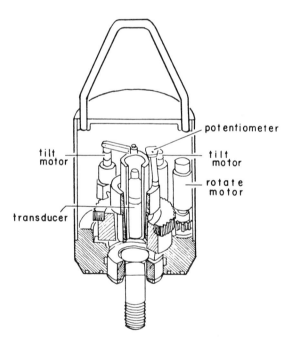

Figure 11.10 Sectional drawing of the transducer and drive motor assembly of the instrument used for in situ measurement of thread strain. (Modified from Ref. 8.)

the elastic region (at least if you know ahead of time the approximate relationship between stress and strain for the body in the plastic region).

This longitudinal-shear (or bircfringent) technique has not been reduced to practice. Some laboratory work has been done, but there are some severe practical problems in building a combination of shear and longitudinal transducer that can be used on a small-diameter bolt. The experimental work also suggests that the technique will not work on a material that exhibits any anisotrophy, indicating that new bolt materials and/or improved control over present materials would be required. Experiments also indicate that the technique would not be usable if one wished to measure preload with an accuracy much better than 10% or 20%. We can do almost that well with torque! Nevertheless, the technique may be perfected some day, and is worth knowing about [10].

III. CALIBRATION OF A TRANSIT TIME INSTRUMENT

There is no way at present to predict the exact relationship between a change in transit time in a fastener and its change in length. Nor, as we saw in Chap. 9, can we predict the exact relationship between preload or tension and the change in length. Unless you're content with control errors approaching those of torque or turn control systems, therefore, you must determine the behavior of your fastener experimentally and use this information to calibrate the ultrasonic instrument.

To "determine its behavior" means that you must measure accurately the change in length of the fastener when it is placed under load. If you are using an instrument which reads out in terms of change of length, you must now adjust it to agree with the actual change in length of the fastener.

With other kinds of bolt gage, of course, it is also necessary to measure the exact amount of load placed on the fastener. You now calibrate the instrument to give you the correct readout in terms of tension or stress.

The crudest instruments available today measure bolt elongation to the nearest 0.001 in., the best ones achieve an accuracy of 0.00001 in. The only way to calibrate an instrument of any sort is to compare it to something that is more accurate than the bolt gage.

Micrometers are often proposed, but, as we saw in Chap. 9, there are many practical problems in making accurate micrometer measurements of a bolt, whether the bolt is in a calibration stand or the actual joint. If you have no other way to measure the "actual" change in length, however, and must use a micrometer, then you should make a large number of micrometer measurements, preferably using a large number of people to do so, and then use linear regression techniques to estimate the true change in length as a function of applied load.

A tensile-testing machine can be used if it is equipped with a mechanical or electronic extensometer. The extensometer on these machines, however, measures the change in length in the gap between jaws on the machine—and this does not necessarily equal the change in length of the fastener under test, especially if there is substantial relaxation in the threads. Relaxation is often encountered with a fresh thread, furthermore, so this can be a source of significant error. The tensile machine, however, is our best bet at the present time for large fasteners.

The clamping force produced by the bolt can also be measured with strain-gaged or piezoelectric load washers or, more crudely, by the Skidmore-Wilhelm equipment shown in Fig. 7.24. The bolt gage can now be adjusted to display the correct readout in terms of tension—or by using the tensile stress area (A_S), the correct stress. Or, if stretch readout is preferred, the change in length of the bolt can be measured, while it remains on the load cell, with electronic micrometers or displacement instruments (an LVDT perhaps) or in an instrument such as a shadowgraph. Bolts calibrated in load cells, incidentally, can be tightened with wrenches; the torsion introduced by the wrench will not affect the results.

Whenever you are calibrating an instrument for force or stress, you must also feed it the geometrical or bolt stiffness information used in that algorithm. When calibrating for stretch no geometrical data are required; merely adjust the calibration constant SF until the instrument gives the correct readout at room temperature. The material velocity (MV) will already have been input, using a "standard" number such as 250,000 in./sec for alloy steel bolts. As explained earlier, the stress factor adjustments compensate for any variations in actual velocity.

If this procedure bothers you, and you'd prefer to use the actual material velocity, the usual practice is to use a fast oscilloscope to display the first and second echoes of the ultrasonic burst. The scope can be used to measure the exact elapsed time between these echoes. The physical length of the bolt is also measured, using the best equipment available (e.g., a shadowgraph). Length divided by *half* the transit time (the scope displays the round trip) gives the average velocity of sound in the unloaded bolt at room temperature. This velocity is affected by the residual manufacturing stresses in the bolts, and so will often vary slightly from bolt to bolt in a batch, and will vary even more between bolts received at different times and/or from different sources.

Because the response of the bolts will be somewhat scattered, and because the equipment we use for calibration is not significantly more accurate than the ultrasonic instrument, it's necessary to use statistical techniques to determine the most accurate calibration factors. Calibrate against several bolts rather than one, whenever possible. Make several tests with each bolt. Calibrate the instrument to mean values rather than to the values obtained from an individual test on a single bolt.

To determine the temperature factor (TF), place an unloaded bolt, and the ultrasonic transducer, in a temperature chamber. Use the bolt gage, previously calibrated at room temperature, to measure the initial length of the bolt. Now raise or lower the temperature of the bolt, letting it "soak" long enough to stabilize at the new temperature. (It's important in all this to measure the temperature of the *bolt* at each step, not the temperature of the environment.)

Once the temperature of the bolt has stabilized, the new temperature is input to the bolt gage (via keyboard, usually) and the length of the bolt is again measured. The length will be found to have changed. A temperature factor (TF) is now entered (again by keyboard) and is adjusted until the instrument displays the same bolt length it did when the bolt was first measured at room temperature. The TF required to correct the length for the temperature change is now recorded.

The temperature of the bolt is now adjusted to a different value and the process described above is repeated. Usually three or four different temperatures are used, including room temperature. The TF values required for the various tests will vary slightly. Their average is used as the final calibration constant.

All of this sounds complicated, and it is. The companies which manufacture the equipment are prepared to perform the calibration tests for you, but this is not always feasible. The complexity of the procedure is determined by two unavoidable facts: The instrument we're calibrating is as accurate as—or more accurate than—the equipment usually available for its calibration, and the response of the bolts on which we must calibrate it varies. There's no way around all this at the present time.

Once a bolt gage has been calibrated on a stretched bolt, it can be used to measure steps in a fixed-gage block. It will see these steps incorrectly, since there is no change in stress level to accompany the change in dimension of the block. If its incorrect interpretation of these steps is recorded, however, the instrument can always be recalibrated—for the *bolt*—by adjusting it to give the same "incorrect" reading on the gage block. This technique is used for recalibration on the production floor or in the field. Calibrations against an actual bolt, therefore, are required only once (or once per batch, or supplier, depending on your accuracy requirements).

IV. ACCURACY OF TRANSIT TIME INSTRUMENTS

A. In General

How accurately can a transit time instrument measure or control the change in length of a bolt or the tension in it? As we consider the answer to this question, we'll see that ultrasonic control or measurement, while often giving us a significant advantage over torque, turn, or some of the other techniques we have already considered, is far from perfect. Many variables still affect the accuracy of the result. Fortunately, none of these variables are subject to as much scatter as, for example, the coefficient of friction between thread surfaces. But they are scattered. One big advantage of the ultrasonics, however, is that we can usually use the calibration exercise to reduce the uncertainties. We can calibrate the instrument using a small sample of our bolts, then use it with reasonable accuracy on the rest of the bolts—at least in the same lot. This sort of thing is not possible with torque control where, by comparison, we'd have to "calibrate" each individual bolt in many cases.

In general, ultrasonic equipment is capable of measuring the change in length of a fastener to the nearest ±1% or so. In some situations it can also measure the tension in a bolt to a similar accuracy. It is not easy or economical, however, to achieve this level of measurement accuracy in most applications, and it's even more difficult to control preloads with this precision. In automotive or aerospace work we will usually settle for preload control of ±5–10% when using ultrasonics. In field applications involving large bolts tightened under difficult circumstances we will often settle for preload (not measurement) accuracies like ±20–30%. This frequently leads to the charge that "ultrasonic control is no more accurate than torque." But that's not true. Ultrasonics, even if used in a "quick and dirty fashion," can give us ±30% of control of *residual* preload or tension in the bolts after such things as embedment relaxation, elastic interactions, thermal effects, and the like. Torque control under ideal conditions may give us ±30% of *initial* preload in individual bolts. But, as we have seen in Chap. 6, the max-to-min scatter in residual tension can be as great as 4:1 or 10:1, or even worse.

To understand why we settle for ±20–30% control when ultrasonics is capable of giving us ±1% measurement accuracy, we must look more closely at how a typical transit time instrument works and at the factors which affect the results. Then we must look at the field or usage problems which usually contribute most of the final variations.

B. Basic Principle of Operation

Four basic steps are always involved when we use an ultrasonic instrument to measure the tension in a bolt.

1. We calibrate the instrument for that particular type of bolt in that particular application, as discussed in Sec. III.
2. The instrument is now used to measure the time required for a burst of ultrasound to make a round trip through the unloaded bolt and the slightly longer time required for a burst of sound to make its round trip in that same bolt after it has been tightened.
3. A microprocessor within the instrument now uses the difference between the unloaded and loaded transit times to estimate the change in length of the fastener as it was tightened. In making this calibration, the instrument also relies on the ''calibration constants'' which we input to the instrument as a result of our calibration exercise.

    Thanks to the calibration exercise there is usually a very good relationship between the measured change in transit time and the computed change in length. The variables involved here are reliable and easy to define. We usually have a good deal of confidence in the change in length displayed by the instrument.
4. The microprocessor or a computer associated with the instrument now uses the change in length of the bolt to estimate the tension in it, or the stresses in the threaded region, whichever we are interested in. To do this it must again use calibration information plus, this time, information about the geometry of the bolt and the joint. More variables are involved here than in the estimation of change in length, and several of these variables are more difficult to pin down than those we used earlier. As a result, the accuracy with which the instrument can compute tension is usually less than the accuracy with which it can compute a change in length.

To understand the accuracy—or the limitations—of ultrasonic measurements we must (1) look at anything which affects the measurement of transit time itself, (2) look at anything which affects the use of a change in transit time to estimate the change in length, and (3) look at anything which affects the use of the estimated change in length to compute tension. Such analysis will give us the ''theoretical'' accuracy of the instrument. In addition, we must recognize that we're living in a real world and so must also look at potential operator and equipment problems, things which could prevent us from realizing the full potential of the instrument.

C. Measurement of Transit Time

Sound travels through an alloy steel bolt at approximately 250,000 in./
sec. If we wish to measure a change in length of that bolt to the nearest
0.0001 in. we'd like to be able to resolve transit time to the equivalent of
the nearest 0.00001 in. From a transit time point of view, we'd like to be
able to resolve

$$2 \times \frac{1 \times 10^{-5}}{250,000} = 0.8 \times 10^{-10} \text{ sec} \qquad (11.9)$$

The "2" is introduced because we're dealing with round-trip transit times.

We're using a digital instrument. To resolve 0.00001 in. in a single
round trip we'd need a system which responded at a 12.5-GHz rate.

$$\frac{1}{0.8 \times 10^{-10}} = 12.5 \times 10^9$$

That's not practical at the present time so, once again, we must resort to
statistical techniques. In one example, the instrument "interrogates" the
bolt 160 times. It measures the length of the bolt "crudely" each time.
Each answer is stored in memory; then 160 answers are averaged to determine the most likely mean of the population. The central limit theorem
tells us how many measurements are required to achieve a desired level
of accuracy. With a basic clock frequency of 100 MHz and 160 measurements we can estimate the actual transit time with a standard deviation
on the order of 10^{-8} sec.

At the end of this process the instrument updates the display. It then
makes another 160 measurements and repeats the computation. The display is updated anywhere from 3 to 400 times a second, depending on the
instrument used, the accuracy required, and the length of the bolt.

The length of the bolt is a factor for two reasons. First, it takes longer
for the ultrasound to make a round trip in a longer bolt, so fewer updates
are possible. Second, longer bolts stretch more. We don't have to measure
stretch with as much precision to achieve a given percentage accuracy.

Incidentally, the other numbers used in this discussion—100 MHz,
160 measurements, etc.—were for example only. The actual values depend, again, on the instrument used, the bolts involved, and the application.

D. Transit Time vs. Change in Length

Several factors affect the relationship between transit time—or, more correctly, a change in transit time—and the estimated change in length of
the bolt. Let's take a look at these factors.

Sonic Velocity

Very simplistically, of course, transit time is related to bolt length by the following relationship:

$$T_T = \frac{2L}{MV} \qquad (11.10)$$

where MV = velocity of sound through the unloaded, unstressed fastener (in./sec, mm/sec)
L = length of the bolt (in., mm)
T_T = transit time (sec)

Depending on the material used and its condition, the velocity of sound in a bolt will range from about 150,000 to over 300,00 in./sec (3810–7260 m/sec). We are primarily interested in how it varies for our bolts. Theoretically, the velocity of sound in a rod or bar is [12]

$$MV = \left(\frac{E}{\rho}\right)^{1/2} \qquad (11.11)$$

where MV = the velocity of sound (in/sec, mm/sec)
ρ = the density of the rod or bar (slugs/in.^{3}, g/mm^3)
E = modulus of elasticity (psi, GPa)

Simplistic handbook data would suggest that E and ρ will be constants for a given material. But this is not true. Bolt manufacturers don't control modulus or density except "by chance." They've had no need to, and can't even provide data. Some experts in such matters say that the modulus of as-received bolts will vary from 29 to 30 × 10^6 psi. Others say that even wider variations are possible. I assume that density will vary as well. Certainly, both modulus and density will be functions of bolt temperature. And the density will be affected by stress, creating the stress-vs.-velocity effect in Fig. 11.7.

As mentioned earlier, we use the estimated room temperature velocity of sound in the unloaded bolt in the algorithm used to estimate change in length or preload. We have to be concerned about variations in as-received modulus or density, but why worry about the effects of stress at this point?

The problem: Manufacturing operations such as thread cutting, thread rolling, cold heading, forging, heat treatment, etc., leave a variety of residual stresses in the bolts. Although the average residual stress level is low, there can be significant concentrations at thread roots, head-to-body fillet, etc. The ultrasound will travel more slowly through the concentrations than in unstressed regions of the bolt. The average velocity, therefore, tends to vary somewhat from bolt to bolt in a given lot, and will often vary substantially between bolts made by different vendors.

As a result, several bolts having identical physical lengths will usually have different "acoustic lengths" if these lengths are measured by a bolt gage set at a given stress factor and material velocity. Variations in average velocity, bolt to bolt, will result in apparent variations in length.

As an example, we once compared physical and acoustic lengths in a group of A193-B7 studs which were nominally 18 in. long. The indicated acoustic lengths differed from actual lengths by a maximum of about 50 mils. This doesn't sound like a large difference, but a bolt with a 14 in. grip length tightened to 50% of yield would only be stretched 21 mils or so. As a result, variations in acoustic length (caused by variations in physical length and in velocity) become very important. We're forced, in practice, to measure and store or log the initial acoustic lengths of individual bolts before tightening them.

There are several procedures for dealing with the variations in room temperature velocity. In the procedure described at length in Sec. IIB we feed the bolt gage a single, assumed velocity, e.g., of 250,000 in./sec. Variations are then accommodated by calibration procedures designed to determine the mean stress factor calibration constant, and the scatter in that constant if of interest. But other procedures or algorithms will require the mean and scatter of the actual sonic velocities in a lot of bolts to be determined and input to the instrument.

Variations in residual stress, and therefore room temperature sonic velocity in the unloaded bolt, are one of the most intangible, unavoidable variables we face when we use ultrasonics on bolts. Field experience in many industries, however, has shown that the variations between bolts in a given lot are almost always small enough to make "reasonable" calibration of the instrument possible. If we're dealing with large bolts, on pressure vessel flanges, for example, we can even get reasonable results on bolts of nominally the same material, in spite of the fact that these have probably been obtained from different sources at different times. In any case, however, we can only determine the actual material velocity and/or stress factor by a calibration exercise, and are well advised to use as large a sample of bolts as practical when performing the tests.

This discussion has focused on "length," not "change in length." Obviously, however, we get the latter by measuring length twice, before and after tightening. Variations in sonic velocity, therefore, affect the change-of-length measurements directly.

Residual stress isn't the only thing which creates uncertainty in velocity, however. Temperature and preload also affect it.

Velocity vs. Temperature

An increase in bolt temperature will lead to a decrease in sonic velocity, as shown in Fig. 11.6. Again, the effect will vary somewhat from bolt to bolt and/or between lots of bolts. We cope with the change by determining an average "temperature factor" on a sample of bolts tested at several temperatures, as described earlier. This is less of a concern in most applications than is the residual stress-velocity factor because most bolts are tightened at a constant temperature. However, even a relatively modest change in temperature (for example, 1° or 2°) must be reported to the instrument and used to correct the change-in-length calculations if we want maximum measurement accuracy. One consideration, therefore, is the accuracy of the instrument and technique used to determine the temperature of the bolt.

Normally we have no problem in determining bolt temperature with acceptable accuracy. A temperature probe, built into the transducer housing, can do that for us. In some applications, however, temperature can vary within the bolt. This can be the case in the bolts on a pressure vessel flange when the system is first started up, or even after it has reached steady-state operating condition. The middle of the bolts may be exposed to air, and therefore be cooler than the ends. Bolts on the upper side of a vertical flange may have different profiles than bolts on the bottom. Rain can cool exposed portions of some bolts, etc. In all such cases the temperature of the bolts will vary from point to point, and the velocity of sound at each point will be a function of the temperature at that point. Programmable bolt gages can be fed thermal profiles, and can properly compensate for such variations; but this is an undesirably complicated thing to do. More commonly the instrument is instructed to base calculations on a "best guess" of the average temperature within the bolt.

Velocity vs. Stress

Stress affects velocity; that's one of the principal attractions of ultrasonic measurements. We use shifts in velocity (determined by measuring transit times) to estimate stress levels. The exact relationship between stress and velocity, however, is also subject to bolt-to-bolt and lot-to-lot scatter. We cope with this variation by using the previously described techniques to determine the mean and scatter of the stress factor calibration constant.

Length Change Not Caused by Tension

If something changes the length of the bolt without changing its stress level, the bolt gage could interpret the change in transit time incorrectly.

Such changes are not uncommon, but, fortunately, they can be accommodated.

Let's assume that our transit time instrument has been calibrated to read, correctly, the change of length in a particular bolt when this bolt is stretched. One-third of the change in transit time is created by an increase in path length (ΔL); the rest of the change is created by the stress-induced reduction in the velocity of sound.

We measure the length of the bolt with the instrument, with the bolt unstressed. Then we remove 0.015 in. of material from one end and use the bolt gage to measure the change in length. The bolt is still unstressed. Since the instrument has been calibrated to interpret only one-third of any change in transit time as a change in dimension, it will tell us that the bolt is now 0.015 ÷ 3 or 0.005 in. shorter. This is the sort of "wrong answer" it gets when it measures steps in gage blocks.

In practice, we can encounter dimensional changes which are unrelated to stress change when the temperature of the bolt changes.

$$\Delta L = \rho(\Delta t)L_T \qquad\qquad (11.12)$$

where ΔL = change in length (in., mm)
ρ = coefficient of linear expansion (in./in./°F, mm/mm/°C)
Δt = change in ambient temperature (°F, °C)
L_T = total (not effective) length of the bolt (in., mm)

The length will increase or decrease, of course, as the temperature goes up or down. The coefficient of linear expansion of steel might be 6.48 × 10^{-6} in./in./°F or 11.7 × 10^{-6} mm/mm/°C (the coefficient varies somewhat with alloy content, etc.).

We can also encounter a change in length unaccompanied by a change in stress if there's a long time between initial and final readings and one end of the bolt has corroded. We can't make good accoustic coupling to the bolt through rust, so we must remove it with a file or sandpaper. If we shorten the bolt by 0.006 in. in this process, the instrument will think it sees a decrease in length of 0.002 in.

The temperature factor calibration constant (TF) allows the instrument to compensate, automatically, for changes in length caused by thermal expansion or contraction. Other dimensional changes, caused by mechanical means, are much less common and must be dealt with "manually." One trick is to measure the reduction in length of a bolt as it is loosened; then to take an unloaded reading of its length and compare this to the length logged before the bolt was first tightened. The log can now be corrected, and the bolt can be retightened. All of this is rarely necessary, however.

Stress Factor Accuracy

The stress factor calibration constant (SF) has entered the discussion repeatedly. By now you must realize that it plays a key role in determining the accuracy of the change-in-length measurements. It defines the basic ratio between velocity and change-in-length effects in the fastener material; and it can be, and usually is, used to tidy up such things as variations in initial velocity caused by residual stress in the bolts. If we don't input the correct value for SF, the instrument will not be able to report the actual change in length.

The accuracy of the SF will depend entirely on the accuracy of the equipment and techniques used to determine the true change in length of or tension in the sample bolts when we calibrate the bolt gage. As we learned in Sec. III, we probably won't have calibration equipment which is substantially more accurate than the bolt gage. Statistical techniques can be used, however, to derive SF with acceptable accuracy. Manufacturers of the ultrasonic equipment will perform these tests for you, or tell you how to do them yourself. In any event, however, the instruments used to calibrate the ultrasonics must, themselves, be properly (and frequently) calibrated, maintained, and used. If this is done, and the statistical work is up to par, all other room temperature variations and questions can be ignored. The ultrasonic equipment can be likened to a "rubber ruler" which can be adjusted to give any reasonable answer. If we program it with the right SF, it will give the correct answer, at least for room temperature measurements. Adding the other important calibration constant, the temperature factor (TF), will lead to correct answers at all temperatures. It's certainly useful for you to know the factors which cause uncertainty in the ultrasonic instrument's estimates of change in length—the factors discussed in Secs. I–IV—because there may be times when you'll have to reduce the bolt-to-bolt uncertainties by controlling or measuring velocities or moduli more carefully. In most cases, however, your measurements will be accurate enough if you follow Sec. III and/or the manufacturer's instructions and calibrate the instruments properly.

Resulting Accuracy in Change-in-Length Measurements

Results depend on calibration techniques, bolt size, the care taken in using the instrument, bolt condition and heat treatment, temperature measurements, and all the other factors we've discussed but, in general, it's probably safe to say that:

> Most present-day transit time instruments are capable of measuring the change in length of fasteners to the nearest 0.0001 in. Special

instruments are also available which can detect change to the nearest 0.00001 in. These absolute accuracies get less as fastener length increases, if for no other reason than that the digital circuits and display can handle only a fixed number of significant digits. The stretch of threaded rods up to 24 ft in length, however, has been measured with accuracies which are believed to have approached the nearest 0.001 in.

The equipment accuracies cited above are not always achieved in practice, especially when the equipment is used under difficult field conditions. Since there are few ways to confirm or challenge measurements of this accuracy in the field, data are generally unavailable. My guess is that bolt stretch can probably be measured to the nearest 5% even under extreme conditions (although we'll often settle for less accuracy than this). Production line or laboratory measurements will more likely approach the accuracy for which the instruments are capable. In any case, the accuracy of the calibration exercise, and the way the instruments are used, rather than the capability of the instruments, determine the accuracy of the results.

E. Accuracy of Preload Estimates

We now use the change-in-length algorithms of Eq. (11.6) and (11.7) (or equivalent) to estimate bolt preload. The problems in doing this, in relating change of length to tension, are described in detail in Chap. 5, and involve uncertainty in the geometry of individual features more than in material properties. The most significant uncertainty, at least in relatively short bolts (L/D of less than 4:1) concerns the way the male threads engage the female threads, since this determines the "effective length" of the fastener, and that, in turn, determines its stiffness. Thread run-out vs. body length is also significant in short bolts. See Chap. 5 for details.

As mentioned earlier, an accurate stress factor tells us what percentage of the total change in transit time measured by the instrument was caused by the stress-related velocity change in the bolt. In fully threaded bolts this can give us a relatively clear picture of tensile stress and preload. All we need to know in addition to the change in transit time is the effective length of the bolt. In a partially threaded bolt we must also know, separately, the effective length of the threads. Both dimensions are subject to bolt-to-bolt and lot-to-lot scatter. They can, and often do, change as a bolt is reused (because of localized plastic deformation and creep in the threads). The length of time a bolt has been in service can make a difference, too, because creep is a time-based phenomenon.

What prevents all this from creating serious problems in most applications is "calibration." In fact, measuring "actual" forces produced by a bolt on a load cell is easier than measuring "actual stretch," partially offsetting the additional variables. And the bolt-to-bolt scatter in stretch vs. preload is usually reasonably small, at least within a given lot. As a result, we feel that the equipment is capable of residual preload measurements with an accuracy approaching ±5% in most field situations, and can approach ±2% in laboratory work. There are even well-documented examples where ±1% was achieved consistently in aerospace applications; but this was accomplished only by calibrating the instrument several times a day, by controlling bolt temperatures very carefully, by careful design and treatment of the bolts (with special attention to end surface flatness and parallelism), etc. I think it's worth noting that the *instruments* are capable of such accuracy, but that economics make this precision impossible in most applications.

As a final note, the accuracies cited above for stretch and preload measurements should be considered typical, but not definitive. Detailed studies have been made of the theoretical precision with which present instruments can measure stress or strain in bolts [13]. Uncertainties were estimated using first-order, second-moment methods, lognormal approximations, and Monte Carlo simulation. Some 11 variables were found to play key roles in the measurements, and it was found that, in most cases, one or two of these variables contribute most of the uncertainty in the final estimates. Different variables, furthermore, dominate different applications. For example, if the bolts are very long, variations in the modulus of elasticity can be one of the main problems. In very short bolts uncertainty in grip and thread lengths can dominate. One value of the study is that it allows us to identify the one or two variables which, if more carefully measured or controlled, can produce a significant improvement in overall accuracy. We needn't control them all. We needn't, in fact "control" any of them; measuring and knowing them is sufficient is the instrument is programmed to accept changes in these factors [14].

V. USAGE FACTORS AFFECTING ACCURACY OF MEASUREMENT AND CONTROL

We've analyzed the accuracy with which current transit instruments are able to measure bolt stretch or tension. Several times we learned that there can be a significant difference between the accuracy of the instrument and the accuracy we actually achieve in the field. We will often settle for cruder measurements; and we will almost never try to use the instrument to *control* preload with instrument precision. Let's look next at some of the factors that lead us to "settle for less."

A. Transducers

The transducer plays an important role. It consists of a crystal mounted in a protective housing with a "wear plate" on the business end, to protect the crystal from mechanical damage or abrasion. Since it's important that the transducer remain in a fixed position during a measurement, the protective housing is usually provided with strong magnets to hold the transducer in place. Transducers used on nonmagnetic bolt materials are held in place by mechanical means. Sometimes, for example, they're threaded onto the end of a stud or bolt.

Selection of a Transducer

Transducers come in a variety of diameters, heights, resonant frequencies, operating temperatures, etc. They range in diameter from about $\frac{1}{4}$ in. to 1 in. In general, one should always use the largest diameter which will fit on the end of the bolt without extending past the root diameter of the threads. One reason for selecting a larger diameter: It will pump more acoustic energy into the bolt than would a smaller diameter. This means stronger echo signals and usually greater reliability and accuracy. A transducer whose diameter basically matches that of the bolt, furthermore, is easy to position or reposition on the bolt (see below).

The resonant frequency of the transducer is another important consideration. The frequencies used for bolt measurements usually range from 2 to 10 MHz, with 5 MHz being the most popular. A lower frequency will penetrate the bolt with less loss, but results in a broader "cone" of transmitted energy—which can mean unwanted echoes from nearby threads, etc. As a result, higher frequencies usually work more reliably on longer, thinner bolts.

Most transducers are unable to function at operating temperatures in excess of 150°F (302°C), but special ones are available which can tolerate temperatures up to 400°F (752°C).

In the past it was often necessary to try several transducers before finding one which would work most reliably on a given application, but advances in the ultrasonic equipment (more "punch power" and ability to respond to weaker echoes, for example) make trial and error much less common at the present time.

Placement of the Transducer

The way the transducer is placed on the bolt can affect results. If the transducer and bolt have the same or similar diameters, placement is simple—the transducer is roughly centered on the end of the bolt, with no part of the transducer overhanging the edge of the bolt.

On large-diameter studs the transducer can also be placed at the center. It is then usually moved around slightly while the operator observes the "initial length" displayed by the instrument. He searches for the minimum displayed length, and leaves the transducer in that spot. If the transducer is left in place as the fastener is tightened, no further action is necessary. If repeat measurements are to be made in the future, with the transducer removed in between, it is also desirable at this point to mark the selected location of the transducer with a prick punch, or spot of paint, or the like, so that it can be replaced in the same location.

This procedure is also required if the transducer is used on a stud having a heater or gage rod hole gun-drilled along the axis. A large-diameter transducer can be centered over a small-diameter hole, but more commonly the transducer will be located to one side of the hole, between the hole and the outer diameter of the bolt. Again the operator tries several spots until he finds the minimum initial length. That location is then marked for future reference, if required.

B. Acoustic Couplant

The Role of the Couplant

The strength of the ultrasonic signal affects the accuracy with which the instrument can measure the bolt. In general, the stronger the signal, the more precise and dependable the measurement. A number of things influence this signal strength.

As one example, the strength of the ultrasonic signal is dependent on the amount of energy which can be pumped across the gap bridging the transducer and the fastener. If this is an air gap, the signal will be almost totally reflected back into the transducer before it reaches the bolt—because there is a poor acoustic impedance match between the transducer and air. In practice, therefore, the gap is always filled with an acoustic coupling material of some sort; glycerine, water, various oils, and greases have all been used. Each provides a better impedance match between transducer and fastener than would air alone. The choice of couplant for a given application depends on such things as roughness of the end of the fastener, the ambient temperature, fastener location and orientation, etc. Sometimes contamination or corrosion problems limit our choice (e.g., in nuclear applications).

Voids in the Couplant

Each spot on the end of the transducer functions as a tiny "antenna" to propagate an acoustic wave. These small waves all combine to create the wavefront which travels down the bolt, echoes off the far end, and returns to the transducer.

If there are voids in the film couplant between transducer and bolt, then some of these tiny, spot antennas will be unable to contribute much energy to the combined wavefront. The wavefront becomes distorted. The shape and/or polarity of various portions of the echo signal is distorted—and this will affect the measurement of round-trip transit time. As a result, it's important to make sure that the gap between transducer and bolt is completely filled with a uniform layer of couplant.

Gap Errors

Transit time instruments will measure the distance between the face of the transducer element and the far end of the fastener. This distance may include not only the length of the bolt—which we are interested in—but the thickness of the gap between the operating face of the transducer and the bolt. Since we are making very precise measurements of the change in length of the fastener, it is important that this gap be controlled. We don't care what it is. We want it to remain the same as the bolt is tightened or be the same before and after tightening. This means that the transducers must be held securely against the bolt by magnetic or mechanical means during the bolt-tightening operation, or must be consistently mounted if used before and after. Changes in thickness of the acoustic couplant must be minimized. There is also a wear plate between the working face of the transducer element and the outer surface of the transducer housing assembly. We must be sure always to use the same transducer on a given bolt, or to calibrate out changes in wear plate thickness if we change transducers on that bolt (and are looking for the ultimate in accuracy).

None of the above is a problem, of course, if the transit time instrument is designed to measure the time between first and second echoes rather than between main bang and first echo (or between first and tenth echoes as with the Harrisonics system). In this case, however, there is the offsetting difficulty that the extensometer must work correctly with weaker signals.

C. End Surfaces

Roughness, Grade Markings, Socket Heads

A very rough surface on either end of the bolt can create problems in two ways. Roughness between the transducer and the input end of the bolt can reduce the amount of ultrasonic energy pumped across the gap, creating weak, uncertain echoes. Extreme roughness on the far or echo surface of the fastener can lead to interference effects and can distort the signal in the same way that near-field interference effects do. This, again, can lead to improper timing of the echo signal. This is not to say that bolt ends must be ground and polished in every application. In most cases, roughness affects accuracy rather than ability to do a job.

Raised grade markings, dished ends, hex sockets, etc., can also cause coupling problems (but are not a problem, usually, if they're on the echo-surface end of the bolt). If the transducer can't be placed on a reasonably flat surface, inside of or beside surface holes or embossings, the bolt cannot be read ultrasonically.

Grade markings can be removed with a file. The resulting surface is usable from an ultrasonic point of view, but removing such markings may not be acceptable to quality control people.

Fasteners with special end treatments can be expensive, forcing many people to decide against ultrasonic measurements for this reason. A few people, however, have decided to machine the ends or remove grade markings on a small sample of the fasteners they use, painting one end red or something, so that these can be readily identified in the final assembly.

These special fasteners are fed into the assembly operation in a normal fashion, and conventional assembly tools—torque or turn, or whatever—use them along with the regular fasteners. Quality control inspectors then watch for "red bolts" in the final assembly. Each one of these is measured ultrasonically to monitor the performance of the basic torque-turn or other control system being used on a production basis. If each of the special fasteners has approximately the same acoustic length to begin with (which generally equates to approximately the same physical length), then this system is a very effective way to monitor the behavior of assembly tooling and/or to follow subsequent working loads or relaxation effects.

Nonparallel End Surfaces

If the opposite ends of the fastener are not sufficiently parallel with each other, the acoustic signal will be weakened or may be lost altogether. This is not usually a problem with as-received fasteners (although the ends are never exactly parallel to each other), but it is a frequent problem when the bolt bends as it is tightened.

Nonparallelism can cause a problem simply because the "length" of the bolt is not uniform from one point to another. This is no problem if the diameter of the transducer is nearly as large or as large as the diameter of the bolt; under these circumstances the ultrasonics will average the length and/or change in length of the fastener.

If we're using a $\frac{1}{2}$-in.-diameter transducer on a 4-in.-diameter bolt, however, the instrument will see a different length when it's mounted in a 12 o'clock position than it would in a 6 o'clock position, or when mounted on the centerline of the fastener. (Mounting on the centerline is not always possible because of heater holes and the like.) Under these conditions the location of the transducer must be marked with dye or a prick punch when the zero stress reading is taken. It must then be returned to approximately the same position for the aftertightening reading.

If the nonparallelism is created by or made worse by bending, an additional problem arises. Remember that the velocity of the ultrasound is inversely proportional to the stress level within the fastener. The tension stress down one side of a bent bolt is, of course, higher than the tension stress down the opposite side. The ultrasound wave front will, therefore, travel more slowly down one side of the bolt than the other. The bolt acts as sort of a lens, gradually warping the ultrasonic wavefront, as suggested in Fig. 11.11.

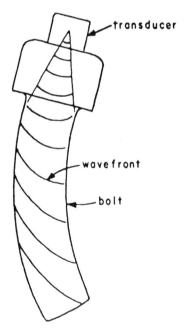

Figure 11.11 Tensile stress is higher on one side of a bent bolt than on the opposite side. This will rotate the wavefront of the acoustic signal as shown.

If the difference in stress level from one side of the fastener to the other is too great, the ultrasonic signal can end up echoing off the side of the bolt rather than the end, and, in effect, will be lost. This is an extreme case, but it does happen, especially in bolts which have heater holes in them. (The hole adds an additional level of complication, since the bolt can no longer act as a waveguide for the signal.)

Nonparallelism, regardless of its sources, can also distort the shape of an echo signal. Each portion of the echo surface acts as a tiny reflector. All such reflections are combined to produce the echo signal. If, because of nonparallelism, these small reflective surfaces are out of phase with each other, they can partially cancel each other. These interference effects distort the signal in the same way that the near-field effect distorts it, as suggested in Fig. 11.12.

D. Controlling Preload

The problems we've discussed above can introduce error in our measurements, but they still don't explain why we sometimes settle for preload scatter of 20% or worse when the ultrasonic instruments are capable of a relatively easy 5%. The reason for the discrepancy: Even though we can *measure* preload, during assembly, or after assembly, with an accuracy of 5% or better, we cannot always afford to *control* it with that precision. The normal example of this is the large, pressure vessel flange. We tighten the bolts one or a few at a time, in a cross-bolting pattern. Ultrasonic measurements are made on a few of the bolts during the early passes—the first two passes, typically. Only on the final routine pass is the tension in each bolt measured. Meanwhile, elastic interactions, embedment, and variations in the torque-preload relationship have led to the results illus-

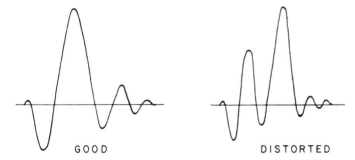

GOOD   DISTORTED

Figure 11.12 Interference patterns produced by nonparallel transmission and echo surfaces can seriously distort—or even destroy—an echo signal.

trated by Fig. 6.25, a "saw-toothed curve" with residual preloads varying by 4:1 or more in most cases. And we'll get these results whether or not the initial preloads achieved in each bolt are monitored ultrasonically as the bolts are tightened individually. Even perfect ultrasonic measurement at this point cannot prevent subsequent scatter.

When we first used ultrasonics on such joints, we assumed that it was only proper to end up with residual preload scatter comparable to the accuracy of the instrument, ±1-2% or so. And we found that this was certainly possible—if one could spend a day or two tightening an individual joint. But, thank goodness, we also found ourselves forced to give up when scatter had been reduced from 4:1 to ±10% or 20%, because every other part of the plant was ready to go back on line. As a result of being forced to stop, we learned that ±20% or ±30% or "worse" was a big improvement over 4:1 and was acceptable on all but the most stubborn leakers.

On hard joints, or on joints where most or all of the bolts are tightened simultaneously, scatter in the ±5-10% range is probably the accepted norm, because interactions are fewer or are nonexistent. Even here, however, the final variations are determined more by the nature of the joint and by the tools used to generate bolt loads than by the accuracy of the ultrasonic measurements.

VI. CASE HISTORIES

The ultrasonic bolt gage is used to control or monitor the tension created in bolts by torque tools or tensioners. Earlier chapters, therefore, have included, indirectly, "case histories" of its use, situations in which ultrasonics was used to reveal the results achieved with a variety of assembly procedures. The additional examples below will help illustrate the fact that ultrasonics allows you to see things you cannot see as conveniently by any other means. In fact, I think it gives the bolt mechanic or engineer the same sort of insight the electrician gets from an oscilloscope. The mechanic can make a change in lubricant, tools, fastener material, etc., and, thanks to the bolt gage, see immediately the results which that change has produced—instant feedback in either a static or a dynamic situation. For the first ime we can go into the field and determine actual results on actual bolted joints—not just on a few laboratory or field samples specially prepared with strain gages or the like.

The things you will see will not always be pleasant or understandable. Bolted joints are a very complex subject, and their behavior is not well known. They are not very stable, either—they flex and shift and relax under environmental changes and/or variations in external load—as well as just "all by themselves." When you start to use ultrasonics, therefore, you should be prepared for a number of unpleasant surprises. I believe that it was the mathematician Laplace who said, "Getting an education is like climbing a flagpole; the higher you go the more horizon you can see." Using ultrasonics on bolted joints gives you such an education. Here's a typical example.

A. Connecting-Rod Bearing Caps on a Large Diesel Engine

The fasteners were $1\frac{3}{4}$–10 × 17, 4140 studs. The problem: Fatigue failure of one stud had created a chain of events which led to the failure of the entire engine. Ultrasonics was used to answer some of the questions raised by the analyst.

First question: What would failure of one stud do to the load distribution in the other seven studs holding down the bearing cap?

To answer this question, eight new studs were used to fasten a bearing cap onto an actual engine. Ultrasonics was used to control the elongation in each. One stud—that corresponding to the one which had initiated the original failure—was now loosened completely. The residual preload in the remaining studs was measured, with the results shown in Fig. 11.13.

This is a hard joint (no gasket). There is a good fit between mating parts, joint surfaces are good, parts are massive; yet each stud experiences

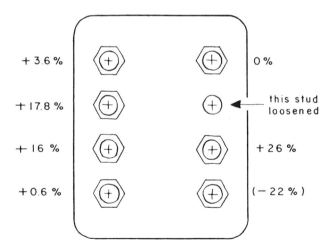

Figure 11.13 Percent growth (more load) or relaxation (less load) in seven studs in a connecting-rod bearing cap when the eighth stud is loosened completely.

a different change in load when one stud is loosened, and some of the changes are substantial.

Second question: How much residual preload is there in a typical bearing cap stud $1\frac{1}{2}$ years after initial tightening? The engine has been running for most of this time.

This question was answered by measuring the loss in length in 16 studs (two caps in another engine) when they were removed for routine maintenance purposes.

As you can see from Fig. 11.14, the mean change in length was 0.0226 in., with a range of 0.0176–0.029 in. (-22% to $+28\%$ of the mean).

These results were somewhat surprising, since the maintenance specifications (and use of the short-form torque-preload equation) would suggest that initial stretch, $1\frac{1}{2}$ years earlier, was—or should have been—0.016 in. to a maximum of 0.022 in. They probably were overtightened originally; but they may have tightened somewhat as a result of vibration, a rare, but not unheard of, possibility.

In any event, use of ultrasonics gave the analyst some joint behavior and joint condition information which would have been difficult or impossible to obtain by other means.

B. Ladle Turret Bearing Bolts

A group of $2\frac{1}{4}$–12 $\times$ 29, 4340 studs were used to preload 14-ft-diameter bearings which supported a large ladle in a steel mill. The company which manufactured the bearings told the steel mill that the bearings must be "properly preloaded" or they would not be covered by the product war-

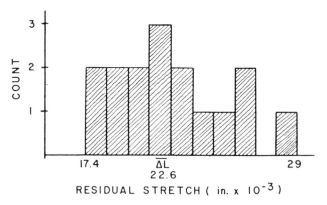

Figure 11.14 Residual elongation (preload) in 16, $1\frac{3}{4}$–10 $\times$ 17, 4140 studs after $1\frac{1}{2}$ years of operation.

ranties and would probably have a short life. The means for achieving proper bolt preload, however, was left up to the steel mill.

Hydraulic tensioners were used to tension the studs, and a uniform run-down torque of 400 lb-ft was applied to each nut. Ultrasonics was used to measure the amount of tension introduced by the tensioners. The ultrasonic equipment was also used to measure the amount of residual tension in each stud, following removal of the tensioners. There turned out to be a surprisingly wide scatter in residual tension, with many studs having no residual tension at all.

The first reaction was, "The ultrasonic equipment isn't working." The tensioners had applied nearly 400,000 lb of force to each stud, and a high-run-down torque had been used. Relaxation and some elastic interaction was expected in this nongasketed joint; but we hadn't expected to loose ALL of the initial preload.

The nuts on the tightened studs were tight, furthermore, even though the bolt gage said there was no preload. This was considered proof that the ultrasonics was wrong. Had ultrasonics NOT been used, the job would have been accepted after the final pass with the tensioners.

The ultrasonic equipment was checked, and found to be working correctly. Further investigation of the studs and joint uncovered the reasons for low or zero preload. There were two reasons.

First, close-fitting (Class 3), fine-pitch threads had been specified for the studs "to do a better job." This was a reaction to the bearing manufacturer's warning about proper preload. When the studs were tensioned, however, the pitch of their threads stretched a little, creating, in some cases, interference between male and female threads. (The nut threads, of course, did not grow when the studs were pulled by the tensioner.) A significant part of the 400 lb-ft of run-down torque was used to overcome this thread interference.

A more serious problem was caused by the fact that the joint surfaces had been spot-faced under each nut so that opposite sides of the joint would be flat and parallel to each other, to minimize bending and, again, "do a better job." The studs were only threaded on the ends, with a solid body or shank in between. The threaded sections were too short and/or the spot faces were too deep. It turned out that many nuts had been tightened down against their own bodies, instead of against the joint surface, as shown in Fig. 11.15. These were the studs which had ended up with zero preload.

Once again, therefore, ultrasonics revealed problems which would not otherwise have been spotted. The circumstances and resulting problems were admittedly unusual, but the ultrasonics reduced the chances of early bearing failure and a subsequent, costly, maintenance shutdown.

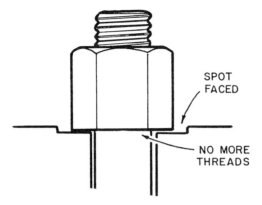

Figure 11.15 In one application, ultrasonics detected zero preload in large studs whose nuts had been tightened down against the unthreaded body section of the studs. The studs were exerting no clamping force on the joint.

C. Automotive Bolting

The automotive industry is the largest user of ultrasonics for bolts at the present time. The equipment is used in place of strain gages for a variety of studies and experiments. It's also used to monitor production tooling, as explained in an earlier chapter. In general, ultrasonics hasn't revealed unexpected behavior, because this industry has historically paid a great deal of attention to bolting, and is fully aware of the problems and variables encountered during assembly operations. In several cases, however, the relative ease with which ultrasonics can be used (as compared to the equally accurate strain gage) has made it easier for the manufacturers to probe and solve suspected problems. A few examples:

Ultrasonics has been used to evaluate quality control procedures for on-line inspection for residual tension in recently tightened bolts, procedures based on "breakaway torque." Ultrasonics showed that breakaway torque was often a poor measure of residual preload, so poor that the procedure was abandoned in several cases.

Ultrasonics was used to monitor the relaxation of head gaskets, including that of proposed asbestos substitute gaskets. Tests were on actual heads, tightened by production tooling. In many cases the relaxation far exceeded that anticipated.

Aluminum is replacing cast iron and other materials in many parts as auto makers struggle to reduce fuel consumption by reducing vehicle weight. One manufacturer used ultrasonics extensively to cope with the extreme relaxation of a tapered aluminum/cast iron joint. Ultrasonic studies led the way to longer, thinner bolts and the use of Belleville springs under bolt heads to compensate for embedment relaxation during thermal cycles.

VII. PROBLEMS REDUCED BY ULTRASONICS

With reference again to Fig. 6.28, ultrasonics allows us to see past many of the groups of variables which determine the outcome when we tighten a group of bolts. Like stretch control, ultrasonics is able to focus on the response of a single part—the bolt—to the assembly tools and procedures. Gone are the variables introduced by interactions between parts, such as the friction between nut and bolt, or the relative stiffnesses of bolt and joint. Because we can read bolts after tightening as well as during, we're able to see and compensate for elastic interaction and relaxation effects, another big plus for residual preload accuracy.

But ultrasonics is not perfect. It can't measure the interface clamping force between joint members, and so can be fooled by hole interference or joint resistance effects. It sees bending stress as an increase in average tension. And a variety of material and geometric properties still affect its accuracy. But it's a big step forward, and its use should become more and more common in the years ahead.

VIII. GETTING THE MOST OUT OF ULTRASONIC CONTROL

In addition to the many generic things we've discussed several times—things like training and supervising the operators, keeping good records, using consistent procedures, etc.—there are a number of special things you can do to maximize the results you get when you use ultrasonics to control or monitor bolting operations.

First, recognize that the more precise an instrument is, the more care and skill are required when you use it. This is true of ultrasonic measurement of bolt stretch or tension. Using the equipment in a "quick and dirty way" may give you better preload control than torque and/or turn procedures, but you must do a lot of things right to get the full accuracy the instruments are capable of giving you. We've already discussed many of these points, but here's a summary list of recommendations for use.

A. Manuals

The equipment and the techniques will be in a state of evolution for some time to come. Study the operating and calibration manuals supplied with the equipment to make sure you're using it correctly.

B. Calibration

Proper calibration is essential. The equipment can only measure transit time. You have to tell it how to interpret transit time data for your application. When calibrating:

1. Calibrate on more than one fastener whenever possible.
2. Make several calibration runs on each fastener, and compute the mean response—for example, the mean slope of the force-elongation curve.
3. Compute the theoretical response (stiffness, stretch, etc.) to check the experimental data. (See Chap. 5 for stiffness equations.) You won't get perfect agreement, but should be within a percent or so by the time the instrument is properly caibrated.
4. Record all calibration data, including bolt information, calibration settings (SF, TF, MV), bolt temperature, instrument and transducer serial numbers, etc.
5. Load the bolt as heavily as possible during these tests. Higher stress levels lead to larger changes in transit time which can be interpreted with greater precision than can small changes.
6. By the same token, use the maximum practical change in temperature when calibrating for temperature factor (but observe the temperature limitations of the transducer).
7. Don't rely on the extensometer on the tensile machine for actual change-in-length data. Thanks to thread embedment the jaws of the machine can creep a bit under load, with no accompanying change in bolt tension or elongation.
8. Monitor and record the temperature of the bolt.
9. For maximum accuracy, recalibrate regularly, especially when starting a new lot of bolts or a new vendor.
10. When calibrating the instrument for an application involving large bolts received from a variety of sources at different times (e.g., petrochemical flanges), it's safe to use generic calibration constants (SF, TF, MV) provided by the manufacturer of the instrument. If you want the ultimate accuracy, however, you must calibrate against a representative sample of your bolts. It may even be necessary to sort and characterize them by ultrasonic behavior. (I've never en-

countered a petrochemical application where this degree of accuracy was necessary, however.)

11. After calibrating the equipment on bolts, use it to read calibration bars and record the results. This will make simple recalibration checks possible.

C. Using the Equipment

1. Make sure that the proper calibration settings have been fed to the instrument.

2. If the instrument has not been used for some time, or has been dropped, for example, recheck the calibration against the calibration bars.

3. Make sure that the ends of the bolts are as flat and parallel as circumstances permit.

4. On small bolts, center the transducer on one end, and make sure that it doesn't overhang the root diameter of the threads. If there's any question, use a smaller-diameter transducer.

5. On large bolts, make preliminary measurements at several locations, and choose that spot which gives the smallest initial length readings. Mark that spot with paint, dye, or a prick punch so that the transducer can be returned to the same location for future measurements.

6. If time allows make several measurements of initial and final bolt tension or stretch. By this process you'll be able to see past early relaxation or interaction effects, and will more likely be dealing with a better coupling film (no voids).
Wipe the end of the bolt before applying new couplant, before each measurement. Work the transducer down into the couplant before recording the readout. After a little practice you should be able to repeat readings within the resolution of the instrument—and without peeking at the readout!

7. Measure the final tension or stretch in all bolts after tightening the last one in the joint. This will reveal elastic interaction effects, and will allow you to compensate for them if desired.

8. Handle the instrument and transducer with care. Store in a clean, dry place. If using them outdoors, in rain, cover them with a sheet of plastic or other material whenever possible.

REFERENCES

1. Couchman, J. C., Acoustic-elastic fastener preload indicator, Report prepared for the Air Force Materials Laboratory, Air Force Systems Command,

Wright-Patterson Air Force Base, under Contract no. F33615-76-C-5151, Final report, June 15, 1978.

2. Gordon, B. E., Jr., and T. Speidel, Stress measurement by ultrasonic techniques, Paper presented at the 1973 SESA Fall Meeting, Indianapolis, IN, October 1973.

3. McFaul, H. J., An ultrasonic device to measure high-strength bolt preloading, *Materials Evaluation*, vol. 32, no. 11, pp. 244–248, November 1974.

4. Erdman, D. C., Measuring fastener strain, Paper presented at the Air Transport Association of America Non-Destructive Testing Meeting, Houston, September 11–13, 1973.

5. Bickford, J. H., Using ultrasonics to measure bolt tension, a state-of-the-art report, SME Technical Paper no. AD76-650, 1976.

6. Heyman, J. S., A self-exciting ultrasonic reflection CW instrument, *1976 Ultrasonic Symposium Proceedings*, IEEE Cat. no. 76, CH1120-5 SU.

7. Sakai, T., T. Makino, and H. Toriyama, Bolt clamping force measurement with ultrasonic waves, *Trans. JSME*, vol. 43, no. 366, pp. 723–729, February 1977.

8. McIntyre, P. J., and N. F. Haines, *The In-Situ Measurement of Thread Strain in Mild Steel Bolts*, Central Electricity Generating Board, Berkeley Nuclear Laboratories, Berkeley, Gloucestershire, England, September 1975.

9. Bobrenko, V. M., et al., Ultrasonic method of measuring stresses in parts of threaded joints, *Sov. J. Non-Destructive Testing*, vol. 10, no. 1, January–February 1974 and November 1974.

10. Bickford, John H., Using ultrasonics to measure the residual tension in bolts, Paper presented at the 1987 SEM Spring Conference on Experimental Mechanics, Society of Experimental Mechanics, Houston, TX, June 14–19, 1987.

11. Gordon, Bennett, Jr., Measurement of applied and residual stresses using an ultrasonic instrumentation system, Paper presented at the 24th International Instrumentation Symposium, Albuquerque, NM, May 1–5, 1978.

12. Sears, Francis Weston, *Mechanics Heat and Sound*, Addison-Wesley, Reading, MA, 1950, p. 497.

13. Veneziano, Danielle, Uncertainty in the estimation of bolt stress, Consulting report to Raymond Engineering, December 15, 1982.

14. Meisterling, Jesse, Accuracy of ultrasonic bolt load determination, Paper presented at CETIM Symposium on Fluid Sealing Application to Bolted Flange Connections, Nantes, France, June 18–20, 1986.

III

THE JOINT IN SERVICE
Importance and Stability of the Clamping Force

12
Theoretical Behavior of the Joint Under Tensile Loads

We've now completed our study of the various ways in which the initial tension in a bolt—its preload—can be controlled during the tightening operation. Preload control has probably received more publicity and attention than any other bolting problem, but it's certainly not the only problem we face.

We're not just interested in the initial tension, we're interested in joints that don't fail in service during the expected life of the product. Correct preload is a critically important factor, but we also need to know and/or control the service or working loads on the bolt and joint, and need to understand the many ways in which joint condition and behavior change with time as a result of relaxation, corrosion, vibration, cyclic loads, etc. We'll consider these and other aspects of the problem in the next group of chapters, starting with an analysis of working loads and then looking at the many ways in which a joint can misbehave or fail—and what we can do about it.

Note that when we turn our attention from assembly to working loads, failure modes, and the like, we start to address a new, major topic. Our first topic, which has occupied most of our attention to date, was establishing the clamping force on the joint. Where does it come from? How can we get the clamping force we want? Etc. Now we begin our studies of an equally important, second topic: How stable will that clamping force be in service? What can change it? How much will it change? Under what conditions will it be lost altogether? Will anything increase it to a dangerous level? We're going to start this study by examining the way the joint responds when exposed to the normal, external loads it has been designed to support, loads created by external pressure, inertia, weight, etc. As we learned in Chap. 2, we often categorize joints by the type of loads they support—calling them "tensile joints" or "shear joints." In this and the next chapter we're going to look at the first of these—joints in which the bolts are loaded in tension. We define tension loads as those which are applied along a line of action more or less parallel to the axes of the bolts. Then, in Chap. 14, we'll study joints loaded in shear.

There are two key questions we must answer whenever we analyze the response of a joint to external loads.

What is the maximum force or tension which, in the worst case, the bolts in the joint will be subjected to?

What will be the minimum clamping force on the joint, again "worst case"?

We're interested in the answer to the first question because we don't want the bolts to break. We're interested in the answer to the second question because we now know, as we learned in Chap. 2, that the life and behavior of a bolted joint will be very short if there's too little clamping force between the joint members.

I. THE JOINT DIAGRAM

To understand the working behavior of a bolted joint—for example, its failure modes—we must first understand in some detail the forces and deflections in the joint—in other words, its elastic behavior. Figure 12.1, greatly exaggerated to ilustrate the effects, shows what happens when we tighten a bolt on a flange or the like. Tightening the bolt sets up stress and strain in both the bolt and flange members. The bolt is placed in tension; it gets longer. The joint compresses, at least in the vicinity of the bolt. It *always* does this, regardless of how stiff it may appear to be.

A. Elastic Curves for Bolt and Joint Members

In Chap. 6 we constructed a "joint diagram," first for a preloaded bolt and then for the preloaded bolt after relaxation. We're now going to see how to extend those diagrams to include the effects of a tensile load on the joint. It's important for us to know where the diagram comes from, however, so let's first repeat the development of the diagram for a pre-loaded joint.

We start with elastic curves for the bolt and for that portion of the joint surrounding a single bolt, as in Fig. 12.2. We then push those two

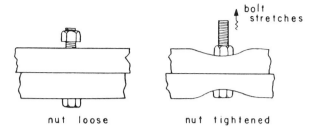

Figure 12.1 Tightening a bolt stretches the bolt and compresses the joint.

curves together against a single, central, preload axis, as in Fig. 12.3 to form the diagram of the preloaded joint. We then reduce the size of this triangle a little to account for embedment relaxation and for elastic interactions between bolts during assembly, as in Fig. 6.27.

B. Determining Max and Min Residual, Assembly Preload

The Equations

Now that we know all there is to know about preload scatter (thanks to Chaps. 6 through 11) we must ask ourselves, "What preload are we talking about: average preload, maximum preload, or minimum preload?" Unfortunately, because it means we must construct two joint diagrams (or a more elaborate and slightly confusing one) we're interested in two things. First, we're interested in the maximum residual preload in individual bolts, because that will help us define the maximum loads seen by individual bolts in service. We don't want any bolts to break under excessive load.

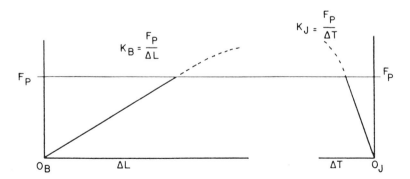

Figure 12.2 Elastic curves for bolt and joint members.

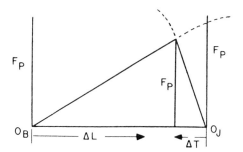

Figure 12.3 The elastic curves for bolt and joint can be combined to construct a joint diagram. O_B is the reference point for bolt length at zero stress. O_J is the reference point for joint thickness at zero stress.

We're also interested in the minimum *average* preload in the group of bolts which form this joint, because that will allow us to estimate the minimum clamping force created on the joint by all of the bolts working together. In a few cases we might be interested in the worst-case minimum residual preload in an individual bolt, but almost always that will be of much less interest than the minimum average because our main concern is for the total clamping force rather than the minimum, per-bolt clamping force. In spite of this, however, we start by considering both the maximum and minimum initial preloads created in individual bolts during assembly.

Let's call average, initial preload at assembly F_{Pa}. If the preload scatter during assembly is called $\pm s\%$ (with s expressed as a decimal), then

$$F_{Pmax} = (1 + s)F_{Pa} \tag{12.1}$$

$$F_{Pmin} = (1 - s)F_{Pa} \tag{12.2}$$

For example, if we're using torque control and expect the initial preloads achieved for a given torque to vary by $\pm 30\%$, then

$$F_{Pmax} = 1.30F_{Pa}$$

$$F_{Pmin} = 0.70F_{Pa}$$

As we saw in Chap. 6, the initial preloads will be decreased, during assembly, by embedment relaxation and by elastic interactions. We can extend Eqs. (12.1) and (12.2) to include these losses, thereby computing the max and min *residual* assembly preloads in individual bolts after relaxation loss. We'll call these Max and Min F_{Pr}.

$$\text{Max } F_{Pr} = (1 + s)F_{Pa} - \Delta F_m - \Delta F_{EI} \tag{12.3}$$

$$\text{Min } F_{Pr} = (1 - s)F_{Pa} - \Delta F_m - \Delta F_{EI} \qquad (12.4)$$

When we use these equations we'll have to assume values for the embedment and elastic interaction losses, and we'll usually find it convenient to express the losses as fractions of the average preload. To do this we substitute the following in Eqs. (12.3) and (12.4):

$$\Delta F_m = e_m F_{Pa} \qquad (12.5)$$

$$\Delta F_{EI} = e_{EI} F_{Pa} \qquad (12.6)$$

where e_{EI} = the percentage of average, initial preload lost as a result of elastic interactions; expressed as a decimal
       e_m = the percentage of average, initial preload lost as a result of embedment relaxation; expressed as a decimal

An Example

Writing the equations is straightforward enough, but we encounter some difficulty when we try to apply them to a given case history or example. The problem comes in deciding what values to assign to the various terms. The many uncertainties associated with bolted joints force us to make a number of assumptions, based on prior experience, or on the published experiences of others.

As we've seen in previous chapters, for example, embedment relaxation might reduce initial preloads by 10%. Elastic interactions might reduce them further by an average of 18% for an ungasketed, two-piece joint; by 30% for a joint containing a sheet gasket; and by 46% for a joint containing a spiral wound gasket. [6]. Let's assume we're dealing with an ungasketed joint. Let's also assume that we're going to use torque to control the buildup of initial preload in these as-received bolts as we assemble the joint, suggesting an initial preload scatter of ±30%.

Theoretically, the last bolt tightened in a given region of the joint ends up with 130% of the average preload, less only some embedment relaxation loss. It wouldn't experience any elastic interaction loss because it is the last bolt tightened in its region. It's highly unlikely, however, at least in my opinion, that a given bolt will see *both* the maximum initial preload *and* the minimum elastic interaction loss (none!). If the failure of an individual bolt in service would compromise a safety-related joint, and if the combination of initial preload plus working loads suggests that failure is possible, then we would have to accept $1.30F_{Pa}$ as the maximum residual preload in an individual bolt. This would mean we'd have to use bolts and joint members large enough to support 30% more than average bolt tension; and that could seriously compromise the cost, weight, and size of the joint.

I'm going to avoid this by using engineering judgment to reduce the maximum anticipated preload. I'm going to assume that, worst case, an individual bolt will see the maximum possible initial preload (130% of average) but will also see average embedment loss of 10% and average elastic interaction loss of 18%. From Eqs. (12.3), (12.5), and (12.6), therefore, I estimate maximum residual assembly preload to be

$$\text{Max } F_{\text{Pr}} = 1.30 F_{\text{Pa}} - 0.1 F_{\text{Pa}} - 0.18 F_{\text{Pa}} = 1.02 F_{\text{Pa}}$$

When the time comes I'll enhance the safety of these assumptions by using no more than 60% of yield as my assembly preload target. Note that "target preload" and "average assembly preload," F_{Pa}, are one and the same.

Now for the low end. As already mentioned, we're interested here in averages, not individual minimums. And once again some engineering judgment will be required to pick values. Theoretically we could algebraically add the average preload, F_{Pa}, to the average embedment loss of 10% and the average elastic interaction loss of 18%. This would suggest an average residual preload of (1.0–0.10–0.18) or $0.72F_{\text{Pa}}$ but I think that this is too optimistic. It's extremely important that our joint end up with enough clamping force, worst case, to survive service loads and conditions. Furthermore, experience shows that there are far more things which result in less clamping force than expected rather than in more, as we saw in Chap. 6. So, I'm going to assume that the average residual will be based on an average initial preload of $0.70F_{\text{Pa}}$ less average embedment and elastic interaction losses of 10% and 18%, respectively. Equations (12.4) through (12.6) now give us

$$\text{Min } F_{\text{Pr}} = 0.70 F_{\text{Pa}} - 0.1 F_{\text{Pa}} - 0.18 F_{\text{Pa}} = 0.42 F_{\text{Pa}}$$

Incidentally, we obtain F_{Pa} by using our old friend the short-form torque-preload equation.

$$F_{\text{Pa}} = \frac{T}{KD}$$

where T = torque in in.-lb (N-m)
       D = nominal diameter of the fastener (in., m)
       K = nut factor (see Table 7.1)

The above gives us the vertical scales for our max and min assembly joint diagram. Now we need to decide how long to make the baselines of our triangles.

We could use the methods of Chap. 4 to actually compute the deflections of bolt and joint members under the residual preloads, but this is unnecessary, and we're not much interested in these deflections. It's sufficient and convenient, therefore, to draw the horizontal lines defining the deflections of bolt and joint members to any convenient scale. It's only necessary that they be in the proper proportions to each other. To determine these proportions we use the methods of Chap. 5 to estimate the stiffness of bolt, and then obtain the joint-to-bolt stiffness ratio from Fig. 5.14 or by computation. If that stiffness ratio is 5:1, then the joint-to-bolt *deflection* ratio will be the inverse of that or 1:5 and the joint's deflection line will be drawn one-fifth the length of the bolt's deflection line. Let's assume this 5:1 ratio for the present example, and draw our joint diagram accordingly.

The resulting "preloaded and relaxed" joint diagrams are shown, combined, in Fig. 12.4. The tensile force in the bolts at this point is called the residual assembly preload (F_{Pr}), and it's assumed to be equal and opposite to the clamping force being exerted by that bolt or those bolts on the joint. We're ignoring complications like weight effect and hole interference in this analysis.

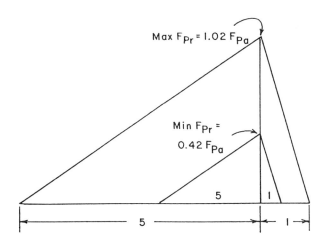

Figure 12.4 Joint diagram showing the anticipated maximum load we expect to see, worst case, in individual bolts (Max F_{Pr}) at the end of the assembly process; taking preload scatter, embedment relaxation, and elastic interactions into account; all with reference to a specific example described in the text. The diagram also shows the worst-case, minimum, average, residual preload (Min F_{Pr}) expected in the same example.

Next we're going to modify this diagram to include an external, tensile load on the joint. Before we do that, however, note that the energy stored in the joint and bolt springs, as a result of the assembly process, is equal to the area enclosed by the joint diagram.

C. Joint Diagram for Simple Tensile Loads

Let's assume that we grab the nut and the head of the bolt with powerful pliers and pull, producing equal and opposite tension forces on each end of the bolt, as in Fig. 12.5. This is obviously an unrealistic method of loading. In fact, we'll never encounter it in practice. But it is the classical way to approach joint behavior—a useful way to further our understanding of the joint and the joint diagram. We'll look at more realistic loading methods later.

Remember, since we tightened the bolt, the joint has been pushing outward on the bolt, keeping the bolt in tension. The new external tension load we have just applied with the pliers helps the joint support the tension in the bolt. In other words, the new external force partially relieves the joint (Fig. 12.6).

Since strain (deformation) is proportional to stress (applied force), the partially relieved joint partially returns to its original thickness, moving back down its elastic curve. Simultaneously, the bolt, under the action of the combined joint force and external force, gets longer—following *its* elastic curve.

Note that the increase in length in the bolt is equal to the increase in thickness (reduction in compression) in the joint. *The joint expands to*

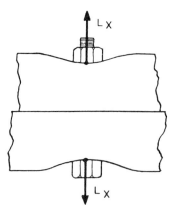

Figure 12.5 Let's assume that an external tensile load (L_X) is applied to the nut and to the head of the bolt as shown.

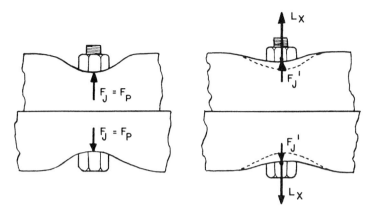

Figure 12.6 Forces on the tightened bolt and joint deflection before and after application of external tension load L_X.

follow the nut as the bolt lengthens. This is an important point; it's a key to understanding joint behavior. Figure 12.7 summarizes the discussion so far.

Now remember: The stiffness of the bolt is only one-fifth that of the joint. This means that, for an *equal* change in deformation (strain), the change in load (force) in the bolt must be only one-fifth of the change of the load in the joint, as noted in Fig. 12.8. The external tension load (L_X) required to produce this change of force and strain in bolt and joint members is equal to the increase in force on the bolt (ΔF_B) *plus* the reduc-

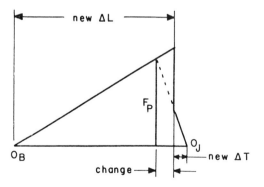

Figure 12.7 When an external tension load is applied, the bolt gets longer and joint compression is reduced. The change in deformation in the bolt equals the change in deformation in the joint.

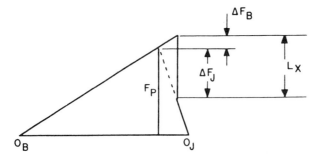

Figure 12.8 Because bolt and joint have different stiffnesses, equal changes in deformation mean an unequal change in force. ΔF_B is the increase in bolt force; ΔF_J is the decrease in clamping force in the joint. L_X is the external load.

tion in force in the joint (ΔF_J), or

$$L_X = \Delta F_B + \Delta F_J \tag{12.7}$$

Many people find this point difficult to accept. After all, we have applied the external load L_X just to the *bolt*, yet the increase in force seen by the bolt is only a small portion of this external load; the rest of the external load is "absorbed" by the joint. We pull the bolt, but it doesn't feel all the pull. This seems to violate a sacred concept, but it really doesn't. We're not, in fact, just "pulling on a bolt." We're applying a tensile load, through the bolt, to a **group** of springs. The bolt spring sees—absorbs—some of this load, the other springs, the joint, sees or absorbs the rest.

It's important to have a clear understanding of all this. The way a bolted joint absorbs external load is another important key to understanding joint behavior. So I think it's worthwhile to look at a crude analogy that may help you see what's happening here.

D. The Parable of the Red Rolls Royce

You're walking up a steep hill in Big City when you see a funny event. A man has just parked his red Rolls Royce on the hill and has gone around back to get a briefcase full of municipal bonds out of the trunk. He's planning to throw them away in an empty lot conveniently provided for trash disposal. The action starts in Fig. 12.9.

Another citizen, carrying a mace, happens by (on his way to collect his unemployment check). He hates guys with red cars, so he seizes a

Figure 12.9   Scene 1 of the parable of the red Rolls Royce.

knife that is lying on the street, and, using his mace as a hammer, nails the first citizen's left foot to the road.

Citizen no. 2 then goes up front and releases the brakes on the car. He locks the doors and exits, stage left, with the key. Our victim must now try to keep the car from rolling downhill, rolling over him on its way (Fig. 12.10).

Being from the country you don't know how to behave in Big City, so you decide to help this guy (whose name is Mr. Joint). You go to the front of the car and start to pull on it—you're applying an external tension force to the car (Fig. 12.11).

But you find that you have a problem. You and Mr. Joint can, indeed, move the car up the hill, but the farther it moves, the less Joint can help

Figure 12.10   Scene 2: The system has been preloaded.

Figure 12.11 Scene 3: An external load (L_X) has been applied, helping the joint (J) support the load applied by the bolt (B).

you, because his foot is still pinned to the road. In order to pull the car away from him, then, you have to pull hard enough not only to add to the force he was originally applying, but also hard enough to *replace* the force he can no longer apply as the car moves away from him.

E. Back to the Joint Diagram—Simple Tensile Load

We have a similar situation in a bolted joint. *Any* external tension load, no matter how small, will be partially absorbed as new, added force in the bolt (ΔF_B), and partially absorbed in replacing the reduction in the force that the joint originally exerted on the bolt (ΔF_J). The force of the joint on the bolt, plus the external load, equals the new total tension force in the bolt—which is greater than the previous total—but the *change* in bolt force is less than the external load applied to the bolt. The joint diagram in Fig. 12.12 shows all this. We'll look at the mathematics of this diagram in a minute. First, let's consider some of the things the diagram can illustrate for us.

F. Changing the Bolt or Joint Stiffness

What happens if we change the stiffness (spring constant) ratio between bolt and joint?

Let's make the bolt a lot stiffer (steeper elastic curve) by using a bolt with a larger diameter. The new joint diagram can be seen in Fig. 12.13. Note that the bolt now absorbs a larger percentage of the same external

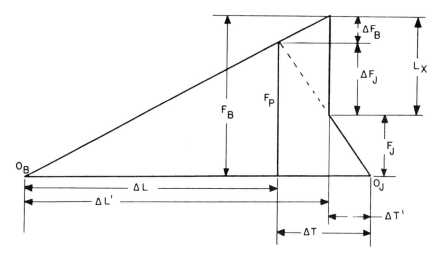

Figure 12.12 Summary of the discussion of the joint diagram. F_P = initial pre-load; F_B = present bolt load; F_J = present joint load; L_X = external tension load applied to bolt.

load. It's as if the owner of that red Rolls were only a child—you work harder as you pull the car away from him, because you're tougher than he is.

If the bolt is made less stiff with respect to the joint, it will see a smaller percentage of a given external load (Fig. 12.14).

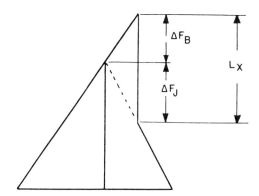

Figure 12.13 Joint diagram when the stiffness of the bolt nearly equals that of the joint.

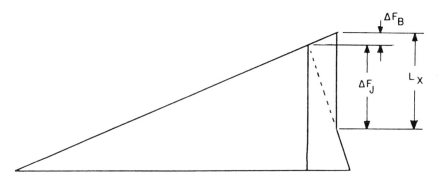

Figure 12.14   Joint diagram with softer bolt and stiff joint.

You might get the same effect if the red Rolls Royce had been owned by a professional football player. You would not have been able to help him as much as you would an ordinary citizen; he's a lot "stiffer" than you are!

The fact that the bolt sees only a *part* of the external load, and the amount it sees depends on the *stiffness ratio* between bolt and joint, has many implications for joint design, joint failure, measurement of residual preloads, etc., as we'll see. But we're not done yet.

G. Critical External Load

If we keep adding external load to the original joint, we reach a point where the joint members are fully unloaded, as in Fig. 12.15. This is called the critical external load. Note that this critical load is *not*, in general,

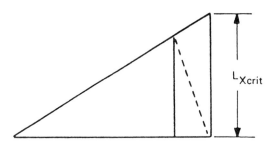

Figure 12.15   A critical external load (L_{Xcrit}) fully unloads the joint (but not the bolt).

equal to the original preload in the bolt, although many people think it is. It's often approximately equal to the preload, however, for several reasons. For example:

1. In many joints the bolt is relatively soft (low spring rate) compared to the joint members. Under these conditions there is a very small difference between the preload in the bolt and the critical external load required to free the joint members.
2. As we'll see later, joints almost always relax after they have first been tightened. Relaxation of 10% or 20% of the initial preload is not at all uncommon. Now, if a bolt has one-fifth the stiffness of the joint (which is also common), then the critical external load required to free the joint members is 20% greater than the residual preload in the bolt when the external load is applied. Under these conditions the difference between the critical external load and the present preload is just about equal and opposite to the loss in preload that was caused by bolt relaxation. Therefore (by coincidence!), the critical external load equals the original preload before bolt relaxation.

H. Very Large External Loads

Any additional external load we add beyond the critical point will *all* be absorbed by the bolt: You're now pulling on the Rolls Royce all by yourself; Mr. Joint has been left behind (Fig. 12.16). Note that if the external load gets still larger, the line describing the action of the bolt becomes nonlinear. We must not forget that all of these joint diagram "triangles" are just portions of the elastic-plastic curves for the bolt and joint members.

Although it's usually ignored in joint calculations, there's another "curve" we should be aware of here. The compressive spring rate of many joint members is not a constant, as discussed in Chap. 2. A more accurate joint diagram would show this, as suggested by Fig. 12.17. More about this in Chap. 13.

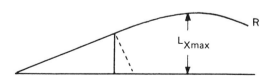

Figure 12.16 L_{Xmax} is the maximum external load the bolt can support before it ruptures (at R).

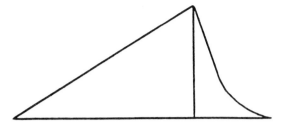

Figure 12.17   The spring rate of the joint is often nonlinear for small deflections of the joint.

So much for our first look at a conventional joint diagram. We'll derive some related mathematics for it later. Before we get into that, however, I want to introduce you to a different type of joint diagram—one we'll find useful when we study fatigue failure. [1]

I. Another Form of Joint Diagram

This time we're going to plot the tension in the bolt and the compression in the joint on different sides of the horizontal axis (which will represent the external load).

Before we apply any external load we have equal and opposite pre-loads in the bolt and in the joint. As we apply external load (see Fig. 12.18), the forces change in both bolt and joint; but the joint, still being

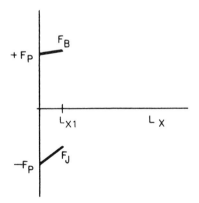

Figure 12.18   The initial preloads in bolt and joint are $\pm F_P$. The bolt load (F_B) increases and the joint load (F_J) decreases when we apply an external load L_{X1}.

five times stiffer than the bolt, sees more change in force for a given change in deformation.

If we apply a critical external load to the assembly, the joint finally becomes completely unloaded. There is no more compressive force in the joint, and the joint members cannot "grow" in thickness any further to follow further elongation of the bolt. You would find, if you increased the external load still further, that there would be an abrupt change in the slope of the bolt curve beyond this point. In fact, you'd find that you were now following the elastic curve for the bolt "alone." The bolt is the only member of the assembly being loaded, now that the joint is fully unloaded and is no longer able to absorb additional load. All of this is shown in Fig. 12.19 [4].

We'll find both types of joint diagram useful later. They are two different ways to look at the same phenomenon.

J. Mathematics of the Joint

Let's return now to the first joint diagram—which we'll use more than the second—and write some equations which will help us analyze and/or design tensile joints. [2] Figures 12.12 and 12.20 shows a completed diagram. The central line, labeled F_P could be representing either maximum or minimum residual "relaxed" preload. We'll call it just F_P at this point

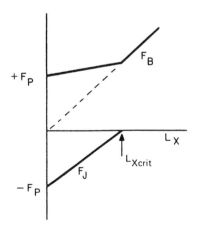

Figure 12.19 If you increase the external load until it exceeds the critical load, you'll see a sudden increase in the rate at which the bolt absorbs further external load.

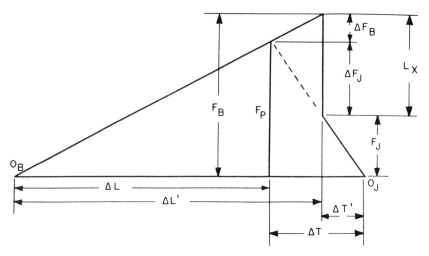

Figure 12.20 Completed joint diagram.

to avoid two sets of essentially identical equations. So—the diagram illustrates:

F_P = initial preload (lb, N)
L_X = external tension load (lb, N)
ΔF_B = change in load in bolt (lb, N)
ΔF_J = change in load in joint (lb, N)
$\Delta L, \Delta L'$ = elongation of bolt before and after application of the external load (in., mm)
$\Delta T, \Delta T'$ = compression of joint members before and after application of the external load (in., mm)
L_{Xcrit} = external load required to completely *un*load joint (lb, N) (not shown in the diagram)

The spring constants or stiffness of the bolt and joint can be defined as follows:

For the bolt: $K_B = \dfrac{F_P}{\Delta L}$ (12.8)

For the joint: $K_J = \dfrac{F_P}{\Delta T}$ (12.9)

By using trigonometry, and by recognizing similar triangles where they occur, we can now derive the following useful expressions:

$$\Delta F_{\mathrm{B}} = \left(\frac{K_{\mathrm{B}}}{K_{\mathrm{B}} + K_{\mathrm{J}}}\right) L_{\mathrm{X}} \qquad (12.10)$$

(until joint separation, after which $\Delta F_{\mathrm{B}} = \Delta L_{\mathrm{X}}$), and

$$L_{\mathrm{Xcrit}} = F_{\mathrm{P}}\left(1 + \frac{K_{\mathrm{B}}}{K_{\mathrm{J}}}\right) \qquad (12.11)$$

The ratio $K_{\mathrm{B}}/(K_{\mathrm{B}} + K_{\mathrm{J}})$ turns out to be so useful that we give it a name and a symbol of its own. Following VDI practice [6] we'll call it the "load factor" (Φ_{K}). It defines that portion of the external tensile load which is seen by the bolt, or:

$$\Phi_{\mathrm{K}} = \frac{\Delta F_{\mathrm{B}}}{L_{\mathrm{X}}} = \frac{K_{\mathrm{B}}}{K_{\mathrm{B}} + K_{\mathrm{J}}} \qquad (12.12)$$

and

$$\Delta F_{\mathrm{B}} = \Phi_{\mathrm{K}} L_{\mathrm{X}} \qquad (12.13)$$

where L_{X} = the external, tensile load (lb, N)

K_{B} and K_{J} are expressed in lb/in. or N/mm

Note that the joint absorbs the rest of the external load, or

$$\Delta F_{\mathrm{J}} = L_{\mathrm{X}} - \Delta F_{\mathrm{B}} = (1 - \Phi_{\mathrm{K}})L_{\mathrm{X}} \qquad (12.14)$$

We're now in position to write more complete equations for the two things we're most interested in: the maximum load seen by the bolts and the minimum clamping force we can expect to see on the joint. To do this we must once again distinguish between max and min residual preloads after relaxation. We're going to write the "room temperature" versions of these equations at this point. In the next chapter we'll extend the equations to include the thermally induced effects of differential expansion.

Maximum anticipated bolt load in service [with reference to Eq. (12.3)]

$$\mathrm{Max}\ F_{\mathrm{B}} = \mathrm{Max}\ F_{\mathrm{Pr}} + \Delta F_{\mathrm{B}}$$
$$\mathrm{Max}\ F_{\mathrm{B}} = \mathrm{Max}\ F_{\mathrm{Pr}} + \Phi_{\mathrm{K}} L_{\mathrm{X}} \qquad (12.15)$$
$$\mathrm{Max}\ F_{\mathrm{B}} = (1 + s)F_{\mathrm{Pa}} - \Delta F_{\mathrm{m}} - \Delta F_{\mathrm{EI}} + \Phi_{\mathrm{K}} L_{\mathrm{X}}$$

Minimum anticipated, per bolt, clamping force in service [with reference to Eq. (12.4)]

$$\mathrm{Min}\ F_{\mathrm{J}} = \mathrm{Min}\ F_{\mathrm{Pr}} - \Delta F_{\mathrm{J}}$$
$$\mathrm{Min}\ F_{\mathrm{J}} = \mathrm{Min}\ F_{\mathrm{Pr}} - (1 - \Phi_{\mathrm{K}})L_{\mathrm{X}} \qquad (12.16)$$
$$\mathrm{Min}\ F_{\mathrm{J}} = (1 - s)F_{\mathrm{Pa}} - \Delta F_{\mathrm{m}} - \Delta F_{\mathrm{EI}} - (1 - \Phi_{\mathrm{K}})L_{\mathrm{X}}$$

As mentioned earlier, we'll usually be more interested in the total clamping force on the joint than in the clamping force created by an individual bolt. To get the total, since we're dealing with averages here, we merely multiply the per bolt force by the number of bolts (N) or:

$$\text{Total, Min } F_J = N \times \text{ per bolt Min } F_J \tag{12.17}$$

We'll extend these equations further in Chap. 13 to include the effects of differential expansion; and further still in Chap. 22 to include the effects of gasket creep; but Eqs. (12.14) through (12.16) will be all we'll need for many (perhaps most) applications. Incidentally, the load factor, Φ_K, is sometimes called the "force ratio" or "the joint stiffness ratio." The latter could be confused with the simpler joint-to-bolt stiffness ratio, so I'll call Φ_K the load factor to avoid confusion.

K. Continuing the Example

In the joint diagram example we started in Sec. IB we computed the maximum and minimum residual assembly preloads; they were $1.02F_{Pa}$ and $0.42F_{Pa}$, respectively, where F_{Pa} was the average, initial, assembly preload. Now let's continue our example by adding the effects of an external load. Let's assume that the external load (L_X) will equal $0.25F_{Pa}$.

First, we must compute a load factor. Remember that the joint-to-bolt stiffness ratio in our example was 5:1. Therefore, from Eq. (12.12):

$$\Phi_K = \frac{1}{1 + 5} = 0.17$$

We can now estimate the changes which will occur in bolt tension (F_B) and the clamping force on the joint (F_J) when L_X is applied. From Eq. (12.13):

$$\Delta F_B = 0.17L_X = 0.17(0.25F_{Pa}) = 0.043F_{Pa}$$

and, from Eq. (12.14),

$$\Delta F_J = (1 - 0.17)L_X = (1 - 0.17)(0.25F_{Pa}) = 0.21F_{Pa}$$

Now we can compute the maximum load to be seen by the bolt, as a result of the assembly process plus the external load (Max F_B) and the minimum clamping force we can expect to achieve on the joint as a result of the same factors (Min F_J). From Eq. (12.15):

$$\text{Max } F_B = \text{Max } F_{Pr} + \Delta F_B$$
$$\text{Max } F_B = 1.02F_{Pa} + 0.043F_{Pa} = 1.06F_{Pa}$$

where F_{Pr} is the residual preload after embedment and elastic interaction loss. Next, from Eq. (12.16):

$$\text{Min } F_{\text{J}} = \text{Min } F_{\text{Pr}} - \Delta F_{\text{J}}$$

$$\text{Min } F_{\text{J}} = 0.42F_{\text{Pa}} - 0.21F_{\text{Pa}} = 0.21F_{\text{Pa}}$$

The results are plotted in Fig. 12.21. Note that these results show that there is an approximately 5:1 ratio between the maximum force which some of the bolts, worst case, must be able to support; and the minimum clamping force we can expect, again worst case. These results are based, of course, on the assumptions I made in Sec. IB and may not be entirely valid, but those assumptions were reasonable and the results are certainly not out of line. The results show us why poor control of the assembly process leads to "overdesign" of the joint. In order to get that minimum clamping force we must use bolts and joint members that are five times more massive than would be necessary if we could control residual, assembly preloads more accurately, perhaps by using something other than torque control to reduce the scatter in initial preload and/or by using ultrasonics or equivalent to allow us to compensate for elastic interactions. Incidentally, the energy contained within an externally loaded joint, in service, is equal to the area enclosed by the joint diagram of Fig. 12.20. If you have computed both deflections and forces you can compute the

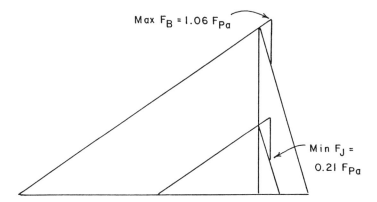

Figure 12.21 Continuing the example first illustrated in Fig. 12.4, this time to include the effects of an external tensile load on the joint. The diagram shows the results of calculations to estimate the maximum, worst-case, load expected in individual bolts (Max F_{B}) and the worst-case, minimum, average, perbolt clamping force on the joint (Min F_{J}). See text for details.

energy stored within the joint from

$$E_J = 0.5 \times \Delta T' \times F_J$$

and the energy stored in the bolt from

$$E_B = 0.5 \times \Delta L' \times F_B$$

These equations are of interest only to those having a morbid interest in bolted joints—like me!—and we won't find much if any use for them. So I'm not going to give them numbers.

So much for simple joint diagrams. They're very basic, and very useful, even though they describe the behavior of the joint under a very *un*common type of loading—a tensile load applied between the bolt head and nut (as with our pliers!) or at least applied to the bolt between the plane of the upper surface of the joint and the plane of its lower surface.

As we're about to see, our equations must be modified and different diagrams must be constructed if the loads are applied at some point other than head-to-nut; but the resulting bolt loads and clamping forces will often be very similar to those we've just computed. Even though head-to-nut loading is almost never encountered in practice, the joint diagram of Fig. 12.12 and its related equations are usually used in design work because they give a worst-case result for the maximum loads—and load excursions—to be seen by individual bolts; and they give reasonable estimates of the minimum clamping force on the joint. They do not, however, give worst-case estimates for the clamping force. If that is the critical factor in our application, then we want to refine our equations and joint diagram to give us a more realistic view of the way tension loads are actually applied to a joint. We do this by introducing the concept of loading planes.

II. LOADING PLANES

We have taken a detailed look at the behavior of a bolt in a joint when tension forces are applied to both ends of the bolt. Now we're going to look at the behavior of the bolt and joint when tension loads are applied at other points.

Note that loads are seldom, if ever, applied to a single "point" in a bolted joint. Loads are created by pressure, weight, shock, inertia, etc., and are transferred to the joint by the connected members. An accurate description of where that load is applied would require a detailed stress analysis (e.g., a finite-element analysis). The people who developed the classical joint diagram, however, have found a simpler way to "place" the load. They define hypothetical "loading planes," parallel to the joint interface, and located somewhere between the outer and contact surfaces of the each joint member. They then assume that the tensile load on the joint is applied to these loading planes. Joint material between the loading planes will then be (theoretically) unloaded by a tensile load; joint material out-board of the planes will be "trapped" between plane of application of load and the head (or nut) of the fastener.

On our first example (last section), these planes coincided with the upper and lower surfaces of the flange or joint. For our next example, loading planes will coincide with the interface between upper and lower joint members, as suggested in Fig. 12.22.

Note that the loading plane is pure fiction. It's a "bugger factor" used to correct a joint diagram analysis, to make the analysis agree, for example, with experimental results (which might show how an external tensile load *actually* affects the tension in a preloaded bolt). In any event, the loading plane is a useful concept. Here's how it works.

A. Tension Applied to Interface of Joint Members

Remember, in our first analysis:

The bolt was treated as a tension spring.

The upper and lower flange members were treated as compression springs.

Tensile loads applied to each end of the bolt stretched (loaded) the bolt and partially relieved (unloaded) the joint.

All of this was analyzed in a joint diagram which showed how one spring was loaded and the other unloaded by the external load.

If the same tensile force, however, were applied to the interface between the upper and lower joint members, then *both* the bolt tension

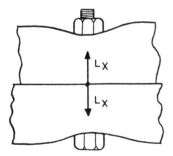

Figure 12.22 In this example the external tension load is applied at the joint interface, as shown here.

spring and the joint compression spring would be *loaded* by the external load. What does the joint diagram look like in this situation.

The two flange pieces are originally exerting equal and opposite forces on each other—forces equal to the preload in the bolt. As we start to apply a small external load at the interface, we partially replace the forces that the two joint members are exerting on each other. We're *relieving* these flange-on-flange forces rather than adding to them, to start with.

Going back to the parable of the red Rolls Royce, this time you're trying to help Mr. Joint by picking him up, reducing the force he is exerting on the road, without changing the amount of force he is exerting on the car, as shown in Fig. 12.23.

In the joint, of course, this means that the external load reduces the flange-on-flange force without increasing the *total* force in either the flange members or the bolt—yet. The joint diagram for this situation is shown in Fig. 12.24 [2]. Note that I have chosen to draw both elastic curves (bolt and joint) on the same side of the common vertical axis (the axis that represents original preload or F_P). I do this because both springs are *loaded* by the external force.

When the external load equals the original preload in the bolt, it will have replaced all of the force that each joint member was exerting on the other. In the red Rolls Royce parable, you've just lifted Mr. Joint completely off the road, but he's still exerting the same amount of force on the car. Neither bolt deformation nor joint deformation has changed to this point.

Increasing the external load beyond this point will now add to the original deformation of both the bolt and the joint members. The bolt gets longer and the joint compresses more. The joint diagram merely "gets larger." (In Fig. 12.25 the dashed lines represent the original joint diagram; the solid lines represent the new joint diagram.) Note that at all times,

Figure 12.23 We're going to relieve Mr. Joint by picking him up and pushing through him.

III. OTHER NONLINEAR FACTORS

A. The Nut-Bolt System

Prying or eccentric action is not the only cause of nonlinear behavior of
a bolted joint. Here's another.

Let's assume that we apply tension to a steel rod of uniform cross
section by pulling on it with our fingers, as shown in Fig. 13.18. As we pull,
we're going to measure the distance between the tips of the fingernails on
our two index fingers; we're also going to measure the change in length
of the rod.

Because of the way in which our fingers are constructed, we would
detect a large and visible change in the distance between our fingernails
even though the balls of our fingertips had only rolled, not slipped, over
the surface of the rod. The simultaneous change in length of the rod itself,
however, would be very small, because we would not be able to exert
much tension this way.

If we plotted the change in length of the rod as a function of the
applied tension, we would find that it would be a straight line. If we plotted
the change in distance between our fingernails as a function of applied
load, however, we would find that it was, in general, not be a straight
line, but depended instead on the load deformation behavior of our flesh
and muscles. I'm not prepared to suggest what the resulting curve would
look like!

A similar situation occurs when we measure the change in spacing
(ΔL_W) between the washers on a bolt, nut, and washer system which is
being subjected to internal pressure load, as in Fig. 13.19.

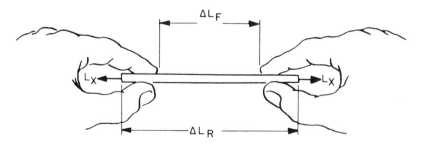

Figure 13.18 We pull on a steel rod to stretch it, measuring as we do the change
in length of the rod (ΔL_R) and the change in the distance between our fingernails
(ΔL_F).

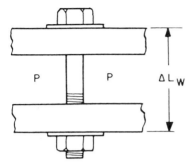

Figure 13.19 Internal pressure (P) applies a tension load to this bolt, nut, and washer system. (Modified from Ref. 8.)

If we also measure the change in length of the total bolt (ΔL_B) as a function of the applied load, we will find that it is a straight line. If we were to compute the stiffness of the bolt (the slope of the line) using the equations of Chap. 5, we would find that our calculations would probably approximate the measured stiffness. The bolt would behave in the anticipated linear elastic fashion, as long as we did not use too much pressure to load it.

If we also plot the change in the distance between washers (ΔL_W) in this situation as a function of applied load, however, we will find that the behavior is very nonlinear, as suggested in Fig. 13.20 [8, 9].

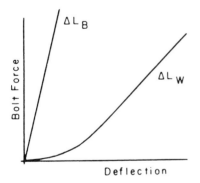

Figure 13.20 Distance between washers (ΔL_W) of the system shown in Fig. 13.19 as a function of the applied load. Simultaneous change in length of the bolt alone (ΔL_B) is also shown. (Modified from Ref. 8.)

The thing we are loading—the bolt—behaves in an elastic fashion, but our method of applying the load—through the nut-and-washer system—introduces nonlinearities if we measure the result at the wrong point. The reason for this nonlinear behavior, of course, is that the nuts and washers have to settle into the threads of the bolt in order to push on the bolt, some embedment occurs, the washers may flatten out a little, etc. It is the nature of the loading mechanism rather than the thing being loaded which determines the apparent behavior.

Who cares? Well, it turns out that the joint designer cares—or should. After all, the joint neither knows nor cares about the behavior of the bolt as an isolated body. It is always loaded by a bolt, nut, and washer assembly. So the force vs. change-in-length behavior which it sees is that reflected by the distance between the two washers which are being used to clamp it together. As a result, the distribution of an external load between bolt and joint, the apparent stiffness of the joint, the apparent stiffness of the bolt as far as the joint is concerned, etc., are all drastically different than would be predicted by calculations based on the assumption that the joint and bolt will both behave as uniformly loaded, linear elastic members.

Some, at least, of this nonlinear behavior is caused by localized plastic yielding in the threads, embedment, etc., so the behavior of a joint which has been preloaded, released, and then reloaded will probably be more linear than the behavior of a fresh joint.

If sufficient load is applied to the bolt, furthermore, it will be operating in a region where its behavior is elastic (the upper or right-hand end of the ΔL_W curve shown in Fig. 13.20). Even here, however, the stiffness of the bolt, as seen by the displacement between washers, is going to be only about half the stiffness computed by our equations (which consider merely the body of the bolt and not the bolt, nut, washer system). This is because the bolt, the nut, and the washers are each springs; they are each loaded, in series, and their combined stiffness will be a function of the stiffness of each one.

$$\frac{1}{K_T} = \frac{1}{K_B} + \frac{1}{K_N} + \frac{2}{K_W} \tag{13.10}$$

where K_T = stiffness of the entire bolt, nut, washer system (lb/in., N/mm)
K_B = stiffness of the bolt (lb/in., N/mm)
K_N = stiffness of the nut (lb/in., N/mm)
K_W = stiffness of the washer (lb/in., N/mm)

Note that in this situation some, at least, of the nonlinear behavior is determined by plastic yielding, etc., within the system.

We could, therefore, expect to find hysteresis effects, etc., if we made a close enough examination.

Note, too, that in this case the apparent stiffness of the bolt (as seen by changes in the distance between the washers) is a function of preload or tension level as well as of the usual dimensions. The system has a very low stiffness at low load levels and a stiffness approaching half that of the bolt alone at high loads. This is similar to the nonlinear behavior of a block under compressive loads, as we saw in Chap. 5, (Fig. 5.8), and is really caused by the same phenomenon: initial plastic deformation of the body under compressive stress.

The situation illustrated in Fig. 13.20 was encountered in some recent experiments with a Superbolt Torquenut shown in Fig. 13.21. This device consists of a cylindrical nut (called a Torquenut) which is run down, by hand, against a heavy, hard washer. A group of "jackbolts" are then tightened, in a cross-bolting pattern, to tension the large stud or bolt on which the Torquenut has been placed. Since only a small wrench, and small amount of torque, is required to tighten the jacking bolts, the technique allows very large fasteners to be tightened in very inaccessible places. It also has other applications, of course.

I describe it here because an attempt was made recently to control the tension built up in the large stud by measuring the change in the gap between the Torquenut and the washer as the jacking bolts were tightened.

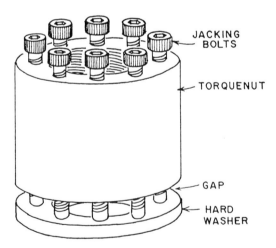

Figure 13.21   A Superbolt Torquenut allows large-diameter fasteners to be tightened with small, low-torque tools. It also illustrates the behavior shown in Fig. 13.20, as explained in the text.

If there were a one-to-one relationship between the change in this gap and the stretch of the bolt, gap measurement would provide a ready means to control the tightening process.

The experiments revealed significant differences between gap change and bolt stretch, however, with the gap change far exceeding bolt stretch. The investigators felt that the difference resulted from the fact that the gap change reflected elastic and plastic deformation of the joint members, washer, and thread surfaces, as well as elastic stretch of the bolt—all as suggested in our earlier discussion [19].

IV. FLANGE ROTATION

Another behavioral factor not accounted for by the classical joint diagrams of Chap. 12 is the phenomenon called flange rotation. It's rarely encountered outside of the pressure vessel and piping world, but it can be a problem there.

Rotation can occur when some form of raised face flange is used, with no contact outside of the bolt circle. As suggested in Fig. 13.22, when the

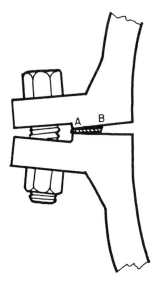

Figure 13.22 Illustration of "flange rotation." The outer edges of the flange are pulled toward each other as the bolts are tightened. The inner diameter of the gasket can be partially unloaded by this process. Thermal expansion and/or contained pressure can also cause flange rotation.

bolts in such a joint are tightened they tend to pull the unsupported outer edges of the joint together. Point A in the illustration acts as a fulcrum in this process. One result is a nonuniform contact pressure between joint members and/or on a gasket. Contact pressure would be greater at point A than at point B, for example.

Rotation can be caused, as mentioned, by the act of tightening the bolts. It can also be caused (or increased) by internal pressure in the system and/or by thermal expansion. Both effects tend to "inflate" the vessel or pipe, with thinner sections deforming (deflecting) more than thicker, again tending to rotate the joints.

When rotation is caused by pressure or temperature, the outer edge of the gasket can be stressed more heavily than it was during initial assembly. The inner edge can be partially or wholly relieved. (One result: As a system comes up to service temperature a leak can open up; or an existing leak can be shut down. More about this in Chap. 19.)

During this process the tension in the bolts decreases as the outer edges of the joint move toward each other. So there can be a simultaneous increase in stress on (a portion of) the gasket and a decrease in bolt tension. That certainly violates the predictions of a classical joint diagram.

I don't know any simple way to predict the amount a given flange will rotate in a given application, but analytical programs which do this are available [10].

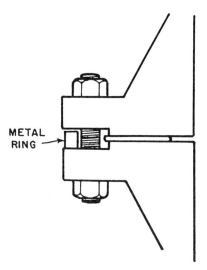

METAL RING →

Figure 13.23 A metal ring, inserted between flange surfaces, can reduce or eliminate objectionable flange rotation.

Petrochemical engineers who must cope with rotation say that it can greatly increase the difficulties of sealing a joint; some even say that rotations as small as 0.10° can make a tight seal almost impossible.

One way to prevent excessive flange rotation is to insert a metal ring between flange surfaces, outboard of the bolts, as shown in Fig. 13.23. Note that the ring must be correctly dimensioned to work properly, however. If it's too thin, the flange may still be able to rotate. If it's too thick, it may prevent the bolts from properly preloading the gasket. If it allows for gasket deflection but prevents rotation, it will provide an outboard support or fulcrum for the flanges, and minimize rotation. Again, however, it won't be easy to dimension a support ring. Designing flanges which are in contact outside the bolt circle is more difficult than designing what the ASME Boiler and Pressure Vessel Code calls Part A flanges: those in contact only inside the bolt circle. [23]. One flange influences the other and the interactions must be considered. This would presumably be true, also, if contact outside the bolt circle were provided by an interface ring.

V. THERMAL EFFECTS

Now let's look at what a change in temperature can do to bolt loads and the interface clamping force. We'll look at five effects: a change in elasticity or stiffness of bolt and joint members; a loss of strength of the bolt; modification of bolt loads and clamping force by differential thermal expansion or contraction; creep relaxation; and stress relaxation. Note that each of these factors acts to change the clamping force in the joint and/or the tension in the bolts. We've called changes of this sort "instability in the clamping force." It's important for us to know how to estimate the amount of change which will occur, and to learn how to reduce it, or compensate for it.

A. Change in Elasticity

The modulus of elasticity of bolting materials decreases as the temperature of the material rises, as we saw in Table 4.6. As a typical example, the modulus of an A193 B7 bolt drops by about 17% when the temperature of the bolt is raised from 70°F (20°C) to 800°F (427°C). Unless there are offsetting differential expansion effects (to be considered soon), the preload in the bolt will decrease in the same ratio, because the bolt gets "less stiff." We can estimate the preload in a bolt at elevated temperature by using the simple expression:

$$F_{P2} = F_{P1} \frac{E_2}{E_1} \qquad\qquad (13.11)$$

where F_{P2} = preload at elevated temperature (lb, N)
F_{P1} = preload at room temperature (lb, N)
E_2 = modulus of elasticity at elevated temperature (psi, GPa)
E_1 = modulus of elasticity at room temperature (psi, GPa)

Note that temperatures below ambient will *increase* the preloads introduced during room temperature assembly. By the same token, preloads introduced during "hot bolting" procedures at elevated temperatures will subsequently increase when the system is shut down and the joint cools (again assuming no differential contraction).

B. Loss of Strength

The tensile strength of most bolts will also decrease as temperature rises, as shown by Table 4.4. The yield strength of an A193 B8 Class 1 bolt, for example, drops from 30 ksi at room temperature to only 17 ksi at 800°F; and B8 is not recommended for use above 800°F. A heavily preloaded bolt could fail, therefore, if exposed to extreme temperatures—by a fire, for example.

This problem has received new attention in recent years. It has been discovered that low-cost suppliers of J429 Grade 8 bolts have often made the bolts of boron steel instead of medium carbon steel. Boron steel is permitted for Grade 8.2, but many suppliers have instead marked and sold boron steel bolts as Grade 8. And this has caused problems at elevated temperatures.

Boron steel and medium carbon alloy steel have similar properties at room temperature. But boron steel bolts can lose as much as 75% of their prestress after 80 hr of exposure to temperatures as low as 700°F (371°C), compared to a 45% loss for the alloy steel bolts [11]. Attempts are currently being made to identify the so-called "counterfeit" bolts in the system, and to prevent further substitutions of this sort.

C. Differential Thermal Expansion

One of the most troublesome thermal effects the bolting engineer must deal with is differential thermal expansion or contraction between joint members and bolts. To illustrate the problem, consider the automotive head joint shown in Fig. 13.24. This is a sketch of an actual joint. Carbon steel bolts are used to bolt an aluminum head to a cast iron engine block, loading a steel and asbestos gasket. The relative coefficients of expansion for these materials might be as follows (all $\times$ 10^{-6} in./in./°F):

Carbon steel   6–7

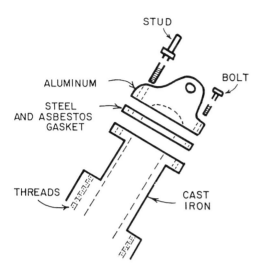

STUD

ALUMINUM

BOLT

STEEL
AND ASBESTOS
GASKET

THREADS

CAST
IRON

Figure 13.24 Automobile engine head. The variety of materials used here results in differential expansion between the aluminum head and the carbon steel bolts. This creates the changes in gasket stress seen in Fig. 13.25.

Cast iron   6
Aluminum   12–13

Figure 13.25 shows the stress on the gasket (or the equivalent tension in the bolts), as the engine of Fig. 13.24 is assembled and then used. A certain amount of initial stress is created when the engine is assembled at room temperature (point A). When the engine is started, gasket stress rises sharply to point B, thanks to the fact that the aluminum head heats up more rapidly than the bolts—and because the coefficient of expansion of the aluminum is approximately double that of the carbon steel bolts. The aluminum tries to expand but is trapped between the cast iron block and carbon steel bolts. The differential expansion increases the tension in the bolts and the clamping force on the gasket.

As the engine continues to run, the temperature of the bolts rises to more nearly approximate (or equal) that of the aluminum head. As a result the bolts expand some more, somewhat reducing the clamping force and stress on the gasket. Because the aluminum still wants to expand more than the other materials, the steady-state stress on the gasket remains higher than the initial assembly stress.

When the engine is turned off, and returns to room temperature, the stress on the gasket returns to the original value. At least it will do this

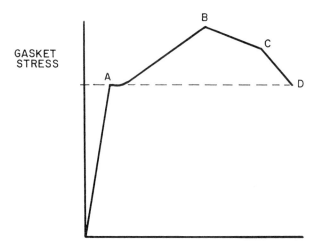

Figure 13.25 The automotive gasket of Fig. 13.24 is initially loaded to point A during assembly at room temperature. As the engine heats up (ahead of the bolts), the stress rises to point B. It falls to point C when the temperature of the bolts "catches up" with the temperature of the head, then returns to the original preload value at point D when the engine is shut down and returns to room temperature.

if the added stress has not caused some irreversible plastic deformation in the gasket. As we'll see in Chap. 19, thermal cycles can "ratchet" all of the clamping force out of a gasketed joint, thanks to hysteresis and creep in the gasket. In the present example, however, the manufacturer says that the stress merely returns to the original value.

It's often useful to be able to estimate the change (increase or decrease) in bolt tension, and/or clamping force on the joint, created by thermal expansion. We can proceed as follows.

The relationship between initial preload in a bolt, and the change of length of that bolt, can be estimated with Hooke's law.

$$F_\mathrm{P} = \frac{A_\mathrm{s}E}{L_\mathrm{E}} \Delta L_\mathrm{B} \tag{13.12}$$

where F_P = preload (lb, N)
  A_s = tensile stress area (in.^{2}, mm^2)
  E = modulus of elasticity of bolt (psi, N/mm$^\mathrm{s}$)
  L_E = effective length of bolt (in., mm)
  ΔL_B = change in length of bolt (in., mm)

The additional tension (or loss of tension) created in the bolt by differential expansion between joint members and bolt (F_T) can be approximated by

$$F_T = \frac{A_s E}{L_E}(\Delta L_J - \Delta L_B) \tag{13.13}$$

where ΔL_J = change in length (thickness) of the joint (in., mm)
ΔL_B = change in length of the bolt (in., mm)
F_T = the additional tension or loss of tension created by differential expansion (lb, N)

Both the tension in the bolt and the clamping force on the joint will be increased if ΔL_J is greater than ΔL_B. They'll be decreased if the bolt expands more than the joint. We can compute these changes in length/thickness as follows.

$$\Delta L_B = \rho_1 L_G(\Delta t) \tag{13.14}$$

$$\Delta L_J = \rho_2 L_G(\Delta t) \tag{13.15}$$

where ρ_1 = coefficient of thermal expansion of the bolt material (in./in./°F, mm/mm/°C)
ρ_2 = coefficient of thermal expansion of the joint material (in./in./°F, mm/mm/°C)
L_G = the grip length of the joint (in., mm)
Δt = the change in temperature (°F, °C)

It's important to note that the length/thickness changes used in Eqs. (13.13) through (13.15) are changes which would caused by a change in temperature if the bolts had not been tightened. For example, if the bolts and joint members were lying on a bench, and the temperature in the room raised or lowered, the parts would experience the changes in dimension calculated by Eqs. (13.14) and (13.15). Note that Eq. (13.13) can also be written in terms of K_B, the stiffness of the bolt, as follows:

$$F_T = K_B(\Delta L_J - \Delta L_B) \tag{13.16}$$

Now, all of the above is "conservative." The increase in tension estimated by Eq. (13.16) will probably be greater than the actual increase, because the equation assumes, in effect, that the total thermal load on the joint will be seen by the bolts. In practice, the stiffness ratio between bolts and joints, and effects such as flange rotation, will modify the tension created by the thermal load. For example, Evert Rodabaugh of Battelle, in the flange design program FLANGE [10], suggests that:

$$F_T = \frac{\Delta L_J - \Delta L_B}{\dfrac{1}{K_B} + \dfrac{1}{K_{J1}} + \dfrac{1}{K_{J2}} + \cdots} \tag{13.17}$$

where the various K_J's define the stiffness of individual joint members, bolts, gasket, joint rotation, etc.

In another, private, reference, an engineer in a large petrochemical company has used the expression

$$F_T = \frac{\Delta L_J - \Delta L_B}{\frac{1}{K_B} + \frac{1}{K_J}} \tag{13.18}$$

This reduces to

$$F_T = \frac{K_B K_J}{K_B + K_J} (\Delta L_J - \Delta L_B) \tag{13.19}$$

which says that the change in force in the bolt is equal to that portion of the external load a joint diagram would predict the bolt would see. (See Fig. 13.26.)

We can rewrite Eq. (13.19) in terms of the load factor (Φ) first defined in Chap. 12 and further discussed in Sec. IIB of this chapter.

$$F_T = \Phi K_J (\Delta L_J - \Delta L_B) \tag{13.20}$$

where Φ = any one of the load factors we've defined; Φ_K, Φ_{en}, etc.
K_J = the joint stiffness associated with that particular load factor (lb/in., N/mm)
F_T = additional tension or loss of tension (lb, N)

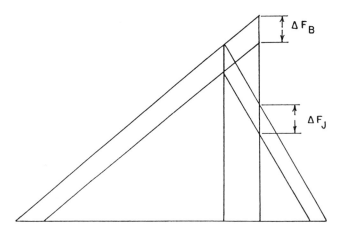

Figure 13.26 Differential thermal expansion between bolts and joint members can simultaneously increase—or simultaneously decrease—the tension in the bolts and the clamping force on the joint, causing the entire joint diagram to grow or shrink as shown here.

For example, if the joint is loaded eccentrically, the bolt is offset from the axis of gyration and the load is applied at some point within the joint members, we'd use the load factor Φ_{en} as defined in part B of Fig. 13.15. We'd also, then, have to use the resilience of that sort of joint, r_J'' (also defined in Fig. 13.15) to define the joint's stiffness ($K_J = 1/r_J$). This force F_T is seen by both the bolt and the joint, because it's created as these two elements fight each other. The force caused by differential thermal expansion, in other words, should not be treated like an "external" load.

We can draw a joint diagram to illustrate this, as in Fig. 13.26. I've included an external load in this diagram because a change in temperature is usually accompanied or preceded by the application of load. We're interested in the changes which affect an in-service, working joint. But it's not the external load which causes the changes shown in Fig. 13.26, it's the differential expansion. This makes the whole joint diagram—with the exception of the external load line—grow or shrink, depending upon whether F_T is positive or negative.

Note that two different bolt lengths must be used in these equations for accuracy. Since we're interested in the forces created by differential expansion between the joint and bolts, we must look first at the relative expansion of the joint—and of that portion of the bolt which "traps" the joint. We consider ΔL_B, in other words, only for the grip length of the bolt.

Consider what we might conclude if we focused, instead, on the overall length or on the effective length of the bolt. Assume, for a moment, that bolts and joint members are made of the same material and experience the same rise in temperature. If we computed the expansion of bolt and joint, using different "lengths" for each—grip length for the joint and effective length for the bolt, for example—we'd conclude that the bolt would expand more than the joint. After all, expansion is an "inch per inch" proposition. So we'd predict a drop in bolt tension. But this would clearly not occur if all parts were of the same material and experienced the same temperature changes. Thermal expansion of such a system would change its dimensions, but not the internal stresses or the forces between subassemblies. The bolt as a free body will expand more than the joint, it's true, but the expansion within and past the nut doesn't create the forces we're concerned about. So we focus on the relative expansion of the joint and of the bolt within the grip length.

When we want to estimate the effect this bolt-joint interference has on bolt tension or clamping force, however, we must use the correct values for bolt and joint stiffness. If we assumed at this point that the bolt was only as long as the grip, we'd conclude that it was stiffer than it really is. So we must now use the effective length of the bolt (grip length plus half the height of the head plus half the thickness of the nut) when computing bolt stiffness K_B. The "length" of the joint, of course, remains unchanged as the grip length. This is illustrated in Fig. 13.27.

D. Stress Relaxation

Two other thermal effects we must be concerned about are the closely related phenomena of creep and stress relaxation. Creep is the more familiar of the two, and can be illustrated as follows.

Let's assume that we've fastened one end of a steel bolt to a ceiling. We now hang a heavy weight from the lower end, and raise the temperature in the room to 1000°F (538°C). Since this temperature would place the bolt in its "creep range," it will slowly stretch, necking down as it does so. Eventually it will get too thin to support the weight, and the bolt will break. The slow increase in length of a material under a heavy, constant load is called creep. It can occur in some materials (e.g., lead) at room temperature, but is more commonly encountered at elevated temperatures.

Stress relaxation is a similar phenomenon. This time, however, we're dealing with the steady loss of stress in a heavily loaded part whose dimensions are fixed. A bolt, for example, is tightened into a joint, which means it is placed under significant stress. But the bolt doesn't get longer. Its length is determined by the joint (and the nut). If exposed to 1000°F, however, the bolt will shed a significant amount of stress—of initial preload, if you will—as the molecules struggle to relieve themselves of the imposed load.

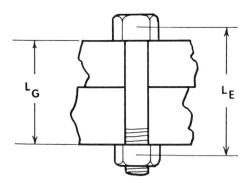

Figure 13.27   When computing the effects of differential expansion, one must use the grip length of the bolt (L_G) to estimate the amount of "interference" between bolt and joint, but use the effective length (L_E) of the bolt when computing the effect this interference will have on bolt length or tension. See text for discussion.

In Fig. 4.3 we saw an estimate of the percentage of initial tensile stress various types of bolts would lose in 1000 hr if exposed to temperatures up to 1472°F (800°C). A carbon steel bolt, for example, would lose about 90% of its initial preload in 1000 hr at 752°F (400°C); while an A193 B8 bolt would lose only 10% or so at that temperature.

These losses, incidentally, would not be repeated. The carbon steel bolt would not lose 90% of the remaining 10% during the second 1000-hr period; it would probably have stabilized at the 10% figure, and might serve indefinitely there, at 400°C, if left in place.

This is because the tendency to relax decreases as the "driving force," the tensile stress in the bolt, decreases. This is illustrated in Fig. 13.28, which shows the relaxation of several A-286 studs in 100 hr at 1200°F (649°C) [12]. The stud whose behavior is illustrated by curve C in Fig. 13.28 appears to have stabilized, at least. The others may lose a little more as time goes by; but won't lose the 50% or so they lost in the first 100 hr.

The fact that a material stabilizes after a certain amount of stress relaxation means that we can compensate for it by overtightening the bolts during initial assembly. If they're going to lose 50%, we put in an extra 50% to start with, assuming that they and the joint can stand this much at room temperature. But we must realize that the loss can be substantial, and can continue for long periods of time. Mayer reports relaxation of

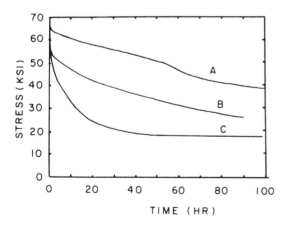

Figure 13.28 Stress relaxation of A-286 bolts exposed to 1200°F for 100 hr. Bolt A was originally stressed to 70 ksi; B and C only to 60 ksi. In the test the bolts were allowed to creep a small amount, then the loads on them were reduced to reverse the creep. Bolt B was relieved in coarser steps than bolt A.

50% in CrNiMo fasteners tested at 500°C; with other materials, tested at 425–480°C, still relaxing after periods as long as 10^4 hr [14]. Markovets says that E110 (25Cr2MoV) steel, normalized at 1000°C and tempered at 650°C, will relax 20% in the first 100 hr at 500°C, 56% (and still relaxing) at 10^4 hr [13]. Other data are reported in Tables 3.7 and 3.8.

Note that there can be a penalty for using better materials to reduce stress relaxation. Bolts made from a material such as Nimonic 80A, which can be used to 1382°F (750°C), can cost five to six times as much as bolts made of a more common material such as A193 B7 [18].

Stress relaxation is affected by the geometry of the bolt as well as by the material from which it's made. A test coupon of a given material, for example, will usually relax much less than a bolt of the same material, because so much of the relaxation occurs in the threads. There are exceptions to this rule, but it's generally true [18].

Whether or not threads are rolled or cut can make a difference at the highest temperatures, but this is said not to matter at temperatures where relaxation properties are generally good (presumably below the service limits). Poor-quality threads, however, which we considered in Chap. 3, will relax much more than good-quality threads [18].

The importance of quality threads is also shown by the fact that Grade B16 bolts used with high-quality stainless steel nuts (Grade 8) relaxed, in one series of tests, much less than the same bolts with carbon or low-alloy nuts [18].

Attempts have been made to use readily available creep data to predict the amount of stress relaxation one might encounter in a given situation (because stress relaxation data are much less common). Creep data are subject to much variation, however, with one investigator reporting data which differ significantly from data reported by another. As one result, short-term stress relaxation estimates cannot be based on creep data. Estimates of long-term loss, however, are more reliable. One can use, for example, a relationship called *Gieske's correlation*, which says that "the residual stress in a bolting material, after 1000 hr at a given temperature may be taken as equal to the stress required to produce a 0.01% creep in 1000 hr at that same temperature" [12].

E. Creep Rupture

Although bolts are loaded under constant strain (constant length) conditions, they can fail by a mechanism known as "creep rupture." We tighten the bolt (i.e., load it heavily) at room temperature, then expose it to temperatures in the creep range. Stress relaxation will partially relieve it. Now we return it to room temperature, to perform maintenance on the system, for example. After completing our work, we retighten it to the original loads and put it back in service at elevated temperatures. It will experience a second cycle of stress relaxation.

During each such cycle the bolt experiences some creep damage, even though its length never exceeds its length when first tensioned. After a certain number of cycles it will develop a crack, and will eventually break. The failure is called creep rupture.

It needn't take many cycles to rupture a bolt or stud in high-temperature service. Seven cycles can do it if initial preloads are high enough. Lower preloads in the same studs might extend this to 25 cycles. With only two maintenance cycles a year, however, 25 cycles could equate to $12\frac{1}{2}$ years of useful life [15]. Another source recommends a maximum of six retightening cycles for bolts used in elevated temperature service and recommends that the accumulated plastic deformation be restricted to a maximum of 2% [18]. In any event, designers are cautioned to consider such things when selecting bolting materials and specifying maintenance procedures [16].

In the discussion so far we've considered only creep or stress relaxation in the bolts. These phenomena can also occur in joint members and gaskets, with creep being the more common. The neck of a pipe flange, in high-temperature service, will creep, as might the flanges themselves [16, 17]. Gasket creep is also common, and will be discussed in Chap. 19.

F. Compensating for Thermal Effects

The behavior of a bolted joint will ultimately depend to a large extent on the clamping force on that joint in service, as opposed, for example, to the clamping force created during assembly. Thermal effects which change the initial clamping force, therefore, can be a real threat to behavior.

Fortunately, all of the changes we've looked at have limits; they don't go on forever (at least not rapidly enough to be a problem). Stress relaxation can be severe in the first few hours or few hundred hours, but stresses will eventually stabilize. Most gasket creep, as we'll see in Chap. 19, occurs in the first few minutes. Differential expansion or contraction ceases when the temperature of the system stabilizes. So we're dealing with transitions from one stable point to another, rather than with continuous change. That's a big help!

We're also dealing, still, with a system of springs which will share loads and changes in load, as suggested by the joint diagram of Fig. 13.26. That, too, can be a big help, because it means we can often adjust bolt-to-joint stiffness ratios to reduce the impact of thermal change.

Specifically, therefore, we overtighten bolts at assembly to compensate for anticipated losses from differential expansion, gasket creep, stress relaxation, and loss in bolt stiffness. Before overtightening them, of course, we must determine whether or not the higher preloads will damage anything during assembly or after temperatures (and therefore loads and strengths) have changed.

If we're designing the joint, using Eqs. (13.21) through (13.23), we'll have anticipated the amount of "overdesign" required to compensate for differential expansion. If we're dealing with a troublesome, existing joint we could use the same equations to estimate the amount of overtightening required. We could then use the bolt strength equations of Chap. 2 to decide if the present bolts will take this much stress. If not, we could consider using bolts made of a stronger material.

In Chap. 1 we learned that we usually want to use the highest clamping forces the parts can stand. That's the goal in the process described above.

One common limitation on assembly preload will be the rotation of a raised face flange. It's not unusual to find yourself trapped between "too little residual bolt load after differential expansion and/or stress relaxation" and "flange rotation great enough to open a leak path." The answer to this dilemma can be more *uniform* residual tensions in the bolts. Don't tighten them above a limit determined by rotation, but make sure that every one is left as tight as that limit allows. More about this in Chap. 22.

If more preload isn't possible, or doesn't solve the problems created by thermal change, you should consider altering the stiffness ratio between bolts and joint members. Assuming that the nominal diameter of your bolts is set by the design of the joint, and that you can't change joint stiffness, you may want to increase bolt length. This will reduce bolt stiffness, which will mean less change in bolt tension for a given expansion-interference forced change in dimension. A less stiff bolt will also absorb a smaller percentage of a load change than will a stiffer one; as the joint diagram teaches us.

The normal way to increase bolt length is to stack Belleville washers under the head and/or nut; or to place elongated collars there. Figure 13.29, for example, shows a very long collar used in a nuclear power application to solve a difficult, differential expansion problem in a normally inaccessible joint [20]. Where longer bolts are unacceptable, bolt stiffness can be reduced by gun-drilling a hole down the bolt axis, and/or by turning down a portion of the bolt.

Another way to reduce thermal effects, at least those involving differential expansion, is to use "more similar" materials for bolts and joint members.

Although you should avoid it if possible, "hot bolting" can also be used to compensate for thermal changes. Retighten the bolts after they've relaxed, gaskets have crept, etc. How much torque should you use for

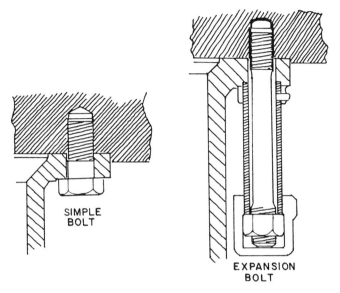

SIMPLE
BOLT

EXPANSION
BOLT

Figure 13.29   The elongated "expansion bolt" on the right replaced the conventional bolt on the left, to solve a severe differential expansion problem in a nuclear power application.

this? It's a frequent question and I've never heard a definitive answer. If the change in temperature didn't affect the lubricity of the parts, then reapplication of the original torque should reestablish the original clamping force. If the lubricity has decreased by 10%, torques should probably be raised a like amount. Perhaps your lubrication supplier can tell you what changes, if any, to expect. If not, an experiment in an oven, testing the force required to slip a heavy block over a plate, might give you a way to estimate the "hot torque." Don't forget, however, that a change in temperature will also affect a lot of other factors which influence the torque-preload relationship—things like fits and clearances, the ease with which operators can reach and tighten the nuts, the calibration of the tools used in proximity to hot parts, etc. But a lubricant analysis should be a reasonable way to get a rough estimate.

Beyond that, keep track of the torques used (and the temperatures involved). Experience will show you whether those torques were "good" or "bad" for the next shutdown or repair.

VI. JOINT EQUATIONS WHICH INCLUDE THE EFFECTS OF ECCENTRICITY AND DIFFERENTIAL EXPANSION

A. The Equations

We can now extend the equations of Chap. 12 to include the effects of differential expansion and joint eccentricity (i.e., prying action). Remember that we're interested in two things: the maximum tensile load which must be supported by an individual bolt, and the minimum clamping force we can expect to find, worst case, in the joint; both of these "in service." Remember, too, that the sign of the differential expansion force is not always the same. If the joint expands more than the bolts, it's positive; if the bolts expand more than the joint, however, there will be a loss of both clamping force and bolt tension, and the sign will be negative. With all that in mind, let's rewrite our equations using Φ_{en} as the load factor and $1/r_J''$ as the joint stiffness (which we'll call K_J'').

Maximum bolt load [extending Eq. (12.15)]:

$$\text{Max } F_B = (1 + s)F_{Pa} - \Delta F_m - \Delta F_{EI} \tag{13.21}$$
$$+ \Phi_{en}L_X \pm \Phi_{en}K_J''(\Delta L_J - \Delta L_B)$$

Minimum per-bolt clamping force on the joint [extending Eq. (12.16)]:

$$\text{Min } F_J = (1 - s)F_{Pa} - \Delta F_m - \Delta F_{EI} \tag{13.22}$$
$$- (1 - \Phi_{en})L_X \pm \Phi_{en}K_J''(\Delta L_J - \Delta L_B)$$

Total minimum clamping force on a joint containing N bolts:

$$\text{Min total } F_J = N \times \text{ per bolt Min } F_J \tag{13.23}$$

In the three equations above,

F_{Pa} = the average or target assembly preload (lb, N)

ΔF_m = the change in preload created by embedment relaxation (lb, N); Eq. (12.5) showed us that $\Delta F_m = e_m F_{Pa}$

e_m = percentage of average, initial preload (F_{Pa}) lost as a result of embedment, expressed as a decimal

ΔF_{EI} = the reduction in average, initial, assembly preload caused by elastic interactions (lb, N); Eq. (12.6) told us that $\Delta F_{EI} = e_{EI}F_{Pa}$

e_{EI} = the percentage of average, initial preload (F_{Pa}) lost as a result of elastic interactions, expressed as a decimal

ΔL_B = the change in length of the grip length portion of a loose bolt created by a change of Δt (°F, °C) in temperature (in., mm); see Eq. (13.14)

ΔL_J = the change in thickness of the joint members, before assembly, if exposed to the same Δt (in., mm); see Eq. (13.15)

s = half the anticipated scatter in preload during assembly, expressed as a decimal fraction of the average preload; see Eq. (12.1)

K_J'' = the stiffness of a joint in which both the axes of the bolts and the line of application of a tensile force are offset from the axis of gyration of the joint, and in which the tensile load is applied along loading planes located within the joint members (lb/in., N/mm); K_J'' is the reciprocal of the resilience of such a joint (see Fig. 13.16)

Φ_{en} = the load factor for the joint whose stiffness is K_J'' (see Fig. 13.16)

N = number of bolts in the joint

Again, these equations can be rewritten using any appropriate combination of Φ and K_J. This nearly completes our main bolted joint design equations, but in Chap. 22 we'll add one more factor—gasket creep—which will be required if we're designing a gasketed joint. Equations (13.21) through (13.23) will, however, be appropriate for all other joints.

B. An Example

Now let's run an example to practice using Eqs. (13.21) through (13.23), extending the example we worked out in the last chapter to include the effects of eccentricity and differential expansion. We'll assume the following values. Some of these are the same as those we used in Chap. 12; others are added to include eccentricity and expansion. This time, for example, we'll need to enter actual data for such things as bolt and joint dimensions and material strengths.

Assembly parameters:

s = tool scatter = 0.30
e_m = 0.1
e_{EI} = 0.18
F_{Pa} = target preload = 50% of bolt yield

Service conditions:

Operating temperature = 400°F
External load, L_X = $0.25F_{Pa}$
n = decimal defining distance between loading planes = 0.5

The joint: dimensions and properties (see Figs. 5.11 and 13.12)

Material: mild steel
E = 27.7 × 10⁶ psi at 400°F (Table 4.6)
a = 0.4 in.
s = 0.2 in.

T = total joint thickness; also = grip length L_G = 1.0 in.
T_{min} = thickness of thinner joint member = 0.5 in.
b = distance between bolts = 0.75 in.
$D_J = W = 2 \times D_B$ = 1.0 in.
ρ_J = coefficient of expansion = 8.3×10^{-6} in./in./°F (Table 4.5)
N = number of bolts in the joint = 8

Bolt: dimensions and properties:

We'll use a $\frac{3}{8}$–16 × $1\frac{1}{2}$ Inconel 600 bolt; identical in dimensions to that used as an example of bolt stiffness in Chap. 5, Section ID (Fig. 5.5)
A_S = tensile stress area of the threads = 0.0775 in.^{2} (Appendix F)
$E = 30 \times 10^6$ psi at 400°F (Table 4.6)
$K_B = 2.124 \times 10^6$ lb/in., as computed in Chap. 5, Section ID
S_y = yield strength at 70°F = 37×10^3 psi (Table 4.4)
D_B = width across flats of the head of the bolt = 0.5 in. [24]
D_H = 0.4 in.
ρ_B = coefficient of expansion = 9.35×10^{-6} in./in./°F (Table 4.5)

Now, using the data listed above, we need to compute several factors required for Eqs. (13.21) through (13.23). First let's compute the "area" of the joint (A_J) and of the "equivalent cylinder" of the joint (A_C). See Figs. 5.10 and 5.11 for reference.

$$A_J = b \times W = 0.75 \times 1.0 = 0.75 \text{ in.}^2 \quad \text{(see Fig. 5.11)}$$

$$A_C = \frac{\pi}{4}(D_B^2 - D_H^2) + \frac{\pi}{8}\left(\frac{D_J}{D_B} - 1\right)\left(\frac{D_B T}{5} + \frac{T^2}{100}\right)$$

$$A_c = \frac{\pi}{4}(0.5^2 - 0.4^2) + \frac{\pi}{8}\left(\frac{1}{0.5} - 1\right)\left(\frac{0.5 \times 1}{5} + \frac{1^2}{100}\right)$$
(5.20)

$$A_C = 0.11 \text{ in.}^2$$

Next we must compute the radius of gyration (R_G) for the rectangular area A_J, using Eq. (5.26), where d = the length of the longer side (in this example $d = W$).

$$R_G = 0.209d = 0.209(1.0) = 0.209 \text{ in.}$$
$$R_G^2 = 0.0437 \text{ in.}^2$$

Next, we compute the stiffness (K_{Jc}) and then the resilience (r_J) of the equivalent cylinder (reference Fig. 5.10).

$$K_{Jc} = \frac{EA_C}{T} = \frac{27.7 \times 10^6(0.11)}{1.0} = 3.05 \times 10^6 \text{ lb/in.}$$

$$r_J = \frac{1}{K_{Jc}} = 0.328 \times 10^{-6} \text{ in./lb}$$

Next we need the factor λ^2 (reference Fig. 13.15)

$$\lambda^2 = \frac{s^2 A_C}{R_G^2 A_J} = \frac{0.2^2(0.11)}{0.0437(0.75)} = 0.134$$

Now we can compute resiliencies r'_J, r''_J, r_B (reference Fig. 13.15) and joint stiffness K''_J.

$$r'_J = r_J(1 + \lambda^2) = 0.328 \times 10^{-6}(1 + 0.134) = 0.372 \times 10^{-6} \text{ in./lb}$$

$$r''_J = r_J\left[1 + \left(\frac{a\lambda^2}{s}\right)\right] = 0.328 \times 10^{-6}\left[1 + \left(\frac{0.4 \times 0.134}{0.2}\right)\right]$$

$$r''_J = 0.416 \times 10^{-6} \text{ in./lb}$$

$$r_B = \frac{1}{K_B} = \frac{1}{2.124 \times 10^6} = 0.471 \times 10^6 \text{ in./lb}$$

$$K''_J = \frac{1}{r''_J} = 2.404 \times 10^6 \text{ lb/in.}$$

We can now compute load factor Φ_{en} (see Fig. 13.16).

$$\Phi_{en} = n\left(\frac{r''_J}{r'_J + r_B}\right) = 0.5 \frac{0.416 \times 10^{-6}}{(0.372 + 0.471) \times 10^{-6}}$$

$$\Phi_{en} = 0.247$$

Next we need to compute the changes in length of bolt and joint when subjected to an increase in temperature of 330°F (from 70°F to 400°F).

$$\Delta L_J = \rho_J(L_G)\Delta t = 8.3 \times 10^{-6}(1.0)\,330 = 0.00274 \text{ in.}$$

$$\Delta L_B = \rho_B(L_G)\Delta t = 9.35 \times 10^{-6}(1.0)\,330 = 0.00309 \text{ in.}$$

Finally, let's select an assembly preload, which becomes our target or average preload (F_{Pa}). We said that we wanted F_{Pa} to equal 50% of the bolt's yield strength. That will be the room temperature yield strength because assembly is done at room temperature.

$$F_{Pa} = 0.5(S_y)A_S = 0.5(37 \times 10^3)0.0775 = 1.43 \times 10^3 \text{ lb}$$

Now we're finally ready to compute the two things we're most interested in, the maximum force the bolts must be able to support (Max F_B) and the minimum clamping force on the joint ($N \times$ Min per-bolt F_J) using Eqs. (13.21) through (13.23). And let's compute each term in Eqs. (13.21) and (13.22) separately before combining them. It's instructive to see how much change in preload each factor contributes to the final result. Starting then with Eq. (13.21):

Max assembly preload $= (1 + s)F_{Pa} = (1 + 0.3)\,1.43 \times 10^3 = 1.86 \times 10^3 \text{ lb}$

Embedment loss $= e_m F_{Pa} = 0.1(1.43 \times 10^3) = 0.143 \times 10^3$ lb
Average elastic interaction loss $= e_{EI} F_{Pa} = 0.18(1.43 \times 10^3) = 0.257 \times 10^3$ lb
Increase in bolt tension caused by the external load $= \Phi_{en} L_X = \Phi_{en}(F_{Pa}/4) = 0.247(1.43 \times 10^3/4) = 0.0883 \times 10^3$ lb
Decrease in bolt tension caused by differential expansion $= \Phi_{en} K_J''(\Delta L_J - \Delta L_B) = 0.247(2.404 \times 10^6)(0.00274 - 0.00309) = 0.208 \times 10^3$ lb

We can now use Eq. (13.21) to combine these to get the Max F_B in service:

Max $F_B = 1.34 \times 10^3$ lb

This shows us that, in this application, the maximum tension in the bolts will occur at room temperature, during assembly, on the most lubricious bolts (maximum positive tool scatter). This is a desirable situation; it means that if the bolts don't break during assembly they won't break in practise.

Now for minimum clamp force [Eq. (13.22)]

Minimum assembly preload $= (1 - s)F_{Pa} = 0.70(1.43 \times 10^3) = 1.00 \times 10^3$ lb

Embedment and elastic interaction losses are the same as computed above.

Loss in clamp created by the external load $= (1 - \Phi_{en})L_X = (1 - \Phi_{en})(F_{Pa}/4) = (1 - 0.247)(1.43/4) \, 10^3 = 0.269 \times 10^3$ lb

Loss in clamp caused by differential expansion is the same as above.

We use Eq. (13.22) to combine these and get the minimum, per-bolt clamp force on the joint in service.

Min $F_J = 0.123 \times 10^3$ lb

The total clamp force on the joint would be N times this, or $8 \times 0.123 \times 10^3$

Total Min $F_J = 0.984 \times 10^3$ lb

Note that there's a 15:1 ratio between the maximum tensile force the bolts must be able to withstand and the minimum force we can count on for interface clamp.

$$\text{Ratio} = \frac{\text{Max } F_P}{\text{Min per-bolt } F_J} = \frac{1.86}{0.123} = 15.1$$

Once again we see that inability to control assembly preload, elastic interactions, and embedment, plus the effects of external loads and differential expansion, have forced a significant "overdesign" of the joint. We could

improve results by using bolt and joint materials having similar coefficients
of expansion and/or by using feedback assembly control to reduce tool
scatter and relaxation losses; but the situation explored in this example
is not uncommon and may be more economical than the "improvements"
just listed.

Note that we've taken a very conservative view in running this example. Some of the factors we've dealt with—such as the effects of external
load or of differential expansion—are presumably unavoidable. Each and
every joint of this sort, used in the hypothetical service, will be exposed
to those effects. But will each joint see a full ±30% scatter in the torque-
preload relationship? I very much doubt it. If we had placed this joint in
production we could assume that an occasional joint would see a full
+30% or a full −30%; but I would suspect never both. Also, if this joint

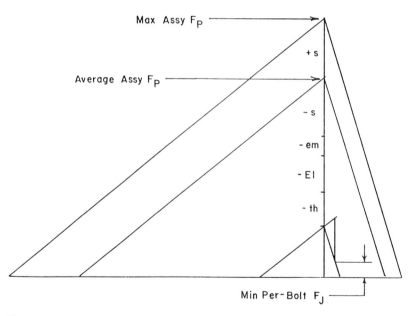

Figure 13.30 This joint diagram is based on the example given in Sec. VI of
this chapter. The maximum bolt load, in this case, is simply the average assembly
preload, F_{Pa}, plus the maximum anticipated scatter in the relationship between
applied torque and achieved preload. The minimum, per-bolt clamping force is
the average assembly preload less the loss in preload caused by embedment relaxation (em), elastic interactions (EI), differential expansion (th), and external tensile
load on the joint. The resulting maximum bolt load, which will define the size of
the bolts, is 15 times the minimum per-bolt clamping force we can count on from
those bolts.

was to be used in a safety-related application we might assume both $\pm 30\%$ to be sure we had covered all possibilities. But if we were only concerned about a few joints, each containing only a few bolts, we could undoubtedly use less than $\pm 30\%$ for the anticipated scatter.

How much less? I'm afraid that I'll have to leave an "accurate" answer to the statisticians of this world. If the joint is not safety related, however, I would suggest that $\pm 10\%$ might be used if our bolts are new and "as-received" (unlubed) but were all obtained at the same time from the same source; and perhaps $\pm 5\%$ for new, single-source, lubed bolts. Higher figures would, of course, be used for old, reused bolts, especially if they show signs of handling and/or are slightly rusty or something. For design purposes, however, the full range illustrated in Fig. 13.30 would apply.

REFERENCES

1. Fisher, J. W., and J. H. A. Struik, *Guide to Design Criteria for Bolted and Riveted Joints*, Wiley, New York, 1974, pp. 260ff.
2. Nair, R. S., Peter C. Birkemoe, and W. H. Munse, High strength bolts subject to tension and prying, *J. Structural Div.*, *ASCE*, vol. 100, no. ST2, proc. paper 10373, pp. 351–372, February 1974.
3. Agerskov, H., Analysis of bolted connections subject to prying, *J. Structural Div.*, *ASCE*, vol. 103, no. ST11, November 1977.
4. Simon, M. W., and J. B. Hengehold, Finite element analysis of prying action in bolted joints, Paper no. 770545, Society of Automotive Engineers, Earthmoving Industry Conference, Central Illinois Section, Peoria, IL, April 18–20, 1977.
5. Junker, G., *Principles of the Calculation of High Duty Bolted Connections—Interpretation of the Guideline VDI 2230*, VDI Berichte no. 220, 1974, an Unbrako technical thesis, published by SPS, Jenkintown, PA.
6. Finkelston, R. J., and P. W. Wallace, Advances in high performance fastening; Paper no. 800451, SAE, Warrendale, PA, February 25–29, 1980.
7. Grotewohl, A., *Calculation of a Dynamically and Eccentrically Loaded Bolted Connecting Rod Connection According to VDI 2230*, VDI Berichte no. 220, 1974.
8. Pindera, J. T., and Y. Sze, Response to loads of flat faced flanged connections and reliability of some design methods, University of Waterloo, Canada, Trans. CSME, March 1972.
9. Pindera, J. T., and Y. Sze, Influence of the bolt system on the response of the face-to-face flanged connections, Reprints of the 2nd International Conference on Structural Mechanics in Reactor Technology, vol. III, part G-H, Berlin, Germany, 10–14 September 1973.
10. Rodabaugh, E. C., FLANGE, a computer program for the analysis of flanged

joints with ring-type gaskets, ORNL-5035, U.S. Department of Commerce, National Technical Information Services, Springfield, VA, January 1985.

11. False Grade 8 engineering performance marks on bolting and improper marking of Grade 8 nuts, Research Report of the Industrial Fastener Institute, Cleveland, OH, April 4, 1986.

12. Halsey, N., J. R. Gieske, and L. Mordfin, Stress relaxation in aeronautical fasteners, National Bureau of Standards Report no. NBS-9485, February 1967.

13. Markovets, M. P., Graphic analytic method of calculating the time to failure of bolts under stress relaxation conditions, *Teploenergetika*, *NG*, pp. 52–54, June 1970; English translation, *Thermal Eng.*, vol. 17, no. 6, pp. 77–79, June 1970.

14. Mayer, K. H., Relaxation tests on high temperature screw fastenings, *Wire*, vol. 23, pp. 1–5, January–February 1973.

15. Valve studs—tightening, inspection and replacement recommendations (TIL-891), General Electric Company, Tampa, FL, December 1979.

16. Criteria for design of elevated temperature Class 1 components in Section III, Division 1, of the ASME Boiler and Pressure Vessel Code, ASME, New York, May 1976.

17. Kraus, H., and W. Rosenkrans, Creep of bolted flanged connections, *Welding Research Council Bulletin*, no. 294, Welding Research Council, New York, May 1984.

18. Sachs, K., and D. G. Evans, The relaxation of bolts at high temperatures, Report C364/73 on work done at the GKN Group Technological Center, Wolverhampton, UK, September 6, 1973.

19. From information received from the manufacturer, Superbolt, Inc., Carnegie, PA, 1988.

20. Recommended practices in elevated temperature design: a compendium of breeder reactor experiences (1970–1987), A. K. Dhalla, Editor, Published by the Pressure Vessel Research Committee of the Welding Research Council, New York, in *Welding Research Council Bulletin* no. 362 (April 1991), no. 363 (May 1991), no. 365 (July 1991), and no. 366 (August 1991).

21. Grosse, I. R., and L. D. Mitchell, Nonlinear axial stiffness characteristics of axisymmetric bolted joint, available from the University of Massachusetts, Amherst, MA, or from the Virginia Polytechnic Institute and State University, Blacksburg, VA, 1988.

22. VDI 2230, Systematic calculation of high duty bolted joints: issued by Verein Deutscher Ingenieure (VDI) VDI Richlinien, Dusseldorf, Germany, October 1977; Translation by Language Services, Knoxville, TN, published by the U.S. Department of Energy as ONRL-tr-5055, p. 44ff.

23. Modern Flange Design, Bulletin 502, Published by Taylor Forge, G&W Taylor Bonney Division, Southfield, MI (no date).

14
In-Service Behavior of a Shear Joint

We've studied axial tension loads at length because they're always present (since we preload—tension—the bolts as we tighten them) and because they will often dominate the behavior of the joint even when other types of load are also present. Tension loads in general are also the most difficult to understand, furthermore, because of the intricate way in which bolt and joint share them.

Before leaving the subject of working loads, we should examine shear loading. This is also a common type of load, especially in structural steel joints; and it demands an entirely different type of joint analysis—and creates a different joint response—than does a tension load. In fact, as mentioned in the Preface, it would require a second book to do justice to the subject of shear joints; I won't attempt to cover them here. Geoffrey Kulak, John Fisher, and John Struik have produced a well-written and comprehensive text on the subject, *Guide to Design Criteria for Bolted and Riveted Joints* [1, 2], which I recommend.

Although we can't examine them at length, it's pertinent for us to take a brief look at shear joints—enough to see why they're different, and enough to know when to read another book!

I. BOLTED JOINTS LOADED IN AXIAL SHEAR

In a shear joint the external loads are applied perpendicular to the axis of the bolt, as in Fig. 14.1.

504

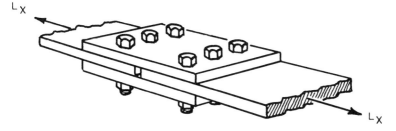

Figure 14.1 Bolted joint loaded in shear.

A. In General

A joint of this sort is called a *shear joint* because the external load tries to slide the joint members past each other and/or to shear the bolts. If the line of action of the external force runs through the centroid of the group of bolts, it's called an *axial shear* load, as shown in Fig. 14.2.

The strength of such a joint depends on (1) the friction developed between the joint surfaces (called *faying surfaces*) and/or (2) the shearing strength of the bolts and plates. Until recently, joints loaded in shear were formally classified as either "friction-type" or "bearing-type" and were specified and analyzed accordingly. Laboratory studies were used to confirm design procedures. Now it is recognized, however, that, while it's possible to construct a pure friction-type or pure bearing-type joint in a laboratory, no such distinction exists in most field joints.

We'll look at the field situation a little later, and at contemporary shear joint classifications. First, though, I think that the original definitions are still a useful place to start learning about the design and behavior of shear joints. I also think that, although pure friction or bearing joints may be

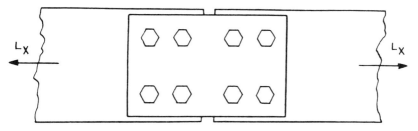

Figure 14.2 This is called an axial shear joint because the line of action of the external load passes through the centerline of the bolt pattern.

difficult or impossible to achieve in structural steel work, they are possible in nonstructural applications, the main requirement being the ability to control joint and hole geometry more closely than is possible in structural steel applications. So, let's start our review by looking at the classical friction-type and bearing-type joints. Then we'll return to the present structural steel viewpoint to see why that differs.

B. Friction-Type Joints

The amount of friction between two mechanical parts is, of course, proportional to the normal force clamping the two parts together, and to the coefficient of friction at the interface. In a joint the normal force is produced by the preload (axial tension load) in one or more bolts. The total friction force developed in the joint is called the *slip resistance* of the joint (RS). In Chap. 23 we'll see how to estimate the slip resistance of a joint. Suffice it to say at this point that it's generally a function of the clamping force between joint members and the interjoint friction forces which can be developed as a result.

Bolt Load in Friction-Type Joints

The bolt holes are always $\frac{1}{16}$ in. or so larger than the bolt diameter in structural steel joints. Until slip occurs, therefore, there are no shearing forces on the bolt (see Fig. 14.3). Under these conditions the bolt is loaded by the pure axial *tension* created when the nuts were tightened. In other words, we're back to the original joint diagram—with zero external loads (Fig. 14.4).

The elastic curve for the bolt will be that of a bolt under axial tension load—for example, the curve shown in Fig. 2.1—even though the external load on the *joint*, in the present case, is perpendicular to the axis of the bolts.

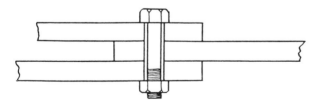

Figure 14.3 Any contact between bolts and joint members in a friction-type joint is accidental, so there is no shear stress on the bolts. The space between bolt and joint members is free.

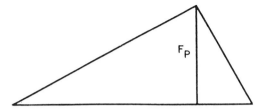

Figure 14.4   The only load on the bolts in a successful friction-type joint is the preload (F_P).

The specifications recommend that the bolts in structural joints be set to a tension at least equal to the proof load of the bolt—i.e., approximately equal to the yield strength of the bolt. This is the *minimum* setting recommended. Through use of turn-of-nut procedures (described in Chap. 8), or tension-indicating fasteners (Chap. 10), most of the bolts are actually set well past this point; they are set into the plastic region of the elastic curve. This means, as we've seen, that every bolt is set to approximately the same tension. It follows, therefore, that until and unless slip occurs, all of the bolts in a friction-type joint are essentially loaded equally. This is not true in a bearing-type joint, as we'll see.

Stresses in Friction-Type Joints

As long as the joint doesn't slip, the tension in one set of plates is transferred to the others as if the joint were cut from a solid block. Lines of principal (tension) stress flow from one to the other without interruption (neglecting for the moment the local stress concentrations caused by the holes or by the clamping forces produced by individual bolts); see Fig. 14.5.

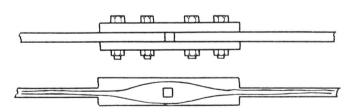

Figure 14.5   As long as there is no slip between joint members, a shear joint acts as if it were a solid block, with a smooth transfer of stress from one input member to the other.

C. Bearing-Type Joints

When the external loads rise high enough to slip a friction-type joint, the joint plates will move over each other until prevented from further motion by the bolts (Fig. 14.6). This joint is now considered to be "in bearing"—or to be (for purposes of analysis) a *bearing-type joint*. As mentioned earlier, some joints are designed to be in bearing from the start.

It's worth noting that the *ultimate* strength of all shear joints is determined by their strength in bearing—not by their frictional slip resistance. Friction-type joints, however, are often considered to have failed if they slip into bearing. Ultimate strength is rarely a good measure of the design strength or useful strength of a joint, any more than it is a good measure of the useful strength of a mechanical part.

Stresses in Bearing-Type Joints

The stress patterns in bearing-type joints are more complex than those in friction-type joints. The tension in one set of plates is transmitted to the others in concentrated bundles through the bolts (Fig. 14.7). What's more, each row of bolts transmits a different amount of load—at least in so-called long joints (those having many rows of bolts) (Fig. 14.8). The outermost fasteners always see the largest shear loads (see p. 93 in Ref. 1; p. 12 in Ref. 3).

We saw a similar phenomenon when studying the stresses in nut and bolt threads. Remember that the inboard (first engaged) threads saw the most load. The outboard ones saw the least, because the inboard had already transferred some of the bolt load to the nut. In the joint case, the outer rows of bolts transfer some of the load on one set of plates to the other plates, reducing the loads seen by both plates and bolts toward the center of the group of bolts.

As a result the outer bolts in a joint see far more than the average stress level—some say as much as five times the average—while the inner

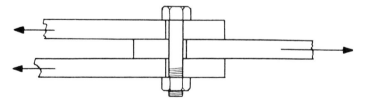

Figure 14.6 In a bearing-type joint the bolts act as shear pins. The spaces between bolt and joint members are offset.

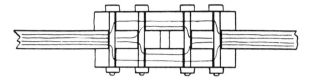

Figure 14.7 In a bearing-type joint the tension in one plate is transmitted to the other plate through the bolts, making for a more complex stress distribution than in a friction-type joint (compare with Fig. 14.5).

ones see less than average. The outer bolts, therefore, usually are loaded plastically rather than elastically. This results in plastic flow, which helps to distribute the load more uniformly between the various rows of bolts.

Before we leave the subject of the stresses in bearing-type joints, it's worth noting that since the bolts do bear on the joint, there are shear stress concentrations in the plates. These can cause local yielding of the plates (the bolt holes become slots), or the ends of the plate can tear out, as we'll see in our discussion of joint failures. These failures would not be seen in a friction-type joint unless it slipped.

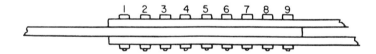

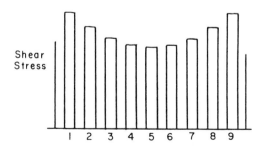

Figure 14.8 Shear stress in individual bolts varies substantially, especially in long joints. (Modified from Ref. 1.)

II. FACTORS WHICH AFFECT CLAMPING FORCE IN SHEAR JOINTS

As far as bolt preload and clamping force are concerned, shear joints are affected by all of the factors which affect those things in tensile joints.

Initial preloads will be *scattered* as a result of variations in geometry, lubricity, condition, etc. of the parts involved, as well as variations in tools, operators, procedures, and all the rest.

The bolts and joint members *embed* in shear joints, just as they do in tensile joints.

Elastic interactions occur as a group of bolts are tightened in a shear joint.

Preloads and clamping forces in a shear joint will be altered by *differential expansion* if the parts are made of different materials and/or are subjected to different temperatures.

As a result we can use Eqs. (13.21) through (13.23) to estimate maximum bolt loads and minimum clamping forces; but with one important difference. The term $\Phi_{en}L_X$ should be omitted from Eq. (13.21), and the term $(1 - \Phi_{en})L_X$ should be omitted from Eq. (13.22), because bolt loads and clamping forces will not be affected this way by external shear loads. Bolt-to-joint stiffness ratios have no influence on the way shear joints absorb loads, although they still affect the way such joints respond to temperature swings and the resulting differential expansion.

We can draw a joint diagram for such a joint. It would look exactly like the diagram of Fig. 13.28 but without the external load effect shown at the bottom of that diagram. The minimum, residual clamping force on the joint in that diagram, Min per-bolt F_J, will be the clamp left after differential expansion (shown as "th" in the diagram).

This is not to say, however, that shear joints don't respond to external loads. They do, but in ways that differ completely from the way their tensile cousins respond. Let's take a look.

III. RESPONSE OF SHEAR JOINTS TO EXTERNAL LOADS

Figure 14.9 diagrams the way in which a shear joint responds to ever-increasing loads. The figure shows the overall deformation and displacement of joint members as a function of applied shear force. To start with (part 1 of Fig. 14.9) there is linear, elastic deformation of joint members under relatively mild loads.

As applied force increases, the friction forces between joint members are overcome, and the joint slips into bearing (part 2 in Fig. 14.9).

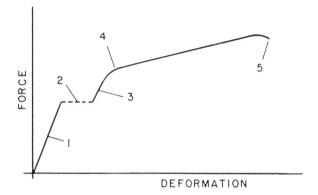

Figure 14.9   This chart illustrates the sequence by which a shear joint fails. See text for a discussion.

Higher loads create more elastic deformation, this time of both bolts and joint members, as in part 3 of Fig. 14.9. When these loads rise still further the parts start to deform plastically (part 4, Fig. 14.9).

Finally—part 5 of Fig. 14.9—something breaks. Either the bolts shear and the joint members are free to pull past each other or the bolts tear out through the sides of the joint members.

Clamping forces and bolt stress are changed when shear loads are applied to the joint. Once the bolts have been brought into bearing shear stress combines with the original tensile stresses created when we tightened the bolts to increase total stress in the bolts. Tensile stress doesn't change; but total stress does. This is true, at least, until the joint nears failure. Under extreme loads the parts have been much deformed, and bolts have probably been pulled sideways at an angle. Some of the shear load may now be seen as an increase in bolt tension. Some experiments indicate, however, that the bolts shed preload tension when abused this way. The bolts in shear joints fail, in shear, at about the same shear stress levels, regardless of how they were originally preloaded [2, p. 49].

None of this, however, can be usefully shown in a joint diagram. The diagram of Fig. 12.4 still stands as our best representation of a joint loaded only in shear.

IV. JOINTS LOADED IN BOTH SHEAR AND TENSION

There are joints which must support both shear and tensile loads in service. Such a joint is subject to all of the variables and potential problems faced by joints loaded solely in tension, as well as all of the problems faced by joints loaded only in shear. The full Eqs. (13.21) through (13.23) can be used to define the effect of the tensile load on bolt tension and clamp force. The tendency of the joint to slip into bearing can be based on the resulting Min per-bolt F_J.

The following equation can be used to determine how much shear stress the bolt can stand if subjected to a given tensile stress, or vice versa (see p. 69 in Ref. 1; p. 14 in Ref. 3; p. 51 in Ref. 2).

$$\frac{S_T^2}{G^2} + T_T^2 = 1.0 \tag{14.1}$$

where S_T = the ratio of shear stress in the shear plane(s) of the bolt to the ultimate tensile strength of the bolt
       T_T = the ratio of the tensile stress in the bolt to the ultimate tensile strength of the bolt
       G = the ratio of shear strength and tensile strength of the bolt (0.5–0.62 typically if computed on the thread stress area)

It is best to compute both S_T and T_T using the equivalent thread stress area formulas of Chap. 2 rather than the shank area.

If one plots the equation above for a given bolt, he will obtain an elliptic curve such as that sketched in Fig. 14.10 (see p. 54 in Ref. 1; p.

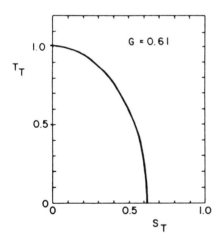

Figure 14.10 Elliptic curve used to relate the tensile capacity of a bolt to the shear stress imposed on a bolt, and vice versa. See Eq. (13.1) for definition of the terms T_T, S_T, and G. This curve is for static loads on the thread stress area of ASTM A325 or A354 BD bolts. (Modified from Ref. 1.)

51 in Ref. 2), which represents static loads on A325 or A354 BD bolts. Similar curves can also be drawn for dynamic loads.

Note that Eq. (13.1) does *not* compute the allowable stress limits for a given bolt. It merely shows the relationship between tensile and shear stresses. However, a number of authors have plotted and published solutions of the equation for stress levels that would constitute failure—either plastic yield or fracture of the bolts in question. Use such curves with caution. Note that some are for static loads, some for cyclic loads; some assume that the shear planes will pass through the shank of the bolt, others assume that shear planes will pass through the threads, etc.

REFERENCES

1. Fisher, J. W., and J. H. A. Struik, *Guide to Design Criteria for Bolted and Riveted Joints*, Wiley, New York, 1974.
2. Kulak, Geoffrey L., John W. Fisher, and John H. A. Struik, *Guide to Design Criteria for Bolted and Riveted Joints*, 2nd ed., Wiley, New York, 1987.
3. *High Strength Bolting for Structural Joints*, Booklet 2867, Bethlehem Steel Company, Bethlehem, PA.

15
Joint Failure

One of our main goals in studying the design and behavior of bolted joints is to avoid joint failure by proper design and/or through the use of effective assembly techniques. Now that we have studied in some detail the assembly process and the effects of external loads on the joint, we are ready to turn our attention to the topic of joint failure.

As we learned in Chap. 1, the main function of a bolt is to clamp two or more joint members together. The main function of the assembly process is to introduce the tension in the bolts which produces the clamping force.

Joint failure occurs when the bolts fail to perform their clamping function properly; for example, if they exert too high a force on the joint.

More commonly, as we will see, joint failure will occur if the bolts provide too little clamping force. In most such situations, the clamping force will probably be insufficient because of deficiencies in the assembly process. Remember that most of the factors included in the block diagram of Fig. 6.28 would result in less than anticipated clamping force rather than in excessive force.

In other cases, however, insufficient clamping force can result from some form of instability in bolt tension. In Chap. 4, for example, we looked at some of the material properties which could lead to instability in service; and in Chap. 13, we considered changes caused by variations in temperature. We will be taking a look at other forms of instability in this and in the next few chapters.

We are going to start with an overview of the whole subject of joint failure; then go on to a closer study of the four principal types of failure: self-loosening, fatigue, corrosion, and leakage. But first, lets examine some of the ways in which a joint and/or bolt can fail.

I. MECHANICAL FAILURE OF BOLTS

Obviously, bolts will fail to exert sufficient clamping force on a joint if they are broken. They can break for a variety of reasons.

Mechanical failure during assembly (the mechanic pulled too hard on the wrench! or the bolts weren't up to par).
Mechanical failure at elevated temperatures (bolt strength dropped as temperature rose).
Corrosion ate through the bolt.
Stress corrosion cracking.
Fatigue failure.

We will look at corrosion, stress corrosion cracking, and fatigue in Chaps. 17 and 18. These are relatively common causes of bolt failure. Mechanical failure during assembly or in service is much less common, but is certainly not unknown. In recent years, in fact, there has been an increase in the number of elevated temperature/mechanical failures reported because of the fact that manufacturers of low-cost bolts have been using boron steel instead of medium carbon steel for Grade 8 fasteners, and boron steel loses strength more rapidly than carbon steel does as the temperature is raised.

There have also been recent reports that heads have snapped off low-cost bolts when they were tightened because of such things as improper heat treat (e.g. creating quench cracks), small fillet radii, or poor material. In a few cases, it was even found that suppliers had welded hexagonal heads onto threaded rod to manufacture the bolts.

If you use the proper material, maintain bolt quality, and dimension the bolts to support the intended loads (see Chap. 2), your bolts should not break because of mechanical failure. They still may break because of corrosion, stress corrosion, or fatigue, however, as we will see in Chaps. 17 and 18.

II. LOST BOLTS

It's also obvious that bolts won't perform their clamping function properly if they're missing. Perhaps the most common reason for missing bolts is a phenomenon called self-loosening, which we will look at in detail in Chap. 16. Self-loosening is most commonly caused by vibration, but can also be caused by such things as temperature or pressure cycles. Anything, in fact, which puts reversing loads on the joint in a direction at right angles to the axis of the bolts may cause the bolts to loosen.

Self-loosening isn't the only cause of missing bolts, however. In a surprising number of situations, the bolts are missing because the mechanic didn't install them. In some cases, involving large, heavy equipment, some of the bolts were not installed because of hole misalignment or the like. In other cases, only carelessness was involved. Because most bolted joints are grossly overdesigned, most of them can get by with a few missing bolts, but obviously, we don't want to do this in critical situations.

III. LOOSE BOLTS

Bolts which "aren't tight enough" are probably the most common cause of joint misbehavior and failure at the present time. Broken and missing bolts could be considered an extreme form of "loose" as far as failure analysis is concerned. Any one of these three problems can lead to such failure modes as:

Joint leakage
Joint slip
Cramping of machine members (for example, bearings can get out of line)
Fatigue failure
Self-loosening

The relationship between loose bolts and self-loosening is a chicken-and-egg proposition. If the bolts are too loose to start with (were improperly tightened during assembly, for example), this will encourage self-loosening. Self-loosening, on the other hand, will progressively loosen the bolts. It's often difficult to tell, therefore, which was the cause and which the effect when one is analyzing a joint failure.

The relationship between loose bolts and fatigue requires some explanation, which you'll find in Chap. 17.

IV. BOLTS TOO TIGHT

It's less obvious and less common, but bolts which are too tight can also contribute to joint failure. Excessive bolt loads can crush gaskets, for example, or damage (gall) joint surfaces.

Excessive bolt loads can also encourage stress corrosion cracking, as we will see in Chap. 18, or can reduce fatigue life, as we will see in Chap. 17.

We learned a minute ago that insufficient preload (loose bolts) can encourage fatigue failure; now, we learn that "too much" can be a problem as well. Fatigue is one of those problems which can only be avoided by just the right amount of tension in the bolts, at least according to some experts. We will learn more about this in Chap. 17.

V. WHICH FAILURE MODES MUST WE WORRY ABOUT?

Which failure modes must the bolting engineer worry about? The answer depends on the job being performed by the bolted joints, on the consequences of failure, on the environment the bolts are working in, and usually on the industry. The petrochemical industry, for example, is primarily concerned with leakage from gasketed joints and with corrosion problems. Fatigue and vibration loosening are usually of little concern.

The automotive industry, on the other hand, would probably name self-loosening and corrosion as the two main problems, but leakage from head gaskets is also a significant concern. Fatigue is, again, a relatively minor issue.

The primary concerns of the structural steel industry are joint slip and corrosion. Fatigue is sometimes an issue, but usually of joint members rather than of the bolts. Self-loosening and leakage are never encountered.

The aerospace industry would probably list fatigue first.

VI. THE CONCEPT OF "ESSENTIAL CONDITIONS"

I think it's useful to recognize that each type of failure is set up by a limited number of "essential conditions," usually three or four in number. For example, to have a corrosion problem, you *must* have:

An anode
A cathode
An electrolyte
A metallic connection between anode and cathode

If you can eliminate any one of these essential conditions, you can completely eliminate the corrosion problem.

There are also essential conditions for the other types of failure we have discussed, as listed below:

Stress corrosion cracking requires:

A susceptible material
Stress levels above a threshold
An electrolyte
An initial flaw

Hydrogen embrittlement requires:

The same conditions as stress corrosion cracking, but with hydrogen instead of an electrolyte

Fatigue failure requires:

Cyclic tensile stress
A susceptible material
Stress levels above an endurance limit
An initial flaw

Mechanical failure requires:

Stress levels exceeding the static strength of the bolts or threads

Self-loosening of the fastener requires:

Cyclic loads at right angles to the bolt axis
Relative motion (slip) between nut, bolt, and joint members

The fact that the essential conditions are limited makes it appear that it would be relatively easy to avoid joint failure. The problem, however, is that dozens—maybe even hundreds—of secondary conditions can establish the essential conditions required for a particular type of failure. We'll look at some of these secondary conditions when we study corrosion, fatigue, self-loosening, etc., in detail in subsequent chapters. To give you an example of the diversity of secondary conditions, however, a few years ago the Nuclear Regulatory Commission (NRC) became concerned about the growing number of reports they were receiving from nuclear operators concerning bolt problems on safety-related joints. No joint failures had been reported, only the failure of individual bolts. These failures included loose bolts, missing bolts, broken bolts, corroded bolts, etc. [1].

Studies were made to assess the extent of the problems and to reduce or prevent them. Tabulations were made of the factors which had contributed to the potential problems which had been reported. These "factors" are what I have called secondary conditions, above.

Remember that, a few minutes ago, we said that the only essential condition for mechanical failure was "stress levels exceeding static strength." A total of 170 safety-related, mechanical failures of bolts were reported to the NRC over a 3-year period. The following secondary conditions were reported as possible causes for these failures.

Bolt material not as specified
Poor choice of material by the designer
Improper heat treat (including quench cracks)
Excessive preload
Shear, bending, and torsion stress
Creep damage
Abnormal loads (water hammer, seismic shock, etc.)
Poor fastener dimensions (e.g., poor thread fit)
Elevated temperatures
Construction procedures

It's easy to think of additional things which might cause mechanical failure of bolts. We've already discussed some of these things. The point is that the single "essential condition" we've defined for mechanical failure can be set up by a large number of problems behind the problem. In many cases, in fact, it is not at all obvious what conditions have led to the failure of the joint and/or what we could do to prevent a recurrence of the problem. In many cases, we can't afford to perform the basic metallurgical, chemical, or analytical work required to reach a correct answer to the questions "How did it fail, and why?"

In many cases, furthermore, altering the conditions which lead to one problem can merely create conditions for another. For example, we know that more preload helps fight vibration or other forms of self-loosening. Excessive bolt stress, however, can encourage stress corrosion cracking. If we are concerned about both problems in a given application, then we're going to be forced to produce exactly the right amount of preload in those bolts.

VII. THE IMPORTANCE OF CORRECT PRELOAD

This is not the only situation in which correct preload is useful, incidentally. A study of the essential conditions of the various failure modes shows that improper preload can be a contributing factor in almost every situation. Here's a tabulation of the relationship between failure mode and preload (which includes, in this discussion, its equivalents "bolt tension" and "clamping force").

Corrosion

Higher stress levels (higher preload) can make a material more anodic, more active in a corrosive environment.
Insufficient preload can allow electrolytes to leak from pressure vessels and piping systems, exposing the bolts to corrosion attack.

Stress Corrosion Cracking (SCC)

Excessive preload can raise stress levels above the SCC threshold.
Insufficient preload can again allow leakage of corrosive materials.

Fatigue Failure

Excessive preloads can raise stress levels above endurance limits.
Excessive preloads can mean an unnecessarily high mean stress in the material.
Insufficient preload can increase the stress excursions seen by the parts.

Mechanical Failure

Excessive preload can add to subsequent service loads, thereby exceeding the strength of the fastener in service.
Insufficient preload can expose the fastener to the full extent of the external load (see Chap. 12 on joint diagrams).

Self-Loosening of Fastener

Insufficient preload can allow transverse slip of the bolt and joint members, an essential condition for self-loosening.

A major weapon against joint failure, therefore, is "the correct clamping force on the joint." This, in turn of course, depends on the correct preload during assembly and then stability of bolt tension and clamping force in service.

VIII. LOAD INTENSIFIERS

A large number of factors can contribute to the failure of bolted joints. In fact, this whole book could be called a discussion of such factors. It's worthwhile listing some of them here, however, to emphasize their effect on joint integrity.

Parts fail when subjected to loads which exceed their strength. That seems straightforward enough. Unfortunately it's often difficult or impossible to predict or control the loads. All sorts of things can make the actual loads worse than we thought they'd be. Other factors can increase the stress levels created by a given load; as far as the fastener is concerned, the load itself has increased.

In Chap. 13, for example, we saw that bolt loads are increased by prying action if the bolt and external load are not coaxial; and that the problem is magnified if joint members aren't "stiff enough." Someone has discovered all this and opened our eyes to it, which is lucky, because this particular load intensifier is anything but obvious.

Table 15.1 lists some other things which can increase (intensify) the stress levels within the loaded bolt, and thereby increase the chance of failure. After each I've listed the chapter in which you'll find more information.

The last two items in Table 15.1 require a brief explanation. "Poor fits" increase stress levels because they reduce the contact areas between nut and bolt threads, or between nut and joint members (e.g., if the bolt hole is oversize) and, therefore, increase contact pressures. The resulting stresses can exceed those anticipated by the designer.

Nonuniform preload in a group of bolts can cause a few of them to carry more than their share of the total load placed on the system.

Table 15.1 Factors Which Increase the Stress Levels Produced in Bolts or Joint by a Given External Load

Factor	Reference chapters
Prying action	13
Eccentric loads	23
Bending	2
Improper bolt/joint stiffness ratio	12, 13, 17
Shock or impact	
Gaskets	12, 19
Perpendicularity of threads and hole-to-joint surfaces	2
Poor fits (e.g., bolt to nut)	
Nonuniform preload	

IX. FAILURE OF JOINT MEMBERS

We've said that a joint fails when the bolts fail to provide a suitable clamping force, and we've looked at several ways in which this may happen. What about failure—rupture—of the joint members themselves?

This is uncommon in joints loaded in tension. Sometimes you'll encounter failure in the neck of a piping flange, but the flanges themselves rarely crack. Automotive or other castings can crack. But these failures are seldom, in my limited experience at least, related to the bolting. Failure of joint members loaded in shear, however, is more common.

We took a brief look at some of the failure modes of joint members under noncyclic shear loads in Chap. 2; for example, in Fig. 2.18. If the bolt holes are too closely spaced, the joint can tear through the so-called "net section," as in Fig. 15.1B. If there are only a few bolts, and they're placed too near the edge of a plate, the bolts can pull their way through the plate, shearing it, as in Fig. 15.1A.

It won't always be the plates which fail, however. In short joints with widely spaced holes located well back from the free edge, failure can

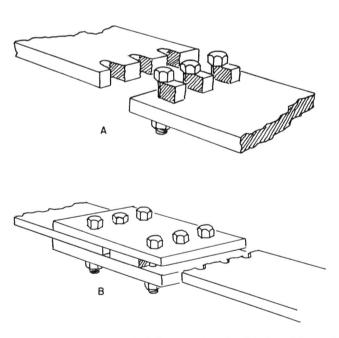

Figure 15.1 Some static failure modes of axial shear joints. (A) Tearout or marginal failure. (B) Failure through the "net section."

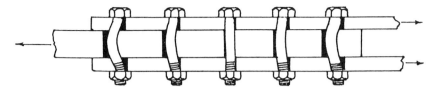

Figure 15.2 In long joints the bolts will often fail first, starting with the outermost ones because these see the greatest loads (see Fig. 14.8).

consist of the simultaneous shearing of all bolts. It will be the bolts which fail in long joints, too, but in a different way.

Here, many rows of bolts are involved, and there is substantial frictional restraint between joint members, even if the joints are not slip-critical and the bolts have not been heavily preloaded.

The outermost bolts transfer the largest loads from plate to plate, and therefore see the largest loads (see Fig. 14.8). As the loads on the joint increase, relative slip between joint members occurs first at the outer ends of the members. They're stretching. This act will distort the outer bolt holes and will eventually shear those bolts. Failure of these bolts can occur before the innermost bolts have suffered at all. Figure 15.2 shows such a joint at this point.

Under still higher loads, the remaining frictional restraint between joint members, in the center of the bolt pattern, will be overcome, and the rest of the bolts will shear [2].

All of this applies to joints under noncyclical loads. Fatigue loads lead to other types of failure, as we'll see in Chap. 17.

X. GALLING

A. Discussion

We have now looked briefly at various ways in which a bolted joint can fail—in most cases because the bolts have failed to clamp the joint members together properly. Before going on to a detailed look at some of the more important failure mechanisms, I think we should take a brief look at another type of bolt "failure" which, while not affecting the performance of the joint in service, can be a real nuisance and expense. I'm referring to the galling of some bolts—especially larger ones—as they are tightened or removed during assembly or maintenance operations.

Galling is a cold welding process which occurs when the surfaces of male and female threads come in such close contact that an atomic bond can form between them. Galling is encouraged by such things as lack of lubrication, lack of oxide film on the metal, high contact pressure, and heat. Stainless steel bolts are particularly likely to gall.

Minor galling can cause minor damage to thread surfaces, but these can often be chased with a tap or die; the bolts can be reused. Major galling, however, can prevent removal of the bolt or nut. If this happens, the bolts have to be drilled out; the nuts have to be cut apart with a nut splitter, or, if they're too large for that, burned off with a torch.

There is no foolproof answer to galling. The following techniques, however, have worked for some people. These tips are listed in no particular order:

Use coarse threads instead of fine, use a Class 2A fit.

Use a good thread lubricant or antiseize compound. Here are some popular choices:

Moly disulfide works well if stresses are below 50% of yield and temperatures are below 750°F (400°C).

FelPro C670 lubricant is popular—again, at temperatures below 750°F (400°C).

Silver-based lubricants or antiseize compounds are especially effective.

Milk of magnesia has been found to be effective in high-temperature petrochemical and refinery operations.

Silicon grease works well on stainless steel bolts tightened into aluminum blocks.

Liquid dish detergent is said to work well if bolts are tightened into aluminum.

Better material combinations:

Use stainless steel nuts on low-alloy bolts (for example, on A193 B7, B16, etc.).

Cold-drawn 316 stainless steel nuts work well on cold-drawn 316 bolts.

400 series stainless steel nuts work well on 316 series bolts.

Grade 2 ("pure") titanium fasteners work better than the higher-strength grades of titanium.

ARMCO Nitronic 60 bolts work well with Nitronic 50 nuts—but not the other way around, for some reason.

Carpenter Gall-tough is a recently introduced austenitic stainless steel which the manufacturer claims provides good galling protection.

B. Removing Galled Studs

If all else fails and the studs gall and you have to remove them, you might try some of the following tricks.

If galling is only minor, or if the studs have been exposed to high-temperature service for a long time, and you are sure that the original lubricants have dried out and are concerned about galling, you might try applying a good penetrating oil. These oils are usually slightly acidic. A product called "Masteroil" is popular.

Iodine, another mild acid, has worked well for some.

In some piping situations, people have removed the entire flange from the system and soaked it in a mild acid bath.

Note that, after removing a stud on which you have used these mild acids, you should clean and relubricate the studs as soon as possible to prevent unnecessary acid attack.

If the oils don't work, rapid heating and cooling of the flange or bolt—but not both—will sometimes work. Some people have gun-drilled a hole through a stud, then filled the hole with weldment to heat it suddenly.

If one of the problems is getting a good enough grip on a stud, apply penetrating oil and then weld a nut onto the end of the stud. Use the nut like the head of the bolt to apply more torque.

If the nut or stud has been threaded into a blind hole and you can reach the bottom of the hole (by drilling through the side of the flange, or block, or vessel, or whatever), drill such a hole, then tap it for a pipe fitting. Pump penetrating oil under relatively high pressure—2000 psi has been suggested—into the blind hole, wait for that oil to start to appear on the surface of the joint, then remove the bolt or stud.

If all else fails and the bolts must be destroyed, drill them out using a magnetic hold-down drill. EDM can also be used to "drill" them out, but the EDM process leaves a hard-surfaced hole which is more difficult to retap. After removing the bolts or studs, drill the joint members for a collar, thread it on both ID and OD (the ID, of course, is tapped for the

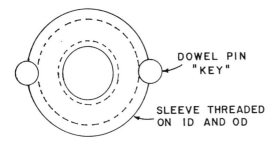

Figure 15.3 If galling forces us to drill out a bolt, the hole in the joint can sometimes be repaired by drilling and tapping a larger hole, then inserting a collar threaded on the ID to accept a new bolt of the original diameter. Dowel pins can be used as keys to retain the collar.

original bolts or studs), thread the collar into the drilled-out joint member, drill a couple of small holes, and insert some pins to retain the collar (as shown in Fig. 15.3), then replace the bolts [3].

REFERENCES

1. Bickford, J. H., Nuclear bolting issues and programs, Pressure Vessel Components Design and Analysis—PVP-Vol. 98-2, American Society of Mechanical Engineers, New York, 1985.
2. Kulak, Geoffrey L., John W. Fisher, and John H. A. Struik, *Guide to Design Criteria for Bolted and Riveted Joints*, 2nd ed., Wiley, New York, 1987, pp. 89ff.
3. From a conversation with Richard Potter of Bolting Services, Inc., Houston, TX, 1985.

16
Self-Loosening

I. THE PROBLEM

When we tighten a fastener, we pump energy into it: tension, torsion, and bending energy. The fastener is a stiff spring, and we stretch, twist, and bend it.

After we let go, this energy is held in the fastener by friction constraints in the threads or between contact faces of the nut and joint. If something overcomes or destroys these friction forces, the energy stored in the fastener will be released; the bolt will return to its original length with the inclined plane of the bolt threads pushing the inclined plane of the nut threads out of the way.

Subjecting the bolted joint to vibration will do this. Under certain circumstances, all preload in the fastener will be lost as a result. In fact, the fastener itself can shake loose and be lost. This can be a severe problem for any product that is bounced around or handled a lot—anything from a vehicle to a toy. Losing all preload and/or losing the fastener can, of course, lead to all sorts of other failures we would rather avoid. So it's useful to know what causes vibration loosening and some of the things we can do to minimize or prevent it.

Note that vibration loosening is a common cause of what we have called "clamping force instability." That force can be significantly reduced, or lost altogether, as a result of vibration.

It can also be lost by other forms of self-loosening, incidentally. Vibration may be the most common, but transverse slip, flexing of joint members, thermal cycles, and other things can also cause a joint to loosen. The self-loosening mechanism is the same in each case, however. We'll assume that vibration is the culprit and focus on it for now.

II. HOW DOES A NUT SELF-LOOSEN?

We probably don't know why a fastener will self-loosen under vibration, shock, thermal cycles, or the like. A number of theories have been advanced, and their authors believe they know, but the theories vary [1, 2, 12, 23–26]. They can't all be right—perhaps none are. A few years ago an ASME committee attempted to establish a working group to resolve the question. It was decided that a substantial amount of money would be required to finance the necessary research; but the attempts to attract financial support drew a blank. The project was, therefore, abandoned.

We'll examine one of the current theories in detail, and will briefly review the others. Before we do that, let's take a look at some basic factors which form a part, at least, of most or all of the existing theories.

Everyone agrees that a threaded fastener will not loosen unless the friction forces existing between male and female threads are either reduced or eliminated by some external mechanism acting on the bolt and joint. The disagreements concern the type of mechanism that does that. Before looking at these mechanisms, let's examine these all important friction forces we're trying to preserve.

They are created, of course, by the preload or tension in the bolt, which creates a "normal force" between male and female threads. The nut will turn with respect to the bolt only if some "antifriction force" (or torque) exceeds the thread-to-thread friction force.

Note that it is not necessary for the antifriction forces to be in the same direction as the forces which tend to loosen the nut. Let's place a block on a table as in Fig. 16.1, for example. We want to move the block from point A to point B by exerting a force on it: force 1. The amount of force required, of course, is equal to that required to overcome the frictional force between block and table μW (where μ is the coefficient of friction and W is the weight of the block) plus the small additional force required to accelerate the block (to get it moving).

If someone else were to apply a second force to the block in any other direction—for example, force 2 in Fig. 16.1—and if this second force were enough to overcome the frictional restraint between block and table, then we would only have to overcome the inertia of the block in order to move it from point A to point B.

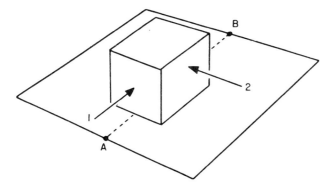

Figure 16.1 If force 2 is large enough to overcome any frictional force which exists between the block and the table, then only a very small force 1 would be required to move the block from point A to point B. Force 1 would only have to overcome the inertia of the block.

As an example, it has been pointed out that it is easier to pull the cork from a wine bottle if one first rotates the cork to break the friction forces, then pulls the cork out. A straight pull must overcome both the friction forces and any suction forces, and is more difficult [1].

The point of all this is that vibration forces in direction 2 would allow some other low-level force to move a relatively heavy block across the surface of a table—if those direction 2 vibration forces were large enough to break most or all frictional restraint between block and table.

The fastener, of course, is not a block on a table. It can better be modeled by the inclined plane and block shown in Fig. 16.2. There is now a small force which wants to move the block down the plane—a force equal to $W \sin \theta$, where W is the weight of the block and θ is the angle of the plane on which the block is resting. This tendency to slide, of course, can be overcome, or more than overcome, by frictional restraints ($\mu W \cos \theta$) between the block and plane. If we were to shake the plane vigorously in the direction shown by the double arrow, however, the block would gradually or rapidly walk its way to the bottom of the plane. This vibration in one direction would destroy the friction forces and allow $W \sin \theta$ to do its job [4].

In the fastener, of course, it is not the weight of the nut, but rather the tension stored in the bolt, which creates the force which pushes the inclined plane of the nut out of the way when the vibration forces are broken. This is suggested in Fig. 16.3.

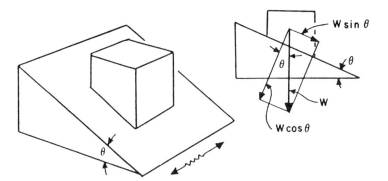

Figure 16.2 Shaking the inclined plane in the direction shown by the double arrows would destroy frictional contraints between block and plane and allow force $W \sin \theta$ to move the block to the foot of the plane. (W is the weight of the block.)

Note that the bolt in Fig. 16.3 must be exerting a force on the nut—or the block in Fig. 16.2 must be exerting a (gravitational) force on the inclined plane—if the reduction of friction force is to result in motion. The theories insist—and experiments appear to confirm—that some "off-torque" is required to create relative motion between nut and bolt, even under severe vibration. The source of this off-torque is generally considered to be the tension in the bolt acting against the inclined plane of the nut threads [1, 12, 24, 26]. Other theories, however, have been proposed [2, 23, 25].

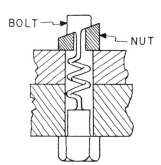

Figure 16.3 Schematic of a nut and bolt. Anything which breaks the friction forces between them (and between the nut and the joint) will allow the tension in the bolt to push the nut out of the way as the bolt attempts to return to its initial length.

One thing that has been agreed upon is that vibration is a far greater problem in a joint loaded in shear than in a joint loaded only in tension. Severe vibration parallel to the axis of the bolt, for example, might succeed in reducing preload by 30% or 40% over a long period of time. But it will usually not result in total loss of preload or in loss of the fastener. Severe transverse vibration, perpendicular to the axis of the bolt, can, and often does, cause complete loss of preload. The theory is that only transverse vibration destroys those frictional restraints in what are basically horizontal or transverse surfaces. I found this a little difficult to understand at first, because instinct tells me that if I tap the inclined plane in Fig. 16.2 with a pencil—in any direction—or bang it up and down, the block will slip to the foot of the plane just as readily as it would if I shake the plane back and forth in the direction shown by the arrows. In this case, therefore, the direction of vibration would be unimportant.

In the bolt, of course, it's easy to see how one could break the frictional restraints by substantial transverse vibration. If hole clearances, etc., allow it, I could certainly slip the joint members with respect to the nut, for example. (Similar slip occurs on a smaller scale between nut and bolt threads.) It's difficult to envision, however, how even severe *axial* vibration would cause slip or separation in these surfaces. Axial vibration would actually increase the contact forces and strains between parts during part of each cycle. In fact, at least one reference [3] says that axial vibration can sometimes tighten a bolt as well as loosen it. Severe axial vibration, however, can cause periodic dilation of the nut, creating the relative motion required to break the friction forces [11].

Some fasteners are subjected to neither pure transverse nor pure axial motion but to a combination, or to *arc slip*, as shown in Fig. 16.4, where the vibratory motion is along a circular path rather than a straight one. Experiments have shown that this type of vibration will sometimes tighten, sometimes loosen—and sometimes "neither"—the fastener [2].

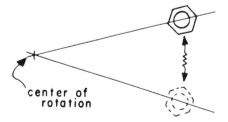

center of
rotation

Figure 16.4 Vibration along the circular path will sometimes tighten, sometimes loosen, and sometimes "neither" a fastener.

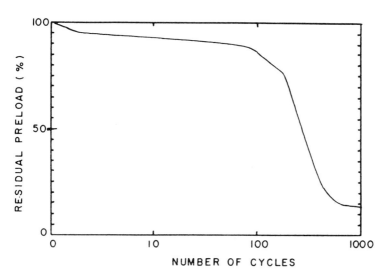

Figure 16.5 Vibration loosening starts with a slow and gradual relaxation of initial preload. Only when preload has fallen below a certain critical value does the nut actually start to back off. Loosening is rapid beyond this point. (Modified from Ref. 1.)

III. LOOSENING SEQUENCE

All agree that a fastener subjected to shock or vibration or thermal cycles will not lose all preload immediately, but will first undergo a relatively slow loss of preload. No one knows for sure why this progressive loss occurs, but it has been well documented [1, 12]. Most seem to think that cyclic forces applied to the thread surfaces by vibration and the like cause additional embedment and the slow destruction—the breakdown—of contact surfaces. Only after sufficient preload has been lost by this process will the friction forces between thread surfaces be low enough to be overcome by subsequent load cycles. At this point the nut will loosen rapidly. Various patterns of loss have been seen, as suggested in Figs. 16.5 and 16.7.

IV. JUNKER'S THEORY OF SELF-LOOSENING

I mentioned earlier that several theories of vibration loosening have been proposed. The best known is probably that of Gerhard Junker [1]. His work has been published in many forms, and has apparently been confirmed by experiments conducted on a "Junker machine," which we'll discuss later. Further confirmation comes from results obtained with special vibration-resistant fasteners whose designs have been based on his theories. His theory is, therefore, a reasonable starting point. Let's look at it in detail and then review some of the alternatives suggested by others.

A. The Equations

Junker's theories are based, in part, on the so-called "long-form torque equation" relating the torque applied to a fastener to the frictional and elastic reactions to that torque. The form of the equation I prefer, because I find it most descriptive, is a simplified version proposed by Nabil Motosh [22]:

$$T_{in} = F_P \left(\frac{P}{2\pi} + \frac{\mu_t r_t}{\cos \beta} + \mu_n r_n \right) \tag{16.1}$$

where
T_{in} = torque applied to the nut (in.-lb, N-mm)
F_P = preload in bolt (lb, mm)
P = thread pitch (in., mm)
μ_t, μ_n = coefficient of friction in thread and nut surfaces, respectively
r_t, r_n = effective contact radii of thread and nut surfaces (in., mm)
β = half-angle of thread tooth (usually 30°)

If a "prevailing torque" fastener is being used, the long-form equation must be modified to include this torque—a reaction torque (T_P) which is *not* proportional to preload.

$$T_{in} = F_P \left(\frac{P}{2\pi} + \frac{\mu_t r_t}{\cos \beta} + \mu_n r_n \right) + T_P \tag{16.2}$$

where T_P = prevailing torque (in.-lb, N-mm)

B. The Long-Form Equation in Practice

The long-form equation is believed to explain, correctly, the basic relationship between input and reaction torques in a threaded fastener. In practice, however, it is virtually useless as a means of predicting the exact relationship between applied torque and achieved preload in a given fastener, since we never know what values to assign to the parameters involved.

In spite of these limitations, the long-form equation can be used to "explain" behavior, even if it's not especially useful for numerical calculations.

C. The Equation When Applied Torque Is Absent

Junker uses the long-form equation with applied torque, $T_{in} = 0$. In other words, he's only concerned with the torque created by the elastic stretch of the bolt and the frictional reaction torques. Note that the elastic torque still wants to rotate the nut in a counterclockwise direction, as it did while the bolt was being tightened. The friction torque, however, has now reversed sign because it's opposing the counterclockwise elastic torque rather than an externally applied clockwise torque. The prevailing torque, if present, will also change sign. Taking all of this into account, and using Motosh, we can write for T_{OFF}, the net torque tending to loosen the fastener,

$$T_{OFF} = F_P \left(\frac{P}{2\pi} - \frac{\mu_t r_t}{\cos \beta} - \mu_n r_n \right) - T_P \tag{16.3}$$

How far would the coefficients of friction have to drop before there would be a net torque to loosen the bolt (assuming nominal geometry and no prevailing torque)?

As an example, in an ASTM A325 fastener, 1-in. in diameter with eight threads per inch tightened to proof load ($F_P = 51{,}500$ lb), the coefficients of friction would have to drop to about 0.015 before there would be a net, internal off torque to loosen the bolt. This is one-fifth to one-tenth what we would expect the coefficients to be under normal conditions. Some other factor must enter the picture if loosening is to occur. Junker attempts to explain this other factor. He suggests that (assuming that no locking or prevailing torque device is present):

1. The elastic stretch in a bolt will create a torque that attempts to loosen the nut, as suggested by the long-form equation.
2. If vibration is severe enough, transverse slip will occur between male and female threads and, *simultaneously*, between the joint surface and the face of the bolt *head*.
3. This slip momentarily overcomes all frictional restraint between parts, frees both ends of the bolt, and allows a portion of the elastic energy stored in the bolt to escape as the *bolt*, not the nut, rotates.
4. Whether or not any energy will escape depends on whether or not *external forces* on the system are great enough to overpower the fric-

tion or other forces which resist slip in the threads and between bolt and joint.

5. How *much* energy will be lost during each cycle depends on the *thread slip distance* involved; the greater the thread clearance, the more energy will be lost each cycle before slip ends and friction forces are reestablished.

6. The amount of energy lost during each cycle also depends on the magnitude of the net "off torque" on the nut during slip. According to the simplified Motosh equation, this would be, simply:

$$T_{OFF} = \frac{F_P P}{2\pi} \tag{16.4}$$

since, in effect, $\mu = 0$ momentarily, and there is no T_P.

D. Why Slip Occurs

An important part of Junker's theory is the explanation of why transverse vibration causes thread and nut/joint slip. In the 1969 paper [1] he suggests the following sequence of events.

1. As a result of the previous cycle, the nut and bolt are in the relative positions suggested by Fig. 16.6A—where thread clearance has been magnified for clarity.

2. Now the top joint member starts to move toward the right; the lower member moves left.

3. At first, the nut and bolt remain in the relative position shown in Fig. 16.6A, locked in that relationship by thread friction. As joint slip continues, however, the bolt bends toward the right because of the relative motion of joint members (Fig. 16.6B).

4. After a while, bending forces overcome the friction forces between male and female threads. The bolt straightens up and moves toward the opposite side of the nut. This action alone is not sufficient to cause loosening since the other end of the bolt is still held by friction forces (Fig. 16.6C).

5. If conditions are severe enough, however, the head of the bolt also slips against the adjacent joint surface—after the bolt threads have started to slip over the nut threads. Under these conditions both ends of the bolt are momentarily free and it will rotate slightly, losing a little of the stored potential energy.

6. As joint slip continues, the bolt is now cocked again to the right. During this period no further preload loss occurs because there is no further thread clearance in the necessary direction (Fig. 16.6D).

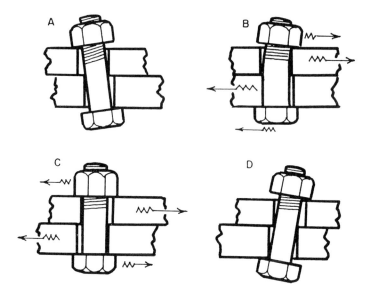

Figure 16.6 Illustration of the process by which fasteners loosen, according to Gerhardt Junker. See text for discussion.

7. The whole process then reverses, dumping a little more energy on the return stroke.

The loosening sequence described above has been supported by experiments made on a "Junker machine," shown in Fig. 16.11 and discussed in Section VI. Careful measurements show the brief instant during which slip occurs, as well as those portions of the cycle during which the bolt is clamping the upper and lower joint members. To my knowledge, none of the other theories for vibration loosening have been demonstrated in so convincing a fashion.

E. Other Reasons for Slip

Although Junker doesn't mention it, slip can presumably be caused by factors other than bending of the bolt. The bolts don't necessarily bend during a MIL-STD-1312 test, for example, but nuts loosen. Inertial forces presumably cause the slip there. There may be other situations where a combination of inertia and bending could do it. Anything which causes simultaneous slip of threads and head will satisfy Junker's theory.

V. OTHER THEORIES OF SELF-LOOSENING

Now let's take a brief look at some of the alternative explanations which have been proposed for self-loosening.

A. ESNA Theory

The Elastic Stop Nut Corporation of America (ESNA) agrees with Junker that a nut will "slide down" the inclined plane of the bolt threads if thread friction gets too low. But they maintain that this is a "near-terminal condition" that occurs after most preload has been lost; that most self-loosening occurs before there is any relative motion between male and female threads [12, 26].

They report an initial embedment loss of preload amounting to 2–10% when the fastener is first tightened, depending upon the initial roughness of the threads. This loss is blamed on embedment plus the loss of the torsional stress induced in the shank of the bolt by wrenching.

If now exposed to vibration or shock the fastener will lose more preload, as shown in Fig. 16.7. They believe that vibration or shock loads

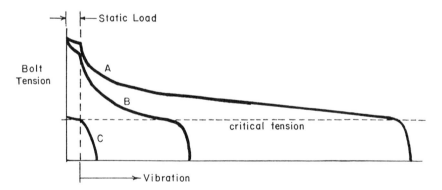

Figure 16.7 A bolt will lose some tension after first being tightened, presumably through embedment and/or loss of the torsional stress introduced when the nut or bolt was torqued. This initial loss occurs under "static" loading conditions as suggested here. When subjected to vibration the bolt will lose further tension. This loss occurs without any relative motion between the nut and the bolt, and, says ESNA, is caused by further embedment, wear, and "hammering" of the threads. Only when residual bolt tension drops below a critical value, determined by the coefficient of friction of contact surfaces, does the nut start to back off the bolt. The amount of vibration required to reach this point depends in part on how heavily the bolt was preloaded to start with.

cause the fastener to vibrate at a resonant frequency that is always much higher than the exciting frequency. Because of the clamping forces involved, and the massiveness of joint members, such vibration is rapidly damped, as suggested by Fig. 16.8, but the vibration of the fastener, they believe, "hammers down" the thread surfaces, causing the progressive loss of preload.

Their theory is supported by experiments which show that most loss of preload occurs before the nut starts to back off. Because of this they claim that prevailing torque (which we'll examine later) can only prevent loss of the nut; it does not prevent substantial loss of preload. Their solution is a nut which contains a nylon ring which further damps the resonant ringing of the fastener, reducing the "hammering" and extending the fastener's resistance to vibration. We'll take a closer look at this nut in a minute. They acknowledge that such factors as initial preload and the bolt's stiffness affect the rate of self-loosening, but suggest that those factors are predetermined by the design of the joint, and that users facing vibration problems are best served by increasing the dampening in the system, which, thanks in part to their nut, is controllable.

B. Kasei and Ishimura

These Japanese research workers believe that as transverse slip occurs in the joint the nut, at first, sticks to the adjacent joint member and that this causes the bolt to bend (in agreement with Junker). They suggest, however, that, as the bolt bends the male threads slide over the female threads and that this motion, thanks to the helix angle of the threads, creates a clockwise torsional stress in the shank of the bolt [23]. When

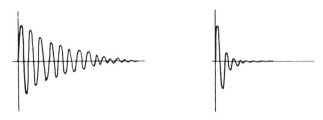

Figure 16.8 Each time a fastener receives a "blow" from the vibrating system in which it is mounted, it will resonate briefly at its own frequency. Damping these internal vibrations can reduce the number of load cycles which the fastener sees, thereby reducing the rate at which it will relax, embed, etc. A well-damped joint will therefore resist vibration longer than a poorly damped one. (Modified from Ref. 12.)

the nut finally looses its grip on the joint the male threads are allowed to lock onto the female threads. The torsional stress in the shank is relieved as the bolt turns the nut counterclockwise.

C. Daadbin and Chow

In England, Daadbin and Chow say that the contact between male and female thread surfaces is damped but elastic. Shock or vibration initially increases contact forces, but since the parts are elastic, they will then rebound. If the rebound is great enough the friction forces will be substantially reduced or even momentarily eliminated—and the nut will "slide down" the inclined plane of the bolt threads [24]. Their theoretical equations suggest that a smaller lead angle, longer duration of applied force, and larger coefficient of friction should all retard loosening; and their experiments confirm that this is so.

D. Yazawa and Hongo

Yazawa and Hongo believe that shear loads cause the bolt head to slip over the joint surface, creating friction forces on the underside of the head, as shown in Fig. 16.9. Because the contact between head and joint plate is rarely uniform, these forces will be greater on one side of the head

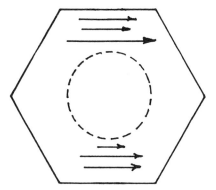

Figure 16.9 According to Yazawa and Hongo, the head of a bolt slips over the surface of the joint exposed to transverse vibration. The relative motion between head and plate creates friction forces in the direction of slip, as suggested here. Since these will not be uniform, a net clockwise or counterclockwise torque will be created on the bolt, either tightening or loosening it. The authors claim that this is the source of the torque which results (in some cases) in self-loosening.

than on the other. This creates torques which will either tend to tighten or to loosen the fastener [25].

E. Sakai

Sakai agrees with Junker and with Kasei and Ishimura that as transverse slip occurs the bolt inclines in the direction of slip. The male threads slip over the female threads, with one side slipping "in the ascending direction of lead angle," the other side in "the descending direction." This creates loosening torques both as the plate slides forward and as it returns [2].

So much for the theories of self-loosening. We may never know why it occurs, but everyone agrees on some of the basic things which retard self-loosening—and all experiments confirm these basics even if disagreeing on the specifics of the loosening mechanism. All agree that more preload, more friction, flatter helix angle, and less severe imposed vibration all retard self-loosening. All agree that transverse forces—and slip—are required. All agree that prevailing torque helps, but they disagree on "how much." In a minute we're going to look at some of the many vibration-resistant fasteners which have been placed on the market, each of which relies on one or more of the basics just listed. First, though, let's look at the way in which such fasteners are evaluated.

VI. TESTING FOR VIBRATION RESISTANCE

In a moment we will look at many ways to improve the vibration resistance of a bolted joint. Since we can't predict vibration loosening mathematically, the usefulness of such techniques must be determined experimentally, in a test machine.

Experts warn that it is not sufficient to conduct a laboratory test on a simulated joint, however. If possible you should always repeat the test on another system—perhaps the one you are having trouble with—to be sure that you have really made a difference. Our knowledge of vibration loosening is entirely empirical, and there are many factors which can make a difference. Some experiments, in fact, have suggested that complex interactions between suspected factors, perhaps more than the factors themselves, determine the rate at which a given system will loosen; and/or that there probably are other factors which we have not been able to pin down as yet, which also make a difference [6]. You could easily be fooled by some of these unknown interactions and factors if you tested only a "test joint."

A. ALMA Test

One popular way to test for vibration resistance is shown in Fig. 16.10. The nut and bolt are tightened onto a small cylinder. This cylinder is placed in the slot of a test block. The cylinder is longer than the block, and washers are used at both ends so that the cylinder is free to bang around in the slot without being able to fall out of it. The block is vibrated in a vertical direction, causing the fastener and its cylinder to bang back and forth between the top and bottom of the slot until the nut and bolt shake apart. Vibration frequency, amplitude, and time are measured to set a numerical value on the vibration resistance of the fastener system under test. This is sometimes called the ALMA test [3, 5]. All pertinent dimensions are defined in the government specification MIL-STD-1312.

It is easy to measure the amplitude and frequency of input vibration in the ALMA test system. Some people, however, believe that it is more pertinent to measure the actual magnitude of the force exerted on the joint under test and/or the actual displacement of joint members. This is not possible with the ALMA test. You can measure the amplitude of the shake table, but it is difficult to tell what, if any, displacement has actually resulted between the fastener and the test cylinder—or to determine the forces on the fastener—because both of these depend on impact, and impact is very difficult to predict or control.

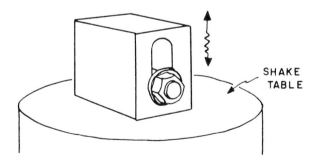

SHAKE
TABLE

Figure 16.10 The ALMA vibration test. The fastener under test is mounted in a cylinder which is free to bang up and down within the slot in the block as the block is vibrated vertically.

B. Junker Test

An alternative vibration test machine is shown in Fig. 16.11. This is called the Junker machine. An eccentric cam generates a controllable amount of transverse displacement on the joint under test. Force cells measure the actual transverse forces exerted on the joint. One can now determine the relationship between residual preload in the fastener under test and external vibratory forces created by the test machine as a function of time. Junker machines can be purchased from SPS, whose address is given in Table 16.1 and Appendix D.

Theoretically, when you test a fastener for vibration resistance, you would like to subject it to the vibration frequencies and magnitudes you expect the joint to encounter in your application. Predicting the vibration environment a given product will see, however, is even more difficult than predicting external loads. You will rarely be able to find good data in the literature for your application or product, but will, instead, have to rely on your own field tests. Vibration frequencies and magnitudes will not be uniform, furthermore, but can vary from moment to moment as well as from user to user. Your only recourse is to provide a fastener system that is immune to the range of frequencies you expect it *might* encounter in practice and then to determine by trial and error whether or not you have been successful.

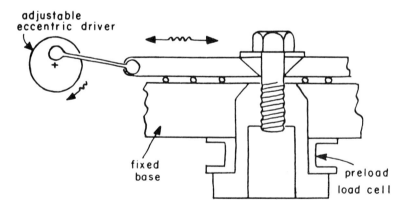

Figure 16.11   The Junker vibration test machine. The forces exerted on the test joint and the displacement of the test joint can both be measured with this device. (Modified from Ref. 17 and from information published by SPS.)

Table 16.1 Trade Names and Manufacturers
Cited in Illustrations

Greer/Smyrna
Spiralock
Smyrna, Tennessee

ESNA Division
Amerace Corporation
Union, New Jersey

SPS Technologies
Aerospace and Industrial Products Division
Jenkintown, Pennsylvania

CrestLock
National Twist Drill
Division of Lear Seigler, Inc.
Rochester, Michigan

Disk-Lock International
Los Angeles, California

Stage 8 Fastening Systems
San Rafael, California

Loctite Corporation
Anaerobic adhesives
Newington, Connecticut

Detroit Tool Industries
Spiralock
Madison Heights, Michigan

TrueLock Co.
Herndon, Virginia

MacLean-Fogg Co.
Fasteners and Assemblies Div.
Mundelein, Illinois

VII. TO RESIST VIBRATION

In Chap. 15 we learned that self-loosening will occur only if two essential conditions are present: cyclic, transverse loads and relative slip between thread and or joint surfaces. Junker and several others suggest various ways in which these conditions might cause self-loosening. According to most theories, we can prevent self-loosening if we can eliminate one or both of these conditions. It's also obvious that we could prevent at least complete loss of preload if we could somehow fasten or lock the nut to the bolt, relying on mechanical or chemical means rather than on friction to guarantee the integrity of the fastener.

We're now about to examine a few of the many ways which have been proposed for doing these things. Most of our options will fall into one of the following categories.

1. Keep the friction forces in thread and joint surfaces from falling below the forces which are trying to loosen the nut.
2. Mechanically prevent slip between nut and bolt or nut and joint surfaces.
3. Reduce the helix angle of threads to reduce the back-off torque component ($W \sin \theta$ in Fig. 16.2).
4. Provide a "prevailing torque" or locking action of some sort which counters the back-off torque created by the inclined planes of the threads, and does so even after friction forces in the system have been overwhelmed by vibration.

Let's look at some of the ways we can accomplish these goals.

A. Maintaining Preload and Friction

Conventional Wisdom

The least expensive and simplest way to fight self-loosening is often by preventing loss of preload in the fastener. High preload or bolt tension provides a high "normal force," which, in turn, creates frictional forces which discourage relative motion between nut and bolt. So we want to insure proper control of preload during assembly and do whatever is possible to reduce or eliminate or compensate for the subsequent relaxation of preload caused by embedment, elastic interactions, and the like.

It is generally agreed that we want to tighten the fasteners to the threshold of yield if we need maximum vibration resistance [7, 8]—always recognizing, however, that we may not be able to tighten them this much if external loads and/or safety factors make this much preload unwise. As far as vibration resistance alone is concerned, however, the more preload the better, as shown in Fig. 16.12 [7].

We can also do things to modify the coefficient of friction of thread or other surfaces, avoiding lubricants, for example, and/or plating parts to increase the coefficient beyond that which we would get with as-received parts [9].

Introducing some form of vibration damping can also help maintain friction, as mentioned earlier, because it reduces the rate at which preload

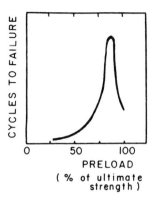

Figure 16.12 Vibration resistance of a Grade 5 fastener as a function of preload (expressed as percentage of yield strength).

will relax under vibration. Nylon inserts in the bolt or nut threads are said to be an effective way to accomplish this [3].

Note that one way to reduce preload loss is to provide a low bolt-to-joint stiffness ratio. As the joint diagrams in Chap. 12 told us, a relatively "stiff" joint and "soft" bolt will reduce the amount of bolt preload lost for a given amount of bolt length change; in this case a given amount of embedment or hammering.

Anything else which reduces the amount and rate of relaxation will also be helpful. If possible, for example, you should avoid gaskets in a joint subjected to severe vibration.

Disc-Lock Products

There's an interesting group of products, produced by Disc-Lock International of Los Angeles, California, that are designed to prevent the loss of preload in a fastener by mechanical means [27]. The original Disc-Lock product was a two-piece washer, shown in Fig. 16.13. This consisted of two mating discs with interposing multitoothed cams, a series of short ramps. Sharp ridges on the upper and lower outer surfaces of the discs dig into the nut and joint surfaces. If the nut tries to back off, it must drag its cam disc along with itself. The cam surfaces on its disc are forced to climb the cam surfaces on the disc which is gripping the joint. The cam angle is greater than the lead angle of the threads, so relative motion between the two cams prevents loosening of the fastener. In fact, if the nut tries to back off the tension in the bolt can actually *increase*, at least

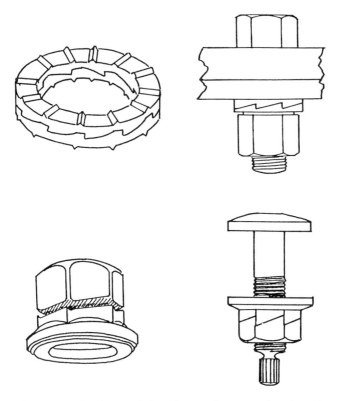

Figure 16.13 Some of the Disc-Lock International products described in the text. The washer illustrated by the two sketches at top was the first Disc-Lock product. The four-piece nut illustrated by the bottom left sketch was introduced only recently. It can be used on a standard bolt or, as shown at bottom right, can be provided with a twist-off, torque-controlled bolt.

if initial preload is low, as shown in Fig. 16.14. If initial preload is high, some of it will be lost, but the loss is minimized by the action of the mating cams.

More recently the Disc-Lock company has introduced a four-piece (but preassembled) nut which acts on the same principle as the washer. The top piece is threaded and has ramped cams on its lower side. The middle piece is unthreaded and has cam surfaces on its upper side which engage the cams on the upper piece. A flat washer and a ring which holds everything together complete the assembly. If the threaded part of the nut tries to back off it must do so by climbing the ramped cams on the

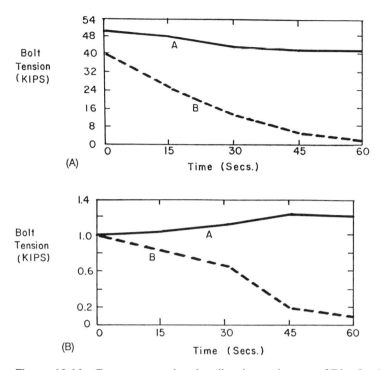

Figure 16.14 Curves comparing the vibration resistance of Disc-Lock nuts (the A curves to conventional automotive wheel nuts (B). If heavily preloaded, as in the top set of curves, the Disc-Lock nut will lose some tension before stabilizing, when subjected to severe vibration in a Junker's machine. When lightly preloaded, bottom curve set, the Disc-Lock nut actually increases the tension in the bolt by a few percent before stabilizing. The conventional nut loses all preload under either preload condition, and does so in only 1 min.

mating part. Again, the cam angle is greater than the lead angle of the threads. Hex flange Disc-Lock nuts are available in sizes ranging from $\frac{5}{16}$ to $1\frac{1}{8}$ in. and in metric sizes from M8 to M27. The company also manufactures a twist-off bolt (like that shown in Fig. 10.4) equipped with a Disc-Lock nut, and a bolt with a multipiece, Disc-Lock head which can be used in blind holes.

Not long ago I witnessed some very severe and impressive tests of several Disc-Lock nuts on a Junker machine, and came away a believer. This product is already widely used as a wheel nut on large military and commercial vehicles both here and abroad, and it is starting to find other applications in the automotive world and elsewhere.

B. Preventing Relative Slip Between Surfaces

Providing and maintaining adequate bolt tension is probably the easiest and least expensive way to combat moderate vibration. In many cases, however, it is impossible to provide a large enough fastener to withstand the vibration present. The preloads required to resist vibration forces would yield or break the fasteners we have to work with—because of limitations on joint size, shape, cost, or the like. Under these circumstances, something else is required.

If you're designing the joint, one way to minimize slip is to orient the bolts and joints so that bolt axes are parallel to the expected direction of vibration. Remember that axial vibration is far less of a problem than transverse vibration.

In many cases a designer can shape the joint so that relative slip between joint members is prevented or at least minimized. Remember that there must be actual slip before we break the friction forces that resist vibration loosening.

Joints such as those shown in Fig. 16.15, therefore, can be very helpful in fighting vibration [19].

The fastener can sometimes be used as a dowel pin—in a tight hole—to reduce joint slip. Or an actual dowel pin can be added to the joint.

Joint members can be tack-welded together. Or adhesives can be used between joint surfaces to minimize slip.

Experiments have indicated that the nuts of long, thin fasteners won't slip over joint surfaces under transverse vibration; instead, the fasteners

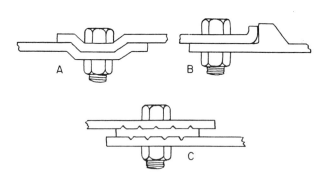

Figure 16.15 Joints can sometimes be designed to resist transverse slip. This can be a very effective way to resist vibration loosening. A toothed "shear washer" has been introduced in joint C.

bend. If the length-to-diameter ratio is greater than 8:1, "you can't shake them loose" [10].

C. Reducing Back-Off Torque

One way to reduce back-off torque on the nut would be to reduce preload. Everyone who has studied the problem, however, says that we always want the maximum possible preload to reduce vibration loosening. The loss in friction forces which would result from less preload, in other words, overwhelms any advantages we would get by reducing the off-torque created by the inclined planes.

We can, however, reduce the off-torque by decreasing the helix angle. The only practical way to do this in most situations is to use a fine-pitch thread instead of a coarse-pitch thread. This can make a useful difference, as shown in Fig. 16.16.

D. Countering Back-Off Torque

When all else fails—and it often does—the only thing we can do to fight vibration loosening is to provide another source of torque to counter the back-off torque produced by the inclined planes of the threads. Now, even if vibration totally destroys all friction forces, some other mechanism prevents that nut from being pushed out of the way by the bolt threads.

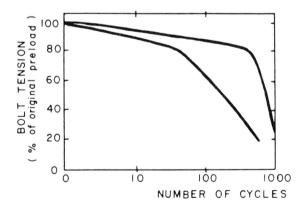

Figure 16.16 Reducing the helix angle of the threads by using a fine-pitch thread instead of a coarse-pitch thread can make a useful difference in the vibration resistance of a fastener. (Modified from Ref. 17.)

Prevailing Torque Fasteners

There are many different types of prevailing torque fasteners, a few of which are shown in Fig. 16.17. In general they can be classed as (1) all metal nuts or bolts whose threads have been purposely distorted or modified to provide some interference with the mating part, (2) nuts or bolts with a plug or patch or insert of nonmetallic material—often nylon—in the threads to create interference, and (3) nuts with a collar or ring of nonmetallic material, again to create interference with the mating bolt and, in this case, as discussed earlier, to dampen the resonant frequency vibrations of the fastener.

   Since the torque required to run down the nut on a prevailing torque fastener must be added to the torque required to achieve a desired preload, it's important to be able to predict the run-down torque. The Industrial Fasteners Institute in Cleveland, Ohio has developed standards which specify acceptable run-down torques. They also specify the minimum torques required to disassemble the fasteners on the first and fifth removals. These torques, of course, are those which would be acting to prevent "self-removal," and so are also of interest. Prevailing torque nuts are covered in IFI Standard 100/107. English and metric series bolts are covered in IFI 124 and 524, respectively [31].

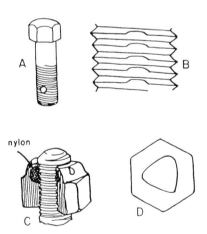

Figure 16.17   A selection of prevailing torque nuts and bolts. Those shown include (A) nylon pellet in bolt threads (Greer); (B) interference fit threads (SPS); (C) nylon locking collar in nut (ESNA); (D) nuts with out-of-round holes.

Free-Spinning Lock Nuts or Bolts

Free-spinning lock nuts or bolts can be run down with normal (very little) torque. As they are tightened against the surface of the joint, however, they dig into it, or distort in some way to create an interference fit with the mating parts. A few of the many choices are sketched in Fig. 16.18 [11, 13–15, 17]. The serrated head bolt shown in Fig. 16.18B is reported to be especially effective. So is the Spiralock thread form of Fig. 16.18A. The bolt threads here are conventional, but the root of the nut threads is a tiny ramp or inclined plane. As the nut is tightened, the tips of the male thread are forced into interference fit with the ramps. This eliminates all clearance between male and female threads. The inventor of this thread

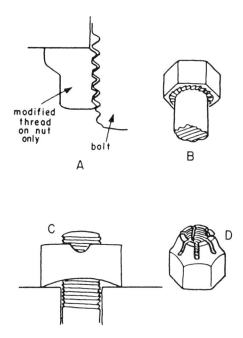

Figure 16.18 Free-spinning lock nuts and bolts which can be run down with normal torque, but which create an interference of some sort of final tightening. Those shown include (A) interference thread nut (Spiralock by Greer/Smyrna or Detroit Tool Industries); (B) serrated head bolt by SPS (serrations dig into joint surface and resist reverse rotation; serrated face nuts are also available); (C) spring head nut, which distorts inwardly to pinch bolt; (D) spring arms on top of nut provide interference fit with bolt threads (called beam-type nut).

form, Harold ("Ace") Holmes of Detroit, is a firm believer in the Junker theory of vibration loosening. He designed the Spiralock thread form to eliminate loosening by eliminating slip clearance—and the results seem to support Junker's theories.

The Spiralock thread form has been tested at MIT [33] and at Lawrence Livermore Laboratories [34]. In the resulting reports we learn that, although the nut is free spinning until seated, this thread form requires 20% more torque than a standard thread form to achieve a given preload. The extra torque, of course, is required to pull the male threads up the root ramps of the female threads.

Another feature: the Spiralock thread form creates a more uniform distribution of load and stress than does a standard thread form. For example, Nayak of MIT says that only 18% of the tensile load in the bolt is transferred to the first engaged thread; versus 34% in a standard thread form [33]. He also says that the Spiralock thread requires three times as much off-torque to start loosening the nut as does a standard form, a measure of the Spiralock's vibration resistance.

Holmes has also designed another interference fit thread form called the CrestLock, shown in Fig. 16.19. Again, the bolt threads are conventional. The innermost threads on the nut are conventional, too, so the nut is free spinning as it is started on the bolt. The outer nut threads are undercut, however, to create an interference fit with the tips of the bolt threads, as shown (exaggerated) in the illustration. Again, the slip clearance between male and female threads has been eliminated.

When selecting a prevailing torque or locking fastener, you should consider these points (in consultation with potential suppliers):

Operating temperature limits
Mating thread accommodation
Effect on mating parts (may damage or Brinell them)
Reusability
Type of installation tools required
Effect on the mechanical properties of mating parts (for example, does it
    reduce the fatigue life of another part by providing a "softness" in
    the joint or by creating stress concentrations?)

E. Mechanically Locked Fasteners

Sometimes it's impossible to provide enough prevailing torque to prevent loosening under severe shock or vibration conditions; other times self-loosening would threaten safety and we want to be absolutely sure the fasteners won't come loose. In such situations we can consider the use

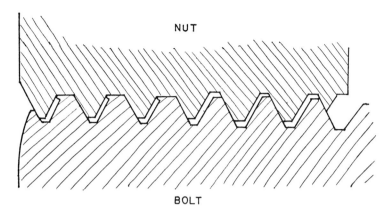

NUT

BOLT

Figure 16.19 The CrestLock thread form includes an interference fit between the outermost threads of the nut and the bolt threads, eliminating clearance for slip.

of fasteners in which the nut and bolt are mechanically locked together. We'll find both new and old options in this category.

Lock Wires and Pins

The earliest attempts to prevent self-loosening of a threaded fastener probably involved lock wires, keys, and cotter pins; and these still find a lot of use. A couple of examples are shown in Fig. 16.20. These can effectively prevent total loss of the nut—which may be extremely important—but they are not very effective in preventing substantial loss of preload within the fastener. It has been reported, for example, that two degrees of rotation in the nut can reduce preload in a hard joint by 27%; six degrees can reduce it 42% [1]. Most lock wires or cotter pins aren't intended to provide tight control of nut motion. Even if they save the nut, the loss of preload may lead to fatigue or another type of failure.

Welding

Nuts can be welded to bolts, at least if they're large enough. The normal procedure is to tack weld the nut to the end of the bolt. Another procedure is to tack weld both the nut and the head of the bolt to joint surfaces. Either procedure makes removal of the nut (for maintenance purposes, for example) very difficult and will probably make it necessary to replace them with new parts if they are removed. A related procedure, which

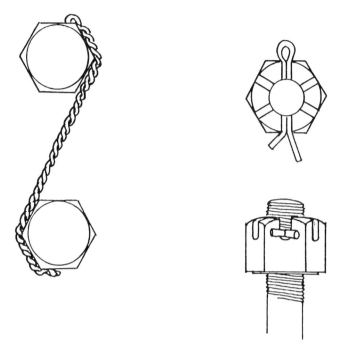

Figure 16.20   Lock wires and cotter pins are ancient ways to mechanically fasten the nut to the bolt. They're both still widely used, but are more awkward to install than more modern locking fasteners. The cotter pin is often used in conjunction with a castellated nut, as shown here.

preserves the parts, is to place a "keeper" over the tightened nut and then weld the keeper to the joint, as shown in Fig. 16.21. This trick has been used to trap the large nuts used on the foundation studs of nuclear reactors; nuts which presumably will not be removed until the reactor is taken out of service.

Stage 8 Fastening System

The Stage 8 fastening system, shown in Fig. 16.22, fights loosening in still another way. The bolt and nut are conventional except for snap-ring grooves, as shown. Loosening is resisted by a retaining arm which is slipped over the nut and/or bolt head. The far end of the arm butts up against an adjacent nut or other reaction surface to prevent counterclockwise rotation. The snap-ring prevents loss of the retaining arm. A wide

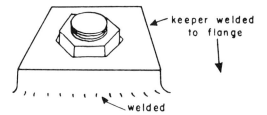

Figure 16.21 A square plate with a hexagonal hole in it is placed over the nut. Then the plate is welded to the top surface of the joint to retain the nut.

variety of retaining arms, collars, rings, etc., are available, depending on the bolt pattern, the reaction surfaces available, etc.

Boydbolt Fastening System

I've never seen one, but as the second edition of this book was being prepared the locking bolt shown in Fig. 16.23 was being offered for sale by Rexnord Specialty Fasteners of Torrance, California. A spring-loaded

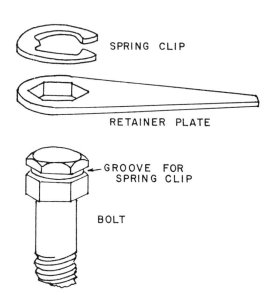

Figure 16.22 The Stage 8 fastening system provides a reaction or retaining arm to prevent reverse rotation of the nut or bolt. A variety of reaction arm configurations are available.

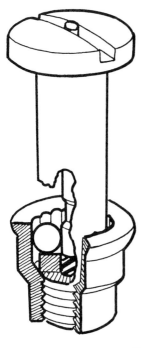

Figure 16.23   The Boydbolt uses steel balls, spring loaded into a detent groove in the bolt, to resist vibration (Rexnord, Torrance, California).

plunger, running down the axis of the bolt, forces two hardened steel balls out into detent grooves in the nut, locking the nut to the bolt. The plunger is depressed, manually, to retract the balls and allow the nut to be removed, if desired. The device is shown in Fig. 16.23.

Huck Lockbolt

Lest we forget, the so-called lockbolts which are described and illustrated (Fig. 10.5) in Chap. 10 will also resist severe vibration. Remember that these bolts are held by collars swaged into annular grooves, or into threads and a keyway instead of by nuts.

TrueLock Bolt

Two new products which provide mechanically locked nuts have recently been placed on the market. One is the TrueLock Bolt, made by the

TrueLock Company of Herndon, Virginia [28]. As shown in Fig. 16.24, a locking pin, parallel to the axis of the bolt, is run into a rounded "keyway," one-half of which is formed by the nut and one-half by a "lock crown" which has been connected to the bolt by a fine-pitch threaded pin run into a hole tapped along the centerline of the bolt. This description makes it sound more complicated than it really is; see the illustration for a better understanding. Because the threaded center pin in the lock crown is elastic, the manufacturer claims that the lock can be achieved at any nut position. Some other mechanically locked nuts lock in only a limited number of positions.

Axilok Nut

Another recently introduced product is the Axilok spindle nut shown in Fig. 16.25, made by the MacLean-Fogg Company of Mundelein, Illinois [30]. It is intended for use on automotive axles. A locking ring is prevented

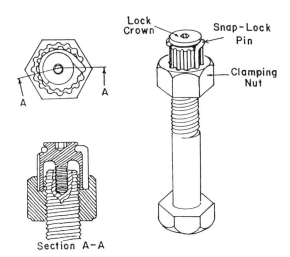

Section A–A

Figure 16.24 A TrueLock fastener, recently offered for sale. A Lock Crown, shown cutaway in section A–A, contains a series of half-round grooves. The fastener's nut also contains a series of half-round grooves. Since one piece contains one more groove than the other, only one pair of grooves can be aligned at any one time. A fine-pitch, threaded pin on the Lock Crown is screwed into a tapped hole in the end of the bolt, and is tightened until one of its grooves is aligned with a groove in the previously tightened nut. A snap-lock pin is now run into the hole formed by the aligned grooves, and is snapped into a retaining ring on the Lock Crown to complete the assembly.

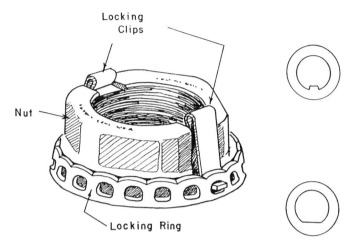

Figure 16.25 The Axilok automotive spindle nut recently introduced by Mac-Lean-Fogg. The perforated ring seen at the base of the nut is prevented from rotating on the end of the axle by a flat or keyway, as shown in the smaller sketches. The nut contains two spring clips which are depressed by a conventional drive socket as the nut is tightened. Tabs on the lower ends of these spring clips find holes in the base ring either when the socket is first removed or after the nut has backed off a degree or two. Further rotation of the nut is impossible, once the tabs have found their holes; but a conventional socket can once again be used to depress the clips and remove the nut for maintenance purposes.

from rotating on the axle by virtue of a milled slot or D-shaped keyway. Springy locking clips on the nut engage holes in the locking ring to prevent the nut from backing off after it has been tightened. These clips are automatically depressed when a conventional drive socket is placed over the nut, so installation and removal of the nut can be done with standard tools.

F. Chemically Bonded Fasteners

Chemically bonded fasteners, while not, perhaps, having the vibration resistance of mechanically locked ones, are more popular. Part of the reason for this may be the fact that the chemicals can be used on small as well as large fasteners. Also contributing to their popularity: the chemicals cost little and can be used on standard fasteners. In many cases the chemicals are applied by the user before assembly; in other cases they're applied by the manufacturer. Several chemicals are used, including acrylics, but I believe that microencapsulated anaerobics are the most popular [32]. Before taking a brief look at them, let's look at another, less common way to bond nuts to bolts.

Rust

The U.S. Marines have found an interesting way to fight vibration loosening in tank tread bolts. After they have assembled the tread, they drive the tank through surf. Saltwater corrosion effectively "welds" the nut and bolt together and/or welds them to the joint members. Petrochemical engineers often find rain helpful.

Anaerobic Adhesives

One very good way to resist off-torques is to "cement" the nut and bolt together. The most common way to do this is with an anaerobic adhesive, a material which is activated (hardens) when subjected to high pressure in the presence of metal [16]. It is applied to fastener threads much as a lubricant would be applied. It "glues" the threads together when they are tightened; it can, however, be overcome if you subsequently wish to take the joint apart. The material does no permanent damage to the threads.

A wide variety of anaerobic adhesives are available. Selection would be based on such things as the size of the gap to be filled, adhesive strength (off-torque requirements), size of the fastener, method of applications, etc. It's best to consult the manufacturers for details.

We learned earlier that it is important to know how much torque will be required to install a prevailing torque fastener, because that torque must be added to the torque required to achieve a desired preload. The same is true of fasteners coated with anaerobic adhesives or other chemicals. We're also interested, of course, in knowing how effective such coatings are in resisting vibration, which means we'd like to know how much breakaway torque is required to back off the nut and how much prevailing torque is involved if we try to loosen it further. The Industrial Fasteners Institute of Cleveland, Ohio has developed standards listing acceptable values for each of these torques when the coating is applied by the manufacturer. Standard IFI-125 covers English series fasteners; IFI-525 covers metric series fasteners [31, 32].

G. Vibration-Resistant Washers

A wide variety of vibration-resistant or locking washers are also available—and are very popular. A few of them are shown in Fig. 16.26.

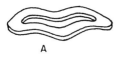 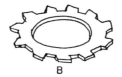 

A

B   C

Figure 16.26 Washers used to fight self-loosening. The wave washer is supposed to provide some spring tension in the bolt after the nut has loosened; but I doubt if it's of much help. The toothed washer digs into both nut and joint surfaces and fights relative motion between them. The Belleville washer is usually used in stacks, on large fasteners in pressure vessel and similar applications. A stack provides a fairly high spring rate, and also allows us to use longer bolts whose lower spring rate will be more comparable to that of the Bellevilles. The lower spring rate also makes the bolt tension less susceptible to fluctuations in external load, thermal change, and the like, thereby fighting self-loosening by retaining bolt preload.

Washers Which "Maintain Tension" in the Fastener

The wave and Belleville washers shown in Fig. 16.26 are intended to push outward on the nut and so maintain some tension in the bolt if the nut loosens. The wave washer is typically used on small fasteners, the Belleville on large.

Obviously, all tension will have been lost in the bolt before the wave washer takes over. Its effectiveness must be questioned, to put it politely. The Belleville washer, however, has an impressive track record, at least in the pressure vessel world. They're usually used in stacks, four or more washers being piled on top of each other. The spring rate of the stack, while probably less than the spring rate of the bolt, is still high enough to provide some clamping force on the joint. The stack also allows us to use a longer bolt, reducing bolt stiffness and making some loss of deflection less significant; as suggested in Fig. 6.17.

Toothed Washer

The toothed washer shown in Fig. 16.26 is designed, I believe, to bite into both joint member and nut, preventing relative motion. They're widely used in appliances and in other applications where some self-loosening is, while annoying, of relatively little concern.

Helical Spring Washer

At first glance the helical spring washer shown in Fig. 16.27 would appear to be of as little value as the wave washer; unless the cut ends manage

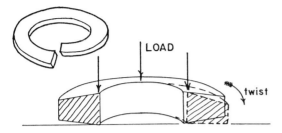

Figure 16.27 A helical spring lock washer would appear to be a fairly inefficient way to resist self-loosening; but recent research—described in the text—shows that this washer twists and rolls when it's fully loaded. Since that requires a clamping force that can equal 65% of the proof load of the bolt, this washer can, indeed, provide significant resistance to self-loosening.

to bite into joint and nut and resist relative motion the way the toothed washer does. Recent research at the Lawrence Technological University in Southfield, Michigan, however, suggests that this device is more effective than it appears to be [29]. Dr. Clarence Chambers has shown that, while this washer is flattened by bolt tension equal to only 5% of its proof load, increasing bolt preload to 70% of proof will cause the trapezoidal cross section of the washer to roll and twist down on the outside diameter, which also grows. This complex action results in a washer spring rate which can approach 65% of the spring rate of the fastener. That spring rate will dominate the behavior of the fastener under load, and will reduce the amount of preload lost under a given applied load. Retaining preload, of course, is an effective way to resist self-loosening.

Table 16.2 Relative Performance of Various Types of $\frac{3}{8}$–16 Locking Fasteners

Type of fastener	Percentage of initial load retained
Serrated locking screw	85
Anaerobic adhesive	85
All-metal locking screw	50
Epoxy locking screw	45
Patch-type locking screw	30
Prevailing torque nut	30

Table 16.3 Typical vibration Performance of Various Vibration-Resistant Nuts (See text concerning the omission from this list of the Disc-Lock nut)

Type of nut	Relative number of cycles required to shake nut from bolt
Nut with locking ring of nylon	100
Beam-type self-locking nut—aircraft	53
Castellated nut and spring pin	38
Distorted thread nut—aircraft	19
Castellated nut and cotter key	18
Beam-type self-locking nut—commercial	4–17
Distorted thread nut—commercial	1–10
Castellated nut and lock wire	8
Plain nut and spring-type lock washer	5
Plain nut, with or without tooth-type lock washer	1

H. Comparison of Options

As you go through the literature on prevailing torque fasteners, lock washers, anaerobic adhesives, etc., you will find many claims and counterclaims about the efficiency and values of various techniques and products. I am not in a position to pass judgment on these claims—and some of them may be obsolete by the time you read this. The Disc-Lock nut, for example, is not included in Table 16.3 and yet I believe might top the list. Table 16.2 rates a number of different possibilities from the point of view of one manufacturer (who provides all of the methods tabulated) and may or may not be pertinent for your own applications [18]. Table 16.3 rates others from the point of view of a different manufacturer [20]. I suggest that you talk to many possible suppliers, giving them full information about your problems, before making a final selection. Even then, it would be best to test several possibilities in a test mechine such as that described in Sec. VI before making a final selection.

REFERENCES

1. Junker, G. H., *New Criteria for Self-Loosening of Fasteners Under Vibration*, Reprinted October 1973 from Trans. SAE, vol. 78, 1969, by the Society of Automotive Engineers.
2. Sakai, Tomotsugu, Investigations of bolt loosening mechanism, *Bulletin of the JSME*, vol. 21, Paper No. 159-9, September 1978.

3. *Maintaining the Tightness of Threaded Fasteners*, ESNA Division, Amerace Corporation, Union, NJ, 1976.
4. Shigley, J. E., *Mechanical Engineering Design*, 3rd ed., McGraw-Hill, New York, 1977, p. 244.
5. Landt, R. C., Vibration loosening—causes and cures, Presented at Using Threaded Fasteners Seminar, University of Wisconsin-Extension, Madison, WI, April 24, 1979.
6. Walker, R. A., The factors which influence the vibration loosening of fasteners, Presented at 1973 Design Engineering Conference, Philadelphia Civic Center, April 10, 1973.
7. Smith, Stanley K., Fastener tension control—what it's all about, *Assembly Eng.*, November 1976.
8. Junker, G., *Principles of the Calculation of High Duty Bolted Connections—Interpretation of Guideline VDI 2230*, VDI Berichte no. 220, 1974. An Unbrako technical thesis, SPS, Jenkintown, PA.
9. McKewan, John, The effects of plating on torque-tension relationships and vibration resistance, Paper no. 800452, Congress and Exposition of the SAE, Cobo Hall, Detroit, MI, February 19, 1980.
10. Hardiman, Russell, Vibration loosening—causes and cures, Presented at Using Threaded Fasteners Seminar, University of Wisconsin-Extension, Madison, WI, May 9, 1978.
11. Aronson, R. B., How locknuts remove the "unknowns" in joint design, *Machine Design*, p. 166, October 1978.
12. *ESNA Visual Index*, ESNA Division, Amerace Corporation, Union, NJ, 1979.
13. *There's Nothing Quite Equal to Spiralock® Nuts*, Greer/Central, a division of the Microdot Corp., Troy, MI, 1980.
14. 1979 fastening and joining reference issue, *Machine Design*, Penton/IPC Publications, Cleveland, OH, November 15, 1979.
15. Through-hardened DURLOK 180® screws lock tight without gouging, Advertisement of SPS Technologies, Jenkintown, PC, published in *Machine Design*, April 1980.
16. *Loctite® Technology*, Loctite Corp., Newington, CT, 1979.
17. Ingenious locking concepts abound, *Product Eng.*, p. 39, March 1979.
18. Finkelston, R. F., and P. W. Wallace, Advances in high performance fastening, Paper no. 800451, Presented at the Congress and Exposition of the SAE, Cobo Hall, Detroit, MI, February 29, 1980.
19. Bonenberger, P. R., A basic approach to joint design, Presented at Using Threaded Fasteners Seminar, University of Wisconsin-Extension, Madison, WI, May 13, 1980.
20. Baubles, R. C., and G. J. McCormick, *Loosening of Fasteners by Vibration*, ESNA, Union, NJ, December 1966.
21. Landt, Richard C., Comments made during presentation of a paper, Vibration loosening—causes and cures, University of Wisconsin-Extension, Madison, WI, Using Threaded Fasteners Seminar, May 1980.
22. Motosh, N., Development of design charts for bolts preloaded up to the plastic range, *Journal of Engineering for Industry*, August 1976.

23. Kasei, Shinji and Mitsutoshi Ishimura, Basic investigation on self-loosening of threaded joints subjected to repetition of transverse loads, no provenance or date given in the document; but S. Kasei is a joint author (with A. Yamamoto) of papers on self-loosening published in the *Bull. Japan Soc. of Prec. Eng.* and so this paper may be a reprint of an article in that journal.

24. Daadbin, A., and Y. M. Chow, A theoretical model to study thread loosening, Department of Mechanical Engineering, University of Newcastle on Tyne, NE1 7RU, UK, paper received February 27, 1990.

25. Yazawa, Shimpachi, and Kaoru Hongo, Loosening of a bolt-nut connection induced by a tangential load applied to a clamped plate. The journal name, date, etc. are given in Japanese, but I believe this came from the *Bull. Jap. Soc. of Mech. Eng.*, No. 86-1296 A.

26. Baubles, R. C., G. J. McCormick, and C. C. Faroni, Loosening of fasteners by vibration, document ER272-2177, Eng. No. 30 published by the Elastic Stop Nut Corp. of America, Union, NJ, December 1966.

27. From information furnished by Alistair McKinlay of Disc-Lock International, Los Angeles, CA, June 1994.

28. From information furnished by Dr. Michael Vassalotti of the TrueLock Company, Herndon, VA, June 1994.

29. Chambers, Clarence, Spring rate of helical spring lock washers, Lawrence Technological University, Southfield, MI, 1991.

30. From information furnished by the MacLean-Fogg Co., Fasteners and Assemblies Division, Mundelein, IL, 1994.

31. Standards published by the Industrial Fasteners Institute of Cleveland, Ohio, as listed below:

    IFI 100/107, Prevailing-torque type steel hex and hex flange nuts, regular and light hex series, revised version published on August 15, 1993.
    IFI 124, Test procedure for the locking ability performance of nonmetallic element type prevailing-torque lock screws, 1987.
    IFI 125, Test procedure for the locking ability performance of chemical-coated lock screws, October 1973.
    IFI 524, Test procedure for the locking ability performance of metric nonmetallic locking element type prevailing-torque lock screws, 1982.
    IFI 525, Test procedure for the locking ability performance of metric chemical-coated lock screws, 1982.

32. Petras, Jeff, Designing with locking fasteners, *Fastener Technology International*, pp. 48–49, December 1993.

33. Nayak, P. Narayan, Spiralock: vibration resistance and stress distribution, prepared at MIT for the Chrysler Corporation, published by Spiralock, Detroit Tool Industries, Madison Heights, MI.

34. Harrel, Alfred, III, Spiralock, finite element analysis, prepared for Lawrence Livermore Laboratories, published by Spiralock, Detroit Tool Industries, Madison Heights, MI.

17
Fatigue Failure

A metallic part subjected to cyclic tensile loads can suddenly and unexpectedly fail—even if those loads are well below the yield strength of the
material. The part has failed in fatigue. Note that the failure occurs under
tensile loads. I've heard that fatigue failure under cyclic compressive loads
is possible—but is rare—so we'll ignore it.

Since failure only occurs under tensile loads, only the bolts (but not
the joint members) in tension joints and only the joint members (but not
the bolts) in shear joints can and do fail in fatigue. We'll devote most of
this chapter to the fatigue failure of bolts; but will also look briefly at
shear joints before we move on.

Fatigue failure of a single bolt means a reduction in clamping force.
This in turn can increase the load excursions seen by the rest of the bolts,
as we'll see later in this chapter, and that can encourage them to fail too.
As a result, fatigue failure often means the complete loss of the joint.

I. THE FATIGUE PROCESS

A. The Sequence of a Fatigue Failure

We learned in Chap. 15 that fatigue will be a potential problem only if
four "essential conditions" are present: cyclic tensile loads, stress levels
above a threshold value (called the endurance limit), a susceptible material, and an initial flaw in that material. If these conditions are all present,

then a natural sequence of events can occur, and can lead to fatigue failure. These events are called

1. Crack initiation
2. Crack growth
3. Crack propagation
4. Final rupture

This sequence of events is shown in Fig. 17.1. A tiny crack grows slowly, then more rapidly, until the bolt is destroyed, as shown in Fig. 17.1. Let's examine this sequence of events one at a time.

Crack Initiation

Many things can produce that first fatal flaw which starts the fatigue process. A tool mark can do it. So can a scratch produced when the part is mishandled. Improper heat treatment can leave cracks. Corrosion can initiate them. Inclusions in the material can do it. It is probably safe to say, in fact, that no part is entirely free from tiny defects of this sort.

Crack Growth

A tiny crack creates stress concentrations. When the part is subjected to cyclic tension loads, these stress concentrations yield and/or tear the material at the root of the crack. Since most of the bolt still remains undamaged to support the load, initial crack growth is fairly slow.

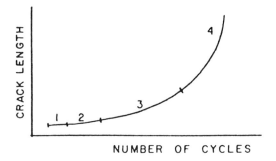

Figure 17.1 Fatigue failure occurs when a tiny crack in the bolt grows under cyclic tension loads until the crack is so large that the next cycle of load breaks the bolt. The stages of failure are (1) initiation, (2) growth, (3) propagation, and (4) rupture. (Modified from Ref. 1.)

Crack Propagation

As the crack grows, stress levels at the end of the crack also increase, since less and less cross section is left to support the loads. The crack grows more rapidly as stress levels increase.

Final Rupture

There comes a time when the crack has destroyed the bolt's capability to withstand additional tension cycles. Failure now occurs very rapidly. As far as the user is concerned, failure has been sudden and unexpected because, until this part of the fatigue process is reached, there is often no visible damage nor change in the behavior of the bolt. Everything appears to be fine until suddenly, with a loud bang, the bolt breaks.

The number of cycles required to break the bolt this way is called its fatigue life. Apparently identical bolts in apparently identical applications can have, of course, substantially different fatigue lives, depending on the location and seriousness of those initial cracks as well as on apparently minor, but important, differences in such things as bolt and joint stiffness, initial preload, alloy content, heat treat, location and magnitude of external tension loads, etc. As a result, there is a lot of scatter in the fatigue life of the bolts used in a given application.

B. Types of Fatigue Failure

Fatigue failures are called high-cycle or low-cycle failures, depending on the number of load cycles required to break the part. High-cycle fatigue requires hundreds of thousands or even millions of cycles before rupture occurs. Low-cycle failure occurs in anything from one to a few ten thousand cycles. You can demonstrate low-cycle fatigue to yourself by bending a paper clip back and forth until it breaks.

The number of cycles required to break a bolt is determined by the magnitudes of mean and alternating stresses imposed on the bolt by external cyclic loads, as we'll see in a minute. Low-cycle failure occurs under very large loads, high-cycle failure under lesser loads. In many applications the bolt can see some of each—lots of relatively mild loads interrupted once in a while by a sudden shock or larger load (perhaps when the tractor hits a rock). In many cases it's difficult to know whether to characterize the failure as a low-cycle or a high-cycle failure. In most well-designed bolted joints, however, fatigue failure, if it occurs at all, will be high cycle.

C. The Appearance of the Break

Close examination of the broken bolt can often tell you whether or not it failed in fatigue. That portion of the break surface which failed slowly, as the crack initiated and grew, will have a relatively smooth and shiny surface. That portion which failed during crack propagation will have a rougher surface; that portion which failed during final rupture will have a very rough surface. If the entire fastener fails suddenly during tightening or the like, the entire break surface will be rough; so these smooth "beach marks" seen on a fatigue surface can be used to distinguish fatigue breaks from breaks which occur under static load. See Fig. 17.2.

You may find more than one crack in a bolt which has failed in fatigue. The initiation and growth of one crack may drastically increase loads in another region of the fastener, causing a second crack to grow and propagate there. Failure can occur in whichever one reduces the strength of the bolt more rapidly.

The most common places to find fatigue cracks and failures in bolts are in the regions of highest stress concentration. These are

Where the head joins the shank of the bolt
The thread run-out point
The first thread or two of engagement in the nut
Any place where there is a change in diameter of the body or shank

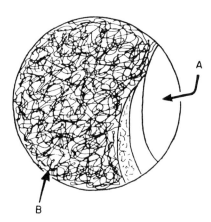

Figure 17.2 Break surface of a bolt which has failed in fatigue. The surface is smooth and shiny in those regions which failed during crack initiation and growth (A). It is rough in those regions which failed rapidly (B).

II. WHAT DETERMINES FATIGUE LIFE?

In general, the higher the cyclic loads seen by the bolt, the sooner it will fail. Whether or not, or how rapidly, a fastener will fail depends on the mean stress level and the variation in stress level under cyclic loads.

There are techniques for estimating what the life of a given material and/or body will be. Accurate prediction, however, is possible only through actual experiments on the body of interest—in our case, on a bolt. Test results are usually presented in the form of *S-N diagrams*, where *S* stands for stress level and *N* for number of cycles of applied load. An examination of these diagrams gives considerable insight into the fatigue process.

A. *S-N* Diagrams

Figure 17.3 shows one possible form of the *S-N* diagram. Alternating tension and compression loads have been applied to the test specimen. Maximum compression stress equals maximum tension stress. Maximum amplitude of either stress is plotted on the vertical axis of the diagram.

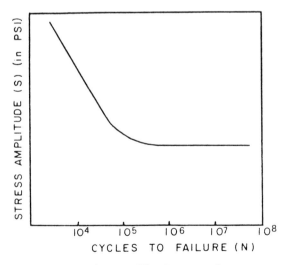

Figure 17.3 The mean life of a group of test coupons subjected to fully alternating stress cycles. When stresses are "full alternating," maximum tension stress equals maximum compression stress and the mean stress on the part is zero. (Modified from Ref. 2.)

The number of cycles required to fail the test coupon is plotted on the horizontal axis. The curve shows the mean life of the test coupons.

Because the fatigue life of one test coupon may differ drastically from that of others, it is necessary to test many coupons before plotting the results shown in Fig. 17.3. The statistical deviations in life can also be determined by such tests. A more complete picture of the tests, therefore, would be shown by a diagram such as that given in Fig. 17.4. Note that many of the test specimens will fail at some number of cycles less than the mean. The remainder will fail at some number of cycles greater than the mean. If the lowest line in Fig. 17.4 represents the minus two standard deviation data, then 95% of the test coupons will survive more cycles before failure than the number of cycles indicated by this line. Only 5% will last longer than the number of cycles indicated by the uppermost line.

Note that either Fig. 17.3 or Fig. 17.4 says that cycle life will be very short when applied alternating stress levels are very high. As alternating stresses are reduced, cycle life increases. Below some stress level, in fact, the curve becomes essentially parallel to the horizontal axis, and fatigue life becomes very large. This stress level is called the *endurance limit* of the material, or part, and is defined as the completely reversing stress level below which fatigue life will be infinite. There is some such limit for

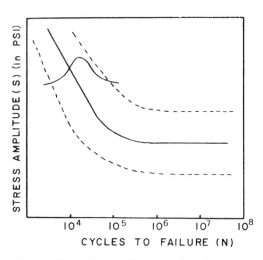

Figure 17.4 There will be considerable scatter in the life achieved in a group of test coupons as a result of a particular stress pattern. Rather than show just the mean life results as in Fig. 17.3, therefore, it is sometimes useful to plot the statistical deviations as well. (Modified from Ref. 2.)

any material and any part. Unfortunately, endurance stress levels are usually only a small fraction of the static yield strength or static ultimate strength of a material or body.

Note, however, that only the *change* in stress must stay below the endurance limit. The total or mean stress can be considerably higher [5]. Incidentally, not all materials have an endurance limit. Aluminum alloys, for example, exhibit finite life even at very low cyclic stress levels [18].

We could now test another large number of test specimens, this time changing the mean tension while leaving the excursion (difference between maximum and minimum tension) the same as it was in the previous tests. This would result in a family of curves such as that shown in Fig. 17.5. For clarity, only the mean curves are shown.

Although all the *S-N* data we have examined are based on tension (and/or compression) loading along the axis of the fastener, it is worth noting that if the fastener is subjected to some other form of stress as well as tension, its fatigue life will be adversely affected.

Shear stress, for example, would rob a portion of the strength of the fastener, making it more susceptible to tension fatigue. Bending stress, which is often present, magnifies the tensile stress on one side of the bolt and can also be a significant problem in a fatigue situation.

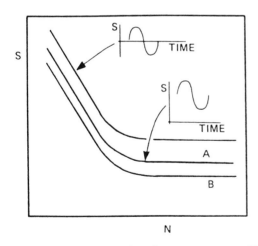

Figure 17.5 Increasing the mean stress will reduce the number of cycles to failure produced by a given magnitude of alternating stress. The uppermost curve here is repeated from Fig. 17.3 for comparison. The mean stress associated with curve B is higher than that associated with curve A. (Modified from Ref. 2.)

B.   Material vs. "the Part"

If we were to test a bunch of coupons made from a different material and/
or subject it to a different heat treatment, we would, in general, generate
a set of curves which would be different from those shown in Figs. 17.3
through 17.5. Fatigue life, in other words, is a function of material and
heat treatment. It is also—and perhaps even more so—a function of the
shape of the part being tested, just as the stress-strain performance of a
body was different from the stress-strain performance of the material from
which it is made (see Chap. 2).

The reason in both cases, of course, is that the shape of the body
determines stress levels. These vary from point to point; the behavior of
the body, therefore, varies from point to point. The gross behavior of the
body is determined by the accumulation of its point-to-point behavior.

A bolt is a very poor shape when it comes to fatigue resistance. Although the average stress levels in the body may be well below the endurance limit of the material, stress levels in unavoidable stress concentration
points such as thread roots, head-to-body fillets, etc., can be well over
the endurance limit. As a result, the apparent endurance limit of the commercial fasteners can be as little as 10% of the endurance limit of the base
material [3]. One source gives the endurance limit of a Grade 8 fastener,
for example, as 18,000 psi [4], well below its proof strength of 120,000
psi.

Another reference says that the fatigue strength of a smooth test bar
of steel is approximately half the ultimate tensile strength (UTS) of the
steel, if the steel has a UTS under 200 ksi (which most bolt materials do)
and if the test is conducted under fully reversing loads (defined as $R =
1$ as we'll see in Sec. III below). The reference goes on to say that the
fatigue limit of the part under test will be less than half the UTS of the
material if any of the following conditions are present:

The part is notched or threaded.
There are residual tensile stresses at the surface of the part.
The part has been electroplated.
The part is corroded.
There has been mechanical damage to the surface.

All of which says "bolt." The reference concludes that if you can't estimate the influence of these factors, then the only way to determine the
fatigue strength of the part is to conduct S-N tests under service conditions
[18]. We'll look at some actual fatigue strength data for fasteners in a
minute. First, though, let's continue our review of fatigue in general.

C. Summary

In summary, the major factors which affect fatigue life are

1. Choice of material
2. Shape of the part
3. Mean stress level
4. Magnitude of stress excursions or variations
5. Condition of the part

Of these, the shape of the part may be the most significant, magnitude of stress excursions the next most significant, and, within reason, material choice the least significant.

III. OTHER TYPES OF DIAGRAM

A. Constant Life Diagram

The *S-N* diagram is only one way to plot the results of a series of fatigue tests. Another more informative diagram is called a constant life diagram (Fig. 17.6). Because of the amount of information on this diagram, it takes a little practice to read it. Let's take an example.

The curved lines marked 10^3 cycles, 10^4 cycles, etc., represent average, "constant," fatigue lives under a variety of conditions.

They intersect the mean stress line at a common point, which is equal to the ultimate tensile strength (S_u) of the material being tested—100 ksi in the example in Fig. 17.6. The data shown, incidentally, are for a group of unnotched, polished test specimens of 100-ksi material. Most, if not all, of the constant life diagrams you'll find are for polished, unnotched test specimens of this sort rather than for bolts or particular shapes.

Working now with the curved line in Fig. 17.6 which represents 10^6 cycles, we learn that the average test coupon will have this life when it sees a maximum tensile stress of 80 ksi (vertical axis of the chart) and a minimum tensile stress of 30 ksi (horizontal axis). A variety of other combinations of max and min stress would also result in a coupon life of 10^6 cycles, but we'll focus on the 80/30 point. This point also represents a load ratio (R) of 0.375—the ratio between the minimum and maximum tensile stresses on the part. The mean stress on the part, at this point, is 55 ksi, which we determine merely by adding the max to the min and dividing by 2.

The total variation in stress on the part is 50 ksi; the difference between the maximum 80 and the minimum 30 ksi. The "alternating stress" for this

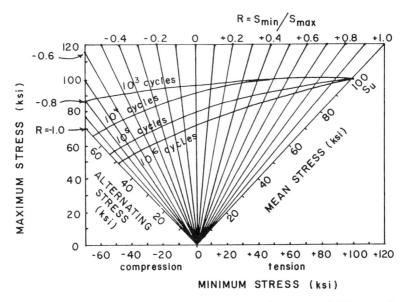

Figure 17.6 A constant life diagram, the most informative of all fatigue diagrams. See text for discussion.

situation is half of 50 or 25 ksi, a fact which is perhaps best illustrated by Fig. 17.7, described next.

B. Center Portion of Constant Life Diagram

Sometimes, only the center portion of the constant life diagram is given, as in Fig. 17.7, with the alternating stress and mean stress lines now forming the axes of the diagram. The only information which is missing from this diagram, maximum and minimum applied stresses, can be computed from the plotted values for mean and alternating stress. The lines representing the various load ratios (R), which I have shown in Fig. 17.7, are often omitted from this type of diagram.

The concept of load ratios, incidentally, is illustrated in Fig. 17.8. In part A, the load is static and there would be no fatigue problems. In part B, the load varies (fluctuates) slightly. The load fluctuations are progressively more severe in parts C, D, E, and F, with F being the worst situation from a fatigue point of view. This situation, where the maximum positive stress equals the maximum negative stress, is called a "completely revers-

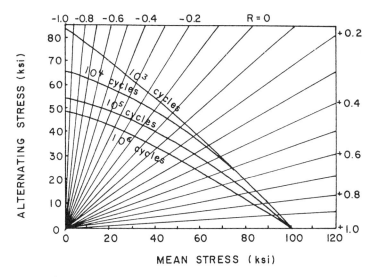

Figure 17.7   A simplified or modified constant life diagram; this one consists of only the center portion of the diagram shown in Fig. 17.6.

ing load'' and represents a load ratio of -1.0. Note that a load ratio of -1.0 is represented by the vertical, alternating stress, axis in the diagram of Fig. 17.7.

C.   Approximate Constant Life Diagram

Note that the constant life lines labeled 10^3, 10^4, etc., in Figs. 17.6 and 17.7 are nearly straight lines. This allows us to construct an approximate but conservative constant life diagram, as illustrated in Fig. 17.9.

To do this, we first make a series of tests in which the mean stress is always zero. Only the magnitudes of the alternating stresses are varied. If we plotted the results on an S-N diagram, it would look like the diagram in the left side of Fig. 17.9. As in all fatigue tests, we would have to test a number of specimens to get a true mean value—there will be a lot of scatter in individual results.

Having done this, we can now plot the average alternating stress for each mean life point of interest on the vertical axis of our modified constant life diagram, as shown on the right of Fig. 17.9. We next draw straight lines between these alternating stress points and the point on the horizontal (mean stress) axis which represents the ultimate tensile strength of the material (S_u). Since these straight lines will always lie below the actual,

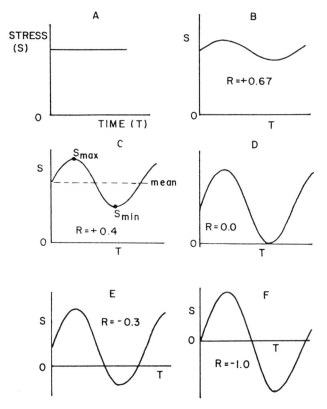

Figure 17.8 Graphical representation of a variety of fatigue loading conditions. In (A), the loads are static, and there will be no fatigue problem. (B) shows a slight cyclic fluctuation in load. (C) through (F) show progressively more severe loads. That in (F) is called a completely reversing load. This is the type of load used to determine an endurance limit.

slightly curved lines, they are safe and conservative approximations of the actual lines. We don't have to make any tests at load ratios other than −1.0 to construct this diagram.

D. Endurance Limit Diagram

The line representing infinite life in the diagram on the right side of Fig. 17.9 would seem to define, fully, the conditions required for infinite product life. After all, fully alternating stress is the worst condition from a

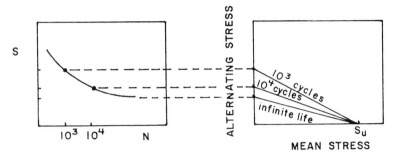

Figure 17.9 An approximate (conservative) constant life diagram is shown on the right. It can be constructed from the data used to draw the mean fatigue life line on an *S-N* diagram, as shown here.

fatigue standpoint. This line, however, omits one factor; namely, that the safe *static* load which can be applied to a part is not its ultimate tensile strength, but is rather its yield strength. We must take this fact into account when predicting true "infinite life."

To do this, we construct the diagram shown in Figure 17.10. We start by repeating the infinite life line, but this time we "lower it" a little to represent the worst-case condition. Remember from Fig. 17.4 that there will always be a considerable scatter in the life achieved in a group of test specimens. For the infinite life diagram, we want to plot the equivalent of the lower dashed line in Fig. 17.4 rather than the mean line we used in Fig. 17.9. Let's assume for the discussion that the worst-case line was 20% below the mean line.

There will also be some scatter, of course, in the yield and ultimate strengths of a material. We'll use the same 20% reduction for these values. We're now ready to construct our final infinite life diagram.

Instead of the original infinite life line, we now connect a point representing 80% of the mean infinite life alternating stress to 80% of the ultimate tensile strength—line A in Fig. 17.10. We now connect points representing 80% of the yield strength of the material on *both* horizontal and vertical axes—line B in Fig. 17.10.

The shaded region in the figure represents the true infinite life of the part. Any combination of mean and alternating stresses which fall within this region will never cause a fatigue failure. Note, however, that we're still dealing with data taken from tests on polished, unnotched test cou-

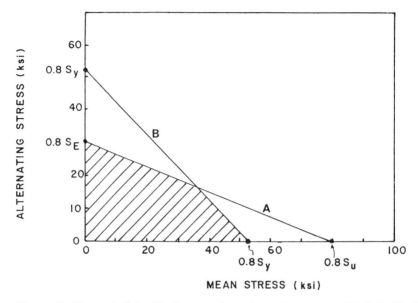

Figure 17.10 An infinite life diagram. Line A is a "worst-case" infinite life line, similar to the one shown on the right side of Fig. 17.9. Line B connects points representing the worst-case yield strength (S_y) of the material on both axes. Any combination of mean and alternating stress which falls within the shaded region would never fail the part.

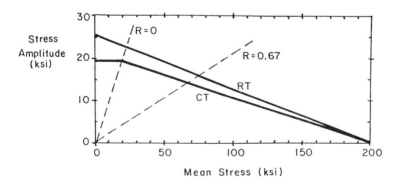

Figure 17.11 A constant life fatigue diagram for SAE Grade 8 bolts at 10 million cycles of life. This chart is based on a literature search conducted by the authors of Ref. 19. Line RT represents fasteners having rolled threads; CT those with cut threads.

pons rather than bolts since all of our data have been based on data obtained from the diagram of Fig. 17.7. Had we instead conducted the original tests on actual bolts, then the final diagrams would have represented the infinite life conditions for the bolt itself.

E. Fatigue Life Data for Fasteners

I have found specific fatigue life or endurance limit data for threaded fasteners hard to come by. I included what little I had until recently in Table 4.11 in Chap. 4. Both endurance limits and fatigue strengths (maximum stress excursions the bolts can stand for a given number of cycles) are included. Figures 17.11 and 17.12 summarize data which have been available for some time but are new to me. Figure 17.11 gives fatigue data for SAE Grade 8 bolts loaded 10×10^6 times [19] and Fig. 17.12 gives data for ASTM A325 and A354-BD bolts used in structural steel applications [20]. The tests reported were stopped after 2×10^6 cycles, but the author says that studies have shown that if heavy structural steel joints survive that many cycles their life can be considered infinite.

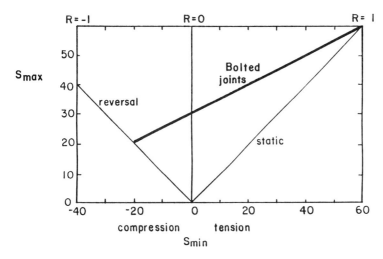

Figure 17.12 A modified Goodman diagram for structural steel bolts ASTM A325 and A354-BD subjected to 2 million load cycles. According to Ref. 20, a heavy structural steel which survives this many load cycles is considered to have infinite life.

IV. THE INFLUENCE OF PRELOAD AND JOINT STIFFNESS

A. Fatigue in a Linear Joint

As we saw in Chap. 12, the bolt will see a portion of any external tension load which is imposed on the joint. The magnitude of the mean load on the bolt depends on the preload in the bolt. The magnitude of the load excursion (ΔF_B) depends on:

The magnitude of the external tension load
The bolt-to-joint stiffness ratio (K_B/K_J)
Whether or not the external tension load exceeds the critical load required to separate the joint (which is determined by the magnitude of the initial preload)

The effect of the first two factors is summarized in Fig. 17.13.

We could also use the triangular joint diagram to show the effects of very large external loads and/or insufficient preload. I think it's more instructive, however, to use the form of the alternative joint diagram given in Figs. 12.18 and 12.19, in which we plotted the bolt load as a function of external load and could readily see what happens to the bolt load when the external load exceeds the critical level required for joint separation.

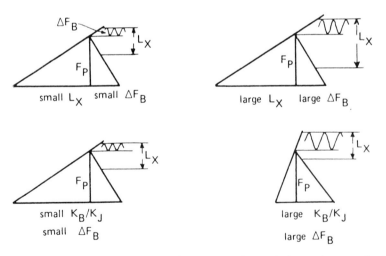

Figure 17.13 The load excursions (ΔF_B) in the bolt are increased with an increase in external load (L_X) and/or an increase in the bolt-to-joint stiffness ratio (K_B/K_J). Note that the initial preload is the same in each case.

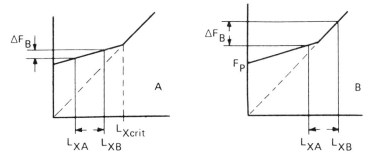

Figure 17.14 The bolt sees a far greater variation in tension (ΔF_B) if the external load exceeds the critical load required for joint separation (as in B) than it does when external loads are less than the critical value (A). Note that initial preload and joint stiffness ratio are the same in both cases. (Modified from Ref. 2.)

In Fig. 17.14, for example, we apply external loads to two joints having the same initial preload and the same stiffness ratios as each other. Furthermore, the excursion (difference between maximum and minimum) of the external load is the same in both cases. Only the values of the maximum and minimum loads have changed. In Fig. 17.14A, the maximum external load is less than that which would be required for joint separation. The resulting excursion in bolt load (ΔF_B) is relatively small.

In Fig. 17.14B, however, the maximum external load exceeds the critical load. The bolt sees 100% of any external load that exceeds the critical level and so, under these circumstances, the excursion in bolt load is greatly increased.

The critical load depends on the initial preload. If we lower preload, we can get into trouble, as shown in Fig. 17.15. Conversely, of course, raising the preload in the joint can get us out of trouble.

B. Nonlinear Joints

The above analysis was based on the assumption that the joint will behave in a linear and fully elastic fashion. As we saw in Chap. 13, this is not always the case—in fact, it may very seldom be the case. If the external load, for example, is not applied along the axis of the bolt and/or the bolt is not located in the center of the joint, prying action can increase the load seen by the bolt. This can make a substantial difference in the load excursions produced in the bolt by a given cyclic external load, as suggested in Fig. 17.16.

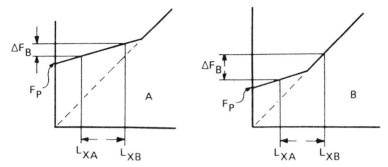

Figure 17.15 The maximum and minimum external loads are the same in both cases here. The maximum load, however, exceeds the critical load required for joint separation in B because of insufficient initial preload (F_P). Note that the joint stiffness ratio is the same in both cases.

With reference to Fig. 17.14, we can sometimes reduce the load excursions seen by the bolt by increasing the preload. This only works, however, if the new, higher preload raises the critical load required for joint separation above the maximum external load seen by the joint. If the maximum external load was already below the critical level, increasing the preload does not reduce the excursion seen by the bolt but merely increases the *mean* stress in the bolt, as shown in Fig. 17.17.

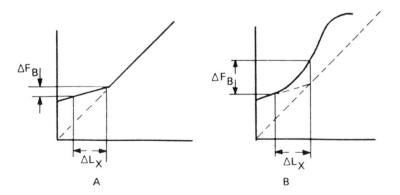

Figure 17.16 Comparison of the loads seen by a bolt in a linear concentric joint (A) and an eccentric joint in which the bolt is subjected to prying action (B). Note that the initial preload, the bolt-joint stiffness ratio, and the maximum-minimum external loads are the same in both cases. At least the *apparent* stiffness is the same. The fact that prying action occurs alters the stiffness of the joint.

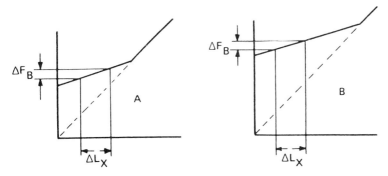

Figure 17.17 If the maximum external load (L_{XB}) is already below the external load required for joint separation, then raising bolt preload will not reduce the load excursion (ΔF_{B}) seen by the bolt; it will merely increase the mean load.

Increasing the preload in an eccentric prying joint will also increase the mean tension seen by the bolt. This time, however, because of the nonlinear nature of the joint, it's likely that increasing the preload will also reduce the load excursion seen by the bolt, as suggested in Fig. 17.18. Under these conditions, increasing the preload can result in a net gain in fatigue life. The increase in mean stress is detrimental but is more than offset by the reduction in excursion.

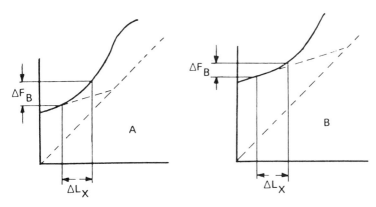

Figure 17.18 Increasing the preload in an eccentric nonlinear joint subjected to prying action will increase the mean load seen by the bolt, but will also reduce the load excursions it sees. Since large excursions are worse than large means, as far as fatigue life is concerned, this change can result in a net gain in fatigue life.

C. What is the Optimum Preload?

We're now in position to answer the question "What preload should we use for maximum fatigue life?"

In general, greater load *excursions* reduce fatigue life more than higher mean load—but neither is helpful. As a result, we can conclude that:

1. A higher preload will help if it reduces bolt load excursions substantially. Higher preload will therefore always help if it raises the critical load required for joint separation above the maximum external load which will be seen by the joint. Using a bolt material having a higher UTS and/or using a bolt with a larger diameter (even though that means a stiffer bolt) will allow you to increase the initial preload and so reduce the alternating stresses on the bolts by preventing the partial opening of the joint under prying action [5, 21]. Higher preload will also help if it reduces the amount of prying experienced by the joint.
2. A higher preload will *reduce* fatigue life if it makes no change in the load excursions seen by the bolt (e.g., Fig. 17.17).
3. A higher preload is probably neither good nor bad if it doesn't reduce the load excursions by very much. I suspect, in other words, that there are gray areas where results could go either way.

SPS has published data which suggest that higher preload, up to the yield point of the fastener, is always desirable in a fatigue situation because they also believe that prying is (almost) always present as well. Some of their results are shown in Fig. 17.19 [5, 22], which shows how prying action can adversely affect the fatigue life of a joint and how an increase in preload can improve fatigue behavior.

In another project SPS tested automotive connecting rod joints initially tightened "well past yield." They subjected the joints to 10^6 load cycles without failure—even with extreme plastic deformation of the bolts. In other tests they obtained acceptable fatigue lives from $\frac{3}{8}$–16 "industrial bolts" which had been yield tightened as much as three times. They report that yield tightening did decrease the fatigue life of the bolts slightly, and that the benefits of rolling the threads after instead of before heat treatment were reduced somewhat by yield tightening; but that, in general, yield tightening was beneficial [23].

Anyway, many people—including those at SPS—insist that higher preload will always improve fatigue behavior. Others, however, argue that since a higher mean will reduce fatigue life, though not as much as a higher excursion will, a higher preload can be helpful but may not be [6] and only a careful analysis will answer the question.

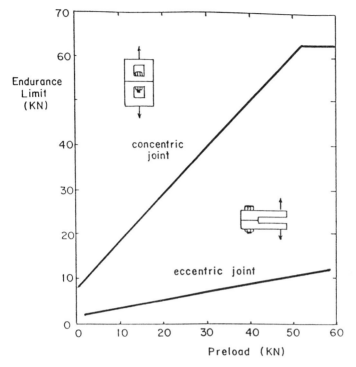

Figure 17.19 This chart, based on tests sponsored and/or conducted by SPS, shows that the endurance limit of a joint is increased if the initial preload applied to the joints is increased. It also shows that concentric loading results in much better fatigue performance than eccentric loading. The fasteners used in these tests were M10 × 65, Grade 12.9 bolts. The length of the concentric joint was 50.8 mm and its contact area was 38 mm². The contact area of the eccentric joint was 38 × 38 mm and the distance between the centerline of the bolts and the line of action of the applied load was 78.9 mm—a substantial eccentricity [22].

D. Fatigue and the VDI Joint Design Equations

We last examined the VDI joint design equations in Chap. 13. Now it's time to see what they tell us about joint failure. At first glance the answer is, "not much." Equation (13.21), for example, tells us how to compute the maximum anticipated tensile stress to be seen by a bolt, as a function of assembly preloads, relaxation factors, thermal change, and the like. Most of the factors included in that equation won't fluctuate, and therefore will only affect the mean preload seen by the bolts. That can influence fatigue life but we're far more interested in fluctuations in load, if any.

There is one term in Eq. (13.21) which addresses a change in tension in the bolt caused by external load, namely the ΔF_B term given as $\Phi_{en}L_X$ where Φ_{en} is the load factor (sometimes called the joint stiffness ratio) for a prying joint (with the axes of the bolt and the line of action of the tensile load both offset from the axis of gyration of the joint) and L_X is the external load placed on the joint. Other Φ's also discussed in Chap. 13—for concentric or nonprying joints etc.—could be multiplied by L_X to compute ΔF_B for other situations, of course.

Now, the ΔF_B used in Eq. (13.21) was described as the result of a *static* L_X exerted on the joint. If, however, the external load fluctuates between some L_X and zero ($R = 0$) as illustrated in Fig. 17.14, then ΦL_X can be taken as the stress excursion seen by the bolt. If the external load fluctuates between an L_{Xmax} and an L_{Xmin} other than zero, then the difference between those values must be multiplied by the appropriate Φ to compute the excursion seen by the bolt.

The "stress amplitude" of this load excursion is now compared to the endurance limit of the bolt (if infinite life is desired) or to its fatigue strength for a desired number-of-cycles life. The stress amplitude is the difference between mean stress and maximum stress, as illustrated in Fig. 17.20, or one-half the *fluctuating* L_X. In VDI terms we have, if we want infinite life,

$$\Delta F_{Ba} = \frac{F_{Bmax} - F_{Bmin})}{2} = \frac{\Phi L_{Xa}}{2} \qquad (17.1)$$

and

$$\Delta F_{Ba} < \sigma A_r \qquad (17.2)$$

where
A_r = the root diameter area of the threads (in.², mm²)
ΔF_{Ba} = stress amplitude seen by the bolt (lb, N)
F_{Bmax} and F_{Bmin} = max and min tensions in the bolt as a result of cyclic changes in the external load on the joint (lb, N)
L_{Xa} = the fluctuating portion of the external joint on the joint (lb, N)
Φ = load factor or joint stiffness ratio.
σ = the maximum stress in the outer fiber of the root of the first, load-bearing thread (psi, Pa). You'll find the worst-case (eccentric prying) expression for σ in Eq. (13.8).

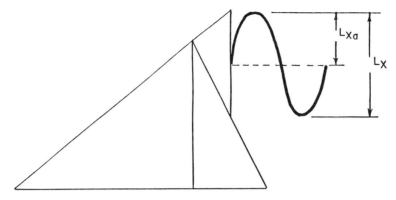

Figure 17.20 This joint diagram shows the relationship between the full excursion of a tensile-tensile (nonreversing) load L_X and the half-excursion L_{Xa} called the "stress amplitude" in the German VDI Directive 2230. Most of the fatigue data I've seen have been based upon tests conducted under fully reversing loads ($R = -1$). But VDI here used half of a fluctuating tensile load in endurance limit calculations. Fatigue data are based on many different types of loading conditions and we're well advised to understand the basis of the data we're using for our own applications.

Note that in most of this chapter the endurance limit has been based on tests in which a fully reversing load ($R = -1$) was applied to the parts. With these loading conditions the maximum stress seen by the bolts is indeed only half the change in stress, since the mean stress is zero. VDI Directive 2230 suggests that we compare the endurance limit to only half of a fluctuating stress, as shown Fig. 17.21; even though the mean stress—thanks to preload—is certainly not zero. Their Directive also contains a table of endurance limits for DIN steels which may be based on the same conditions. Certainly most bolts don't see fully reversing stress cycles. But all of this should alert us to the fact that fatigue test data can be based on many different kinds of tests and should be applied with caution to our own applications if conditions differ.

In any event, the VDI 2230 aims at infinite life for every application, even when prying is present. A commendable goal, but, obviously, it won't always be economically practical to design for infinite life, especially under prying conditions. Of course, we'll always *want* to do it if practical and always *must* do it if joint failure would have serious consequences.

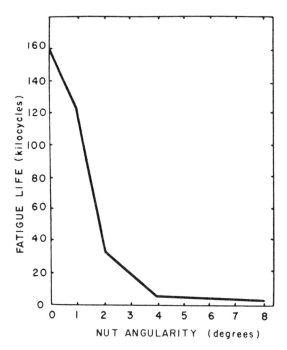

Figure 17.21 If the face of the nut is not exactly parallel to the surface of the joint, fatigue life can be seriously affected, as shown by this study made by Viglione [10]. Bolts were ⅜–24 MIL-B-7838 with 2-in. grip length.

V. MINIMIZING FATIGUE PROBLEMS

You should realize that each of the things we can do to reduce or eliminate a fatigue problem is an attempt to overcome one or more of the four essential conditions without which failure would not occur. Remember that these conditions were cyclic tensile loads, stresses above an endurance limit, a susceptible material, and an initial flaw. We can rarely eliminate any one of these completely; but if we can at least reduce one or more of these factors we can usually improve the fatigue life of our bolted joints.

In general, most of the steps we can take are intended to reduce stress levels (including stress concentrations) and/or to reduce the load excursions seen by the bolt. Surface flaws and the susceptibility of the material will usually be the concern of the bolt manufacturer rather than the user. Let's look at some of our options.

A. Minimizing Stress Levels

The following are not listed in order of importance. They merely describe some of the many things which can be and are done to fasteners to limit stress concentrations and/or general stress levels. Some of them are relatively obvious; others are subtle. Many are incorporated in so-called fatigue-resistant fasteners which are available from some manufacturers. In any event, here are some of the things which work.

Increased Thread Root Radius

Sharp, internal corners are natural places for fatigue cracks to start, so using threads with radiused roots can increase fatigue life. For example, going from a flat root thread to one with a radius equal to 0.268 times the pitch (a so-called 55% thread) increased the fatigue lives of various specimens from 80 to as much as 2800% even though the change increased the tensile strength of the thread by only 1 to 12% [22]. For illustrations of flat and rounded radius roots see Fig. 3.2.

Rolled Threads

Rolling the threads instead of cutting them provides a smoother thread finish (fewer initial cracks). Rolling provides an unbroken flow of the grain of the material in the region of the threads, partially overcoming their notch effect, and it builds compressive stress into the surface of the bolt. This compressive "preload" must be overcome by tension force before the thread roots will be in net tension. A given tension load on the bolt, therefore, will result in a smaller tension excursion at this critical point (point to stress concentration).

Threads can be rolled either before or after heat treating. After is better but is also more difficult. Rolling before heat treating is possible on larger diameters.

In any event, one authority says that cold-rolled threads have double the fatigue strength of cut threads, although he doesn't specify whether the rolling occurs before or after heat treatment [24]. Others report that the higher the basic strength of the material, the greater the benefit of rolling after heat treat [25].

Fillets

A generous fillet between head and shank will reduce stress concentrations at this critical point. The exact shape of the fillet is also important; an elliptical fillet, for example, is better than a circular one [4].

Increasing the radius of a circular fillet will help. So will prestressing the fillet (akin to thread rolling) [7].

Perpendicularity

If the face of the nut, the underside of the bolt head, and/or joint surfaces are not perpendicular to thread axes and bolt holes, the fatigue life of the bolt can be seriously affected. Fig. 17.21, for example, shows the effect of a few degrees of nut angularity (lack of perpendicularity) on fatigue life. A two-degree error reduces fatigue life by 79% [8].

Overlapping Stress Concentrations

Bolts normally see stress concentrations at thread run-out, first threads to engage the nut, and head-to-shank fillet. Anything which imposes additional load or concentration of load at these points is particularly damaging. Some such factors are shown in Fig. 17.22. For best performance, for example, there should be at least two full bolt threads above and below the nut. Thread run-out should not coincide with the joint interface (where shear loads exist), etc.

Thread Run-Out

Thread run-out should be gradual rather than abrupt, as suggested in Fig. 17.23.

Thread Stress Distribution

As we saw in Chap. 2, most of the tension in a conventional bolt is supported by the first two or three nut threads. Anything which increases the number of active threads will reduce stress concentrations and increase fatigue life. Some possibilities are suggested in Figs. 2.9 through 2.11 in Chap. 2. The so-called tension nuts of Fig. 2.9 create nearly uniform stress in all threads, as shown in Fig. 2.10, for example.

Modifying the nut pitch so that it's slightly different from the pitch of the bolt threads can also make a substantial improvement in fatigue life. One authority [1] suggests that a nut with 11.85 threads per inch be used with a bolt having 12 threads per inch. He points out that this not only provides more uniform distribution of stress in the threads, but also reduces the stiffness of the bolt with respect to the joint by making the effective length of the bolt a little greater. Reducing the stiffness ratio helps, as we saw in Fig. 17.13.

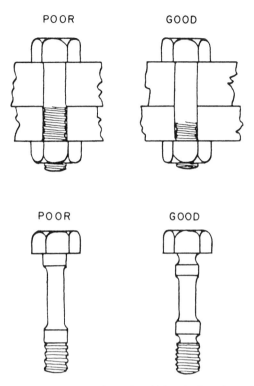

Figure 17.22 Joints should be designed so that maximum loads do not fall on stress concentration points of the fastener. Several points of good and bad practice are suggested.

Another way to smooth stress distribution in the threads is to use a nut that is slightly softer than the bolt. The nut can now conform to the bolt more readily. Standard nuts are softer than the bolts they're used with, for this reason; still softer nuts are possible if you can stand the loss in proof load capability.

A helical thread insert in a tapped hole also "conforms" to the male threads, because the insert is flexible; but the insert doesn't reduce static strength the way a soft nut will.

A jam nut improves thread stress distribution too, by preloading the threads in a direction opposite to that of the final load.

A final way to improve the distribution of stress in the threads is to taper them slightly, as shown in Fig. 17.24. Tapering the lower threads

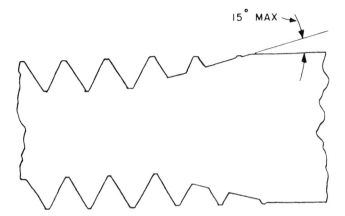

Figure 17.23 Thread run-out should be gradual—some people suggest a maximum of 15°—to maximize stress concentrations at this critical point in the fastener.

of a nut at a 15° angle, until the first thread had been removed, for example, improved the fatigue life of one fastener by 20% [1]. You must be sure to put such a nut on in the right direction, however, or it will increase stress concentrations and reduce fatigue life over that obtained with conventional nut.

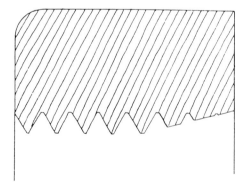

Figure 17.24 Tapering the input threads of the nut can distribute stresses more uniformly and increase fatigue life. The taper is 15° and is sufficient to just remove the first thread.

Bending

Bending increases the stress levels on one side of the fastener. This is one of the reasons why nut angularity hurts fatigue life. One way to reduce bending is to use a spherical washer.

Corrosion

Anything we do to minimize corrosion will reduce the possibilities of crack initiation and/or crack growth and will therefore extend fatigue life. This is confirmed by the fact that running bolts in a hard vacuum results in an order-of-magnitude improvement in fatigue life [1] because it completely eliminates corrosion. Corrosion, as we'll see in Chap. 18, can be more rapid at points of high stress concentration. Since this is also the point at which fatigue failure is most apt to occur, fatigue and corrosion aid each other, and it is often difficult or impossible to tell which mechanism initiated and/or resulted in failure.

Flanged Head and Nut

At one time the fastener shown in Fig. 17.25 was being proposed as a fatigue-resistant, ISO standard configuration. To my knowledge, such a standard has not yet been published. The proposal is informative, however, because it shows that details of fastener geometry can have a significant effect on fatigue life. All of the refinements shown here are intended to reduce stress concentrations. In addition to the flanges, the design includes details we previously examined in Figs. 2.9A and 17.24. Dishing

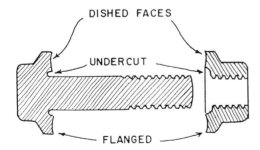

Figure 17.25 Flanged, dished, and undercut nut and bolt head improve stress distribution and therefore fatigue life. (Modified from Ref. 11.)

the flanges slightly, incidentally, creates more uniform distribution of stress between flange and joint surface.

Surface Condition

Any surface treatment which reduces the number and size of incipient cracks can improve fatigue life substantially. Polished surfaces, for example, will make a big difference. Shot peening the surfaces also helps—not only because it smooths out beginning cracks, but also because it puts the surfaces in compressive stress (much as thread rolling does).

B. Reducing Load Excursions

Nothing can help extend the fatigue life of a bolt or joint more dramatically than a reduction in load excursions. We have discussed this at some length in an earlier section; I repeat it now simply because it is the single most important thing you can do. Your means of doing it include the following.

Prevent Prying

As we've seen, prying action greatly increases the load excursions seen by the bolts, and so should be avoided by proper design of the joint if at all possible. This, however, may mean economically unattractive, massive joint members.

Proper Selection of Preload

Correctly identify the maximum safe preload that your joint can stand, estimating fastener strength, joint strength, and external loads, analyzing them carefully with the help of a suitable joint diagram.

Control of Bolt-to-Joint Stiffness Ratios

Conventional wisdom says that we should try to minimize the bolt-to-joint stiffness ratio so that most of the excursion and external load will be seen by the joint and not by the bolt. Use long, thin bolts, for example, instead of short, stubby ones, even if it means using *more* bolts in a given joint. Eliminate gaskets wherever possible and/or use stiffer gaskets. (This may not, however, be helpful if you have leak problems, as we'll see in Chap. 19.)

Against all this conventional wisdom, however, we have those who argue—as we learned earlier—that using stiffer bolts of a larger diameter will *increase* not decrease fatigue life because this will allow you to increase initial preloads and therefore reduce prying action. Again I think the stiff bolt vs. soft bolt argument will be won only on a case-by-case basis.

Achieving the Correct Preload

Poor-quality tools and/or controls will increase the preload scatter and force you to work to a lower mean preload. Use the best you can afford, as discussed in Chaps. 7–11.

VI. PREDICTING FATIGUE LIFE OR ENDURANCE LIMIT

Techniques for theoretically predicting endurance limit or fatigue life of bolts are beyond the scope of this text. You will find some data, however, to Table 4.11 in Chap. 4. From these data you can see that the endurance limit of most bolts is significantly less than the endurance limit of the base materials. We've already learned that one expert [3] says that the endurance limit of bolts is only about one-tenth the endurance limit of the base materials. Others say that the cyclic loads imposed on a joint should be kept below 4% of the ultimate tensile strength of the fasteners if infinite life is desired [14]. A third source says that we can guesstimate the endurance limit of a bolt by experimentally determining the endurance limit of a polished, notch-free specimen of bolt material, then dividing that limit by a suitable stress concentration factor [15]. As an example, stress concentration factors for $\frac{1}{2}$–13 × 6, SAE J429 Grade 2 fasteners in pure tension were found to range from 1.57 to 2.11 [16]. So here we have three experts saying that the endurance limit of a bolt is $\frac{1}{10}$, $\frac{1}{25}$, and $\frac{1}{2}$ of the endurance limit of a test coupon. Take your pick! And accept this confirmation that fatigue test data are often "scattered."

Here's a more carefully thought-out way to estimate the endurance limit of a bolt. An automative company estimates the endurance limit (S_n^1) by multiplying the endurance limit of a standard test specimen by a series of "correction factors" [17] using the following equation:

$$S_n^1 = S_n(C_1 \times C_2 \times C_3) \qquad (17.3)$$

where S_n = the endurance limit of a standard test coupon (they say that this limit is one-half the ultimate tensile strength for wrought ferrous metals or 0.4 of the ultimate tensile for stainless steels)

C_1 = the loading factor (0.85 for axial loading, 0.58 for torsional loading)

C_2 = the size vs. type of stress effect factor (0.85 for bending or torsional loads in fasteners 0.5–2 in. in diameter, 1.0 for axial loads of any diameter)

C_3 = the stress concentration factor (0.3 for rolled threads in quenched and tempered fasteners)

Other correction factors are added if the fastener is to be exposed to a corrosive environment or if the consequences of failure are great and they want to add a safety or reliability factor.

To guarantee that 98% of the fasteners will exceed the predicted life, for example, a reliability factor $C_4 = 0.8$ is included in Eq. (17.3).

The multipliers need not all be less than 1.0, incidentally. If the fastener has been cold-worked, or surface-hardened and plated, correction factor C_5, greater than 1.0, is also included. The reference, however, doesn't suggest how MUCH greater. The use of special thread, nut, and head geometry—as, for example, in Figs. 17.23–17.25—might also allow use of a C_5 greater than one.

I'm sure that Eq. (17.3), and the proposed correction factors C_1 through C_5, are reasonable and appropriate for the fasteners used by the auto manufacturer who published this procedure for estimating endurance limits. The procedure would presumably work for other types of fastener in other industries as well; but it would be best to base your correction factors on fatigue tests or experiences of your own, rather than on data published by others.

VII. THE FATIGUE OF SHEAR JOINT MEMBERS

As I mentioned at the beginning of this chapter, it's the bolts which fail in joints loaded in tension, but it's the joint members which fail under shear loads. Such failures—especially of symmetric butt splice joints—are described at length in the text by Kulak et al. [26] and so I'll only touch on a few highlights here. All of the following comments are derived from that text.

In a properly preloaded, slip-resistant shear joint the fatigue failure will occur through the gross cross section of the joint member. (See Fig. 2.18.) If the joint is a bearing type, or is supposed to be slip resistant but was improperly preloaded and has slipped into bearing, then failure will occur through the net cross section which intersects a line or group of holes. In general, bearing-type joints have less fatigue resistance than slip-resistant joints of comparable size. In fact, just increasing the slip resistance in a slip-resistant joint (presumably by increasing the coefficient

of friction between faying surfaces and/or by increasing the initial clamping force) improved the fatigue behavior of the joints.

Kulak et al. summarize the results of fatigue tests conducted by many
different workers. In most of these tests the maximum applied stresses
exceeded the yield strength of the net sections of the test specimens and
often approached or sometimes exceeded the yield strength of the gross
sections. These tests show that stress excursion is the dominant factor in
determining crack growth. In fact, the fatigue strength of these structural
steel members was relatively independent of the grade of steel tested, or
of its strength. The materials tested had yield strengths ranging all the
way from 36 to 120 ksi; but this variation in strength had negligible effect

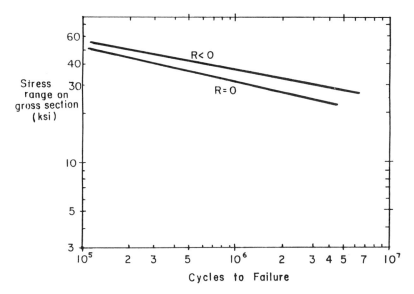

Figure 17.26 Mean *S-N* curves summarizing many tests of slip-resistant structural steel joints. The tests were conducted by many workers under a wide variety
of conditions. Interestingly, they included steels having yield strengths ranging
from 34 to 120 ksi, and show that yield strength has little influence the fatigue
behavior of structural steel joint members. The upper line represents tests conducted with reversing loads fluctuating between tension and compression (i.e., *R*
< 0). The lower line represents the mean *S-N* data for tests conducted under
tensile loads varying from some maximum value to zero. (i.e., *R* = 0) Note that
these *S-N* curves, unlike those we studied earlier, are plotted on log-log scales,
which converts the "curves" we saw earlier to straight lines. This is a very common way to present *S-N* data. I used the "curved" versions because I thought
they were more informative for teaching purposes.

on fatigue life. (This is not to say that strength will never affect fatigue life of a metal part—a bolt, for example. It merely says that strength variation in the tested range did not affect the lives of structural steel members.) The joint members tested under reversing loads ($R < 0$) had better fatigue lives than those tested under cyclic tensile loads only ($R = 0$). This was true for both slip-resistant and bearing-type joints. The authors suggested that this result from the fact that crack growth is inhibited by compressive loads.

Kulak et al. point out that it is theoretically possible to predict fatigue life using the techniques of fracture mechanics, but to do so one must know the shape and size of the initial flaw—and the stress gradient. Since this is not practical for structural steel design, all of the fatigue data they report, such as the data shown in Fig. 17.26, must be—and have been—obtained by laboratory tests. I've been fortunate enough to see some of the machines used for these tests—at Lehigh University in Bethlehem, Pennsylvania, at the University of Toronto, and at the University of Texas at Austin—and they are very large and very impressive. The failure of a structural steel building is to be avoided at all costs, so a great deal of work has gone into the design, codification, and testing of this type of bolted joint. Those interested in this should attend meetings of—or at least follow the activities of—the Research Council on Bolted Joints, sometimes called the "Bolting Council." They sponsored the preparation of the Kulak text, and they have sponsored many of the tests reported therein as well. Much of their knowledge is summarized in the AISC document describing the proper use in structural applications of ASTM A325 and A490 fasteners [27].

VIII. CASE HISTORIES

A. Transmission Towers

A Midwestern power company installed 128, two-pole steel H-shaped towers to carry two 345-kV circuits in a horizontal configuration. The poles were fabricated in sections and the sections were connected by flanged, bolted joints. In some cases the flanges weren't pulled together completely, but were left open on one side.

The bolts used were $1\frac{1}{4}$ and $1\frac{1}{2}$ in. in diameter, varied in length from $4\frac{1}{2}$ to $10\frac{1}{2}$ in., and most were made of ASTM A354 (Grade BD) or A325 steel.

Several years after the towers were erected, 15 bolt heads were found lying on the ground. An investigation showed that wind blowing on the towers had subjected the bolts to dynamic loads which had led to fatigue failure. The fact that some joints were essentially loose on one side undoubtedly contributed to the problem; since preload was low and the joint very "springy," the critical load was low and the bolts had to absorb a larger share of any external load than intended. It was also felt that the bolts, being "high strength" (especially the A354), were too brittle to be acceptable in a fatigue situation.

The joints—there were 688 of them involving 15,000 bolts—were retightened, and many bolts were replaced. The problem was discovered before failure of any joint [12].

B. Gas Compressor Distance Piece

The studs originally used in this application were ASTM A193 B-7 with cut threads, used with standard A194, 2H heavy hex nuts, installed with a nominal torque of 385 lb-ft. Since molydisulfide lubricant was used, this torque produced bolt tension ranging from 25,000 to 45,000 lb [1]. Studs were $1\frac{1}{8}$–7 × 7 in size.

The operator of this compressor reported daily failure of these studs when the equipment was first put into operation. An analysis of the failed bolts showed that they had failed in fatigue, and that bending stresses had contributed substantially to the problem. The following steps were taken:

1. Stud material was changed from B-7 (basically SAE 4140) to 4340, heat-treated to 37–43 R_C. This provided a minimum tensile strength of 160 ksi and a minimum yield of 145 ksi. The cleanliness of the 4340 was controlled.
2. Force washers were installed under one-quarter of the nuts to monitor installation torque and relaxation on a sample basis.
3. Standcote SC-1 PTFE coating (baked on) was used on all studs, nuts, and washers to reduce torque-preload scatter.
4. The threads were rolled after heat treating.
5. Nuts with washer faces were used. The faces were carefully machined perpendicular to thread axes, and seat correction spacers were added.
6. Preload was increased to the range 50,000–55,000 lb.
7. The studs were inspected ultrasonically for fatigue cracks from time to time.

The results were dramatic. The first stud failure didn't occur until the studs had been in service for 6 months. The second and third failures occurred after the seventh and eight months; and the fourth after 18 months. Nine additional bolts were found to be broken (out of 16) after 23 months. (All earlier failures had, of course, been replaced.)

No additional changes were made. All studs were subsequently re-placed after each "first" failure, or after 18 months of service. This is considered an excellent service life for this very demanding application.

REFERENCES

1. Keith Grigg, Standco Industries, Inc., Houston, Texas, Personal communication, 1979.
2. Wayne D. Milestone, Fatigue design considerations in bolted joints, Presented at Using Threaded Fasteners Seminar, University of Wisconsin-Extension, Madison, WI, April 1979.
3. Robert Finkelston, SPS Laboratories, Jenkintown, PA, Personal communication, March 1980.
4. Thread forms and torque systems boost reliability of bolted joints, *Product End.,* December 1977.
5. G. H. Junker, *Principle of the Calculation of High-Duty Bolted Joints,* Unbrako-SPS, Jenkintown, PA.
6. G. A. Fazekas, On optimal bolt preload, *Trans. ASME, J. Eng. Ind.,* pp. 779–782, August 1976.
7. R. J. Lingscheid, *Environmental Conditions for Advanced Fastener Systems,* Rocketdyne Div., North American Aviation, Inc., Society of Automotive Engineers, New York, September 1967.
8. High-strength bolted joints, *SPS Fastener Facts,* Standard Pressed Steel Co., Jenkintown, PA, Sec. IV-C-4.
9. Carl C. Osgood, *Fatigue Design,* Wiley, New York, 1970, p. 180.
10. J. Viglione, The effects of nut design on the fatigue life of internal wrenching bolts, Naval Air Engineering Center, Aeronautical Materials Lab., Report no. NAEC-AML-1910, AD 446 247, June 17, 1964.
11. Friesth, E. R., Modern metric hardware—simpler and better, *Assembly Eng.,* pp. 36ff, October 1977.
12. Ficken, L. A., Bolt failures on 345 kV pole type structures, Paper presented to the Transmission and Distribution Committee of the Edison Electric Institute, St. Petersburg, FL, January 18, 1980.
13. Bickford, John H., Fatigue failure of threaded fasteners, *J. Japan Res. Inst. for Screw Threads and Fasteners,* Vol. 14, No. 10, 1983.
14. Private conversion with Paul Bonenberger, General Motors Corp., Detroit, MI, 1981.
15. Considerations for the design of bolted joints, Industrial Fastener Institute, Cleveland, OH, Undated.
16. Walker, R. A., and R. J. Finkelston, Effect of basic thread parameters on fatigue life, Society of Automotive Engineers, Paper no. 770851, October 1970.
17. *Fastener Engineering Handbook,* Vol. 3, Ford Motor Co., Detroit, MI, June 1983.

18. Krams, Williams E., and Louis Raymond, Ph.D., Installation torque and fatigue performance of threaded fasteners, part III of a series, *American Fastener Journal,* May/June 1991.

19. Krams, William E., and Louis Raymond, Ph.D., Installation torque and fatigue performance of threaded fasteners, last of a series, *American Fastener Journal,* September/October 1991.

20. Osgood, Carl C., Saving weight in bolted joints, *Machine Design,* pp. 128ff, October 25, 1979.

21. Junker, G. H., and P. W. Wallace, The bolted joint: economy of design through improved analysis and assembly methods, *Proc. Inst. Mech. Eng.,* vol. 198B, no. 14, 1984.

22. Hood, A. Craig, Factors affecting the fatigue of fasteners, Attachment 7, Presented at a meeting of the Bolting Technology Council, in Cleveland, OH, April 19, 1993.

23. McNeill, W., A. Heston, and J. Shuetz, A study of factors entering into the calculations for connecting rod joints according to VDI 2230 guidelines, *VDI Berichte* (Germany), no. 766, 1989.

24. Eccles, Williams, MIED, bolted joint design, *Engineering Designer* (UK), pp. 10ff, November 1984.

25. Nassar, Sayed, Importance of initial tension on bolt fatigue load capacity, material passed out at a Bolted Joint Seminar presented by Lawrence Technological University, Southfield, MI, May 1992.

26. Kulak, Geoffrey L., John W. Fisher, and John H. A. Struik, *Guide to Design Criteria for Bolted and Riveted Joints,* 2nd ed., Wiley, New York, 1987.

27. *Specification for Structural Joints Using ASTM A325 or A490 Bolts,* prepared by the Research Council on Structural Connections, published by the American Institute of Steel Construction, Chicago, 1988.

18
Corrosion

One of the most common problems we face when dealing with bolted joints is corrosion. It can take many forms, and can affect the stability of the clamping force and the useful life of the bolts or a joint in many ways.

For example, such mechanical failures as thread stripping and fatigue can be accelerated or made more likely by corrosion. Alternatively, initial buildup of rust can increase the tension in the bolt and the clamping force on the joint, because rust IS a buildup, increasing dimensions.

Excessive corrosion, of course, can eventually lead to a reduction in preload as parts weaken; or to the total loss of clamping force through corrosion wastage, or, more unexpectedly and suddenly, through the mechanisms of hydrogen embrittlement or stress corrosion cracking.

Even if corrosion doesn't proceed far enough to affect the clamping force or life of the joint, it can cause problems. It can spoil the appearance of a product, or make assembly/disassembly difficult or impossible. So corrosion, which has been defined as "the deterioration of a material because of a reaction to its environment" [1], is a problem we can't ignore. Let's look at some of the factors which cause corrosion, and at some of the things we can do about it.

I. THE CORROSION MECHANISM

A. The Galvanic Series

Every metal has a characteristic electrical potential, determined by its atomic structure and based on the ease with which the material can produce or absorb electrons. Those materials which will provide electrons more readily are called *anodic*; those which absorb electrons more readily are called *cathodic*. Anodes and cathodes are called electrodes. If properly interconnected they create "batteries."

No material is just an anode or just a cathode. Any material can serve either function, depending on the other materials to which it is connected. Steel, for example, is anodic in the presence of stainless steel or brass. It is cathodic in the presence of such materials as zinc or aluminum.

The relative anodic-cathodic potential of metals is defined by a table called the galvanic series. Materials listed toward the beginning of the table are anodic compared to those listed nearer the end of the table. The following list shows the relative anodic-cathodic relationship of many of the materials we will encounter in bolted joints [10].

Anodic end
(least noble—most likely to corrode)
Magnesium
Zinc
Aluminum 1100
Cadmium
Aluminum 2024-T4
Steel or iron
Cast iron
Chromium iron (active)
Nickel-resist cast iron
Types 304 and 316 stainless steel (active)
Tin
Nickel (active)
Inconel (active
Hastelloy Alloy C (active)
Brasses
Copper
Bronzes
Monel nickel copper alloy
Nickel (passive)
Inconel (passive)

Types 304 and 316 stainless steels (passive)
Hastelloy Alloy C (passive)
Silver
Titanium
Graphite
Gold

Cathodic end
(most noble—least likely to corrode)

Note that some materials appear more than once in the table. Their electric potential depends on whether or not they are in an "active" or "passive" condition. Although passivity is not fully understood, it is believed to be caused by the presence of a very thin oxide layer on the surface of the material—a layer which, in effect, partially insulates the material and so reduces the ease with which it can give off electrons.

Although each metal will have a characteristic electric potential, this potential can vary, depending, as one example, on whether or not the metal is in an active or passive condition. There are other things which can alter the electric potential of a particular material. An increase in stress level, for example, can make a material more anodic. So can high temperature. Some materials are more anodic at grain boundaries, or in the vicinity of impurities [2].

Reducing the amount of oxygen in the electrolyte (the solution which forms the battery) will also make it easier for metals to give up electrons—it will make them more anodic.

Mechanisms such as these make it possible for a single body to act as both anode and cathode and so, alone, form its own battery. We'll take a detailed look in a moment.

B. The Corrosion Cell

We learned in Chap. 15 that the four "essential conditions" for corrosion were an anode, a cathode, an electrolyte, and a metallic connection. A body will not corrode until it is immersed in, or wetted by, a solution of some sort, and provided with an electrical connection to another body having a different potential which is also immersed in, or wetted by, the same solution. The two bodies of different potential, electrically connected together, and in the presence of a liquid, form a miniature battery, as suggested in Fig. 18.1. The anode in the battery will provide electrons which flow to the cathode. In this process the anode is gradually destroyed—in other words, it corrodes. The cathode, on the other hand, collects material which plates out on its surface. Remember:

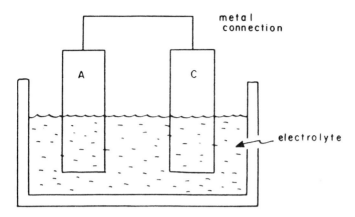

Figure 18.1 An electrical battery is formed whenever two metals having different electrical potentials are connected together by a piece of metal and by a liquid of some sort. Under these conditions the more anodic of the two materials will corrode.

*A*nodes *A*way
*C*athodes *C*ollect

Corrosion "batteries" can be formed in many different ways and by many different combinations of material. Some examples are given below. Corrosion chemists have distinguished among various classes of corrosion depending on the basic nature of the battery at work and/or the appearance of the results. Names given to various types of corrosion include:

General corrosion or uniform attack
Galvanic or two-metal corrosion
Concentration cell corrosion
Stress corrosion cracking
Pitting

And there are others. In *every* case, however, the basic process is that described above—a relatively anodic material is connected to a relatively cathodic material in the presence of a solution, creating a miniature battery, destroying the anode as it produces electrons.

C. Types of Cells

To minimize corrosion problems we must either prevent the formation of "batteries" or reduce their size and effectiveness. There are many ways to do this, as we'll see later. It will be helpful first, however, to look in detail at how some of these corrosion cells are formed in practice.

Two-Metal Corrosion

The most obvious way to encourage corrosion is to connect two different metals together, electrically, in the presence of a fluid. The farther apart the metals are on the galvanic series, the greater the potential difference between them, and the more apt the anode is to corrode. Using steel bolts on an aluminum tower that will be exposed to seawater, for example, is not a good idea, although it has been done (the tower collapsed!) [6].

In a less obvious fashion, dissimilar metals can be coupled through wet earth, puddles of rainwater, etc. It's surprising, in fact, how often we can inadvertently design batteries this way, or how often we must live with them because the design demands dissimilar materials.

Batteries involving two materials are relatively easy to spot and understand. There are many less obvious ways to create corrosion cells, however.

Broken Oxide Film

Let's take a piece of steel which has rusted slightly. We will scrape the rust off one portion of the steel and wet the entire surface with water. That portion of the surface which is still protected by an oxide film (rust) will be cathodic with respect to that portion which is not so protected. There will, therefore, be an electrical potential difference between two adjacent portions of the surface (connected together inside the body), so the anode will corrode—forming rust on its surface, as suggested in Fig. 18.2 [2].

Note that the amount of oxide film can also determine the relative anode-cathode relationship of various portions of the surface of a single body. As rust builds up on one portion of the body, it becomes less anodic and will eventually become cathodic with respect to a previously rusted area. In this way batteries form and re-form across the surface until the entire body has been destroyed.

Stress Corrosion Cracking

Stress makes a body more anodic. Stress concentrations at the root of a tiny crack, therefore, will make that portion of the body anodic with respect to adjacent portions, creating a tiny battery which corrodes and enlarges the crack (Fig. 18.3). This process often aids the growth of fatigue cracks—in fact, it's often difficult to tell whether or not a part has failed

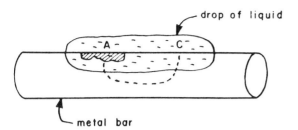

Figure 18.2 A single body can serve as both anode and cathode if a portion of it is protected by an oxide film (such as rust) while another portion is not. Variations in amount of oxide film can also make a difference. As a result, single bodies can and will rust when exposed to moisture.

in fatigue or in corrosion cracking, or in a combination thereof. Hydrogen embrittlement may be a form of stress corrosion cracking, according to some experts [3].

Stress corrosion and hydrogen embrittlement are serious problems for bolting engineers. They're relatively common in bolts, and they lead

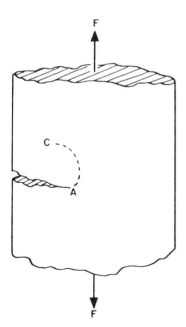

Figure 18.3 Stress concentrations make the tip of a crack more anodic than adjacent regions, leading to stress corrosion cracking.

to sudden and unexpected failure. As a result they deserve special attention. We'll take a closer look in Secs. II and III.

Crevice Corrosion

We said earlier that reducing the oxygen content of the electrolyte will make it easier for adjacent metals to produce electrons. Water trapped under the head of the fastener, for example, as in Fig. 18.4 will be oxygen starved. Being trapped in a crevice, the oxygen cannot be replenished by the oxygen in nearby air. The oxygen starved water becomes acidic. As a result, the crevice becomes more anodic than adjacent regions of the joint, and corrodes.

Fretting Corrosion

If two oxide-coated bodies are rubbed together, the oxide film can be mechanically removed from high spots between contacting surfaces, as shown in Fig. 18.5. These exposed points will now be active (anodic) compared to nearby portions of the surface which are still protected by an oxide film (more passive or cathodic). So the exposed regions will rust. Further relative motion will knock off the next high spots which will rust, etc.

Over a period of time, this combination of electrochemical corrosion and mechanical motion will produce a very fine rust powder in the joint, called the products of fretting corrosion.

There are undoubtedly many other ways in which a corrosion cell can be formed, but these examples should suffice to convince you that there *are* many ways, they are not all obvious, and it can be difficult to eliminate

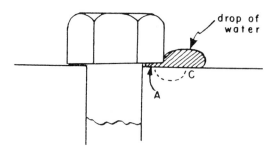

Figure 18.4   The oxygen content of water trapped in a crevice is less than that of water which is exposed to air. As a result, the crevice is anodic with respect to surrounding joint material.

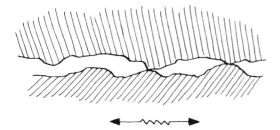

Figure 18.5 Mechanical action (fretting) removes corrosion products from surface high spots. Exposed areas will be anodic with respect to other areas which are still covered with rust or the like.

them all. We'll consider some ways of doing this in Sec. III. First, however, let's take a closer look at one of the bolting engineer's main corrosion concerns—stress corrosion.

II. HYDROGEN EMBRITTLEMENT

A. Stress Cracking Failure Modes

The fastener industry—both suppliers and users—face two, major types of delayed failure by which an apparently good bolt or joint will suddenly and unexpectedly fail after hours or months or even years of satisfactory service. The two are fatigue failure, which we discussed in the last chapter, and stress cracking, which we'll consider now. Of the two, stress cracking is probably the more troublesome because it's more common. It is caused by a combination of tensile stress and corrosion, or tensile stress plus absorbed hydrogen. Unlike fatigue, stress cracking can cause failure even if the applied loads on the fastener are static, i.e., noncyclic.

Four different mechanisms have been identified which can cause stress cracking:

Hydrogen embrittlement
Stress embrittlement
Stress corrosion cracking
Hydrogen-assisted stress corrosion

We will discuss each of these but will concentrate on those which cause bolting engineers the most problems, namely, hydrogen embrittlement and stress corrosion cracking.

The time required for a bolt to fail under stress corrosion can be anything from a few hours after tension has been developed in the fastener to many years. As we'll see later, sophisticated ultrasonic techniques have been developed for detecting beginning stress corrosion cracks in in-service bolts, but these techniques are not readily available. Fasteners can be removed and early cracks may be found with magnetic particle or dye penetrant inspection, but stud or bolt removal and replacement is not always possible. As a result, many cracks go undetected until failure, which is sudden and unexpected.

Since failure is always complete—the bolt breaks—stress corrosion in its various forms can be a very serious problem.

Let's start by taking a close look at hydrogen embrittlement.

B. The Hydrogen Embrittlement Mechanism of Failure

As far as the number of people affected are concerned, the most troublesome form of stress cracking is probably hydrogen embrittlement because it can and often does cause failure of common bolt materials being used in common applications. It's a frequent problem for the auto industry, for example, but also concerns the aerospace, structural steel, pressure vessel, and every other industry where bolt or joint failure has safety implications.

It can occur any time atomic hydrogen has been absorbed and retained by the fastener, and there are many ways that this can happen. For example, atomic hydrogen can be absorbed into the surface of the fastener during cleaning, descaling, pickling, or electroplating operations. If the fasteners are not baked properly as part of the plating process, the hydrogen will remain trapped by the plating.

Although electroplating is the most common source of entrapped hydrogen, the fastener can acquire it from other sources as well. Since hydrogen can be created at the cathode of a corrosion cell, embrittlement is sometimes caused when a sacrificial anode is used to protect a structure (a procedure described later in this chapter) [31]. Aluminum alloys coupled to steel will generate hydrogen at the steel electrode [33]. Lubricants can produce hydrogen if they break down during drilling, machining, or forming operations. Conversion coating operations such as phosphating or black oxiding can cause problems. Hydrogen present in the service environment can also be absorbed by the fastener. And these effects can be accumulative. Incidentally, highly stressed parts absorb more hydrogen than lower stressed ones [29].

The fact that properly plated fasteners can still acquire hydrogen and fail through embrittlement has caused a lot of liability problems for fas-

tener suppliers, since most users automatically blame "poor plating practices" for any embrittlement failure. As the problems often don't become
apparent until the user's product has been assembled and/or put in use, the
liability claims can far exceed the cost of the supposedly faulty fasteners.
Fortunately—at least if they're called upon—there are experts who can
usually determine whether or not the failure was caused by improper manufacturing procedures or by service conditions [28,31].

The hydrogen absorbed by nonplated fasteners may or may not cause
problems, because it can and often will diffuse out of the parts under
certain conditions. Bolts given a phos-oil treatment, for example, shed
the absorbed hydrogen if left on the shelf for 30 days or more at room
temperature, or if baked for 34 hours at 199°F (93°C). Plated fasteners,
on the other hand will not readily shed absorbed hydrogen if the plating
is more than 0.06 ml (2.5 μm) in thickness [37]. In summary then—even
unplated fasteners can and do fail, but electroplated ones are most likely
to give us problems.

Failure occurs when the fastener is stressed above a "threshold"
level—by preloading, for example. Curiously enough, the entrapped hydrogen will tend to migrate to points of stress concentration within the
fastener. The pressure created by the hydrogen creates and/or extends a
crack; which grows until the bolt breaks. That, at least, is one of several
models for hydrogen embrittlement which have been proposed [10]. The
main point for our purposes is that all of the proposed mechanisms start
with the absorption of hydrogen by the base metal—usually during electroplating operations [3].

Note that although hydrogen embrittlement is usually included in a
discussion of corrosion it is not really a corrosion failure. It usually occurs
in "corrosion-resistant" (i.e., plated) bolts, however, and it's often difficult to distinguish this type of failure from others we're about to consider
are *are* corrosion related. So, I'll follow the conventional path and include
it here.

C. Susceptible and Safe Materials

Traditionally, hydrogen embrittlement has been most commonly encountered in cadmium-plated, "high-hardness" steels. In general, common
experience says that if the hardness of the fastener is less than 35 HRC,
you'll probably have no problems; if it's above 40 HRC, problems are
almost certain; in between you may or may not have a problem [10].
Socket head cap screws made of medium carbon alloy steel and hardened
to over 45 HRC are so apt to fail if plated that plating is considered "risky
business." Plated SAE J429 Grade 8 fasteners, hardened to 39 HRC, are

susceptible, but Grades 5 (34 HRC) or less are said to be immune [37]. There is evidence, however, that lower-hardness, plated steels are *not* immune, but simply take longer to fail—years instead of hours—because crack growth is slowed by a decrease in strength and the toughness of the steel increases, so that the crack must grow over a larger distance before the bolt will fail [28].

Although we'll usually encounter hydrogen embrittlement in plated steels, it can and does occur in other fastener materials such as austenitic stainless steels [3], aluminum, and titanium [29]. In fact, embrittlement was a major problem when titanium fasteners were first introduced, but improved manufacturing procedures have made this type of failure rare. The alloy Ti-6A1-4V is said to be especially insensitive to embrittlement [35].

The company Weserchemie GmbH Brueder Mlody in Germany has recently introduced fasteners hardened to 40 HRC and above which are coated with a 0.05-mil (2-μm) layer of copper before being plated with zinc or nickel. The copper provides a barrier which inhibits the absorption of hydrogen. The coating process, a combination of mechanical plating and chemical action, also "polishes" the surface of the fastener and thereby reduces the "reaction area" for hydrogen [30].

Certain exotic, high-strength materials are also said to be relatively immune to hydrogen embrittlement. These include Inconel 718 and MP35N, for example. Unfortunately, their relatively high cost makes them unacceptable for many of the applications troubled by embrittlement.

D. Testing for Embrittlement

Several documents include test procedures designed to detect hydrogen embrittlement. Specification ASTM F606-90, Section 7, for example, describes a procedure for the hydrogen embrittlement testing of commercial grade, through hardened fasteners having diameters ranging from $\frac{1}{4}$ to $1\frac{1}{2}$ in. A sample group of fasteners is taken from a production run. The fasteners are inserted into a test fixture. A wedge with a 4° to 6° taper is placed under the head of each fastener, which is then tightened to 75% of its minimum ultimate tensile strength. The fasteners are left under this stress for 48 hr, after which a breakaway torque (in the tightening direction) is applied. If that torque is less than 90% of the initial torque a fastener is assumed to have relaxed and its test is aborted. Those which pass the breakaway torque test are examined under 20 power magnification. If any cracks are found the lot of fasteners is rejected [27].

Although the torque test and search for cracks are specified as the F606-90 inspection criteria, if hydrogen embrittlement is present the heads of the fasteners will probably break off when the breakaway torque is applied, or will have popped off during the 48 hours under the high stresses created by the tapered washer. And failure can be violent: the flying head can cause serious injury, such as loss of an eye or worse [27, 28]. For example, if you were testing a $\frac{3}{4}$-in.-diameter, 2-in.-long, Grade 8 bolt loaded to 75% of its UTS or 113 ksi—and it broke—the potential energy stored in the bolt would be sufficient to project the fastener up onto the roof of a building six stories high. So—be careful when you conduct these tests!

Another test for hydrogen embrittlement is described in MIL-STD-1315-5A. The fasteners tested here are preloaded then left under stress for 200 hr. Magnetic particle inspection is then used to inspect them for cracks. If any are found, the sampled lot is rejected [38].

ASTM subcommittee E08.06 has proposed—and may by now have published—a test procedure in which fasteners hardened to 39 HRC (175 ksi UTS) or higher are left under load for 7 months before being inspected. Those hardened to less than 39 HRC are to be loaded for 14 months before inspection; confirming the earlier statement that softer fasteners can fail but will take longer to do so [31].

And there are still other recommendations. The Industrial Fasteners Institute recommends a 24-hr test for fasteners in the 32–38 HRC range. The U.S. Navy requires tests of up to 4 years duration in some cases [31]. In each case visual or other crack detection methods are used at the end of the test if the fasteners haven't already failed. And failure can occur quite rapidly. One authority says that if a fastener breaks 1 to 48 hr after initial preloading the problem is almost certain to be hydrogen embrittlement. If it breaks during assembly it's almost certain to be something else [31].

A current incentive to conduct a test of some sort is provided by the new Public Law 101-592, which requires that all fasteners having a minimum tensile strength of 150 ksi or more shall be "recertified" after they are electroplated. Failure to do so can lead to a 5-year jail term and/or a fine of $25,000 [27].

It would obviously be desirable to find an effective way of testing for embrittlement in a shorter time than those cited above. A "rising step load" test has recently been developed by Dr. Louis Raymond. Computer-controlled equipment is used to monitor the onset of crack growth with great precision. Dr. Raymond says that with this procedure dependable hydrogen embrittlement tests can be run in 24 hr [28, 31].

E. Fighting Hydrogen Embrittlement

There are several ways to combat hydrogen embrittlement. The most common is to try to prevent hydrogen from being absorbed by the fastener. In most situations this involves the use of correct plating procedures: baking the fasteners properly, keeping the baths clean, etc. The preparation and cleaning steps are special problems, with overloaded or underloaded barrels playing a major role [29]. As far as baking is concerned, a bake time of at least 3 hr at a temperature of 350–400°F (178–204°C) within 4 hr of plating used to be recommended for steels of hardness 32 HRC (150 ksi UTS) or higher. If the hardness was 40 HRC or more, a minimum of 23 hr was recommended; above 50 HRC the recommendation was "don't electroplate" [38]. It's my understanding that the military, at least, now specify a minimum of 23.5 hr for hardness of 32 HRC or more.

The requirement that baking start within 4 hr of plating is based on the fact that if baking is delayed the absorbed hydrogen may already have concentrated and have started cracks. No amount of baking will eliminate these cracks. The baking, incidentally, doesn't drive out the hydrogen, it merely forces it into "traps" where it loses its mobility and therefore prevents it from concentrating [29].

A second, popular way to avoid entrapped hydrogen is to use fastener coatings which do not involve electroplating. We'll look at some of these in a minute, and will see that they include such things as mechanically plated cadmium or zinc as well as bonded coatings of molybdenum disulfide or Teflon. Ceramic and aluminum coatings are also finding some applications. Although these all reduce the chances for embrittlement they don't always eliminate it, as we've seen—and many of the new coatings can be expensive.

A third way to avoid hydrogen embrittlement would be to minimize the stresses placed on the bolts. This would work—but it's not a popular solution because high clamping force is usually desirable to prevent other types of bolt or joint failure. We'll see that stress control is the most popular way to reduce stress corrosion problems—but only because it's virtually our only option. With hydrogen embrittlement we have other, more attractive options such as those just cited.

A fourth—and also common—way to minimize embrittlement problems is to use a less susceptible bolt material. For the budget conscious this usually means using Grade 5 instead of Grade 8, or Grade 5 equivalents such as ASTM A325 or Metric 8.8. For those with critical applications, safety concerns, and money to spend, this means such high-class materials as Inconel 718 or MP35N.

III. STRESS CORROSION CRACKING

A. The Mechanism of Failure

We learned earlier that stress tends to make a metal more anodic. The tip of a crack in a bolt under tension will, therefore, be more anodic than the materials surrounding it. If an electrolyte is added, we have produced a small "battery" which will provide the energy needed to eat away the anode at the tip of the crack. The crack, therefore, will grow again until the bolt breaks.

Having said all that, I should confess that it is only one of several theories which attempt to explain the mechanism of stress corrosion cracking [3]. Although not universally accepted, it certainly illustrates the apparent process.

There is, I think, no dispute about the essential conditions required for stress corrosion, whatever the mechanism may be. These conditions are:

A susceptible material
Tensile stress above a threshold limit
An electrolyte

As far as susceptible material is concerned, all metallic bolting materials are susceptible to some extent, though some of them much less than others. As far as the electrolyte is concerned, we do not need to immerse the part or constantly drip on it, as is usually required for general corrosion wastage. Only a tiny amount of electrolyte is required for stress corrosion cracking (SCC). I once heard a "dirty fingerprint" had caused a failure. The story may have been apocryphal, but the message was valid. Such normally benign electrolytes as humid air can cause SCC. We'll look at some other possibilities in Sec. V.

The importance of "stress above a threshold level" is more involved, and is illustrated in Fig. 18.6. It shows the time to failure of a number of AISI H-11 bolts subjected to a variety of tensile loads [40]. The higher the load, the shorter the life of the bolts, at least if the loads exceeded 3500 lb. Bolt life was essentially infinite, however, if the applied load was 3500 lb or less. This load defined the "threshold stress level" for that material in that environment.

The threshold stress level for a given material can sometimes be predicted, using the techniques of fracture mechanics and a material property called K_{ISCC}. Let's take a detailed look at this important concept.

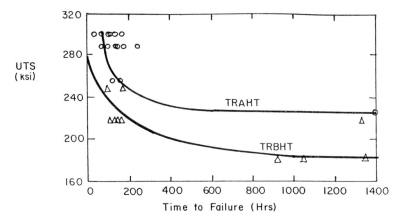

Figure 18.6 Rolling fastener threads after heat treatment (TRAHT) gives the fastener greater resistance to SCC than rolling them before heat treatment (TRBHT). The data shown here are based on tests of $\frac{1}{4}$–28 × 2 AISI H-11 bolts heat treated to a strength of 260 ksi UTS. (From Ref. 40.)

B. The Concept of K_{ISCC}

Experts in failure analysis and fracture mechanics have given us a linear elastic fracture mechanics (LEFM) equation with which we can estimate the amount of stress that can safely be applied to a part which might otherwise fail through SCC. The equation is [17]:

$$K_{ISCC} = C\sigma\sqrt{\pi a} \qquad (18.1)$$

where K_{ISCC} = the threshold stress intensity factor for SCC (ksi $\sqrt{\text{in.}}$)

C = the shape factor (1.5 has been used for threads) [17]

σ = nominal stress (ksi)

a = crack depth (in.)

Note that Eq. (18.1) does not "explain" the mechanism of stress corrosion cracking. It just provides a means for characterizing the tendency of the body—in this case a bolt—to crack, given an existing flaw and various levels of applied stress. The equation tells us that if the product $C\sigma\sqrt{\pi a}$ exceeds a certain critical value K_{ISCC}, then a crack (the initial flaw, for example) will grow and the part will break. The initial flaw can be a tool mark, a corrosion pit, a crack caused by heat treatment, etc.

Even though the equation doesn't give us new knowledge about the SCC process, it has physical meaning—the all-important threshold stress intensity factor, K_{ISCC} is as much of a material property as yield strength or coefficient of expansion. It can (and must) be determined experimentally for a given material, in a given condition (heat treat, etc.) in a given environment. If we know K_{ISCC} for out bolting application, then we can relate an anticipated flaw size (e.g., crack depth) to an allowable (safe) stress or preload in the fastener.

Many factors affect K_{ISCC}. We'll look at some of these next. It's important to realize, however, that the number of variables involved is so large that experimental results tend to be scattered.

Different investigators get different results because one or more variables—often unidentified—vary between one set of experiments and another. Even local variations within a bolt can affect SCC properties [19]. As with vibration and fatigue problems, therefore, we're dealing with conflicting data, and if avoiding failure is essential, we must set limits on stress or preload in accordance with the "worst-case" results which have been reported.

If the calculations are based on the worst-case (i.e., lowest) value of K_{ISCC}, the results are often very conservative. If an occasional failure is acceptable and the consequences of failure are of no great concern, then applied stresses can be higher than suggested by Eq. (18.1) [17].

Unfortunately, there is no common source of information for K_{ISCC} for bolting materials. The literature is scattered, and sometimes contradictory [10]. In critical applications, you should make your own tests, using bolts and conditions which closely reflect your own application.

C. Factors Affecting K_{ISCC}

Bolt Material

As already mentioned, K_{ISCC} is a material property, like the modulus of elasticity or the coefficient of thermal expansion. It's sometimes called a measure of the strength of a material in a corrosive environment [34]. Like most (all?) other material properties, however, it's not a constant, but is a function of a number of variables.

The Environment

As mentioned earlier, K_{ISCC} is not a single-valued material property, and therefore there is no single threshold stress level. These things depend very much on the environment [37]. And, unfortunately, K_{ISCC} data for many corrosive environments are just plain not available [32].

Thread-Forming Procedure

The way the threads are formed also affects the resistance of the bolt to SCC. Threads rolled after heat treat have greater resistance than those rolled before heat treat, as suggested by Fig. 18.6 [40].

Bolt Strength or Hardness

The hardness of the bolt—which relates to its strength—affects the stress corrosion behavior of the bolt and, therefore, its K_{ISCC} value. Figure 18.7, for example, shows K_{ISCC} vs. yield strength results (small circles) for a variety of tests made on a variety of low-alloy quenched and tempered (LAQT) steels. Examples of such materials would be ASTM A193 B7 or B16, SAE J 429 Grade 8, AISI 4340, ASTM A490 or A307 or A540, etc. The environments involved included humid air, seawater, aqueous solutions of sodium chloride, and distilled water [17].

The solid line on the graph shows the worst-case relationship between K_{ISCC} and hardness. If our bolts had a yield strength of 150 ksi, for exam-

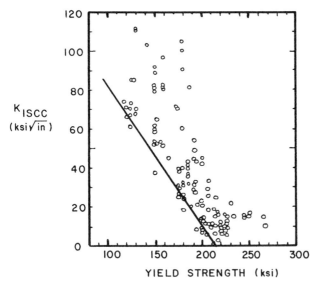

Figure 18.7 K_{ISCC} vs. yield strength for low-alloy quenched and tempered (LAQT) steel in humid air and chloride-containing aqueous environments. Circles show results of test made by a variety of investigators under a variety of conditions. The solid line defines the so-called ''lower bound'' relationship between K_{ISCC} and yield strength. Assuming a higher K_{ISCC} for a given yield strength would introduce some probability of SCC failure.

ple, then, using Fig. 18.7, we could safely assume a K_{ISCC} value of 50 ksi $\sqrt{in.}$ or less. From this we could determine an acceptable preload. Let's take an example.

Let's assume that our bolts are A540, Grade B21's, $\frac{7}{16}$ in. in diameter, with a yield strength of 140 ksi and a K_{ISCC} of 55 ksi $\sqrt{in.}$ For the shape factor (C) we'll use 1.5 [17].

We next have to make some assumptions about the depth of the cracks or flaws we can expect to find in the bolts before we introduce them to the SCC environment. This is a tough problem, of course, with no absolute answers, but we'll assume that the maximum flaw size (a) is 20 mils. Now we are prepared to answer the question "How much preload can I develop in this bolt without risking an SCC failure?" We start by solving Eq. (18.1) for nominal stress.

$$\sigma = K_{ISCC}/C \sqrt{\pi a}$$

$$\sigma = 55/1.5 \sqrt{\pi (0.02)}$$

$$\sigma = 146 \text{ ksi}$$

If our bolts have a $\frac{7}{16}$–16 UN thread, then the tensile stress area will be 0.114 in². We can now compute the safe tension which can be developed in this bolt when we preload it.

$$F_p = \sigma As$$

$$F_p + 146(0.114) = 16.6 \text{ kips}$$

That calculation is correct, however, only if—and this is a big if—the environment with which we are involved is similar to those used during the tests which created the data on which Fig. 18.7 is based. Those tests were made in "humid air" or salt water environments or equivalent. The K_{ISCC} value, and safe preload, would be reduced substantially if our environment included more aggressive electrolytes, such as a strong acid or hydrogen sulfide, as we'll see later.

If we wish, we could now go on and use the short-form torque-tension equation $\tau = KDF_p$ to compute the assembly torque.

Note that Fig. 18.7 teaches us that the higher the yield strength of our bolts, the less the preload we can introduce, safely, at assembly. This suggests that "stronger" bolts can, in fact, be less dependable than "weaker" bolts in an SCC situation, at least as far as LAQT steels are concerned.

Fortunately, not all "strong" bolts are prone to stress corrosion cracking. Socket head screws, for example, with a hardness of 39–44 HRC, are resistant to SCC if they are made with a low-alloy steel having sufficient alloying content to gain a high as-quenched hardness and, therefore, a lower yield-to-ultimate-strength ratio (i.e., good ductility). A generous chromium content in the steel helps too [10]. We'll look at some other exceptions in a minute.

Type of Electrolyte

The aggressiveness of the electrolyte has a significant impact on the probability of an SCC failure. Petrochemical plants manufacturing acids, for example, have had SCC failures in A193 B7 bolts tempered to values as low as 22 HRC (80 ksi yield). In a humid air environment, the same bolt material can be tempered safely to 38 HRC. (See also Fig. 18.13).

As mentioned earlier, even a small amount of electrolyte can cause problems. In recent years, for example, the nuclear power industry has been concerned about a number of SCC failures in A193 B7 and other LAQT steels, where the electrolyte was formed of a combination of humid air and molydisulfide thread lubricant. The molydisulfide decomposes (hydrolyzes) at modestly elevated temperatures to form corrosive hydrogen sulfide [21].

Moly isn't the only lubricant which can cause such problems. In one study, sulfur-based, copper-based, and lead-based lubricants also contributed to the cracking of such materials as 17-4PH, cold-worked 304, and even annealed 304 stainless steels, as well as Inconel and Inconel-X. Only graphite-based lubricants led to crack-free behavior [22].

It's interesting to note, I think, that some environments which cause no corrosion in unstressed parts can cause SCC when tensile stress is present. Conversely, other electrolytes which lead to rapid general corrosion may *not* cause SCC [10]. This, apparently, is one of the mysteries which tends to refute the "mechanism" discussion I gave in Sec. I.

Temperature

Cracking susceptibility is also a function of temperature. For example, the resistance of a 132-ksi-yield steel in aqueous hydrogen sulfide solution was halved when the temperature was raised from room temperature to 300°F (150°C) [19]. The reverse can be true, however, if the bolts are entirely immersed in the electrolyte. Here, the higher temperatures may drive off some of the oxygen present in the electrolyte, slowing the rate at which SCC cracks will develop.

Bolt Diameter and Thread Pitch

Apparently there's a relationship between the depth of the threads on a bolt and its sensitivity to SCC. The presence of a thread makes it difficult to choose an equivalent crack depth, a. Furthermore, the shape factor, C, is affected by such things as thread engagement [17]. So, although the thread itself is not a "crack," it affects SCC sensitivity.

As a result, larger-diameter bolts (deeper threads) of a given material (given K_{ISCC}) have lower threshold stress (σ) levels than smaller bolts, at least up to a point (see Fig. 18.8). For the same reason (thread depth) bolts with fine-pitch threads are less sensitive than those with coarse threads [17, 20].

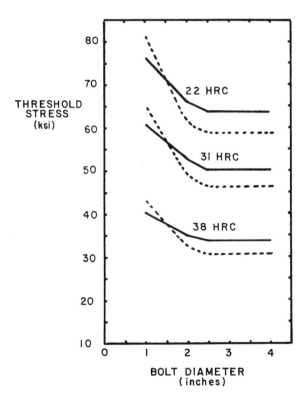

Figure 18.8 Threshold stress as a function of bolt diameter for LAQT steels with UNC threads. Initial crack depth is assumed to equal the thread depth plus 0.1 in. in each case (solid line) or twice thread depth (dotted line).

D. Combating SCC

To fight SCC, we must try to find a way to eliminate or minimize one of the three essential conditions: (1) a susceptible material; (2) a stress level above a threshold limit; or (3) the presence of an electrolyte. Here are some of the steps commonly taken to accomplish these things.

Susceptibility of the Material

Although every metallic bolting material is susceptible to SCC under certain conditions, most of them can be made resistant to it if properly heat-treated, except in the most aggressive of environments. For example, although carbon steel and LAQT fasteners can have SCC problems at all strength levels, they are usually safe to use unless hardened to an ultimate strength in excess of 160 ksi (40 HRC or higher). They should also be used with some caution in a hardness range 35–39 HRC. Below 35 HRC, they are generally considered immune to SCC; all of this, again, is for "normal" environments (humid air, aqueous chloride, etc.). As an example, steels with yield strengths below 100 ksi are highly resistant to SCC [19]. ASTM A325 bolts are considered safe from SCC because their hardness (strength) is not high enough to cause a problem—at least in the sort of environments they are most likely to encounter in structural steel and similar applications.

If the environment involves hydrogen sulfide, then such materials as A193, B7M, and A320 L7M should be considered. They have higher threshold stress levels than the more common B7 or L7 grades; in fact, they are intended specifically to resist SCC and have carefully limited hardnesses.

Austenitic stainless steels (such as A193 B8 and AISI 316) give better SCC service than martensitic (such as AISI 410, 17-4PH or ASTM A449) stainless steels because the martensitic materials have a propensity for pit formation and crevice corrosion which apparently encourage the entry of hydrogen into the material [19].

As mentioned earlier, socket head screw materials give exceptionally good service.

Many aerospace materials are essentially immune to stress corrosion under normal conditions. MP35N, for example, has excellent resistance to SCC; so do some titanium alloys, although these may be susceptible to SCC at elevated temperatures unless properly processed [24].

In general, however, SCC failure of aerospace bolts is limited to alloy steels. These can still be used successfully, but only after being coated with combinations of cadmium and nickel or other inorganic materials [26]. More about coating in Sec. IV.

Aluminum 7075-T73 (a proprietary Alcoa heat treatment) is fairly impervious to SCC and is stronger than 2024 T4 aluminum but there is a significant cost differential as well [24].

NASA's George C. Marshall Space Flight Center has published extensive lists of resistant and susceptible materials which are to be used or are proposed for use in space vehicles, other flight hardware, ground support equipment, and test facilities. This means exposure to seacoast or mild industrial environments according to their report [37]. The materials they evaluated include those used in structural or joint applications as well as for bolting. Their list of alloys having high resistance to SCC includes: carbon (1000 series) and low-alloy (4130, 4340, etc.) steels having ultimate tensile strengths below 180 ksi and Custom 455 stainless steel in condition H1000 and above. The list also included the following materials "in all conditions" (i.e., any hardness): A286 stainless steel, Inconel 718, Inconel X-750, Rene 41, Unitemp 212, Waspaloy, MP35N, and several titanium alloys including Ti-6Al-4V.

NASA's list also includes many materials used for joints, structures, or other purposes as well as for bolting. This includes a number of stainless steels, wrought and cast aluminum alloys, copper alloys, beryllium, and magnesium.

Another long list of resistant materials can be found in Ref. 40. In addition to the materials just cited, they list many titanium alloys including Ti-7Al-12Zr, Ti-8Al-1Mo-1V, and Ti-5Al-5Sn-5Zr. They also list resistant joint materials such as 7075-T6 aluminum, type 321 stainless steel, titanium Ti-6Al-4V, and aluminum alloys 2219-287 and 2014-T6.

Materials to avoid include carbon and alloy steels with hardnesses over 40 HRC and high-strength maraging steels such as Vascomax 250 or Marage 300. Materials such as these are so sensitive to SCC that a tiny flaw can be fatal. NASA's list of susceptible materials included carbon, H-11, and alloy steels having UTS above 200 ksi and various maraging steels aged at 900°F.

Eliminating the Electrolyte

One common way to "eliminate" the electrolyte is to coat the bolts to prevent electrolyte from contacting them. Materials such as aluminum, ceramics, and graphite, for example, can be very effective against SCC. You'll find more details in the discussion of fastener coatings. Other than coating them, it is difficult to isolate the bolts, fully, from environments which can produce electrolytes sufficiently aggressive to cause SCC. As mentioned several times, humid air can do it. A dirty fingerprint may cause a problem. Bolts completely embedded in concrete, and therefore apparently isolated from corrosive liquids, have failed by SCC because the concrete leached chlorines which formed the electrolyte [20]. Normal thread lubricants, as already mentioned, can also lead to SCC problems. Joint sealants (chemical gaskets) have also been identified as the source of leachable sulfur, flourine, and chlorine materials which led to cracks [25]—although there is some debate about these findings. But properly applied coatings will, in effect, "eliminate the electrolyte."

Although complete elimination of possible electrolytes is very difficult, without coating protection, reducing the exposure to electrolytes can extend SCC life. Tightening gasketed joints so that they don't leak electrolytes onto the bolts is an obvious step for a bolting engineer to take. This, of course, can also reduce other types of corrosive attack on the bolts as well as SCC.

Keeping Stress Levels Below a Threshold Limit

Although many of the charts in this chapter imply that only preloads affect a bolt's resistance to SCC, in fact any source of stress can contribute to the problem. Preload is usually a major factor, but we mustn't forget residual manufacturing stresses, bending stress, the stresses created by hole interference or press fits, etc. In spite of this complexity, probably the most common way to combat SCC is to keep stress in the fasteners below a threshold limit defined (or at least computed) by the K_{ISCC} value. As already discussed, the acceptable stress limit will be a function of K_{ISCC}, the crack shape factor, and the size of the initial flaws which must be tolerated. With K_{ISCC} data in hand, we can compute acceptable maximum stress levels from a rearrangement of Eq. (18.1).

$$\sigma = K_{ISCC}/C\sqrt{\pi a}$$

Figure 18.9 gives the resulting data, as a function of bolt hardness, for LAQT steels used in aqueous or mildly chlorine environments, assuming initial crack depths of one and two times the thread depth [17]. Note that this author reported that these plotted stress limits, suggested by use of the worst-case (lower bound) stress intensity factor K_{ISCC}, were "very conservative." This seems to be confirmed by the theoretical calculations of acceptable preload stress vs. hardness which resulted in the plot shown in Fig. 18.10, one of a series of 36 such plots (for a variety of bolting materials, threads, etc.) reported in Ref. 20. He suggests that a preload stress as high as 68 ksi would be acceptable for a 4-in.-diameter LAQT 4340 bolt hardened to 42 HRC. Figure 18.9 suggests a maximum preload stress of 32.5 ksi for a $1\frac{1}{2}$-in.-diameter bolt of the same material hardened to 42 HRC, in spite of the fact that larger-diameter bolts are supposed to

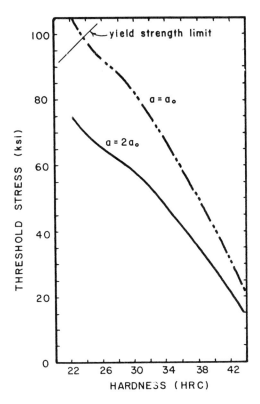

Figure 18.9 Plot of threshold stress limits vs. hardness for $1\frac{1}{2}$-in.-diameter bolts of LAQT steel, used in humid air or chloride-containing aqueous environments. The K_{ISCC} values used to compute threshold stresses were taken from the solid line of Fig. 18.7. Crack depths equal to the thread depth (dashed line above) and twice that (solid line) were assumed.

be more sensitive to SCC than smaller ones. Both references were using the same K_{ISCC} data; but the authors of Ref. 20 assumed different crack depths and crack shape factors than the authors of Ref. 17.

There's another difference here as well. The solid line in Fig. 18.10 is a linear regression line representing the average results of the stress vs. hardness tests. The line in Fig. 18.9 is based on the lower-bound, worst-case K_{ISCC} values shown in Fig. 18.7. Although the people who developed Fig. 18.10 don't say so, they imply that maximum preload (i.e., threshold stress) decisions can be based on the average response of the

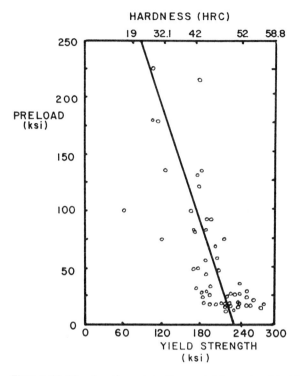

Figure 18.10   Another plot of acceptable preload (which equates to "threshold stress") for an LAQT steel with a UNC thread. The differences between the recommendations of this plot and that of Fig. 18.9 are discussed in the text.

material, rather than on the worst case. So, how we use the data can be another variable.

Note that this plot shows threshold stress (maximum safe preloads) as a function of both hardness, as in Fig. 18.9, and yield strength. We'll use yield strength in the remaining SCC plots. This is a common practice. We should keep in mind, however, that yield strength alone doesn't determine SCC behavior. Heat treatment, microstructure, and the composition of the material also play a role [19].

Figure 18.11 repeats the regression line (called here the "average response" line) of Fig. 18.10. This time, however, I've also included lower-bound and upper-bound response lines which fully encompass the individual test points "spotted" throughout Fig. 18.10. In Fig. 18.11 I've also replotted the lower-bound threshold stress vs. yield strength line for a 4-

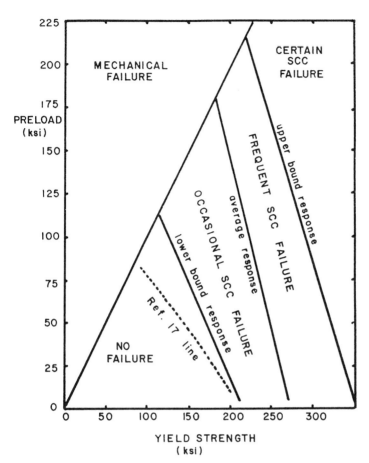

Figure 18.11 A more complete presentation of the data plotted in Fig. 18.10 (from Ref. 20) plus some comparison data from Ref. 17. See the text for details.

in.-diameter bolt, from Ref. 17. As you can see, there's reasonable agreement between the lower-bound lines of the two references. Our choice of lower-bound or average data will presumably be based on the consequences of bolt failure in our own application.

Another aspect of Fig. 18.11 is worth noting. In addition to the lines derived from K_{ISCC} data and calculations, I've plotted a yield line representing all the points at which the recommended preload (threshold stress) equals the yield strength of the fasteners. It's important to realize that Eq. (18.1) will recommend preloads in excess of yield, since the K_{ISCC}

values on which it depends are blind to the yield strength of the material. Your calculations, therefore, must always be checked against yield.

Note, too, that the type of bolting material and the character of the electrolyte also play roles in the selection of the maximum or threshold preload for a given situation. Figure 18.12 shows the threshold stress vs. yield strength for six different bolting materials. (The response of LAQT steels in general is also shown here, by implication, since it's essentially identical to that of 4340 steel.) Figure 18.12, like Figure 18.11, is for 4-in.-diameter bolts in a variety of aqueous and chloride environments.

As you can see from Fig. 18.12, which is taken from various plots in Ref. 20, different materials exhibit different degrees of sensitivity to SCC

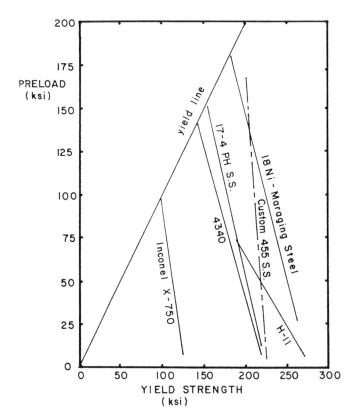

Figure 18.12 The average response, yield strength vs. threshold stress (max safe preload), for six different bolting materials. The data are for 4-in.-diameter bolts in aqueous environments. The response of miscellaneous LAQT steels would be essentially identical to that of the 4340 steel plotted here.

in aqueous environments. I'm a little surprised by the relatively high standing of maraging steel in this group, since it generally gets low marks in the literature on SCC. These are all linear regression (average response) lines, however, which may explain things. The lower-bound line for the maraging steel, not shown in the figure, is well to the left of the average 4340 line.

Figure 18.13 shows the threshold stress vs. yield strength response of miscellaneous low-alloy steels to three different environments: aqueous, including humid air, hydrogen gas, and hydrogen sulfide. As you can see, the preloads which can be applied safely to a low-alloy steel bolt in humid

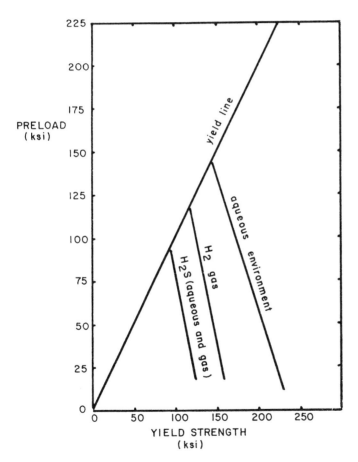

Figure 18.13 The average response, yield strength vs. threshold stress lines, for LAQT steels in three different environments: aqueous or humid air, hydrogen gas, and hydrogen sulfide.

air are many times greater than those which can be applied if the bolt is exposed to hydrogen or H_2S. All of which emphasizes that knowing an experimentally determined K_{ISCC} doesn't eliminate the uncertainties for our predictions of SCC life, safe preload levels, etc. Reported data are too scattered, and the assumptions we must make on C and a and the many other factors which affect SCC are too uncertain to guarantee our predictions. As in so many aspects of bolting, your own tests and prior experience should count for more than the data and conclusions published by others.

If the tests and better data aren't available, try to keep preloads and working loads in the bolts as low as possible, consistent with any need to fight leaks or vibration loosening or other problems suggesting high tension. The Industrial Fastener Institute suggests that you should "be alert" if bolts are to be tightened to 50% of yield or higher; so you might try to keep them below that [10].

Another approach is to use very soft bolts—below 22 HRC, for example—to maximize K_{ISCC}, then tighten them to yield. This maximizes the clamping force available from the most SCC-resistant bolts, and has worked for petrochemical plants dealing with very aggressive electrolytes.

One final way to reduce stresses in the fastener is to shot-peen or pressure-roll the fastener in production to build up a compressive stress on the thread and other surfaces. This reduces the tensile stress when the fastener is placed under load.

Figure 18.14 gives us a final look at the relationship between SCC resistance and tensile stress in the fastener. I include these data because they relate to high-strength H-11 steel instead of to the LAQT steels covered in the previous several figures. The corrosive environment here is still the same—a sodium chloride solution. Note that even such a high-strength material as H-11 still has a threshold stress level below which it exhibits long—probably infinite—life in this environment [40].

E. Surface Coatings or Treatment

The K_{ISCC} value for a fastener can be affected by surface coatings or treatment. Figure 18.15 shows the effect various plating have on low-alloy steel fasteners, for example [34]. A coat of nickel under conventional cadmium led to satisfactory SCC resistance in H-11 bolts heat treated to a UTS of 260 ksi and then loaded to 90% of the proportional limit [40]. The same reference says that nickel and nickel-cadmium coatings in general result in a significant improvement in SCC resistance, as does electroplated cadmium with a chromate conversion coating. Vapor-deposited cadmium and a zinc chromate primer give less but some protection. But

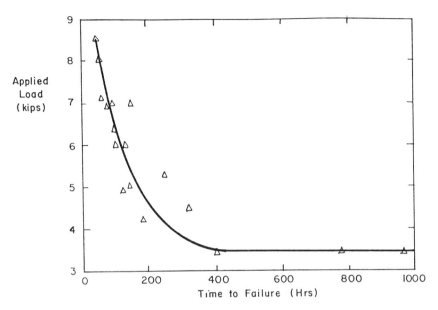

Figure 18.14 Stress corrosion cracking is similar to fatigue failure in many ways. For example, SCC won't occur if tensile stress within the bolt is kept below a threshold value which is a function of the environment. The tests reported here involved exposure to a $3\frac{1}{2}$% NaCl bath. The fasteners were immersed for 10 min, then dried for 50 min. This cycle was repeated until failure. The $\frac{1}{4}$-28 × 2 uncoated AISI H-11 bolts, whose threads had been rolled after heat treatment, had infinite SCC life if subjected to a tensile force of 3500 lb or less in this environment. The 3500 lb load would correspond to an average tensile stress of 96,154 psi or about 37% of the 260 ksi UTS of these bolts. (From Ref. 40.)

plating or surface treatment is not an automatic cure. Gold plating makes little difference [40]. Nor can you eliminate SCC by chrome plating 440 C stainless steel, for example, or by anodizing 2024-T3 aluminum. And carburizing a low-strength carbon steel to a surface hardness corresponding to a UTS of 200 ksi can make that usually-resistant material susceptible [37].

F. Detecting Early SCC Cracks

As mentioned earlier, SCC cracks can usually be detected by magnetic particle or dye penetrant techniques if bolts or studs are removed from the joint for inspection. It would obviously be desirable, however, to inspect them in place.

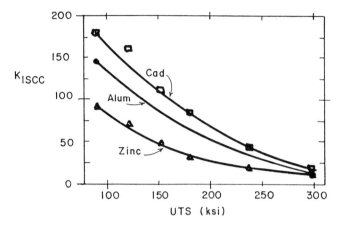

Figure 18.15 K_{ISCC} vs. tensile strength for low-alloy steel fasteners electroplated with cadmium, aluminum, or zinc. As the chart shows, some platings are more beneficial than others in fighting SCC, but none help very much if the bolts are too hard. (From Ref. 34.)

Conventional ultrasonic flaw detection equipment can be used on short studs (perhaps up to a foot in length), especially if the cracks are relatively large and are oriented more or less at right angles to the axis of the bolts. A crack which has penetrated through one-quarter or one-half the diameter of the bolt, for example, would be relatively easy to spot (but might also be propagating so rapidly that it would be detected too late).

More sophisticated ultrasonic techniques have been developed for detecting cracks with depths of 50 mils or larger in the threaded regions of bolts and studs which are up to 112 in. (285 cm) in length [18]. A cylindrically guided wave technique was used. The investigation was sponsored by the Electric Power Research Institute of California.

IV. OTHER TYPES OF STRESS CRACKING

Hydrogen embrittlement and stress corrosion cracking are our main concerns. We should, however, take a brief look at the other two stress cracking mechanisms before we finish the discussion.

A. Stress Embrittlement

Stress embrittlement is the same as hydrogen embrittlement except that it starts with a chemical reaction between a noncoated fastener and the atmosphere; for example, a reaction between a high-carbon steel and hydrogen sulfide. The hydrogen is introduced by the environment rather than by a plating process.

High-carbon and high-strength steels in general are the most susceptible to stress embrittlement [10]. High-strength martensitic stainless steels are also very susceptible, whereas austentitic and ferritic stainless materials will rarely cause problems [24].

Other bolting materials which are very resistant to stress embrittlement include ASTM A193/A193M Grade B7M bolts and A1942M nuts. The bolts, however, should not be hardened above 99 HRB; the nuts should be 22 HCR or less [10].

B. Hydrogen-Assisted Cracking

Hydrogen-assisted cracking is basically hydrogen embrittlement or stress embrittlement combined with stress corrosion cracking. A corrosion cracking process, in other words, is aided by a buildup of hydrogen pressure, with the hydrogen coming either from a plating process or from a chemical reaction with the environment.

Hydrogen-assisted cracking, for example, was encountered a few years ago in some low-carbon martensite bolts tempered to 40 HRC and higher (metric Class 12.8). Used in automotive rear suspension applications, the bolts began to fail unexpectedly about 3 years after they were installed. The failure mechanism was identified as hydrogen-assisted cracking which had been delayed but not prevented by the fact that the fasteners also had a decarburized (i.e., softer) surface [23].

V. MINIMIZING CORROSION PROBLEMS

A. In General

To fight corrosion we must find a way to reduce or eliminate one or more of the four essential conditions: the anode, the cathode, the electrolyte, or the metallic connection between anode and cathode. Anything we can do to reduce the efficiency of this corrosion "battery" will be helpful.

Note that we're now talking about reducing corrosion in general. We've already considered the steps required to combat the special case of stress corrosion in its various forms.

B. Detailed Techniques

Some of the ways in which we can accomplish our basic goals of destroying the corrosion battery and/or of minimizing its effectiveness are fairly obvious. If we can keep a bolted joint dry, for example—perhaps providing a roof, or drainage holes—we "remove the electrolyte." Simple cures of this sort, of course, aren't always possible. Some of the other things we can do are far from obvious. Here are some of the things which are commonly suggested.

1. When designing fastener, joint, and structure, select materials as close together as possible in the galvanic series, minimizing electrical potential differences. The best solution of all, of course, is to use identical materials (although we would still be faced with differences in potential created by differences in stress level, temperature, oxide film, etc.).

2. Since the anode is destroyed, it is desirable to have a large anode and a small cathode. It now becomes relatively easy for the anode to supply the electrons demanded by the cathode; only a relatively small percentage of the anode is destroyed in a given time. By comparison, it would be very bad to have small aluminum fasteners (small anode) holding together a large steel structure (large cathode). The anode would be destroyed very rapidly by the demands of the cathode. In general, therefore, the fasteners should be the most noble, most cathodic element in the joint.

3. Break the metallic circuit connecting anode to cathode by electrically insulating one or the other with paint or other coatings, or with spacers and the like. If you use coatings, you must keep them in good repair, however. A small break in the material coating a large anode will produce a small anode—that portion of the body which is no longer coated. This can lead to very rapid corrosion at that point.

   Coatings, in general, are such an effective way to fight corrosion they deserve special treatment. We'll take a detailed look at them in Sec. VI.

4. Introduce a third electrode—a sacrifical anode—to reverse the flow of current in the battery. As an example, let's assume that you must use steel bolts to clamp brass joint members together, and are troubled by rapid corrosion of the bolts. If blocks of aluminum are placed near the bolts, they will act as a sacrificial anode. Both steel and brass now become cathodes, absorbing electrons from the aluminum, which is rapidly destroyed. The sacrificial aluminum anode is replaced from time to time to protect the steel bolts. As we'll see, some types of coating provide "galvanic protection" of this sort, too.

5. In some critical applications, an actual battery—such as an automobile battery—is physically connected, in the reverse direction, be-

tween the natural anode and cathode. Since the potential of the battery exceeds that of the corrosion battery, the current is reversed and that part which would have been an anode is protected.

6. Seal crevices and the like to prevent accumulation of oxygen-starved moisture. Sealing materials can include paint, putty, nonwicking plastic washers (such as nylon), and the like.

7. Minimize stresses and/or stress concentrations by providing fillets, polishing or shot peening surfaces, designing for uniform distribution of external loads, preloading bolts uniformly, using conical washers to minimize bending stress, etc.—all the things we have talked about in previous chapters for minimizing stress variations, fatigue damage, and the like.

8. If you have some control over the electrolyte, you can sometimes add inhibitors which reduce its capacity for transporting electrons (ions). We use such materials to protect the radiators in automobiles, for example.

9. Use materials which resist the electrolyte. The Industrial Fastener Institute (IFI) publishes a long table showing how much resistance such materials as nylon, brass, and stainless steel have to various chemicals and solutions [4]. Type 304 stainless steel, for example, has excellent resistance to such materials as turpentine, sulfur, fresh water, or wine; but it has poor resistance to sulfuric acid or zinc chloride. Some very general information on corrosion-resistant materials is given in Table 4.12, but you'll find the very long and detailed IFI presentation more helpful, I'm sure.

10. If all else fails, or is impractical for your application, you might consider replacing the bolts periodically, before they fail. This can be a less expensive solution to a corrosion problem than a more technical response. No need to decide exactly what type of corrosion you're facing; no need to search for that perfect bolt material or coating; just throw them away once in awhile. Fresh anodes! It's a valid response.

VI. FASTENER COATINGS

A. In General

One of the most common ways to combat corrosion in bolts is to make the bolts of a corrosion-resistant material, the type of material usually being a function of the industry you are in. Petrochemical people, for example, favor various kinds of stainless steel. Aerospace favors Inconel and titanium. Automotive users favor such things as aluminum and plastics. Marine users like silicon bronze, Monel, and titanium.

In spite of the popularity of corrosion-resistant base materials, a more popular way to protect bolts is to coat them with a protective layer of some sort. One source, for example, says that 90% of all carbon steel bolts are coated with something or other [10].

Coatings can resist corrosion in one of three ways [15].

They can provide barrier protection, isolating the bolt from the corrosive environment and/or breaking the metallic circuit which connects the anode to the cathode.

They can provide "passivation" or "inhibition" slowing down the corrosion, making the "battery" less effective.

They can provide "galvanic" or sacrificial protection, reversing the direction of current in the "battery" to protect the more important electrode (in this case the bolt).

We'll look at coating examples in a minute. Note first, however, that barrier protection—involving such things as paint, cad plating, etc.—requires a perfect coating. A small break in the coating can create a tine anode which will erode more rapidly than would the bolt as a whole. Sacrificial coatings on the other hand—such as aluminum or zinc—do not have to be perfect. As long as some of what remains of the coating is near the material to be protected—and is immersed in the electrolyte—the coating will provide some protection.

Coatings in general have an enormous impact on fastener performance. Before we look specifically at their role in the war against corrosion, it's worth noting that they also provide other important features.

Certain coatings, for example, will have a very desirable effect on the coefficient of friction, reducing the drag between male and female threads and/or between nut and workpiece. Perhaps more important, good coatings also reduce the amount of variation or scatter in friction, improving the accuracy of the torque-tension relationship. (See Table 7.1.)

A less important, but also popular, use for coatings is to change the appearance of the fastener—matching colors to hide the fasteners, or preventing rust for appearance sake rather than for any structural reasons, or even to make the fasteners stand out as design accents.

Our main concern at the moment, however, is with corrosion protection.

Coatings are often divided into three groups: organic, metallic, and composite. Let's look at some examples.

B. Organic Coatings

Organic materials are derived from plant or animal matter and contain compounds of carbon.

They can provide as much as twice the corrosion protection of such things as cadmium or zinc plating and have the additional advantage that they can be provided in a wide variety of colors. Like the other types of coatings we will consider later, organics can be applied by many different methods—dip/spin, spraying, painting, etc. The resulting layers are not always uniform in thickness; materials applied by these techniques tend to build up in the cracks and crevices of a fastener, as shown in Fig. 18.16.

One of the advantages of organic coatings is that they eliminate the hydrogen embrittlement problems sometimes caused by the electroplating process. Also, they involve no heavy metals, such as zinc or cadmium, which are of concern to environmentalists. Organics such as the fluorocarbons, furthermore, can provide more resistance to salt spray than can some of the more popular metallic coatings such as cadmium or zinc.

Typical organic coatings include the following:

Paints

A few years ago, alkyd and phenolic patients were very popular, but better, more lubricious materials such as the fluorocarbons and other polymers (to be discussed below) have generally replaced them [9]. Zinc-rich paint is still popular in structural steel work, but it is generally the entire joint, including the bolts, which is painted rather than the bolts alone.

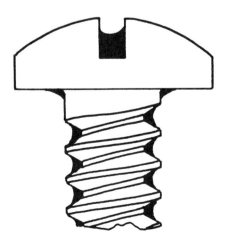

Figure 18.16 Organic coatings tend to build up in thread roots and other crevices as shown here (exaggerated).

Phos-Oil Coatings

Zinc phosphate and manganese phosphate are mild acids. If fasteners are placed in a solution of one of these materials and then tumbled, their surfaces will become slightly porous; a "chemical conversion coating" has been created. Such surfaces provide an excellent base for the retention of oils, waxes, or other organic lubricants [10]. Such coatings as phosphate plus oil, phosphate plus paint plus oil, phosphate plus zinc-rich paint, etc., have all been very popular [15].

Solid Film Organic Coatings

A bonded, solid-film lubricant provides a coating which might be described as "a thin layer of slippery paint." The films generally consist of an air-dried or oven-dried resin binder in which are embedded tiny particles of one or more lubricating or corrosion-resistant materials. These can include such things as molybdenum disulfide, graphite, or polytetrafluoroethylene (PTFE)—at least as far as the organic coatings are concerned [11]. We'll consider composite solid film coatings later. (A composite coating has more than one "active" lubricating and/or corrosion-resistant component.)

Solid-film coatings are available in a wide variety of proprietary formulations. For example, the fluorocarbons (which are used on carbon steel, stainless steel, and aluminum fasteners) are sold under such trade names as Teflon-S, Stalgard, Xylan, Emralon, and Everlube. In general, the fluorocarbons can be used in applications which involve temperatures ranging from $-450°F$ ($-268°C$) to $+400°F$ ($+204°C$).

C. Inorganic or Metallic Coatings

Inorganic materials are any materials *not* containing plant or animal matter; hence inanimate. The class can include such things as ceramic coatings, but we're going to concentrate on the more common metallic materials.

Metallic coatings can be applied to fasteners by a variety of processes, including electroplating, hot dipping, vacuum deposition, and so-called "mechanical plating" techniques.

If electroplating is used, the inorganics tend to build up on the sharp edges of fastener surfaces rather than in the cracks and crevices, as shown in Fig. 18.17.

Let's look at some examples.

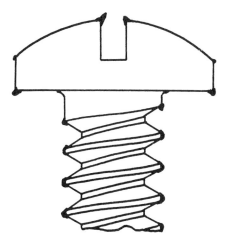

Figure 18.17 Metallic coatings build up on the tips of threads and other sharp edges and corners.

Electroplated Coatings

Cadmium and zinc are the two most common electrodeposited coatings, although more expensive materials, such as nickel, chromium, and silver, can also be applied this way and are used in special applications.

Cadmium protects fasteners more effectively than zinc does in marine environments; but zinc is a better choice in most industrial environments (a combination of cadmium and zinc does better than either in both environments) [13].

A few years ago, cadmium came under attack from the general public because a cyanide rinse used in the plating process can create a dangerous effluent [7]. The final coating was not dangerous; but the process was considered environmentally unsound; and so steps were taken to find substitute coatings.

This search is still going on. One popular "solution" to the problem is to use zinc instead of cadmium. One of the attractions of zinc, incidentally, is that it is a cheaper coating. But zinc is less lubricious than cadmium, so a given preload requires a higher assembly torque. Zinc also tends to double the scatter in the torque-tension relationship, which has created a number of problems in automated assembly operations. Zinc, furthermore, can provide significantly less corrosion protection than cadmium in certain environments.

Another disadvantage is that zinc will develop a dull, white corrosion product called "white rust" unless protected by a clear or colored chromate coating [10]. One advantage of zinc is that it gives galvanic protection as well as barrier protection. Cadmium provides only barrier protection.

In recent years, techniques have been developed to process cadmium's cyanide rinse effluent more effectively. Although these increase the cost of cadmium plating, the lubricity and corrosion problems encountered when zinc is substituted for cadmium have led to a rebirth of interest in cadmium. The search for other substitutes was partially successful, however, so cadmium will probably never regain its previous popularity.

We'll look at some of the coatings selected as cadmium substitutes in a minute. But first, let's continue our survey of coating types.

Hot Dip Coatings

In general, two materials are applied by hot dip techniques: aluminum and zinc. Fasteners coated with aluminum are said to be "aluminized." Zinc-coated fasteners have been "galvanized."

These are both low-cost coatings and are generally used on relatively inexpensive, high-strength fasteners; ASTM A325 structural steel bolts are often galvanized, for example.

The hot dip process is difficult to control, so the resulting coatings tend to vary quite a bit in thickness. Threads are generally undercut or overcut to provide room for the coating—and then should be recut with a tap or die after the coating has been applied. Although this process is relatively common, it can result in a significant reduction in the stripping strength of the threads, a problem that is generally avoided if the fasteners are mechanically galvanized (see below) rather than hot dip galvanized.

Hot-dip-galvanized fasteners can have more corrosion resistance than mechanically galvanized ones, however, because the coating thickness is greater. On the other hand, because of the difficulties of controlling the hot dip process, corrosion resistance can vary more for the hot dip products. Galvanizing in general, incidentally, tends to give greater corrosion protection than electroplating, again because of the greater thickness.

Mechanical Plating

Fasteners are said to be mechanically plated when a ductile metal such as cadmium, zinc, or tin is cold-welded onto the metal substrate by mechanical energy. Glass beads are usually used to do the "welding."

Fasteners which have been mechanically plated with zinc are said to be mechanically galvanized. As already mentioned, coating thicknesses are much more uniform than they are with the hot dip galvanizing process, so it is not necessary to chase the coated threads with a die.

One big advantage of mechanical plating, as opposed to electroplating, is that no baths are involved, eliminating hydrogen embrittlement and detempering concerns.

In one recently developed process, combinations of aluminum and zinc can be uniformly deposited on fastener surfaces without buildup in thread roots, etc. These combinations are said to give the durability of aluminum coatings plus the galvanic protection of zinc (aluminum alone does not always give sufficient galvanic protection because it will form oxide films which partially or wholly isolate it from the electrolyte and other metals).

The mixed aluminum-zinc coating thickness run about half a mil, making overtapping unnecessary, and, therefore, preserving thread strength. The coatings are applied at room temperature so that detemper and hydrogen embrittlement are not problems. Costs are comparable to those associated with galvanizing.

A phosphate coating is usually applied on top of the aluminum-zinc coat for better corrosion resistance and lubricity. The resulting coating is said to provide better corrosion resistance than galvanizing [13].

Miscellaneous Coating Processes

A large number of other coating processes are available. High-strength steels and titanium, for example, are sometimes coated with aluminum in a process called ion vapor deposition (IVD). The resulting coating is said to have excellent resistance to stress corrosion cracking and to be usable to temperatures as high as 950°F (510°C) [7].

In another process, a nonporous layer of high-purity nickel is actually alloyed with the surfaces of carbon or alloy steel fasteners. The process gives good resistance to severe acids and alkalines and creates no buildup. Additional coats of cadmium or zinc are sometimes added for additional corrosion protection and lubricity [8].

D. Composite Coatings

It's not entirely clear to me when a multiple-component coating should be classed as organic or metallic rather than composite, but, in general, composite coatings consist of a wide variety of combinations of "active" organic and/or inorganic materials applied in separate layers and/or in various mixtures. As one example, using a dip/spin process, one might apply zinc to a fastener for corrosion protection, cover this with an organic paint for color, and finish up with a PTFE coating for lubricity [16]. Another available combination is inorganic aluminum coated with phosphate coated with a chromate. Aluminum with an inorganic ceramic binder is another offering [15]. In fact, various combinations of aluminum and inorganic coatings are said to have replaced cadmium across the board in airframe applications, and are also recommended for electrical connections. Such coatings are said to give torque-tension characteristics virtually identical to those of cadmium [14].

Note that stainless steel fasteners are often used in aerospace applications, which usually involve aluminum structural members. A combination of stainless steel and aluminum leads to galvanic attack which can be prevented by coating the stainless steel fasteners with an aluminum/inorganic coating [14].

The number of composite (and other) coatings which are available is nearly endless, but the preceding should be sufficient introduction. Further details on a number of coatings are given in Table 18.1. I don't mean to "recommend" those listed by including them; nor do I mean NOT to recommend others by not listing them. The table merely lists typical offerings that are available, in most cases, from many sources. (Addresses for the suppliers listed in Table 18.1 are given in Table 18.2.)

The buildup of composites is shown in Fig. 18.18. There's some buildup in thread roots, but much less than is (sometimes) the case with organic coatings. There's also some buildup on sharp edges but, unlike the edge buildup of metallic plating, the composite material accumulated on edges will break off easily.

E. Rating Corrosion Resistance

One of the columns in Table 18.1 gives the resistance of various coatings to salt spray. Many different types (concentrations, etc.) of salt spray are possible; I believe that most, if not all, of the data given in Table 18.1 came from use of the ASTM B 117 procedure. Resistance is given in hours of successful exposure (before excessive damage has occurred).

Obviously resistance to salt spray doesn't completely define the corrosion resistance of a coating. You may be primarily interested in resistance to something else, like a specific acid or fuel or sulfur or wine. Salt spray is commonly used to rate coatings, however, and so I've used it here. Suppliers can presumably tell you how well their coatings will stand up to other types of environment, although the information they'll give you is often fairly general (good, fair, poor, etc.).

ASTM Standard B-117 is not the only possible way to test for corrosion, of course. Many manufacturers have developed special procedures appropriate for the environments seen by their products. Another published standard is the German DIN 50018 (called the "Kesternich test"). These tests, which are also specified in the Factory Mutual Standard FM 4470, are conducted in a special cabinet and simulate the effects of acid rain [45].

Predicting corrosion results is not easy. As is so often the case with bolted joints, laboratory tests can rarely predict field results. If tests must be made in a laboratory environment, try to duplicate field conditions as accurately as possible. Test actual joints, not chunks of metal with bolts stuck in them. Better yet, find a way to conduct tests in the field, on the actual job sites, if possible. And remember that past experience, recorded in maintenance records or the like, may be a more accurate way to evaluate a corrosion-resistant material or coating than a quick and dirty lab test.

F. Substitutes for Cadmium Plate

As mentioned earlier, many people have been actively seeking acceptable substitutes for cadmium plating. Let's take a brief look at some of the resulting choices.

Ideally, a perfect substitute would equal or exceed cadmium plate in corrosion resistance, lubricity, and cost. Substitutes which avoided the hydrogen embrittlement problems sometimes associated with cadmium plate would be especially attractive. Although none of the substitutes found so far match all of cadmium's characteristics, they're close enough to be acceptable.

Here are some of the coatings which have been adopted as substitutes. (See Table 18.1 for trade names and sources.)

Solid film organics. Fluorocarbons or molybdenum disulfide in resin binders has been used by some of the automotive companies [8].

Electrodeposited, alloyed coatings of zinc plus tin have been found to be a good substitute for cadmium when magnesium auto blocks are involved (e.g., European auto manufacturers) [8].

Aluminum with inorganic binders (e.g., ceramics) provides torque-tension relationships which are virtually identical to cadmium, and have replaced cadmium in most airframe applications [14].

Ion vapor deposited (IVD) aluminum is being used in place of cadmium in many high-performance applications [8].

Tin flash over electroplated zinc. The zinc provides galvanic protection; the tin prevents galling.

Table 18.1 Fastener Coatings

Type	Trade name[a]	Manufacturer[b]	Corrosion resistance (hr)[c]	Temp. range (°F)	Ref. and notes
Aluminum in ceramic binder	SermaGard	SermaGard/Teleflex	1500	+1200	e
Alum. filled, corrosion inhibitive, organic binder	Alumazite 'Z'	Tiodize Company	—	+500	8
Alum., inorganic binder and chromates	SermaGard	SermaGard/Teleflex	Many thousand	+900	e
Ion vapor deposition alum.	IVD aluminum	SPS Technologies	1000	+925	e
Cadmium, plain	—	3M	50–320[d]	—	8
Cadmium plus chromate	—	3M	200–2000[d]	—	8
Cad-tin (50/50)	—	3M	50–320[d]	—	8
Cad-tin-chromate	—	3M	200–2000+[d]	—	8
Cad-zinc(25/75)-chromate	—	3M	2100–6700[d]	—	8
Cad-zinc(50/50)-chromate	—	3M	5900–6720[d]	—	8
Fluorocarbon (PTFE)	Teflon	DuPont	—	-450 to +400	7
Fluorocarbon + resin binder	Reflon-S	DuPont	1100->3000	-450 to +400	7, 44
PTFE plus binding resin	Emralon 305	Acheson Colloids Co.	500	+500	7
PTFE in phenolic binder	Everlube 6108	E/M Corporation	500	-100 to +400	11
Fluoropolymers with high-temp. organic polymers plus corrosion inhibitors	Xylan	Whitford Corporation	96–2000[d]	—	7
Graphite plus binder	Dag	Acheson Colloids Co.	Poor	+850	7
Graphite plus binder	Pepcoat	G*Chemical Company	1200	-350 to +650	e
Molydisulfide + epoxy binder	Ecoalube 642	E/M Corporation	500	-365 to +500	11
Molydis. + silicone binder	Everlube WL-135	E/M Corporation	100	-365 to +1300	11

Description	Trade name	Manufacturer		Temperature (°F)	
Molydis. in bonding resin	Lube-Lok 2109	Electrofilm Inc.	—	−459 to +450	e
MoS$_2$/graphite/phenolic resin	Electrolube E-40	Electrofilm Inc.	—	−459 to +450	e
Molydis. plus binder	Molydag	Acheson Colloids Co.	—	+650	7
Surface alloy of nickel	Sanbond	AMCA International	2000	+600	8
Surf. alloy Ni-Cad-chromate	Sanbond-Cad	AMCA International	600	+600	8
Surface alloy Ni + zinc	Sanbond-Z	AMCA International	—	+600	8
Silver + indium + binder	Lube-Loc 4253	Electrofilm Inc.	—	−459 to +450	e
Zinc, plain	—	3M	36–192[d]	+500	8
Zinc + clear chromate	—	3M	24–192[d]	—	8
Zinc plus yellow chromate	—	3M	100–650[d]	—	8
Zn-tin(75/25)-chromate	—	3M	400–1000[d]	—	8
Zn-tin(50/50)-chromate	—	3M	600–1300[d]	—	8
Alloyed coatings of electro-deposited zinc and nickel	Ni-Alloy	Deveco Corporation			
Yellow bronze chromate over zinc plating	Macrocor 250	Nucor Fastener	250	0	41
Thermosetting epoxy coating containing aluminum flake over zinc plating	Magnigard Silver 17	Magni Industries, Inc.	1000–10,000	—	42
Phenoxy topcoat over organic, zinc-rich film over zinc phosphated steel	Magnigard-Black	Magni Industries, Inc.	100–1000	—	42
Aqueous coating dispersion containing chromium, proprietary organics, and zinc flake	Dracromet 320	Metal Coatings International	f	700	43

[a] Most, if not all, of these trade names are registered.
[b] Addresses for these manufacturers are given in Table 18.2.
[c] Usually tested per ASTM B117.
[d] Depending on the thickness of the coating.
[e] From the manufacturer's literature.
[f] Said to have "three times the resistance of zinc chromate coatings."

Table 18.2 Fastener Coating Suppliers

Acheson Colloids Company
Port Huron, Michigan (Dag, Molydag, Emralon)

AMCA International
New Bedford, Massachusetts (Sanbond)

Deveco Corporation
Addison, Illinois

DuPont
Wilmington, Delaware (Teflon)

Elco Industries
Rockford, Illinois

Electrofilm Inc.
Valencia, California (Lub-Lok, Electrolube)

E/M Corporation
West Lafayette, Indiana (Ecoalube, Everlube)

Fel-Pro
Skokie, Illinois (N5000, C5A)

G* Chemical Corporation
Wayne, New Jersey (Pepcoat)

MacDermid Inc.
Waterbury, Connecticut (mechanical Al plus Zn)

Magni Industries, Inc. (Magnigard)
Birminham, Michigan

Metal Coatings International (Dacrotizing, Dacrosealing)
Chardon, Ohio

Never Seez Compound Corp.
Broadview, Illinois (Never-Seez)

Nucor Fastener Corp.
St. Joe, Indiana

Serma Gard
Division of Teleflex Incorporated
Limerick, Pennsylvania (Sermatel, SermaGard)

SPS Technologies
Jenkintown, Pennsylvania (IVD Aluminum)

3M Plating Systems Dept.
Commercial Chemicals Division
St. Paul, Minnesota (mics. plates)

Tiodize Company
Huntington Beach, California (Alumazite)

Whitford Corporation
West Chester, Pennsylvania (Xylan)

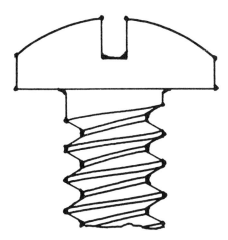

Figure 18.18 Composite coatings build up in crevices and on sharp edges. The crevice buildup, however, is usually less than that encountered with organic coatings; and the edge buildup breaks off easily.

Yellow bronze chromate over zinc. This coating has been adopted by the Defense Industrial Supply Center (DISC) as a substitute for cadmium, especially on Grade 8 fasteners [41].

REFERENCES

1. Myers, J. R., *Fundamentals and Forms of Corrosion*, University of Wisconsin-Extension, Madison, WI, 1980.
2. Neill, W. K., Corrosion prevention—basic theory and practical design pointers, Presented at the Using Threaded Fasteners seminar, University of Wisconsin-Extension, Madison, WI, May 9, 1978.
3. Latanision, R., O. H. Gastine, and C. R. Compeau, Stress corrosion cracking and hydrogen embrittlement—differences and similarities, Paper presented at Symposium on Environment Sensitive Fracture of Engineering Materials, October 1965.
4. Fastener Standards, Industrial Fastener Institute, Cleveland, OH, 1988, pp. B40–B43.
5. Brochure, Standco/Bolt Division of Standco Industries, Inc., Houston, TX.
6. Aimone, M. A., Fundamentals and the forms of corrosion, Presented at the Using Threaded Fasteners seminar, University of Wisconsin-Extension, Madison, WI, May 1980.
7. Emrich, Mary, Corrosion protection for fasteners, part 1, *Assembly Eng.*, pp. 36–40, October 1982.

8. Emrich, Mary, Corrosion protection for fasteners, part 2, *Assembly Eng.*, pp. 17–21, November 1982.
9. 1984 fastening and joining Reference issue, *Machine Design*, pp. 50, 52–53, November 15, 1984.
10. *Fastener Standards*, Industrial Fastener Institute, Cleveland, OH, 1988, pp. B30–B49.
11. Gresham, Robert M., Bonded solid film lubricants for fastener coatings, *Fastener Technol. Int.*, April/May 1987.
12. Taylor, Edward, SermeTel W aluminum coating for aerospace fasteners, SPS Laboratory Report no. 5392, SPS Technologies, Jenkintown, PA, March 24, 1972, Reprinted October 30, 1980.
13. From a speech given by an employee of MacDermid, Inc., Waterbury, CT, in a meeting of the Research Council on Structural Connections, Miami, Fl, June 8, 1988.
14. From a speech given by A. E. Simmons, Jr., of the SermaGard Division of Teleflex, Inc. (USA) at the 1988 Annual Meeting of the Industrial Fastener Institute, Laguna Niguel, CA, February 22, 1988.
15. From a speech given by T. Dorsett of Metal Coatings International Inc. at the 1988 Annual Meeting of the Industrial Fastener Institute, Laguna Niguel, Ca, February 22, 1988.
16. From a speech given by Roger Kelly of the Man-Gill Chemical Company at the 1988 Annual Meeting of the Industrial Fasteners Institute, Laguna Niguel, CA, February 22, 1988.
17. Chung, Y., Threshold preload levels for avoiding stress corrosion cracking in high strength bolts, Tech Report no. 0284-03 EV, Bechtel Group, San Francisco, April 1984.
18. Light, Glenn M., Harayan R. Joshi, and Soung-Nan Liu, Detection of stress corrosion cracks in reactor pressure vessel and primary coolant system anchor studs, Southwest Research Institute, San Antonio, TX, about 1985.
19. Rungta, Ravi, and Bhaskar S. Majumdar, Stress Corrosion cracking of alternative bolting alloys, Report on research project 2058-12, Battelle Memorial Institute, Columbus, OH, December 1985.
20. Czajkowski, C. J., Bolting applications, NUREG/CR-3604 BNL-NUREG-51735, U.S. Nuclear Regulatory Commission, Washington, DC, May 1984.
21. Czajkowski, Carl J., Testing of nuclear grade lubricants and their effect on A540 B24 and A193 B7 bolting materials, Report NRC FIN A-3011, Brookhaven National Laboratories, Upton, NY, March 1984.
22. Rowland, M. C., and T. C. Rose, Tests on thread lubricants for use in nuclear reactors, Report APED-4422, R63APE 29, General Electric Co., San Jose, CA, December 1963.
23. Hughel, Thomas J., Delayed fracture of Class 12.8 bolts in automotive rear suspensions, *Fastener Technol.*, pp. 36–38, April 1982.
24. Blake, Alexander, *What Every Engineer Should Know About Threaded Fasteners*, Marcel-Dekker, New York, 1986.
25. Degradation of threaded fasteners in the reactor coolant pressure boundary

of PWR plants, IE bulletin no. 82-02, U.S. Nuclear Regulatory Commission, Office of Inspection and Enforcement, Washington, DC, June 2, 1982.

26. Taylor, Edward, Stress corrosion crack protection from coatings on high strength H-11 steel aerospace bolts, Special Technical Publication 518, ASTM, Philadelphia, 1972.

27. Greenslade, Joe, Hydrogen embrittlement testing requirements for bolts per ASTM F606-90, section 7, *American Fastener Journal*, pp. 8ff, January/February 1993.

28. Raymond, Dr. Louis, Hydrogen embrittlement of fasteners: problems and solutions, *American Fastener Journal*, pp. 85ff, May/June 1993.

29. Grobin, Allen W., Jr., Hydrogen embrittlement problems, *ASTM Standardization News*, March 1990.

30. Mlody, Martina, New method for the prevention of hydrogen embrittlement, *Fastener Technology International*, pp. 74ff, February 1994.

31. Raymond, Dr. Louis, Test methods for hydrogen embrittlement of fasteners: meaning and use, *American Fastener Journal*, May/June 1994.

32. Raymond, Louis, Ph.D., Titanium alloys, *American Fastener Journal*, pp. 8ff, May/June 1992.

33. Raymond, Louis, Ph.D., Aluminum alloys, *American Fastener Journal*, pp. 40ff, January/February 1993.

34. Raymond, Louis, Ph.D., Copper alloys/steels, *American Fastener Journal*, pp. 12ff, March/April 1993.

35. Faller, Kurt, and William A. Edmonds, The incentives for titanium alloy fasteners in the Navy/marine environment, *American Fastener Journal*, pp. 44ff, September/October 1992.

36. Notes prepared for a bolting seminar by Raymond Tootsky of McDonnell Douglas Corp., Long Beach, CA, December 1991.

37. Greenslade, Joe, Zinc plating socket head cap screws is a risky business, *American Fastener Journal*, pp. 33ff, March/April 1993.

38. Greenslade, Joe, Military hydrogen embrittlement requirements, *American Fastener Journal*, pp. 9ff, February/March 1989.

39. Raymond, Louis, Ph.D., Use of a hard plate in hydrogen embrittlement testing is not everything it's cracked up to be, *American Fastener Journal*, pp. 49ff, March/April 1994.

40. Hood, A. Craig, Preventing stress-corrosion cracking in threaded fasteners, *Metal Progress*, September 1967.

41. Irving, Robert R., DISC gives its approval to new coating, *Metalworking News*, pp. 6ff, December 18, 1989.

42. From a brochure published by Magni Industries, Inc., Birmingham, MI, 1993.

43. From information provided by John Dutton of Metal Coatings International, Chardon, OH, September 1993.

44. From a brochure provided by Prosper Engineering, Ltd., Irvine, Ayshire, UK, 1993.

45. Willis, David P. Jr., Engineered coatings for fasteners, *Product Finishing*, May 1987.

19
Gasketed Joints and Leaks

Preventing liquid or gas leaks is one of the most important and most difficult jobs faced by bolted joints. It is important because the contained fluids are often important from an economic standpoint. Many of these fluids are also dangerous—toxic or flammable, for example. A leak is always a nuisance; it will be an expense; it can even lead to a fire, explosion, or some other disaster.

This is a difficult job for two reasons. The requirement that a joint be liquid or gas tight means that it must be assembled and tightened with an extra measure of knowledge—an almost intolerable burden to place on what is already a very difficult control problem. Liquid or gas joints, furthermore, usually contain a gasket of some sort, and this always adds major uncertainties to the behavior of the joint. In spite of their widespread use, we still don't know as much as we'd like to about how a gasket behaves or why. Attempts are being made by the engineering societies to correct this situation, but there are so many variables, and the task is such a large one, that it may be years before the final uncertainties are resolved.

Incidentally, there's no such thing as a truly leak-free gaseous joint. They *all* leak to some degree; but the best ones leak so slowly that it doesn't matter and/or may even be extremely difficult to detect. One current problem facing investigators, in fact, is how to define "leak free." So we not only face uncertainties when dealing with gasketed joints, we face a theoretical impossibility, at least as far as gas-containing joints are concerned (liquid-containing joints can be leak free).

We have to start somewhere, however. We have to go on designing and using things even if we don't know all of the theoretical answers. Let's look, therefore, at the mechanical properties of a gasket. We'll consider first the basic mechanical properties, then look specifically at the elevated-temperature behavior of the gasket. Next we'll consider the gasket's leakage behavior and the new gasket factors being developed by the Pressure Vessel Research Committee (PVRC) to quantify such behavior. We'll go on to review means of selecting a gasket, selecting the assembly stress (torques, for example) to be used, and will end with some advice for and an example from the field.

Most of the material which follows is based on the types of gasket used in pressure vessels, pipelines, process systems, and similar equipment because that is where my own experience lies. Obviously, this means that the joints to be considered (primarily the raised-face pressure vessel flange) will differ from the types of joint found elsewhere—on an automotive engine head, for example. The basic behavior of the gasket materials, however, is the same in all industries and the construction of the gaskets tends to be similar (though not identical). I believe that the discussion will be of use to people in most industries using gaskets.

I. WHY DOES A JOINT LEAK?

A joint leaks when the material being contained escapes through pores or gaps in the gasket, or escapes around the gasket. Persistent leaks through a gasket can often be reduced by a change in gasket material or type. Sometimes it helps to coat the gasket with a gasket compound of some sort. Soaking a gasket in water can reduce nitrogen leaks, for example.

Eliminating leaks around the gasket is a more difficult job. It's not just a question of eliminating major openings or gaps between gasket and flange surface—this would be relatively simple to do in most cases. Instead we must maintain "sufficient" contact pressure—usually called *gasket stress*—between the flange and gasket surfaces, and we must maintain this as bolts or gasket relax, and as the temperature and pressure of the contained fluid or gas change.

We generate the contact pressure by tightening the bolts or studs which hold the joint together, tightening them enough to create intimate (near perfect) contact between gasket and joint surfaces. Nonuniform tightening can distort the joint members and/or the gasket, opening leak paths. Rough or damaged flange surfaces will also cause problems. Small leaks allowed by these conditions will often erode and/or corrode the leak path, greatly increasing the leak rate.

II. MECHANICAL BEHAVIOR OF A GASKET

Let's start with a review of the mechanical characteristics of gaskets. This is critically important information for joint designers, because a gasket dominates the behavior of a joint. Understanding its mechanical behavior will also help us understand the equally important subject to follow: why do gasketed joints leak and what can we do about it? So first—mechanical characteristics.

A. Load-Deflection Behavior

The load-deflection characteristics of a gasket are not reported in terms of force and deflection, they're reported in terms of compressive stress on the gasket vs. its deflection under that stress. The symbols we'll use for these terms are stress, S_G; deflection, D_G. These parameters are usually measured by loading a small square of gasket material, or by compressing a small-diameter ring between two hydraulically loaded platens. By these means a uniform stress can be applied over the entire are of the material being tested. The actual gasket, in service, does *not* see uniform stress, as we'll see in Sec. C.

The gasket can be considered as another spring in series with the bolt, nut, washer, and flange springs. Since the gasket is part of the joint, we can combine its spring constant with that of the flange members as before:

$$\frac{1}{K_J} = \frac{1}{K_G} + \frac{1}{K_F} \tag{19.1}$$

where K_G = spring constant of the gasket (lb/in., N/m)
$\quad\quad\quad K_F$ = spring constant of the combined flange members (lb/in., N/m)
$\quad\quad\quad K_J$ = the combined spring constant of the joint (lb/in., N/m)

And we can use the resulting K_J to construct a joint diagram and/or make joint calculations, as in Chap. 12.

The spring constant of the gasket, however, has some very special and unpleasant characteristics. For example:

1. The spring constant of the gasket in use is often smaller than the spring constant of the bolt and/or joint members. This "spring," therefore, will dominate the elastic behavior of the joint, as suggested in Fig. 5.16.
2. The spring constant of the gasket is nonlinear, as suggested in Fig. 19.1. Thus we cannot represent its behavior by a single spring constant but rather must use some function of the compression of the gasket.

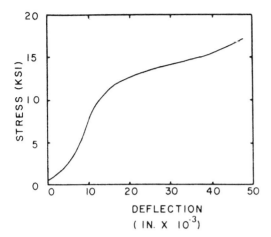

Figure 19.1. The force-deflection behavior of gaskets is usually nonlinear, making the spring constant of the gasket a function of applied force. Data shown are for asbestos-filled, stainless steel, spiral-wound gasket, $12\frac{13}{16}$ in. OD, $11\frac{9}{16}$ in. ID, and 0.175 in. thick. (Modified from Ref. 3.)

3. The gasket is not a fully elastic spring. It exhibits a great deal of hysteresis and will take a permanent set. Sometimes some of this "permanent" deformation will disappear over a period of time, so that the behavior of a particular type of gasket may be very different in a dynamic load situation than in a static situation [6]. A typical hysteresis curve is shown in Fig. 19.2 [7].

4. The temperature coefficient of expansion or contraction of the gasket is always substantially different from that of the flange members or bolt, so the behavior is always affected by severe temperature changes.

5. In some types of joint configuration, for example a raised-face flange, stiffness of joint members is going to be determined primarily by their stiffness in rotation rather than their compressive stiffness. There are computer programs for estimating rotation and rotational stiffness [10] and the data plotted in Figs. 5.19 and 5.20 give us a new rule of thumb for estimating overall joint stiffness. Rotation, however, complicates our efforts to define joint member/gasket stiffness ratios.

If we take a spring of the sort described and introduce it to a joint diagram, we will see something like that suggested in Fig. 19.3. When the liquid or gas within the system is pressurized, a working or external load

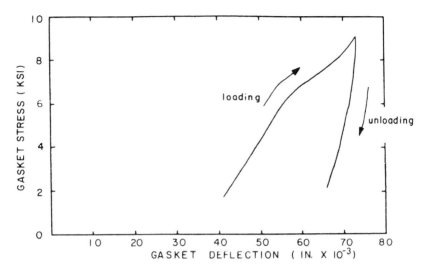

Figure 19.2 Load-deformation curve for a spiral-wound gasket, made of 304 stainless steel with asbestos filler, as it is loaded initially and then unloaded in use. In use the gasket operates as a very rigid, nearly elastic body. Gasket dimensions: OD = 5.75 in.; ID = 4.75 in.; thickness = 0.177 in. Reloading, in use, will occur essentially along the "unloading" curve above. (Modified from Ref. 7.)

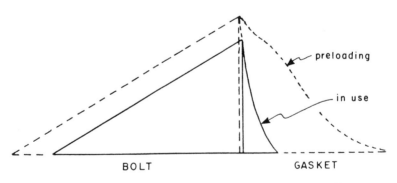

Figure 19.3 Joint diagram for a gasketed joint, showing the effects of nonlinearity and hysteresis of the gasket. The dashed lines show the initial, preloaded state; the solid lines show the possible working state, after some initial relaxation of the joint.

is placed on the joint. The exact influence such a load will have on the members of the joint, including the gasket, depends on the configuration of the joint members—which determine the so-called loading plane as we saw in Chap. 12. In the worse-case situation the loading plane will coincide with the upper and lower surfaces of the flange members, so we will assume that the load is placed there.

Under the influence of this load, the bolt increases in length somewhat and the gasket is partially unloaded. This is the situation we encounter when trying to make predictions, using such a joint diagram. We must know the stiffness of the gasket/joint system as it is unloaded and re-loaded—the slope of the "in use" portion of the curve shown in Fig. 19.3. If a relatively stiff flange is involved, the stiffness of the joint system will be dominated by the lesser stiffness of the gasket. Until recently that didn't help much—very little gasket stiffness information was available. Now, however, data are emerging from studies sponsored by the Pressure Vessel Research Committee of the Welding Research Council. You'll find a compilation of the available data in Table 5.1. As just mentioned, the gasket-driven stiffness of standard raised-face flanges is plotted in Figs. 5.19 and 5.20.

The stress-deflection curves shown in Figs. 19.1 and 19.2 define the behavior of a spiral-wound gasket; not the behavior of gaskets in general. Other curves are possible, as shown in Fig. 19.4, with that shown for flexible graphite (the dotted curve in Fig. 19.4) perhaps being a more "typical" curve than the others.

B. Creep Relaxation of Gaskets

Another mechanical characteristic of gaskets which has an important effect on joint behavior is creep relaxation. As bolt preload is being developed, during assembly of a joint, the gasket is compressed—becomes thinner—the change in its thickness depending upon its "compressibility" and upon the amount of bolt load created during assembly. When the compressive stress is maintained for a long time the gasket slowly but surely becomes still thinner (and slightly wider), allowing the bolts clamping the joint to relax. This is reflected in Fig. 19.3. The dashed line shows the shape of the joint diagram when the gasket is first loaded during assembly; the solid line shows the shape of the joint diagram after creep relaxation and internal pressure have taken their toll. The odd shape of the dashed line in Fig. 19.3 is merely the curve of Fig. 19.2 reversed to accommodate normal joint diagram usage.

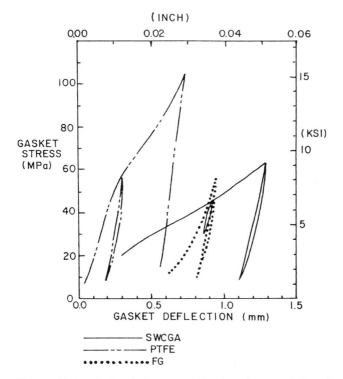

Figure 19.4 Mechanical stress-deflection characteristics of spiral-wound, chlorite-graphite-asbestos-filled (SWCGA), sheet PTFE, and flexible graphite (FG) sheet gaskets.

Definitions

Gasket creep is measured and described in several different ways in the literature. Here's a brief glossary of the terms used and their meaning [17, 40, 47, 48, 51].

Pure Creep

Pure creep is the loss of thickness of the gasket under a *constant* compressive stress load. This loss can usually be measured only in a test rig, since the clamping force exerted by the bolts in a conventional flange will decrease as the gasket creeps. This type of creep is usually reported as a percentage change in the thickness of the preloaded gasket. That is usually not equivalent to the loss in stress on the gasket.

Pure Relaxation

Pure relaxation is the loss of compressive stress on a gasket loaded under constant deflection conditions (e.g., by using it in a flange which has metal-to-metal contact). It would take very sophisticated equipment to measure this. Relaxation is reported as the percentage change in the compressive stress on the gasket versus the initial, preloaded stress [17].

Creep Relaxation

Creep relaxation is a combination of the above. The gasket, for example, is loaded by tightening bolts. When the gasket creeps, becomes thinner, the bolts can relax and so there's a simultaneous and corresponding loss of bolt elongation, bolt tension, clamping force, and, therefore, stress on the gasket. Since the torque required to restart the bolts is a function of bolt elongation or tension, there's also a "torque loss" and creep-relaxation data are often reported in those terms. The ASTM F 38 B and DIN 52913 creep tests are really creep-relaxation tests, but many of the PVRC tests are not. They're conducted under constant stress or constant deflection because that's easier to arrange with their automated, computer-controlled test rigs. But they've also sponsored tests under bolt load.

Here's an example of creep relaxation as measured by the change in length of the bolts. Let's assume that preloading the bolts created an average elongation in them of 5×10^{-3}. The gasket has been compressed during assembly, but we usually won't know or care by how much. After a few hours, however, we find that the average elongation of the bolts is now only 4×10^{-3}. The loss of 1 mil represents creep relaxation of 20%. The same loss of 1 mil has also occurred in the gasket, but will probably be a smaller percentage of this compressed thickness. Since we're primarily interested in the change in compressive stress of the gasket, the percentage change in bolt load is of more interest to us than the percentage change in gasket thickness. The ASTM Standard F 38B test is reported in terms of the bolt change, so the C-R creep specifications reported in Table 19.1 reflect the true loss of stress on the gasket during that standard test.

Because creep relaxation occurs under a steadily reduced compressive load on the gasket it should typically be less than the pure creep of that same gasket if the preload stress were maintained constant during the test. As one reported example, a $\frac{1}{16}$-in. compressed asbestos gasket lost 6% during a pure creep test of 300 hr at 200°F (93°C) but lost only 3% when loaded by bolts under the same conditions [52].

Table 19.1 Gasket Creep-Relaxation Data

Type of gasket	Thickness (in.)	Temp. (°F)	Type of test[a]	Time (hr)	Loss[b] (%)	Ref.
Compr. asbestos	NA	75	C-R	24	21	37
	c	75	c	1000	28	c
		750		24	22	
		750		1000	35	
Compr. aramid fiber	NA	75	C-R	24	22	37
		75		1000	24	
		750		24	80	
		750		1000	94	
Flexible graphite	NA	75	C-R	24	11	37
		75		1000	12	
		750		24	3	
		750		1000	37	
Cork plus rubber	0.062	200	C-R	18	82	52
				200	90	
Cork	0.062	200	C-R	18	68	52
				200	82	
Compr. asbestos	0.062	200	C-R	18	28	52
				200	42	
Compr. asbestos	0.062	200	C	300	6	52
Spiral-wound, asbestos-filled with outer gage ring	0.175	70	C	1	0.5–8	17
			Cy-C	1	5–64	
			R	1	14–23	
Spiral-wound, asbestos-filled, no gage ring	0.125	70	C	1	3–19	17
			Cy-C	1	7–67	
			R	1	17–23	
Double-jacketed, asbestos-filled	0.125	70	C	1	1.2–2.4	17
			Cy-C	1	2–2.5	
			R	1	6–10	
Compr. asbestos	0.125	70	C	1	3–8	17
			Cy-C	1	4–10	
			R	1	5–17	
Spiral-wound, mica-graphite–filled, with outer gage ring	0.125	800	C	1.5	20	57

[a] C = pure creep: Cy-C = cyclic creep: R = relaxation: C-R = creep-relaxation. Creep and cyclic creep = ΔD_G/initial D_G; relaxation = ΔS_G/assy. S_G; creep-relaxation = $\Delta(\Delta L_B)$/assy. ΔL_B, where D_G = gasket deflection, S_G = gasket stress, L_B = effective length of the bolt.
[b] Loss of clamping force or one of its equivalents (such as bolt stretch, bolt tension, or gasket thickness).
[c] A blank means "same as above."

Cyclic Creep

The PVRC found that many gaskets will creep more under cyclic loads than under static loads, so they designed and sometimes report cyclic tests under pure creep (constant stress) conditions [17].

Some Factors Affecting Creep

Gasket creep is unavoidable because a good gasket material must have some plasticity to allow it to mate intimately with all the imperfections of the flange surfaces. The amount of creep will depend on the materials of which it's made, and its construction. It will also depend on a number of other factors:

Initial Thickness

Creep relaxation is proportional to the thickness of the gasket. A $\frac{1}{4}$-in.-thick compressed aramid fiber gasket will relax about four times as much as a $\frac{1}{16}$-in.-thick gasket made of the same material. This is why gasket manufacturers tell us to use a gasket which is "as thick as it needs to be but as thin as possible." The flange condition may force us to use thicker gaskets than we'd like to. Watch out for the thickness effect: most gasket manufacturers base published creep data on tests of single thickness; usually $\frac{1}{32}$ in. because that's the default thickness for the ASTM F 38 B test. You can get into trouble if you apply that data as-is to a thicker gasket.

Time

Most creep relaxation occurs in the first 15–20 minutes after preloading a gasket, or after 20–25 load cycles [17], but can and probably will continue at measured rates for several hours, as suggested by Fig. 19.5 [46]. Further creep will occur after the gasket's temperature is raised, but, again, this will usually be a short-time affair. Gaskets will continue to creep slowly, perhaps forever, but the additional change in deflection usually isn't enough to cause problems. Because they stabilize after a few hours in service, for all practical purposes, many gasket manufacturers recommend retightening the bolts after the first 18–24 hr to recover clamping force lost during this time. The stabilized gasket won't shed the same percentage of clamping force in the next 18 hr. Retightening, however, often requires "hot bolting," which is a dangerous practice.

One group of investigators who have studied long-term creep effects say that creep rates become so slow after 18 hr as to be negligible. As an example they say that a compressed asbestos gasket which shed 37% of its bolt load in 18 hr would require 100 years to shed the rest [52].

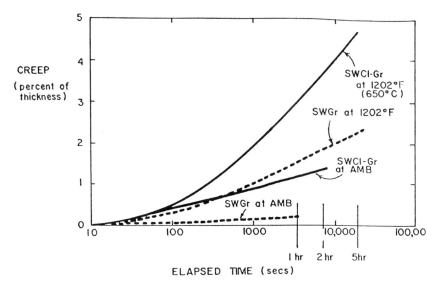

Figure 19.5 Creep of chloride-graphite and asbestos-filled spiral-wound gaskets at room temperature and at 1202°F (650°C). This is a linear-log plot. The asbestos-filled gasket exhibits less creep at all times and at both temperatures. (Adopted from Ref. 46.)

Temperature

Temperature is a major factor. High temperature can increase the amount of creep by a factor of 10 or more, as shown by some of the data in Table 19.3. The creep of an aramid-elastomer gasket increases significantly with the time of exposure at elevated temperatures; with compressed asbestos, however, the effect is relatively small [33]. In fact, one of the main attractions of compressed asbestos is its ability to function at temperatures of 750°F (399°C) or more. We'll take a closer look at elevated-temperature creep later. But be warned: most of the creep data published by gasket manufacturers are based on room-temperature tests.

Applied Loads

The amount of initial stress, preload, placed on the gasket affects its creep rate. Higher loads tend to produce higher amounts of creep; but sometimes, paradoxically, they can also result in lower percentage losses [17].

Table 19.2 Mechanical Properties of Sheet Gaskets

Type of gasket	Creep-relax[a] (%)	Tens. str.[b] (psi)	Compress. (%)	Recovery (%)	$P \times T$ × 1000	Max P (psi)	Max T (°F)
Compr. asbestos	16–30	2600–5000	7–17	40–50	350	1200–3700	650–1100
Compr. aramid fiber	18–25	2200–2800	7–17	40–50	350	900–1200	700
Compr. carbon fiber	14–15	1500–1800	7–17	50	700	2000	900
Pure PTFE	10–55	1625–2000	6–55	20–40	50–350	100	500
PTFE/st. st. core	20	5000	4–9	45–50	NA	2500	500
Rubber and synthetic rubbers	NA	1000–2600	25–75	NA	15–20	100–150	250–350
Flexible graphite	<5–5	900	40	20	700	2000	932–5432[c]
Inorganic-filled PTFE	30	2000	8–16	40	NA	1200	500
Vegetable fiber sheet	NA	1000–2000	25–55	40	40	200	212–250

Reported range of properties

[a] At room temperature.
[b] Across the grain (i.e., weak direction).
[c] Highest in a neutral or reducing atmosphere; lowest in an oxidizing atmosphere.
Source: All data taken from manufacturers' catalogs: Refs. 40, 47, 48, 49, and 59.

Bolt Stiffness

The stiffness of a bolt won't affect a gasket's creep-relaxation properties, but it will affect the relationship between the loss of thickness of the gasket and the simultaneous loss of clamping force or compressive stress on the gasket. Long, thin bolts will allow less loss of clamping force, for a given change in gasket thickness, than will short, stubby bolts. This is one of the reasons why collars or stacks of Belleville springs are so often found on bolts in process industry joints.

Creep can be a major problem. It can reduce clamping force to the point where a leak opens up, or, even worse, reduce clamping force to the point where the gasket will be blown out of the joint if its tensile strength alone is not sufficient to resist the contained pressure. For this reason an asbestos substitute gasket which relaxes twice as much as a compressed asbestos gasket is considered four times poorer in performance capability [30]. Gasket leakage is bad news and gasket blowout is a danger to be avoided! So we want to know by how much a proposed gasket will creep in our application.

Gasket Creep at Room Temperature

Some years ago the PVRC investigators studied the room-temperature creep behavior of several types of gasket, because trustworthy data were lacking, especially for fabricated gaskets such as double-jacketed or spiral-wound types. In fact, manufacturers only report creep data for sheet materials such as compressed asbestos, compressed aramid fiber, etc.—the materials included in Table 19.3. The PVRC was also encouraged to study creep because engineers often blame it for many of their pressure vessel and piping problems. The initial PVRC testes involved four types of gasket [16, 17].

Spiral wound, asbestos filled, with outer compression ring
Spiral wound, asbestos filled, without compression ring
Stainless steel, double jacketed, asbestos filled
Compressed asbestos with rubber binding

Three types of test were made:

Creep under constant gasket stress (maintained by hydraulically loaded platen)
Creep under cyclic stress
Gasket stress relaxation at constant gasket deflection

None of these tests imitate the true, bolted joint situation, which would involve simultaneous creep and relaxation with neither constant stress nor constant deflection. The tests can be used, however, to estimate the amount of creep and/or relaxation which might be encountered in practice. Here are some of the results of these tests [17]. (Incidentally, these were all room-temperature tests. We'll look at elevated-temperature creep in Sec. III.)

1. Seventy-five percent of the creep occurred in the first 10–15 min after stress was applied to the gasket.
2. Relaxation under constant stress continued for a longer period of time, with perhaps half of it taking place in the first 10–15 min.
3. The cyclic stress tests involved continuous and fairly rapid changes in stress (100 psi/sec or 0.7 MPa/sec). Under these conditions, most of the creep occurred within 20–25 cycles.
4. Constant and cyclic stress creep tests were conducted at a variety of maximum stress levels. The amount of creep which occurred was found to decrease as initial preloads were increased. For example, the deflection of the spiral-wound gasket without a gage ring increased 1.85 mils (0.047 mm) in 1 hr under a full-surface-area stress of 20 ksi; but increased by 3.46 mils (0.088 mm) under a stress of only 5 ksi (35 MPa). The latter stress, incidentally, corresponds to the seating stress (y factor) recommended for such a gasket by the ASME Boiler and Pressure Vessel Code, as we'll see in a later section. As a percentage, deflection increased 2.8% under 20 ksi; 19.1% under 5 ksi.
5. The outer gage ring reduced the amount of creep in a spiral-wound gasket, the increase in deflection being 0.5% and 9.7% at 20 ksi and 5 ksi, respectively.
6. Relaxation under constant deflection followed a different pattern, with relaxation increasing as initial stress increased—at least for spiral-wound and double-jacketed gaskets. Relaxation of the compressed asbestos gasket was independent of initial stress.
7. Leakage rates tended to decrease with time, i.e., as creep occurred. Apparently creep helps the gasket mate with flange surfaces.
8. A general conclusion was that it took the gaskets tested about 10 min to "stabilize." This fact was used to design subsequent leak tests; it could help people in the field analyze the assembly behavior of their bolted joints.

Relatively modest amounts of creep were observed during these tests. Typical losses might correspond to 1–5% of initial preloads, rising to 15–20% for lower stress levels. Bolting engineers in the field, however, have often encountered much greater "torque loss" than this. Why?

There are three possible explanations. In some cases the relaxation or loss of initial preload is caused by the elastic interactions discussed in

Chap. 6 rather than by (or in addition to) creep. Remember that interactions could leave some bolts with only a fraction of their initial preloads.

A second possibility: the gaskets being used may not have been of the types tested by the PVRC. Gaskets vary drastically in their creep behavior. One investigators found that creep of a compressed asbestos gasket resulted in a 28% loss of clamping force in the joint after 1000 hr at room temperature. The loss with a compressed organic fiber gasket was 24%; the loss with a flexible graphite gasket was only 12%.

When the same gaskets were tested at 750°F, however, the compressed asbestos and flexible graphite gaskets each allowed about 35% relaxation; but the organic fiber gasket allowed 94%—essentially total loss of clamping force [37]. Which brings us to our third possibility: users were troubled by creep at elevated, not room, temperatures. More about elevated-temperature creep in a moment.

Room-temperature creep relaxation data for sheet gaskets have become more available since the first edition of this book appeared (or at least I've become more aware of it). Most of the published data are based on tests made in accordance with ASTM Standard F38. Some typical numbers are given in Tables 19.1 and 19.2. Creep data for fabricated gaskets are still not available.

C. Other Mechanical Properties of Gaskets

Tensile Strength and Blowout Resistance

Another mechanical property we're interested in is the tensile strength of the gasket. A gasket with zero tensile strength could theoretically be held in the joint if the clamping force on the gasket were high enough, but this much clamp is rarely possible—or desirable (see the discussion of crushing strength below). As a result the gasket must count, at least in part, on its tensile strength to keep it from being torn apart and blown out of the joint.

Most gaskets have relatively modest tensile strengths, as shown by the data in Table 19.2. The strength, furthermore, is often different "with the grain" and "across the grain," at least in mass-produced, composite sheet materials such as compressed asbestos or compressed aramid fiber. The data published by reputable manufacturers are always for the worst case, across the grain. The tensile strengths of fabricated (spiral wound, double jacketed, etc.) or solid metal gaskets are presumably considerably higher than those of the sheet materials reported in Table 19.2 but are not reported by gasket manufacturers. Because of their strength, solid metal gaskets are often preferred when contained, in-service pressures exceed 2500 psi (17.2 MPa).

Whether or not a gasket will be blown out of the joint depends not only on its tensile strength and the clamping force exerted on it by the bolts, but also on service temperatures and pressures. The dependence on contained pressure is obvious; more pressure will create higher tensile stress in the gasket. Temperature is a factor because the tensile strength of a gasket is a function of temperature; higher temperatures mean less strength. Most of the tensile strength data currently published are based on the room-temperature ASTM F 152 test; but the AHOTT, EHOTT, and other PVRC tests described in Sec. V of this chapter will, if adopted by the ASTM or by gasket manufacturers, change that.

Meanwhile, blowout resistance of a given material is measured by clamping a test gasket between a pair of flanges and then subjecting it to elevated pressure and temperature. Typically the gasket and flange are heated and the contained pressure is then raised until blowout occurs. After many crush tests a combined "$P \times T$" rating can be assigned to the gasket. Let's take a quick look at this important rating factor.

The $P \times T$ Factor

For every gasket material there's a maximum service temperature above which it cannot be used, safely. Each also has a limiting service pressure. For example, a proposed material may have a temperature limit of 300°F and a contained pressure limit of 1000 psi. It could not be used at 300°F when the contained pressure is 1000 psi, however; no gasket material can stand maximum temperature and pressure simultaneously. So the material will also be assigned a $P \times T$ factor: a product of pressure and temperature. This $P \times T$ factor can be used to find any acceptable combination of maximum temperature and maximum pressure. For example, let's say that the $P \times T$ value for the hypothetical material being considered is 250,000. If the service pressure in a proposed application is 1000 psi, then the maximum acceptable service temperature for that application will be 250,000 divided by 1000 or 250°F. If the service temperature will be 350°F, then the maximum acceptable service pressure will be 250,000/350 or 714 psi. Similar calculations can be made for any other pressure or temperature.

Crushing Strength

As we'll learn later in this chapter, a gasket will leak less if the clamping force on the joint is increased—at least up to a point. If the clamp is too severe, the gasket will be crushed, may actually break up. This strength, like so many other gasket properties, is a function of temperature and of gasket thickness—and is not usually reported by gasket manufacturers.

Results of one independent test give the room-temperature crushing strength of 1.5 mm (60 × 10^{-3} in.) thick compressed asbestos (CA) as 15,950 psi (110 MPa) and that of 3-mm-thick CA material as only 8700 psi (60 MPa). The crushing strength of the same gaskets at 508°F (300°C) was 10,440 psi (72 MPa) and 5510 psi (38 MPa), respectively.

The same investigator also tested the crushing strength of compressed synthetic fiber (polyaramid) gaskets of the same thicknesses and at the same temperatures. The room-temperature crushing strengths of these gaskets were comparable to those of the compressed asbestos, but their crushing strengths at 508°F were only 6,525 psi (45 MPa) for the 1.5-mm material and 4350 psi (30 MPa) for the 3-mm material [41].

Another reference reports that the crush resistance of some "high performance" fibrous materials may be as much as 100,000 psi (689 MPa) or more, but doesn't name the materials [48].

I remember being told that the crushing strength of a spiral-wound asbestos-filled gasket was "above 35,000 psi (240 MPa)," but you had better check that with a manufacturer before accepting it. In fact, one current reference says that the maximum safe seating stress for a spiral-wound, asbestos-filled gasket is 25,000 psi; that of a PTFE-filled spiral-wound gasket is only 13,000 psi; and that for a grafoil-filled one is 20,000 psi [49].

Compressibility and Recovery

The last mechanical properties we're going to look at are compressibility and recovery. Compressibility defines the amount by which a gasket will change thickness—be compressed—when first preloaded. It tells us something about the stiffness of a gasket, but is usually not of prime interest to us. Recovery, however, is more important. It's a measure of the "spring-back" of a gasket, how much thickness it recovers when the loads on it are reduced or removed. This is compared to the total compression of the gasket from the as-received thickness to the end-of-test-after-creep-relaxation thickness to compute a percentage recovery. This can be a major factor if we're dealing with cyclic loads. We don't want the compressive stress on the gasket to fall too far when the contained pressure is reduced, for example, or when a rainstorm temporarily cools the flange and reduces the thermal expansion of bolts and joint members. Manufacturers report both compressibility and recovery of their products, but—as usual—the data are based on room-temperature tests of a given thickness of material, often $\frac{1}{32}$ or $\frac{1}{16}$ in. Some manufacturer's data are given in Table 19.2.

Gasket Stress Not Uniform

The compressive stress seen by the gasket when we assemble the joint and when we put it in service is not a "mechanical property" of the gasket, but is certainly a mechanical factor which affects its behavior and is worth a brief discussion. As we'll see in a minute, the amount a gasket leaks is a function of the assembly and working stresses on it. Thanks to the work of the PVRC and others were now able, in many situations, to predict the amount of leakage, based on a prediction of stress. But we have a problem: those stresses are not uniform. They vary for a number of reasons, as follows.

Nonuniform bolt load: Scatter in the torque-preload relationship, elastic interactions, embedment, and all of the other factors we've discussed will, of course, create some point-to-point variations in the loads created in the gasket by individual bolts.
Flange rotation: Raised-face, pressure vessel flanges will, if thin enough, and if the bolt forces are high enough, pull together slightly along their outer diameters. This partially unloads the inner diameter of the gasket, creating a radial variation in stress.
Flange friction: The friction between mating surfaces can affect the stresses seen by both flanges and gasket, I assume because friction fights things like thermal expansion and flange rotation [44].
Gasket width: It has been reported recently that the contact pressure on wide, pressure vessel gaskets is less uniform than that on narrow ones [34]. Presumably this is because the bolts are farther from the inner diameter (ID) of the gasket than from the outer diameter (OD).
Gasket stiffness: The same reference reports that thinner gaskets are more affected by the contained pressure than are thicker gaskets. Thinner ones will be stiffer than thicker ones, so this makes good joint diagram sense.
Bending loads on the flange: External bending loads on the flanges affect both the amount and distribution of stress on the gasket. The PVRC committees were just starting an attempt to quantify this effect in 1993 [45].

One more gasket stress comment is in order. Because of the possibility of flange rotation the ASME Boiler and Pressure Vessel Code, which we'll take a look at a little later, assumes that only the outer half (approximately) of most gaskets is loaded or stressed during assembly or in service. Their *m* and *y* gasket factors (again to be discussed later) are used with this assumption in mind. All of the studies made in the last two decades by the PVRC, CETIM in France, the British Hydromechanics Research Agency in the UK, and others, however, report the results in terms of full area contact; so be careful if you try to compare PVRC results to *y* factors, for example.

Flange Surface Finish

Here's another mechanical factor which, though not a material property of the gasket, can and does affect the gasket's leak behavior. "Everyone knows" that damaged flange surfaces, radial gouges, etc. can cause leaks. Everyone also seems to have a theory or rule that equates surface roughness and/or pattern of tool marks with leakage behavior. Some say, for example, that only circular tools marks are acceptable; others insist that "phonograph" marks are also acceptable.

Acceptable surface roughness is also a hotly debated issue. Many years ago the PVRC sponsored roughness tests, checking the leakage of several type of gasket against surfaces whose roughness varied from 32 to 1000 μin [60]. The results suggested that, although roughness influenced leak rate, it was a minor factor compared to such things as gasket type, gasket stress, and contained pressure. (Solid metal gaskets were an exception: they were seriously affected by surface finish.) The results were challenged and so the tests were repeated, with the same results.

Recently, however, the subject has been raised again. There's a feeling that surface roughness will be important for meeting fugitive emission standards. So the PVRC Subcommittee on Gasket Testing and Elevated Temperature Behavior has agreed to sponsor further research [58]. In the meantime, gasket manufacturers generally publish recommendations regarding surface finish. Contact your suppliers for advice.

III. MECHANICAL BEHAVIOR AT ELEVATED TEMPERATURES

Elevated temperatures can affect the behavior of gasketed joints in several ways, most of them unpleasant. Thermal cycles can cause even more problems for us. Let's look at some of these effects.

A. Physical Degradation of the Gasket

High temperature can degrade a gasket in many ways, changing such things as tensile strength, resilience, modulus of elasticity, and the ability to support compressive loads. Figure 19.6 shows a general decrease in maximum-service stress for a variety of materials, for example [21]. Only

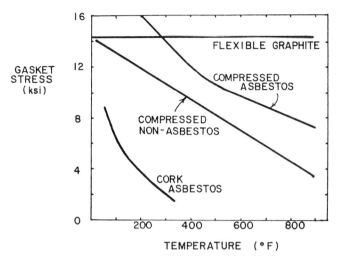

GASKET STRESS (ksi)

TEMPERATURE (°F)

Figure 19.6 Curves representing the maximum-service, pressure-service temperature points for four different types of gasket. If these pressure-temperature limits are exceeded, the gaskets would suffer irreversible damage.

flexible graphite remains unaffected by the temperatures shown there (but also see Fig. 19.7). Curves for the other materials represent the points at which the combination of stress and temperature plotted would create irreversible distortion and/or decomposition of the gasket.

Many of the proposed asbestos substitute materials fail by thermal degradation; the clays or other inorganics used as binders turn to powder when temperatures are too high.

Another form of physical degradation is the inward buckling of gaskets confined by outer gage rings. The gaskets try to expand as the temperature rises. The rings confine them. They are forced to buckle to accommodate their increased length.

B. Creep and Relaxation

Elevated temperatures can also increase the amount of creep or relaxation a gasket experiences under load. The PVRC reported room-temperature relaxation under light loads (5 ksi) of a compressed asbestos gasket to be about 17% [17]. Others have reported relaxation of 60% of similar gaskets under similar loads at 390°F (200°C) [21].

In another project, PVRC investigators studied room- and elevated-temperature relaxation for spiral-wound gaskets with asbestos and with a proprietary mica-graphite filler. Behavior was similar when the two gaskets were tested at room temperature. The asbestos-filled gasket relaxed much less than the mica-graphite–filled one at 800°F (427°C), however. Both, incidentally, deflected (crept) under high temperature until the flange surfaces were in contact with the gasket's outer gage ring. The gaskets then relaxed, with the gage ring taking more and more of the load exerted by the flanges. As one result, the flanges used with the mica-graphite gasket showed a heavy imprint of the ring at the end of the test [22].

Other PVRC creep results are shown in Table 19.3 [24].

Elevated-temperature creep data are surprisingly rare, considering the importance of creep to joint behavior, plus the fact that many gasketed joints must operate at high service temperatures. Some of the original PVRC test results are given in Table 19.3. Other data are given in Table 19.1 and still other are plotted in Fig. 19.7. The latter shows the results of many tests on one of the most popular "substitutes for compressed asbestos," flexible graphite. Although none was used here, most flexible graphite gaskets include a solid or perforated metal core of steel because flexible graphite has a lower tensile strength and lower flexural strength than compressed asbestos, and is, therefore, more difficult to handle if it's not reinforced with a laminated or solid sheet of some other material.

Flexible graphite's popularity is due in part to its ability to function at temperatures of 750°F (399°C) or more, just like compressed asbestos. As Fig. 19.7 shows, however, flexible graphite does creep rather substan-

Table 19.3 Gasket Creep at Room and Elevated Temperatures

Gasket[a]	Approximate decrease in gasket thickness in 2 hr (%)	
	At room temp.	At 1200°F
DJ, mica filled	0.4	4.0
DJ, asbestos filled	0.4	4.0
SW, chlor/graph. filled	2.0	2.5
SW, asbestos	1.3	3.5
SW, mica-graphite filled	3.0	2.0[b]
SW, mica filled	0.1	0.2
SW, graphite filled	0.1	1.8

[a] DJ = double jacketed; SW = spiral wound.
[b] This material "hardens" at elevated temperature.

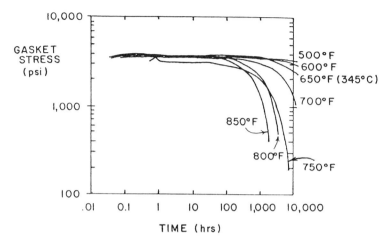

Figure 19.7 A log-log plot of gasket stress vs. time for gaskets made of flexible graphite with no appreciable amount of binder subjected to temperatures ranging from 500°F (260°C) to 850°F (454°C) in an oxidizing atmosphere. On the basis of these tests it was decided that this material could not be used for continuous exposure above 650°F (345°C).

tially, at least in an oxidizing atmosphere, after an initial delay, if the temperature is high enough [42]. PVRC investigators at École Polytechnique report that the creep relaxation of flexible graphite increases above 700°F (371°C) because it softens and the oxidation process accelerates [50]. Tests at École Polytechnique have also suggested, however, that flexible graphite's creep relaxation at 800°F (427°C), while greater than that of compressed asbestos, has surprisingly little effect on its sealing ability [31]. These tests, however, were made in a nonoxidizing atmosphere and were primarily relaxation tests; gasket deflection was held constant during most of the test. Many say, in fact, that flexible graphite gaskets should not be counted on for continuous service above 650°F (343°C) in an oxidizing atmosphere [37, 42].

C. Thermal Expansion or Contraction

One-shot differential thermal expansion or contraction can increase or decrease the loads on a gasket, as we saw in Chap. 13. This phenomenon, then, can raise or lower gasket stress along the unloading-reloading curves shown in Figs. 19.2 and 19.4. Sealability can be improved or degraded, depending on whether or not gasket stress is increased or decreased. Such changes, however, are relatively easy to estimate, by the techniques presented in Chap.13.

D. The Effect of Thermal Cycles

More complex and difficult-to-predict changes occur if repeated thermal cycles are involved. Two types of behavior have been reported, and are illustrated in Fig. 19.8 [23]. These observations were made during studies sponsored by the Task Group on Gasket Testing of the PVRC.

If the load (contact stress) on the gasket was cycled between constant max-min values, the deflection of the gasket "ratcheted," with the gasket becoming thinner with each cycle.

If the gasket was cycled between max and min deflections, the stresses on the gasket progressively decreased, as shown in the lower part of Fig. 19.8.

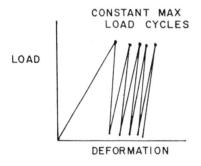

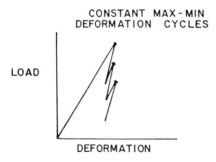

Figure 19.8 A gasket subjected to repeated load cycles, between given max and min values, becomes thinner and thinner, as shown in the upper plot. Repeated deformation cycles (lower plot) will progressively "ratchet" the contact stress on the gasket to lower and lower values, eventually opening a leak.

The gaskets were tested between hydraulically loaded platens in the tests just described. As a result, stress and deflection could be controlled independently. In an actual bolted joint the ratcheting which occurs is expected to be a combination of those shown in Fig. 19.8. A more realistic representation of the real-world situation might be as shown in Fig. 19.9. The relative amounts of stress ratcheting and deflection ratcheting will be determined by the configuration of the joint, the characteristics of the gasket, the thermal cycle involved, etc.

The ratcheting is caused by a combination of factors, including time-dependent plasticity (creep), time-independent plasticity (hysteresis), the general nonlinearity in the stress-stain response of the gasket, and the fact that these effects (as well as such things as gasket stiffness) change as the temperature changes. The attempt to analyze or predict ratcheting is complicated by the fact that most gaskets are not constructed of "a" material, but are combinations or systems of materials, and by the fact that present, commonly available analytical codes do not describe materials having this complex a behavior.

The PVRC has established a Task Group to deal with elevated-temperature behavior. Analytical tools are being developed but probably won't be available until the late 1990s.

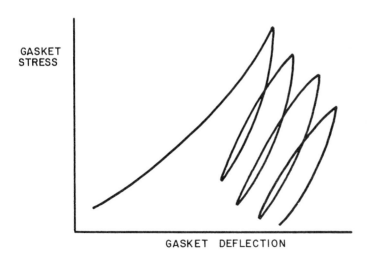

GASKET
STRESS

GASKET DEFLECTION

Figure 19.9 Most gaskets in field applications will experience a combination of stress ratcheting and deflection ratcheting, as suggested here, rather than one or the other, as suggested by Fig. 19.8.

IV. LEAKAGE BEHAVIOR OF A GASKET

A. The Classical View: *m* and *y* Factors

The leakage behavior of pressure vessel gaskets is defined, by most engineers, in terms of two "gasket factors" found in Section VIII of the ASME Boiler and Pressure Vessel Code. These are called the *m* or maintenance factor and the *y* or seating stress factor. They were and are intended by Code authors to be used solely for the design of flanged connections; they are not supposed to be used to predict or explain leakage, nor to define assembly bolt loads. They are, however, seen by most engineers as at least a clue to the relationship between assembly or working stress on the gasket and "leak-free" behavior. If I offer you a new type of gasket, perhaps a proposed substitute for compressed asbestos (CA), and give you data on such things as creep rate, tensile strength, $P \times T$ factor, compressibility, recovery, etc., it may help you compare that gasket to asbestos; but it won't give you a clue as to whether this new material will require more or less seating stress—more or less torque—than does CA. You'll get a feeling for that only if I also give you suggested *m* and *y* factors for the new material. So it's inevitable that most users include *m* and *y* in their deliberations when they try to analyze a leakage problem or pick an assembly torque.

Research in recent years has shown that the classical gasket factors give a very simplistic view of the relationship between gasket stress and joint behavior. New gasket factors have been proposed and work is currently under way to incorporate these in the code. All of this will be discussed at length in Sec. V. Before looking at the new factors, however, let's take a closer look at *m* and *y*, to see why new factors are necessary.

The *y* Factor

The *y* factor is the initial gasket stress or surface pressure required to preload or seat the gasket to prevent leaks in the joint as the system is pressurized. Some typical published values are given in Table 19.4.

The *m* Factor

As already mentioned, when the system is pressurized, the contact pressure on the gasket is reduced to some residual value, depending on the elastoplastic behavior of the gasket and its relationship to the elasticity of the joint. Experiments show that the liquid or gaseous pressure a joint will contain is proportional to the amount of residual contact pressure exerted by the joint surfaces on the gasket—and that the contact pressure

Table 19.4 Typical Gasket Factors from ASME Code

Type of gasket	Maintenance factor, m	Min. seating stress, y (psi)
$\frac{1}{8}$-in. asbestos with suitable binder for operating conditions	2.00	1,600
Spiral-wound metal, asbestos filled	2.50–3.0	10,000
Solid, flat, aluminum (soft)	3.25	8,800
Elastomers with cotton fabric insertion	1.25	400

Source: From ASME Boiler and Pressure Vessel Code, Section VIII, Division 1, Appendix II, Table UA-49.1, p. 317, 1977.

on the gasket must usually be larger than the pressure being contained. (Some gasket materials will stick to the joint; now contact pressure can be less than contained pressure and still seal the joint.)

The ratio of contact pressure to contained pressure is called the m factor, and may be different for different types of gaskets, as suggested by Table 19.4.

Problems with the m and y Factors

Note that the y and m factors, respectively, define the amount of assembly stress which must be placed on the gasket *and* the amount of residual stress that must be present to prevent a leak after the system has been pressurized. Those not familiar with gasket behavior might think that only the second really matters, but decades of experience and experiment have shown clearly that both of these gasket stresses—initial seating stress and in-service stress—are equally important. Failure to achieve a given seating stress, for example, can lead to a leak even if a theoretically correct maintenance stress is present. This fact has puzzled many people. As we'll see in Sec. B, however, an explanation is now available.

I have said that the m and y gasket factors are experimentally determined. Those published in the ASME code, however, were originally presented as "suggestions" and the experiments from which they were derived were never published. Many attempts have been made to confirm or "improve" them with often conflicting results [3]. Here are some of the claims made about m and y factors.

Most people believe that the y factor for a thick gasket should be greater than the y factor for a thinner sheet of the same material, yet the Code y for $\frac{1}{8}$-in. compressed asbestos is smaller than the Code y for thinner CA.

One investigator found that while both *m* and *y* should be increased with an increase of thickness of a synthetic fiber gasket, thickness didn't affect *m* and *y* for compressed asbestos very much [41]. The same investigator also concluded that *m* and *y* depend on whether the contained fluid will be a liquid or a gas.

There are many other examples of the challenges laid on *m* and *y*, but those will give you an idea. To see why these challenges are valid we're going to take a look, next, at the early leakage tests run by PVRC sponsored engineers. Then, in Sect. V, we'll see how the problems are resolved.

B. Early PVRC Gasket Tests

The following discussion is based on the behavior of a spiral-wound, asbestos-filled gasket. We'll consider other kinds of gaskets later. As we'll see, the behavior of many types differs in detail from that of the spiral-wound gasket, but their basic behavior is the same. If we understand one, we will understand many, including automotive and other nonpressure vessel gaskets too, incidentally. We'll also look at some exceptions, gaskets whose behavior differs from this norm.

In the PVRC tests, data are obtained simultaneously on the mechanical and leak behavior of the gasket [7, 12]. A typical plot of the mechanical (force-deformation) behavior, expressed in terms of stress on the gasket and gasket deflection, is shown in Fig. 19.10.

In the PVRC tests, a gasket of a given type would be loaded and unloaded progressively. As it was first loaded, for example, the investigator would apply a partial seating stress to the gasket and would then pause to make a leak test at that point. He would then apply additional seating stress and make another leak test, etc.

After loading the gasket to some maximum seating stress, he would then partially unload the gasket and again check leak rate. The gasket would be almost fully unloaded and then reloaded to the initial assembly stress. By the end of a given test series the leak rate might have been checked at al the points shown by small circles in Fig. 19.10. Contained pressure was constant during a given test series.

In some tests, the investigator loaded and unloaded the gasket from several different initial stresses, as suggested in Fig. 19.11. The first unloading test series was run from initial stress A. After unloading and reloading cycles the seating stress was raised to point B, followed by further unloading and reloading tests. The leak test points are still defined by the small circles.

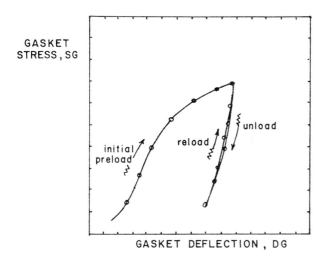

Figure 19.10 Typical stress-deflection curve for a spiral-wound gasket as it is loaded during assembly, unloaded when the system is pressurized, and reloaded when the bolts are retightened or the system is depressurized. During PVRC studies, leak tests were made at each point marked with a small circle. Contained pressure remained constant during all leak tests.

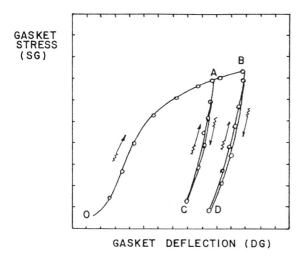

Figure 19.11 Similar to Fig. 19.10 but the gasket was first unloaded and reloaded between points A and C, then taken to stress B, and unloaded/reloaded between B and D.

As the tests progressed the investigators discovered that the leak rate of a gasket at any particular point was influenced more by the *deflection* of the gasket than by the stress on the gasket at that time. In fact, they could represent a given leak rate by a nearly vertical line superimposed on the gasket stress–gasket deflection curve, as suggested in Fig. 19.12.

If the current stress and deflection or the gasket was represented by any point to the right of the constant 10^{-3} ml/sec leak rate line shown in Fig. 19.12, then the actual leak rate at that point would be less than that represented by the constant leak rate line. Conversely, any combination of gasket stress and deflection which placed the gasket to the left of the constant rate line would mean a greater leak rate than represented by the line.

As the gasket in Fig. 19.12 is first loaded, for example, the leak rate will drop steadily until it reaches 10^{-3} ml/sec at an initial stress of 9.9 ksi. If loaded further, to 10.5 ksi, it will leak less than 10^{-3} ml/sec. As it is unloaded, the leak rate increases, finally reaching 10^{-3} ml/sec, again at a gasket stress of 6.5 ksi.

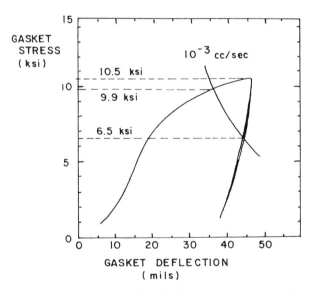

Figure 19.12 Same as Fig. 19.10, but with a constant leak rate line superimposed on the stress-deflection curve. The leak rate would drop to 10^{-3} ml/sec when initial seating stress reached 9.9 ksi. If further loaded to 10.5 ksi, it could then be unloaded to 6.5 ksi before the leak rate would climb to 10^{-3} ml/sec again. See text for details.

Note that more than one constant leak rate line can be defined and superimposed on our gasket curve, as suggested in Fig. 19.13. Lines toward the right end of the diagram represent lower leak rates than those to the left end of the diagram.

The combination of gasket behavior curve and constant leak rate curve, furthermore, can be used to shed light on the m and y gasket factors listed in the ASME Code. In Fig. 19.14, for example, let's assume that we've loaded the gasket to point A, representing an initial seating stress of 10 ksi. This assembly stress can be compared to the Code-recommended y factor for this type of gasket. In effect, it becomes "our" y-related factor for this assembly. The PVRC investigators, incidentally, have always computed gasket stress based on the full width of the gasket, not on the approximately half-width used in the ASME Code.

As the gasket is unloaded from this initial assembly point, the stress on it decreases and the deflection also decreases by a lesser amount. Let us assume that we can detect a leak—or are otherwise bothered by a leak—when the unloading curve intersects the 10^{-3} ml/sec constant leak rate line we have drawn on this figure. The intersection between this line and the unloading line will define a minimum acceptable in-service gasket stress and, therefore, can be compared to the minimum or maintenance stress defined by the code m factor. In Fig. 19.14 the minimum stress is

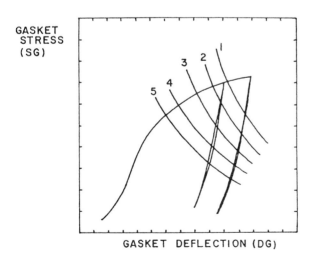

Figure 19.13 Same as Fig. 19.11 but with a family of constant leak rate lines superimposed on the stress-deflection curve. Line 1 represents the smallest leak rate; line 5 the greatest.

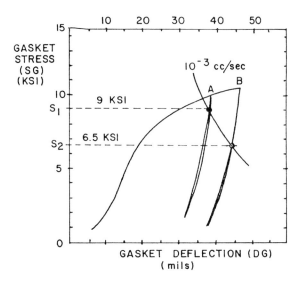

Figure 19.14 Curve set showing the relationship between initial seating stress and the maintenance stress required to keep leak rates below an acceptable maximum of (in this case) 10^{-3} ml/sec. If initially loaded to point A, this spiral-wound gasket could be unloaded only to 9 ksi. If loaded to point B, however, it could be unloaded to 6.5 ksi before the leak rate would reach 10^{-3} ml/sec.

9 ksi. If we assume that the contained pressure in this test was 1 ksi, then we have determined that "our" m-related factor (which I'll call m_1) is 9 divided by 1, or 9. Note that the Code m for a spiral-wound, asbestos-filled gasket is 3. Part of the difference here, but only part, can be accounted for by the fact that the Code m is based on effective rather than full gasket width.

We can now begin to see why previous gasket leak tests made by different investigators under different conditions often produced different results. Figure 19.14, for example, shows what would happen if we loaded the gasket to a slightly higher assembly or y-related stress of 10.5 ksi (point B). We could unload the gasket from this point all the way down to a stress of 6.5 ksi before intersecting the 10^{-3} ml/sec constant leak rate line, giving us an effective m-related factor (m_2) of only 6.5. This suggests that the gasket factors m and y are not unique constants for a given material, but depend on each other—and on that leak rate we have defined as "acceptable."

The tests also have shown repeatedly that a small increase in initial seating stress can result in a joint which remains tight even under significantly lower in-service stress. This is illustrated below. Increasing the initial seating stress on a gasket by about 23% lowered the maintenance factor (for a given contained pressure and acceptable leak rate) from over 21 to 5.9.

Initial seating stress	Corresponding maintenance factor
21.1 ksi	21.2
25.9	5.9

The implications of all this are significant for people who design or assemble bolted flanged joints: The current Code m and y gasket factors do not really define the characteristics or behavior of gaskets. As a result, they do not provide a firm foundation for either design or assembly decisions.

The PVRC investigators have sought, and appear to have found, an improved way to define the gasket, to improve the performance of Code-designed flanges. Their goal has been a design procedure which takes our new knowledge of gasket behavior into account, yet is similar in complexity to the present design procedure outlined. Since the actual behavior of the gasket is far more complex than that implied by m and y factors, this has been a real challenge.

As of this writing (mid-1994) a new design procedure has been defined and work is under way to incorporate it in the ASME Code. What follows is based on preliminary suggestions and findings emerging from the work of the PVRC investigators. It's "accurate" as far as it goes, but it may well appear in modified form by the time it enters the Code and becomes a legal document in most states. It should be used with some caution, therefore, and checked against more recent publications of the PVRC and/ or ASME before being applied to critical joints.

I am about to describe a new way of defining gasket characteristics and behavior, based on the concept of gasket "tightness," and on three new gasket constants, called G_b, a, and G_s, which, unlike m and y, ARE "constant" for a given type of gasket (at least at a given temperature and from a given manufacturer!). Let's start by looking at tightness.

C. The Tightness Parameter

The investigators discovered, by testing gaskets under a variety of conditions and with a variety of contained fluids, that test data could best be summarized by use of a dimensionless parameter called a "tightness parameter," defined as follows: [13, 18]

$$T_P = \frac{P}{P^*}\left[\frac{L^*_{RM}}{L_{rm}D_t}\right]^a \tag{19.2}$$

where T_P = the tightness parameter (dimensionless)
        P = the contained pressure (psi or kPa)
        P^* = reference atmospheric pressure (14.7 psi or 101.3 kPa)
        L_{rm} = mass leak rate (lb/hr·in., mg/sec·mm)
        L^*_{RM} = reference mask leak rate (0.008 lb/hr·in., 1 mg/sec·mm)
            which is keyed to a normalized reference gasket of 5.9
            in. (150 mm) outside diameter.
        a = experimentally determined exponent (e.g., 0.5 if the con-
            tained fluid is a gas, 1.0 if it's a liquid)
        D_t = gasket OD (in., mm)

A tightness parameter of 100, therefore, would mean that it takes a contained pressure of 100 atmospheres (1470 psi or 10.1 MPa) to create a
total leak rate of about 1 mg/sec from a gasket having a 6-in. (150-mm)
outer diameter.

Note that the tightness parameter is expressed in terms of *mass* leak
rate rather than volumetric leak rate. This allows the investigators to lump
together test data representing a variety of test fluids; some gaseous, some
liquid. Since mass leak rates are difficult for the uninitiated to visualize,
Table 19.5 lists some equivalents in terms of volumetric leak rates, and
Table 19.6 gives some bubble equivalents, for gaseous leaks [11].

Figure 19.15 is a log-log plot of gasket stress vs. tightness parameter
for a spiral-wound gasket with flexible graphite filler and stainless steel
windings [12]. A plot of this sort summarizes a large mass of test data
involving a variety of contained pressures, test fluids, gasket stresses,
and leak rates. Note that two different unloading loops are shown in the
figure. The test fluid was helium.

Table 19.5 Volumetric Equivalents for a
Mass Leak Rate of 1 mg/sec (0.008 lb/hr)

Contained fluid	Volumetric equivalents	
	ml/sec	pt/hr
Water	1×10^{-3}	0.008
Nitrogen	0.89	7.1
Helium	6.15	49

Table 19.6 Bubble Equivalents for Volumetric Leak Rates

Leak rate (ml/sec)	Volume equivalent	Bubble equivalent
10^{-1}	1 ml/10 sec	Steady stream
10^{-2}	1 ml/100 sec	10/sec
10^{-3}	3 ml/hr	1/sec
10^{-4}	1 ml/3 hr	1 in 10 sec

D. "Acceptable Tightness"

I've mentioned that there's no such thing as a perfectly tight joint, at least if the contained fluid is a light gas. Even if the system contains a liquid, "perfectly tight" can be expensive, requiring heavy flanges, lots of bolts, accurate preloads, etc. As a result, PVRC investigators felt it necessary to define "acceptable leak rates" in terms of tightness. Three levels of tightness were suggested, called economy, standard, and tight, as listed in Table 19.7. Economy might be used for a low-pressure water line in

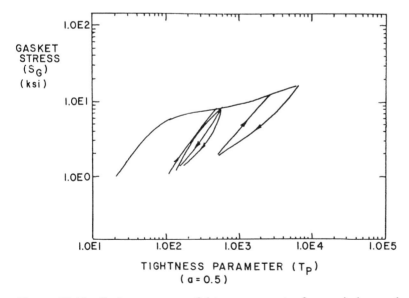

Figure 19.15 Gasket stress vs. tightness parameter for a spiral-wound gasket with flexible graphite filler. The dimensionless tightness parameter allows the PVRC investigators to summarize the results of leakage tests under different contained pressures.

Table 19.7 Tightness Criteria for Flange Design

Tightness classification	Mass leak rate per unit diameter (L_{rm})		Constant C	Tightness parameter pressure ratio (T_{pmin}/P_r)
	(mg/sec·mm)	(lbm/hr·in.)		
T1—economy	1/5	1/25	1/10	0.18257
T2—standard	1/500	1/2480	1	1.8257
T3—tight	1/50,000	1/248,000	10	18.257

the back lot. Standard would presumably be the most common choice. Tight might often be required to combat fugitive emissions.

At this point I believe that the ASME is planning to include only the standard tightness in revised flange design rules. They may, however, provide a new appendix to Section VIII of the Code, describing how the other levels of tightness can be used as alternates.

The Code committee is also, I've heard, planning to use a simplified form of the tightness equation. The full expression, Eq. (19.2), is useful for our purposes because it reveals the definition and meaning of the tightness parameter. The equation can be simplified, however, by combining terms and taking advantage of the various constants [46]. This gives us:

$$T_{Pmin} = 1.82574(C)P_r = 1.83(C)P_r \qquad (19.3)$$

or

$$T_{Pmin} = 0.124(C)P \qquad (19.4)$$

where T_{Pmin} = the dimensionless minimum tightness required to keep leakage equal to or less than a preselected value
   C = a constant defining an acceptable leak rate (see Table 19.7)
   P_r = the pressure ratio, e.g., P, psi/14.7 psi
   P = the service pressure, psi

E. The Effect of Elevated Temperature on Tightness

The elevated-temperature tests of the PVRC have revealed that the basic sealability of some gasket materials changes as the temperature of the gasket is increased. The gasket becomes far more sensitive to reductions in clamping force or compressive stress, as suggested by Fig. 19.16. The

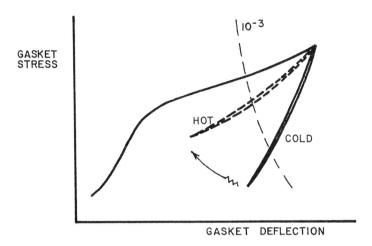

Figure 19.16 Elevated temperatures can make some gaskets more sensitive to a subsequent loss of gasket stress, as shown here.

amount by which gasket stress can be reduced (S_H in Fig. 19.17) before the leak rate exceeds an acceptable value has decreased drastically.

A spiral-wound, mica-graphite gasket exhibited this behavior. This gasket outperformed even a spiral-wound, asbestos-filled gasket at room temperature and other low temperatures, but the SWCA gasket is better at high temperatures [both were tested at temperatures of 800°F (427°C), for example] because its unloading curve is far more stable than that of the SWMG gasket [46].

Examination following the test revealed that the mica-graphite–filled gasket had partially buckled inward, reducing the contact area between gasket and flange at one point, and thereby affecting sealability by "rotating" the unloading portion of the gasket curve.

A different change in sealability has been noted during tests of aramid-fiber, asbestos-substitute gaskets. This time the slope of the unloading curve remained the same (or perhaps steepened a little) but the entire unloading surve moved to the left (toward reduced tightness, reduced sealability).

In these same tests, incidentally, elevated temperature created an initial *improvement* in sealability as increased gasket creep created a more perfect mating of gasket and flange surfaces. The subsequent reduction in sealability may have been caused by degradation of the binder during prolonged exposure to elevated temperatures. There is, in fact, a very definite time factor in all of these tests. A gasket which maintains a given sealability after several weeks at 650°F, for example, may exhibit that same sealability after only a few hours at 800°F. The loss of strength, of integrity, and of sealability takes less time at higher temperatures.

Studies continue on the high-temperature behavior of various gaskets, especially those which might be used as substitutes for compressed asbestos. This effort was discussed in Sec. III, where we looked at the elevated-temperature behavior of gaskets, so I won't repeat it here. Remember that flexible graphite, usually reinforced with a steel core of some sort, has emerged as the most likely replacement for CA, at least for service temperatures up to 650°F (343°C).

F. New Gasket Factors

Figure 19.17 is an idealized approximation of the sort of chart shown in Fig. 19.15, this time with linear regression lines replacing the actual loading and unloading curves. It is the present intention of the Subcommittee on Bolted Flanged Connections of the PVRC to summarize and report years of gasket test data with the aid of such plots, by reporting three new gasket factors which define such plots.

These factors are:

G_b, the intercept of the upper portion of the loading curve with the gasket
     stress axis
a, the slope of that same line
G_s, the intercept of all unloading lines with the gasket stress axis

The PVRC investigators originally labeled these parameters B, d, and S^*; and they were so reported in the first two editions of this book. The ASME Code committee requested different terms because B, d, and S had already been used for other flange design parameters in the existing Code. Hence the terms listed above.

A word of caution. Since these are log-log plots the gasket stress axis can be drawn through many different tightness parameter values. In Fig. 19.15, for example, it's located at a tightness parameter of 1.0E1. If you refer to the PVRC literature, you'll find that both 1.0E1 and 1.0E0 are frequently used.

The true values of G_b and G_s, however, are determined by the intercepts of the loading and unloading lines with $T_P = 1.0E0$, as in Fig. 19.17.

The three constants define the leak behavior of a gasket. We'll look at some specific values in Table 19.8. Note at this point, however, that the unloading lines all intersect the gasket stress axis in the vicinity of a common point, defined as G_s and called "the point of ambient tightness."

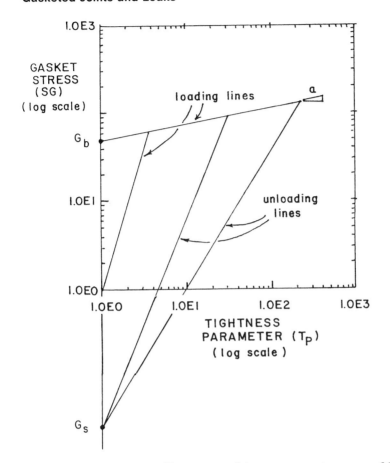

Figure 19.17 The "actual" stress vs. tightness parameter curve of Fig. 19.15 can be replaced by the set of linear regression lines shown here. The upper portion of the initial loading curve is described by a line having intercept G_b and slope a. Unloading and reloading lines have intercepts near G_s. These three constants, G_b, a, and G_s, can, therefore, be used to describe the behavior of a given type of gasket under a wide variety of conditions.

Take some time to study Fig. 19.17. The three new constants which define this chart—or its mathematical equivalent—are the proposed replacements for the classical gasket constants m and y. Like m and y, furthermore, the three new constants are intended primarily for the use of flange designers. They are idealized constants, the summation of many

Table 19.8 PVRC Gasket Constants G_b, a, and G_s

Gasket description	G_b (psi/MPa)	a	G_s (psi/MPa)
Beater fiber, premium, $\frac{1}{16}$ in. tk $\times$ $\frac{1}{2}$ in. $\times$ $5\frac{7}{8}$ in. OD, 250 m-in., fl pltns	900/6	0.45	0.3/0.00207
Compr asbestos, premium, $\frac{1}{16}$ in. tk $\times$ $\frac{3}{4}$ in., 8 in. OD, fl pltns	2500/17	0.15	117/0.8069
Compr abestos, premium, $\frac{1}{8}$ in. tk $\times$ $\frac{1}{2}$ in., $5\frac{7}{8}$ in. OD, 250 m-in., fl pltns	400/2.759	0.38	15/0.10345
Compr fiber, economy, $\frac{1}{8}$ in. tk $\times$ $\frac{1}{2}$ $\times$ $5\frac{7}{8}$ in. OD, 250 m-in., fl pltns	200/1	0.33	15/0.10345
Compr fiber, service, $\frac{1}{16}$ in. tk $\times$ $\frac{1}{2}$ in. $\times$ $5\frac{7}{8}$ in. OD, 250 m-in., fl pltns	1200/8	0.23	56/0.38621
Compr glass fiber, premium, $\frac{1}{32}$ in. tk $\times$ $\frac{1}{2}$ in. $\times$ $5\frac{7}{8}$ in. OD, 250 m-in., fl pltns	285/1	0.45	117/0.8069
Expanded PTFE, flat, .110 in. $\times$ $\frac{3}{8}$ in., $5\frac{7}{8}$ in., 255 m-in., fl pltns	250/1	0.25	1E-06/6.9E-09
Flt soft aluminum, $\frac{1}{16}$ in. tk $\times$ $\frac{3}{8}$ in., $5\frac{7}{8}$ in. OD, 125 m-in., fl pltns	1525/10	0.24	200/1.37931
Flt soft Cu, $\frac{1}{16}$ in. tk $\times$ $\frac{3}{8}$ in., $5\frac{7}{8}$ in., 125 m-in., fl pltns	5000/33	0.133	258/1.7793
Jktd Cu w/asb millbd, $\frac{1}{8}$ in., $5\frac{7}{8}$ in. $\times$ $4\frac{7}{8}$ in., 125 m-in., flat platens	1800/12	0.35	15/0.10345
Jktd low C stl w/mica fill, $\frac{1}{8}$ in. tk, $5\frac{7}{8}$ in. $\times$ $4\frac{7}{8}$ in., 125 m-in., fl pltns	2900/20	0.23	15/0.10345
Jktd st stl w/asb millbd, $\frac{1}{8}$ in. tk, $5\frac{7}{8}$ in. $\times$ $4\frac{7}{8}$ in., 125 m-in., fl pltns	2900/20	0.23	15/0.10345
Jktd low C stl, corr, pipe, w/mica fill, $\frac{1}{8}$ in. tk, 125 m-in., fl pltns	4200/28	0.25	200/1.37931
Lam flex grph w/chem bond to 316SS sheet (ASTM 6FM F2) $4\frac{1}{2}$ in. $\times$ $7\frac{1}{8}$ in. $\times$ $\frac{1}{16}$ in.	816/5	0.377	0.066/0.0005
Lam flex grph w/mech bond to 316SS sheet (ASTM 6FM F1) $4\frac{1}{2}$ in. $\times$ $7\frac{1}{8}$ in. $\times$ $\frac{1}{16}$ in.	450/3	0.45	0.009/6.2E-05
Lam flex grph w/316SS wireless (ASTM 6FM F2) $d\frac{1}{2}$ in. $\times$ $7\frac{1}{8}$ in. $\times$ $\frac{1}{16}$ in.	1700/11	0.26	15/0.10345
Nbin soft Cu, fl, $\frac{1}{8}$ in. $\times$ $\frac{1}{2}$ in. $\times$ $5\frac{7}{8}$ in., 125 m-in., w/one $\frac{1}{64}$ in. nubbin	2400/16	0.2	0.250/1.72414
Nbin L C stl, fl, $\frac{1}{8}$ in. $\times$ $\frac{1}{2}$ in. $\times$ $5\frac{7}{8}$ in., 125 m-in., w/one $\frac{1}{64}$ in. nubbin	12000/82	0.11	65/0.44828
SpWnd 6 in. CL600 SS304/Abs fill w/ext gage rings, 125 m-in. fl pln	3400/23	0.3	7/0.04828
SpWnd 4 in. CL600 SS304/mica-graph fill and ext gage ring	2600/17	0.23	15/0.10345

Table 19.8 Continued

Gasket description	G_b (psi/MPa)	a	G_s (psi/MPa)
SpWnd 4 in. CL600 SS304/graph fill and ext gage ring	2300/14	0.237	13/0.08966
SpWnd 4 in. CL150 SS304/PTFE fill and ext gage ring, 125 m-in. fl pln	4500/31	0.14	70/0.48276
Corrugated soft Cu, .02 in. $\times$ $\frac{1}{2}$ $\times$ $5\frac{7}{8}$ in., 125 m-in. fl pltns	1500/10	0.24	430/2.96552
Corrugated low C stl, .02 in. $\times$ $\frac{1}{2}$ in. $\times$ $5\frac{7}{8}$ in., 125 m-in. fl pltns	3000/20	0.16	115/0.7931
Corrugated 18Cr-SN8 stl, .02 in. $\times$ $\frac{1}{2}$ in. $\times$ $5\frac{7}{8}$ in., 125 m-in. fl pltns	4700/32	0.15	130/0.89655
Compr glass fiber $\frac{1}{16}$ in. thick	1150/8	0.3	117/0.8069
SpWnd, st. st. inner gage ring, PTFE filled	2250/16	0.19	67/0.4621
Flat metal jacketed with soft copper or brass insert	1800/12	0.35	15/0.10345
Flat metal jacketed with soft steel or iron insert	2900/20	0.23	15/0.10345
Corr metal jacketed with soft steel or iron insert	8500/59	0.134	230/1.5862
Expanded PTFE form in place chord 0.110 $\times$ $\frac{3}{8}$ in., $\frac{1}{8}$ in. thick	800/4	0.25	0.005/3.45E5

Source: Preliminary values prepared by James R. Payne of the PVRC Committee on Bolted Flanged Connections. From Ref. 46 and 62.

tests and much scatter in test results. They are not specifically intended for use by people trying to analyze an existing flange or trying to decide how much torque to apply to the flange bolts, but, like m and y, they are already being used for such purposes and will continue to be so used until the gasket world finds some other way to relate assembly preloads and/or working loads on the gasket to gasket sealability. But at least one engineer, who has been involved with the PVRC effort from almost the beginning, says that G_b, a, and G_s can't be used to estimate leak rate during operation; that we must use actual operating temperatures and pressures, and an actual tightness graph (such as Fig. 19.15) to do this [35]. It's still my opinion, however, that since most engineers won't have access to such graphs, the three new constants will, by default, be used for such purposes. And, at least, will give more insight than can now be gained by the use of m and y factors.

V. TESTING AND EVALUATING GASKETS

This book is written for those who use gaskets, not those who make them, so few of you will ever need to conduct the tests described below. Instead you will rely on tables of data provided by gasket suppliers, tables giving the results of tests which they have conducted. It is useful, however, for the users to have a general awareness of the types of test used to evaluate a gasket material, because these quantify the properties which the manufacturers believe will affect the behavior and serviceability of a gasket when it is put to work. So let's take a quick look at gasket tests.

A. ASTM and DIN Tests

The ASTM, DIN, and other agencies have long published a variety of standard gasket tests. Some of these test the gasket for its mechanical properties; such as tensile strength, compressibility under load, ability to recover its initial shape (primarily thickness) when the load is removed, and the way it will relax or creep if squeezed by a compressive load for a long time. Other tests evaluate the gasket's resistance to various chemicals, and, of course, its ability to seal a joint. The latter, usually called "leak tests," have normally been conducted with a convenient, safe, inexpensive, standard test fluid such as water or nitrogen. Here's a check list of a few of the more important (for our purposes) current test standards.

ASTM F 36—test for short-term compressibility (not creep) and recovery.
ASTM F 37 B—a room-temperature leakage test, using liquid or gas as the test fluid.
ASTM F 38 B—measures the loss of gasket thickness caused by creep, at room temperature, usually using a $\frac{1}{32}$-in. thick gasket. The gasket is loaded by a bolt, so this is a creep-relaxation test rather than a "pure" creep test as explained in Sec. III B.
ASTM F 146–tests the physical properties of the gasket after immersion in various fluids.
ASTM F 152–measures tensile strength of most, but not all (e.g., not rubber), gasket materials.
ASTM F 65586—standard method for leak rate vs. m and y gasket factors. Room-temperature test.
DIN 3535—a gas permeability test (i.e., leakage through the gasket).
DIN 52913—torque retention test. Gasket loaded by a bolt. The temperature is raised. Residual gasket compression measured after 16 or 100 hr.

B. Early PVRC Tests

As already mentioned, several groups have, in recent years, sponsored or conducted extensive research designed to identify improved methods of evaluating a gasket's behavior in service. The initial incentive for such work, at least in the United States, was to define better "gasket factors" for use by engineers who design pressure vessels, pipe flanges, and related equipment; and thereby improve our ability to design better flanged joints. For nearly a decade literally thousands of mechanical and leak tests were conducted on a variety of gasket materials in a largely successful attempt to learn why, how, and by how much a given gasket will leak under specific sets of service conditions. These tests were all conducted at room temperature in order to simplify a very complex research problem, and because the sought-for new gasket factors were to be used in the ASME Boiler and Pressure Vessel Code, whose flange design procedure generally ignores the effects of temperature on gasket behavior. The Pressure Vessel Research Committee (PVRC) in the United States was the principal sponsor for these tests, which were conducted in laboratories in Canada, France, and England, as well as here.

C. Elevated-Temperature PVRC Tests

Once the basic, room-temperature behavior of a gasket was thoroughly understood, tests at elevated temperature were initiated. The original purpose was the same as that of the room-temperature tests: to gain a better understanding of flange behavior, gasket deformation, and leakage at elevated temperatures, and thereby improve our ability to design successful joints. Almost immediately, however, another concern was added to the original definition of the high-temperature problem. The U.S. Environmental Protection Agency (EPA) announced, in July 1989, that the manufacture, importation, and processing of asbestos-containing gaskets would be prohibited as of August 25, 1993; and the distribution of them would be further prohibited as of August 25, 1994 [38]. Gaskets exposed to service temperatures of 750°F (400°C) or above, and those used in corrosive environments, were exempted from the prohibition. We'll consider some of the results of this later on in this chapter, but are concerned at this point only with its impact on gasket tests.

Before leaving the subject of the EPA ban, however, I should complete the story by saying that the search for suitable asbestos substitutes was largely unsuccessful; and the use of some of the proposed substitutes caused problems; including accident and injury. As a result, the EPA ban was overthrown in October 1991 [40]. It is now legal to make, sell, and use asbestos gaskets in the United States, as long as certain precautions are taken (e.g., gaskets must be wetted before and during removal). Elimination of the ban has not lessened the search for substitutes, however. A large number of lawsuits involving asbestos in other products make gasket manufacturers reluctant to continue to supply it. Other countries have or are planning to ban asbestos gaskets, furthermore. Italy banned them as of April 1994, for example [36]. So an aggressive search for substitutes is still going on.

One of the principal characteristics of asbestos is its ability to function under high temperatures. A truly satisfactory substitute, therefore, had to do the same; and so the elevated-temperature tests being sponsored by the PVRC and others became focused—for awhile at least—on tests of possible substitutes.

Most recently another concern has been raised—government regulators have announced that "fugitive emissions" must be controlled by the petrochemical and process industries. In the past leakage from bolted joints had been a matter for regulation only in the nuclear power industry. The new law has further increased the interest in better gasket tests. Since the plants creating the emissions are usually operating at elevated temperatures, these investigations in many ways dovetail with studies of elevated-temperature leakage in general, as well as with the search for asbestos substitutes.

D. The Search for Better Tests

Traditional ASTM and DIN tests can be quite time-consuming, but that's not a problem when you're using them for routine tests of production samples, or to test an occasional new material. Most ASTM tests are also conducted at room temperature. Because of the urgency of some of the new concerns, however, because large numbers of new materials had to be tested, and because the search for asbestos substitutes necessitated elevated-temperature tests—it became obvious to those doing the research that "better" gasket tests were required and that they must, for one thing, be faster. Many gasket materials will provide satisfactory service when first exposed to elevated temperature, for example, but their behavior worsens upon long exposure. They may crumble, or creep excessively, or lose tensile strength, or lose their ability to seal a joint. We need to know if they're going to do these things, but we can't wait months or years for the results of a test, at least not if a fast test is possible.

Users also face the bewildering task of choosing new gasket materials for a variety of applications where they've always used asbestos and/or always accepted a certain amount of leakage. This job would be made easier if gasket properties could be defined by standard quality factors of some sort. So—new ways both to test and to rate or compare gaskets were sought—and are still being sought. Those most interested in gasketed joints, in fact, should continue to monitor the work of the various bolted joint and gasket subcommittees and task groups of the PVRC because the work is ongoing and will apparently continue for many years in the future.

As of this writing a number of new tests have been designed by PVRC investigators, and some of these are being codified as ASTM standards. These measure such things as leakage, relaxation, and tensile strength after exposure to high temperatures. Each of these has been defined by an acronym which you may be faced with some day, so here's a check list of some of them [30, 31].

ARLA—Aged Relaxation Leakage Adhesion test. Measures the weight loss of the gasket, creep relaxation, leakage, and adhesion to the flange surfaces under thermal exposure in an air oven. Like the ATRS test but uses ring gaskets so leakage can be measured.

ATRS—Aged Tensile/Relaxation Screening test. Dumbbell-shaped test specimens are tested for creep during and tensile strength after up to 42 days of exposure to 750°F (400°C).

EHOT—Emissions HOTT test. A room-temperature leakage test followed by a 3-day HOTT test, followed by an elevated-temperature leakage test.

FIRS—FIre Simulation screening test. Specimen subjected to 1200°F (649°C) for 30 min, then tested for tensile strength and relaxation properties.

FITT—Simulated Fire Tightness Test. Gasket subjected to 1200°F for 15 min. The leak rate is measured during and after the test.

HATR—High-Temperature Tensile/Relaxation screening test. Conducted at 1100°F (593°C).

HOTT—Hot Operational Tightness Test. Gasket tested for leak tightness and blowout resistance under contained pressure, gasket stress, and temperatures (up to 800°F) (450°C) used in process industries.

ROTT—Room-Temperature Tightness Test.

E. Measuring Gasket Stress and Deflection

Most users won't make the gasket tests described above, but some of us may want to analyze a flange or gasket problem by measuring the in-service deflection of a gasket and/or the compressive stress on it. Here's a quick check list of some of the ways to make such measurements.

Solder plug deflection measurements: Fat plugs of solder are placed in holes punched through the gasket during assembly of the joint. The final thickness of the gasket is measured by measuring the thickness of the solder plugs, when the joint is next disassembled. A time-honored procedure [52].

Bolt stretch: If the flange has through bolts or studs, their lengths can be measured before, during, and after the flange is put in service. The change in elongation of the bolt will equal the change in thickness of the gasket unless thermal effects also affect bolt and flange dimensions. The change in elongation of the bolts will also correspond to a change in compressive stress on the gasket, unless there's metal-to-metal contact of flange surfaces (and/or a gage ring on the gasket). The change in elongation can be related to the change in gasket stress by applying Hooke's law to the bolts, as in Chap. 5, with the same exceptions.

Flange gap: You can sometimes get a rough idea of gasket deflection by measuring the gap, and change in gap between flanges, but the gap is too often affected by rotation and thermal effects to be much use as a measuring tool. Lines have been scribed on the sides of the flat-faced "flanges" used in test rigs, however, and used successfully to measure gasket deflections.

Fuji prescale film: This pressure-measuring film is placed between flange surfaces. Its final color intensity is a function of the pressure applied to it. The film reveals variations in the contact stress between flange surfaces as well as the magnitude of stress at each point. A densitometer is used to quantify color intensity [53, 54].

Uniforce sensors: These thin-film sensors contain crossing grids of metallic conductors separated by an insulator. The greater the pressure applied to a given point, the greater the conductivity between grid wires at that point. A computer reads the conductivity at each grid intersection and produces a colored picture of the contact stress magnitude and distribution. Standard sensors are available to measure the contact forces around individual bolts. Custom-made sensors can read the contact between larger areas. One advantage of this sensor is that it can monitor dynamic changes in stress in service, as well as static stress before and after or during service [55, 56].

VI. GASKET QUALITY FACTORS

The results of the PVRC tests have also been used to define a number of new gasket quality factors which are designed to help users compare proposed new gaskets to compressed asbestos gaskets [30, 31]. Some of these are now being considered by the ASTM and others. Here's a sampling of those which may find use.

A. Proposed Quality Factors

A_e—the equivalent aged exposure factor. A dimensionless factor which relates the product of exposure time and exposure temperature to a standard based on tests of a "typical" compressed asbestos gasket. Using this factor, the gasket manufacturer can estimate the probable maximum service temperature of a proposed new gasket after a relatively short test at high temperature

$$A_e = K(T - 300) \times H^{0.2} \tag{19.5}$$

where T = exposure temperature (°F)
$\quad\quad$ 300 = reference temperature below which new materials tested to date see no damage (°F)
$\quad\quad\ H$ = exposure time (hr)
$\quad\quad\ K$ = 0.0502 or 1/19.9054, a constant based on the fact that A_e typically equals 100 for the substitute materials tested to date
$\quad\quad$ 0.2 = standard exponent based on ATRS tests of elastomer-bound materials

The anticipated service temperature limit (T_s in °F) of a given length of exposure, H, can be computed from:

$$T_s = \frac{19.9054 A_e}{(H^{0.2}) + 300} \tag{19.6}$$

Q_r—the load-retaining capability factor. A dimensionless factor which compares the percentage of stud load retained by the new gasket material to the percentage retained by a "typical" compressed asbestos gasket.

$$Q_r = \left(\frac{\text{ATRS percent stud bolt load retained}}{75}\right)^2 \tag{19.7}$$

This factor is squared to emphasize the importance of the retention of bolt load.

Q_t—the tightness quality factor. A dimensionless factor comparing the sealability of a new material to compressed asbestos.

$$Q_t = \frac{1}{2}\left[\frac{\log(T_{pmin})}{1.5} + \frac{(\text{slope})}{1.5}\right] \tag{19.8}$$

$$\text{Slope} = \frac{1}{\log(T_{pmax}) - \log(T_{pmin})} \tag{19.9}$$

where T_p = the tightness parameter; see Secs. IVB and IXC.

Q_m—the tensile strength factor. The cross-grain tensile strength of the new material is compared to that of compressed asbestos.

$$Q = \frac{\text{ATRS cross-grain posttest tensile strength, psi}}{1000 \text{ psi}} \tag{19.10}$$

Q_p—the overall quality parameter; the product of Q_x and Q_r; acknowledge that stud bolt retention and tensile strength are both extremely important in evaluating a gasket.

$$Q_p = Q_x \times Q_r \tag{19.11}$$

A_p—equivalent exposure parameter based on the weight loss of the gasket. I've left this until last because it deserves a little more discussion than the rest. First, let's look at the present equations for A_p, based on the temperatures and times used to evaluate the elastomeric bound gaskets tested to date. These constants may change, so you should follow the ongoing activities of the PVRC Subcommittee on Gasket Leakage and Elevated Temperature Behavior if you wish to make use of A_p [31]. In the following equations, H is the time of exposure.

Compressed asbestos gaskets
If the exposure temperature, $T < 647°F$

$$A_p = 1000 \times H^{0.6} \times \left(\frac{1}{366} - \frac{1}{T}\right) \tag{19.12}$$

If $T > 647°F$

$$A_p = 17{,}000 \times H^{0.6} \times \left(\frac{1}{626} - \frac{1}{T}\right) \tag{19.13}$$

Compressed aramid fiber gaskets
If $T < 625°F$

$$A_p = 1000 \times H^{0.6} \times \left(\frac{1}{364} - \frac{1}{T}\right) \tag{19.14}$$

If $T > 625°F$

$$A_p = 17{,}000 \times H^{0.6} \times \left(\frac{1}{606} - \frac{1}{T}\right) \tag{19.15}$$

Now, for a brief discussion of A_p. The PVRC investigators have discovered that there's a strong correlation between the weight loss of an elastomeric bound sheet gasket material (such as compressed asbestos or compressed aramid fiber) and the gasket's mechanical properties, such as loss of tensile strength and an increase in creep relaxation. Weight loss also correlates to an increase in leak rate. This allowed them to develop the A_p rating factor, which can be used to predict the long-term elevated-temperature behavior of a proposed gasket material, using relatively short-term mechanical tests. As a result they believe that A_p will be more useful than A_e.

The investigators have also found, however, that thermal degradation of an elastomeric bound gasket depends on such things as the width of the gasket, the initial seating stress on the gasket, and whether or not an oxidizing fluid is contained by the gasket. This means that weight loss tests must take such factors into account.

B. Some Examples of the Use of Quality Factors

Let's see how these quality factors might be used. Remember their purpose: to help us compare proposed asbestos substitute gaskets to compressed asbestos itself. Figure 19.18 presents data taken during ATRS

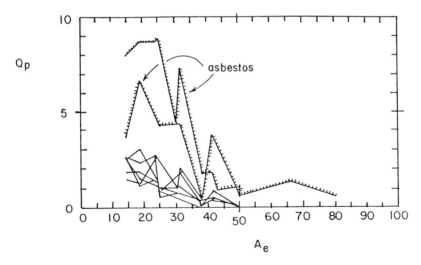

Figure 19.18 Results of Aged Tensile/Relaxation Screening (ATRS) tests on elastomer bound gaskets reinforced with asbestos, glass, or aramid fibers. The glass and aramid results are shown by the solid lines, and are grouped below the compressed asbestos results. The data are presented as a plot between the overall quality parameter, Q_p, and the equivalent aged exposure parameter, A_e. See the text for the meanings of these terms. Note that only the compressed asbestos gasket maintains positive values of Q_p past an A_e of 50, and that it ranges between an A_e of 0.5 and 1.5 at these higher A_e values.

(Aged Tensile/Relaxation Screening) tests on compressed asbestos gaskets, as well as on elastomer-bound gaskets reinforced with glass or aramid fibers. Quality factors Q_x and Q_r were determined by test results and were then combined to compute a Q_p value for each test. The exposure times and temperatures used for each test were used to compute an A_e value as well. The various pairs of Q_p and A_e values were then used to plot the lines shown in Fig. 19.18.Note how Q_p for compressed asbestos levels off at an A_e of about 50 and stays within a band of about $1 \pm \frac{1}{2}$, while Q_p for the other materials becomes zero at $A_e = 50$ [30].

Figure 19.19 is another plot of the Q_p and A_e data for just one of the aramid fiber gaskets tested. Test results for this material were, as usual, scattered. All results leading to a Q_p of $1 = \frac{1}{2}$, the stable, long-term Q_p of the compressed asbestos tested, are included in the shaded area. This range of Q_p data corresponds to A_e values ranging from 18 to 37. The PVRC investigators now used Eq. (19.6) to compute a probable maximum service temperature (T_s) range for this material, for an anticipated exposure time was to be 3 years (27,600 hr). An A_e of 18 led to a T_s of 347°F (175°C). An A_e of 37 suggests a T_s of 396°F (202°C) for the same exposure time [30].

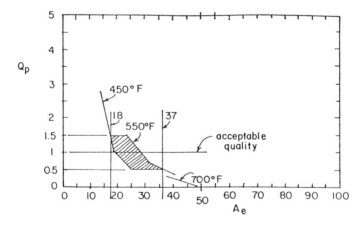

Figure 19.19 Another graph of some of the data plotted in Fig. 19.18. This time a "summary" of the aramid fiber data is given, still as a function of the overall quality parameter, Q_p, and the equivalent aged exposure parameter A_e. An acceptable Q_p quality level of $1 \pm \frac{1}{2}$ (based on the performance of compressed asbestos as plotted in Fig. 19.18) was used to determine the acceptable range of A_e values for the aramid fiber gaskets tested. The acceptable A_e range of 18 to 37 was later converted to a range of maximum service temperatures for this material, as described in the text.

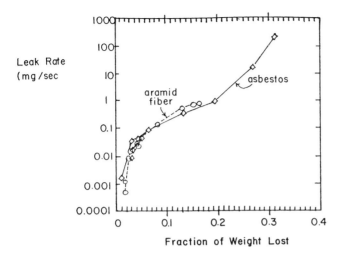

Figure 19.20 Results of a series of PVRC tests in which the weight lost by elastomer-bound asbestos and aramid fiber gaskets was correlated against leakage through the gaskets after various elevated-temperature tests. These were conducted at a variety of exposure times and temperatures. The close correlation between these factors leads the PVRC investigators to believe that the elevated-temperature leakage behavior of elastomer-bound materials can be predicted on the basis of relatively short-term elevated-temperature studies involving only weight loss.

Figure 19.20 is the closest I've come to A_p data, which it isn't. But it does show that the concept of A_p, that a gasket's leakage behavior can be predicted on the basis of short-term weight loss tests, is valid. It shows mass leak rate vs. fraction of weight lost for compressed asbestos and aramid fiber gaskets [31]. As mentioned earlier, this weight loss work is continuing. It will be reported in a Welding Research Council Bulletin someday.

VII. SELECTING A GASKET

A. In General

I have said that G_b, a, G_s, and the concept of tightness fully define the behavior of a gasket under many different conditions. Can we use this information, then, to select the optimum gasket for a given application and/or to distinguish between a "good" gasket and a "bad" one?

The new gasket constants can help us select an appropriate gasket, but they provide only a small part of the information we need for this task. Selecting a suitable gasket is a complex task involving a great deal of knowledge about gasket materials, configuration, strength, etc., plus a great deal of knowledge about the proposed application. Good references are available from many gasket manufacturers, including those organizations listed in Table 19.9. In general, these tell us that the selection of a gasket should be based on [19, 20]:

1.   The chemical resistance of the gasket
2.   The heat resistance of a gasket
3.   Tensile and crush strengths of the gasket
4.   The seating stress required by the gasket
5.   The cost of the gasket
6.   The resilience (stiffness/recovery) of the gasket
7.   The pressure of the contained liquid or gas
8.   The configuration of the joint to be sealed
9.   The nature of the contained fluid (explosive, carcinogenic, benign, etc.)
10.   Permeability of the gasket

Table 19.9   Sources of Information for the Selection of Gaskets

Fel-Pro, Inc.	Richard Klinger Inc.
7450 N. McCormick Blvd.	2350 Campbell Rd.
P.O. Box 1103	Sidney, Ohio 45365
Skokie, Illnois 60076-8103	(513) 498-1181
Flexitallic Gasket Co.	Lemons Metal Gasket Co.
P.O. Box 760-T	P.O. Box 947
Deer Park, Texas 77536	Houston, Texas 77001
(713) 478-3491	(713) 222-0284
Fluid Sealing Association, Inc.	Union Carbide Corp.
2017 Walnut St.	P.O. Box 94637
Philadelphia, Pennsylvania 19103	Cleveland, Ohio 44101
(215) 569-3650	(216) 529-3900
Garlock Inc.	
608 N. 10th St.	
Camden, New Jersey 08101	
(609) 964-0370	

Note: I do not intend to endorse the organizations listed here, or to belittle the many gasket manufacturers NOT included. These are just the few with which I have some familiarity.

11. Lubricity of the gasket (for easy release)
12. Creep characteristics of the gasket
13. Thermal conductivity of the gasket
14. The configuration of the gasket

plus a number of lesser factors.

Upon reflection we see that G_b, a, G_s, and T_P can only help us with items 4, 6, 7, and 10 above. At least that's all we can expect from the room-temperature gasket factors published to date. Work is being done on the elevated-temperature behavior of gaskets, which will result in high-temperature values for G_s. (G_b and a remain the same because initial loading is always at room temperature). G_s can change significantly as temperatures rise. That information, when available, will help us deal with item 2 above—one of the factors which dominate gasket selection. The gasket factors discussed to date, however, can only help with the four items mentioned earlier.

Nevertheless, they ARE a help there, and so are worth reporting. You'll find a tabulation of the factors published to date in Table 19.8.

You'll note, I'm sure, that the constants given for one type of gasket are similar to those for the others in this list. As far as leak rate vs. gasket stress is concerned, furthermore, the constants are even more similar than they appear in Table 19.8. This should come as no surprise. The m and y factors listed in the ASME Code are also similar to each other for most types of commercially acceptable pressure vessel gasket.

There ARE gasket types whose behavior differs significantly from those reported in Table 19.8, however. Solid metal gaskets, for example, tend to have ambient tightness factors (G_s) fairly close to their G_b values. In other words, they are very sensitive to loss of gasket stress, and will lose "tightness" quickly when they're unloaded. They also tend to be difficult to evaluate. They exhibit a lot of scatter in stress vs. leak rate; they behave differently on reuse than on initial use; etc. As one result, new gasket constants for these gaskets have not been published as yet.

Note that even though the new gasket factors can help us deal with some of the parameters which guide our choice of gasket, the new factors do not fully resolve any of these points all by themselves. Take item 7, contained pressure, for example. Contained pressure is included in the calculation of the tightness parameter (T_P) and so helps to define leak rate in relationship to G_b, a, and G_s. The gasket factors help us relate pressure to our application, therefore. But there are other pressure considerations which are not addressed by the gasket factors or tightness parameter. A higher contained pressure, for example, means greater bolt loads—and a gasket which can resist them. Higher pressure also increases the tendency of a gasket to blow out—and so influences our consideration of such things as gasket tensile strength and flange surface finish. So, the gasket factors tell us only part of what we need to know when introducing "pressure" to the gasket selection process.

One critically important consideration in the selection of a gasket is its response to temperature. Tensile strength (and therefore blowout resistance), creep, recovery, and general sealability are all affected by an increase in temperature, as we've seen. This has become painfully clear in recent years as more people than usual are selecting gaskets—to replace the old standby, compressed asbestos, they've used for many decades. Even though CA gaskets are once again allowed in the United States, they are still outlawed in many countries, and the handling or use of asbestos in any form opens manufacturers and others to the possibilities of lawsuits, so the search for replacements goes on.

Many of the proposed substitutes consist of a clay or other inorganic base material to which has been added substitute fibers for strength, flexibility, and improved processing. Aramid, acrylic, glass, and cellulose fibers have been used, but, of these, only glass has the heat resistance of asbestos (but it lacks tensile strength). Some of the proposed substitutes, incidentally, equal or better asbestos in tightness at room or moderately elevated temperatures, but fail mechanically when temperatures are raised.

As already mentioned several times, flexible graphite sandwiched with a steel core is often chosen as a CA substitute [21] but can't be used in an oxidizing atmosphere if service temperatures exceed 650°F (343°C). In an inert or reducing atmosphere it can be used at temperatures up to 5432°F (3000°C) [59].

As suggested earlier, a full treatment of gasket types and the selection of an optimum gasket is beyond the scope of this text. Those interested in a full treatment would do well to acquire a copy of Ref. 46. Other help is available for those who have personal computers.

B. Gasket Selection by Computer

I have complained earlier that information concerning such things as the crushing strength of a gasket, or the amount of seating stress it requires, has been difficult to obtain, since it was not published by gasket manufacturers. That's now starting to change. Several companies can now provide their customers with floppy discs containing full information about their products and about the standard flanges they're usually used in. The user inputs data about his application: flange size and pressure rating, gasket contact area, bolt lengths, service temperature and pressure, the type of fluid to be contained by the gasket, the type of bolts to be used (i.e., the material), etc. Program outputs include a recommendation for the type of proprietary gasket which should be used in the application, plus such things as the recommended seating stress (including assembly torques), information about the total clamping force those torques will generate on the flange, the working stress on the gasket, the maximum allowable gasket stress, sometimes the crushing strength, and other useful information. In some programs the user can also input special flange dimensions if a standard flange won't be used.

Such programs have been created by Klinger [32], the Anchor Packing Division of Robco, Inc. [41], Armstrong World Industries [43], and perhaps by others.

VIII. SELECTING ASSEMBLY STRESS FOR A GASKET

Perhaps the most common question asked by gasket users, at least in the pressure vessel world, is, "What torque should I use on this flange?" If you've plowed through the book from the beginning to this point you'll know that it's not an easy question to answer. But each user has to answer it. We're going to look at several possible ways to do so.

Our natural inclination is to turn to our gasket suppliers for advice. Many product brochures include tables of seating stress vs. torque, for standard sizes of flange and/or bolts [47, 49]. Such tables appear to be torque recommendations, but they're not. They simply tell us how much stress we'll create on a gasket in a given flange if we apply the tabulated torque to the bolts. But each of the many types and thicknesses of gasket we might use in our application will presumably require a different amount of seating stress, not just the "standard" stress made possible by those bolts. The torque tables are blind to the requirements of the gasket.

Some gasket manufacturers do have application engineers who, if asked, will provide recommendations for assembly torque, based upon information provided by the customer; often in a telephone conversation. Communication between customer and gasket manufacturer can be imperfect, however, and time constraints can prevent the application engineer from considering all the factors which should enter into such a recommendation. The resulting "answers" have not always been the right ones. But things are changing.

As just mentioned in the last section, some gasket manufacturers can now provide their customers—or will use in response to a potential customer's inquiry—sophisticated computer programs whose outputs include, among other things, recommended seating stresses and/or assembly torques tailored for your application. The programs require all the inputs necessary for a more accurate recommendation than is possible with a brief phone call to an application engineer. I urge you to take advantage of these programs if you can: the answers they give you will be far better than those you could get by using the procedures to be described next. This assistance is only available for those planning to use the products of the companies who have designed these programs, however, so many engineers will still have to struggle to answer the "torque question" themselves. Hence the detailed discussions which follow.

One of the simplest and easiest to apply answers I ever heard goes like this: "I determine the bolt's yield strength and the crushing strength of the gasket (in terms of bolt stress). I then use the smaller of those two to compute assembly torque" [36]. That's fine, as long as things like thermal expansion, flange rotation, or stress corrosion cracking—to mention only three—don't concern you.

Most pressure vessel people I've talked to tend to rely on a y factor obtained from the ASME Code or from a gasket manufacturer to estimate a minimum required gasket stress. Others base their torques on a table of "allowable stress" also found in the Code. We'll examine the Code and what it says about assembly stress in Sec. A, below. Then, having shown that the Code tries to ignore the subject of assembly stress, I'll suggest, in Sec. B, a lengthy procedure in which y factors might be used to compute an assembly torque, taking a lot of assembly and service factors we've already discussed into consideration. Finally, in Sec. C and D, I'll discuss two ways in which the new gasket factors, G_b, a, and G_s, might be used to pick an assembly stress. Be warned again that m, y, G_b, a, and G_s aren't supposed to be used for this purpose. You'll probably have to use them, however, unless you have access to full PVRC or other test results—or to answers provided by one of the gasket selection computer programs mentioned earlier. So use them—but cautiously!

A. Assembly Stress and the ASME Code

The design procedure defined by the ASME Code is widely used by designers and, in general, has been a successful design tool. The m and y gasket factors published in the Code, however, have often been challenged. Even though these factors "work" within the framework of the Code procedure (those many joints and plants are, after all, functioning), experiments have raised considerable doubt as to the validity of the gasket factors [5]. A study published in 1973 in Canada, for example, suggested that the m and y factors should be several times as large as those published in the 1973 Code [3]. Findings such as these have led to the general recognition that Code practice won't guarantee leak-free joints. Does this mean that the ASME Code is no good? Not at all. It means that its role is frequently misunderstood.

The Code was never intended to define leak-free behavior of a gasketed joint. It was intended, instead, to force designers to design pressure systems which would not "blow up" in service, to eliminate the frequent boiler explosions which occurred—with often fatal results—before the Code was written. Leaks were of no real concern to the original authors.

As operating pressures and temperatures have risen, however, we have become more sensitive to the impact which a leak, or multiple leaks, can have on the environment and on our safety. As a result there has been an increasing interest in modifying the Code so that flanges designed by its procedures would more often operate in a leak-free manner.

To support the statement that the Code is not intended to define leak-free behavior, note that the gasket factors listed in the Code are suggestions only. They are not mandatory and—this is less obvious—they are not intended to define *assembly* limits on gasket stress.

While we're on the subject of "assembly limits," it should also be pointed out that the "allowable stresses" listed several places in the Code are not intended to limit assembly stress in the bolts either. These allowables are intended to force flange designers to overdesign the joint, to use more and/or larger bolts and thicker flange members than they might otherwise be inclined to use. This is one of the many ways in which the Code introduces significant factors of safety into the design of the system—again, aiming at elimination of serious accidents.

Only in nonmandatory Appendix S does Section VIII of the Code deal with assembly stress. And it deals with this topic only in relatively general terms.

Appendix S, for example, suggests that for assembly purposes you apply at least one-and-one-half times the allowable stress to the bolts in order to support the external loads they will see during a hydro test. Appendix S also goes on to suggest, however, that if one-and-one-half times allowable isn't enough—if the system still leaks—you should feel free to go to still higher stress levels.

The closest Appendix S comes to quantifying assembly stresses in the bolts, as a matter of fact, is in Eq. (19.16), which suggests the amount of stress you might expect to produce in the bolts at assembly.

$$S_A = \frac{45,000}{\sqrt{D}} \tag{19.16}$$

where S_A = stress created in the bolts at assembly (psi)
D = nominal diameter of the fasteners (in.)

In short, then, the ASME Code is intended to be a designer's document, not an assembler's document.

The comments above are based on the provisions of Section VIII of the Code, the section most commonly used. Things are somewhat different in Section III, which is used by the nuclear power industry. Assembly stresses in bolts are dealt with in two different ways. A mandatory Appendix XII, identical to nonmandatory Appendix S in Section VIII, suggests that bolts can be tightened as much as necessary to prevent leaks. In this respect Section III agrees with Section VIII.

In several other places, however, Section III takes a different tack. Consider subsection NB, for example, which deals with the design and integrity of Class 1 components whose failure would imperil the primary pressure boundary of the plant. Paragraphs under NB-3232 describe the "actual service stresses in bolts, such as those produced by a combination of preload, pressure and differential thermal expansion." Service stresses allowed in this section can sometimes reach the yield point of the bolting materials, and commonly will be two-thirds of yield. A similar discussion will be found in subsection NE (NE-3230).

Note that these discussions don't recommend loading bolts during assembly to yield, or to two-thirds yield, or to any other point—they merely say that service loads of this sort are acceptable. To repeat my earlier assertion: Nowhere does the Code specify or recommend assembly preloads.

B. Using *y* to Pick an Assembly Preload

Now that we know that the *y* factor is not supposed to be used to pick an assembly stress or torque, let's see how we might use it to do just that. I warn you that the procedure I'm about to suggest is rather involved; it must be to compensate for the shortcomings in the *y* factor. The process involves many of the topics we've looked at in previous chapters, too; forms a sort of summary of how to use the information presented so far to do a particular job; to pick an assembly torque for a pressure vessel. Sorry if some of the following seems redundant. I think it's necessary—and you might find it useful. So, here goes.

The bolts on a gasketed pressure vessel joint must be tightened enough to mate the gasket surfaces "skintight" with the flange surfaces and further tightened to compensate for losses in clamping force which occur during tightening and when the joint is put in service. These losses can be caused by one or more of the following factors.

Variations in the amount of preload created by a given torque. We call this "tool scatter."

Loss in average preload created by elastic interactions between bolts as they are tightened one by one.

Embedment of thread and joint surfaces.

The loss in preload caused by the hydrostatic end load when the vessel is pressurized.

The loss in preload caused by creep of the gasket at room or elevated temperatures.

The change in preload caused by differential thermal expansion between bolts and joint members as the temperature of the vessel changes. This change can cause an increase or decrease in preload and clamping force, depending upon the change in temperature of each part, their coefficients of expansion, etc.

Let's look at each of these factors and see how to estimate them.

Mating Gasket and Flange Surfaces

The amount of preload required to mate the gasket and flange surfaces is, at the present time, best defined by the "y factor" in the ASME Boiler and Pressure Vessel Code. This factor is defined as the minimum seating stress required for acceptable behavior. It is intended to be used only by flange designers (who use it to size the flange members and bolts) but it's usually the only clue a user has for the seating requirements of a given type of gasket.

Let's take an example. The Code says that a $\frac{1}{16}$-inch-thick compressed asbestos gasket has a y factor of 3700 psi. This apparently tells us that the bolts should be tightened to create this much stress on the full seating surface of the gasket—but that's not the case. The design rules in the Code tell the designer—and us—that the bolts must be tightened to create this much seating stress on the "effective area" of the gasket. For our purpose as users we can take this to be half the actual contact area. Table UA49.1 in Section VIII of the Code will tell you how to compute this area more exactly if you wish.

Assuming the half area, proceed as follows:

Multiply it by the y stress to get the total bolt load required (lb or N)

Divide that load by the number of bolts in the joint to get the force required for each bolt (lb, N). This is the minimum bolt preload required. Now we must add the additional bolt load required to replace the anticipated losses.

Preload Scatter

Conventional wisdom says that we'll get a scatter of $\pm 30\%$ in the preload we achieve when we tighten a group of as-received bolts with a given torque. If the bolts and nut faces have been lubricated with a good thread lubricant the scatter might be half that, or ±15%. With a large group of bolts, therefore, the average preload should be the middle of these ranges. Most flanged joints, however, don't have enough bolts in them to achieve a dependable average. It's probably safer to assume we'll need perhaps 10% more torque than we might think we need, to achieve a desired average, if we're using as-received bolts; and perhaps 5% more if using lubricated bolts. Obviously these numbers will vary depending upon the condition of the bolts, their size (larger ones probably require more overtorquing), etc. but they're probably a reasonable starting point. So the minimum preload computed above must be increased 5 or 10% to accommodate preload scatter.

Embedment

When the bolts are first tightened the thread and joint surfaces contact each other only on the "high spots" of their always irregular surfaces. These contact areas will creep; the surfaces will settle into each other, until enough surface has been brought into contact to stabilize the process. This embedment can cause a loss of 10% of average preload if the nuts and bolts are new, perhaps only a third or half as much if they're reused. This 5 to 10% must be added to our growing bolt tension figure.

Elastic Interaction Loss

When a bolt is tightened in a given location on the flange—let's say the 12 o'clock position—it is stretched a little and it partially compresses the joint at that point. When neighboring bolts—at the 11 o'clock and 1 o'clock positions—are tightened later in the torquing procedure, they further compress the joint in the 12 o'clock region. That allows the 12 o'clock bolt to relax a little; to lose some of its initial preload.

   Work done by Dr. George Bibel of the University of North Dakota suggests that the average loss of preload in a joint containing sheet gasket will be something like 30%. If a spiral-wound gasket is used, Dr. Bibel says the average loss will be as much as 48%. So our bolt load target must be increased another 30–48% for this factor. Note that this loss occurs during assembly, before we put the joint in service.

Hydrostatic End Load

Now we turn on the vessel. The internal pressure attempts to blow the flange apart. We call the force created by the pressure the *hydrostatic end load.* It will increase the tension in the bolts and simultaneously decrease the clamping force on the gasket. The initial, assembly bolt load must be enough to compensate for this loss. The bolts must also be strong enough to support the increase in tension caused by the pressure load. We'll assume that they are (since Code-designed joints have at least a 4:1 safety factor built into the selection of bolt size and material strength). Let's concentrate on the loss; which we compute as follows:

Determine the area enclosed by the ID of the gasket. (The Code uses a slightly greater area but this one should suffice for assembly calculation purposes.)

Multiply that area by the contained pressure. This gives you the total pressure force being exerted on the joint.

Divide that force by the number of bolts in the joint, and add it to the previous total.

The number you get by this procedure will be a very conservative one, and is one used in the ASME Code design rules. In actual fact only about two-thirds of the hydro load will be seen as a loss in stress on the gasket. (The other third is seen as increase in bolt tension.)

Gasket Creep Loss

Gaskets must be partially plastic in order to seal a joint. As a result they'll creep when first preloaded. The amount of creep will vary and will be, in part, a function of service temperatures (higher temperature means more creep). Thicker gaskets creep more than thin ones, too. You'll have to contact your gasket supplier for creep rates of the gaskets you're planning to use. *Be sure to tell him your operating temperatures and the gasket thicknesses required in your applications.* If you're taking the data from a manufacturer's catalog, be sure to read the fine print concerning temperature and thickness before accepting a creep rate figure.

Creep amounts can range from 10% or so to almost 80%. Small amounts can be compensated for by increasing assembly preloads by the same percentage. Large amounts of creep can only be compensated for by rebolting (retorquing, for example) after the system has been on line for awhile. Most creep occurs soon after system start-up; or at least soon after service temperatures are reached. Retorquing within an hour or two should compensate for the bulk of the creep. Fortunately, a gasket which causes a clamping force loss of 50%, because of creep, will not cause a further loss of 50% if reloaded. It will allow some further loss, but much less than it caused when first loaded.

Differential Thermal Expansion

To determine the effects of a thermal change on preload you must:

Compute the change in length which would be created in the grip length
of the bolts by the change in temperature, assuming the bolts are not
in the joint.

$$\Delta L_B = \rho_B (\Delta t) L_G \qquad\qquad (19.17)$$

Compute the change in thickness of the joint members, assuming they're
not clamped by the bolts.

$$\Delta L_J = \rho_J (\Delta t) L_G \qquad\qquad (19.18)$$

If the bolts and joint members change by a different amount, because
they're at different temperatures and/or are made of different materials,
you must next compute the force created by their differential expansion.

$$F_T = (A_S E)(\Delta L_J - \Delta L_B)/L_E \qquad\qquad (19.19)$$

where ΔL_B = change in length of a free bolt (in.)
$\quad\quad\ \Delta L_J$ = change in thickness of the joint (in.)
$\quad\quad\ L_E$ = effective length of the bolt (grip length plus one nominal
$\quad\quad\qquad$ diameter) (in.)
$\quad\quad\ L_G$ = grip length (distance between face of nut and underface
$\quad\quad\qquad$ of head) (in.)
$\quad\quad\ \rho_B$ = coefficient of thermal expansion of the bolt material (in./
$\quad\quad\qquad$ in./°F)
$\quad\quad\ \rho_J$ = coefficient of expansion of the joint material (in./in./°F)
$\quad\quad\ \Delta t$ = change in temperature (°F) (may be different for bolts and
$\quad\quad\qquad$ joins members)

This force (F_T) must be added to our earlier total. We now have a target
preload for these bolts. It consists of:

The per bolt force required to create a y stress on half the contact surface
of the gasket, plus
5 to 10% more to compensate for tool scatter, plus
5 to 10% more to compensate for embedment, plus
30 to 48% more to compensate for elastic interactions, plus
Whatever is required to compensate for the hydrostatic end load, plus
Whatever is required to compensate for the loss caused by gasket creep,
plus
Whatever is required to compensate for differential expansion, if any.

Safety Checks

Before using the resulting total to pick a torque or assembly preload, you should:

Divide the desired force by the "tensile stress area" of the bolt and compare the result to the yield strength of the bolt at service temperatures. You'd like your choice of preload to be less than 60% of yield, though higher percentages are not uncommon.

Discuss the planned seating stress—based on the full, planned preload not just on the y factor—with your gasket supplier. All gaskets have crushing strengths which must not be exceeded.

Picking a Torque

If your desired preload passes the safety checks, you're now ready to use it. If assembly will be controlled ultrasonically, you'll use it "as is." In this case you can compensate for preload scatter, embedment, and elastic interactions during assembly, and will need an assembly target which is much less than you'll need if you're going to use torque control. If torque is your choice, use your target preload to compute torque with the following equation:

$$T = KDF/12 \tag{19.20}$$

where T = torque in ft-lb
D = nominal diameter of the bolt (in.)
F = target preload (lb)
K = dimensionless "nut factor"

If you're using:

As-received steel on steel bolts (not stainless steel) $K = 0.2$
Stainless steel bolts $K = 0.3$
Lubricated steel bolts $K = 0.17$

I told you it wouldn't be easy! But that's what you should go through when using a y factor to pick an assembly torque for a gasketed joint. It's an involved procedure because y is a design factor; it is *not* intended to be used to define an assembly torque and it ignores all of the other stuff we've just considered. But, used as suggested above, I think that this is a safe and reasonable procedure. Now let's look at one more way to pick an assembly stress and torque, this time using the proposed, new gasket factors G_b, a, and G_s.

C. Using the New Gasket Constants to Select an Assembly Stress

Before attempting to use G_b, a, G_s, and T_P we should take a moment to think about a fundamental difference between the new and old gasket constants. The present (old) constants, m and y, provide specific design recommendations for maintenance and seating stress for a variety of gaskets. One set of numbers for each type of gasket. Concrete, tangible, easy to use.

The new constants, G_b, a, and G_s are not "recommendations" at all. Instead, they help define the behavior of the gasket under all possible stress conditions. The only design guidance emerging from the PVRC work is in the concept of "tightness classifications," and even here the designer may be given, not a set of numbers, but a choice of three possible tightness levels. Each of those, as we'll see in a minute, can be achieved by a wide variety of seating and maintenance stress combinations; and some trial and error and/or past experience is required to pick an appropriate set. Once the designer has learned how to convert G_b, a, G_s, and a selected tightness level to specific stress targets, he can presumably design "better" flanges than he can with the more tangible, present recommendations. But his job has admittedly been made more complicated.

In one way the new and old constants are the same; neither is intended to guide the assembler. Since G_b, a, and G_s do genuinely define gasket behavior, however, I believe that they CAN be useful to the assembler. In fact, I suggest we start our discussion of use by using them to select an assembly stress target for a gasket. Then, in Chap. 22, we can go on to consider their use in design.

Before describing a detailed procedure, let me say that there are many different ways in which the PVRC information might be used to select an assembly stress for an existing gasket. I believe that the procedure suggested below is a reasonable one, and will help me explain to you the relationships between such factors as leak rate, seating and maintenance stress on the gasket, assembly bolt loads, anticipated scatter in bolt loads, etc. Other procedures are possible, however, and may be more efficient or "safer." Specifically, Jim Payne, of J. Payne Associates, Long Valley, New Jersey, feels that my procedure might inadvertently lead an engineer to overstress a flange, in order to obtain a lower leak rate, even though the engineer is warned against this in step 6 below. I will start, however, by describing my procedure because it illustrates the relationship of PVRC data to other assembly factors, as already mentioned, and should be a safe procedure if my warnings are heeded.

1. First we must select an acceptable leak rate, guided by Tables 19.5–19.7. Let's assume that we're dealing with a standard 10-in. ANSI B16.5 flange. The gasket, therefore, will have an outside diameter of 12.75 in. We'll also assume that the system contains a noncarcinogenic liquid with a density similar to water. Perhaps a "standard" (T_2) leak rate would be acceptable.

    A standard leak rate would be 0.002 mg/sec·mm (0.0004 lb/hr·in.) or, in easier to visualize volumetric terms, 0.0004 pt of liquid per hour per inch of gasket diameter (from Table 19.5). For a 12.75-in. gasket that would be:

    $$L_{RM} = 0.0004 \times 12.75 = 0.0051 \text{ pt/hr}$$

    I should mention that these calculations are only approximate because they're based on the assumption that *mass* leak rate is identical for various fluids. Our starting point is a standard leak rate based on tests with helium. In fact, conditions allowing a standard mass leak rate with helium would probably allow a higher leak rate for some alternate fluids, and a lower rate for still others. To a first approximation, however, the procedure I'm describing is considered valid. Note that the leak rate we've computed is so low it is unlikely that we'd be troubled by slight variations in either direction. It's so low, in fact, that at this point many designers would be tempted to start over, using an "economy" leak rate instead of the standard. We'll stick with the standard for our example, however.

2. Next we use Eq. (19.2) to convert the maximum acceptable leak rate to a minimum acceptable tightness parameter. If contained, (operating) pressure will be 235 psi.

    $$T_{Pn} = \frac{235}{14.7} \left[\frac{0.008}{0.0004 \times 12.75} \right]^{0.5} = 20$$

    Note that for L_{RM} we use, not the total volumetric leak rate from the 10-in. flange, but the selected mass leak rate per inch of diameter (i.e., a mass leak rate of 0.0004 lb/hr·in. rather than 0.005 pt/hr). We computed the total volumetric leak rate earlier merely to evaluate our selection of an acceptable level of tightness.

3. Now we use published G_b, a, and G_s data, plus the T_{Pn} calculated above, to draw the graph shown in Fig. 19.21. Let's assume that we're trying a new gasket material, and have determined that G_b, a, and G_s are 7 ksi, 0.33, and 55 psi, respectively. (See Table 19.8 for some actual gasket factors.)

Since this is a log-log plot, we use a ruler to develop slope a (a rise of 0.33 in. in 1 in. on the graph, for example).

4. Next we draw one of many possible unloading lines, such as that shown in Fig. 19.21. This line must intersect the gasket stress axis at G_s, and should intersect the loading line (defined by G_b and a) at some tightness parameter greater than the minimum T_{Pn} computed above.

Trial and error, or past experience, is involved at this point. Note, however, that the intersection of this unloading line and the T_{Pn} line defines the lowest gasket stress the gasket can tolerate without exceeding the maximum leak rate we've selected (and defined by T_{Pn}). I've called this intersection S_{ml} in Fig. 19.21.

5. How do we know if the assembly stress we've arbitrarily selected, labeled $S_A$min in Fig. 19.21, is truly acceptable?

One relatively simple way is to assume that the full hydrostatic end load created in service is seen as a reduction in stress by the gasket.

$$\Delta S_G = S_H \frac{A_v}{A_G} \tag{19.21}$$

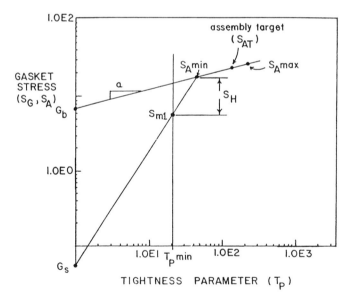

Figure 19.21 Plot illustrating the use of G_b, a, G_s, and the tightness parameter to select an assembly stress target (S_{AT}) for, in this case, a double-jacketed, mica-filled gasket. See text for details.

where ΔS_G = reduction in gasket stress (psi, MPa)
          S_H = contained pressure (psi, MPa)
          A_v = cross-sectional area of vessel (in.^{2}, mm^2)
          A_G = area of contact between gasket and flange
               (in.^{2}, mm^2)

In a 12.75-in.-OD gasket the ratio between vessel area and gasket contact area will probably be about 5:1. The contained pressure is 235 psi. We could expect the reduction in gasket stress, therefore, to be 5 × 235 or 1175 psi, assuming, again, that the entire external load is "seen" by the gasket. The joint diagram would tell us that only a portion of this load will actually be seen by the gasket, so our assumption is conservative.

Looking at Fig. 19.21 we see that the $S_A$min we've selected as a first cut is about 18 ksi. Reducing that by 1175 psi would lower gasket stress to 16.8 ksi, still comfortably above the level of S_{m1}, which is 6 ksi. So, our choice of $S_A$min has passed its first test.

6. Next we must decide whether or not our choice of $S_A$min is too high. As we saw in the last step, it could presumably be lowered by 10 ksi or so, but gasket tests show that more stress is better, means less leakage. Why not leave it at 18 ksi—or even raise it?

To answer these questions we must take a closer look at $S_A$min.

$S_A$min is not "the" assembly stress; it is the minimum expected assembly stress. Somewhere above it will be an average assembly stress (S_{AT}) and somewhere above that a maximum assembly stress ($S_A$max). Our tools and procedures will not be perfect; assembly stresses will be scattered over a range between $S_A$min and $S_A$max. So, when we address the question "Is $S_A$min too high?" we must locate the upper end of the preload scatter band defined by $S_A$min, and ask, instead, whether or not $S_A$max will be too high. ($S_A$max and min are not equidistant from S_{AT} in Fig. 19.21 because this is a log-log plot.)

The PVRC studies do not attempt to define upper limits on gasket stress; their goal was merely to define gasket behavior. So we must look elsewhere for the things which could limit the assembly stresses applied to the gasket. A number of factors must be considered.

Too much stress can crush a gasket, but the gasket isn't the only issue. Gasket stress is created by stress in the bolts and is accompanied by stress in the flanges as well. We can accept assembly targets which would raise the stresses in these components past the *design* allowables listed in the Code; but we usually don't want to take them past yield. More about this later.

Excessive bolt loads might also lead to excessive flange rotation, a situation in which apparently better (higher) gasket stress can actually make a leak more likely.

Further complicating our lives is the fact that when we analyze maximum stresses on an operating flange we should, in critical applications at least, consider not only the stresses (preloads) introduced at assembly, but possible increases in these stress levels when pressure loads and/or differential thermal expansion is encountered. Section VIII Code rules don't include such factors, but they include several factors-of-safety which lead a designer to a safe design. We, on the other hand, are trying to estimate a safe, maximum assembly stress, taking advantage of Appendix S, which implies that we can stress the bolts as high as we need to, to avoid leaks; but it does *not* give us license to use infinite stress. By combining assembly stresses with subsequent, anticipated, stresses introduced by in-service loads we can presumably avoid overstressing flanges or bolts.

All of which brings us back to the concerns voiced by Jim Payne of J. Payne Associates, mentioned earlier. Mr. Payne states, correctly, that it is imperative that the flanges not be stressed too high by the selection of an assembly, gasket stress based simply on "less leak rate." The *designer* should start by selecting a leak rate. He then proportions the flange to keep stress levels within Code rules (as discussed in the next section). The assembler, however, cannot "proportion" his flanges; they already exist. Maximum stresses, not a preselected leak rate, must dominate his decisions, So, Mr. Payne suggests that the assembler start with a bolt preload, plus an assembly tolerance, which will take the flange to near yield. The assembler can use Code rules, and some trial and error, to do this.

He then estimates the minimum stress that gasket will see, based on target preload less an assembly tolerance—and less pressure relief, etc., if he wants to carry his analysis this far. He then uses Eq. (19.2) to predict the leak rate. This will be the best (lowest) leak rate he can achieve, because it's based on the highest tolerance flange stress.

I don't disagree with all this; far from it. My procedure starts with a leak rate, ends with flange stress estimates, and recycles if the stresses are too high. I think the procedure illustrates use of the PVRC data, and that was my primary goal. But I acknowledge that Mr. Payne's procedure may be more direct; starting with the highest tolerable stress and then computing leak rate. No recycling necessary if done properly.

Regardless of just how we use it, the PVRC data can give us significant insight as we struggle to pick an assembly stress. It is mainly useful in picking a target, or minimum stress; other considerations must be used to set upper limits—and may even force reductions in desired minimums.

A final comment about all this: the procedure I've just outlined is, I think, a reasonable one, but is incomplete. It ignores several of the complicating factors included in the previous section,where we used a y factor to compute an acceptable assembly stress. It does not specifically address gasket creep or differential expansion effects, for example, because I didn't want to make you go through all those considerations a second time. But you should go through them—modify your first answers to the assembly torque question—if you believe that they'll affect the behavior of your joint in a significant way.

Understand, too, that the procedure I've suggested is my idea; it is not something which is emerging from the work of the PVRC and ASME. The PVRC committees, as already mentioned, are working with ASME Code committees on modifications of the flange design rules. The PVRC investigators have included an "assembly efficiency" factor in their recommendations, but this is included only to force the flange designer to create heavier flanges if he expects their bolts to be tightened by quick and dirty means than if he expects them to be tightened by sophisticated preload control equipment. No attempt is being made to suggest ways in which appropriate assembly stresses or torques can be computed with the aid of the new gasket constants. That will come someday, I'm sure, but not in the foreseeable future.

We're not done yet. Before leaving the topic of how to select an assembly bolt load or torque for a gasketed, pressure vessel joint I want to describe one more possibility. If nothing else it should convince you that all of the procedures we've discussed so far are just "possibilities."

D. An Alternate Procedure, Again Using the New Gasket Constants

Here, in outline form, is an alternate way to use the new gasket constants to pick an assembly torque. It is one of several procedures defined by J. Ronald Winter of Tennessee Eastman, who warns that the procedure used for a given application will depend on the level of calculation detail required for that application. Joints containing carcinogens or chemicals that the EPA says should be monitored will obviously receive more attention than those of lesser concern [35]. Anyway, here's one of the procedures he has considered [61].

1. Pick an acceptable, maximum leak rate: economy, standard, or tight. He believes that standard will be acceptable for most situations, even

those involving fugitive emissions (except, again, for those joints containing carcinogens etc.).

2. Use Table 19.7 plus your operating pressure (P) and Eq. (19.4) to compute an acceptable minimum tightness (T_{pmin}).
3. Multiply T_{pmin} by 1.5 for hydrotest purposes. Call this value T_{pn}.
4. Get gasket constants, G_b, a, and G_s from Table 19.8 and use them, plus T_{pn}, to compute a possible seating stress.
5. Compute the bolt preload required to create this seating stress. Call this F_p.
6. Assume that the bolts will also see the full load created by the contained pressure (i.e., assume that the joint is infinitely stiff and will absorb none of the pressure load). Call the per-bolt pressure load L_x.
7. If $F_p \geq 2L_x$ stop! You're done.
8. If not, add a safety factor to F_p.
9. In either case you might also want to add an assembly tool scatter factor to F_p.
10. Final things to check before using F_p to compute an assembly torque:
    (a) Don't load bolts > 50% of yield using this procedure (because we've ignored things like thermal effects or bending).
    (b) The selected F_p should correspond to a bolt stress > 10 ksi. (He's assuming that ASTM A 193 or equivalent bolts are used; 10 ksi represents too little a percentage of their yield strength.)
    (c) Don't pick an F_p which will crush the gaskets.
    (d) Don't pick an F_p which will rotate the flanges excessively. (This could be a tough one to estimate. Past experience may have to guide you—or trial and error.)
    (e) Don't pick an F_p which will encourage stress corrosion cracking.

That's it. The procedure comes from a man with a great deal of field experience, as well as long association with the PVRC. It's probably the best procedure described; but, again, is only one of several Mr. Winter uses. I think that we've spent enough time on this subject and should move on. After, however, a final comment.

I've never used any of the computer-program gasket selection routines mentioned earlier and now provided by several gasket manufacturers [32, 41, 43]. I'm impressed, however, by the fact that the outputs of these programs include the gasket manufacturers' recommendations for assembly bolt loads and torques, and that these recommendations are based on a lot of application data provided by the user. If the programs are as good as they appear to be, and if they deal with the kinds of gaskets you've going to use, then I would think that they would give more accurate assembly torque recommendations than would *any* of the procedures described in this section of the chapter.

IX. SUGGESTIONS FOR THE FIELD

In spite of the uncertainties, many people use many gaskets and most joints remain leak free for practical purposes. Here are some suggestions for reducing the possibility of a leak.

1. Follow Code practice. In spite of the questions now being raised, the Code defines the best-known way to design liquid or gas joints at the present time. They have safely and repeatedly "done the job" even if they don't answer all of the theoretical questions. Used with the suggestions given below, they should solve your problems as well—until a well-analyzed and well-tested alternative becomes available. Now, let's look at some ways to improve the odds.

2. Flange surfaces should be in good condition. Any radial grooves or erosion should be removed by machining, or the flange should be replaced. Annular grooves are usually not a problem. Ask the gasket manufacturer for surface finish recommendations, however.

3. Tighten flange bolts in a cross or star pattern such as that suggested in Fig. 19.22. Starting at one bolt and working your way around the flange in sequence 1, 2, 3, etc., can seriously warp the flange and/ or produce a wave in the gasket which will make sealing almost impossible.

4. Tighten several bolts simultaneously, if you can afford duplicate sets of tools, extra operators, etc. Hydraulic tensioners can be ganged and synchronized, for example. So, to a lesser degree of pull-down accuracy, can hydraulic wrenches. In Fig. 19.22, for example, you might tighten 1, 11, 6, and 16 together; then move on to 2, 12, 7, and 17. Or do them two or three at a time (e.g., 1, 11, and then 6, 16).

5. Use the best bolt control means you can afford, to minimize preload scatter, Experience indicates that *uniform* tightening is sometimes as useful as a particular value of preload. Ultrasonic control, for example, has been used successfully to control the tightening of many joints which have been notorious leakers in the past.

6. Uniform preload is important, but it often isn't sufficient. A high preload is always better than a low preload unless it damages joint members or the gasket. In fact, the higher the preload you can safely

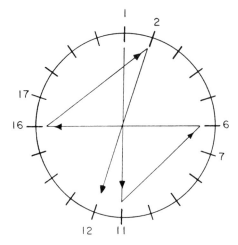

Figure 19.22 Tighten bolts on a gasketed joint in a star or cross pattern, such as the 1, 11, 6, 16, 2, 12, etc., sequence shown above. Such a sequence will minimize the danger of damaging the gasket during assembly.

achieve, the more closely the behavior of a gasketed joint will approximate the theoretical behavior, in spite of all the uncertainties and nonlinearities involved. High preload means that a given internal pressure will result in the least change in contact pressure on the gasket, for example. It is probably this fact—that "higher is better"—that has resulted in the degree of success found today. Mechanics, in desperation, ignore recommended torque values and just keep tightening those bolts until the leak stops, often taking the flange bolts past yield. The design theories are now beginning to catch up with this almost universal practice. Listen to Appendix S of the Code: if you have leaks, it's probably safe for you to do what most others do—tighten the bolts some more.

7. Use a gasket that is in good condition—new, if possible. And don't damage it as you assemble and tighten the joint.

8. Follow the engineering literature. Design recommendations and gasket practices are going to change drastically in the next few years.

9. Keep records which will allow you to recover critical information about your tools, procedures, crews, results. As mentioned several times, the large number of variables we have to deal with makes bolting an empirical art rather than an exact science at the present time. You'll learn more from your own experience if you can recover it accurately.

10. Be as consistent as possible. Define your procedures, then follow them. Avoid changes in lubricants, tools, bolts, etc., unless previous results have been unsatisfactory. Every such change can alter the set of variables you're dealing with, increasing the scatter in results.
11. Train and supervise your bolting personnel. Welding isn't the only assembly technology which requires skill and care! Tell the guy with a wrench what you're trying to accomplish; why you want to accomplish it; why good results aren't automatically achieved. (Training videotapes for pressure vessel and piping engineers and mechanics are available from the Electric Power Research Institute, P.O. Box 217097, Charlotte, NC 28221.)
12. Remember that thin flanges require more care and assembly accuracy than thick ones; and tailor your training, supervision, selection of tools, etc. accordingly.
13. If the gasket manufacturer recommends retightening the bolts after 18 or 24 hr to compensate for initial gasket creep, do so. If this involves hot bolting and/or dangerous fluids take suitable precautions.

X. A CASE HISTORY

Let's look at an actual example of a joint which has misbehaved. We'll look at the response of this joint to a change in operating temperature, and we'll then see what the owners did to eliminate leaks [25, 26, 46].

The joint in question is shown in Fig. 19.23. It is a horizontal heat exchanger used in 400°F, 600-psig steam service. Although it was designed some years ago by Code rules, it has an unusual configuration and involves four different materials. The design is not recommended!

The raised-faced flanges in this joint are made of carbon steel. The tubesheet is 304 stainless steel. ASTM A193 B7 alloy steel bolts are used. The gasket is spiral-wound; stainless steel and asbestos.

The coefficients of thermal expansion of the basic materials are shown below:

Material	Coefficient of expansion ($\times 10^{-6}$ in./in./°F)
Carbon steel	6–7
Alloy steel	6–7
Stainless steel	9–10

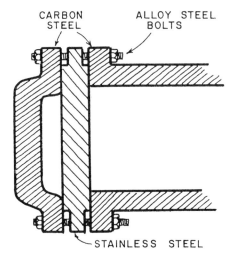

Figure 19.23 Simplified, cutaway view of a poorly designed "problem" heat exchanger discussed in the text.

Figure 19.21 shows the changes in stress and tempearture which preliminary calculations suggest occur in this joint when it is put in service.

As a first step the bolts are tightened at room temperature, creating an estimated gasket stress of 20 ksi (Part A, Fig. 19.21). Next, 100-psi steam is slowly backed into the heater over a 1-hr period. As this process starts, and the joint responds to the increase in contained pressure, bolt loads increase slightly and gasket stress decreases—as shown in part B of Fig. 19.24.

As this process continues, the flange and bolts start to heat up. Thermal expansion increases gasket and bolt stresses, as shown in part C of Fig. 19.24. Note that in parts B through E the stress on the OD of the gasket is greater than that on the ID. The difference is caused by the fact that this relatively light flange is rotating as it is loaded by pressure and by thermal effects.

The next step in the start-up process is to allow 600-psi steam to gradually force the condensate out of the unit, as suggested in parts D and E of the figure. During the first part (D) of this period the gasket temperature is approximately 400°F, but the bolts only reach 100°F—as shown at the top of Fig. 19.24. The gasket is hotter than the bolts during this half-hour period simply because it is in more intimate contact with the steam being introduced to the system.

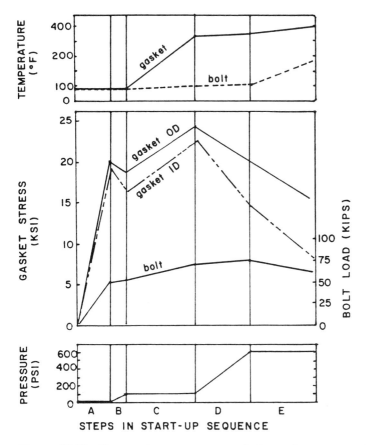

Figure 19.24 Temperature, pressure, and stress changes experienced by the exchanger shown in Fig. 19.23 during start-up. The start-up sequence is described in the text.

During phases C through E the stainless steel tubesheet—which is also in intimate contact with the steam—expands more rapidly than the B7 bolts, both because it has a greater coefficient of expansion and because it is heating up sooner than the bolts. This tends to raise the stress on the gasket.

The joint responds first, however, to the increase in contained pressure, which reduces the stress on the gasket (part D of Fig. 19.24). This reduction is partially offset by the differential expansion described above, but the pressure effect dominates this phase.

As the system continues to heat up, the bolt temperature eventually reaches approximately 250°F. This expands the bolts so that they partially catch up with the expansion of the stainless steel tubesheet. That process, of course, further reduces the clamping force which the bolts are exerting on the joint, so that the gasket stress drops some more, falling to 8 ksi on the ID and approximately 17 ksi on the OD in steady-state operation. In the steady-state condition, therefore, the average stress on the gasket is not much more than half of what was originally introduced at assembly. This amount of change is not uncommon, incidentally, and a residual stress on the gasket equal to half the assembly stress is usually acceptable.

This heat exchanger was a problem for its owner. It never stayed sealed for more than 3 months of operation and often opened up within a few days or weeks of plant start-up.

I should mention that the 8-ksi and 17-ksi stresses reported above are computed, average values. If such stresses had really been present, the system probably would not have leaked. Factors such as uneven bolt-up, circumferential temperature gradients, flange surface imperfections (the flanges had been repaired several times after leakage), the effects of rain on temperature distribution throughout the flange, etc., were not included in the analysis. It is suspected that actual gasket stresses were considerably lower than those reported above, at least in some regions of the flange.

The solution to this problem, in fact, was improved control of assembly preloads. The average preload was increased and the bolt-to-bolt scatter in assembly preload was reduced.

Ultrasonic equipment was used to monitor the change in length of the bolts in this joint as they were tightened with hydraulic torque tools. At first the torque and procedure which had "always been used" were used again, but now with ultrasonic equipment monitoring results. Resulting assembly tensions in the bolts are shown in line A of Fig. 19.25. Since these bolts were lubed with Moly, the engineers who had specified the assembly torque had assumed a nut factor of 0.17 and had, therefore, believed that the minimum preload created in any of the bolts in this joint would correspond to a change in length of the bolt of 16 mils, as shown by the dashed line in the upper part of the figure.

In fact, the actual preloads being achieved turned out to be an average of 12 mils ranging from as little as 6 to as much as 23 mils in individual bolts. This was presumably the condition of the joint when it was normally put into service and normally misbehaved after subsequent thermal cycles, rain showers, or the like.

With the results of the ultrasonic study in hand, the engineers increased the assembly torques and, in fact, used the ultrasonic equipment

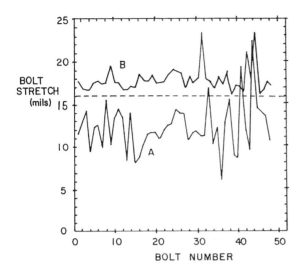

Figure 19.25 Recurring leakage problems with the exchanger shown in Fig. 19.23 were eliminated by improving the pattern of bolt preload, at assembly, from that shown in line A to that shown in line B. The mean preload was increased, and the scatter reduced, with the help of ultrasonic measurements of bolt tension.

to control final tensions. Minimum tension was now raised to 16 mils; maximum was still held at 23 mils. The mean preload now was at a level corresponding to an average bolt stretch of 18 mils, 50% higher than that obtained earlier. Final preloads are shown in line B of Fig. 19.25.

The result: No further leaks after 3 years of continuous service. The residual tensions in these bolts, incidentally, were measured after 1 year of service during a routine shutdown (and without loosening any of the bolts). The average bolt had lost approximately 30% of its assembly pre-load during this year of operation. But again, that had not opened up a leak, nor did one open up in the next 2 years of operation. All of this would appear to confirm the earlier discussion which suggested that an improvement in the assembly or y stress created in a joint would lead to a substantial improvement in the maintenance or m stress.

Several years later this flange, which was $5\frac{3}{8}$ in. thick, was replaced with an $8\frac{1}{8}$-in.-thick one. The thicker flange was much less sensitive to preload variations, and a return to simple three-pass torque control was possible [46].

REFERENCES

1. Swick, Robert H., Designing the leak free joint, *Machine Design,* pp. 100ff, January 22, 1976.
2. *Modern Flange Design,* Taylor-Bonney Division of Gulf and Western, Southfield, MI, 1979.
3. Stevens-Guille, P. D., *Successful Application of Spiral Wound Gaskets Requies ASME Code Revisions,* Atomic Energy of Canada, Ltd., Chalk River Nuclear Laboratories, Chalk River, Ontario, August 1973.
4. Raut, H. D., and G. F. Leon, Report of gasket factor tests, *Welding Res. Council Bull.,* no. 233, pp. 1–35, December 1977.
5. Roberts, I., Gaskets and bolted joints, *J. Appl. Mech.,* vol. 17, pp. 169–179, June 1950.
6. Stahl, I. G., An analysis of gasketed joints, *Automotive Des. Eng. (Great Britain),* vol. 12, pp. 75, 77, 79–81, October 1973.
7. Bazergui, A., Luc Marchand, and Gerald Lague, Analysis of exploratory gasket tests, Interim Report presented to Subcommittee on Bolted Flanged Connections of the Pressure Vessel Research Committee, May 1980.
8. Payne, J. R., The PVRC's gasket test program II—A Status Report, Reprint no. 19-80 of a paper presented at the 45th Midyear Refining Meeting of the American Petroleum Institute Tank Update, Houston, TX, May 14, 1980.
9. Bazergui, Andre, Oral report to Subcommittee on Bolted Flanged Connections of the Pressure Vessel Research Committee, at United Engineering Center, New York, May 1980.
10. Rodabaugh, E. C., F. M. O'Hara, Jr., and S. E. Moore, FLANGE—A computer program for the analysis of a flanged joint with ring type gaskets, Report ORNL-5035, U.S. Dept. of Commerce, National Technical Information Services, Washington, DC, January 1976.
11. NUREG/CR-1312 UCRL-51738, Table 1, p. 2, Nuclear Regulatory Commission, Washington, DC, March 1980.
12. Bazergui, A., L. Marchand, and H. D. Raut, Development of a production test procedure for gaskets, Bulletin 309, Welding Research Council, New York, Novermber 1985.
13. Payne, James R., George F. Leon, and Andre Bazergui, Getting new gasket design constants from gasket tightness data, Experimental Techniques, Society for Experimental Mechanics, vol. 12, no. 11, pp. 22s–27s, November 1988.
14. Payne, James R., New gasket constants from the PVRC gasket test program, Report to Subcommittee of Bolted Flanged Connections, PVRC of the WRC, New York, January 22, 1988.
15. Payne, J. R., Gasket description and typical tightness constants, Data presented at a meeting of the PVRC's Subcommittee on Bolted Flanged Connections, New York, May 17, 1988.
16. Kraus, H., and W. Rosenkrans, Creep of bolted flanged connections, Welding Research Council Bulletin no. 294, WRC, New York, May 1984.

Gasketed Joints and Leaks 727

17. Bazergui, A., Short term creep and relaxation behavior of gaskets, WRC Bulletin no. 294, WRC, New York, May 1984.
18. Payne, J. R., A. Bazergui, and G. F. Leon, New gasket factors—A proposed procedure, Proceedings of the 1985 Pressure Vessels and Piping Conference, PVP, vol. 98-2, pp. 85–93, ASME, New York, 1985.
19. Rhine, John, A. B. Owen, and Richard S. Owen, eds., Gasket Handbook, Lamons Metal Gasket Co., Houston, TX, 1980.
20. Technical Handbook, 2nd ed., Metallic Gaskets, Fluid Sealing Association, Philadelphia, PA, 1979.
21. Petrunich, P. S., Gasket designs and applications using flexible graphite, SAE Technical Paper no. 830214, SAE, Warendale, MI, 1983.
22. Bazergui, Andre, and Luc Marchand, Development of a tightness test procedure for gaskets in elevated temperature service, WRC Bulletin 339, December 1988.
23. Payne, James, J. Payne Associates, Inc., Long Valley, NJ, Private communication, 1988.
24. Marchand, L., M. Derenne, and A. Bazergui, Short term mechanical tests at elevated temperatrures, Report to the Joint Task Group of Elevated Temperature Behavior of Bolted Joints, PVRC, New York, Ocrober 1986.
25. Bickford, J. H., K. Hayashi, A. T. Chang, and J. R. Winter, A preliminary evaluation of the elevated temperature behavior of a bolted flanged connection, *WRC Bull.*, 341, February 1989.
26. Winter, J. Ronald, Use of an ultrasonic extensometer to determine variations in the assembly bolt loads of a problem industrial flange, *Experimental Techniques,* Society for Experimental Mechanics, vol. 12, no. 11, pp. 6s–11s, November 1988.
27. Bazargui, A., and G. Louis, Tests with varoius gases in gasketed joints, *Experimental Techniques,* Society for Experimental Mechanics, vol. 12, no. 11, pp. 17s–21s, November 1988.
28. Bazergui, A., L. Marchand, and J. R. Payne, Effect of fluid on sealing behavior of gaskets, Proceedings, 10th International Conference on Fluid Sealing, Innsbruck, Austria, BHRA, Cranfield, Bedford, England, 1985, pp. 365–385.
29. Bazergui, A., L. Marchand, and H. D. Raut, Further gasket leakage behavior trends, *WRC Bull.*, 325, pp. 1–10, July 1987.
30. Payne, James R., Andre Bazergui, et al., Summary report on elevated temperature tests for asbestos free gasket materials, MTI Publication No. 35, published by the National Association of Corrosion Engineers, 1990.
31. Derenne, Michel, Luc Marchand, Jim Payne, and Andre Bazergui, Elevated temperature gasket test: Final report, submitted January 1992 to the Subcommittee on Gasket Testing and Elevated Temperature Behavior of the PVRC, for publication in *WRC Bull.*
32. Softwear associated with the Klingerexpert gasket design, application, and analysis computer program, published by Rich. Klinger GmbH, Idstein, Switzerland.
33. Marchand, Luc, Andre Bazergui, and Michel Derenne, The influence of ther-

mal degradation on the sealing performance of compressed sheet gasket materials with elastomer binder, part 1, experimental methods, published by École Polytechnique de Montréal, 1992.

34. Comments by Yves Birembaut of CETIM, France, contained in the minutes of a meeting of the Committee on Bolted Flanged Connections of the PVRC, San Antonio, Texas, February 1, 1994.

35. Comments made in a letter to me from J. Ronald Winter of Tennessee Eastman Co., March 24, 1993.

36. Comments made by Max Ghirlanda of Guarco, Italy and reported in the minutes of an Information Meeting on Gasket Testing arranged by the PVRC, San Antonio, Texas, January 31 and February 1, 1994.

37. Jones, W. F., and B. B. Seth, Evaluation of asbestos free gasket materials, presented at the Joint ASME/IEEE Power Generation Conference, Boston, MA, October 21–25, 1990.

38. Cheng, Robert T., and Henry J. McDermott, Exposure to asbestos from asbestos gaskets, *Applied Ocupational and Environmental Hygiene,* vol. 6, no. 7, pp. 588 ff, July 1991.

39. Information distributed to its customers by the Flexitallic Gasket Co., Houston, TX, on March 6, 1990.

40. Brochure published by the Durabla Manufacturing Co., Lionville, PA, received January 1994.

41. Corol, Walter, Get better asbestos free gaskets, *Power,* pp. 31 ff, December 1990.

42. Jones, William F., and Brij B. Seth, Asbestos free gasket materials for turbines, *ASTM Journal of Testing and Evaluation,* pp. 94 ff, January 1993.

43. Moser, Mark A., and Brian C, Lehr, GMX: An expert system that selects the optimum gasket material for an application, technical paper from Armstrong World Industries, Inc., Lancaster, PA, received in 1992.

44. Minutes of the meeting of the Flange Parameter Study Task Group of the PVRC, Penn State University, College Park, PA, May 24, 1993.

45. Draft on request for proposal on bolted flanged joints subjected to externally applied loads, issued by the PVRC Task Group on Flange Parameter Studies, June 18, 1993.

46. Winter, J. Ronald, Gasket selection: A flowchart approach, presented at the 1990 ASME Pressure Vessel and Piping Conference, Nashville, TN, June 18–21, 1990.

47. Engineered gasket products, brochure published by the Mechnical Packing Division of Garlock, Palmyra, NY, 1990.

48. Brochure published by Armstrong World Industries, Inc., Lancaster, PA, received 1992.

49. Brochures published by the Lamons Gasket Co., Houston, TX, received 1993.

50. Derenne, Michel, and Luc Marchand, Elevated temperature tests on metal reinforced flexible graphite material: Progress report to the Subcommittee on Gasket Testing and Elevated Temperature Behavior of the PVRC, May 1993.

51. Marchand, Luc, Jerome Daoust, and Andre Bazergui, Effect of creep relaxation on gasket behavior at room temperature, preliminary report to the Subcommittee on Bolted Flanged Connections of the PVRC, October 1986.
52. Smoley, E. M., F. J. Kessler, R. E. Kottmeyer, and R. G. Tweed, The creep relaxation properties of a flat faced gasketed joint assembly and their relation to gasket and flange design, published by the SAE, reprint received from Armstrong Cork Co., no date given.
53. Fuji Prescale Film: An Instruction Manual, published by the Fuji Photo Film Co., Tokyo, Japan, no date given.
54. Cavicchioli, James A., Fuji prescale—A picture of pressure, SAE technical paper 850186, delivered at the SAE International Conference and Exhibition, Detroit, MI, February 25–March 1, 1985.
55. Czernik, Daniel E., and Frank L. Miszczak, A new technique to measure real time static and dynamic gasket stresses, SAE paper 910205, delivered at the SAE International Congress and Exposition, Detroit, MI, February 25–March 1, 1991.
56. Brochure published by Force Imaging Technologies, Chicago, IL, 1994.
57. Bazergui, Andre, and Luc Marchand, Comparison of elevated temperature tightness test procedures on two spiral wound gaskets, draft of a report to the Subcommittee on Bolted Flanged Connections of the PVRC, October 1987.
58. Minutes of the meeting of the Subcommittee on Gasket Testing and Elevated Temperature Behavior of the PVRC, Columbus, OH, May 23, 1994.
59. Brochure published by Marine and Petroleum Manufacturing, Inc., Houston, TX, received 1994.
60. Payne, James R., Effect of flange surface finish on spiral wound gasket constants, presented at the 2nd International Symposium on Fluid Sealing of Static Gasketed Joints, La Baule, France, Sepbember 18–20, 1990.
61. Private correspondence with J. Ronald Winter of Tennessee Eastman Co., 1992.
62. Payne, James R., Progress report for the PVRC Project: ASTM Standard/ASME Gasket Factors, under PVRC grants 91-5 and 92-16, presented to the Committee on Bolted Flanged Connections of the PVRC, October 13, 1992.

IV
USING THE INFORMATION

20
Selecting Preload for an Existing Joint

We have reviewed all of the main topics which concern the "beginner" who wants to learn about the design and/or behavior of bolted joints. Even though this was only an "introduction," it has been a long and sometimes complex story. It's often difficult for the novice to decide how much of this is pertinent for a given application. In most situations, he or she can't afford—or won't need—to address all of the issues we've discussed. We'll end our studies, therefore, with several chapters which show how to put it all together; how to focus on the factors of importance for a given application; how to identify those which can be ignored; and how to estimate the combined impact of the chosen factors in order to make better design or assembly decisions.

This first chapter will help you decide how much preload you should use in an existing joint; or "what torque?" as the question is usually stated. To answer this question we must briefly review many of the topics we've covered. Hopefully this summary will help you decide which of the many factors we've looked at will affect your results. And it will give you a reasonable way, I think, to make acceptable assembly decisions without having all the hard and fast data we'd like to have. Such data cost money—often a lot of money—and are rarely available to those who must deal with existing joints. In fact, in many (most?) situations you won't even need most of the procedures to be described in this chapter. You'll be able to use the "simple ways to select preload" described in Sec. II. If you've had problems with a joint, however, or have failure or economic concerns, then you'll need something practical but a little more elaborate; you'll need the procedures which make up the bulk of this chapter. For safety-related joints you may want to start with these procedures, but use the information in prior chapters and/or the many references to go well beyond them. So—here's how we would start our search for a "better torque" if we had had problems or wanted to do a better job with an existing joint.

In picking preload or torque for such a joint we should answer two questions: "How much initial, assembly clamping force do we *want* in this joint, considering the service loads and conditions the joint will face?" and "How much clamping force—and scatter in clamping force—can we *expect* from the assembly torques, tools, procedures, etc. we plan to use?" Let's see how we might answer these questions in an economically acceptable way.

I. HOW MUCH CLAMPING FORCE DO WE WANT?

A. Factors to Consider

We start by considering the in-service clamping force *needs* of the joint. What loads and service conditions must the clamping force between joint members resist? Note carefully that this is the in-service clamping force we're talking about; not just that theoretically created by the initial assembly preloads. That initial clamp must be high enough to compensate for all of the mechanisms which may reduce the clamping force to the in-service level, including:

Embedment relaxation
Elastic interactions
Creep of metal parts, gaskets, etc.
External tensile loads
Hole interference
Resistance of joint members to being pulled together
Prevailing torque
Differential thermal expansion

We must start by deciding how much clamp the joint will need in service; then try to estimate how much each of the above factors will have robbed from the initial clamp. Only then can we decide how much initial clamp—and therefore initial, assembly preload—we need. OK—here's a check list of the main factors which the in-service clamping force might have to resist.

Joint Slip

This is a key one because slip can cause a lot of problems. It can cause unfortunate stress concentrations in a slip-distorted structure. It can cause fretting corrosion or fatigue of joint members. It can cause self-loosening, or misalign and cramp bearings, etc. It's relatively easy to calculate the clamping force to reduce slip, however, if we know the magnitude of the shear loads imposed on the joint.

Let's assume that a shear load of Lx is to be imposed on this joint. To avoid slip the frictional forces created by the clamping force must exceed the external load as follows:

$$F \geqslant Lx\mu \tag{20.1}$$

where F = clamping force on the joint (1b, N)
$\quad\quad Lx$ = external load (1b, N)
$\quad\quad \mu$ = coefficient of friction (typically 0.15–0.30)

Self-Loosening

We learned in Chap. 16 that self-loosening will occur when transverse loads cause slip between joint members and thread surfaces. Although it is often difficult to quantify the vibratory or other forces creating self-loosening, the joint slip equation above could be used to estimate the point at which the joint members will slip if the external loads are known.

If external loads cannot be estimated and the application is likely to involve vibration or other forms of self-loosening, then a good practice would be to plan for the "maximum clamping force the parts can stand." The more the better, up to, but probably not exceeding, yield.

Pressure Loads

The influence of pressure loads on bolted joints is explained in some detail in Chap. 19. One way to quantify the required initial and in-service clamping forces is to use the equations of the Boiler and Pressure Vessel Code. These are currently (early 1995) being modified to utilize the new PVRC gasket factors (see Chaps. 19 and 22) but the equations found in the Code at present are:

$$W_{M1} = \frac{\pi G^2}{4} P + 2\pi GbmP \tag{20.2}$$

$$W_{M2} = \pi bGy \tag{20.3}$$

where W_{M1} = tension in the bolts in service (1b, N)
$\quad\quad\quad W_{M2}$ = initial tension in the bolts at assembly (1b, N)

G = diameter of gasket (in., mm)
P = contained pressure (psi, Pa)
b = effective width of the gasket (in., mm)
m = gasket maintenance factor (see Chap. 19)
y = initial gasket stress at assembly (see Chap. 19)

As suggested in Chap. 19, these equations assume that the entire pressure load will be seen by the bolts (i.e., ignoring the implications of the joint diagram), but these equations or their equivalent could presumably be used for an approximate (and conservative) estimate of the clamping force required in most gasketed joint situations.

Joint Separation

In some applications, a noncritical foundation bolt is an example, gravity holds the joint in place and it is sufficient for the bolts merely to maintain alignment. In this situation, clamping force could be as low as 0 without risk of joint failure. Even here, however, some clamp would be useful merely to retain the nuts. This little clamp would also be acceptable in those structural steel joints which are not slip critical. In most tension joints, however, it's as important to avoid separation as it is to avoid joint slip—maybe even more important. Separation can lead to such horrors as gross leakage and low-cycle fatigue of bolts. To avoid separation we must be sure that the initial preloads are high enough to compensate for all of the clamp loss factors listed a minute ago and still leave some residual in-service clamp. A margin of safety, if you will.

Fatigue

Although joint separation can reduce the fatigue life of bolts by a substantial amount it is not, as some handbooks imply, the only cause of fatigue failure. "Too little clamping force" or, less commonly, "too high a mean bolt tension" can also cause problems. This time, however, we have no simple equation to compute the amount of in-service bolt tension or the related interface clamping force required. A fairly complex analysis is required, as described in Chap. 17 and in the references cited at the end of that chapter.

B. Placing an Upper Limit on the Clamping Force

When determining the amount of clamping force required to combat separation, self-loosening, slip, or a leak, we are interested in establishing the essential minimum of force. In each of those situations, additional clamping force is usually desirable (for added safety) or is at least acceptable. You might remember that another early theme we addressed was "We always want the maximum clamping force the parts can stand." There is, however, always some upper limit on that clamping force. If that weren't the case, we could simply "tighten them a lot more" to avoid failures. Instead, we must define an upper limit for our application.

In fact, this is one of two key issues we first addressed in Chaps. 12 and 13. Remember our attempts to identify "the maximum tension which will be seen by some of the bolts" and "the minimum clamping force we can expect in some joints"? Well, in Sec. A above we discussed the minimum clamping force requirements for a joint; now we're about to address the equally important issue of maximum tension or stress in the bolts because this will usually—though not always—be the thing which places an upper limit on the amount of clamp we want. We might *like* more; but not if it means broken or threatened bolts. So—here are some bolt factors which limit clamping force.

The Yield Strength of the Bolt

There is a good deal of debate about this in the bolting world at the present time, but most people feel that it is unwise to tighten bolts past yield in most applications. There are many exceptions, with structural steel being the most obvious. As we saw in Chap. 8, torque-turn equipment which tightens the fastener past yield, or to yield, is growing in popularity for automotive and similar applications. In general, however, we usually won't want to tighten them past yield during initial assembly. Bolt yield, then, is one easy-to-estimate, "worst-case," upper limit on the in-service clamp force.

Thread-Stripping Strength

Obviously, we will never want to tighten the fasteners past the point at which their threads will strip. This then provides another, simplistic, worst-case upper limit. If we didn't consider this limit when selecting the bolts, we should do so now (unless one of the limiting factors listed below will obviously dominate our decision).

Design Allowable Bolt Stress and Assembly Stress Limits

We always want to identify any limits placed on bolt stress by codes, company policies, standard practices, personal bias, or the like. Both structural steel and pressure vessel codes, for example, define maximum design allowable stresses for bolts. It's necessary, however, to distinguish between a maximum design stress and the maximum stress which may be allowed in the fastener during assembly. This will differ from the maximum design allowables if a design safety factor is involved. In the structural steel world, for example, bolts are frequently tightened well past yield, even though design allowables are only 35–58% of yield. Pressure vessel bolts are commonly tightened to twice the design allowable, as explained in Chap. 19. Aerospace, auto, and other industries may also impose more stringent limits on design stresses than on actual stresses to force the designer to use more or larger bolts than he might otherwise select (and, therefore, to introduce safety factors in the design).

Torsional Stress Factor

If the bolts are to be tightened by turning the nut or the head, then they will experience some torsional stress as well as tensile stress during assembly. If tightened to yield, they will yield under a combination of tensile and torsional stress. If we plan to tighten them to or near yield, it's pertinent to reduce the maximum tensile stresses allowed at assembly by a "torquing factor" which makes room for the torsional stress. If as-received steel-on-steel bolts are used, then a reduction in the allowable tensile stress of 10% is probably reasonable. If the fasteners are to be lubricated, you might use 5%. The true amount of strength absorbed by torsion will be determined by all of the variables which affect the torque-preload relationships, so the torsional effect on the tensile capacity of the bolts may be much greater than 5 or 10% (and is as difficult to predict as the specific torque-tension relationship for a given set of parts). Anyway, 5–10% will probably be acceptable in most applications. Most people don't tighten bolts to the yield point anyway.

Shear Stress Allowance

If the bolts will also be exposed to shear stress we must take that into account in defining maximum assembly preloads and the resulting in-service clamping force, since shear stress will reduce the amount of bolt strength capacity available for the tensile stress.

The shear reduction is explained and illustrated in Chap. 14 [Eq. (14.1)].

Stress Cracking

As we learned in Chap. 18, stress cracking is encouraged by excessive tension in the bolts and we are told to "be alert if service loads exceed 50% of yield," at least for low-alloy quenched and tempered steels. See the many bolt stress vs. yield strength curves in Chap. 18 for a more complete definition of this important upper limit.

Combined Loads

As we try to determine the maximum stress the parts can stand, to avoid damage to the parts, SCC, or other problems, we must evaluate our estimates in light of the anticipated service loads on the bolts. Most of these service loads would add to the bolt stresses introduced during assembly (the preloads) even as they reduce interface clamping force.

The joint diagrams of Chaps. 12 and 13 can be used to add external loads to preloads. The effects of a temperature change can be estimated using the procedure described in Chap. 13. The bolts must be able to withstand worst-case combinations of these pre- and service loads.

All of the limiting factors discussed so far deal with bolt strength, which, in turn, limits the clamping force available for the joints the bolts are used in. There are also a few joint factors which can limit clamping force. Here are some of the most common ones.

Damage to Joint Members

Too much tension in a bolt can cause its head and nut to embed themselves into the surfaces of the joint; not just by a "normal" amount of a few mils, but enough to cause visible damage to the joint surfaces. The VDI Directive 2230, for example, says that the "boundary surface pressure" (P_G) of joint materials is usually slightly greater than the yield strength of the material, and they recommend that bolt tension not exceed the value suggested by this equation.

$$P_G A_P \geq 0.9 \text{ Max } F_B \tag{20.4}$$

where A_P = contact area (e.g., between nut face and joint) (in.^{2}, mm^2)
F_B = tension in the bolt (lb, N)
P_G = boundary surface pressure (psi, N/mm^2)

The 0.9 is added as a safety factor.

Distortion of Joint Members

Joint members can sometimes be distorted by excessive bolt loads. For example, the outer ends of raised face flanges can be pulled toward each other—perhaps bent—by too much preload. This can unload the ID of the gasket, opening a leak path.

Gasket Crush

Excessive preload can so compress a gasket that it will not be able to recover when internal pressure or a thermal cycle partially unloads it. Contact the gasket manufacturer for upper limits. Note that these will be a function of service temperatures (see Fig. 19.6 for example).

C. Summarizing Clamping Force Limits

Let's try to summarize the range of clamping force we want. Not too high—that might cause joint problems or, more commonly, bolt problems. Not too low—that might cause joint slip or separation or leakage.

Figure 20.1 illustrates one possible scenario based on some of the factors we've just discussed. Each factor is described as a percentage of the ultimate strength of the bolt, to give us a common, vertical axis. The figure shows that 100% of ultimate would break the bolts. It suggests that bolt tension above about 80% of ultimate would take the bolts past yield, and assumes that we don't want to do that in our application: this region is considered "unusable."

The chart suggests that we subtract about 5% from the tensile yield strength of the bolts to account for torsional stress introduced when torque is applied. It suggests that we might have reduced the maximum allowable bolt stress still further to accommodate a Code or other specification limit. And it further reduces the limit by a 5% "safety factor" to acknowledge that our data are limited and/or that we haven't included such factors as external loads or differential expansion which can add to bolt tension. The final result is a suggested upper limit on desired clamp force equal to 62% of the ultimate strength of the bolts.

The chart also defines an acceptable lower limit on clamping force; suggesting that we want a clamp equivalent to at least 30% of ultimate to make sure the nuts don't fall off and that, more seriously, we want a clamp equivalent to as much as 42% of ultimate to prevent that failure mode (or modes) we're most concerned about, whether that be self-loosening, leakage, slip, fatigue, or combination of these. Since the exact amount of load or vibration to be seen by a joint is often hard to calculate, and since "more is generally better," we throw in a larger safety factor on the low end than we did at the high. The result: a lower limit on desired clamp which is equivalent to 48% of the bolt's ultimate strength (or about 60% of yield if the bolt yields at 80% of ultimate).

All of which is very broad-brush and seat-of-the-pants, but that's the way most assembly preload or torque decisions are made in the field. The principal advantage of the procedure is that it forces us to consider separately, and assign at least gut-feel values to, those factors we think

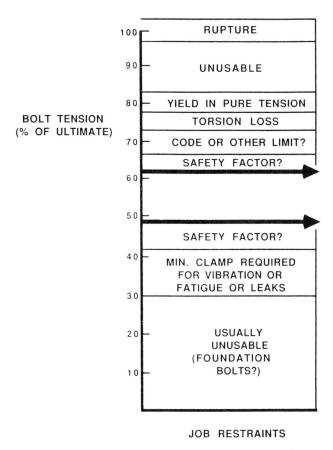

Figure 20.1   Chart summarizing the clamp load decisions made for a hypothetical joint. See the text for a detailed discussion.

might affect our joint. The chart shown in Fig. 20.1, incidentally, can also be constructed in terms of yield strength or torque, for single bolts or the whole group.

OK—we have now roughly defined the range of clamping force we'd like to see in our joint, a range defined in part by limitations in the strength of the bolts. Now we want to pick a target preload to specify for assembly purposes. If dealing with a problem joint we'll also want to estimate the range of clamping force we can expect to achieve during assembly, taking our choice of preload plus our assembly tools and procedures into account. It will often be necessary, too, to estimate how much that range might be increased by postassembly and in-service conditions. With all this information in hand we can then compare the desired range of clamping force to the anticipated range. If they agree, we're done. If they don't agree we'll have to take steps to correct our assembly procedures—perhaps "pick a better torque"—to make them agree.

Before proceeding, let's acknowledge, once again, that it usually won't be necessary for us to go through the "full" procedure I'll describe in a minute. If service conditions are reasonable and we have no safety concerns, we can choose a preload or torque by some of the very simple ways described next. We'll want to go beyond these things only if we're concerned about results, or—more commonly—have had problems with this joint in the past.

II. SIMPLE WAYS TO SELECT ASSEMBLY PRELOADS

In most situations, as the bulk of prior experience tells us, we can pretend we never read this book, and can "pick a preload" by one of several simple, time-proven methods that have worked on most joints in the past. And we don't want to forget this point because, if we do, we can waste time and money solving problems which don't exist. So, before considering a procedure to use when it is necessary to "do it right," let's look quickly at some of the simpler ways that will often be sufficient.

A. Best Guide: Past Experience

If you've had previous experience assembling this joint, and the results have been acceptable, don't change your procedures or tools. As we've seen, bolting involves more variables than we can cope with. We can *never* predict results with perfect accuracy; so prior experience, if satisfactory, is better than all of our theories. Leave it alone! Note that the tools and procedures you've used have, by default, "selected" a preload range for you. It worked. It's acceptable.

Note that you won't usually be able to identify a successful preload per se in situations such as those we're discussing. You'll know what torque the mechanics have used; or which size of tool; or which procedure.

Note that even if we know exactly which torque has been used on this joint we don't know "the" preload created by that torque. A given torque—*any* torque—will create a range of preload in a group of bolts. Variations in friction can scatter the preload by as much as ±30%. Elastic interactions can leave residual preloads scattered by max:min ratios of 20:1 or more. Embedment, tool error, operator problems, etc. etc. can also enter the picture as we have seen. In spite of all this, in spite of the fact that we haven't measured, or even been aware of, such factors, if the tools, procedures, torques, etc. we've used in the past have *worked*—we should keep using them.

B. Second Best: Ask the Designer

If you've had no prior experience with this joint, or have had some problems, your best source of information is the man who designed the joint. As we've seen, many factors must be considered in the design of a successful joint. It's a rare customer or user or assembler who knows enough about the materials and configurations of—or loads on—a joint to duplicate the designer's expectations when it comes to specifying the correct clamp force. If the selection of preload matters, the designer should know what the desired range of values should be. Presumably he will have gone through a design procedure similar to that discussed in Chap. 21.

There will be many cases, however, in which the designer has done no such thing—but even here he may have been guided by the past experience of other customers, or by general experience with this type of product. Even if he hasn't made an analysis, therefore, he may be able to recommend a preload—again, specified as a torque in most situations. Note that the designer's recommendations can sometimes be found in the operating or maintenance manuals which came with the equipment. If not—give the designer a call!

C. Unimportant Joint: No Prior Experience

In another common situation, you'll have no prior experience with this joint, and no way to reach the designer (or he can't help) and so must pick the preload (probably a torque) yourself. If the joint is a "common" one, and you have no real concern about the consequences of failure, then the normal procedure is to pick a torque from a suitable table. You might use that in Appendix F of this book, for example. Picking a torque (and a tool and a procedure) determines the range of preload you'll achieve at assembly, even if you won't be able to quantify this range with any precision. Most joints are overdesigned; most joints will behave themselves when assembled with a torque selected this way.

Again, picking a torque will, in effect, pick a range of preload, as suggested in Sec. A above.

D. When More Care is Indicated

If your past experience with the joint could be better and/or you're a little concerned about the consequences of failure, you might want to compute an appropriate torque instead of picking one from a table. One way to do this:

Determine the yield strength (S_y) of the bolt material at the operating temperature of the joint (data on yield strengths will be found in Chap. 4).
Pick a target "percentage of yield" (P) from Table 20.1.
Determine the tensile stress area of the bolt (A_s) (see Appendix F or H)
Estimate the lubricity of the fastener by picking a nut factor (K) from Table 7.1 in Chap. 7.
Now use the following version of the short-form, torque-preload equation to compute an assembly torque (T) in in.-lb.

$$T = KDPS_yA_s \tag{20.5}$$

(D is the nominal diameter of the bolt in inches.)

Why is a computed torque "better" than a torque taken from a table? The computation takes specific job constraints into consideration; a table is based on "normal" assumptions which may or may not be valid for your application. In the procedure above, for example, we made application-specific decisions based on:

Percentage of yield strength as a preload (not torque) target
The lubricant we're planning to use
The operating temperature of the joint

In addition, or instead, you could base the desired preload on such things as the fatigue endurance limit of the bolts or on the limitations suggested by stress corrosion cracking concerns instead of basing it on a percentage of yield. But the end result at this level of the preload selection process is still a torque, with all the subsequent uncertainties of the torque-tension relationship. Neither this process nor a torque table should be used on critical joints.

And again—we end up here with a specified torque and with the un-specified but unavoidable range of preload it will create.

E. If Improvements Are Required

If you've used a torque based on past experience, or picked from a table, or computed with the short-form equation, and have had nuisance prob-

Table 20.1 Typical Target Preloads as a Percentage of the Yield Strength of the Bolts

This percentage of yield	Might be appropriate for applications such as these
25	Unimportant nongasketed joints exposed to static loads, foundation and anchor bolts under static load, also joints where there have been serious stress corrosion cracking problems
40	Gasketed joints in routine service, including those covered by the ASME Code, which have not given problems
50–60	Average nongastketed joint, with "normal" safety or performance concerns, where past experience does not suggest higher or lower preloads; a good place to start a search for the optimum preload when some trial and error is acceptable
	Probably the maximum acceptable preload for gasketed joints designed to ASME Code rules (although there will be a few exceptions)
70–75	Upper limit for nongasketed joints with which you've had "low preload" problems in the past (leaks, self-loosening, fatigue, etc.) and where torque control will be used at assembly
85–95	Joints which have had consistent "low preload" problems in the past; where the need to avoid failure dictates the use of special techniques such as stretch or ultrasonic control; and where service loads (or ignorance of service loads) make it unwise to take the fasteners any closer to the yield point
100	Structural steel bolts tightened by turn-of-nut procedures; also high-performance or problem joints facing self-loosening, fatigue, or other "low preload" problems, but where service loads can be predicted with sufficient accuracy to guarantee that the fasteners won't be "ratcheted" to rupture later on; sometimes used in gasketed joints in automotive applications, for example; requires special assembly techniques

lems, and can't get any help from the people who designed the equipment, you might want to try a different torque. Eureka! Seriously, "trying a different torque" is valid. Bolting is an empirical art at present; "experiments" are often the most cost-effective way to make improvements. They may, in fact, be your only choice. (Again, you shouldn't use this or any other "simple" preload selection technique if the joint is critically important and failure would be disastrous.)

What torque should you try? If the problems you've experienced included a leak, self-loosening, joint slip, joint separation under load or fatigue, more preload is indicated. Try increasing it (e.g., try increasing assembly torques) by 10% or so.

If the problem has involved stress corrosion, stripped threads, crushed gaskets, or excessive flange rotation, try 10% less preload or torque the next time.

In either case, keep good records so that you can benefit fully from your "experiments" and ultimately find the optimum preload, torque, procedure, etc.

Note that "picking a better torque" probably won't reduce the range of preload created in the bolts; it will merely move the average or mean preload to a higher or lower level. But that will frequently solve a problem.

F. Selecting Preload for Critical Joints

The procedures described above can be used in most applications. Obviously they can't be applied—except perhaps as a first step—to critical or safety-related joints. A good design engineer must then be called upon to select a preload, and an assembly procedure (to control the range). Either he—or a well-trained bolting engineer—might be able to use the more rigorous procedures described in Sec. IV of this chapter; hopefully with the help of well-founded data on the factors which can affect results in a given application.

Let's assume that we've used the quick-and-dirty procedures described above to pick a target preload which we'll use to specify an assembly torque. We know that this preload will result in a range of residual preload, and that that range will be further modified by service conditions to give us a still wider in-service range. We're very interested in estimating the size and limits of this range, because that range in preload will determine whether or not we have created at least the minimum clamping force we decided we needed in Sec. I. Let's see how we might estimate the in-service range of clamp force.

III. ESTIMATING THE IN-SERVICE CLAMPING FORCE

Placing exact limits on the in-service clamping force will usually be impossible. There are too many variables, and we'll rarely have good data with which to define their exact effect on the overall results. We can, however, often use the procedure I'm about to describe, plus general knowledge of "typical" values, to give us some idea as to the possible range. The answers we get will be good enough for most applications.

The procedure we'll follow is this:

We'll list the variables, or groups of variables, which we think will affect the preload achieved at assembly and/or the stability of that preload in service. Most, if not all, of the possible assembly problem groups are listed in Fig. 6.28. It would also be useful to consult the chapter on the assembly technique (torque or turn, etc.) you're planning to use. Or you might identify possible variables by looking at the table of contents or the index to this book. In most situations, incidentally, you'll be primarily concerned with a half-dozen or so key factors. You don't have to consider every little thing which could affect the outcome.

Next we'll estimate the contribution which each key variable, or group, might make to the preload scatter experienced during assembly, or might contribute to the subsequent instability. We'll deal with the key variables one at a time, or one group at a time, as we make these estimates.

Then we'll use the chart shown in Fig. 20.2 to combine these estimates and so arrive at final answers to our questions.

The procedure is simple and is easy to use, in part because it's heavily based on assumptions, estimates, and "typical values." In most situations this is all we'll have. Accurate data will be rare, and only in the most critical situations will we be able to afford the time and money to make the experiments or analyses required to obtain hard data. Nevertheless, the proposed process will take us well beyond the "usual procedure" for estimating in-service bolt loads or clamping force on the joint.

What is the "usual procedure"? Most people seem to base their estimates of assembly accuracy—or in-service preload scatter—entirely on the process used to tighten individual bolts. If torque control is used at assembly, for example, it's usually assumed that final assembly preloads and in-service bolt tensions will be scattered $\pm 30\%$ around a mean or target value of tension. But the group of variables which create the uncer-

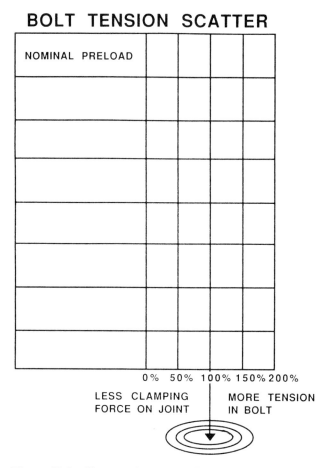

BOLT TENSION SCATTER

Figure 20.2 Chart used to summarize the results of scatter in the major variables which affect the amount of bolt tension and interface clamping force expected during assembly and when the joint is put into service.

tainty in the torque-tension relationship for single bolts is only one of the many groups of variables which can affect assembly or in-service results. Elastic interactions, relaxation effects, thermal effects, and many other things can also play a significant role. The procedure I'm about to describe allows us to consider all or most of the factors that will make a significant difference in a given application.

As mentioned, Fig. 20.2 is the worksheet that we'll use to estimate the combined effect of the factors we think will affect the outcome in a particular application. Let's work an example to see how it's used.

A. Basic Assumptions

Let's assume that we'll use a torque wrench to tighten the bolts on a straightforward, nongasketed joint. We'll also assume that we've used the procedure described in Sec. I to define an acceptable range or preload (48–62% of the bolt's ultimate tensile strength as in Fig. 20.1). Now, after reviewing Fig. 6.28 and Chaps. 6 and 7 (on torque-controlled assembly), let's further assume that we've decided that five factors, each of which involves a different group of variables, will determine assembly results and subsequent changes of bolt tension in service. At the end of the example we'll briefly review some of the other factors we might have considered.

We've picked the five factors listed below. We could have picked more—or less. There's nothing magic about five. The following list, however, includes several different types of problem or variables; some assembly, some in-service; some introducing plus or minus scatter, others biased in only one direction; some uncertain, others inescapable; etc. So this is a good list with which to study the procedure.

Tool accuracy
Operator accuracy
Control accuracy
Short-term relaxation
External loads

B. Combining the Scatter Effects

Now let's use the worksheet to combine these assembly and service factors. Then we'll be able to compare anticipated results with the desired results established in Sec. I.

We start with the concept that there is an "ideal" in-service tension for this joint. If we could always introduce that tension at assembly, and if subsequent load and environmental factors never altered it, we'd have achieved perfection. We would have "hit the bull's-eye" 100% of the time. We represent this ideal tension by a vertical line drawn from the top of our worksheet to the center of the target at the bottom. As a first cut, we'll assume that this tension is at the midpoint of the 48–62% range established earlier—i.e., our target, in-service bolt tension corresponds to 55% of the ultimate tensile strength of the bolt. This ideal bolt tension, of course, will create an ideal clamping force between joint members.

Now we introduce our first "factor" of concern—the accuracy of the tools to be used at assembly. Note that anything which reduces the preload or service tension in the bolts also reduces the all-important clamping force on the joint, making it less than our ideal or target value. Anything which increases the preload or tension means more tensile stress in the bolt than is "ideal." Let's see what our first variable, tool accuracy, does to us.

By "tool accuracy" I mean the accuracy with which the tool produces the thing it's supposed to produce. In this example, the tool produces "torque" and we'll assume that it does this with an accuracy of ±5%. (We'll consider the accuracy with which that torque creates preload in a minute.) We'll assume that this 5% (and the other percentages to follow) are "3 sigma" values, and represent the worst-case errors we can expect to encounter in nearly all such assemblies. The worksheet could, instead, be based on one standard deviation (68% of the population), or on 2 sigma values (95%), to be less conservative. But we'll assume 3 sigma (99%) for now.

In the real world, would we really consider tool accuracy to be a key variable? Sure. It's not uncommon to believe (sometimes correctly) that a more accurate tool can make a significant difference. For our example, let's pretend that a torque wrench salesman has heard that we've had vibration loosening problems, and he says that our problems will be solved if we use his 5% wrench instead of the 15% multiplier arrangement we've been using. We can evaluate his claims as we make our analysis; can decide if a better, more accurate, torque tool would be likely, or unlikely, to reduce our problems.

We said that we expected the torque accuracy of this tool to be ±5%. Since torque is linearly related to preload (see the short-form equation in Chap. 7), this means that tool inaccuracies, worst case, will introduce as much as 5% more tension in some of our bolts than we'd like to see; or create, in other assemblies, as much as 5% less clamping force in the joints. These are both undesirable results, and therefore are of concern to us. We don't want to break the bolts (too much tension) and we don't want the clamping force to be too low (remember that we always want the maximum clamping force those parts can stand). So, as we plot tool accuracy on our worksheet, as in Fig. 20.3, we label plus errors as "more tension in the bolt" and negative errors as "less clamping force on the joint." The accumulated 3 sigma scatter at this point is, of course, just ±5%.

BOLT TENSION SCATTER

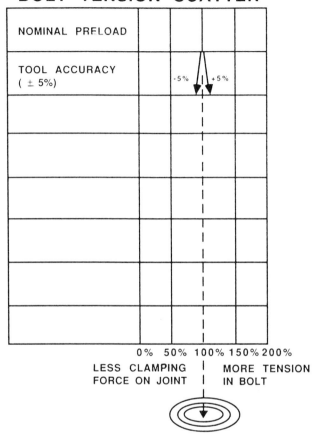

Figure 20.3 The first variable, tool accuracy, is assumed to contribute ±5% of scatter to the bolt tension or clamping force.

Now we introduce the second factor of concern: operator accuracy. (By "operators" I mean the mechanics or assemblies or others who "operate" the tools; use the wrenches; tighten the bolts.) This factor defines the amount of scatter introduced by operator errors, operator carelessness, poor accessibility (which makes it difficult for the operator to do a "perfect" or "ideal" job), etc. Let's assume that this factor will contribute, alone, ±10% of scatter.

Do operators contribute this much uncertainty to bolted joint results? They certainly do. Some people claim, in fact, that *all* in-service problems result from improper assembly or maintenance practices: the result of operators who don't care or lack skill. We'll assume that we're dealing with well-trained mechanics who know that bolting is important. They'll "contribute" ± 10% to the scatter primarily because accessibility is poor and working conditions are difficult. Even with a perfectly accurate torque wrench they won't always be able to apply the "ideal" or target torque to each nut.

Note that this contribution to the scatter will be especially difficult to quantify. Our procedure, however, makes our guess more realistic by isolating it from the other variables we must deal with. Anyway, let's assume ± 10% for the example.

We don't just add this ± 10% to the tool's ± 5%. Probability theory tells us that it's unlikely that we'll experience a worst-case tool error and a worst-case operator error in the same assembly. We may see that once in awhile, but in 99% of the assemblies (3 sigma again) the combined effect of these two errors will be less. We can compute the probable combined error by taking the square root of the sum of the squares of the two error percentages, as follows:

$$\pm V_T = \pm [V_{TL}^2 + V_{OP}^2]^{1/2} \tag{20.6}$$

where V_T = the total 3 sigma scatter
          V_{TL} = the 3 sigma scatter contributed by the tool
          V_{OP} = the 3 sigma scatter contributed by the operator

So the combined effect of these two errors is only ± 11%, as shown in Fig. 20.4.

Now we add the third factor, control accuracy. This is the accuracy with which the selected control variable—torque in this example—produces the thing we're interested in—which is always bolt tension. Table 20.2, incidentally, lists typical control accuracies for a variety of assembly tools and procedures. In our example, we'll use the conventional wisdom and say that the torque-tension scatter will be ± 30%. We could introduce elastic interactions as an additional variable affecting the bolt tension achieved at assembly, but won't for this first example. We'll assume a fairly rigid joint with metal-to-metal contact and insignificant interactions.

We take the square root of the sum of the squares again, taking three variables into account this time, and now find a cumulative scatter of ± 32% (Fig. 20.5).

Our next variable, short-term relaxation, is handled the same way; but now we have to compute positive and negative errors separately,

BOLT TENSION SCATTER

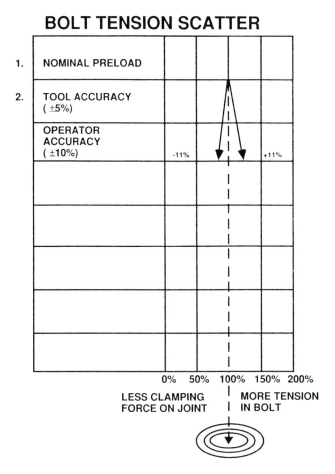

Figure 20.4 The second variable studied in the example given in the text is operator accuracy; and is assumed to contribute ± 10% of scatter. The accumulated scatter at this point is ± 11%. (See text for that calculation.)

since they're different. Let's assume embedment relaxation of − 10%. Relaxation will never increased bolt tension, so the plus scatter is zero. Cumulative errors now stand at, worst case, + 32% and − 34%. This tells us that, over a long period of time and many assembly operations (i.e., a large statistical sample), we could expect to see, worst case, 32% more tension in some bolts than the "ideal" hoped for; and that, in other cases (not simultaneously!), we'd see 34% less than the ideal clamping force in some of the joints (Fig. 20.6).

Table 20.2 Preload Scatter Reported for a Variety of Bolting Tools or Procedures

Torque control with hand wrench	±30
Stall torque air tool	±35
Click type torque wrench	±60 to 80
Torque wrench plus multiplier	−70 to +150
Turn-of-nut (structural steel)	±15
Computer-controlled air tool to yield point	±3 to 10
Rockwell International's LRM (torque-angle) system	±3 to 10
Strain gaged load washers	±15
Strain gaged bolts	±1
Swaged lockbolts	±5
Air-powered impact wrench	−100 to +150
Hydraulic tensioners with vernier gage readout	±20
Operator feel	±35
Bolt stretch (micrometers)	±3 to 15
Ultrasonic control	±1 to 10

All values are in "%" and come from the author's own experiences or from the *Standard Handbook of Machine Design*, edited by Joseph R. Shigley and Charles R. Mischke, McGraw-Hill, New York, 1986, p. 23.23.

This summarizes the situation at the end of assembly. Note that the ±30% torque-tension variable has dominated the results so far (justifying those who often assume that it's the only variable!). It wouldn't have dominated results if we had included elastic interactions in our analysis, or had assigned a large scatter to the operators.

The final variable we've selected involves an in-service condition. We'll handle it in the same way, however. When we put the joint into service it sees an external tensile load. This load will both increase the tension in the bolts and decrease the clamping force on the joint (remember the joint diagrams in Chaps. 12 and 13). If we assume a 5:1 joint-to-bolt stiffness ratio, and an external tensile load equal to 25% of preload, we might see a change of +7% in bolt tension and −20% in clamping force on the joint when the load is applied.

This factor differs from the others we've considered in several ways. First, it simultaneously affects both bolt tension and clamping force. Most of the previous variables would affect both, but not in the same assembly.

Second, the external load may be relatively easy to quantify, compared at least to such factors as operator accuracy or torque-tension scatter.

BOLT TENSION SCATTER

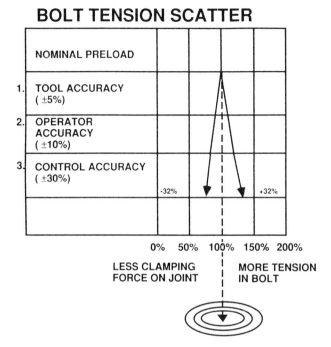

Figure 20.5 Control accuracy (torque vs. tension) scatter of ±30% raised the accumulated scatter to ±32%.

Third, this effect will probably be unavoidable. Tools and operators may perform perfectly, in some assemblies at least, but all assemblies have been designed to carry an external load. When we combine this "certain" variation with the earlier "probabilistic" ones, therefore, we're adding apples to oranges and will undoubtedly distress our neighborhood expert on probability. But I think the procedure is useful.

We're trying to estimate—roughly, simply, inexpensively—the combined effect of many variables. Most of the "data" we're using are "soft," to say the least. This latest violation of probability mathematics (if that's what it is) won't, in my opinion, make things much worse.

Adding the first service factor of concern, the effect of tensile loads, doesn't change the picture much, thanks to the "smoothing" effect of the square root of the sum of squares equation. The accumulative worst-case errors now stand at +34% and −39%, as in Fig. 20.7.

Let's pause at this point and see what those errors suggest. Plus 34% means that some of our bolts will experience 134% of ideal or optimum

BOLT TENSION SCATTER

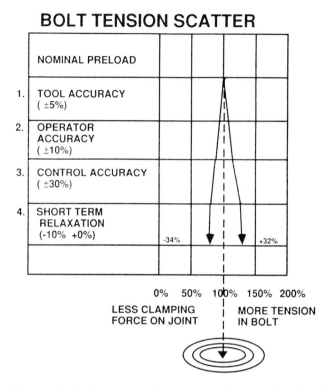

Figure 20.6 Short-term relaxation can decrease the clamping force but not increase bolt tension. The accumulated scatter is now $+32\%$ and -34%.

tension because of the combined assembly and service factors considered so far. Minus 39% tells us that, worst case, we can only count on 61% (100% − 39%) of the desired clamping force in some of those joints. The ratio, therefore, between the tension the bolts must be able to support without breaking and the clamping force we can count on in 99% (3 sigma) of our assemblies is 134%/61%; in other words, the ratio is greater than 2:1.

Let's assume that past experience has suggested to the designer that a $1\frac{1}{8}$-in.-diameter bolt will probably be required for this joint. This means, presumably, that a $1\frac{1}{8}$-in. bolt can withstand the maximum 134% of ideal stress experienced by bolts in this application. But we can only count on 61% of the ideal clamping force. We could get that much clamp from a perfectly tightened $\frac{3}{4}$-in. bolt.

BOLT TENSION SCATTER

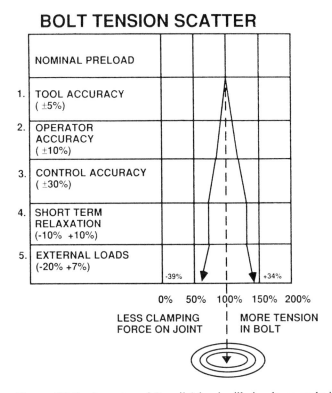

Figure 20.7   An external (tensile) load will simultaneously increase bolt tension by 7% and reduce clamping force by 20% if the bolt/joint stiffness ratio is 5:1 and the external load is 25% of preload. Note that the accumulated scatter can now be defined as −39% to +34% of the desired or target tension; or it can be defined as 61–134% of the desired tension.

This demonstrates a phenomenon that we've mentioned several times. Bolted joints are usually "overdesigned" to compensate for uncertainties in assembly results and/or in-service conditions. Designers know that they have to use lots of big bolts to avoid problems. If the assembly and service variations could be reduced, less overdesign would be required. Smaller bolts, lighter joint members, smaller tools could do the job now done by the heavy joints and large bolts. All of this would mean economic advantages—which would be partially offset by the added costs of better control at assembly. At present the benefits offset the costs only in applications where weight, size, efficiency, etc., are also significant factors.

Table 20.3 Joint Overdesign Factor

Ratio between the amount of tension the bolts must support and the amount of clamping force we can count on (worse case)	That ratio would require a bolt of the diameter shown below (a perfectly tightened $\frac{3}{4}$-in.-diameter bolt could provide the same clamping force)
2:1	$1\frac{1}{8}$ in.
5:1	$1\frac{3}{4}$ in.
10:1	$2\frac{1}{2}$ in.
16:1	3 in.

Even though we've considered only five variables, the ratio between the force we can count on and the force the parts must support is over 2:1. Many additional variables will have to be included in some applications, often increasing the ratio well beyond 2:1. Table 20.3 shows how the ratio affects the amount of overdesign required in a joint.

In Sec. V we'll try to answer the question, "Which variables should we include in our analysis?" The answer will have a large impact on the results we get, and on the estimated ratio between the maximum bolt tension and the minimum, per-bolt clamping force in the joint. The ratio between "must support" and "can count on" will often be much greater than 2:1. Before deciding which variables to include, however, let's see how we might use the resulting estimate.

IV. RELATING DESIRED TO ANTICIPATED BOLT TENSION

We estimated the range of bolt tension we wanted in Sec. I and the range we can expect above. Now we will combine our two analyses. To do this we turn the target of our last worksheet on its side and hold it up against the "desired tension" summary of Fig. 20.1. The results are shown in Fig. 20.8. In doing this we have to convert the scatter percentages of Fig. 20.7 to percentages of ultimate strength. The target or ideal preload was 55% of ultimate. On the downside, then, we'd expect a worst-case 61% of 55%, or 34% of ultimate. On the upper end, we'd expect 134% of 55%, or 74% of ultimate.

We have said that we want bolt tension to lie between 48% and 62% of ultimate tensile strength. The assembly/behavior analysis suggests the actual range will be 34–74%, exceeding both the high and low ends of the desired range. How can we respond to this?

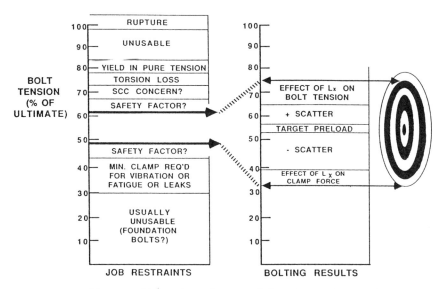

Figure 20.8 Chart combining the estimates of Figs. 20.1 and 20.7. The range in bolt tension desired is shown on the left; the range expected is summarized on the right.

Remember that the anticipated range was based on a preselected type of assembly tool, a torque wrench in our example. Note that we were using a "better" torque wrench with ±5% scatter instead of ±15%. This undoubtedly got us closer to the desired results, but many assemblies will still lie outside the desired range. Going to a still better torque tool (±2% is also available) obviously wouldn't provide sufficient improvement.

The scatter in anticipated results could be narrowed considerably by going to a computerized torque-turn system; and reduced even further by ultrasonic control of assembly. A second analysis would show whether or not such refinements would make the anticipated range fall within the desired range. I'm sure, in this example, that they would. But this much improvement in assembly accuracy may be economically unattractive for you. Fortunately, there are other options.

If the consequences of failure aren't too great, you can shrug your shoulders and accept the fact that some assembled joints will fail; will require field repair or maintenance, for example. The ranges we've estimated are 3 sigma ranges—covering the worst case with 99% probability. Most of the preloads introduced at assembly will be nearer the ideal than the extremes we've estimated. And, of course, we ARE making estimates, and not very accurate ones at that. We may luck out! So, let some fail. Keep good records. Build on your experience to modify your original selection of the target preload and/or assembly procedures to minimize failure, even if you can't eliminate it.

Note that your estimates of the bolt tension desired may involve as much uncertainty as your estimates of expected tension. In our example we estimated max and min tensions desired—then a couple of safety factors. Maybe that was unnecessarily conservative. So, all is not lost if "expected" exceeds "desired."

If the analysis is correct, however, and it matters, another option would be to use a different fastener. A stronger one would increase ultimate strength, raising the upper end of the desired range. The target preload could now be shifted upward as well. The lower end of the desired range would fall (would become a smaller percentage of the higher ultimate strength of the new bolt). So a stronger bolt will open up the desired range.

The desired range can also be opened up by other fastener-related changes. You may have selected a relatively high minimum tension to fight vibration loosening. Perhaps you can use an anaerobic adhesive or a prevailing torque fastener to reduce minimum tension requirements.

The maximum desired (acceptable) tension might be increased by choosing a different fastener material, if stress corrosion cracking is the concern, etc. There are many other possibilities of this sort which can be used to make the bolts, and the joint, better able to tolerate a wide variation in assembly and/or behavioral results. Think "overdesign"!

In some situations you'll find it easy to fit the anticipated range into the required range. In this case you could ask yourself whether or not LESS accurate assembly procedures would be acceptable. Presumably that would reduce assembly costs; and there's no economic sense in keeping the range of anticipated results under tighter control than required by the joint. Such results might also indicate excessive overdesign.

All of this will sound too casual to the purists reading this book. But we're dealing with a "chaotic" number of variables. A rigorously accurate analysis may be impossible and is often going to be prohibitively expensive. It will rarely be worth the effort. But the estimating procedure I've described, even though crude, will take you well beyond the "accuracy" implied by codes and other documents which merely suggest that you "tighten them correctly." And in some cases, the accuracy of our rough estimates can be quite good.

That accuracy, of course, will depend on the accuracy with which we estimate the scatter of the individual key variables. Sometimes an experiment can be made to refine our guess for one or a couple of the variables which contribute the most to the overall scatter. The larger the number of variables you pin down this way, the more accurate your estimate, of course. But you'll never—or rarely—be able to pin them all down.

In any event, I think that you'll find the analysis summarized in Fig. 20.8 to be an efficient and relatively easy way to get an overview of the many different aspects of bolted joints which we have discussed in this book.

V. WHICH VARIABLES TO INCLUDE IN THE ANALYSIS?

A. In General

Which variables should we include when making an analysis of this sort to pick preload or torque for an existing joint? The answer to this question will obviously have a major impact on the results we get, and the answer depends upon what it is we are trying to do. Why are we trying to pick preload or torque or other control variables for this particular joint? Picking preload or a torque is something the designer should have done—if a particular value is important. If we're doing it, it can only be because we think a choice is more important than he did, or his choice wasn't communicated to us, or because—most commonly—we've been having troubles with this joint (whether or not the designer specified a preload).

B. Possible Factors to Include

The VDI analysis, which we looked at in Chap. 13 and will return to in Sec. VII, includes a long list of factors with which to start our own lists when picking a torque or other type of preload control. In their procedure they consider:

Magnitude of the external load
Stiffness ratio ("load factor")
Eccentricity, if present
Embedment
Bending stress in the bolts
Tool scatter (e.g., torque-to-preload variations)
The effect of torsional stress on the tensile stress capacity of the bolt
The fatigue endurance limit of the bolts
The bearing stresses between nut or bolt head and joint surface
The clamping loads required to combat self-loosening, fatigue, or leakage

As far as I know, however, the VDI Directive 2230 doesn't include any of the following effects.

Differential expansion
Gasket creep
Elastic interactions
Hole interference
Resistance of joint members to being pulled together ("weight effect")
Operator errors caused by lack of training, carelessness, or accessibility
    problems
Tool calibration errors
Safety factors

C. Which Should We Include?

Almost all of these factors are present and influence results when we tighten a group of bolts. Only differential expansion, gasket creep, hole interference, and the weight effect are application specific and may be absent. And we won't care about the fatigue endurance limit if the applied loads are noncyclical. But the others, including such major factors as tool scatter and elastic interactions (and often operator problems) are essentially always there. If so—why leave any of them out?

One big reason most of these factors have always been left out is that we have been generally unaware that they exist. Another reason: even if we know they exist we have had little or no data with which to support specific values. The VDI Directive is the only publicly available document I'm aware of which includes things like embedment relaxation and tool scatter in a formal design procedure, and gives tables of typical values. Gasket manufacturers publish some gasket creep data, but as we learned in the last chapter it is usually limited to one thickness of a given material tested only at room temperature. In this book I've included whatever further data I could find on other variables, but, considering the number of variables involved in bolting, the data will often not apply to your application.

How can we leave them out and still design or assemble most joints successfully? Easy: we have traditionally overdesigned bolted joints—and included healthy "safety factors"—to cover our ignorance. The overdesign gives us room to include the factors now, whenever we want a clearer understanding of why a joint has caused problems and want to pick a better torque or equivalent. If this is a safety-related joint, however, in an industry where designers *do* consider many of these factors, we should only use the analysis to probe our problems. Only the designer should pick a better torque.

Even though we'll usually include factors the designer overlooked, and/or use different assumptions than he did about things like the stiffness ratio and the magnitude of external loads, it's still useful to make the analysis described earlier—or the more rigorous one to be described in Sec. VII. Either analysis is relatively easy to make, so we can repeat them, using several different sets of factors, and using the "typical" data found in this book if we have no application-specific data of our own. Such analyses can often reveal design flaws and/or suggest root causes for a troublesome problem.

But don't overdo it. Start by including only the three or four factors you believe will have the most impact on the behavior of your joint, based on previous experience with that joint whenever possible. Include a safety factor or "tolerance" to cover uncertainties. Revise your list—or the values you're using—if the results don't seem reasonable, or don't agree with prior experience. Add or subtract percentages from some of the variables to see if results are more realistic. Try adding or subtracting 10% from the target torque, as suggested earlier in this chapter, and see what effect that has on your estimated results. *Play* with this stuff: don't let it dominate your life. We know too little to lock it in concrete. Remember: past experience is a better guide than *any* theory or procedure or equation. If it ain't broke—don't fix it!

VI. THE BOLTING TECHNOLOGY COUNCIL

I have a final suggestion for those of your who are frustrated by the fact that bolted joint assembly is still dominated by uncertainty, trial-and-error, and chance; and by the fact that most standards organizations have ignored this. Slow, but steady progress has been made on bolting materials, configurations, and design procedures (finite-element codes, for example) but very little progress has been made in assembly technology. As a result a group of us, now representing nearly 100 organizations, have formed the Bolting Technology Council (BTC), sponsored by the Materials Properties Council of New York (address in Appendix C). The goal of the BTC is to sponsor research, provide recommendations, act as a clearinghouse for information, and, in general, further the technology of bolted joint assembly and (because it affects assembly) of the behavior of bolted joints in service. We have sponsored several small research programs, but expect that it will be years before we can make a significant contribution to this complex technology. Nevertheless, a start has been made. Anyone interested in participating in this important effort should contact the Materials Properties Council for details.

VII. A MORE RIGOROUS PROCEDURE

The procedure we've looked at is very useful for making seat-of-the-pants decisions about preload and other matters; using our best guess about which factors might most influence the results—and our guesses about the possible magnitude of each effect. We use rough statistics to make the results less intimidating than they might be if we assumed the worst case for each effect. We ignored many factors we might have included and assumed typical values for such difficult-to-estimate factors as the magnitude of the external load or the joint stiffness ratio. In spite of all this, and although we certainly wouldn't want to use it to make final decisions on safety-related or critical joints, I think that the procedure is valid for most situations.

A. Experiments Required for True Accuracy

When truly accurate answers are required, we'd have to make carefully controlled experiments on the actual joint. Finite-element analysis might help, but at the present time experiments are the only way to get the accurate application-specific data we need about such things as elastic interactions, stiffness ratios, and the response of the joint to service loads. Just picking a torque or other assembly control parameter isn't enough, either. We'd also have to impose strict quality standards on such things as assembly procedures, bolt quality, and lubricity to improve the chances that results of our experiments accurately and consistently reflect production assembly results. The auto industry, for one, does all of this now and has for years.

We could use Eqs. (13.21) through (13.23) to utilize the experimentally collected data. Those equations are repeated below, with some examples, for your convenience.

B. The Equations

Whether or not we have experimental data with which to make our analysis, we will often want a more rigorous way to analyze the possibilities than the seat-of-the-pants method described earlier. Here's a set of equations roughly based on those found in the VDI Directive 2230, but including differential expansion and elastic interactions. The equations can easily be extended to include more factors, or reduced to include less, to fit your own applications. The discussion in Sec. V is just as pertinent for this rigorous analysis as for the simpler one described earlier.

Maximum bolt load [extending Eq. (12.15)]:

$$\text{Max } F_\text{B} = (1 + s)F_\text{Pa} - \Delta F_\text{m} - \Delta F_\text{EI} + \Phi_\text{en}L_\text{X} \pm$$
$$\Phi_\text{en}K_\text{J}'' (\Delta L_\text{J} - \Delta L_\text{B}) \quad (13.21)$$

Minimum per-bolt clamping force on the joint [extending Eq. (12.16)]:

$$\text{Min } F_\text{J} = (1 - s)F_\text{Pa} - \Delta F_\text{m} - \Delta F_\text{EI} - (1 - \Phi_\text{en})L_\text{X} \pm$$
$$\Phi_\text{en}K_\text{J}'' (\Delta L_\text{J} - \Delta L_\text{B}) \quad (13.22)$$

Total minimum clamping force on a joint containing N *bolts:*

$$\text{Min total } F_\text{J} = N \times \text{per-bolt Min } F_\text{J} \quad (13.23)$$

where F_Pa = the average or target assembly preload (lb,N)
ΔF_m = the change in preload created by embedment relaxation (lb,N); Eq. (12.5) showed us that $\Delta F_\text{m} = e_\text{m}F_\text{Pa}$
e_m = percentage of average, initial preload (F_Pa) lost as a result of embedment, expressed as a decimal
ΔF_EI = the reduction in average, initial, assembly preload caused by elastic interactions (lb,N); Eq. (12.6) told us that $\Delta F_\text{EI} = e_\text{EI}F_\text{Pa}$
e_EI = the percentage of average, initial preload (F_Pa) lost as a result of elastic interactions, expressed as a decimal
ΔL_B = the change in length of the grip length portion of a loose bolt created by a change of Δt (°F,°C) in temperature (in.,mm); see Eq. (13.14)
ΔL_J = the change in thickness of the joint members, before assembly, if exposed to the same Δt (in.,mm); see Eq. (13.15)
s = half the anticipated scatter in preload during assembly, expressed as a decimal fraction of the average preload; see Eq. (12.1)
K_J'' = the stiffness of a joint in which both the axes of the bolts and the line of application of a tensile force are offset from the axis of gyration of the joint; and in which the tensile load is applied along loading planes located within the joint members (lb/in., N/mm); K_J'' is the reciprocal of the resilience of such a joint (see Fig. 13.15)
Φ_en = the load factor for the joint whose stiffness is K_J''; see Fig. 13.16
N = number of bolts in the joint
L_x = external tensile load (lb,N)

Again, these equations can be rewritten using any appropriate combination of Φ and K_J.

C. Minimum Clamping Force—Some Examples

First Example—Using Worst-Case Values

Let's feed these equations some numbers to see how important it is to use the best available numbers. Let's start with Eq. (13.22) and compute the anticipated minimum clamping force on the joint. For a first round we'll assume the parametric values used in the example given in Chap. 13, Sec. VIB. To repeat, these were:

Average assembly preload (F_{Pa}) = 1,430 lb.
Tool scatter (s): we assumed a full 30% below target preload, so s = 0.3.
Embedment: assumed a "typical" loss of 10%, so e_m = 0.1.
Elastic interactions: assumed an "average" loss of 18% for a two-piece, metal-on-metal joint, using results obtained in Dr. George Bibel's experiments, so e_{EI} = 0.18. (Note that Dr. Bibel tells us that the average loss would be 30% if a sheet gasket were included in the joint; or 46% if the gasket is a spiral wound type. See Chap. 12, Sec. IB.)
External load effect: the tensile load reduced the per-bolt clamping force on the joint by 269 lb, based on the assumption that the external load was equal to 25% of the preload we were trying to compute (talk about winging it!) and on the "eccentrically and internally loaded" load factor Φ_{en} computed in Chap. 13. Instead, of course, we could use any of the other Φ's listed in Chap. 13, Sec. IIB.
Differential expansion loss: a per-bolt clamping force loss of 208 lb based on the assumptions that Inconel bolts were being used in a mild steel joint and that the operating temperature would be 400°F.

These assumptions give us the results shown in Fig. 20.9, a minimum, per-bolt clamping force on the joint of 123 lb. The ratio between average, initial preload and residual clamping force is about 12:1. A ratio of 15:1 was calculated in Chap. 13 but was based on maximum, not average, preload and also included a gasket creep factor which I've omitted here.

Whether the ratio is 12:1 or 15:1, this is not a pretty picture! And it doesn't include many factors which might have made it worse, such as the gasket creep, which was included in the Chap. 13 example, or hole interference or the weight effect, both described in Chap. 8. But how valid is this calculation?

It purports to define the minimum tension we can expect to see in some of the bolts. To do that, however, we would have to assume a worst-case elastic interaction loss of 36% instead of an average 18%. I tried

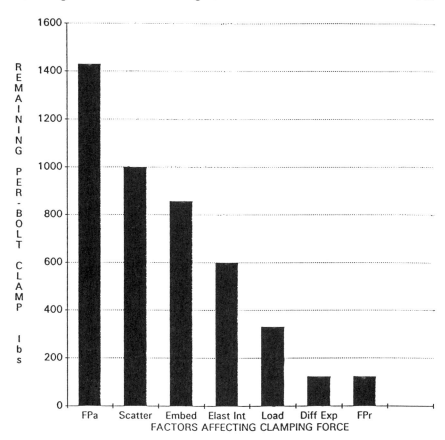

Figure 20.9 Chart showing the large difference expected between the average, initial, per-bolt clamping force applied to a joint during assembly (F_{Pa}) and the residual, in-service clamping force (F_{Pr}) when we assume worst-case tool scatter and embedment losses in bolt tension, plus an average elastic interaction loss, plus the unavoidable losses in clamping force caused by the external tensile load and differential expansion between bolts and joints. This is a plot of the calculations made in Chap. 13, Sec. VI.

that—and computed a residual, minimum, per-bolt clamping force of minus 135 lb. Since my goal was to teach you how to use Eq. (13.22), not to define and analyze an actual example, I used 18%. But it means that the computed 123 lb is an "almost worst case" answer.

Equation (13.23) says that we can compute the minimum anticipated clamping force on a joint containing N bolts by multiplying 123 by N, which we did in Chap. 13. But this assumes that *every one* of the N bolts in some joints will only contribute an "almost worst case" amount of clamping force to the joint. I think that that's highly unlikely. I think that Eq. (13.22) can reasonably predict the worst-case clamp contributed by an occasional bolt, but not by each bolt in a given joint.

Second Example—Using Statistically Combined Values

In the quick-and-dirty procedure we discussed earlier in this chapter we used the square root of the sum of the squares of the variances to reduce the combined effect of tool scatter, embedment, and elastic interactions; to get a more cheerful—and more realistic—picture. This introduces the assumption that no one bolt will simultaneously be exposed to the worst-case minimum assembly preload, maximum embedment, and maximum interaction loss. The results, shown in Fig. 20.10, are based, therefore, on the square root of the sum of squares of the following assumptions:

Preload scatter: $\pm 30\%$
Embedment: minus 10% max
Elastic interactions: minus 36% max

The load and differential expansion effects remain unchanged at 269 and 208 lb, respectively. Combining the three factors tabulated above suggests a 3-sigma maximum loss of only 47.9% of average assembly preload (from which we further subtract the load and differential losses). The resulting per-bolt clamping force is now 268 lb, more than twice what it was under our earlier assumptions. The preload-to-residual-clamp ratio is about 5:1; not something to celebrate, but a definite improvement.

But again, I think it would be very conservative to assume that each of the N bolts in a given joint ended up with 47.9% less tension than anticipated by the target preload. What we really need is a series of experiments, a resulting list of "combined losses" and the mean and scatter for the combined data. Then we could use Eq. (13.23) with more confidence. Such data, unfortunately, do not exist.

Third Example—Using Average Values

Could we resolve this by feeding only "average" data into Eq. (13.22)? That would suggest the following:

Tool scatter: 0% average loss!
Embedment: 5% average
Interaction loss: 18% average
Load and differential expansion losses unchanged

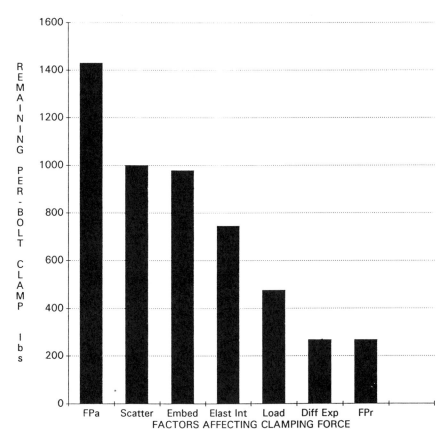

Figure 20.10 Chart showing the anticipated difference between initial and residual clamping force when we assume that worst-case effects will never (or rarely) occur simultaneously; that the expected deviations in individual factors can be combined by the square root of squares Eq. (20.6).

These assumptions would result in a *big* improvement in predicted clamping force, thanks to the fact that the average of a ±30% tool scatter factor is 0%. Although the ±30% figure appears again and again in the literature, I think that it's unwise to assume that the plus scatter will equal the minus scatter on a given joint. We know that more factors lead to less preload than to more preload. Therefore, although your own experience may well lead you to use a different value, I'm going to use a minus 10% for the average, single-joint, per-bolt, tool scatter.

The results are shown in Fig. 20.11, an average worst-case (if there is such a thing) per-bolt, in-service, clamping force of 495 lb; and a preload-to-clamp-force ratio of only 2.9:1. I would feel comfortable multiplying this number by N to estimate the total force on the joint. To repeat, however, every answer we get depends upon which factors we've included. Only experiments on the actual joint will give us a truly accurate answer.

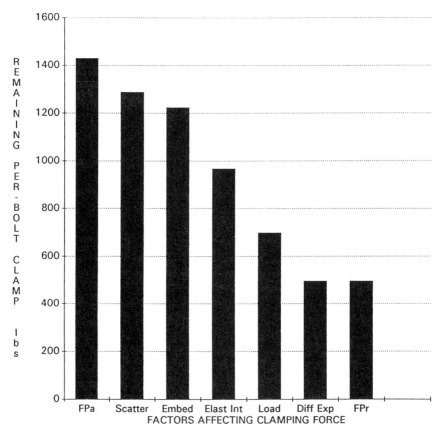

Figure 20.11 Chart showing the difference between initial and "average" residual per-bolt clamping force when we assume an average deviation for the various loss factors. This is probably a more realistic view of the results than those shown in Figs. 20.9 and 20.10; especially if our goal is to compute the total clamping force applied to the joint by N bolts. See the text for a detailed discussion.

Fourth Example—Using Feedback Control Values

If we use strain gage, ultrasonic, or other "feedback" control of final, assembly preload we can reduce the combined effects of tool scatter, embedment, and elastic interactions to ±10%. The results are shown in Fig. 20.12; a ratio between target preload and final, per-bolt clamping force of only 1.8 to 1. Since the load effect and differential expansion losses can be accurately predicted if we know joint materials and geome-

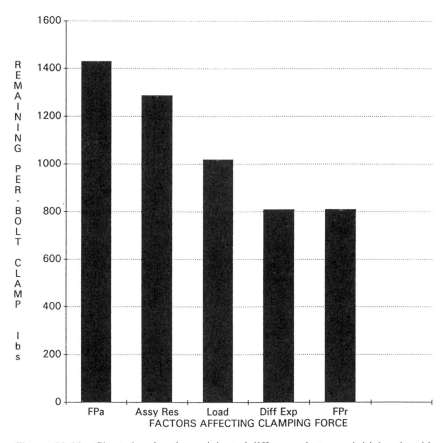

Figure 20.12 Chart showing the anticipated difference between initial and residual clamping force when feedback control is used during assembly to correct for tool scatter and compensate for embedment and elastic interaction loss; giving a practical, worst-case deviation of assembly preload of only 10% of average assembly preload.

try—and, the tough part, the magnitude of the loads we will see in service—this would also eliminate the need for joint experiments. The feedback gives us the "experimental" data we need as we assemble each joint. No more need to estimate assembly loss factors or even to decide which to include in our analysis. But this type of control is expensive and we'll usually have to settle for estimates. Which we'll use with either the quick-and-dirty procedures described first in this chapter, or in a more rigorous analysis such as that described in this section.

D. Maximum Bolt Tension

We have focused this "more rigorous" discussion entirely on the use of Eqs. (13.22) and (13.23). We should never accept the assembly preloads suggested by use of these equations without also using Eq. (13.21) to estimate the worst-case stresses our choice would create in some of the bolts. These were calculated, correctly for the worst-case example, in Chap. 13, and so I won't repeat it here. Note that it's perfectly reasonable to use worst-case deviations when estimating maximum bolt loads, because we don't want any bolt to break and we don't multiply the results by N or something. Even here, however, our inputs to the equation can distort the results and penalize our design. If we're planning to use torque is it reasonable to assume a worst case $+30\%$ scatter, for example, or will that be reduced when we lubricate the bolts?

Regardless of which procedure you use, I think that each of those we've examined in this chapter has its place and can be useful—can help you to estimate, however roughly, the assembly and in-service clamping forces in your joints and thereby can help you to analyze and solve many bolting problems. They can certainly give you the insight you need to use good engineering judgment when picking a better torque—or when picking a more appropriate assembly control variable. They've also given us a chance to review many of the topics we've considered in this book. And so, with this review under our belts, the time has come to move on to the subject of the design of bolted joints.

21
Design of Joints Loaded in Tension

This is the first of three chapters on the design of bolted joints. In this first chapter we are going to discuss, in general, the design of joints loaded in tension, first examining a "typical" procedure, then the more rigorous, VDI procedure. In Chap. 22 we'll consider the design of a specific type of tension joint, a gasketed pressure vessel or piping joint. Then, in Chap. 23, we'll take a brief look at the design of a joint loaded in shear.

To repeat one of the earliest points we covered, the most common purpose of the bolted joint is to clamp two or more things together. When we design a bolted joint, therefore, we are usually designing a clamp and will be concerned about such things as the strength of the clamp, the integrity or reliability of the clamp, the stability of the clamp in service, and the life of the clamp.

The specific factors a given designer must consider will depend, of course, on the application of his particular joint. Designers of joints for nuclear power plants or aircraft will obviously have to consider many more factors and make better decisions than will designers of less critical joints.

I. A MAJOR GOAL: RELIABLE JOINTS

To the extent that he can afford it, every designer will be interested in "reliable" joints. This will mean different things in different applications, of course, and will require joint sizes and configurations appropriate for specific service conditions. In spite of the fact that reliability can mean many different things, I think that the following checklist will be a useful place to start our summary of the design process. It tabulates the main issues which determine reliability. Few applications will involve *all* of the points listed here, but each topic is a valid concern somewhere.

When you start a new design, review this list. Check the items as you read them, to force yourself to focus, if only briefly, on each issue. Double-check, or make a separate list of, those items which you feel will affect reliability in your particular application.

Checklist for Reliable Bolted Joints

I recognize that a bolt is a clamp; that an "unreliable joint" is synonymous with "an inadequate clamp."

I realize that reliability requires the following:

A joint that is strong enough ("rigid" enough) to provide sufficient structural integrity to prevent slip, separation, vibration, misalignment, wear, etc., of the interconnected parts of the product or system.

> Enough bolts
> Adequate bolt diameter
> Appropriate material strength
> Sturdy joint members

A joint (clamp) which is also stable under service conditions: i.e., won't degrade or weaken. (Factors such as corrosion, vibration, fatigue, and elevated temperatures can change the clamp.) Stability requires the things tabulated below.

Materials which are stable in the service environment.

> Won't corrode excessively
> Have an acceptable resistance to stress corrosion cracking
> Have an acceptable resistance to fatigue
> Won't lose too much strength at operating temperatures
> Won't relax too much at elevated temperatures (because of stress relaxation, creep, or reduction in the modulus of elasticity)

Bolt and joint geometry (shape which encourages stability of the clamp.

> Properly designed to minimize stress concentrations (e.g., bolt head-to-shank fillets and thread run-out details which reduce the chances of fatigue failure; another example—joint shear planes which coincide with thread run-out encourage failure of the bolt in shear).

Bolt-to-joint stiffness ratios which direct external loads and/or loads created by differential expansion to the components (bolts or joint members) best able to support them. (Usually means stiff joint members, flexible bolts.)

An in-service clamping force which is able to enforce stability. (The in-service clamping force will be determined by the preloads introduced at assembly, but modified by service loads and/or thermal effects.)

Enough clamp:
The clamping force must be high enough to minimize self-loosening of the bolts under vibration, thermal cycles, joint flexing, etc.
It must be high enough to minimize load changes in the bolts (thereby improving fatigue life).
It must be high enough to prevent joint slip (which encourages self-loosening, wear, unexpected stress concentrations in joint members and connected parts, etc.).
It must be high enough to prevent leaks, a problem in itself, but which also lead to corrosion and further degradation of the clamp.
But not too much:
The clamping force must also be low enough to avoid excessive stress in bolts; this could lead to stress corrosion cracking, for example, or tensile failure under unexpected loads or thermal effects.
The clamping force must be low enough to avoid damaging or distorting joint members (e.g., excessive flange rotation).
The clamping force must be low enough to avoid crushing a gasket (which can also cause leakage).

II. TYPICAL DESIGN STEPS

The checklist above helps us set our sights. Now, how do we proceed?

Most bolted joints in this world probably are, and will continue to be, designed by gut feel based on past experience with similar joints. This is perfectly acceptable. A complete design analysis can be very expensive, thanks to the large number of variables and factors involved. Analysis is usually justified if one is starting from scratch with no prior experience in a particular application, and if the consequences of failure are severe; or if the joint is to be mass-produced and overdesign would be uneconomical.

Even with critical joints, past experience will be involved and pertinent, of course, again because of the large number of variables and the inevitable uncertainty in each which will always make the outcome less than certain.

No matter what the specific application for a bolted joint, certain common design steps will usually be included in the procedure. The amount of attention devoted to each step will depend on the importance of the joint, but the step will be there in some form or other in most situations. The steps I'm about to describe are summarized in Fig. 21.1.

A. Initial Definitions and Specifications

The obvious first step is to define the purpose of the joint and to rough out some preliminary specifications concerning the use of the equipment or system in which the joint is located, defining such things as operating speed, temperature, desired life, estimated cost targets, etc.

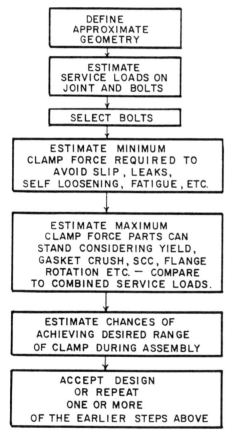

Figure 21.1 Block diagram which summarizes the activities involved in the design of bolted joints.

B. Preliminary Design

Preliminary geometric layouts are usually next. These roughly define the size and shape of the various parts involved, including the joint members.

C. Load Estimates

Once we know how big this thing is, roughly how much it weighs, and the intended use, we can start to guesstimate the possible loads on the bolted joints. Except for obviously critical joints, this is an oft-neglected step, but for a good reason. It's often very difficult to estimate service loads. If you're serious about the analytical design of bolted joints, however, you can't ignore this step.

As far as the joints are concerned, these loads will presumably include such things as weight, inertial affects, thermal affects (see Chap. 13), pressure, shock, etc. Both static and dynamic loads must be estimated. Load intensifiers (prying, eccentricity, etc.) should be acknowledged if present (see Chaps. 13 and 23).

D. Review Preliminary Layouts; Define the Bolts

The load estimates will, of course, affect the size and shape of parts, and vice versa. In fact, in most situations, the geometric layout and load-estimating steps will be performed simultaneously. In any event, with approximate joint geometry and loads established we can now make a preliminary selection of bolt size and number. Important parameters to be defined at this point include nominal diameter, grip length, number of threads per inch, tensile stress area, the bolt material, and the strength of the material.

At this point it would also be useful to estimate the stiffness of the bolt using Eq. (5.10) and then go on to estimate the stiffness of the joint using the procedure described in Sec. II of Chap. 5. We'll need these estimates to predict the effects of external loads on the joint, the danger of fatigue failure, etc.

It's also useful, when selecting the bolts, to estimate their static strength to set an upper bound on the stress or tension which they can support. We will rarely design the joint to impose such stresses on the bolt, but it useful to define this upper limit. At this point, the designer should also estimate the stripping strength of the threads of bolt unless an off-the-shelf bolt is to be used with an off-the-shelf and recommended nut. Stripping strength should always be estimated if the length of thread engagement is abnormally short or if the bolts are to be tightened into tapped holes in a soft material such as aluminum.

E. Clamping Force Required

Minimum Clamp

We have now roughed out the joint configuration, picked the bolts, and estimated the maximum tensile strength (and, therefore, clamping force) available from these bolts. Our next step (which many designers will place first!) is to estimate the minimum amount of clamping force the joint must have to avoid failure in this application. We'll design and assemble for more than this minimum, to be safe, but start by asking, "What's the least required here?"

If the joint must only face static or slowly moving tensile loads, then it will probably be sufficient to design for clamping forces which are somewhat greater than the maximum anticipated tensile load. If some overdesign is acceptable, the designer can ignore such things as the joint diagram and assume that the bolts will see any external tensile load in its entirety.

The amount by which the assembly and in-service clamping forces selected by the designer should exceed the external load will depend on such things as the accuracy with which the bolts are to be tightened, the accuracy with which service loads can be estimated, the consequences of failure, etc. As far as assembly accuracy is concerned, the less accurate the tool, the greater the design clamping force. The larger design value, of course, means that the joint will have "enough" clamping force even if grossly "undertorqued" during assembly. This is one of the factors which cause most joints to be "overdesigned"—heavier than they need to be to compensate for assembly uncertainties. As an example, many people want the nominal clamping force to be three to four times the anticipated service loads.

In critical situations, the designer must carefully estimate the service loads on the joint and the required resistance to those loads when selecting a clamping force. The goal, of course, is to eliminate any chance of bolted joint failure. The analysis, therefore, can be defined in terms of failure modes, as follows. The designer need only concern himself with one or a couple of these possibilities in most applications. Few joints are threatened by all possible modes of failure. The possibilities were listed in the last chapter and included joint slip or separation, self-loosening, and fatigue. We'll take a closer look at this in Sec. IV of this chapter.

Maximum Clamp

As mentioned several times in this and previous chapters, it's not enough to define a minimum clamping force. We must also satisfy ourselves that the joint—and especially the bolts—are never exposed to too much stress.

As we saw in the last chapter, the clamping force is usually limited by a bolt strength factor; including the bolt's fatigue strength, thread-stripping strength, or susceptibility to stress corrosion cracking. Sometimes, however, it's the crushing strength of a gasket or the bearing strength of the joint member which determines the upper limit. See the last chapter, or Sec. IV of this chapter, for more specifics.

III. JOINT DESIGN IN THE REAL WORLD

Figure 21.1 implies that joint design is a well-organized, step-by-step logical procedure. It can be—it has been reduced to this by codified procedures in several industries. But, in most situations, I suspect it's governed more by impulse and intuition than logic. The designer starts at this point in the process which most interests or concerns him, leapfrogs to his next concern, backs up to repeat an earlier step, balances one concern against another, circles and cycles until a joint is born.

Design is, after all, a creative process, and too many "rules" can be counterproductive.

The complex technology of the bolted joint, however, makes "rules," "shopping lists," "guidelines," etc., often helpful and sometimes essential. If the joint is "important" the several topics listed in Fig. 21.1 should be addressed by the designer, even if he chooses to do so in a different order. For truly critical joints, where the consequences of failure are severe, the designer will want to go well past the relatively simplistic approach described in Sec. II. Perhaps he'll resort to a finite-element analysis. In other situations he'll want to adhere rigorously to a codified procedure, many of which are mandated by law. Or, he may want to use the procedure published some years ago by the German engineering society VDI. This is the most detailed, publicly available, general-purpose, joint design procedure I've encountered. We first used a version of it in Chaps. 12 and 13 when we were studying the behavior of a joint under tensile loads. We returned to it in the last chapter; using it to define a more rigorous procedure for selecting preload for an existing joint. Now we're going to use it for the purpose for which it was intended, to design a joint.

IV. THE VDI JOINT DESIGN PROCEDURE

I'm basing the following discussion primarily on the G. H. Junker paper [1] listed in the references at the end of this chapter. Although my copy isn't dated I know that this paper predates the first edition of this book, and therefore is probably about 15 years old. The VDI procedure was first published in 1977 [2] and was modified later. The most recent version I'm aware of was published in 1986 [3]. I'm sure that the procedure described by Junker is still valid, however, and we'll follow it step by step, with only a few modifications to accommodate some personal judgments and/or new factors such as elastic interactions.

A. Terms and Units

We'll use the following terms and units while following the VDI procedure:

A_P = contact area between bolt head, or nut face, and the joint (in.^{2},mm^2).

A_r = root diameter area of the threads (in.^{2},mm^2); see Table 3.3.

A_s = tensile stress area of the threads (in.^{2},mm^2); see Appendix F.

F_B = tension in bolt, in general (lb,N).

ΔF_B = change in bolt tension caused by external load, L_X (lb,N).

Max F_B = maximum estimated bolt tension, e.g., as a result of Max F_P plus the effects of external load, L_X etc. (lb,N); equals the maximum per-bolt clamping force on the joint under the present assumptions.

Min F_B = minimum estimated bolt tension and per-bolt clamping force (lb,N).

F_{Krqd} = the minimum preload (or clamping force) required to prevent separation of an eccentrically loaded joint (lb,N).

F_P = preload in general (lb,N).

ΔF_P = loss of preload during or immediately following assembly because of embedment and elastic interaction effects (lb,N).

Max F_P = maximum anticipated per-bolt preload during assembly (lb,N).

Min F_P = minimum anticipated per-bolt preload during assembly (lb,N).

F_{Pa} = "target" preload used to compute the torque or other control parameter to be used at assembly (lb,N).

F_{Prqd} = minimum preload (or per-bolt clamping force) required to prevent slip, separation, or leakage of a concentrically loaded joint (lb,N).

F_J = per-bolt clamping force on the joint, generally assumed to equal the existing per-bolt preload (lb,N).

ΔF_J = the change in per-bolt clamping force created by external load, L_X (lb,N).

Max F_J = maximum bolt preload and clamping force created during assembly; before the joint is put in service (lb,N).

Min F_J = minimum bolt preload and clamping force created during assembly, before the joint is put in service (lb,N).

Γ_y = the tensile force required to yield the bolt (lb,N).

K'_J = stiffness of a concentric joint loaded at internal loading planes (lb/in., N/mm).

ΔL_B = increase or decrease in the grip length section of the bolt because of a temperature change (in., mm).

ΔL_J = increase or decrease in thickness of the joint because of a temperature change (in., mm).

L_S = external shear load (lb,N).

L_X = external tensile load applied to joint (lb,N); also equals the maximum external load experienced during a load cycle if the load varies.

L_{Xmin} = minimum external load experienced during a load cycle if the load varies (lb,N).

P_G = maximum allowable pressure which can be exerted on the joint by the bolt head or nut without damaging the joint; P_G is usually a little greater than the yield strength of the joint material (psi, MPa).

R_G = radius of gyration of the joint (in., mm).

a, s, u = important dimensions of an eccentrically loaded joint, illustrated in Fig. 13.12 (in., mm).

s = scatter in preload caused by the assembly tools and procedures; e.g., if torque is used for control, with a resulting scatter of $\pm 30\%$, then $s = 0.30$. A few values, based on a VDI table, are given in Table 21.1.

α_A = the scatter in preload caused by the assembly tools and procedures as defined by VDI. α_A = Max F_P/Min F_P. If we use a $\pm 30\%$ torquing procedure, $\alpha_A = 1.30/0.70 = 1.86$. See Table 21.1 for other VDI values.

Φ_{Kn} = load factor for a concentric joint, loaded internally at loading planes. $\Phi_{Kn} = \Delta F_B/L_X$.

μ = coefficient of friction between joint surfaces.

σ_A = the bolt stress at its endurance limit (psi, MPa).

α_y = the yield stress of the bolt (psi, MPa).

B. Design Goals

We have two main goals.

1. An in-service clamping force great enough to prevent slip, separation, or leakage (partial unloading).

Table 21.1 Preload Scatter Factors

Factor used by VDI[a] α_A	Factor used in this book $\pm$ s	Type of control used during assy
1	0%	Yield control[b]
1.2	10%	Elongation control
1.35–1.86	15–30%	Torque control[c]
1.5–2.64	20–45%	Airstall power wrench
1.67–4.0	25–60%	Impact power wrenches[d]

[a] I computed these values using the data trabulated in Fig. 7 of Ref. 5 and shown in the next column. I believe that the Ref. 5 numbers are based on Table 17 of Ref. 2 (which is in German).

[b] VDI, influenced by Junker and SPS, say that yield control is nearly perfect: that any scatter will be caused only by variations in the yield points of individual bolts. This view can, of course, be disputed.

[c] VDI says that the lower values can be achieved by using an experiment on the actual bolts to pick the torque. The possible reduction in scatter caused by a good lubricant is not considered.

[d] VDI says that the higher values are for soft connections and/or rough surface finishes.

2. Bolts strong enough to survive and support the maximum assembly preload plus the maximum service loads, including those caused by tensile and/or shear forces and thermal effects.

C. General Procedure

We'll assume that we have roughed out the configuration and size of this joint and its bolts, starting basically as we did in the procedure described in Sec. II of this chapter. We have also estimated the type and magnitude of the external loads to be placed on this joint; and have decided how much clamping force will be required to combat anticipated failure modes and service conditions. Now we can use the VDI procedure to analyze, refine, and/or correct our initial decision, as follows.

1. We'll make a preliminary estimate of the assembly preload requirements, taking things like tool scatter and relaxation effects into account. We'll assume that preload equals the per-bolt clamping force, too (i.e., no hole interference or weight effect).
2. Then we'll revise our estimate of the assembly preload requirements to add the effects of the external loads on the joint. As we take this step we'll introduce a value for the minimum clamp load required in service.

3. Next we'll add the effects of some other types of load the joint might see, if we think that they might be present. These include:

Shear loads
Eccentric loads
Bending loads on the bolts and/or on the joint
Load changes caused by differential expansion
Dynamic (i.e., fatigue) loads
Combinations of the above

4. Finally we'll check our results against several limiting conditions to make sure that neither the bolts nor joint members are overstressed.

We'll do all this to determine:

The maximum tension seen by some bolts
The minimum clamp force seen by some joints
Specifications for the optimum bolt for this joint

D. Estimating Assembly Preloads: Preliminary Estimate of Minimum and Maximum Assembly Preloads

VDI says that we should use the following expressions for preliminary estimates of the minimum and maximum preloads to be created during assembly.

$$\text{Min } F_P = F_P + \Delta F_P \tag{21.1}$$

We express this minimum as the sum of an as-yet unidentified preload F_P and the reduction in that preload which will be caused by relaxation effects during assembly. We do this in order not to overlook the fact that the assembly preload must be great enough to compensate for the relaxation.

The ΔF_P term gives us our first problem, or rather, our first need for some engineering judgment. VDI equates ΔF_P to embedment relaxation, but we now know that elastic interactions will also be present and can cause far more loss than will embedment. In my opinion, therefore, we should start by including these interactions. If the results suggest an unacceptable spread between maximum bolt tension and minimum clamping force, then we must consider using an assembly procedure that compensates for relaxation.

The expression we use for a preliminary estimate of the maximum assembly preload is simply α_A times the minimum preload or

$$\text{Max } F_P = \alpha_A(F_P + \Delta F_P) \tag{21.2}$$

It's important to note at this point that VDI assumes that these max and min assembly preloads are equal to and opposite to the max and min clamping forces which the bolt will create on the joint. Things like hole interference and weight effect, in other words, are ignored (as they should be for most joints).

E. Adding the Effects of the External Load

We have not yet fully defined the max and min preloads we want to create during assembly, because we haven't taken another important clamp load loss factor into account. We do so now. The per-bolt clamping force on the joint will be reduced, by the external load, by an amount equal to

$$\Delta F_J = (1 - \Phi_{Kn})L_X \tag{21.3}$$

The external load will also add to the tension in the bolt, as follows:

$$\Delta F_B = \Phi_{Kn}L_X \tag{21.4}$$

We're going to ignore ΔF_B for the moment and concentrate on ΔF_J, the loss in clamp load caused by the external load. We must compensate for this loss in our selection of assembly preload.

The joint must be able to function, in service, with the minimum clamping force which remains in the joint after relaxation and external load effects have done their worst. I'm going to follow the VDI procedure and call this remaining clamping force the "force that's required" to prevent the joint from misbehaving or failing, or F_{Prqd}. Junker calls F_{Prqd} "the decisive factor in the design equation." This is the clamping force which must be able to resist slip, separation, or leakage. I said in an earlier chapter that the purpose of the bolted joint was to generate a force: this is that force.

We can now rewrite the expression for the minimum and maximum preloads we want to create during assembly—taking tool scatter, relaxation, and load effects into account, as illustrated in Fig. 21.2.

$$\text{Min } F_J = \Delta F_J + F_{Prqd} + \Delta F_P \tag{21.5}$$

and

$$\text{Max } F_J = \alpha_A[\Delta F_J + F_{Prqd} + \Delta F_P] \tag{21.6a}$$

We can rewrite that last expression in terms of the external load or

$$\text{Max } F_J = \alpha_A[(1 - \Phi_{Kn})L_X + F_{Prqd} + \Delta F_P] \tag{21.6b}$$

Now, VDI says that we can use Eq. (21.6) to size the bolts. They do this by using the relationship

$$\text{Max } F_J \leq 0.9F_y \tag{21.7}$$

Why $0.9F_y$? VDI assumes that the bolts will be tightened by turning a

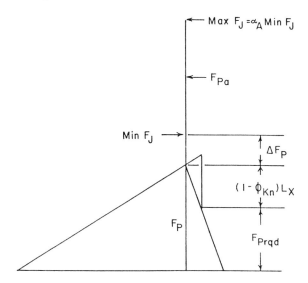

Figure 21.2   This joint diagram illustrates Eqs. (21.5) and (21.6). We start by applying an assembly preload equal to F_P plus $\triangle F_P$ to the joint. Elastic interactions and embedment reduce this to just F_P. This is equal and opposite to the clamping force on the joint, F_J. When the joint is placed in service the external tensile load will further reduce the clamping force by a factor equal to $(1 - \Phi_{Kn})L_X$, leaving a residual clamping force called the "force required to combat joint failure," F_{Prqd}. The minimum required assembly preload, Min F_J, is the sum of F_{Prqd} plus the two loss effects. The maximum assembly preload is α_A times that. The target preload, F_{Pa}, to be used to compute an assembly torque lies halfway between the max and min F_J's.

nut, whether or not torque is selected as the control variable, and that some of the input energy will be turned into torsional stress, reducing the bolt's capacity to support tensile stress by 10%.

Note that these equations for max and min joint clamping force, F_J, are the values we want to or expect to introduce at assembly, before the external load is actually applied. They are, therefore, still equal and opposite to the preloads created when we tighten the bolts. The Max F_J does *not* define the maximum tension the bolts will see when actually placed in service. We'll get to that in Eq. (21.9).

F.   Is the Required Force "Good Enough"?

At the beginning of this design procedure we defined the failure modes we fear, and the minimum clamping force we think will be required to prevent them. We can now enter that clamping force, F_{Prqd}, in Eq. (21.6). Is the value we picked appropriate?

In answering this question we are, in effect, repeating the "how much clamping force do we need?" question raised in Secs. II and IV of the last chapter. If you skipped those, you might want to review them now. Or review the various chapters devoted to fatigue, self-loosening, and the like.

At this point in the procedure, however, VDI gives us the following suggestions, with which to evaluate our estimate. To combat

Separation: The residual clamping force, F_{Prqd} must be greater than zero, and the "more greater" the better! Limited by the forces/stresses the parts can stand.

Transverse slip: $F_{\text{Prqd}} \geq F_S/\mu$ (21.8)

Leakage: F_{Prqd} must be large enough to create the residual gasket stress currently defined by the ASME Code's "m factor"—or the residual gasket stress defined by the new PVRC gasket factors. See Chap. 19 for details.

Fatigue: Clamp load F_{Prqd} must be "as large as possible"—presumably limited by how much force or stress the parts can stand. If this is an eccentric joint, subjected to cyclic loads, F_{Prqd} will be replaced by F_{Krqd} as discussed under Eccentric Loading below.

This ends Mr. Junker's list of possible clamp force requirements. We might want to add other possibilities, again based on the discussion in Sec. IV of Chap. 20. For example:

Creep relaxation: If this is a gasketed joint we'll certainly want to increase the assembly preload to compensate for the subsequent loss due to gasket creep by adding a term covering this loss to those within the brackets in Eq. (21.6). We might call this term ΔF_{Cr}. We want to do this—but sometimes may not be able to. Sometimes, as mentioned in Chap. 19, the loss is so great the initial preload required to compensate for it would crush the gasket, and we have to compensate by retightening the bolts after creep has occurred. Nevertheless, it's a factor we mustn't forget.

Differential thermal expansion: A change in temperature can cause a simultaneous increase or decrease in bolt tension and clamping force. If a decrease is indicated we'll want to add a term covering that loss to Eq. (21.6) as well. The expression for this term would be

$$\Delta F_{\text{th}} = \pm(1 - \Phi_{\text{Kn}})\, K'_{\text{J}}\, (\Delta L_{\text{J}} - \Delta l_{\text{B}})$$

If an increase is indicated we might be able to *subtract* the change from the terms within the brackets in Eq. (21.6). We'd certainly want to include the increase on the left side of Eq. (21.12), below.

G. Further Considerations

VDI adds some further considerations: things to check before we specify the size and material of the bolts to be used in our newly designed joint.

Static Strength of the Bolt

We don't want the stress in the bolts to exceed their yield strength, so:

$$\text{Max } F_B = \text{Max } F_J + \Delta F_B \leq F_y \tag{21.9a}$$

or

$$\text{Max } F_B = \text{Max } F_J + \Phi_{Kn} L_X \leq F_y \tag{21.9b}$$

These equations define the true, maximum, in-service tension the bolts will see.

VDI says that another way of expressing this is by the following: [Eq. (21.7) justifies this alternate way to define the maximum bolt load.]

$$\Delta F_B \leq 0.1 \sigma_y A_S \tag{21.10}$$

Fatigue

If the joint is loaded concentrically, and the load cycles between L_X to zero, we can express the allowable limits on the excursion in load mathematically by:

$$\Delta F_B/2 = \Phi_{Kn} L_X/2 \leq \sigma_A A_r \tag{21.11}$$

We could, of course, substitute a specific fatigue life stress for the endurance limit stress if something less than infinite life were acceptable for the application.

If the load cycles between L_X and an L_{Xmin} which is greater than zero, then:

$$\Phi_{Kn}(L_X - L_{Xmin})/2 \leq \sigma_A A_r \tag{21.12}$$

Or, for a "push-pull" load:

$$\Phi_{Kn}(L_{X \text{ tensile}} - L_{X \text{ compressive}})/2 \leq \sigma_A A_r \tag{21.13}$$

See Table 4.11 for some endurance limit data if you have none of your own, or see the discussion in Chap. 17, Sec. VI.

The presence of fatigue loading will usually have a major impact on the size of the bolt chosen, its material, and/or its design. As a partial response we might want to use a fatigue-resistant fastener and/or take some of the other steps described in Chap. 17.

Bearing Stress

VDI says that the bearing stress under the head of the bolt or between nut and joint should not exceed the "boundary surface pressure" of the joint material, or

$$0.9 \text{ Max } F_J \leq P_G A_p \tag{21.14}$$

where P_G = boundary surface pressure (psi, MPa) and is usually slightly higher than the yield strength of the joint material
A_P = contact area under head or nut (in.^{2}, mm^2)

Why Max F_J instead of Max F_B here? As we say in Eq. (21.9), Max $F_B >$ Max F_J. The *tensile stress* in the bolt increases under an external load, but the clamping force between joint members, and therefore the contact stress between bolt and joint, decreases because of this same load. Equation (21.14) acknowledges this by basing the limit on Max F_J, the highest bolt load created during assembly.

Shear Stress

If the bolt will also be loaded in shear its tensile capacity must be reduced as illustrated in Fig. 14.10. Now we must compare Max F_B not to the yield strength of the bolt, but to remaining tensile capacity of the bolt.

Bending Stress

In critical situations we might want to estimate any bending stresses to be seen by the bolt; perhaps using the VDI expression given in Eq. (13.8). Tensile or combined stress limits would then be based upon the loads seen by the more highly stressed convex side of the bolt rather than on the average tensile stress.

Eccentric Loading

With eccentric loading and the resulting prying action, things get a little more complicated. The required force, which we have previously called F_{Prqd}, is now called F_{Krqd} and is the force required to prevent even partial separation of the joint. The equation for this force is:

$$F_{\text{Krqd}} = \frac{(a - s)u}{(R_G^2 + su)} \qquad (21.15)$$

There are potential problems here. We're told that a can be statically indeterminate in some joints (e.g., for an automotive conrod) [3] and that s and R_G are not easy to quantify [1]. See Fig. 13.12 for an illustration defining a, s, and u.

If the eccentric joint is subjected to cyclical, fatigue loading, then we must calculate a fatigue strength based upon the maximum fiber stress. We use Eq. (13.8) to compute the change in fiber stress and use that change instead of $\triangle F_B$ in Eq. (21.11).

H. Revised Bolt Specifications

At the beginning of this design procedure we tentatively selected a bolt size and material. The analysis which followed was intended to confirm or refute that choice. VDI tells us that we can now use Eqs. (21.6) through (21.15) to revise our choice of size and material, if revisions are indicated. Note that the new selection will be based, not just on gut feel or past experience, but on an impressive list of design considerations, including:

Yield strengths of the bolt and the joint material
The external loads to be placed on the joint (magnitude, static or dynamic, tensile or shear, or a combination)
Maximum fiber stress in the bolt
Under head or nut contact stress on joint
Expected relaxation (embedment and elastic interactions)
Tool scatter (which means that we have decided upon the type of tool and assembly procedure to be used)
The joint stiffness ratio or load factor, with a consideration of how the joint is going to be loaded (internally or externally; eccentrically or concentrically)
The amount of bolt strength which will be absorbed by torsional stress
The clamping force required to prevent slip, separation, or leakage

V. AN EXAMPLE

Let's try an example to see how we might use the VDI equations.

A. Inputs

We must first decide whether we want to assign maximum, minimum, or average values to the various design parameters, in effect facing the same

decisions we dealt with in Chap. 20, Sec. VIIC. My preference at that
time was for average values, and that's what we'll use here. If this were
a safety-related joint we'd also want to run the calculations for worst-
case values. The VDI Directive 2230, incidentally, does not specifically
address this issue, but that document contains several tables of data defining possible inputs; and I'm sure that these are all average (i.e., "typical")
values. Only tool scatter is treated as a statistical variable by VDI; input
loads may also be treated as variable in fatigue situations, but always with
the implication that the max/min loads are known. In any event—here
are the "average" inputs we're going to use in this example.

Let's assume that we have roughed out the design of our product and
have need for a joint which will be subjected to a concentric, per-bolt
tensile load which will cycle between 3000 and 4000 lb. We'd like the
minimum clamping force to be at least 1000 lb per bolt, to prevent joint
separation. Based on past experience we've decided to use $\frac{1}{2}$–13 UNC
SAE J429 Grade 8 bolts. Joint material will be a low-carbon steel. We've
computed the load factor, using the procedure in Chap. 12, and estimate
it to be 0.2. The equipment will be assembled using manual torque
wrenches, and it will be used at room temperatures only. Here, then, are
our inputs:

A_S = 0.1419 in.^{2} (from Appendix F)
$\triangle F_P$ = 28% of F_P (10% from embedment plus 18% average relaxation
    through elastic interaction)
L_X = 4000 lb
L_{Xmin} = 3000 lb
s = ±30%
Φ_{Kn} = 0.2
σ_A = 18 ksi (from Table 4.11)
σ_y of bolts = 130 ksi (from Table 4.1)
σ_y of joint material = 34 ksi (from Table 4.16)

B. Calculations

We start with some preliminary computations.

α_A = $(1 + s)/(1 - s)$ = 1.86
F_y = $\sigma_y A_S$ = 130,000 (0.1419) = 18,447 lb
$\triangle F_P$: To compute $\triangle F_P$ we must assume or estimate an F_P; let's say that
    will be
        50% of the yield strength of the bolt, or
        F_p = 0.5 F_y = 0.5(18,447) = 9224 lb
        Therefore: $\triangle F_P$ = 0.28F_P = 2583 lb
$\triangle F_J$ = $(1 - \Phi_{Kn})L_X$ = 0.8 (4000) = 3200 lb [from Eq. (21.3)]

Max and Min Assembly Preloads

Now we're ready to compute min and max assembly preloads [Eqs. (21.5) and (21.6)]

$$\text{Min}F_J = (\triangle F_J + \triangle F_{Prqd} + \triangle F_P) = (3200 + 1000 + 2583) = 6783 \text{ lb}$$
$$\text{Max}F_J = \alpha_A \text{ Min } F_J = 1.86(6783) = 12,616 \text{ lb}$$

Static Strength of the Bolts

Are the bolts strong enough? From Eq. (21.7):

$$0.9F_y = 0.9(18,447) = 16,602 \text{ lb}$$

This is greater than Max F_J, so the latter seems OK, but a second test is required.
From Eqs. (21.3) and (21.6):

$$\triangle F_B = \Phi_{Kn}L_X = 0.2(4000) = 800 \text{ lb}$$
$$\text{Max } F_J + \triangle F_B = 12,616 + 800 = 13,416 \text{ lb}$$

That's only 73% of the yield strength (F_y) of the bolt, leaving a 13% safety factor. Our choice of a $\frac{1}{2}$-in. Grade 8 bolt may not be optimum, but it certainly appears to be acceptable.

Fatigue Strength

Next we analyze the fatigue situation; using Eq. (21.9).

$$\frac{\Phi_{Kn}(L_X - L_{Xmin})}{2} \leq \sigma_A$$
$$\frac{0.2(4000 - 3000)}{2} = 100 \text{ psi}$$

which is well below the endurance limit of 18 ksi, so our choice of bolt still seems acceptable.

Contact Stress

Finally, we'll check the contact stress between bolt head and joint. We obtain a sample of the bolt and measure the across-flats distance of its hex head to be 0.75 in. The bolt, of course, has a nominal diameter of 0.5 in. We use the familiar $\pi(D^2/4)$ to compute the areas described by these diameters; then we compute the difference between them.

$$\text{Contact area } A_P = A_{flats} - A_{diam}$$
$$A_P = 0.442 - 0.196 = 0.246 \text{ in.}^2$$

The yield strength of the joint material is 30 ksi, so we'll call that the maximum allowable contact pressure, P_G [from Eq. (21.14)]

$$A_P P_G = 0.246(34,000) = 8364 \text{ lb}$$
$$0.9 \text{ Max } F_J = 0.9(12,616) = 11,354 \text{ lb}$$

$A_P P_G$ is *less* than 0.9 Max F_J and so the contact stress exceeds the limits imposed by Eq. (21.14). We could respond by specifying less preload; but that should be avoided if possible because it would mean a minimum clamping force of less than 1000 lb per bolt. We could also specify a stronger joint material, but that sounds expensive. (I'm assuming that our other joint stress computations say that the material is acceptable.) The simplest thing to do is to use washers under the bolt head and nut to reduce contact stress.

A quick calculation shows that a 1-in.-diameter washer would raise $A_P P_G$ to 20,040 lb, well above the 0.9 Max F_J value of 11,354 lb.

No gasket or temperature change or shear loads or eccentric loads are involved here, so we're done. As you can see, the procedure is easy to use. But estimating the values to be used for the input data—loads and things—can require some effort and some engineering judgment.

VI. OTHER FACTORS TO CONSIDER WHEN DESIGNING A JOINT

That concludes the discussion of the VDI procedure. There are, however, many other factors we'll often want to consider when designing a joint. These include the following, in no particular order.

A. Thread Strength

We want the body of the bolt to break before the threads strip. That will be the case if we've selected standard bolts and nuts; but we'll want to use the thread strength equations of Chap. 3 to check length of engagement etc. if we're planning to use the bolts in tapped holes, especially in soft materials. For example, one rule of thumb I've heard suggests that the length of engagement should be at least two times the bolt diameter if a steel bolt is to be used in a hole tapped in an aluminum joint.

B. Flexible Bolts

Although some say use of stronger bolts with the resulting higher preload is more important than flexibility, conventional wisdom says that we'll usually prefer flexible bolts. They are much better energy storage devices and are, therefore, less sensitive to thermal change, vibration, embedment or other relaxation loss, etc. Although there are no absolutes in bolting, previous chapters have included statements like this: "If the bolt's length to diameter ratio is 8:1 or more, it will never self-loosen" or "We always want the joint-to-bolt stiffness ratio to be 10:1 or greater." In general, we want to avoid such poor energy storage parts as short, stiff bolts; composite or soft joint materials; etc.

C. Accessibility

We want a minimum of 60° in which to swing a wrench, and would like to be able to see the bolts. Remember the guy who wrote that "the preload in most bolts in this world is directly proportional to their accessibility." Try to put yourself in the mechanic's shoes when locating those bolts.

D. Shear vs. Tensile Loads

In general, shear joints are subjected to fewer failure modes than are tensile joints. If you have a choice, therefore, you'd be well advised to connect your parts with well-designed shear joints rather than with tension joints. See Chap. 23 for more on shear joint design.

E. Load Magnifiers

We want to avoid things which magnify the loads seen by the bolts. The most common sources of this problem are prying, as illustrated in Fig. 13.4, and eccentric shear loads, which will be discussed in Chap. 23. (See Fig. 23.8.)

F. Minimizing Embedment

We can minimize embedment relaxation by chamfering holes, by insisting on flat and parallel joint surfaces, by specifying that holes should be drilled perpendicular to joint surfaces, and/or by specifying hard washers.

G. Differential Expansion

Differential expansion can create a significant and simultaneous increase—or decrease—in both bolt tension and the clamping force of the joint. We can at least reduce the effects by using bolts and joint members made of materials having similar coefficients of thermal expansion. It helps if bolts and joint members are exposed to the same changes in temperature, though this is often difficult to achieve. Using bolts with a generous length-to-diameter ratio, perhaps with the help of Belleville springs or cylindrical collars, can be a big help, too.

H. Other Stresses in Joint Members

The general-purpose VDI procedure doesn't cover all of the stresses a joint might be exposed to, of course, so we must be careful not to overlook them. Pressure vessel and piping designers, for example, estimate the stresses in flange fillets. Studies have been made of ways to combat bending stresses placed on certain types of aerospace joints [4]. And the PVRC is trying to quantify the effects of bending moments on gasketed, flanged joints. Excess stress can cause shear joints to tear out, etc. These are only a few examples, but presumably they'll give you the message, "Don't base your designs solely on the VDI procedure."

I. Locking Devices

If the joints are to be exposed to extreme shock or vibration, then we should consider using one of the locking devices or fasteners described in Chap. 16. It would also help to design the joints so that the axes of the bolts are more or less parallel to the vibratory or shock loads.

J. Hole Interference

We want to avoid inadvertent interference between the bolts and clearance holes in joint members, by careful dimensioning of hole locations, sizes, etc., because hole interference can absorb a substantial portion of assembly preload. If interference is desired or unavoidable we should specify that the bolts be forced through the holes by other means before being tightened.

K. Safety Factors

Many of the values we'll assign to design parameters will be subject to considerable variation and/or based on an outright guess. The safety and reliability of our designs, therefore, can be enhanced by the judicial use of safety factors, applied either to the individual values we use or to final results. For example, if the analysis suggests that a $\frac{1}{2}$-in.-diameter bolt will do the job we might well decide to use a $\frac{5}{8}$-in. or $\frac{3}{4}$-in. one to cover the uncertainties.

L. Selecting a Torque to Be Used at Assembly

VDI doesn't tell us, specifically, how to pick an assembly torque, but they give us the necessary information to do this. The "target" preload would be the midpoint between Max F_J and Min F_J as defined in Eqs. (21.5) and (21.6). Averaging these gives us

$$F_{Pa} = (1 + \alpha_A)(\text{Min } F_J)/2 \tag{21.16}$$

We'd then use a suitable nut factor (K, from Table 7.1) and the short-form, torque-preload equation to compute a torque value.

$$T = KDF_{Pa}$$

where
T = torque (in.-lb, mm-N)
K = nut factor (dimensionless); typically 0.2 for as-received steel
D = nominal diameter of the fastener (in., mm)
F_{Pa} = preload to be used at assembly (lb, N)

Continuing the example of Sec. V:

$$F_{Pa} = (1 + 1.86)(6783)/2 = 9700 \text{ lb}$$

$$T = 0.2(0.5)9700 = 970 \text{ in.-lb}$$

And that's it! If we do all of the things described in this section and in Sec. IV, we'll end up with better designed bolted joints than *most* of those which are already out there. Who could ask for anything more?

REFERENCES

1. Junker, G. H., Principle of the calculation of high-duty bolted joints: interpretation of Directive VDI 2230, published by SPS Technologies, Jenkintown, PA, no date given (but probably about 1978).
2. VDI 2230: Systematic Calculation of High Duty Bolted Joints, Translated from the 1977 German text by Language Services, Knoxville, TN, reproduced by the U.S. Department of Energy, Office of Scientific and Technical Information, Oak Ridge, TN, no date given.
3. McNeill, W., A. Heston, and J. Schuetz, A study of factors entering into the calculation for connecting rod joints according to the VDI 2230 guidelines, VDI Berichte No. 766, 1989.
4. Baumann, Theodore R., Designing safer prestressed joints, *Machine Design*, pp. 39ff, April 25, 1991.
5. Eccles, William, Bolted joint design, *Engineering Designer* (UK), pp. 10ff, November 1984.

BIBLIOGRAPHY ON BOLTED JOINT DESIGN

You might find the following works useful, in addition to the references cited at the end of this and the next two chapters.

A. General (Includes Auto, Aero, and Miscellaneous Joints)

Blake, Alexander, *Design of Mechanical Joints,* Marcel Dekker, New York, 1985.

Joint design, Prepared by the Technical Information Staff of the Industrial Fastener Institute, *Machine Design,* p. 12ff, September 11, 1969.

Junker, G. H., Principle of the calculation of high-duty bolted joints, SPS, Jenkintown, PA.

Osgood, Carl C., How elasticity influences bolted joint design—part 1, *Machine Design,* p. 92ff, February 24, 1972.

Osgood, Carl C., How elasticity influences bolted joint design—part 2, *Machine Design,* p. 104ff, March 9, 1972.

Osgood, Carl C., *Fatigue Design,* Wiley-Interscience, New York, 1970.

Shigley, Joseph E., *Mechanical Engineering Design,* McGraw-Hill, New York, 1977.

Shigley, Joseph E., and Charles R. Mischke, eds., *Standard Handbook of Machine Design,* McGraw-Hill, New York, 1986.

Systematische Berechnung hochbeanspruchter Schraubenverbindungen, Verein Deutscher Ingenieure (VDI), Berlin, West Germany, October 1977.

B. Pressure Vessel and Piping Joints

ASME Boiler and Pressure Vessel Code, ASME, New York, 1983 (e.g.). Section III—Division 1, Subsection NB-Design of Class 1 components (those contained in the reactor coolant boundary). See especially sections NB3231 and Appendix E.

Section III—Division 1, Subsection NC—Class 2 components (pressure retaining). See especially Appendix XI.

Section III—Division 1, Subsection NF (component supports). See especially sections NF 3324.6 and NF 3332.5.

Section VIII—Division 1 (pressure vessels). See especially Appendix 2 and Appendix Y.

Section VIII—Division 2 (design of pressure vessels "by analysis" instead of "by rules"). See especially Appendix 3.

Blach, A. E., A. Bazergui, and R. Baldur, Bolted flanged connections with full faced gaskets, *WRC Bulletin* 314, Welding Research Council, New York, May 1986.

Petrie, E. C., The ring joint; its relative merit and application, *Heating, Piping and Air Conditioning,* vol. 9, no. 4, April 1937.

Reddy, M. D., and B. S. Nau, Factors affecting the design of large flanged joints and joint seals for chemical plant, British Hydromechanics Research Association (BHRA), Cranfield, Bedford, England, December 1977.

Rodabaugh, E. C., F. M. O'Hara, Jr., and S. E. Moore, FLANGE, a computer program for analysis of flanged joints with ring-type gaskets, U.S. Department of Commerce, National Technical Information Service, Springfield VA, January 1976.

Rodabaugh, E. C., and S. E. Moore, Evaluation of the bolting and flanges of ANSI B16.5 flanged joints—ASME Part A Design Rules, ORNL/Sub/2913-3, NRC-5, PP 115–7h, Oak Ridge National Laboratory, Oak Ridge, TN, September 30, 1976.

Wesstrom, D. B., and S. E. Bergh, Effect of internal pressure on stresses and strains in bolted-flanged connections, *ASME Transactions*, p. 121ff, 1951.

Webjorn, Jan, The bolted joint, a series of problems, Dissertation No. 130, Institute of Technology, Linkoping, Sweden, 1985.

C. Structural Steel Joints

Bibliography on Bolted and Riveted Joints, American Society of Civil Engineers, New York, 1967.

Frank, K. H., and J. A. Yura, An experimental study of bolted shear connections, Report FHWA/RD-81-148, U.S. Department of Transportation, Federal Highway Administration, Published by the National Technical Information Service, Springfield, VA, December 1981.

Kulak, Geoffrey L., John W. Fisher, and John H. A. Struik, *Guide to Design Criteria for Bolted and Riveted Joints*, 2nd ed., Wiley, New York, 1987.

Manual of Steel Construction, 8th ed., American Institute of Steel Construction (AISC), Chicago, 1980.

Specification for structural steel joints using ASTM A325 or A490 bolts, American Institute of Steel Construction (AISC), Chicago, November 13, 1985. (Note: This specification is now available in two versions, one describing allowable stress design procedures, the other load and resistance factor design procedures.)

Merritt, Frederick S., ed., *Structural Steel Designers' Handbook*, McGraw-Hill, New York, 1972.

22
The Design of Gasketed Joints

We've studied, in general, the design of joints loaded in tension. Now we're going to look at the design of a special class of tension joint; the gasketed joint. We'll find that adding a gasket tends to complicate things and that, at least for pressure vessel and piping systems, the engineering community has responded to the complexity by reducing the design process to a series of simplified steps. These steps, furthermore, incorporate a number of safety factors to cover the fact that they ignore many of the factors which affect the behavior of such joints.

We're going to examine two of these codified procedures: that found in the current ASME Boiler and Pressure Vessel Code, and the revised Code procedure recommended by the Pressure Vessel Research Committee (PVRC). We're also going to look at three other, "unofficial" (at least in the United States) design procedures proposed by engineers in the UK and in Germany. We'll start, in fact, with that defined by VDI in Germany, because we spent much of the last chapter studying this procedure, and it makes sense to extend it to gasketed joints while it's still fresh in our minds. It's also a sophisticated procedure which takes far more design factors into account than do the simplified and/or codified procedures to be discussed in the rest of this chapter. As a result it gives us a more accurate look at the gasketed joint than do these other procedures, which also makes it a good place to start. You should review the discussion in Secs. IV through VI of Chap. 21 if you skipped them or need a refresher.

I. THE VDI PROCEDURE APPLIED TO GASKETED JOINTS

Section 5.6 of VDI 2230 shows how the equations we used in the last chapter can be applied to a raised face, pressure vessel, flanged joint. Drawing on another German document, DIN 2505, which I don't have, the authors introduce values for such things as gasket creep and stiffness ratios [1]. In the discussion which follows, I'll attempt to show how the same equations can be used with data published by the PVRC and others, rather than by DIN.

We would start the design procedure exactly as we would for any joint, with a preliminary layout probably based on past experience or custom. We'd estimate the loads to be placed on the joint, and would probably make preliminary decisions about joint and bolt dimensions and materials. Then we'd use the VDI equations to evaluate and refine these initial decisions.

For convenience, let's repeat the VDI equations. We called these Eqs. (21.5) and (21.6). They are general-purpose equations which can be used for the design of any type of bolted joint: we've already seen how to use them to design a joint loaded in tension. They define the minimum and maximum preloads we want to create in the bolts during assembly; but they define these bolt loads in terms of the minimum and maximum clamping forces the bolts will exert on the joint before the joint is placed in service, and they take into account the anticipated relaxation of the bolts during assembly, the external loads the joint must support in service, and the residual clamping force required to avoid failure. So, although they're "initial assembly preload" equations, they encompass the major factors which will affect the behavior and life—and therefore the design—of that joint. Here are those equations.

$$\text{Min } F_J = \Delta F_J + F_{Prqd} + \Delta F_P \qquad\qquad (21.5)$$

Since $\Delta F_J = (1 - \Phi_{Kn})L_X$ this can also be expressed as

$$\text{Min } F_J = (1 - \Phi_{Kn})L_X + F_{Prqd} + \Delta F_P$$

We also have

$$\text{Max } F_J = \alpha_A [(1 - \Phi_{Kn})L_X + F_{Prqd} + \Delta F_P] \qquad\qquad (21.6)$$

where ΔF_P = loss of preload during or soon after assembly (lb, N).

$\text{Max } F_J$ = maximum anticipated bolt tension created during assembly; equals the clamping force that bolt applies to the joint before the joint is put in service (lb, N).

$\text{Min } F_J$ = minimum anticipated bolt tension and clamping force created during assembly, before the joint is put in service (lb, N).

F_{Prqd} = minimum preload (and per-bolt clamping force) required to prevent joint failure (lb, N).

L_X = external tensile load applied to the joint (lb, N).

α_A = the scatter in preload caused by the assembly tools and procedures; α_A = Max F_P/Min F_P.

Φ_{Kn} = load factor for a concentric joint, loaded internally at loading planes. $\Phi_{Kn} = \Delta F_B / L_X$ where ΔF_B = the change in tension created in the bolts by the external load. This is the load factor we used when discussing tension joints in general. As we saw in previous chapters—and will see again below—there are other load factors we might use instead.

Equations (21.5) and (21.6) apply to gasketed joints as well as to any other kind of joint loaded in tension (or in shear, as we'll see in Chap. 23). To use them for gasketed joints we must merely define each factor in an appropriate way. Here's how we do that:

ΔF_P: In a gasketed joint this loss of preload will be caused, in part, by embedment relaxation and by elastic interactions between bolts during assembly; as in any joint. Elastic interactions, however, will usually be greater for a gasketed than for a nongasketed joint. The only specific data we have on this come from a limited number of tests run by Dr. George Bibel of the University of North Dakota [2]. He tells us that the loss will average 30% if the joint contains a sheet gasket, and as much as 46% if it contains a spiral-wound gasket.

That's not the end of the story for a gasketed joint, however. Most gaskets creep when first loaded, and often creep more when put in service and exposed to elevated temperatures. This loss can be substantial, and must be included when we estimate ΔF_P. VDI 2230 apparently lumps gasket creep with true embedment. In any event, they call all gasketed joint relaxation embedment and quote data taken from German Standard DIN 2505.

Gasketed joints often contain dissimilar materials. For example, we might be faced with aluminum joint members, alloy steel bolts, and gaskets made of a combination of materials. Each will have a different coefficient of expansion. Service temperatures can, therefore, either increase or decrease bolt loads and clamping forces.

Putting all this together for a gasketed joint, therefore:

$$\Delta F_P = \Delta F_{em} + \Delta F_{EI} + \Delta F_{CR} \pm \Delta F_{TH} \qquad (22.1)$$

where em, EI, CR, and TH refer to embedment, elastic interactions, creep, and thermal effects, respectively.

F_{Prqd}: This term will define the residual stress on the gasket required to prevent excessive leakage. We might be able to obtain this from the gasket manufacturer. Or we could use the ASME Code "m" factor to define it as a multiple of the contained pressure. (The ramifications of doing this were explained in Chap. 19.) Or, we could do it by using the new PVRC gasket factors, also discussed in Chap. 19. Or, most accurate of all, we could base F_{Prqd} on actual PVRC or other leak test data. We'll be taking another look at the ASME and PVRC procedures later on in this chapter, so I won't elaborate further at this time. VDI 2230 again refers to German Standard DIN 2505 for the residual stress required to prevent leakage. According to the translation I have [1], the gasket they use in their example is made of high-quality "It-material," whatever that may be!?

Max F_J, Min F_J: The definitions of these terms remain unchanged for gasketed joints. It's worth noting, however, that the VDI procedure is the only one we're going to look at which bases the design upon assembly preload limits, and which includes significant consideration of the scatter in assembly preload.

L_X: This is the hydrostatic end load created by the contained pressure. All the procedures discussed here estimate it the same way; i.e., they assume that flange rotation partially unloads the gasket, exposing its inner portions to the contained pressure. This point will be clarified when we run numerical examples.

α_A: the definition of this term is unchanged for a gasketed joint. If we use torque control at assembly and expect a preload scatter of $\pm 30\%$, then $\alpha_A = 1.3/0.7 = 1.86$.

Φ_{Kn}: the definition of this term is unchanged, too, but we must recognize that the gasket will have a major influence on the effective stiffness of the joint; will dominate it, in fact. VDI 2230 has equations with which to compute the stiffness of a raised-face flange, and the resulting load factor for such a joint. I can't repeat those equations here because my copy of the translation is poor, and some of the terms they use are apparently not defined. But I prefer experimental data to theoretical equations when dealing with bolting, and so I suggest that if we're dealing with a raised-face joint we can use the results of studies made by Dr. Bernard Nau of the British Hydromechanics Research Agency to estimate the bolt-to-joint stiffness ratio, as explained in Sec. III-B of Chap. 5. The table of gasket stiffnesses, derived from PVRC test data, Table 5.1, can also be useful. In many cases, however, it will be necessary to conduct an experiment to determine the gasket stiffness. As we learned in Chap. 5, we'll be interested in the stiffness characteristics revealed by a previously preloaded gasket as it is unloaded and reloaded. We're not interested in its stiffness during initial loading.

Is Φ_{Kn} the correct load factor to use for a gasketed joint? Remember that it is only one of several possibilities, and is supposed to be used only for joints which are loaded concentrically along internal loading planes. I think it will be appropriate for the relatively stiff flanged joint to be used as an example in this chapter, but it might not be appropriate for other gasketed joints. We have *very* little information on stiffness ratios for gasketed joints, however, so this is really a moot point. We'll rely more on Dr. Nau than on VDI for the actual values used for this term.

As a final point, only VDI takes stiffness ratios into account. The other procedures we're going to use ignore them.

Having estimated each of the above factors, we can compute Min F_J and Max F_J. We then use these limits to select bolt size and material; and we subject our selection to the same cross checks and tests we used in the last chapter. Again, you should review that discussion if you missed it.

II. AN EXAMPLE, USING THE VDI PROCEDURE

A. The Inputs

We're going to use several of the procedures described in this chapter to evaluate the same proposed design, so that we can see how each procedure works and can compare results. The joint to be designed—more properly the preliminary design to be evaluated—is shown in Fig. 22.1. Here are the initial design parameters:

Service pressure (P): 1000 psi
Service temperature (T); 800°F
Bolt material: ASTM A193-B7
Bolt yield strength (σ_{yB}): 105 ksi at room temperature (Table 4.1); 64 ksi at 800°F (Table 4.4)
Bolt size: $\frac{7}{8}$—16 × 6
Tensile stress area (A_s): 0.521 in.^{2} [7]
Number of bolts: 12
Joint material: low-carbon steel A36
Joint material yield strength (σ_{yJ}): 34 ksi (Table 4.16)

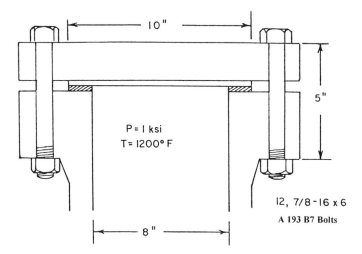

Figure 22.1   Tentative design of a blind flange which will be used as an example in several analyses in this chapter. Further details will be found in the text.

Gasket: double-jacketed; stainless steel, asbestos filled
   OD: 10 in.
   ID: 8 in.
   Width: 1 in.
   Thickness: $\frac{1}{8}$ in.

B.   The Calculations

F_{Prqd}: How much residual clamping force do we need here? VDI 2230 gives a value for this, but takes it from DIN 2505 and doesn't tell us enough about the type of gasket they're using in their example. We need to use U.S. data of some sort to estimate this factor.

   For this example, I'm going to take the common view that the required residual clamping force is adequately defined by the ASME Code m factor, which is 3.75 for this type of gasket, even though I know that the Code's m and y factors are suspect. An m of 3.75 would suggest an F_{Prqd} capable of creating a residual stress on the gasket of $m \times P = 3.75 \times 1000 = 3750$ psi on what the Code calls the "effective area" of the gasket. There are various Code formulas for computing this area, depending upon flange type and gasket width, but I'm going to assume that the effective area will be approximately half the apparent contact area. I know that m is only a rough approximation, so making an engineering judgment about the contact area seems reasonable. (See Sec. III-B of this chapter for further information on this effective area business.) Let's also note in passing that the Code y factor for a double-jacketed asbestos-filled gasket is 9000 psi, which I'll also assume is based on about half the contact area of the gasket. We'll want to compare our Min and Max F_J decisions to this number.

The full contact area of this gasket is:

$$\frac{\Pi(OD^2 - ID^2)}{4} = \frac{3.14159(100 - 64)}{4} = 28.3 \text{ in.}^2$$

So half the area is 14.2 in.^{2}

As a result, I'll assume that the total, residual clamping force on the joint must be:

$$3750 \times 14.2 = 53,250 \text{ lb}$$

or

$$\frac{53,250}{12} = 4438 \text{ lb per bolt}$$

I emphasize that this is only a ballpark figure, based roughly on Code data; but it's the kind of information we have been forced to use in the past. The PVRC tests give us a more accurate look at gasket stress requirements, and we'll deal with them in Sec. IV.

L_X: VDI says that the gasket is exposed to its midradius, presumably by flange rotation. They base their calculation of end load, therefore, on diameter d_D:

$$d_D = \frac{OD + ID}{2}$$

In our example, $d_D = 9$ in.
Therefore, the hydrostatic end load will be:

$$\frac{P \times \Pi \times d^2_D}{4} = \frac{1000 \times 3.14159 \times 9^2}{4} = 63,617 \text{ lb}$$

or

$$\frac{63,617}{12} = 5301 \text{ lb per bolt.}$$

Φ_{Kn}: As already mentioned, our best information concerning the bolt-to-joint stiffness ratio of a raised-face, gasketed flange is found in Dr. Nau's work as reported in Chap. 5 (e.g., in Figs. 5.19 and 5.20). Although that data are not for a double-jacketed gasket, they suggest that the stiffness ratio is the same for gaskets having quite different stiffness characteristics; so I'm going to use those numbers as is. They say that, for a joint this large, joint stiffness K_J is typically twice the bolt stiffness K_B. Therefore:

$$\Phi_{Kn} = \frac{1}{(1 + 2)} = 0.33$$

This, incidentally, is much greater than the value computed in the VDI 2230 example, but, as I've said before, we have no information concerning the type of gasket they're using.

ΔF_P: This is the tough one. We're going to have to make some engineering judgments to estimate it. We must estimate each term in Eq. (22.1). Here goes:

ΔF_{em}: Let's assume 5% of the initial preload F_P. To estimate it we need a preliminary, minimum value for F_P. As a first guess let's say that we want the minimum, initial F_P to be 50% of the room-temperature yield strength of the bolt, or:

$F_P = A_S\sigma_{yB} = (0.521)105,000 = 54,705$ lb per bolt.
Half of that would be 27,353 lb per bolt, and
5% of this would be 1368 lb per bolt.

ΔF_{EI}: We'll use Dr. Bibel's figure of 30% of preload (average loss) for a sheet gasket. So: $\Delta F_{EI} = 0.3 \times 27,353 = 8206$ lb per bolt.

ΔF_{CR}: VDI says that "embedment" will amount to 5% of the thickness of the gasket at room temperature and 10% of thickness at steam temperatures, but, once again, we don't know enough about the gasket used in their example to compare it to our double-jacketed, asbestos-filled one. As we learn from Table 19.3, our gasket will lose, through creep, only 0.4% of its initial thickness at room temperature; but it would lose as much as 4% of its thickness at 1200°F. What can we assume for its loss at 800°F? Let's assume $800/1200 \times 4\%$ or about 3%.

Its initial thickness is $\frac{1}{8}$ in. (0.125 in.). Three percent of that is 3.75 mils. What does that mean in terms of preload? According to Table 9.2 a "typical" ASTM A193, B7 bolt will stretch about 1.6 mils per inch of grip length if tightened to 50% of yield. [Alternatively, we could compute the stretch using Eq. (5.8), but Table 9.2 is accurate enough for these estimates.]

The grip length here is 5 in.; so the bolt will stretch 5 ×
1.6 = 8 mils when tightened to a preload of 27,353 lb. The
3.75 mils loss of thickness in the gasket, therefore, will reduce
preload by:

$$\left(\frac{3.75}{8}\right) \times 27,353 = 12,822 \text{ lb}$$

ΔF_{TH}: Even though the operating temperature here will be 800°F, I
think we can ignore any change in preload caused by differential
expansion, since bolts, joint members, and gasket are all made
of steel. There probably would be some change, because these
are different steels, but the effect, I think, would be minor compared to those we've already considered. See Chap. 13, Sec.
IV-C if you need to estimate this effect.

Putting all of this together we get:

$\Delta F_P = 1368 + 8206 + 12,822 = 22,396$ lb per bolt

ΔF_J: $\Delta F_J = (1 - \Phi_{Kn})L_X = (1 - 0.33)5301 = 3552$ lb per bolt

Min F_J: We can now compute Min and Max F_J.

Min $F_J = \Delta F_J + F_{Prqd} + \Delta F_P$

Min $F_J = 3552 + 4438 + 22,396 = 30,386$ lb (56% of yield)

Max $F_J = \alpha_A$ Min F_J

Max $F_J = 1.86(30,386) = 56,517$ lb (103% of yield)

Max F_B: Max F_J defines the maximum assembly load the bolt will see,
but not the maximum in-service load. That will be:

Max F_B = Max $F_J + (\Phi_{Kn}L_X)$

Max $F_B = 56,517 + (0.33)(5301) = 58,266$ lb (107% of yield)

Looks like we've got some problems. First of all, we said we were
going to tighten the bolts to an initial preload of 50% of yield, but find
that the VDI equations tell us we need at least 56% of yield. Since some
of our calculations were based upon a percentage of preload, it looks like
we should redo the calculations. If we did this, though, we'd get a still
higher Min F_J and theoretically have to repeat the calculations again—and
again.

Remember that the data we're inputting here are all based on typical
or assumed values. The tool scatter of ±30% is worst case and applies

to some bolts but not, probably, to each bolt in a joint. As a result, I think we can accept 30,386 lb as being essentially equal to the initial guess of 27,353 lb, and move on.

We don't want to tighten any of the bolts past yield, however, and in fact VDI tells us not to exceed 90% of the yield strength of the bolts, so I suggest that we use engineering judgment to set Max F_J equal to 90% of the room-temperature yield strength of the bolts, or $0.9 \times 54{,}705 = 49{,}235$ lb, and use α_A to define the Min F_J as follows:

$$\text{Min } F_J = \frac{\text{Max } F_J}{\alpha_A} = \frac{49{,}235}{1.86} = 26{,}470 \text{ lb (48\% of yield)}$$

We now have Max and Min F_J's which are in the ballpark of our first estimates and are, therefore, reasonable in light of our many assumptions. The target assembly preload, F_{Pa}, would be halfway between Max and Min F_J, or

$$F_{Pa} = \frac{49{,}235 + 26{,}470}{2} = 37{,}853 \text{ lb (69\% of yield)}$$

The maximum in-service bolt load in this case will be:

$$\text{Max } F_B = \text{Max } F_J + (\Phi_{Kn}L_X)$$
$$\text{Max } F_B = 49{,}235 + (0.33)(5301) = 50{,}984 \text{ lb}$$

which is only 93% of yield. All of which suggests that our preliminary choice of a $\frac{7}{8}$, B7 bolt appears to be confirmed. If we were nervous about all these assumptions and readjustments we could do one or more of several things. We could

Redo the calculations using larger bolts, perhaps 1 in. in diameter.
Insist that the bolts be lubed at assembly, to reduce the $\pm 30\%$ scatter.
Provide instructions that the bolts be retightened after brief service at 800°F to compensate for gasket creep, the largest loss factor.

C. Discussion

How does our proposed range of assembly preload compare to the Code's suggestion that the joint should be designed so that it is capable of exerting a seating stress of 9000 psi on the effective area of the gasket? The Code only suggest this, it doesn't mandate it, but presumably it's a reasonable suggestion.

At the risk of getting ahead of ourselves: in Sec. IV we'll learn that the Code calculations suggest that a per-bolt load of 7747 lb is required to create 9000 psi on the effective area of this particular gasket. The Code calculations also show us, however, that the joint should be designed for a seating bolt load of over twice this—of 17,290 lb per bolt—in order to satisfy the in-service m factor requirements of this gasket.

We've decided to apply an average preload of over twice this, and a maximum of about three times it! Would that crush the gasket? It might; we'd certainly want to check that possibility with the gasket manufacturer. Of more interest to us at the moment, though, is the question, "Has VDI led us astray here?"

Probably not. The last calculation certainly seems to confirm our bolt choice. Our bolt can exert five times as much initial clamping force on the gasket as that suggested by the Code. No need for a bigger bolt if clamping force is the only consideration. (As we'll see, the Code orders us to consider other factors as well.) But it looks as if these VDI calculations are going to suggest a design capable of producing far more preload than the Code suggests is required by the gasket. Who's right?

First of all, let's acknowledge that I've used a modified version of the VDI equations. As we saw in the last chapter they equate ΔF_P to "embedment" only; they don't add the effects of elastic interactions or, as near as I can tell from the documents in hand, with thermal expansion. (German Standard Din 2505 or other may well consider thermal effects.) If we also eliminate interactions from our calculations we get:

$$\text{Min } F_J = 3352 + 4438 + 14,190 = 22,180 \text{ lb per bolt (41\% of yield)}$$

and

$$\text{Max } F_J = 1.86(22,180) = 41,255 \text{ lb per bolt (75\% of yield)}$$

$$F_{Pa} = 31,718 \text{ lb per bolt (58\% of yield)}$$

So, even this change leaves us recommending much higher preloads than does the present Code, because the Code ignores gasket creep. Does the fact that the Code ignores creep and that VDI ignores interactions mean that we can ignore the inevitable and substantial effects of elastic interactions and gasket creep? The Code at least includes substantial safety factors to cover their assumptions and omissions. Experience shows us, however, that the present procedure creates joints that sometimes have to be retightened to compensate for creep and/or interactions. More massive joints and bigger bolts could provide enough initial preload to compensate, but the gaskets may not be able to support such loads. The VDI procedure assumes that the 90% of the room-temperature yield strength of the bolt determines the upper limit for Max F_J (unless fatigue or the other considerations covered in the last chapter further limit the capability of the *bolt*). But that's for tension joints in general. The gasket—or flange rotation—may well define the upper limit for a gasketed joint.

Anyway—we've used a modified VDI procedure to evaluate our proposed design of a raised-face joint. The design appears to be acceptable, but our choice of assembly preload looks questionable when compared to the Code or in light of our concerns about crushing the gasket. Nevertheless, let's leave it here for now and go on to look at the present Code procedure, which, incidentally, makes no attempt to define assembly preloads. It merely suggests how much seating stress and residual stress will be required to prevent the gasket from leaking excessively; but almost nothing about how to achieve those stresses, nor how they might vary in practice.

III. ASME CODE FLANGED JOINT DESIGN RULES

Two design procedures are allowed by the ASME Boiler and Pressure Vessel Code. That outlined below is the simpler of the two—a "cookbook" approach in which there are relatively conservative limits on design configurations, stress levels, etc. This procedure is covered in Section VIII, Division 1, Appendix II of the Code. The more complex procedure, described in Section VIII, Division 2, of the Code, allows the designer greater freedom of choice, but requires a detailed design analysis to prove the safety of the proposed configuration. In the simpler analysis, the designer takes the following steps, with reference to Fig. 22.2 [3, 4]

1. The designer selects the general size and type of flange to be used.
2. He selects or determines the following design conditions and materials:

    Operating temperatures
    Operating pressures
    Flange, bolt, and gasket materials
    Allowable stress levels

3. He computes the loads which will be placed on the joint by the internal pressure. This consists of two components: the hydrostatic end force, H_D, exerted on the closed end of the vessel or pipe system; and the pressure force, H_T, which acts directly on the exposed inner face of the flange surface (see Fig. 22.2). The end force reaches the flange through the pipe and the hub.
4. He computes two bolt loads: that required for initial seating of the gasket, W_{m2}, and that required to prevent leaks when the system is pressurized, W_{m1}. In so doing he relies on the experimentally determined and published y and m factors discussed above. These calcula-

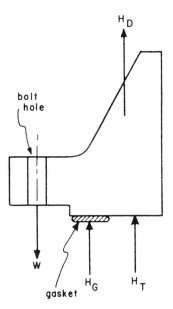

Figure 22.2 Forces involved in the design of a bolted flange joint, and in the selection of gaskets and bolts. See text for definition of terms.

tions also require the designer to estimate the effective width of the gasket, b, using rules given in the Code. Because of such things as flange rotation (discussed later) and nonuniform loading of the gasket, there can be substantial uncertainty about its effective width. Note that the loaded gasket exerts a reaction force, H_G, on the flange, and that this force is in the same direction as the previously calculated pressure forces, H_D and H_T. In any event, the expressions for seating (W_{m2}) and maintenance (W_{m1}) bolt loads are [12]

$$W_{m2} = \pi b G y \qquad\qquad\qquad\qquad\qquad\qquad (22.2)$$

$$W_{m1} = \frac{\pi G^2}{4} P + 2\pi G b m P \qquad\qquad\qquad\qquad (22.3)$$

where G = diameter of the pressure vessel to the midpoint of the gasket (in.)

P = contained pressure (psi)

b = effective width of the gasket (usually about half the actual width) (in.)

m, y = gasket constants discussed earlier

Note that W_{m2} is simply the effective contact surface area of the gasket (πbG) times the recommended seating stress (y, in psi). This defines the total clamping force the bolts are expected to exert on the joint upon initial, room-temperature assembly. The bolt load suggested for in-service W_{m2} is the sum of two terms:

a. The force required to resist the tendency for the internal pressure to pull the flanges apart ($\pi G^2/4$) $\times$ P, plus
b. The additional bolt force required to maintain the suggested, minimum gasket stress on the full-contact surface of the gasket $2b\pi GmP$, with gasket factor m defining this added force as a multiple of the contained pressure P.

Note that, to introduce a safety factor, the expression for W_{m1} ignores the teachings of the joint diagram. It assumes that the bolts will see the entire external end load created by the contained pressure (item "a" above) and not merely some fraction of that load as determined by the bolt-to-joint stiffness ratio.

5. He computes the minimum (total) cross-sectional area of the bolts required for initial seating of the gaskets, the area required to produce the initial bolt load of W_{m2}. Hopefully this amount of bolting will also be sufficient to provide the residual gasket pressure required to contain leaks under later internal pressure loads. He doesn't want to have to retighten the bolts after starting up the system. At this point, therefore, he also computes the minimum bolt area required to produce W_{m1}, the final design choices for actual bolt area, and the flange design bolt load W.

6. Having computed the four load or reaction forces operating on the flange, he can compute the bending moments exerted by these forces on the flange. These moments are taken about the centerline of the bolt hole. From these moments, the designer can compute the stresses in the flange. If these are within allowable limits, he's done; if not, he must go back to the beginning, select different parameters, and try again. The Code gives all of the necessary equations for these calculations, plus a large collection of tables and graphs to aid and guide the designer.

The procedure described above is used for raised-face flanges, the type shown in Fig. 22.2. Some other types require modified or different procedures, and some, such as a flat-faced flange having contact outside of the bolt circle, are very different—and more difficult [4,5]. In my opinion, therefore, it would be misleading—and possibly dangerous—for me to describe the Code procedure in greater detail. The failure of a pressurized joint can be a very serious proposition, and you should design one only with the full set of conditions, procedures, specifications, and equations provided by the carefully developed and well-tested Code. This can be obtained from the ASME, whose address you'll find in Appendix C.

IV. EXAMPLE, USING ASME CODE RULES

Now let's use the present ASME Code rules to analyze the joint illustrated in Fig. 22.1. This time we'll use the terms and symbols found in the Code, rather than those I've been using in this book. As before we have roughed out a design, based on prior habit, and have made preliminary decisions concerning dimensions, materials, bolt number and size, etc. We have:

A. The Inputs

Design pressure (P): 1,000 psi
Operating temperature (T): 800°F
Bolting material: ASTM A193-B7
    From a table of Design Allowable Stresses in the Code we learn that:
        Allowable stress at room temperature (S_a): 26 ksi
        Allowable stress at 800°F (S_b): 19 ksi
Gasket: $\frac{1}{8}$-in.-thick, double-jacketed, stainless steel, asbestos filled
    OD: 10 in.
    ID: 8 in.
    Width (N): 1 in.
    Gasket factors: $m = 3.75$; $y = 9,000$ psi

These are basically the same numbers we started with when using the VDI procedure to analyze this same proposed design. The big difference here, of course, is the "allowable stress" for the bolts. The VDI procedure allowed stresses up to the 90% of the room-temperature yield point of the B7 bolts, while the ASME Code imposes a safety factor of 4 on the maximum stress allowed; reducing it from 105 ksi to 26 ksi. They do this to force the designer to create pressure vessels which won't burst, that being one of the main goals of the Code.

B. The Calculations

We now use these inputs to compute the minimum required bolt loads for assembly and for operating conditions. The Code assumes that flange rotation will prevent the joint from loading the entire surface of the gasket; so it has us start by computing a reduced "effective area" based on an "effective width" (b). Specifically, the procedure is as follows. We compute a b_0 as determined by the flange type (Table 2-5.2 in the 1983 version of the Code)

$$b_0 = \frac{N}{2} \qquad (22.4)$$

and then an effective width, b, using a formula based on the width of the gasket. Since the width of our gasket exceeds $\frac{1}{4}$ in., we use (from the same Code table):

$$b = 0.5(b_0^{1/2})$$
$$b = 0.5(0.5^{1/2}) = 0.354 \text{ in.} \qquad (22.5)$$

Gasket diameter (G) at the "gasket load reaction point" (the effective ID of the gasket)

$$G = OD - 2b$$
$$G = 10 - 2(0.354) = 9.292 \text{ in.} \qquad (22.6)$$

The effective contact area of the gasket is now estimated as the product of the circumference of the gasket at reaction diameter G times the effective width b, or

$$\text{Area} = \Pi \times b \times G$$
$$\text{Area} = 3.14(0.354)(9.292) = 10.3 \text{ in.}^2 \qquad (22.7)$$

Note that this is less than the half-area of 14.2 in.^{2} we assumed when making the VDI calculations. Anyway, we can now compute the minimum bolt load required for seating the gasket, to achieve the y stress on the effective area. The Code calls this bolt load W_{m2}.

$$W_{m2} = (\Pi\, bG)y$$
$$W_{m2} = 10.3(9000) = 92,700 \text{ lb} \qquad (22.8)$$

This, of course, is the total force to be produced by the 12 bolts in the joint. The Code doesn't bother to compute the per-bolt force, but we'll want it for comparison with our VDI calculations.

$$\text{Per-bolt assembly force} = \frac{92,700}{12} = 7725 \text{ lb}$$

There's not a one-to-one relationship, but the per-bolt W_{m2} is roughly equivalent in concept to the Min F_J of our VDI calculations.

We next compute the required bolt load for operating conditions, called W_{m1}.

$$W_{m1} = H_p + H \qquad (22.9)$$

where H_p = the force required to produce minimum stress required on the gasket in service (lb)

         H = the hydrostatic end load (lb)

Again, although a direct comparison is not valid, we could express W_{m1} in VDI terms as follows:

$$W_{m1} = F_{Prqd} + L_X$$

To continue, the Code teaches us that

$$H_p = 2b\Pi GmP$$
$$H_p = 2(0.354)\,3.14(9.292)\,3.75(1000) = 77{,}465 \text{ lb} \tag{22.10}$$

and

$$H = \frac{G^2 \Pi P}{4} \tag{22.11}$$

$$H = \frac{(9.292^2)\,3.14(1000)}{4} = 67{,}778 \text{ lb}$$

Note that Eqs. (22.9) through (22.11) introduce several assumptions which differ from those used in our VDI calculations. First, in the VDI calculations we used the ID of the gasket to compute the hydrostatic end load. The ASME Code assumes that flange rotation will expose the inner part of the gasket to the contained pressure, increasing the estimated end load from 50,265 lb to 67,778 lb. Second, VDI assumes that the clamping force on the gasket will be relieved by only part of the hydrostatic end load, namely $(1 - \Phi_{Kn})L_X$ as determined by the stiffness ratio and load factor, Φ_{Kn}. To introduce another safety factor the Code assumes that the clamping force will be reduced by a full L_X. Anyway, W_{m1} can now be computed:

$$W_{m1} = 77{,}465 + 67{,}778 = 145{,}243 \text{ lb}$$

or

12,103 lb per bolt.

The Code now has us determine the minimum, total, cross-sectional area of all the bolts (A_m) by dividing the larger of W_{m1} and W_{m2} by the design allowable bolt stress at the operating temperature. W_{m1} is larger in this example, so:

$$A_m = \frac{W_{m1}\ (\text{or } W_{m2})}{S_b} \tag{22.12}$$

$$A_m = \frac{145{,}243}{19{,}000} = 7.64 \text{ in.}^2$$

We must now select a group of bolts which will give us slightly more than this much cross-sectional area. We have tentatively decided to use 12 bolts in our design, so the cross-sectional area per bolt will be:

$$\frac{7.64}{12} = 0.637 \text{ in.}^2$$

The Code orders us to see this as the *root* area of the thread, not the tensile stress area, to introduce still another safety factor. Table 3.3 lists the root areas of a few threads, but it's best to consult ASME B1.1 (Unified Inch Screw Thread Standard) for a larger selection. We would have several choices, depending on the number of threads per inch we accept. As one example, p. 84 of the 1989 version of that standard lists root areas for UNC threads [4]. The $1\frac{1}{8}$–7 UNC thread has a root area of 0.693 in.^{2}, so that's the one we'll use. This is much larger than the $\frac{7}{8}$-in. bolt we had tentatively selected for this joint, and which was apparently confirmed by the VDI analysis. The difference, of course, has been forced by the 4 to 1 safety factor the Code imposes on allowable bolt stress, plus the fact that we're required to take the loss of strength created by an elevated service temperature into account.

Next, still following the Code procedure, we compute the "design bolt load for seating" (i.e., for assembly). Note that this is based upon the average of the area we computed, A_m, and the combined area of the nearest, larger, bolts, A_b.

$$W = 0.5(A_m + A_b)S_a \tag{22.13}$$

where A_b = number of bolts × root area
$$A_b = 12 \times (0.693) = 8.316 \text{ in.}^2 \tag{22.14}$$

The Code uses S_a instead of S_b here because the flanges are always seated (assembled) at room temperature.

$$W = 0.5(7.644 + 8.316)(26,000) = 207,480 \text{ lb}$$

or

$$207,480/12 = 17,290 \text{ lb per bolt.}$$

If we were really designing a flange we'd now use this information, plus further Code rules, to check the dimensions of the flange, compute moments on it and stresses in it, etc. See the latest version of the Code for the full procedure.

C. Discussion

ASME vs. VDI Results

This design bolt load is well below the maximum 41,255 lb per bolt we computed using the VDI procedure, when we ignored the effects of elastic interactions—which the Code also ignores. The 17,290 lb represents only 24% of the room-temperature yield strength of these $1\frac{1}{8}$-in. bolts, while the 41,255 lb represented about 75% of yield for the $\frac{7}{8}$-in. bolt, so these results differ significantly from that point of view as well. The VDI procedure is flexible enough to allow us to include creep and interaction losses, and when we did so we presumably planned to load some of the bolts even further; to 90% of the room-temperature yield as approved by that procedure.

One of the major differences between the VDI and ASME procedures, of course, is the fact that the ASME introduces safety factors and VDI does not. Pressure vessel design in Germany, however, is not fully defined by VDI 2230. We've seen references to DIN 2505, and there are presumably other pertinent documents as well. I'm sure that German designs are every bit as safe as ours, however they achieve that goal. The example given in VDI 2230, however, does not include safety factors—at least I haven't spotted them.

ASME Seating Load: Design Requirements

Note that we seem to have computed two "seating loads" for this ASME joint: using Eqs. (22.8) and (22.13). The first, however, defined the *minimum* seating stress required by the gasket; the second defines the higher seating stress required to provide sufficient residual stress on the gasket in service. W_{m1} dominated our choice of bolt diameter and therefore dominated the design bolt load for assembly (which in turn will determine flange stresses and dimensions, etc.). In other situations W_{m2} will dominate and the results of Eqs. (22.8) and (22.13) would be closer even if not identical.

Assembly Preloads per ASME

Neither W_{m2} nor W defines the actual stress to be applied to the bolts at assembly. Unlike VDI, the Code generally avoids this issue, although nonmandatory Appendix S of Section VIII of the Code discusses assembly stress. It suggests that at least 1.5 times S_a be applied to accommodate the hydrotest procedure. This would raise assembly bolt loads to 25,935 lb per bolt in our example. Appendix S also says, however, that it would expect assembly stresses to be:

$$S = \frac{45,000}{D^{1/2}}$$ (22.15)

or, in our example,

$$S = \frac{45,000}{0.875^{1/2}} = 42,094 \text{ psi}$$

This would create a load in our bolts of:

$$F_P = S \times A_R$$

where A_R = root area of the threads = 0.693 in.^{2}, so

$$F_P = 42,094(0.693) = 29,171 \text{ lb}$$

Very close to the Min F_J we computed when using VDI and including creep and interactions—if that's significant. It probably isn't. Anyway, Appendix S also says that assembly loads can be increased still further if that's necessary to prevent leakage. In fact, if the gasket won't be crushed and the flange members can stand it, the implication is that assembly preloads could be taken to the room-temperature yield strength of the $1\frac{1}{8}$-inch bolts:

$$F_P = A_R \sigma_{yB} = 0.693(105,000) = 72,765 \text{ lb per bolt}$$

That's higher than our maximum Max F_J per VDI, because these bolts are so much larger than those we selected when using the VDI procedure. As before, I'd wonder what this much load would do to the gasket. Flange rotation and damage could also be problems.

Nevertheless, all of this seems to suggest that the two procedures, while very different in detail and in intent, are not incompatible with each other. Both have excellent track records. Either can be used to design successful joints. The Code, however, is a legal document in this country and is, therefore, our preferred choice in the United States.

V. EXAMPLE, USING THE PROPOSED PVRC PROCEDURE

Now we're going to run a third and final analysis of the flange shown in Fig. 22.1, this time using the proposed PVRC gasket constants. We studied these in some depth in Chap. 19, and I suggest that you review Secs. IV-C, D and F and Secs. VIII-C and D of that chapter if you skipped them or don't recall that discussion.

In Chap. 19 we used the new gasket constants to estimate assembly preloads. Like the Code's present m and y factors, they're not supposed to be used that way, but until someone gives us more appropriate information relating gasket stress requirements to assembly preloads that's one of the ways in which most people *will* use them. In this chapter, however, we're going to use them as intended, to select the bolts for a raised-face flange.

PVRC and ASME committees are currently working to define the proper use of the PVRC gasket factors. Their goal is to use them to replace the present m and y factors, but to avoid any other, unnecessary changes in the current ASME Code rules for flange design. The procedure I'm about to describe achieves that. The procedure I'm about to describe is also based on recent, but not final, recommendations of the pertinent committees. I doubt if it will be altered in any serious way before being incorporated in the Code, but some alteration is certainly possible. The discussion which follows, therefore, should be taken as an educational guide only. You should use the procedure officially described in the Code when that appears, possibly as early as 1995.

Before proceeding I want to flag one important difference between the present m and y factors and the new PVRC factors. The m and y factors are used in conjunction with an "effective area" of the gasket. The recommended seating stress (y) for our double-jacketed gasket is 9000 psi, applied to only 10.3 in.^{2} of the apparent contact area, as we saw in Sec. IV-B of this chapter. The full surface area of this gasket is 28.3 in.^{2} so a bolt load capable of producing 9000 psi on that area would be able to produce a stress of only 3276 psi on the full area.

The PVRC gasket factors, by comparison, are based on full contact area, because most of their many tests were made using hydraulically loaded test platens rather than rotatable, raised-face, bolted flanges. So if we use the PVRC factors and decide that we want a seating stress of 9000 psi, we'll have to provide almost three times as much seating bolt load than we would for a y factor of 9000. To confuse things, however, the proposed new rules do use an "effective width" of the gasket to compute an "effective diameter," which is used to compute the hydrostatic end load. I'll point these things out again as we proceed.

A. Inputs

We start, once again, with the same preliminary design and design requirements, repeated here for convenience.

Design pressure (P): 1000 psi
Operating temperature (T): 800°F
Bolting material: ASTM A193-B7

Allowable stress at room temperature (S_a): 26 ksi
Allowable stress at 800°F (S_b): 19 ksi
Gasket: $\frac{1}{8}$-in.-thick, double-jacketed, stainless steel, asbestos filled
  OD (G_o): 10 in.
  ID: 8 in.
  Gasket width (N): 1 in.
Gasket constants: This time, instead of the current m and y factors, we
  use the new PVRC factors (see Table 22.1 and Fig. 19.17)
  $G_b = 2,900$ psi
  $a = 0.23$
  $G_s = 15$ psi
We also require two new inputs this time.

1. Assembly efficiency (e), defining the type of tools to be used for
   assembly. See Table 22.1 for the official choices [6, 8] and see the
   end of Sec. C for a discussion of this factor. We're going to assume
   almost no torque-preload scatter here, at least to start with; later we'll
   see how much difference it would make if we assumed that an impact
   wrench would be used.
     $e = 1.0$ for nearly perfect control ($\pm 10\%$ scatter)
2. Tightness factor (c), defining an acceptable, maximum leak rate, as
   listed in Table 19.7. The official choices are:
     $c = 0.1$ for "economy" flanges, where a fair amount of leakage is
   acceptable
     $c = 1.0$ for "standard tightness" (we'll use this in our first PVRC
   example)
     $c = 10.0$ for "tight" flanges, perhaps those containing carcinogens
   or explosive fluids
   At this writing it appears as if the new Code rules will include only
   standard tightness. An appendix to the Code may describe how the
   alternates can be used if desired and/or necessary.

B. The Calculations

We start with some preliminary calculations as follows:

b_o = effective width of the gasket = $N/2$ = 0.5 in.

$b = 0.5 (b_o^{1/2}) = 0.354$ in. $\qquad$ (22.16)

G = effective diameter = $G_0 - 2b$

$G = 10 - 2(0.354) = 9.292$ in. $\qquad$ (22.17)

Now we get serious. The terms I use below are those currently proposed.

$$\text{Stress ratio} = \frac{S_a}{S_b}$$

(22.18)

$$\text{Stress ratio} = \frac{26}{19} = 1.37$$

A_i = pressure load area (i.e., the area used to compute the hydrostatic end load; based on the effective diameter.)

$$A_i = \frac{\Pi G^2}{4}$$

(22.19)

$$A_i = \frac{3.14(9.292^2)}{4} = 67.81 \text{ in.}^2$$

A_g = gasket contact area (the *full* area, as discussed above)

$$A_g = \Pi(G_0 - N)N$$

$$A_g = 3.14(10 - 1)\, 1 = 28.3 \text{ in.}^2$$

(22.20)

$$\frac{A_i}{A_g} \text{ ratio} = \frac{67.81}{28.3} = 2.396$$

Now we compute several tightness factors. See Chap. 19 for the explanation of tightness, and the derivation of the first equation here.

$$T_{pmin} = 0.1243 \text{ cP}$$

(22.21)

$$T_{pmin} = .1243(1.0)1000 = 124$$

$$T_{pn} = X(T_{pmin})$$

(22.22)

where

$$X = 1.5\left(\frac{S_a}{S_b}\right)$$

(22.23)

The 1.5 is included to accommodate hydrotest.

$$T_{pn} = 1.5(1.37)(124) = 255$$

I'm following what the PVRC investigators have called the "convenient method" here. At the moment at least, this is the method favored for the Code. There's an alternate method which uses iteration to arrive at a better multiplier than X and which results in an optimum specification for bolt size. An "optimum specification" is one in which the size of bolt required for seating equals the size required for sealing the joint in service. We'll stick with the convenient method, however.

$$\text{Tightness ratio } T_r = \frac{\text{Log } T_{pn}}{\text{Log } T_{pmin}} \tag{22.24}$$

$$T_r = \frac{2.407}{2.093} = 1.150$$

We're also going to need $1/T_r$.

$$\frac{1}{T_r} = \frac{1}{1.150} = 0.870$$

We can now compute gasket design stresses. The procedure calculates a "design seating stress," called S_{ya}, and two other possible design stresses, S_{m1} and S_{m2}.

$$S_{ya} = \left(\frac{G_b}{e}\right)(T_{pn})^a \tag{22.25}$$

$$S_{ya} = \left(\frac{2900}{1}\right)(255)^{0.23} = 10,373 \text{ psi}$$

$$S_{m1} = G_s\left(\frac{eS_{ya}}{G_s}\right)^{1/T_r} \tag{22.26}$$

$$S_{m1} = 15\left(\frac{1 \times 10,373}{15}\right)^{0.870} = 4433 \text{ psi}$$

$$S_{m2} = \frac{S_{ya}}{(1.5S_{as}/S_b)} - P\left(\frac{A_i}{A_g}\right) \tag{22.27}$$

$$S_{m2} = \frac{10,373}{(1.5 \times 1.37)} - 1000(2.396) = 2652 \text{ psi}$$

The procedure now selects a "gasket design stress," S_{mo}, as the largest of three possibilities:

$2P = 2000 \text{ psi}$
$S_{m1} = 4433 \text{ psi}$
$S_{m2} = 2652 \text{ psi}$

Since S_{m1} is the largest, $S_{mo} = 4433$ psi in this example. Incidentally, the final Code rules may be written in terms of M_o instead of S_{mo}, where

$$M_o = \frac{S_{mo}}{P} \qquad (22.28)$$

The gasket design stress is now used to compute the design bolt load, W_{mo}, which all the bolts together must exert on the gasket. This value will be used in place of the W of the present Code procedure. Note that this is the design load and not necessarily—or even likely—the actual assembly load.

$$W_{mo} = S_{mo}(A_g) + P(A_i)$$
$$W_{mo} = 4433(28.3) + 1000(67.81) = 193,264 \text{ lb} \qquad (22.29)$$

If M_o is used instead of S_{mo} the equation will be:

$$W_{mo} = P(A_g M_o + A_i) \qquad (22.30)$$

There are 12 bolts in our flange, so the individual bolt load, at the design allowable stress level, is

$$\frac{193,264}{12} = 16,105 \text{ lb}$$

The allowable stress for B7 material at 800°F is 19,000 psi, so the indicated thread stress area of an individual bolt would be:

$$A_R = \frac{(W_{mo}/n)}{S_b} \qquad (22.31)$$

$$A_R = \frac{16,105}{19,000} = 0.848 \text{ in.}^2$$

Following present Code practice, we must pick a bolt whose thread *root* area (not the tensile stress area) exceeds 0.848 in.^{2} The thread root area for a $1\frac{1}{4}$–8 bolt is 0.929 in.^{2}, and that's the one we'll pick this time [7]. Our proposed design has $\frac{7}{8}$-in. bolts which appeared capable of sealing this joint, according to our VDI analysis, but they don't come close to satisfying the Code's 4:1 safety factor.

The rest of the new design procedure will be the same as that found in the current Code, as far as I know, but with W_{mo} replacing the present W. Again, however, you should not take my word for this. See the latest edition of the ASME Code for the official procedure. In any event, the choice of bolt will determine flange dimensions.

Assembly Preloads

The new Code design rules, like the present ones, won't tell us anything about the assembly preloads to use on the joint when it's placed in service. For that we'll still have to rely on Appendix S in Section VIII of the Code. This would suggest that:

The minimum assembly preload should probably be 1.5 times the per-bolt design bolt load to accommodate hydrotest. So:

$$\text{Min } F_P = 1.5\left(\frac{W_{mo}}{n}\right)$$

$$\text{Min } F_P = 1.5 \times 16{,}105 = 24{,}158 \text{ lb}$$

The maximum assembly preload could be as high as the room-temperature yield strength of the bolts, if the gasket and flanges can stand that (which they usually couldn't, I suspect). Anyway, that would give a maximum of:

$$\text{Max } F_P = S_{yB}(A_R) \tag{22.32}$$

$$\text{Max } F_P = 105{,}000(0.929) = 97{,}545 \text{ lb}$$

C. Some Optional Choices

Tight Joint

We picked "standard tightness" for the example above. Here's what we would get if we chose "tight" tightness instead: a "design leak rate" of 1/100 the standard rate. We'll still assume perfect preload control, so

$e = 1.0$
$c = 10$
$T_{pmin} = 0.1243(10)1000 = 1243$
$T_{pn} = 1.5(137)1243 = 2550$
$T_r = 3.407/3.094 = 1.101$
$S_{ya} = (2900/1)(2550^{0.23}) = 17{,}616 \text{ psi}$
$S_{m1} = 15\,(1 \times 17{,}616/15)^{0.9083} = 9{,}213 \text{ psi}$
$S_{m2} = 17{,}616/(1.5 \times 1.37) - 2652 = 5920 \text{ psi}$

S_{m1} dominates this time, so

$S_{mo} = 9213 \text{ psi}$
$W_{mo} = 9213\,(28.3) + 67{,}810 = 328{,}538 \text{ lb}$
$328{,}538/12 = 27{,}378 \text{ lb per bolt}$
$A_R = 27{,}378/19{,}000 = 1.441 \text{ in}^2$

So this time we'll use a $1\frac{9}{16}$ –8 bolt which has an A_R of 1.54 in^2 [7]. That makes sense: we'd expect to need a larger bolt to create a tighter flange. Note that the flange dimensions would increase to accommodate the larger bolt while keeping flange stress levels within design allowable limits.

Assembly Preloads

Once again we use Appendix S to estimate possible assembly preloads. This time:

Min F_P = 1.5 (27,378) = 41,067 lb

Max F_P = yield of a $1\frac{9}{16}$ bolt = 105,000 (1.54) = 161,700 lb

Manual Wrench—Lower Assembly Efficiency

Let's check another option. We've assumed so far that we'd have perfect preload control at assembly, giving us an assembly efficiency, e, of 1.0. Now let's see what penalty we'd pay if we decide instead to use a slugging wrench. We'll return to standard tightness in order to compare this option to our first example, so

e = 0.75
c = 1.0
S_{ya} = 10,373/0.75 = 13,831 psi
S_{m1} = 15 $[0.75(13,831)/15]^{0.870}$ = 4433 psi

So the higher S_{ya} has been neatly offset by the lower e and we get the same S_{m1} as before, but

$$S_{m2} = \frac{13,831}{1.5 \times 1.37} - 2652 = 4078 \text{ psi}$$

The factor S_{m2} is now almost double what it was when we assumed $\pm 10\%$ control of assembly preload, but it hasn't overtaken S_{m1}, which still dominates. We'll end up with the same W_{mo} and the same $1\frac{1}{4}$–8 bolt as in our first example. In other situations the design would be penalized—the designer will have to provide larger bolts and flange members—if he chooses an e of 0.75. But it's a little disturbing that he's not forced to do this in *every* case.

As a matter of fact, assembly efficiency e has received relatively little attention from those writing the new Code rules. The present values are shown in Table 22.1 and (like the present Code m and y factors!) are "suggested only," with the hope that more definitive values, supported by experiment, can be found some day [6]. The intent is to include the effects of both tool scatter and elastic interactions. The expression used

Table 22.1 Assembly Efficiencies

Preload control method	Typical tools used	Anticipated preload scatter	Assembly efficiency (e)
None	Impact or slug wrench	Over $\pm$ 50%	0.75
Torque	Torque or hydraulic wrench	From $\pm$ 30 to $\pm$ 50%	0.85
Tension	Multiple stud tensioners	From $\pm$ 10 to $\pm$ 30%	0.95
Stress or strain	Ultrasonics, strain gages, or calipers	Max $\pm$ 10%	1.00

to compute e is

$$e = \frac{100 - 0.5 \text{ (min. initial \% scatter)}}{100} \tag{22.33}$$

I can't explain the 0.5. Anyway, including an assembly factor is a step in the right direction. VDI has long since included tool scatter in their design procedure; it's high time the ASME Code did too. I believe, however, that "more definitive values" would result in a much larger range for e than the present 0.75 to 1.0.

D. The PVRC Designed Joint in Service

Although the Code rules won't, I believe, do this, it's informative to see how the joint we've just analyzed might function when placed in service. As an example we'll consider the two analyses in which we assumed $\pm 10\%$ control of preload ($e = 1.0$). In the first analysis we accepted "standard" tightness; in the second we tried for "tight."

The results of these two analyses are plotted in Fig. 22.3, but with several additions. In our first analysis we computed a T_{pmin} of 124 and a T_{pn} of 255. Vertical lines on the graph illustrate those values, as well as the G_b, a, and G_s values which define the performance of the double-jacketed gasket. (Again, see Chap. 19 if you're uncertain about this.)

The graph also shows a suggested assembly seating stress S_{ya1} for this first set of numbers; the seating stress I have chosen to plot is 10 ksi. This is less than the S_{ya} of 10,373 psi which we computed as part of the proposed PVRC procedure; but you'll recall, I hope, that the Code rules are not intended to define actual assembly preloads. They're used only to design the joint. We're free to choose an assembly stress which is less than—or greater than—the design S_{ya}. The important thing is to choose

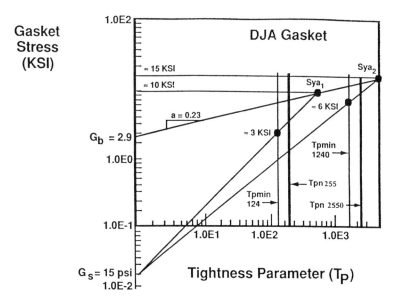

Figure 22.3 The flange shown in Fig. 22.1 was analyzed using a procedure being developed by the PVRC for inclusion in the ASME Boiler and Pressure Vessel Code. Some of the results of the analysis are plotted here. A minimum tightness parameter (T_{pmin}) of 124, combined with S_{ya1}, defines the assembly and in-service leakage behavior of the joint when designed for "standard" tightness. The T_{pmin} of 1240 and S_{ya2} define the behavior for "tight" tightness. Again, see the text for many additional details.

one associated with a tightness greater than T_{pn}, whose value includes a consideration of the 1.5 hydrotest multiplier. Any tightness value greater than T_{pn}, therefore, also provides for hydrotest.

We now draw a line from S_{ya1} to G_s. This defines the leakage behavior of the gasket as the stress on it is relaxed by any means: by service pressure, by gasket creep, by differential expansion, by elastic interactions, etc. The intersection of this line with the T_{pmin} line defines the minimum gasket stress the joint can tolerate without exceeding the "standard" leak rate. That minimum stress turns out to be about 3 ksi. It would be up to us, as users, to estimate the losses listed just above to decide if they would result in a residual stress below 3 ksi. If we decided that they would, we'd increase S_{ya} and try again. Remember, however, that the stress-tightness graph we're using defines average behavior of many double-jacketed gaskets, and won't necessarily define the behavior of the ones we're using. The data on which the graph is based are intended for design use only.

But, as I've said several times before, it's the only information most of us will have when trying to pick and assembly preload.

Figure 22.3 also includes a plot of the results of our second analysis, when we decided that we wanted a "tight" joint. Now $T_{pmin} = 1240$ and $T_{pn} = 2550$. This time I've suggested an S_{ya2} of 15 ksi, which is also less than the calculated design S_{ya} of 17,616 psi. And this time we find that the leak rate will exceed the selected T_{pmin} if gasket stress falls below 6 ksi.

There's nothing magic about the S_{ya}'s I've used in these examples. Many other possibilities could be plotted. Again, this is one of the advantages—or difficulties?—of the new gasket constants. Unlike the old m and y factors, they don't just define a single outcome.

E. Discussion

Table 22.2 summarizes the results of our several calculations. The various procedures we've explored have all led to similar results, which is reassuring and desirable. Although the gasket factors listed in the present Code are suspect, Code designed flanges have a long and impressive track record. A modified procedure which resulted in substantial changes in design requirements would be suspect; and might cause unwarranted concern about the safety of the many Code-designed flanges now in service.

Although the results should be similar, they should also be "better," because they will now be based on a much more accurate understanding of the leakage behavior of gaskets and, therefore, of the seating and residual stresses required to minimize leakage. In the example we've run, the new procedure suggests that a slightly larger bolt will give us better results than would a bolt whose size was influenced by m and y. In other cases we'd find that m and y had forced us to overdesign the flange; the new constants will provide some economic relief.

The bolts we've selected using the present or anticipated Code procedures are also, of course, larger than the $\frac{7}{8}$-inch bolt our first, VDI analysis suggested would be capable of sealing this flange. The Code, however, forces us to base our decision on the "design allowable" bolt stress at service temperature, which introduces a $4:1$ safety factor. Similar precautions are undoubtedly covered in German regulations, if not specifically in VDI 2230. The advantage of VDI 2230, for our purposes, is that it gives us a more accurate look at the actual behavior of a gasketed joint than do the present or proposed Code rules, which are designed to be used by flange designers, not by those who wish to analyze behavior.

The present Code procedure does not deal "openly" with such factors as gasket creep or elastic interactions or tool scatter or thermal expansion,

Table 22.2 Summary of Design Calculations (all tabulated values are in kips per bolt, i.e., lb/bolt × 1000)

	Based on VDI equations		Present Code rules	Code rules being proposed by the PVRC		
Assumed conditions						
Elastic interaction loss	Aver. 30%	None	None	None	None	None
Assumed tool scatter $(+/\pm)$	30%	30%	NA	10%	10%	50%
Assy. efficiency	NA	NA	NA	1	1	0.75
Tightness class	NA	NA	NA	Standard	Tight	Standard
Gasket factors used	NA	NA	m and y	G_b, a, G_s	G_b, a, G_s	G_b, a, G_s
Assembly preload						
Maximum	49.2	41.3	72.8[a]	97.5[a]	161.7[a]	97.5[a]
Minimum	26.5	22.2	25.9[b]	24.2[b]	41.1[a]	24.2[b]
Max bolt tension	51	43	72.8[a]	97.5[a]	161.7[a]	97.5[a]
Design bolt load	49.2	41.3	17.3	16.1	27.4	16.1
Design bolt load × 4[c]	NA	NA	69.2	64.4	109.6	64.4
Selected bolt diameter (inches)	0.875	0.875	1.125	1.25	1.5625	1.25

[a] From an interpretation of Appendix S (can take bolts to yield if the flanges and gasket can stand this).
[b] Also from Appendix S (which says apply 1.5 × design bolt load for hydrotest).
[c] Added to make comparison between VDI and Code results more understandable.
(4 × design allowable = true yield strength at service temperature.)

but these things are provided for by generous safety factors, and by Appendix S (in Section VIII of the Code), which allows us to tighten the bolts well beyond the design stress levels. The proposed Code procedure imitates the present one to some degree but does attempt to include tool scatter (with scatter factors—''assembly efficiencies''—which I think require further work).

VI. SOME ALTERNATE PROCEDURES

Gasketed joints are designed by many people in many countries using a wide variety of procedures. It's useful to take a brief look at these, lest we become too impressed by those we've looked at so far. Unlike the ASME Code procedure, these others are not ''legal standards'' in the

United States, but some of them, at least, have been used enough to prove their worth. We'll start with one which has a very long and successful history.

A. The Klinger Procedure

The Procedure

Some years ago the European gasket manufacturer Klinger suggested a design procedure based on recommended in-service and seating assembly stresses, and therefore similar to that found in the Code [9]. As I mentioned in Chap. 19, they have also—more recently—published a computerized gasket selection procedure which appears to be based on the same philosophy. Their procedure is based upon experimentally determined m and y gasket factors; but the Klinger factors have different meanings than the ASME Code factors, and therefore have different values. For example, the Code y defines a recommended seating stress which will be equal to or greater than the pressure effect. The Klinger y, on the other hand, is a minimum seating stress to which is added the pressure effect. This will become clearer as we proceed.

It's unfortunate that Klinger and the ASME both use m and y, but define them differently. To distinguish between them I will use m' and y' for the Klinger factors.

Klinger starts with an experimentally derived curve of in-service gasket stress vs. contained pressure (curve A in Fig. 22.4). The data shown are for a compressed asbestos gasket, and the test fluid was nitrogen.

Since there is no such thing as a zero leak rate, when the contained fluid is a low-molecular-weight gas such as nitrogen, Klinger assumes that a leakage of 0.25 ml/min (4.2×10^{-3} mg/sec) constitutes a technically tight seal. Any combination of gasket stress and internal pressure located below the curved line (A) of Fig. 22.4 would result in leakage greater than 0.25 ml/min. Any point above the curved line would result in less than 0.25 ml/min leakage.

Klinger next generates the straight line also shown in Fig. 22.4. They call the slope of this line m and its intercept y. (I'll call them m' and y'.) They define the minimum acceptable in-service gasket stress as:

$$L = y' + m'P \qquad (22.34)$$

where L = minimum gasket service stress (psi)
y' = minimum design seating stress (psi)
m' = maintenance factor
P = internal pressure (psi)

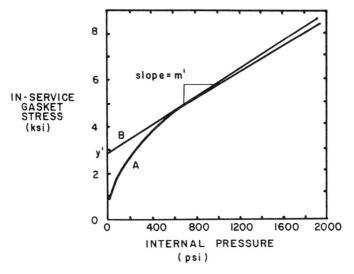

Figure 22.4 Curved line A defines the experimentally determined relationship between in-service gasket stress and contained pressure for a constant leak rate of 0.25 ml/min, for a compressed asbestos gasket. Straight line B is used to derive gasket factors m' and y'.

The recommended initial seating stress on the gasket (SG) is simply the recommended in-service stress (L) plus the reduction in gasket stress which will result from the hydrostatic end load (H).

Presumably the designer can use joint diagrams or equivalent techniques to decide how much of the hydrostatic load will actually result in a reduction in gasket stress. In any event, the seating stress equation becomes:

$$SG = y' + m'P + H \qquad\qquad (22.35)$$

The Klinger equations seem simpler and more graphic than the ASME code equations, but they become similar in complexity by the time the Klinger gasket factors are expressed in terms of bolt load, and the hydrostatic end load is expressed in terms of gasket contact surface area.

The Klinger design procedure is interesting because it's based on available experimental data, and because of its interpretation of the m and y gasket factors. As we saw in Sec. IV, furthermore, the "new" gasket factors emerging from studies sponsored by the Pressure Vessel Research Committee bear a strong resemblance to those suggested by Klinger.

In any event, Klinger gives us an alternate way to select in-service and seating stresses on the gasket. The rest of the procedure for a flange joint could, of course, imitate the ASME Code procedure.

Additional "Klinger lines," defining m' and y' factors for several other types of gasket, can be seen in Fig. 22.5 [9].

The data shown in Fig. 22.5, incidentally, are also based on "standard" leak rate of 0.25 ml/min, with nitrogen as the contained fluid.

Discussion

One important difference between the Klinger procedure and those found in or proposed for the ASME Code is the fact that Klinger defines the amount of stress to be created in the gasket at assembly; not just a design seating stress, but an actual assembly stress. In this Klinger and VDI agree.

Like all the others, however, Klinger appears to avoid such factors as gasket creep or differential expansion effects; at least as far as Ref. 9

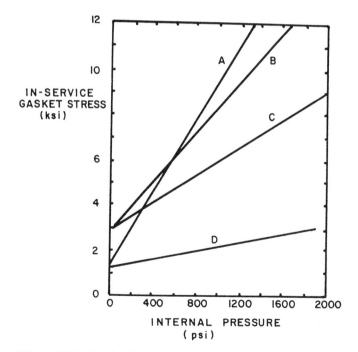

Figure 22.5 Gasket factor lines for (A) expanded graphite, (B) compressed non-asbestos, (C) compressed asbestos (repeated from Fig. 22.4), and (D) cork-asbestos composition gaskets.

is concerned. They may—I think they must—include these factors in their computer softwear, but neither creep nor thermal effects are visible as specific inputs or outputs.

Like the proposed PVRC procedure, Klinger works with the concept of "acceptable leak rates." These are dealt with in terms of volumetric leak rates, ml/min of nitrogen, rather than as the PVRC's *mass* leak rates (which can theoretically be related to any contained fluid). I've lost Ref. 9 but I don't remember their standard leak rate of 0.25 ml/min being related to any particular flange or gasket diameter; but it must have been. Their computer program reports anticipated leak rate in either ml/min or ml/min × m units; so diameter is taken into account. They also refer to a "standard gasket ring measuring 90 × 50 mm," and say that a leak rate of 1 ml/min (measured with nitrogen) is considered "very tight." A leak rate of 1 ml/min of nitrogen would be about what we'd expect to see in a 4-in.-diameter flange with a PVRC "standard" leak rate of 1/500 mg/sec × mm.

B. Procedure Proposed by the BHRA

Dr. Bernard Nau of the British Hydromechanics Research Agency (BHRA) has proposed a simplified design procedure based upon the leakage test data published by the PVRC [10]. Dr. Nau has long been a member of the PVRC committees investigating gasket behavior and has conducted some of the PVRC leakage tests, so he's intimately acquainted with the data and with the attempts to use them in a codified design procedure. He proposes a simpler procedure.

First of all he computes the loss in gasket stress to be caused by the hydrostatic end load, *taking bolt and joint stiffness into account*. This is a significant departure from the present or proposed Code procedures. (As we'll see in a minute, however, Dr. Nau also suggests an optional and simpler procedure which ignores these stiffnesses.) In any event, he defines the hydrostatically created loss as follows. (I'm going to use my terms for these things, not his, to avoid—I hope—confusing my readers.)

$$\Delta S_{\text{G}} = -\left(\frac{K_{\text{G}}}{K_{\text{G}} + K_{\text{B}}}\right) \times \left(\frac{L_{\text{X}}}{A_{\text{G}}}\right) \tag{22.36}$$

where ΔS_{G} = "gasket stress decrement" (stress loss) (psi)
       K_{G} = stiffness of the gasket (lb/in.)
       K_{B} = stiffness of the bolts (lb/in.)
       L_{X} = force created by the contained pressure (lb)
       A_{G} = contact area of the gasket (in.^{2})

This equation is really another version of Eq. (12.14)

$$\Delta F_{\mathrm{J}} = (1 - \Phi)L_{\mathrm{X}}$$

because

$$\frac{\Delta F_{\mathrm{J}}}{A_{\mathrm{G}}} = \Delta S_{\mathrm{G}}$$

and

$$(1 - \Phi) = 1 - \frac{K_{\mathrm{B}}}{K_{\mathrm{B}} + K_{\mathrm{J}}} = \frac{K_{\mathrm{J}}}{K_{\mathrm{B}} + K_{\mathrm{J}}}$$

He equates the stiffness of the joint (K_{J}) to the stiffness of the gasket (K_{G}) and the change in clamping force or gasket stress is, indeed, negative. Anyway, he says that the gasket will dominate joint stiffness and has made studies which confirm this. Some of his results are shown in Figs. 5.19 and 5.20.

He computes L_{X} from:

$$L_{\mathrm{X}} = \frac{\pi P D_{\mathrm{r}}^2}{4} \qquad\qquad (22.37)$$

where D_{r} = diameter of the area affected by the contained pressure (in.)
$\quad\quad\quad P$ = contained pressure (psi)

The affected diameter, in PVRC terms, would be at midwidth of the gasket. Dr. Nau next relates assembly stress to the residual, gasket working stress.

$$S_{\mathrm{m}} = S_{\mathrm{ya}} - \Delta S_{\mathrm{G}}$$

or

$$S_{\mathrm{m}} = S_{\mathrm{ya}} - \left[\left(\frac{K_{\mathrm{G}}}{K_{\mathrm{B}} + K_{\mathrm{G}}}\right)\left(\frac{L_{\mathrm{X}}}{A_{\mathrm{G}}}\right)\right] \qquad\qquad (22.38)$$

where S_{m} = gasket working stress (psi)
$\quad\quad\quad S_{\mathrm{ya}}$ = assembly stress (psi)

We can rewrite this, of course, as

$$S_{\mathrm{ya}} = S_{\mathrm{m}} + \Delta S_{\mathrm{G}}$$

Dr. Nau then replots the PVRC leak test data in lognormal form, as shown for example in Fig. 22.6. Note that his expression for "tightness" is different from that proposed by the PVRC [compare it to Eq. (19.2)] and he plots it on the vertical axis. This allows him to plot the three factors in

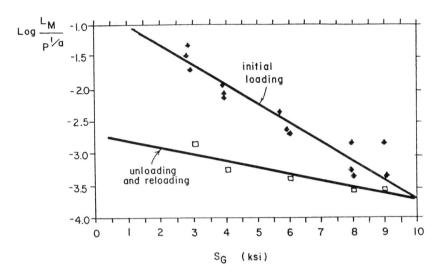

Figure 22.6 Dr. Bernard Nau of the British Hydromechanics Research Agency (BHRA) has replotted some of the PVRC gasket test data using the conventions shown here. Tightness is plotted on the vertical scale as a function of mass leak rate (L_M in Dr. Nau's terms), contained pressure (P), and the PVRC tightness exponent, a. See Chap. 19, Sec. IV-C for the PVRC expression for tightness. Gasket stress (S) is plotted against a linear scale on the horizontal. This plot can be compared to the log-log plot used by the PVRC, which is shown, for example, in Fig. 22.3. Dr. Nau finds the one shown here more convenient for the flange design procedure discussed in the text.

Eq. (22.38) in the manner shown in Fig. 22.7. Here he has added a line, A–B, defining a maximum acceptable leak rate, and has located S_m on that line, thereby defining the minimum acceptable gasket working stress.

Note that Fig. 22.7 contains a number of possible unloading-reloading lines, and, unlike those shown in the PVRC-based graph of Fig. 22.5, these do *not* all pass through a common point on the vertical axis. Dr. Nau acknowledges that PVRC data suggest that they should pass through such a point; but he says that treating them as parallel lines simplifies the design procedure and experimental scatter obscures any differences.

And that's about it. He suggests that the flange design be based on S_{ya}. He says that, in practice, tool scatter and flange rotation (including that induced by thermal expansion) might have to be taken into account.

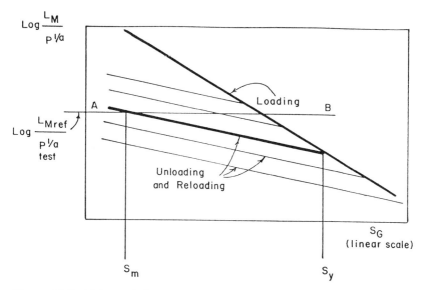

Figure 22.7 This graph illustrates the way Dr. Nau of BHRA uses PVRC gasket leak test data to select assembly stress (S_y) for a proposed flange. The flange would be sized to create this stress.

In common with the PVRC, his equations ignore such things as gasket creep, elastic interactions, differential expansion, etc.

He does include the VDI load factor, but then goes on to say that his studies of bolt-to-joint stiffness ratios show him that "most" of the hydrostatic end load will be seen as a loss of gasket stress. To simplify things, therefore, he suggests that the $K_G/(K_B + K_G)$ factor can be ig- nored; implying that *all* of the end load is seen as a reduction in gasket stress; reducing Eq. (22.38) to:

$$S_m = S_{ya} - \frac{L_X}{A_G} \tag{22.39}$$

He points out that this is a conservative move which would result in slightly more massive flanges and higher S_{ya}'s. It's also the point of view adopted by the Code, of course.

Dr. Nau's approach is similar in simplicity to the Klinger approach. Like Klinger's (and the emerging Code procedure) Nau's procedure is firmly based on actual gasket leakage test data.

C. Flanges in Metal-to-Metal Contact

Dr. Nau's Proposal

The discussion so far has involved only raised-face flanges. These undoubtedly are in the majority, in process industries at any rate, but other joint configurations are possible and may be preferred in many applications. Most of these alternates can be analyzed using the VDI procedure. Some are also covered by the ASME Code. I'm not going to enumerate them but, lest we forget they exist, I think that the following is worth considering.

Dr. Nau of BHRA drafted a paper on what he called controlled compression bolted gasketed joints [11]. The flanges are in metal-to-metal contact both within and outside the bolt circle, as shown in Fig. 22.8b. Any type of "flat" gasket lies in a pocket at the ID of the upper flange. The arrangement can be compared to the conventional, raised-face flange,

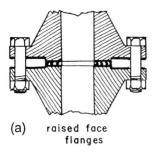

(a)   raised face
       flanges

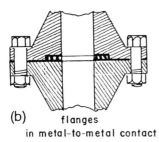

(b)   flanges
    in metal-to-metal contact

Figure 22.8 Sketch comparing a conventional raised-face flange with the controlled compression, metal-to-metal-contact flange suggested by Dr. Nau (BHRA). The conventional flange, in his terms, "floats" upon the spiral-wound or double-jacketed or compressed asbestos or other gasket. In the proposed design the compression of the same types of gasket is controlled by contact between the flanges, contact both within and without the bolt circle.

shown in Fig. 22.8a, where, in Dr. Nau's terms, the flanges "float" on the gasket. It could also be compared to the relatively common (but not illustrated) metal-to-metal flange which is sealed by an O-ring gasket mounted in a midradius pocket. As far as I know, Dr. Nau is the first to suggest using a flat (spiral-wound, double-jacketed, compressed asbestos, etc.) gasket in a metal-to-metal joint.

He says that the bolts in such a flange could be tightened to near their yield point. This won't damage the gasket because its compression is limited by the depth of the pocket it rides in. Nau suggests that the clamping force on the gasket would remain constant unless and until the hydrostatic end load exceeds the preload in the bolts. This would mean no change in gasket stress or leak rate between assembly and in-service conditions.

We learned a few minutes ago that, in a raised-face joint, the joint usually "sees" most of the external load. This means that the clamping force between flange members would decrease by some amount when the system was pressurized. Won't that change the stress on the gasket? No. Surely the stress on that gasket will be determined by its stiffness, not by the forces exerted by one flange on the other. The interflange force might change considerably without affecting the force the gasket exerts on the flanges or vice versa.

As Dr. Nau points out, however, there's a possible problem here. Gasket creep or differential expansion could alter the gasket stress. How much will a gasket creep in a constant-deflection application? In some PVRC tests a gasket was cycled repeatedly between two deflections (Fig. 19.8B) and, in the process, gradually shed stress (i.e., took a permanent deflection). But Nau is talking about loading it only once. It would lose some stress because of creep, but not as much as it would if under a "floating" bolt load where it could get progressively flatter without limit.

A rise in flange temperature would deepen the pocket in which the gasket sits. This would also reduce stress on the gasket unless it grew by a similar amount.

To my knowledge Nau's design has been tested only once. Further tests would be desirable. Perhaps some controlled compression tests should be made first on spiral-wound or other flat gaskets. If they suggest that creep or thermal expansion will not cause problems, full flange tests should be conducted.

Jan Webjorn's Flange Design

Jan Webjorn of Sweden has tried for some years to convince pressure vessel and piping people on several continents to replace raised-face

flanges with an interesting metal-to-metal contact flange he designed while
an adult student at the Institute of Technology in Linkoping, Sweden [12,
13]. Webjorn's design is shown in Fig. 22.9. Here are some of its features.

This is a "rigid, compact" flange, significantly smaller in outer diameter than the raised face flange it would replace.

There is no gasket if an "economy" leak rate is acceptable. A thin
layer of silicone RTV is suggested if a "standard" leak rate is required.
A metallic seal may be required to create a "tight" joint.

The bolts have a minimum length-to-diameter ratio of 6:1. The combination of thick flange, metal-to-metal contact within and without the bolt
circle, and relatively flexible bolts gives this system a large joint-to-bolt-
stiffness ratio. This results in a small load factor (Φ) and so the hydrostatic
end pressure will create little or no change in bolt tension. Almost all of
the hydrostatic load will be absorbed as a loss of interface contact stress.

This distribution of the hydrostatic load is further guaranteed if the
loading planes of this joint fall at or near the joint interface. That is made
more likely by the fact that the bolt circle is placed as close to the centerline of the hydrostatic load as possible; and by the fact that there is very
little flange outboard of the bolt circle to create eccentric leverage. As a
result, there is no eccentric magnification of the pressure load. And, of
course, there's no flange rotation.

Because there's little or no gasket flexibility or creep to contend with,
and no danger of crushing a gasket, the bolts can be loaded to a stress
near their yield point. In spite of the fact that all of the hydrostatic end
load will act to reduce contact stress, the assembly preload can be scattered without having too adverse an effect on performance. (Webjorn
does, however, recommend that properly trained personnel be used to
tighten the bolts, using hydraulic tensioners.)

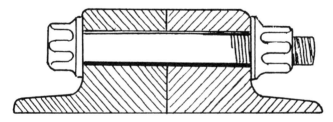

Figure 22.9 Compact, metal-to-metal-contact flange designed by Jan Webjorn
of Sweden. No gasket is used in this flange if "economy" tightness is sufficient.
"Standard" tightness will require the use of a thin film of silicone RTV. The flange
is produced by SPO in Norway.

Webjorn's design is produced by SPO of Drammen in Norway. These flanges, incidentally, are stronger than the pipe to which they are attached.

REFERENCES

1. Systematic Calculation of High Duty Bolted Joints, VDI 2230, Society for Design and Development, Committee on Bolted Joints, Verein Deutscher Ingenieure, VDI Richtlinien, Dusseldorf, Germany, October 1977; Translated by Language Services, Knoxville, TN; translation published by the U.S. Department of Energy as ONRL-tr-5055.
2. Bibel, G. D., and D. L. Goddard, Preload variation of torqued fasteners, a comparison of frictional and elastic interaction effects, *Fastener Technology International*, pp. 31 ff, June/July 1994.
3. Swick, Robert H., Designing the leak free joint, *Machine Design*, pp. 100ff, January 22, 1976.
4. *Modern Flange Design*, Taylor-Bonney Division of Gulf and Western, Southfield, MI, 1979.
5. Rodabaugh, E. C., and S. E. Moore, *Flanged Joints with Contact outside the Bolt Circle—ASME Part B Design Rules*, Performed by Battelle, Columbus, OH, for Oak Ridge National Laboratory, Oak Ridge, TN, May 1976.
6. Payne, Jim, Andre Bazergui, and George Leon, New gasket factors, a proposed procedure, Paper presented at the ASME Pressure Vessel and Piping Conference, New Orleans, LA, June 1985.
7. ANSI B1.1-1989, Unified Inch Screw Thread Standards, Published by the ASME, New York.
8. Winter, J. Ronald, Gasket selection—a flowchart approach, Paper presented at the ASME Pressure Vessel and Piping Conference, Nashville, TN, June 18–21, 1990.
9. Sauter, Ernst M., Current and future gasket materials; methods of evaluating some of their functional properties, Research Laboratories for the KLINGER group of companies, Switzerland, January 1982, (U.S. branch: Sidney, OH).
10. Nau, B. S., On the design of bolted gasketed joints, Presented at the 12th International Conference on Fluid Sealing, Brighton, UK, May 10–12, 1989.
11. Nau, B. S., Controlled compression bolted joints, Draft of a paper written at the British Hydromechanics Research Agency (BHRA) Cranfield, Bedford, England, March 1990.
12. Webjorn, Jan, New look at bolted joint design, *Machine Design*, pp. 81–84, June 20, 1985.
13. The VERAX Compact Flange System, Brochure of SPO (Steel Products Offshore), Drammen, Norway.

23
The Design of Joints Loaded in Shear

I. AN OVERVIEW

Twenty-two chapters ago I described two kinds of bolted joint: those loaded in tension and those loaded in shear. With the exception of Chap. 14 the discussion since then has been focused on tension joints because they're more common, their behavior is more complex, and analyzing them is more difficult. In this last chapter, however, we're going to take another look at the shear joint. To be specific, we'll study the design of such joints and will see the ways in which the design process is the same as that for tension joints, and the many ways in which the two differ.

There are many different types of shear joint, but most can be defined as either a lap joint or a butt joint, as shown in Fig. 23.1. Historically, joints of either type were further classified as "friction type" or "bearing type." The structural steel industry has now abandoned the friction and bearing classifications, as we'll see, but the distinction is still handy for a preliminary review of shear joint design, and so I'll continue to use it.

Shear joints are most commonly encountered in "structures," such as airframes or buildings or bridges. Most of the bolted joints found in structures, in fact, are shear joints. In part this is because of the way loads are applied to structures, but I suspect that part of the reason is that shear joints are more forgiving of assembly errors or preload scatter; they can operate successfully under a much wider range of clamping force than can many tension joints. One of the main reasons for this is that

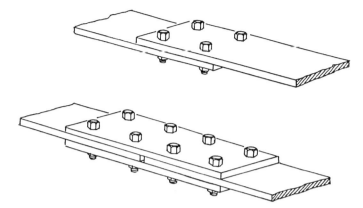

Figure 23.1 Two basic types of shear joint. The upper is called a lap joint; two joint members are bolted to each other. The lower is called a butt joint; the joint members are connected by upper and lower "splice plates."

shear loads don't change bolt tension or clamping force the way tension loads do.

One new problem we do have to be concerned about, however, when dealing with shear joints, is the possible mechanical failure of the joint members themselves. Tension joint failure can usually be blamed on the bolts; either they have created the wrong clamping force or they have themselves failed. Improper clamp can cause a shear joint to loosen under vibration, but most shear joint failures involve the rupture of the joint members.

We're going to start our study of shear joint design with our old friends, the VDI equations. We'll see what they have to teach us about shear joints, and will find that it's useful but not enough. So we'll go on to look at the way the bolts and joint members see and resist shear loads. All of this will be pertinent for the design of shear joints in general. As we go along we'll take an occasional look at some of the codified design procedures which have been developed by the structural steel industry. Those procedures are described and explained in detail in the definitive text by Kulak et al. [7]. Structural steel designers should rely on that text, and/or on the "bolt specs" written by the Research Council on Structural Connections and published by the AISC [8, 13], rather than on my text; so I'm not going to repeat those procedures here. But some comments are certainly in order.

II. THE VDI PROCEDURE APPLIED TO SHEAR JOINTS

Let's return, for a final time, to the VDI joint design equations first encountered in Chap. 21 [11]. These raise two important issues we must address when designing any joint: the minimum clamping force we can expect to see in the joint; and the maximum tension the bolts will have to support. Since clamping force is assumed to equal bolt tension, these equations are expressed in terms of clamp force (F_J) but really define bolt preload limits. I repeat those equations here, for your convenience.

$$\text{Min } F_J = \Delta F_J + F_{Prqd} + \Delta F_P \qquad (21.5)$$

Since $\Delta F_J = (1 - \Phi_{Kn})L_X$ this can also be expressed as

$$\text{Min } F_J = (1 - \Phi_{Kn})L_X + F_{Prqd} + \Delta F_P$$

We also have

$$\text{Max } F_J = \alpha_A[(1 - \Phi_{Kn})L_X + F_{Prqd} + \Delta F_P] \qquad (21.6)$$

where
ΔF_P = loss of preload during or soon after assembly (lb, N).

Max F_J = maximum anticipated bolt tension created during assembly. Equals the clamping force that bolt applies to the joint before the joint is put in service (lb, N).

Min F_J = minimum anticipated bolt tension and clamping force created during assembly, before the joint is put in service (lb, N).

F_{Prqd} = minimum preload (and per-bolt clamping force) required to prevent joint failure (lb, N).

L_X = external tensile load applied to the joint (lb, N).

α_A = the scatter in preload caused by the assembly tools and procedures; $\alpha_A = \text{Max } F_P/\text{Min } F_P$.

Φ_{Kn} = load factor for a concentric joint, loaded internally at loading planes. $\Phi_{Kn} = \Delta F_B/L_X$ where ΔF_B = the change in tension created in the bolts by the external load. This is the load factor we used when discussing tension joints in general.

ΔF_J = the change in clamping force (or bolt tension) created by an external tensile load on the joint (lb, N).

Let's look at each of these terms and see how they apply to shear joints.

ΔF_J: Since there is no tensile load on a joint loaded only in shear, $\Delta F_J = 0$.

ΔF_P: When designing a tensile joint we had to consider four ways in which initial preload might be lost during or after assembly, giving us this expression:

$$\Delta F_P = \Delta F_{em} \mid \Delta F_{EI} + \Delta F_{CR} \pm \Delta F_{TH} \qquad (22.1)$$

where em = embedment relaxation
       EI = elastic interaction loss
       CR = creep loss
       TH = gain or loss because of thermally induced differential expansion

Each of these effects can cause a loss of assembly preload in a shear joint as well as in a tension joint, but the effects are usually smaller; with the exception of embedment. Let's look at each effect as it applies to a shear joint.

> *Embedment:* We can expect to see a typical embedment loss; perhaps 5–10%.
> *Elastic interactions:* Dr. Bibel tells us that the average elastic interaction loss in an ungasketed, metal on metal joint is 18%, well below the values for gasketed joints [10]. This figure, however, is based on limited tests on 24-in.-diameter, raised-face, pressure vessel joints, where the bolts are unsupported by metal-to-metal contact. I would expect to see less loss in most shear joints: but that's just a guess.
> *Creep:* Some structural steel joints are given a thin coat of paint, to control interface friction and/or to provide some corrosion protection, and so we might see some creep loss after assembly. I would expect the loss to be negligible, however; certainly nothing like the major loss created by a gasket.
> *Thermal:* In most of the (few) structural steel joints I've studied the bolts and joint members have both been made of steel, presumably with similar coefficients of expansion. Unlike, say, a pressure vessel joint, both bolts and joint members would experience the same change in temperature at the same time. And the temperature change would be modest; again unlike pressure vessel applications, where temperatures of 1000°F or more are not uncommon. All of which suggests that differential expansion can be ignored in structural steel joints.
> Airframe structures, on the other hand, often involve several materials, including aluminum joint members and bolts of ferrous metals and/or exotic alloys. Temperature changes, how-

ever, are still modest. Grip lengths, furthermore, tend to be small in a structure which must be light enough to fly. I'm sure that airframe designers take thermal change into account; but I doubt if differential expansion is large enough to be a concern.

All of which suggests that in "most" shear joints, ΔF_P is probably going to be less than 30% of initial preload; perhaps much less.

α_A: Shear joint bolts will be subject to the same preload scatter as tension joint bolts. If a torque wrench is used on unlubricated bolts, for example, scatter might be $\pm 30\%$. Following VDI's lead, this would create an α_A of $(1 + 0.3)/(1 - 0.3)$ or 1.86. There are some factors which make tool scatter "different" when we're dealing with a shear joint, however. First of all, preload control is less important in most shear joints, so scatter doesn't matter as much. Second, in structural steel work, at least, the bolts are often tightened well past yield, on purpose, and this reduces scatter to $\pm 5\%$ or so [1] and an α_A of only 1.11. (VDI says that yield control will result in an $\alpha_A = 1.0$, as you'll see in Table 21.1, but I think that that's overoptimistic.)

F_{Prqd}: This is the "force required" to guarantee satisfactory operation of the joint, and it will dominate the selection of the bolts when we design a shear joint. We'll consider it in some detail in the next section of this chapter.

Final equations: As a result of all this we end up with the following VDI equations when we tackle a "typical" shear joint.

$$\text{Min } F_J = 1.3 F_{Prqd} \tag{23.1}$$

$$\text{Max } F_J = \alpha_A \text{ Min } F_J \tag{23.2}$$

The 1.3 comes from my conclusion that we need to create an initial clamping force 30% greater than the in-service force required to accommodate elastic interactions and embedment. I suspect that that's conservative, but I'll use it for this example.

Do we now have what we need to design a shear joint? Hardly. The VDI equations allow us to define the bolts required in a joint. That information helps us size the joint members and pick joint materials, but it isn't sufficient to configure a joint. Most configurations will be based on past experience or custom in a particular application; and shear joints are no exception to the rule. Most structural steel joints are designed "to Code," and I'm sure that most airframe joints are equally well defined by past experience. Both are supported, furthermore, by theoretical studies, exhaustive tests, and by analysis of past failures. The VDI equations give us a first cut only at one important input: bolt tensions.

As Eqs. (23.1) and (23.2) show us, furthermore, we can't even define the bolts until we determine F_{Prqd}. Just as the leakage properties of a gasket determined F_{Prqd} and therefore dominated the design in pressure vessel applications, so does something other than the bolt dominate the design of a shear joint. We'll take a close look, therefore, at F_{Prqd}. First, however, let's look at the way a shear joint resists being torn apart by applied shear loads.

III. HOW SHEAR JOINTS RESIST SHEAR LOADS

A. In General

Shear joints resist applied loads in two ways. First, the interface clamping force generated by the bolts creates friction forces which resist joint slip. Second, the bolts act as shear pins to prevent slip. Because of this, structural steel joints used to be divided into two categories, "friction type" and "bearing type," and shear joints in other industries were presumably seen in the same light. The structural steel industry, however, no longer considers friction type and bearing type to be valid models for building or bridge joints. Instead, they now classify shear joints as "slip-critical" or "not slip-critical." To some extent their concern would apply to shear joints in other industries as well. Here's what they say.

B. The Concept of Slip-Critical Joints

Why are friction-type and bearing-type joints no longer considered valid models for structural steel joints, and what has replaced them?

Steel erectors must work with large, cumbersome, relatively crude components (compared to those found on an assembly line, for example). Holes in joint members must be oversized or slotted to compensate for misalignment and to make it possible to insert the bolts. In virtually every joint, whether it's intended to develop its strength through friction or bearing, some bolts are "in bearing" from the start. Pure friction-type, no-hole-bolt-contact joints are almost nonexistent.

Every joint loaded in shear, furthermore, derives some of its initial strength from friction between the joint members, even if the bolts were only snugged. Furthermore, every joint under unidirectional axial shear loads fails by first slipping into bearing. Final failure of such a joint is by shear of bolts and/or joint members, and the loads required for such failure always exceed the loads required to overcome any friction forces between faying surfaces, even if the bolts were tightened past yield. This is illustrated in Fig. 23.2 [3; p. 89, Ref. 1; p. 94, Ref. 7].

Tests suggest, furthermore, that the shear load at which the joint finally fails is independent of the initial preload in the bolts and of the coefficient of friction between faying surfaces [p. 34, Ref. 8].

As a result of all this, structural steel designers currently use the shear and bearing strengths of bolts and joint members, rather than the friction forces, to estimate the ultimate strength of all joints. They design for friction-dependent, no-slip behavior only in applications which the engineer of record considers to be "slip-critical."

One such situation, for example, could be a joint where the bolt holes are slots and the shear loads on the joint are parallel to the axis of the slots. Since the slots would allow significant motion, joint slip could result in dangerous geometric distortions of the structure, leading to failure even if the joints themselves did not fail in shear or bearing [p. 38, Ref. 8].

Joints subjected to significant load reversals are also considered slip-critical, as are some joints under fatigue loading. Again, it's up to the engineer of record to specify which joints are slip-critical, and should, as a result, be heavily preloaded. (Structural joints loaded in tension are also heavily preloaded, to make sure that each bolt carries a share of the load.)

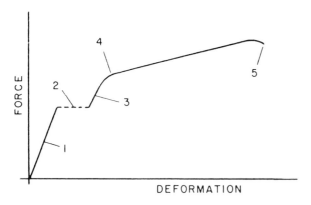

DEFORMATION

Figure 23.2 This graph illustrates the way in which a joint subjected to ever increasing shear loads will fail. First, section (1) on the curve, joint members, splice plates, and perhaps bolts will deform elastically; assuming that all are locked together by interface friction. Second (2), joint members and splice plates slip past one another until the bolts are brought into bearing. Next (3), there's additional elastic deformation of the parts. This is followed (4) by plastic yielding of bolts and/or joint members. Finally (5), something breaks.

So, joints whose integrity depends on resistance to slip are still designed, even though they are no longer referred to as "friction-type" joints. And, to repeat an earlier suggestion, shear joints in manufactured products, where close hole tolerances and alignment are economically feasible, could presumably be designed for friction or bearing strength.

The concept of slip-critical joints was first codified in the November 13, 1985 version of the AISC Specification for Structural Steel Joints Using A325 and A490 Bolts. This was one of several significant differences between the 1985 edition of this specification and earlier editions. After completing the 1985 specification, which defines "allowable stress" design procedures, the "authors" (The Research Council for Structural Connections) developed an optional, alternate Load and Resistance Factor Design Specification—again for A325 and A490 bolts. This so-called LRFD spec was approved for publication on June 8, 1988. People involved in structural steel work should obtain and study these two documents. They are significant departures from previous editions of this important specification. Slip-critical joints, and other contemporary topics, are also covered in Ref. 7, an updated version of Ref. 1.

Although the definition of slip-critical comes from the structural steel industry, the concept is valid for most if not all shear joints. If any amount of slip will place the structure at risk, the designer must specify enough bolting to create sufficient friction to prevent slip. If some slip is acceptable, he can assume and design for strength-through-bearing. There's also a third option, which combines zero slip with strength-through-bearing. Airframe designers frequently specify interference fit bolt holes. The bolts are forced through the holes by an insertion tool of some sort before the nuts are tightened. Slip is prevented by zero bolt-to-hole clearance. It's difficult to create a known amount of interface clamping force under these conditions, however, as illustrated in Figs. 6.6 through 6.8, so the joints must be designed to resist applied loads in bearing. Interference fit holes are possible in structural steel, too, using special bolts which have ribs running parallel to the axis of the bolts, but designers in other industries are more likely to take advantage of this combination of zero slip and bearing strength. One of the attractions of using interference fit holes, I'm told, is that it increases the fatigue resistance of joint members.

When we apply the VDI equations to shear joints we see that F_{Prqd}, the clamping force required to prevent joint failure, can be a key issue. In order to quantify F_{Prqd} we need to relate bolt tension to joint strength. The relationship will depend, obviously, on whether we're designing for strength-through-friction or strength-through-bearing. We're going to look at some joint strength details in the next two sections of this chapter and will see that they do indeed define the F_{Prqd} requirements.

IV. THE STRENGTH OF FRICTION-TYPE JOINTS

A. In General

Frictional resistance to joint slip is created by the normal force which clamps the joint members together; and that force is created by the bolts. In the structural steel world that frictional force is called the "slip resistance" of the joint (R_S) and we'll use the same terminology.

The slip resistance is also a function of the number of slip surfaces (M) involved in the joint, Fig. 23.3. Three surfaces produce three times as much slip resistance as one surface, etc. The equation for slip resistance, therefore, is (see p. 71 in Ref. 1 or p. 75 in Ref. 7)

$$R_S = \mu_S F_P N M \tag{23.3}$$

where R_S = slip resistance of the joint (lb, N)
  μ_S = slip coefficient of the joint
  F_P = preload per bolt (lb, N)
  N = number of bolts holding the joint together
  M = number of slip surfaces (see Fig. 23.3)

This means that F_P in Eq. (23.3) is our friend F_{Prqd}. The slip resistance of the joint, of course, R_S, must be greater than the *maximum* external load which the joint will have to resist in service. By definition F_{Prqd} is the *minimum* bolt load required for a successful joint; which, in this case means to prevent slip.

B. The Allowable Stress Procedure

We can use Eqs. (23.1) through (23.3) to design a friction-type shear joint. We'll look at an example in a minute. But first: I said in the introduction

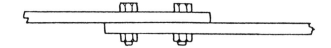

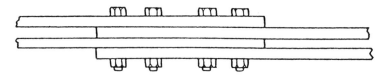

Figure 23.3 Shear joints having one and three slip surfaces.

to this chapter that we'd take an occasional look at the codified design procedures developed by and for the structural steel industry. If we were designing a shear joint for a bridge or building we wouldn't use Eqs. (23.1) through (23.3). Instead we'd use equations found in the text by Kulak et al. [7] or in the AISC bolt specs [8, 13]. One of these specifications presents what is called the "allowable stress" procedure [8]. Using that document, we find that the allowable slip load which can be placed on a slip-critical joint is defined as

$$L_{Xmax} = F_s A_b NM \qquad (23.4)$$

where L_{Xmax} = the maximum load which can be placed on the joint (lb, N)

A_b = nominal body area of a bolt (in.^{2}, mm^2)

N = number of bolts on one side of the joint

M = number of slip planes

F_s = allowable slip load per unit of bolt area (psi, MPa)

A table of allowed values of F_s is given in the reference. These values range from a low of 10 ksi for use with joints having slip coefficients of 0.33 and which contain A325 bolts mounted in long slots which are parallel to the direction of the applied load, to a maximum of 34 ksi for joints having slip coefficients of 0.50 and which contain A490 bolts mounted in standard holes.

This equation does not include any reference to bolt preload, which might seem to be a strange omission. But it's only one of many provisions in the bolt spec. In Table 4 of that document we'll find specifications for the minimum tensions which must be created in bolts of various sizes during assembly; and elsewhere in the same document we'll even find instructions concerning the way in which the bolts should be tightened. Only when we combine these and other Ref. 8 instructions and specifications with Eq. (23.4) will we see a complete picture and have an acceptable, allowable stress design.

C. Other Factors to Consider

Equations (23.1) through (23.3) seem to define everything required to prevent the failure of a general-purpose, friction-type joint. But, as usual, there some other factors to be considered—including some substantial uncertainties.

One factor is the strength of the joint member itself. It's certainly possible to design a joint in which the friction forces will exceed the tensile strength of the joint. That's also easy to avoid. Cyclic loads on a friction joint, however, can cause it to suffer a fatigue failure through the gross cross section as illustrated in Fig. 23.4, and that's much less easy to pre-

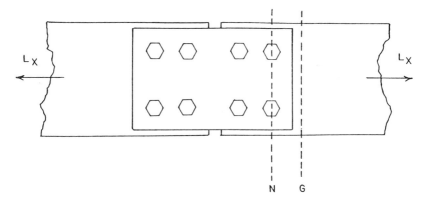

Figure 23.4 A plane passing through a row of bolt holes defines what is called
the "net section" of the joint. A plane passing through an uninterrupted section
of the joint plate defines the "gross section." See also Fig. 15.1. Note that an
external load L_X whose line of action passes through the geometric center of a
symmetrical bolt pattern such as this is called an "axial shear load."

dict. Structural steel and airframe industries spend a lot of time and money
doing joint fatigue tests, and other industries which must deal with cyclic
shear loads will have to do the same. Standard procedures for designing
structural steel joints subject to fatigue loading can be found the references
[7, 8, 13].

As far as uncertainties are concerned, it will often be difficult to predict the applied load. In critical situations we may have to build, instrument, and test a prototype or model. Finite-element analysis is also becoming more popular as a way to do this; but the FEA model should be
confirmed by physical tests whenever possible. As far as structural steel
is concerned, the cited references deal with these uncertainties in a couple
of different ways. More about this in Sec. VII. of this chapter.

In most situations it will also be difficult—perhaps impossible is a
better word—to predict and/or control the coefficient of friction between
joint members. Laboratory tests can give us a clue, but real-world conditions will affect the friction between joint members as much as they do
between nut and bolt or between nut and joint member. The ±30% scatter
in preload we expect to see when we tighten a group of as-received bolts
with a given torque will probably be duplicated by a ±30% variation in
the slip resistance in a group of joints clamped together by a given amount
of preload. To my knowledge, only the structural steel industry has published slip coefficient data. Let's take a quick look at their findings. They

won't apply directly to shear joints in other industries, but they will illustrate the concern and raise issues which must be faced whenever we deal with shear joints.

D. Slip Coefficients in Structural Steel

The slip resistance of the joint is proportional to the coefficient of friction in the joint, the so-called slip coefficient, so surface treatment of the joint members is very important. A number of different treatments have been studied experimentally. The current practice is to allow the use of any coating which has been tested by procedures specified in Ref. 8, and which has been so certified.

Note that the treatment of the faying surfaces and coating, and not just the type of coating, affects the slip coefficient of the joint. For example, a hot-dip galvanized coating has a friction coefficient of 0.18 (average). If that same coating is wire-brushed or grit-blasted, however, the average coefficient becomes 0.4 [p. 212, Ref. 7].

Note, too, that generic classifications for coatings are not possible. The actual slip coefficient of an inorganic, zinc-rich paint, for example, will vary from one paint manufacturer to another, and sometimes even from lot to lot. Hence the current requirement for coating "certification."

In general, faying surfaces should be clean and dry at assembly. Loose scale, dirt, etc., should be removed by wire brushing, but clean mill scale should *not* be removed unless the joint is going to be grit-blasted, or the equivalent, to roughen the surfaces (which should *never* be polished or buffed or "smoothed").

If a coating is required for corrosion protection, it should be one that has been thoroughly tested. The four listed below, for example, have at one time or other been tested by the Research Council. Less expensive vinyl washes and paints are a recent addition to the list [p. 206, Ref. 7].

1. Metallized aluminum
2. Metallized zinc
3. Hot-dip galvanized
4. Inorganic zinc-rich paint

Miscellaneous paints, platings, etc., are not recommended for friction-type joints unless previous tests show that they create an adequate slip coefficient for the joint.

The slip coefficients given above are average figures. In critical situations it's going to be necessary to use minimum figures or (as is the codified practice in structural steel, allowable stress procedures) provide enough safety factor in the bolting requirements to cover such contingencies. Several tables in Chap. 12 in the text by Kulak, Fisher, and Struik [7] list the results of tests on a variety of joints. Surfaces were prepared and/or coated in many different ways. Of most importance for the present discussion, each of these tables lists the average coefficient and the standard deviation. Since the text deals exclusively with structural steel joints, the treatments and coatings are limited to those found in or proposed for that industry. As mentioned above, coatings include hot-dip galvanizing, zinc-rich paint, vinyl coatings, and "metallized" surfaces which have been sprayed with such materials as zinc or aluminum. Some data are also given for uncoated joint members whose surfaces are coated with clean mill scale. Surface treatment—before coating—involves such "rough" procedures as sandblasting or wire brushing; which, as mentioned earlier, made a big difference but which would not be appropriate in the shear joints of many other industries.

Although most of this data won't apply directly to joints in other industries, a sampling may be of interest. You'll find these in Table 23.1. Note that a standard deviation equal to 10% of the average slip coefficient is not hard to find. That would suggest a three sigma deviation of $\pm 30\%$, confirming my statements above. Standard deviations greater than 10%, unfortunately, are also easy to find in the Ref. 7 data.

In any event, Eqs. (23.1) through (23.3) give us much of what we need to know in order to design a shear joint which derives its strength from friction. Here's an example.

Table 23.1 Slip Coefficients

Joint preparation	Type of coating	Coating thickness (mils)	Average coefficient	Standard deviation
Clean mill scale	None	NA	0.33	0.07
Grit-blasted	Zinc spray	0.6 to 1.0	0.42	0.04
Grit-blasted	Aluminum spray	1.6 to 2.2	0.74	0.08
Sandblasted	Zinc dust paint	0.8	0.39	0.02
Sandblasted	Zinc silicate paint	1.0	0.53	0.01
Sandblasted	Vinyl wash	0.3 to 0.5	0.27	0.01
Sandblasted	Vinyl wash; exposed 2 months	0.3 to 0.5	0.27	0.05
Acid pickling bath	Hot-dip galvanized	Not given	0.21	0.08
Acid pickling bath	Hot-dip galvanized	2.4 to 5.0	0.23	0.023

Source: All data taken from Ref. 7.

E. An Example

We'll use the joint shown in Fig. 23.5 as an example. Input data include:

Number of rows of bolts; one on each side of the joint $= 1$
Number of bolts in each row $= N = 2$
Bolts are $\frac{3}{8}$–20 UN
Bolt material: ASTM A325
Cross-sectional areas of the bolt:
  Body $= A_B = \Pi D^2/4 = 0.110$ in.^{2}
  Tensile stress area of threads $= A_S = 0.0836$ in.^{2} (see Appendix F)

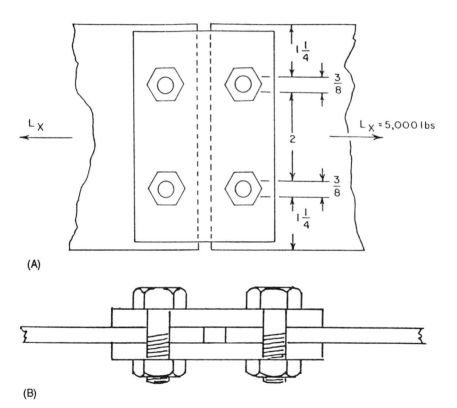

(A)

(B)

Figure 23.5 The joint shown here is used as a design example in the text. Load L_X is also an "axial shear load" because its line of action passes through the centroid of the joint. Note there will be no tendency for the bolt group to rotate under such a load. The joint and splice plates in this joint are $\frac{1}{4}$ in. thick; the bolts have a nominal diameter of $\frac{3}{8}$ in. There are two shear planes here. One passes through the bodies of the bolts, the other through the threads.

Yield strength of bolt material: = σ_y = 92 ksi (see Table 5.1)
Shear strength of bolt material = σ_s = 79 ksi (see Table 5.2)
Joint material: A36
Number of slip surfaces (or shear planes) = M = 2
Average tensile strength of joint material = σ_{tj} = 70 ksi (see Table 5.14)
Average shear strength of joint material = σ_{sj} = 48 ksi (Table 5.14)
Thickness of joint and splice plates: $\frac{1}{4}$ in.
Grip length = L_G = 3 × $\frac{1}{4}$ = $\frac{3}{4}$ in.
Other joint dimensions are shown in Fig. 23.5
Load on joint = L_X = 5000 lb
Joint surfaces: clean mill scale

Minimum Preload Required to Prevent Slip

First we'll use Eq. (23.3) to determine the minimum preload required to prevent slip. Then we'll return to the VDI equations to see what information they can add. So first, from Eq. 23.3, recognizing that the F_P in this equation is our first cut at the VDI F_{Prqd}:

$$R_S = \mu_S F_{Prqd} NM$$

The frictional forces must be large enough to resist the 5000-lb load, so R_S = 5000. The average slip coefficient for clean mill scale is 0.33 (Table 23.1). So we have:

$$5000 = 0.33(F_{Prqd})2(2)$$
$$F_{Prqd} = 3788 \text{ lb}$$

Are the bolts capable of producing this much clamping force? The preload they could create if tightened to yield would be:

$$F_{Py} = A_S \sigma_y = 0.0836(92,000) = 7691 \text{ lb}$$

so everything seems OK so far.
Now, from VDI Eqs. (23.1) and (23.2)

$$\text{Min } F_J = 1.3 F_{Prqd} = 1.3(3788) = 4924 \text{ lb}$$

and

$$\text{Max } F_J = \alpha_A(4924)$$

What value shall we use for α_A? If we plan to use torque control at assembly we'll have to use 1.86, but that would mean a Max F_J greater than the yield strength of the bolts, and it would be impossible to develop that much preload in these $\frac{3}{8}$-in., A325 bolts. If we accept the VDI equations, therefore, we'd have to redesign the joint to use larger and/or more bolts. Or we could specify joint surface treatment giving a larger slip coefficient.

There are other ways of looking at this joint, however. Let's see what happens if we specify that the bolts shall be tightened to or past their yield point, a common structural steel practice. Now we use the VDI equations "backward" and set Max F_J equal to bolt tension at yield. The act of yielding will reduce preload scatter to $\pm 5\%$ [1] so

$$\alpha_A = \frac{1.05}{0.95} = 1.11.$$

Now, from Eq. (23.2)

$$7691 = 1.11 \text{ Min } F_J$$
$$\text{Min } F_J = 6,929 \text{ lb}$$

and, from Eq. (23.1)

$$6929 = 1.3 F_{\text{Prqd}}$$
$$F_{\text{Prqd}} = 5330 \text{ lb}$$

This value of F_{Prqd} is greater than the F_{Prqd} determined earlier from Eq. (23.3). Which is correct? The modified VDI Eqs. (23.1) and (23.2) take into account such factors as tool scatter and elastic interaction loss, which are ignored in Eq. (23.3). That equation allows us to compute the minimum preload required to resist the 5000-lb load, but doesn't tell us how to achieve that preload. VDI tells us we need more preload than we think we do to compensate for preload loss and assembly uncertainties.

Another possible concern here would be the fact that we've assumed an average slip coefficient for clean mill scale. What if the coefficient is, say, two standard deviations less than average? (See Table 23.1.) Now

$$\mu_S = 0.33 - 2(0.07) = 0.19$$

From Eq. (23.3)

$$F_{\text{Prqd}} = \frac{5000}{0.19(2)2} = 6579 \text{ lb}$$

Plugging this value into Eqs. (23.1) and (23.2) would again suggest a Max F_J above the tensile capacity of the bolt. One way or another, the cold, hard, realistic approach of VDI will force us to use larger and/or more numerous bolts, or higher friction joint surfaces, to guarantee that the joint won't slip, even if we decide to use yield control.

Alternate Using the Allowable Stress Procedure

Let's repeat the example, using Eq. (23.4) to compute the maximum slip load which could be placed on this joint per the AISC "bolt spec" [8].

$$L_{Xmax} = F_s A_b NM$$

Table 3 in the specification says that $F_s = 17$ ksi if the slip coefficient is 0.33, the bolts are A325's, and holes are "standard" (i.e., neither slotted nor oversize). So:

$$L_{Xmax} = 17,000(0.110)2(2) = 7480 \text{ lb}$$

Since the anticipated load is only 5000 lb, our design is acceptable. Do we now use Eq. (23.3) to find the preload required to support this load? For this example we'd have to. Normally we'd use Table 4 in Ref. 8 to determine the preload requirements, but that table doesn't include any bolts less than $\frac{1}{2}$ in. in diameter.

If we were using larger bolts, however, and were designing a structural steel or similar joint, we'd be foolish *not* to use the AISC specifications. The design equations and allowed stress or load in these documents are based upon many decades of analysis, test, and experience. Although such factors as tool scatter and elastic interaction are invisible in the design equations, they were always present in practice and are covered by implication in the allowable load figures and/or in the specified procedures for the handling, installation, and tightening of the bolts. If you are designing structural steel joints, therefore, you should follow the AISC design procedures and ignore the complications created by use of the modified VDI equations. If you're designing any other kind of slip-resistant shear joint, however, you'd be wise to use Eqs. (23.1) through (23.3) to get a more detailed look at a proposed design.

In any event, the joint illustrated in Fig. 23.5 would appear to be an acceptable slip-resistant joint, but only if the bolts are properly tightened and an acceptable slip coefficient can be guaranteed.

V. THE STRENGTH OF BEARING-TYPE JOINTS

Several factors determine the load-carrying capability of a bearing-type shear joint. These include:

The shear strength of the bolts
The tensile strength of the joint members (called "plates" in structural steel)

The bearing stress created in the plates by the bolts
The tearout strength of the plates

Let's examine each of these.

A. The Shear Strength of Bolts

The Distribution of Load Among the Bolts

The general shear strength of the bolts in a shear joint can be expressed as:

$$R_B = A_{blt}N\sigma_s \tag{23.5}$$

where R_B = force required to shear all of the bolts on one side of a
joint (lb, N)
A_{blt} = total cross-sectional area of the bolt which must be
sheared (in.^{2}, mm^2)
N = number of bolts on one side of the joint
σ_s = shear strength of the bolt material (psi, MPa)

These terms will become clearer when we look at an example in a minute. But first, some general comments about shear loads on bolts. Equation (23.5) implies that every bolt will bear an equal share of the load in a shear joint, and, for simplicity, we'll make this assumption when selecting the number, size, and type of bolts to be used in our joint. But, in fact, the bolts will *not* share the load equally, a fact which was illustrated in Fig. 14.8. Those bolts closest to the leading and trailing edges of the joint will see far more load than those nearer the center of the bolt pattern. Unfortunately, not even the bolts in a given row will share the load absorbed by that row equally. The placement of the bolts in their holes, minor variations in hole and bolt diameter, and the way the load is applied to the joint will all affect the loads seen by individual bolts. The situation, in fact, is statically indeterminate.

The answer to these uncertainties is our old friend the safety factor. We decide how much shear stress the "average" bolt will see, and then use a bolt capable of supporting many times that stress. All of this applies to what we call an "axial shear load," in which the line of action of the applied load passes through the centroid of the bolt group as it does in Figs 23.4 and 23.5. The safety factor must be increased still further if the applied load is "eccentric." Just as prying significantly magnified the loads seen by some of the bolts in a tension joint, so does eccentricity magnify them in a shear joint. We'll take a look at that in Sec. VI.

For now, however, we're going to adopt the procedures long used by the designers of bearing-type joints and assume that each bolt in the joint sees the same load.

Shear Strength Calculations

Once again we'll use the joint shown in Fig. 23.5 for our example. The design data given at the beginning of the earlier example, Sec. IV-D, still apply.

The shear stress within a single bolt will be:

$$\sigma = \frac{F}{A_{blt}} \tag{23.6}$$

where σ = shear stress (psi, MPa)
   F = shearing force applied to that one bolt (lb, N)
   A_{blt} = total cross-sectional shear area of that one bolt (in.^{2}, mm^2)

If there are N bolts *on one side* of the joint, then

$$F = \frac{L_X}{N} \tag{23.7}$$

For example, in Fig. 23.5 upper and lower splice plates hold two joint members together. There are two bolts to the left of the gap between joint members and two to the right of the gap; so $N = 2$.

The total shear area of a bolt depends upon how many shear planes pass through it and whether they pass through the body of the bolt, the threaded region, or both.

$$A_{blt} = n_b A_B + n_s A_S \tag{23.8}$$

where n_b = number of shear planes which pass through the body
   n_s = number of shear planes which pass through the threads
   A_B = cross-sectional area of the body (in.^{2}, mm^2)
   A_S = tensile stress area of the threads (in., mm) (see Appendix F)

In Fig. 23.5 two shear planes pass through each bolt, one plane through the body and one through the threads, so

$$A_{blt} = 1 (A_B) + 1 (A_S)$$

These are $\frac{3}{8}$–20 UN bolts, so $A_B = 0.110$ in.^{2} and $A_S = 0.0836$ in.^{2}

$$A_{blt} = 0.110 + 0.0836 = 0.194 \text{ in.}^2$$

We can now compute the stress within the bolt and compare it to an allowable stress. The load on this joint is 5000 lb. From Eqs. (23.6) and (23.7)

$$\sigma = \frac{5000}{2(0.194)} = 12{,}887 \text{ psi } (90 \text{ MPa})$$

The allowable shear stress for Grade A 325 bolts used in structural steel applications is 21 ksi (145 MPa) [8], so we're probably OK. The allowable stress in other industries might be higher, closer perhaps to the 79 ksi shear strength of A325 bolt material (Table 5.2). In fact, as an alternative we can compute the total shear strength of the bolts in the joint and compare it to the applied load. From Eq. (23.5)

$$R_B = A_{blt} N \sigma_s$$
$$R_B = 0.194 \ (2) \ 79{,}000 = 30{,}652 \text{ lb}$$

So the joint shown in Fig. 23.5 should be able to support the 5000-lb load with a 6:1 safety factor. But we still have several other factors to consider.

B. Tensile Strength of Joint Plates

We compute the tensile strength of the joint with respect to the net section which includes the most bolts. (All rows may not have the same number of bolts.) In our example, of course, there are only two bolts per row. With reference to the dimensions shown in Fig. 23.5, and to the fact that the joint plates are 0.250 in. thick, the area of the net section is

$$A_J = 0.250(1.25 + 2 + 1.25) = 1.125 \text{ in.}^2$$

Since the tensile strength of A36 is 70 ksi, the tensile strength of this joint will be

$$R_T = A_J \sigma_u = 1.125(70{,}000) = 78{,}750 \text{ lb}$$

Obviously not a concern in our example, where the applied load is only 5000 lb.

C. Bearing Stress

The bearing stress the bolts create in the joint plates is found from

$$\sigma_B = \frac{L_X}{NDL_G} \tag{23.9}$$

where σ_B = bearing stress (psi, MPa)
$\quad\quad\quad L_X$ = load applied to joint (lb, N)
$\quad\quad\quad N$ = number of bolts on one side of the joint

$$D = \text{nominal diameter of the bolts (in., mm)}$$
$$L_G = \text{grip length of the joint (in., mm)}$$

The area in bearing is shown in Fig. 23.6.
   In our example:

$$\sigma_B = \frac{5000}{2(0.375)0.750} = 8{,}889 \text{ psi}$$

Is that an acceptable amount? We're concerned here, for example, with the possibility that the bolt holes will be elongated by the bearing load. According to Ref. 8, bearing stresses of 1.0 to 1.2 times the ultimate tensile strength of the plate material are allowed in structural steel if the bolts are A325 or A490. The exact amount allowed depends upon whether or not the holes are slotted, on the number of bolts, and on the distance between the row of bolt holes and the edge of the plate. If we accept 1.0 times the *minimum* ultimate strength of A36 we can accept a bearing stress of up to 58,000 psi, well below the amount computed above.

D. Tearout Strength

To compute the tearout strength we must first compute the minimum plate area which would have to be sheared. This area is illustrated in Fig. 23.7. Note that there will be two such areas per bolt, or four areas in the present example. The pieces torn out of the plate are sometimes wedge shaped,

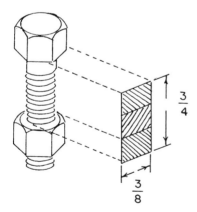

Figure 23.6 We compute the bearing stresses created by the bolt on the joint plates using the crosshatched area shown here, which is equal to the nominal diameter of the bolt times the joint's grip length. In the joint shown in Fig. 23.5, therefore, the bearing area would be $0.375 \times 0.750 = 0.281 \text{ in}^2$.

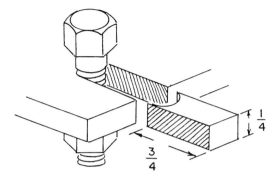

Figure 23.7 If tearout is to occur each bolt must shear at least twice the amount of plate area shown crosshatched here. This area is equal to the thickness of the joint plate times the distance between the centerline of the bolt hole and the edge of the plate. In the joint shown in Fig. 23.5, therefore, the total area to be sheared by the two bolts would be would be $4 \times 0.25 \times 0.75$ in.^{2} = 0.75 in.^{2} (See also Fig. 15.1.)

with larger shear areas, but we're only concerned with the minimum force required. So

$$A_{sh} = 2N_R DL_{eg} \tag{23.10}$$

where A_{sh} = total area which must be sheared if tearout is to occur (in.^{2}, mm^2)

N_R = number of bolts in the row nearest the edge
D = nominal diameter of the bolts (in., mm)
L_{eg} = distance from bolt hole to edge of plate (in., mm)

In our example

$$A_{sh} = 2(2)\,0.375(0.750) = 1.125 \text{ in.}^2$$

And the tearout strength would be

$$R_{TO} = A_{sh}\sigma_s \tag{23.11}$$

where R_{TO} = tearout strength of joint (lb, N)
σ_s = shear strength of plate material (psi, MPa)

In our example:

$$R_{TO} = 1.125(48,000) = 54,000 \text{ lb}$$

Note that computation of tearout strength becomes far more complicated if there are several rows of bolts on each side of the joint. In this situation tearout would involve every row—and would be highly unlikely if not impossible.

E. Summary

We have now decided that the following loads would be required to pull this joint apart.

By shearing the bolts: 30,652 lb
Through tensile failure of the joint member: 78,750 lb
By tearout of the bolts: 54,000 lb

This means that the shear strength of the bolts determines the strength of this joint. Since the applied load is only a sixth of the shear strength, joint failure will not occur.

We also estimated the bearing stress the bolts will create on the joint and found it to be much less than that we can allow. So the design of our bearing-type joint is confirmed.

F. Clamping Force Required by a Bearing-Type Joint

OK so far, but remember that one of our goals is to determine how much F_{Prqd} is required to prevent joint failure, so that we could solve VDI Eqs. (23.1) and (23.2). What do Eqs. (23.5) through (23.11) tell us about F_{Prqd}? They tell us that no specific F_{Prqd} is required. The bearing-type shear joint will theoretically resist shear loads even if all the bolts are dead loose. In practice, we'll tighten them at least enough to retain the nuts—and therefore to retain the bolts. In fact, we may tighten them more to resist self-loosening (which, as we saw in Chap. 16, is caused by transverse slip). Self-loosening requires cyclic loads, however, and they can also cause fatigue failure of joint members. If we're concerned about self-loosening we should probably design the joint to resist shear loads through friction rather than in bearing.

VI. ECCENTRICALLY LOADED SHEAR JOINTS

A. Rotation About an Instant Center

Axial shear load means that the centerline of the external load passes through the centroid of the group of fasteners holding the shear joint together, as suggested in Figs. 23.4 and 23.5. In many cases shear joints are subjected to external loads that don't pass through the centroid of the fastener group, as in Fig. 23.8. Such joints are called *eccentrically loaded shear joints*. The external load is applied to the bolts through some sort

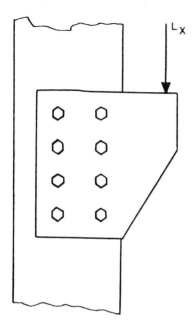

Figure 23.8 A load on a shear joint is said to be eccentric when its resultant does not pass through the centroid of the bolt pattern.

of leverage determined by the geometry of the joint; so "eccentric load" is the shear joint equivalent of prying load in a tension joint (see Chap. 13). In both cases leverage makes a big difference in the load seen by some of the fasteners [1, 5–7].

Under an eccentric load, the entire group of bolts will tend to rotate about an instant center that is determined by the bolt pattern and by the direction of the applied load (Fig. 23.9). Computing the exact load on each bolt under these conditions is time consuming and involves eccentricity and safety factors that must be determined by experiment. Such factors are listed in structural design handbooks, which also list precomputed solutions for a number of standard joints. (See also Refs. 7 and 9.)

For our purposes, it is sufficient to say that those bolts located farthest from the theoretical center of rotation will carry the greatest load and are most apt to fail.

B. Rotation About the Centroid of the Bolt Group

Although the joint will actually attempt to rotate about an instant center, it is easier to estimate the shear loads on the bolts if we assume that it

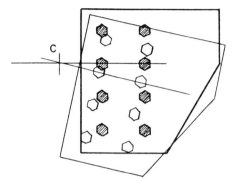

Figure 23.9 Eccentrically loaded joints want to rotate about an instant center as shown here, magnifying the shear loads on the bolts.

tries to rotate about the centroid of the bolt group. This historically earlier and simpler approach results in conservative estimates of shear loads; i.e., it overestimates them, and so is safe to use. Here's an example of this procedure.

Find the Centroid of the Bolt Group

We can usually find the centroid of a bolt group by inspection, as in Fig. 23.8. If the bolts are not in neat rows, however, as in Fig. 23.10, we can use the following equations to locate it [9, 12].

$$\overline{X} = \frac{\sum A_n x_n}{\sum A_n} \tag{23.12}$$

$$\overline{Y} = \frac{\sum A_n y_n}{\sum A_n} \tag{23.13}$$

where A_n = cross sectional area of the body of bolt n (in.^{2}, mm^2)
        x_n = distance of bolt n from an arbitrarily located y axis (in., mm)
        y_n = distance of bolt n from an arbitrarily located x axis (in., mm)
        $\overline{X}$ = distance of the centroid from the y axis (in., mm)
        $\overline{Y}$ = distance of the centroid from the x axis (in., mm)

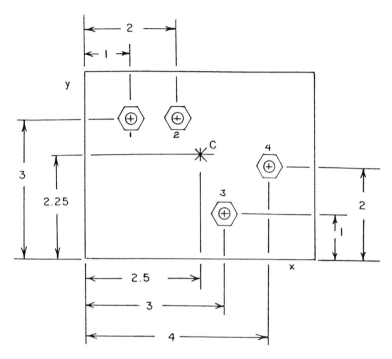

Figure 23.10 This joint is used as an example in the text, to compute the location of the centroid (C) of the bolt group. The bolts are $\frac{1}{4}$ in. in diameter. The x and y axes are arbitrarily located along the left and bottom sides of the plate.

With reference to Fig. 23.10: let's assume that the nominal diameter of the bolts is $\frac{1}{4}$ in. and that we've arbitrarily located x and y axes along the bottom and left edge of the splice plate as shown. The distance of bolt 1 from the y axis is x_1; its distance from the x axis is y_1, etc.

$$A_n = \frac{\Pi\,(0.25^2)}{4} = 0.0491 \text{ in.}^2$$

$$\overline{X} = \frac{A_1 x_1 + A_2 x_2 + A_3 x_3 + A_4 x_4}{A_1 + A_2 + A_3 + A_4}$$

Since each bolt has the same cross-sectional area in this example:

$$\overline{X} = \frac{0.0491(1 + 2 + 3 + 4)}{4(0.0491)} = 2.5 \text{ in.}$$

Similarly,

$$\overline{Y} = \frac{0.0491(1 + 2 + 3 + 3)}{4(0.0491)} = 2.25 \text{ in.}$$

The centroid so located is labeled C in Fig. 23.10.

Estimating the Shear Stress on the Most Remote Bolt

Now, with reference to Figs. 23.11 and 23.12 we're going to estimate the shear load on an eccentrically loaded fastener. Let's assume that the joint shown is a bearing type. First we compute the *primary shear load* on each fastener.

$$P_P = \frac{L_X}{N} \tag{23.14}$$

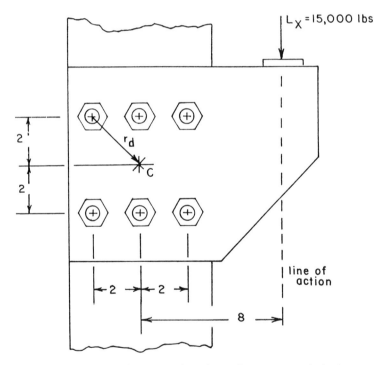

Figure 23.11 The joint shown here is used as an example in the text to compute the effects of an eccentric load on a shear joint. Note that the line of action of the applied force, L_X, is 8 in. from a parallel line passing through the centroid (C) of the bolt group.

where P_P = the primary shear force on each bolt (lb, N)
$\quad\quad L_X$ = the load applied to the joint (lb, N)
$\quad\quad N$ = the number of bolts in the group

In this example

$$P_P = \frac{15,000}{6} = 2500 \text{ lb}$$

Next we must compute the *secondary shear load* on the fastener. To do this we first compute the *distance between the centroid and the most distant bolt* (See Fig. 23.11)

$$r_d = \frac{2}{\sin 45°} = \frac{2}{0.707} = 2.828 \text{ in.}$$

And the reaction moment on the joint

$$M = L_X U \tag{23.15}$$

where M = reaction moment (in.-lb, mm-N)
$\quad\quad L_X$ = load on the joint (lb, N) = 15,000 lb
$\quad\quad U$ = perpendicular distance between the centroid and the line of action of L_X (in., mm) = 8 in.

In our example:

$$M = 15,000(8) = 120,000 \text{ in.-lb}$$

Next we estimate the *reaction force on the most distant bolt*

$$P_d = \frac{M r_d}{\sum r_n^2} \tag{23.16}$$

where P_d = reaction force on most distant bolt (lb, N)
$\quad\quad r_n$ = distance of bolt n from the centroid (in., mm)

From Fig. 23.11 we see that four bolts are each a distance 2.828 in. from the centroid, and two bolts are 2 in. from it, so:

$$P_d = \frac{120,000(2.828)}{4(2.828^2) + 2(2^2)}$$

$$P_d = 8484 \text{ lb}$$

The line of action of the force P_d will be perpendicular to r_d, as shown in Fig. 23.12. Combining the P_s and P_d vectors we get the *resultant shear force (R)* exerted on the most distant bolt by the eccentric load L_X. In our example the resultant is 10,402 lb. This same force would be seen by

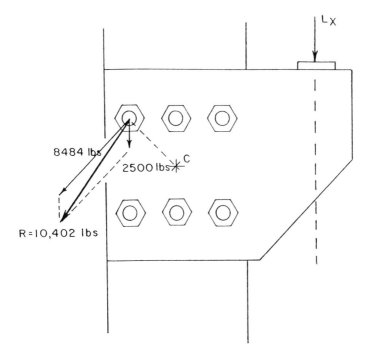

Figure 23.12 The shear forces acting on the bolt which is most distant from the centroid are shown here, along with the resultant force, R. The forces consist of a primary shear force of 2500 lb and a secondary shear force of 8484 lb created by the fact that the load is eccentric. The resultant, total force on the bolt is 10,402 lb, well in excess of the 2500 lb the bolt would have to support if the same external load had a line of action which passed through the centroid of the joint. The same force would be exerted on each of the bolts in the four corners of the group, but each force would be in a different direction.

each of the bolts in the four corners of the bolt pattern, but each force would be in a different direction. We now use Eq. (23.6) to estimate the *shear stress* in the most distant bolts.

$$\sigma_s = \frac{R}{A_{blt}}$$

If these are $\frac{5}{8}$–20 UN, A325 bolts, and there's only one shear plane which passes through the body of the bolt, then:

$$A_{blt} = \frac{\Pi(0.625^2)}{4} = 0.307 \text{ in.}^2$$

and

$$\sigma_s = \frac{10,402}{0.307} = 33,883 \text{ psi}$$

This is less than the average shear strength of an A325 bolt, which is 79 ksi, but it exceeds the AISC allowable stress limit of 21 ksi [8]. Note that the stress would be well under the allowable if the bolt were exposed only to primary shear force. Eccentricity has magnified the loads seen by the most distant bolts.

If we were to analyze this joint assuming rotation about an instant center, we might conclude that the worst-case shear stress in the most distant bolt was within the allowable limit; but, based on our conservative rotation-about-the-centroid procedure we have decided that the farthest bolts will be overstressed. We'd have to go to a $\frac{7}{8}$-in. bolt to bring stress within the allowable limit (or go to a higher-strength bolt material, with less increase in size).

This would not end the analysis, of course. We must still check the bearing stress created by the bolts on the joint plates. Tearout and tensile failure would be unlikely here, but in some designs would have to be checked. We could also design this joint to be slip resistant by determining the resultant forces on each bolt, combining these to compute the total force to be resisted, then choosing bolts and a slip coefficient to provide sufficient slip resistance. I'll leave that as an exercise for the reader.

VII. ALLOWABLE STRESS VS. LOAD AND RESISTANCE FACTOR DESIGN

The structural steel examples I've given in this chapter have all been based upon what that industry calls the "allowable stress design" procedure [7,8]. Until recently this was the only procedure employed by that industry.

Any safe design procedure must take into account the fact that estimates of the strength of a structure are not absolute; that the actual strength will vary because of inevitable variations in dimensions, material properties, and the like. Any safe procedure must also take into account the fact that the exact loads a structure must support are also unknown, thanks to variations in field or service environments, or design assumptions or errors which result in unexpected stresses or stress concentrations. The allowable stress design procedure is based on the worst-case assumption that a structure will have minimum strength but must support the maximum possible load. Uncertainties are covered by a variety of safety factors. The designer, for example, must assume that the strength of the bolts is less than the minimum strength specified by ASTM.

In the last decade an alternate procedure has been developed by the industry, led by the Research Council on Structural Connections. This is called the "load and resistance factor" procedure [7, 13]. Anticipated loads and anticipated strength (resistance) are modified by factors which reflect the probabilistic uncertainties in those estimates. The general expression for relating structural strength to anticipated loads is [7]

$$\Phi R = \alpha D + \gamma(L + I) \tag{23.17}$$

where Φ = resistance factor reflecting the uncertainty in strength
         R = the average strength (resistance)
         α, γ = load factors reflecting the probability of an increase in load
         D = the anticipated dead load on the structure
         L = the anticipated live load on a structure
         I = the anticipated impact load on a structure

The load and resistance factors would, of course, be a function of the application, of the type of joint involved. Those used by the structural steel industry would presumably not be appropriate for the designer of airframes or autos; so I'm not going to go into more detail here. Those involved with structural steel design should consult the cited references.

I will add, however, that the two AISC specifications [8, 13] will lead to similar results. Load and resistance factor design is an alternate procedure; it is not a mandated replacement for the allowable stress procedure. Designers are free to use the one they prefer.

REFERENCES

1. Fisher, J. W., and J. H. A. Struik, *Guide to Design Criteria for Bolted and Riveted Joints*, Wiley, New York, 1974.
2. Munse, W. H., Final report on riveted and bolted structural joints, Division of Highways, Federal Highway Administration, NTIS, PB-201297, 1C34022427.
3. Carter, J. W., J. C. McCally, and L. T. Wyly, Comparative Test of a structural joint connected with high-strength bolts and a structural joint connected with rivets and high strength bolts, *AREA Bull.*, vol. 56, Proceedings, pp. 217–267.
4. *High Strength Bolting for Structural Joints*, Booklet 2867, Bethlehem Steel Company, Bethlehem, PA.

5. Fazekas, G. A., On eccentrically loaded fasteners, *Trans. ASME*, pp. 776–778, August 1976.
6. Fazekas, G. A., Eccentric limit load on fasteners, *Trans. ASME*, pp. 773–775, August 1976.
7. Kulak, Geoffrey L., John W. Fisher, and John H. A. Struik, *Guide to Design Criteria for Bolted and Riveted Joints*, 2nd ed., Wiley, New York, 1987.
8. *Specification for Structural Joints Using A325 and A490 Bolts: Allowable Stress Design*, American Institute of Steel Construction, Chicago, November 13, 1985.
9. Section 23, Bolted and Riveted Joints, *Standard Handbook of Machine Design*, Ed. by Joseph E. Shigley and Charles R. Mischke, McGraw-Hill Book Company, New York, 1986, pp. 23.12–23.16.
10. Bibel, G. D., and D. L. Goddard, Preload variations of torqued fasteners, a comparison of friction and elastic interaction effect, *Fastener Technology International*, pp. 31ff, June/July 1994.
11. *Systematic Calculation of High Duty Bolted Joints*, VDI 223, Society for Design and Development, Committee on Bolted Joints, Verein Deutscher Ingenieure, VDI Richtlinien, Dusseldorf, Germany, October 1977; translated by Language Services, Knoxville, TN; translation published by the U.S. Department of Energy as ONRL-tr-5055.
12. Shigley, Joseph E., *Mechanical Engineering Design*, 3rd ed., McGraw-Hill, New York, 1977, p. 258.
13. *Specification for Structural Joints Using A325 and A490 Bolts: Load and Resistance Factor Design*, American Institute of Steel Construction, Chicago, June 8, 1988.

APPENDICES

A
Units and Symbol Log

Symbol	Uses	English units	Metric units
A	Area	in.^{2}	mm^2
a	Slope of the S_G-T_P line on a log-log plot: one of three new PVRC gasket constants	psi	MPa
a	Distance between centerline of an eccentric joint and the point of application of an external tension load (or resultant of several loads)	in.	mm
a	Experimentally derived exponent used to compute T_P	None	
a	Depth of crack in K_{ISCC} calculations	in.	mm
A_B	Cross-sectional area of the body of a fastener	in.^{2}	mm^2
A_{blt}	Total, cross-sectional shear area of a bolt (sum of the shear areas through body and through threads)	in.^{2}	mm^2
A_C	Cross-sectional area of the equivalent cylinder used to compute joint stiffness	in.^{2}	mm^2
A_G, A_g	Gasket contact area	in.^{2}	mm^2

Symbol	Uses	English units	Metric units
A_i	Area used to compute the hydrostatic end load on a gasketed joint	in.^{2}	mm^2
A_J	Cross-sectional area of the joint	in.^{2}	mm^2
A_p	Contact area between bolt head or nut face and joint	in.^{2}	mm^2
A_r	Cross-sectional area of the minimum minor diameter of bolt threads	in.^{2}	mm^2
A_S	Effective cross-sectional area of the threaded section of a fastener (an assumed area based on the mean of pitch and minor diameters)	in.^{2}	mm^2
A_{sh}	Area which must be sheared to "tear out" a shear joint	in.^{2}	mm^2
A_{TS}	Cross-sectional area of shear of a thread	in.^{2}	mm^2
A_{tot}	Total loaded cross-sectional area of a fastener in shear	in.^{2}	mm^2
A_v	Cross-sectional area of a pressure vessel	in.^{2}	mm^2
b	Effective width of gasket per ASME Code rules	in.	mm
C	Shape factor in K_{ISCC} calculations	None	
c	A tightness factor defining an acceptable, minimum leak rate	None	
C_1–C_5	Correction factors for fatigue calculations	None	
D	Nominal diameter of fastener	in.	mm
D	The anticipated dead load on a shear joint	lb	N
d	Diameter of hole through or into fastener	in.	mm
D_B	Diameter of bolt head or washer (diameter of contact with joint members)	in.	mm
D_H	Diameter of bolt hole	in.	mm
D_J	Diameter of a cylindrical joint, or length of one side of a square joint, or length of short side of rectangular joint	in.	mm
d_m	Minor diameter of male thread	in.	mm

Symbol	Uses	English units	Metric units
D_r	Root diameter of the threads	in.	mm
D_{smin}	Minimum nominal diameter	in.	mm
E	Young's modulus (modulus of elasticity)	psi	GPa
e	Assembly efficiency	None	
E_A	Average loss of preload because of elastic interactions	lb	N
E_B	Modulus of elasticity of the bolt material	psi	GPa
e_{EI}	Percentage of preload lost because of elastic interactions, expressed as a decimal	None	
e_m	Percentage of preload lost because of embedment, expressed as a decimal	None	
E_m	Young's modulus for alternate joint material	psi	GPa
E_{min}	Minimum pitch diameter	in.	mm
E_{nmax}	Maximum pitch diameter of internal thread (nut)	in.	mm
E_p	Basic (nominal) pitch diameter of fastener	in.	mm
E_{Smin}	Minimum pitch diameter of external thread (bolt)	in.	mm
F	Force	lb	N
ΔF	Change in force	lb	N
f	Frequency (Δf = change in frequency)	Hz	Hz
F_B	Tension force in bolt	lb	N
ΔF_B	Change in bolt tension	lb	N
F_{CL}	Clamping force	lb	N
ΔF_{EI}	Loss of preload caused by elastic interactions	lb	N
F_J	Per-bolt clamping force on the joint	lb	N
F_j	Resonant ultrasonic frequency in a bolt	Hz	Hz
F_J	Compression force in joint	lb	N
ΔF_J	Change in per-bolt clamping force created by the external load	lb	N
F_{Krqd}	Minimum preload (or clamping force) required to prevent separation of an eccentrically loaded joint	lb	N

Symbol	Uses	English units	Metric units
ΔF_m	Loss of preload caused by embedment	lb	N
F_P	Preload	lb	N
F_{Pa}	Average, initial assembly preload; also called the "target" preload	lb	N
F_{Prqd}	Minimum preload required to prevent slip, separation, or leakage of a concentrically loaded joint	lb	N
F_S	Force required to shear a fastener	lb	N
F_T	Tension created in bolt by differential thermal expansion	lb	N
ΔF_{th}, ΔF_{TH}	Change in preload caused by a change in temperature	lb	N
F_y	Force required to create a 0.2% permanent set in the bolt (i.e., to yield it)	lb	N
f_z	Amount by which a fastener embeds	in.	N
G	Diameter of a pressure vessel to the midpoint of the gasket	in.	mm
G	Grip length	in.	mm
G	Ratio between shear strength and tensile strength	None	
G	Shear modulus	psi	MPa
G_b	Intercept of the S_G-T_P line with the S_G axis: one of the three new PVRC gasket constants	psi	MPa
G_o	Outer diameter of a gasket	in.	mm
G_S	Intercept of the gasket's unloading-reloading line with the S_G axis: one of the three new PVRC gasket constants	psi	MPa
H	Height of basic thread triangle ($H = 0.86603 - n$, if $\beta = 60°$)	in.	mm
H_D	End force exerted by internal pressure (e.g., in a pressure vessel) on a flange; reaches the flange through the wall of the vessel (or pipe) and through the hub of the flange	lb	N
H_G	Reaction force exerted by a gasket on the flange	lb	N

Symbol	Uses	English units	Metric units
H_T	Force on a flange created by pressure exerted by contained fluid on that portion of the flange which lies between the ID of the gasket and the ID of the pipe or vessel	lb	N
I	The anticipated impact load on a shear joint	lb	N
I	Moment of inertia	in.^{4}	mm^4
K	Nut factor	None	
K	Spring constant or stiffness	lb/in.	N/mm
K_B, K_b	Spring constant of bolt	lb/in.	N/mm
K_F	Spring constant of flange	lb/in.	N/mm
K_G	Spring constant of gasket	lb/in.	N/mm
K_{ISCC}	Threshold stress intensity factor for SCC	psi·in.$^{1/2}$	Pa·mm$^{1/2}$
kip	Force in thousands of pounds	Is an English unit	No equivalent
K_J	Spring constant of joint	lb/in.	N/mm
K_J'	Stiffness of a concentric joint loaded at internal loading planes	lb/in.	N/mm
K_J''	Stiffness of a joint in which both the axes of the bolts and the line of application of a tensile force are offset from the axis of gyration of the joint, and in which the tensile load is applied along loading planes within the joint members	lb/in.	N/mm
K_N	Spring constant of nut	lb/in.	N/mm
K_{nmax}	Maximum minor diameter of nut threads	in.	mm
KSI	English unit for thousands of pounds per square inch	Is an English unit	No equivalent
K_{Smin}	Minimum minor diameter of bolt thread	in.	mm
K_T	Spring constant of a group of springs in series	lb/in.	N/mm
K_T	Stiffness of a nut-bolt-washer system	lb/in.	N/mm
K_{TG}	Stiffness of a gasketed joint	lb/in.	N/mm
K_{tor}	"Torsional spring constant," defined as $T_{tor} = K_{tor} \times \Delta L$	lb	N

Symbol	Uses	English units	Metric units
K_W	Spring constant of washer	lb/in.	N/mm
L	Length (ΔL = change in length)	in.	mm
L	The anticipated live load on a shear joint	lb	N
$\Delta L'$	Bolt stretch after application of external load	in.	mm
l	Length	in.	mm
L_B	Length of fastener (e.g., body plus threads)	in.	mm
ΔL, ΔL_B	Bolt stretch (same as ΔL_C)	in.	mm
L_{be}	Effective length of body of fastener (includes a portion of the head of the fastener)	in.	mm
L_C	Total length of fastener, including head	in.	mm
ΔL_C	Combined or total change in length of bolt	in.	mm
L_E	Effective length of fastener (total length under load)	in.	mm
L_e	Minimum length of thread engagement required to develop maximum strength	in.	mm
L_e, L_{eff}	Effective length of the bolt	in.	mm
L_{eg}	Distance from bolt hole to tearout edge of joint plate in a shear joint	in.	mm
L_G	Grip length	in.	mm
L_J	Length of joint members involved in thermal expansion	in.	mm
ΔL_J	Increase or decrease in the thickness of a joint because of a temperature change	in.	mm
L_S	External shear load	lb	N
L_X	External tensile or shear load applied to joint. Also the maximum tensile load experienced during a load cycle	lb	N
L_{Xmin}	Minimum tensile load experienced during a load cycle	lb	N
L_{max}	Maximum external load a fastener can support	lb	N
L_{RM}	Mass leak rate	lb/hr·in.	mg/sec·mm

Symbol	Uses	English units	Metric units
L^*_{RM}	Reference mass leak rate	lb/hr·in.	mg/sec·mm
L_{ro}	Length of thread run-out per ANSI or other fastener specifications ($L_T + L_{ro} = L_t$)	in.	mm
L_{se}	Effective length of threads of fastener (exposed threads plus a portion, e.g., half, of the threads within the nut)	in.	mm
L_T	Minimum thread length per ANSI or other fastener specifications	in.	mm
L_t	Total length of threads on fastener	in.	mm
ΔL_W	Change in distance between the washers of a bolt as it is loaded	in.	mm
L_X	External load on joint	lb	N
L_{XC}	External compression load	lb	N
L_{Xcrit}	Critical external load (load required to free the joint completely)	lb	N
M	Reaction moment on an eccentrically loaded shear joint	in.-lb	mm-N
M	Number of slip surfaces in a shear joint	None	
m	Gasket factor; ratio of residual (under load) clamping pressure on a gasket required to prevent a leak, to the contained pressure	None	
M_b	Bending moment acting on fastener	in.-lb	N-mm
MV	Material velocity	in./sec	mm/sec
N	Number of bolts in the joint; also the number of threads per inch	None	
N	Full width of a gasket	in.	mm
N	Number of bolts on one side of a shear joint	None	
N	Number of cycles; also number of bolts in a shear joint	None	
n	Number of tests in a statistical sample	None	
n	Number of threads per inch	None	
n	Fraction of joint thickness which lies between the loading planes; also number of bolts in a joint	None	

Symbol	Uses	English units	Metric units
n_b	Number of shear planes which pass through the body of a bolt	None	
N_R	Number of bolts in the row nearest the edge of a shear joint	None	
n_s	Number of shear planes which pass through bolt threads	None	
P	Percentage, expressed as a decimal (e.g., 65% = 0.65)	None	
P	Number of shear planes in a bolt in a bearing-type shear joint	None	
P	Pressure	psi	Pa
P^*	Atmospheric pressure	psi	Pa
P, P_i, p	Pitch of threads	in.	mm
PE_B	Potential energy stored in the bolt	in.-lb	mm-N
P_G	Maximum allowable pressure which can be exerted on the joint by the nut or bolt head	psi	MPa
P_p	Primary shear load on the fasteners in an eccentrically loaded shear joint	lb	N
Q	Prying force	lb	N
R	Radius of curvature of bent bolt	in.	mm
R	Average strength (resistance) of a shear joint	lb	N
R_B	Combined shear strength of a shear joint of all of the bolts in a shear joint	lb	N
R_B	Total strength (in shear) of bearing-type shear joint	lb	N
R_G	Radius of gyration of cross-sectional area of joint	in.	mm
R_S	Slip resistance of shear joint	lb	N
r_B	Resilience of bolt (reciprocal of bolt stiffness K_B)	in./lb	mm/N
r_J	Resilience of concentric joint (reciprocal of joint stiffness K_J)	in./lb	mm/N
r_J'	Resilience of eccentric joint in which external load and bolt are coaxial	in./lb	mm/N
r_J''	Resilience of eccentric joint in which external load and bolt lie along different but parallel axes	in./lb	mm/N

Symbol	Uses	English units	Metric units
r_n	Effective radius of contact between nut and joint or washer	in.	mm
r_t	Effective radius of contact between nut and bolt threads	in.	mm
R_T	Tensile strength of the joint members in a shear joint	lb	N
R_{TO}	Tearout strength of a shear joint	lb	N
r_s	Resilience of the screw (i.e. bolt)	in./lb	mm/N
S	Sample standard deviation	Any	Any
S	Stress		
S_1; $S_t'max$	Maximum principal tensile stress, under combined tension and torsion, at root diameter of bolt thread	psi	Pa
S_2; $S_t'min$	Minimum principal tensile stress, under combined tension and torsion, at root diameter of bolt thread	psi	Pa
s	Scatter in preload caused by assembly tools and/or procedures	None	
s	Distance between centerline of eccentric joint and centerline of bolts	in.	mm
S_a	Allowable stress at room temperature (per ASME Code)	psi	MPa
S_B	Maximum fiber stress created by pure bending	psi	Pa
S_b	Allowable stress at service temperature (per ASME Code)	psi	MPa
SCC	Stress corrosion cracking	None	
SF	Stress factor	None	
S_H	Contained pressure	psi	MPa
S_{m1}, S_{m2}	Design in-service stresses for a gasketed joint	psi	MPa
S_{mo}	Gasket design stress: the largest of three possibilities; $2P$, S_{m1}, or S_{m2}	psi	MPa
S_{ns}	Unit shear strength of nut material	psi	Pa
S_s	Direct torsional stress for combined tensile and shear	psi	Pa

Symbol	Uses	English units	Metric units
S_{ss}	Unit shear strength of bolt material	psi	Pa
S_S'	Maximum shear stress at root diameter of bolt thread under combined tension and torsion	psi	Pa
S_{st}	Unit tensile strength of bolt material	psi	Pa
S_t	Direct tensile stress component of combined tensile and shear stresses (shear resulting from torsion stress)	psi	Pa
S_U	Ultimate shear strength	psi	Pa
S_y	Yield strength	psi	Pa
S_{YP}	Value of tensile stress component (S_t) of total stress when bolt yields under combined tension and torsion	psi	Pa
S_T	Ratio of shear stress in shear plane of bolt to its ultimate tensile strength	None	
S_U	Ultimate shear strength	psi	GPa
S_{ya}	Design seating stress for a gasketed joint	psi	MPa
T	Time (ΔT = change in time or transit time)	sec	sec
T	Thickness of joint	in.	mm
T_1, T_2	Thickness of joint members 1 and 2	in.	mm
ΔT	Compression of joint members before application of an external load	in.	mm
$\Delta T'$	Compression of joint members after application of external load	in.	mm
Δt	Change in temperature	°F	°C
nT	Distance between loading planes in a joint of thickness T	in.	mm
T_1, T_{tor}	Wrench torque which gets converted to frictionally generated torsion from nut to bolt	lb-in.	N-mm
Δt	Change in temperature	°F	°C
ΔT_ϵ	Change in transit time created by bolt strain	sec	sec

Symbol	Uses	English units	Metric units
TF	Temperature factor	per °F	per °C
T_H	Height of head of bolt	in.	mm
T_{in}	Torque applied to nut	lb-in.	N-mm
T_J	Thickness of joint	in.	mm
T_N	Thickness of nut	in.	mm
T_{os}	Threshold or offset torque used in LRM tool control	lb-in.	N-mm
T_P	Prevailing torque	lb-in.	N-mm
TP	Tightness parameter	None	
T_{pmin}	Minimum acceptable tightness parameter for a gasketed joint	None	
T_{pn}	Minimum acceptable tightness parameter for a gasketed joint taking hydrotest pressure and service temperature into account	None	
T_r	Tightness ratio (Log T_{pn}/Log T_{pmin})	None	
T_T	Ratio between tensile stress in bolt and its ultimate tensile strength	None	
T_{tf}	Torque required to overcome thread friction	lb-in.	N-mm
u	Distance from centerline of eccentric joint to edge nearest the point of application of an external load	in.	mm
v	Velocity of ultrasonic wave in bolt, in general	in./sec	cm/sec
v_0	Velocity of ultrasonic wave in unstressed bolt	in./sec	cm/sec
v_t	Velocity of ultrasonic wave in stressed bolt	in./sec	cm/sec
W	Symbol used for bolt tension ASME Code	lb	N
W_{bn}	Work done in bending bolt	ft-lb, in.-lb	N-M, N-mm
W_{in}	Input work done in tightening nut	ft-lb, in.-lb	N-M, N-mm
W_{jc}	Work done in compressing joint	ft-lb, in.-lb	N-M, N-mm
W_{M1}	Maintenance bolt load (ASME Code)	lb	N
W_{M2}	Gasket seating bolt load (ASME Code)	lb	N
W_{nc}	Work done in compressing nut	ft-lb, in.-lb	N-M, N-mm

Symbol	Uses	English units	Metric units
W_{nf}	Work done against nut to joint friction	ft-lb, in.-lb	N-M, N-mm
W_{ten}	Work done in stretching bolt	ft-lb	N-M
W_{tf}	Work done against thread friction	ft-lb, in.-lb	N-M, N-mm
W_{tor}	Work done in twisting bolt	ft-lb, in.-lb	N-M, N-mm
W_{mo}	Design bolt load	lb	N
y	Recommended initial seating stress on a gasket (ASME Code)	psi	Pa
Z_p	Polar section modulus of bolt	in.^{3}	mm^3
α	Y intercept of linear regression line	Any	Any
α_A	Scatter in preload caused by assembly tools and/or procedures as defined by VDI	None	
β	Slope of linear regression line	Any	Any (always a ratio)
β	Half-angle of thread root (30° for UN or ISO metric threads)	deg	deg
Δ	"Change in," e.g., ΔL is "change in length"	None	
ϵ	Strain	Dimensionless	
θ	Turn of nut	deg	rad
$\Delta\theta$	Turn of nut created by a change in preload (ΔF_P)	deg	rad
θ_0	Starting turn used in LRM tool control	rad	rad
θ_G	Relative angle of turn between nut and "ground"	deg or rad	deg or rad
θ_{in}	Input turn of nut	deg or rad	deg or rad
θ_R	Relative angle of turn between nut and bolt threads	deg or rad	deg or rad
θ_{tw}	Angle of twist of bolt under torsion	rad	rad
μ	Coefficient of friction	None	
μ_n	Coefficient of friction between nut and joint surfaces or washer	None	
μ_S	Slip coefficient (of friction) of a shear joint	None	
μ_t	Coefficient of friction between male and female thread surfaces	None	

Symbol	Uses	English units	Metric units
μ, λ, l, m	Second- and third-order elastic constants		
π	3.14159	None	
ρ	Density	slugs/in.^{3}	g/mm^3
ρ	Linear coefficient of expansion	in./in./ΔF	mm/mm/$^\circ$C
ρ_1	Coefficient of thermal expansion of the bolt material	in./in./$^\circ$F	mm/mm/$^\circ$C
ρ_2	Coefficient of thermal expansion of the joint material	in./in./$^\circ$F	mm/mm/$^\circ$C
σ	Stress	lb/in.^{2}	Pa
σ	Standard deviation of a population	Any	Any
σ_A	Fatigue endurance limit stress in the bolt	psi	MPa
σ_{av}	Average stress level in threaded section of fastener	lb/in.^{2}, psi	N/mm^2
σ_B	Maximum fiber stress caused by bending bolt	psi	Pa
σ_B	Bearing stress	psi	MPa
σ_P	Proof strength of fastener expressed as stress	psi	Pa
σ_S	Ultimate shear strength of bolt	psi	N/mm^2
σ_{SA}	Change in bolt stress caused by an external load	psi	MPa
σ_{SAb}	Change in the outer fibre of a bolt caused by bending the bolt	psi	MPa
σ_{ult}	Ultimate tensile strength	psi	MPa
σ_y	Stress required to yield the bolt	psi	MPa
Φ	Resistance factor reflecting uncertainty in strength of a shear joint	None	
Φ_e, Φ_{ek}	Load factor for an eccentrically loaded tensile joint; loaded at the joint surfaces	None	
Φ_{en}	Load factor for an eccentrically loaded tensile joint: loaded internally along loading planes	None	
Φ_k	Load factor for concentrically loaded tensile joint; loaded at the joint surfaces	None	
Φ_{kn}, Φ_n	Load factor for concentrically loaded tensile joint: loaded internally along loading planes	None	

B

Glossary of Fastener and Bolted Joint Terms

Accuracy   See *Preload accuracy*.

AISC   The American Institute of Steel Construction.

Allowable stress   The maximum stress a designer can assume that the parts will stand. Is always less than the minimum strength of the material. For example, the ASME Boiler and Pressure Vessel Code typically specifies an allowable stress that is one-quarter of the service temperature yield strength of the material. This introduces a four-to-one safety factor into the design process and is intended to compensate for uncertainties in estimates of strength, service loads, etc.

Allowable stress design   A design procedure developed for the AISC by the Research Council on Structural Connections. Purposely underestimates the strengths of bolts and joint materials to introduce safety factors into the design of structural steel joints. Is an alternative to the more recently defined load and resistance factor design procedure.

Angularity   The underfaces of the nut and the bolt head should be exactly perpendicular to the thread or shank axes. If the angle between the face and the axis is, for example, 86° or 94°, the fastener is said to have an angularity of 4° (sometimes called *Perpendicularity*).

Anode   That electrode in a battery or corrosion cell which produces electrons. It is the electrode which is destroyed (corrodes).

Area stress or tensile stress   See *Stress area*.

ASME   The American Society of Mechanical Engineers.

ASME Code   See *Boiler and Pressure Vessel Code*.

Barrier protection   The coating on a fastener is said to provide barrier protection if it merely isolates the fastener from the environment. Paint, for example, provides barrier protection.

Body See Figure B.1.

Boiler and Pressure Vessel Code A large and complex document, maintained and published by the American Society of Mechanical Engineers (ASME—see Appendix C for address). The code describes design rules, material properties, inspection techniques, fabrication techniques, etc., for boilers and pressure vessels. The recommendations of the Code have been adopted by most states, and have influenced similar Codes in many countries.

Bolt Officially, a threaded fastener designed to be used with a nut. In this book, the word is often used interchangeably with *Threaded fastener* for convenience.

Bolt gage An ultrasonic instrument, manufactured by Bidwell Industrial Group (see Appendix D for address) and used to measure the stress or strain in bolts.

Bolt, parts of See Fig. B.1.

Brittle A bolt is said to be brittle if it will break when stretched only a small amount past its yield point. (Compare *Ductile*.)

Cathode That electrode in a battery or corrosion cell which attracts electrons.

Clamping force The equal and opposite forces which exist at the interface between two joint members. The clamping force is created by tightening the bolts, but is not always equal to the combined tension in the bolts. Hole interference problems, for example, can create a difference between clamping force and bolt loads.

Constant life diagram A plot of experimentally derived fatigue-life data; perhaps the most complex and complete of the popular charts used to represent such data. See Fig. 17.6 for an example.

Code The Boiler and Pressure Vessel Code published by the American Society of Mechanical Engineers (ASME).

Corrosion cell A natural "battery" formed when two metals having different electrical potentials (an *Anode* and a *Cathode*) are connected together in the presence of a liquid (the *Electrolyte*).

Creep The slow, plastic deformation of a body under heavy loads. Time-dependent plasticity.

Dimensions of bolt See Fig. B.2.

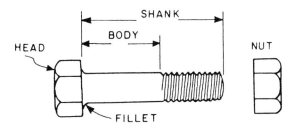

Figure B.1 Parts of a bolt.

LENGTH

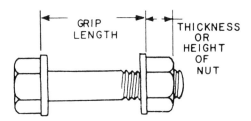

Figure B.2 Bolt dimensions.

DTI Direct tension indicator. A fastener used primarily in the structural steel industry, designed to indicate that a certain minimum amount of tension has been developed in the fastener during assembly. See Figs. 10.1 and 10.3 for examples.

Ductile If a bolt can be stretched well past its yield point before breaking, it is said to be *ductile*. (See also *Brittle*.)

Eccentric load The *external load* on a fastener or groups of fasteners is said to be eccentric if the resultant of that load does not pass through the centroid of the group of fasteners (*eccentric shear load*) or does not coincide with the bolt axis (*eccentric tensile load*).

Effective length of a bolt The *grip length* plus some portion of the bolt (often one-half of the thickness of the nuts) which lies within the nut(s) plus some portion (often one-half the thickness) of the head. Used in stiffness and stretch calculations. (See Fig. 5.3.)

Effective radius of nut, or bolt head, or threads Distance between the geometric center of the part and the circle of points through which the resultant contact forces between mating parts passes. Must be determined by integration.

Elastic interactions When a bolt is tightened it partially compresses the joint members "in its own neighborhood." When nearby bolts are tightened later, they further compress the joint in this region. This allows the first bolt to relax a little (lose a little preload). Tightening bolts on the opposite side of the joint, however, might INCREASE preload in some of the earlier bolts tightened on the near side. These shifts and changes in the elastic energy stored in individual bolts, during assembly, are called elastic interactions. Details can be found in Chap. 6.

Electrode The two metallic bodies in a battery or *Corrosion cell* which give up electrons (the *Anode*) or which attract them (the *Cathode*).

Electrolyte The liquid with which the *Electrodes* of a battery or *Corrosion cell* are wetted.

Embedment Localized plastic deformation in heavily loaded fasteners allows one part to sink into, or smooth the surface of, a softer or more heavily loaded second part. Nuts embed themselves in joint surfaces. Bolt threads embed themselves in nut threads, etc.

Endurance limit That completely reversing stress limit below which a bolt or joint member will have an essentially infinite life under cyclic fatigue loads. Note that the mean stress on the bolts here is zero.

Equation, long form An equation which relates the torque applied to a bolt to the preload created in it, and involves fastener geometry and the coefficient of friction between mating surfaces. A theoretical equation based on rigid body mechanics and the assumption that the geometry of the fastener is perfectly described by blueprint dimensions. [See Eqs. (7.2) and (7.3).]

Equation, short form An empirical equation which relates the torque applied to the bolt to the preload created in it, and which depends mainly on an experimentally derived factor called the *Nut factor*. [See Eq. (7.4).]

Essential conditions Each type of failure to which bolted joints are subject is set up by three or four conditions. The conditions vary, depending on the mode of failure, but never number more than four. Eliminating any one of the essential conditions for a particular type of failure can prevent that type of failure. See Chap. 15 for specifics.

Extensometer Any instrument which measures the change in length of a part as the part is loaded.

External load Forces exerted on fastener and/or joint members by such external factors as weight, wind, inertia, vibration, temperature expansion, pressure, etc. Does *not* equal the *Working load* in the fastener.

Failure of the bolt Term implying that the bolt has broken or the threads have stripped. There can be many reasons for this.

Failure of the joint Failure of a bolted joint to behave as intended by the designer. Failure can be caused or accompanied by broken or lost bolts, but can also mean joint slip or leakage from a gasketed joint even if all bolts still remain whole and in place. Common reasons for joint failure include vibration loosening, poor assembly practices, improper design, unexpected service loads or conditions, etc.

Fastener dimensions See Figure B.2, p. 890.

Fillet Transition region between bolt head and shank, or between other changes in diameter. (See Figure B.1, p. 889.)

Flange rotation Angular distortion of a flange under the influence of bolt and reaction forces. Measured with respect to the center of the cross section of the flange. See Fig. B.3.

Galling An extreme form of adhesive wear, in which large chunks of one part stick to the mating part (during sliding contact).

Galvanic protection The coating on a fastener is said to provide galvanic protection if it is more anodic than the fastener and will, therefore, be destroyed instead of the fastener. Zinc plate (galvanizing) provides galvanic protection to steel fasteners, for example.

Gasket factors Experimentally derived "constants" used to define the behavior of a gasket and/or the assembly and in-service conditions required for accept-

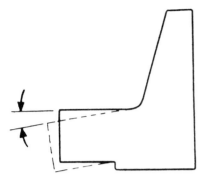

Figure B.3 Flange rotation.

able behavior. The term "gasket factor" comes from the *Boiler and Pressure Vessel Code*, which contains a tabulation of *m* and *y* factors defining the recommended *Gasket stress* in-service and at assembly—for design purposes only. (Actual assembly and in-service stresses will usually be greater, as explained in Chap. 19.) New factors, called G_b, *a*, and G_s, have recently been proposed for the Code. These factors are not design recommendations, but instead, define the behavior of the gasket.

Gasket stress   The contact stress exerted on the gasket by the joint members.

Grip length   Combined thickness of all the things clamped together by the bolt and nut, including washers, gaskets, and joint members. (See *Dimensions of bolt.*)

Head of bolt   See Figure B.1, p. 889.

Height of head or nut   See *Thickness of head or nut* under *Dimensions of bolt*.

Hydrogen embrittlement   A common and troublesome form of *Stress cracking*. Several theories have been proposed to explain hydrogen embrittlement, but, at present, the exact mechanism is still unknown. What is known, however, is the fact that if hydrogen is trapped in a bolt by poor electroplating practices, it can encourage stress cracking. Bolts can fail, suddenly and unexpectedly, under normal loads. See Chap. 18 for a more complete discussion.

Impact wrench   An air- or electric-powered wrench in which multiple blows from tiny hammers are used to produce output torque to tighten fasteners.

Inclusions   Small pieces of nonmetallic impurities trapped within the base metal of, for example, a bolt.

Infinite life diagram   A simple plot experimentally derived fatigue-life data, showing the conditions required for infinite life. See Fig. 17.10 for an example.

Initial preload   The tension created in a single bolt as it is tightened. Will usually be modified by subsequent assembly operations (see *Elastic interactions*) and/ or by in-service loads and conditions.

Joint diagrams   Mathematical diagrams which illustrate the forces on and deflections of fasteners and joint members. (See Chaps. 12 and 13.)

Junker machine   A test machine, first proposed by Gerhard Junker, for testing the vibration resistance of fasteners. See Fig. 16.11.

Length, effective   See *Effective length of a bolt*.

Length of bolt   See Figure B.?, p. 890.

Load and resistance factor design   A design procedure developed for the AISC by the Research Council on Structural Connections. Assigns uncertainties in the strength of (i.e., resistance of) and in the service loads to be placed on a shear joint to estimate the probable strength of the joint. Is a recently defined alternative to the *Allowable stress design* procedure.

Load factor (Φ)   The ratio between an increase in bolt tension and the external load which has caused the increase (i.e., $\Phi = \Delta F_p/L_X$).

Load factors (α, γ)   Factors reflecting the probability of an increase in load in a shear joint. Used in load and resistant factor design.

Lockbolt   A fastener which bears a superficial resemblance to a bolt, but which engages a collar (instead of a nut) with annular grooves (instead of threads). The collar is swaged over the grooves on the male fastener to develop preload. See Fig. 10.5.

Lock nut   A nut which provides extra resistance to vibration loosening (beyond that produced by proper *Preload*), either by providing some form of *Prevailing torque*, or, in free-spinning lock nuts, by deforming, cramping, and/or biting into mating parts when fully tightened.

Material velocity   The velocity of sound in a body (e.g., a bolt). A term used in the ultrasonic measurement of bolt stress or strain.

Mean value   The average value of a number of data points. Computed by dividing the sum of all data by the number of data points.

Monitor, torque   See *Torque monitor*.

Nominal diameter   The "catalog diameter" of a fastener. Usually roughly equal to the diameter of the body, or the outer diameter of the threads.

Nonlinear behavior   A fastener or joint system is said to exhibit nonlinear behavior when the relationship between the *External load* on the joint and deformation of the parts is nonlinear, or when the relationship between increasing *Preload* and deformation is nonlinear. (See Chap. 13.)

Nut factor   An experimental constant used to evaluate or describe the ratio between the torque applied to a fastener and the *Preload* achieved as a result. [See Eq. (7.4).]

Perpendicularity   See *Angularity*.

Pitch   The nominal distance between two adjacent thread roots or crests. (See *Thread nomenclature*.)

Preload   The tension created in a threaded fastener when the nut is first tightened. Often used interchangeably, but incorrectly, with *Working load* or bolt force or bolt tension. (See also *Clamping force*.)

Preload accuracy   A measure of the precision with which a given tool or procedure creates preload in a bolt when the bolt is tightened. A common torque wrench, for example, is said to produce preload with an accuracy of ±30%. The mean preload, however, may not be that which the designer intended, or may not be what he *should* have intended. Accuracy as used here, in other words, is synonymous with *Scatter*.

Preload, initial See *Initial Preload*.

Preload, residual See *Residual preload*.

Prevailing torque Torque required to run a nut down against the joint when some obstruction, such as a plastic insert in the threads, or a noncircular thread, or other, has been introduced to help the fastener resist vibration loosening. Prevailing torque, unlike "normal" torque on a nut or bolt, is *not* proportional to the *Preload* in the fastener.

Proof load The maximum, safe, static, tensile load which can be placed on a fastener without yielding it. Sometimes given as a force (in lb or N) sometimes as a stress (in psi or MPa).

Prying The magnification of an *External load* by a pseudolever action when that load is an *Eccentric tensile load*.

Radius, effective See *Effective radius*.

Raised-face flange A flange which contacts its mating joint member only in the region in which the gasket is located. The flanges do *not* contact each other at the bolt circle. Figure B.3 shows a raised-face flange.

Relaxation The loss of tension, and therefore *Clamping force*, in a bolt and joint as a result of *Embedment*, vibration loosening, gasket creep, differential thermal expansion, etc.

Residual preload The tension which remains in an unloaded bolted joint after *Relaxation*.

Resistance factor Probabilistic factor representing the uncertainties in the designer's estimate of the strength of a shear joint. Used in *Load and resistance factor design*.

Rolled thread A thread formed by plastically deforming the surface of the blank rather than by cutting operations. Increases fatigue life and thread strength, but is not possible (or perhaps economical) on larger sizes.

Rotation of flange See *Flange rotation*.

Sacrificial coating See *Galvanic protection*.

Scatter Data points or calculations are said to be scattered when they are not all the same. A "lot of scatter in preload" means wide variation in the preloads found in individual bolts.

Screw Threaded fastener designed to be used in a tapped or untapped (e.g., wood screw) hole, but not with a nut.

Self-loosening The process by which a supposedly tightened fastener becomes loose, as a result of vibration, thermal cycles, shock, or anything else which cause transverse slip between joint members and/or male and female threads. *Vibration loosening* is a common, but special, case of self-loosening.

Shank That portion of a bolt which lies under the head. (See *Bolts, parts of*.)

Shear joint A joint which is subjected primarily to loads acting more or less perpendicular to the axes of the bolts.

Slug wrench A box wrench with an anvil on the end of the handle. Torque is produced by striking the anvil with a sledge hammer. Called a flogging wrench in England.

Sonic velocity   See *Material velocity.*

Spherical washer   A washer whose upper surface is semispherical. Used with a nut whose contact face is also semispherical. Reduces bending stress in a bolt or stud, by allowing some self-alignment and/or some compensation for nonparallel joint surfaces or *Angularity.*

Spring constant   The ratio between the forces exerted on a spring (or a bolt) and the deflection thereof. Has the dimensions of force per unit change in length (e.g., lb/in.). Also called *Stiffness.*

Standard deviation   A statistical term used to quantify the *Scatter* in a set of data points. If the standard deviation is small, most of the data points are "nearly equal." A large deviation means less agreement.

Stiffness   See *Spring constant.*

Strength of bolt   An ambiguous term which can mean *Ultimate strength* or *Proof load* or *Endurance limit* or *Yield strength.*

Stress area   The effective cross-sectional area of the threaded section of a fastener. Used to compute average stress levels in that section. Based on the mean of pitch and minor diameters. (See *Thread nomenclature.*)

Stress corrosion cracking (SCC)   A common form of *Stress cracking* in which an *Electrolyte* encourages the growth of a crack in a highly stressed bolt. Only a tiny quantity of electrolyte need be present, at the tip or face of the crack.

Stress cracking   A family of failure modes, each of which involves high stress and chemical action. The family includes *Hydrogen embrittlement*, *Stress corrosion cracking*, stress embrittlement, and hydrogen-assisted stress corrosion. See Chap. 18 for details.

Stress factor   A calibration constant used in ultrasonic measurement of bolt stress or strain. It is the ratio between the change in ultrasonic transit time caused by the change in length of the fastener, under load, to the total change in transit time (which is also affected by a change in the stress level).

Stress relaxation   The slow decrease in stress level within a part (e.g., a bolt) which is heavily loaded under constant deflection conditions. A "cousin" to creep, which is a slow change in geometry under constant stress conditions.

Stud   A headless threaded fastener, threaded on both ends, with an unthreaded body in the middle section, or threaded from end to end. Used with two nuts, or with one nut and a tapped hole.

Temperature factor   A calibration constant used in ultrasonic measurement of bolt stress or strain. Accounts for the effects of thermal expansion and the temperature-induced change in the velocity of sound.

Tensile strength   See *Ultimate strength.*

Tensile stress area   See *Stress area.*

Tension, bolt   Tension (tensile stress) created in the bolt by assembly preloads and/or such things as thermal expansion, service loads, etc.

Tension joint   A joint which is primarily subjected to loads acting more or less parallel to the axes of the bolts.

Tensioner   A hydraulic tool used to tighten a fastener by stretching it rather than by applying a substantial torque to the nut. After the tensioner has stretched the bolt or stud, the nut is run down against the joint with a modest torque, and the tensioner is disengaged from the fastener. The nut holds the stretch produced by the tensioner.

Thickness of nut or of bolt head See *Dimensions of bolt* (also called *Height*).

Thread form The cross-sectional shape of the threads, defining thread angle, root, and crest profiles, etc.

Thread length Length of that portion of the fastener which contains threads cut or rolled to full depth. (See Figure B.2, p. 890.)

Thread nomenclature See Fig. B.4.

Thread run-out That portion of the threads which are not cut or rolled full depth, but which provide the transition between full-depth threads and the body or head. (See Figure B.1, p. 889.) Officially called thread washout or vanish, although the term run-out is more popular. (Run-out is officially reserved for rotational eccentricity, as defined by total indicator readings or the like.)

Threaded fastener Studs, bolts, and screws of all sorts, with associated nuts. One of the most interesting, complex, useful—and frustrating—components yet devised.

Tightness A measure of the mass leak rate from a gasketed joint.

Tightness, acceptable Wholly leak-free joints are impossible, at least if the contained fluid is a gas, so it has been proposed that the design of a gasketed joint should start with the selection of an "acceptable" leak rate. The designer would dimension bolts and joint members so that the actual leak rate would never exceed this. Three standard levels of tightness have been proposed as well. See Chap. 19.

Tightness parameter A dimensionless parameter which defines the mass leakage of a gasket as a function of contained pressure and a contained fluid constant. See Eq. (19.2).

Torque The twisting moment, product of force and wrench length, applied to a nut or bolt (for example).

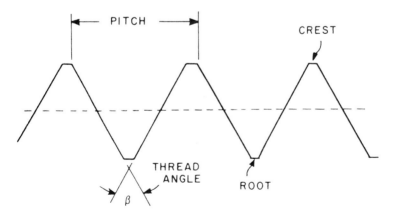

Figure B.4 Thread nomenclature.

Torque monitor A torque tool control system which monitors the amount of torque being developed by the tool during use, but does not control the tool or the torque produced.

Torque multiplier A gearbox used to multiply the torque produced by a small hand wrench (usually a *Torque wrench*). The output of the multiplier drives the nut or bolt with a torque that is higher, and a speed that is lower, than input torque and speed. There is no torque gage or readout on the multiplier.

Torque pack A geared wrench which multiplies input torque and provides a read-out of output torque. In effect, a combination of a *Torque wrench* and a *Torque multiplier.*

Torque wrench A manual wrench which incorporates a gage or measuring apparatus of some sort to measure and display the amount of torque being delivered to the nut or bolt. All wrenches produce torque. Only a torque wrench tells how much torque.

Transducer A device which converts one form of energy into another. An ultrasonic transducer, for example, converts electrical energy into acoustic energy (at ultrasonic frequencies) and vice versa.

Turn-of-nut Sometimes used to describe the general rotation of the nut (or bolt head) as the fastener is tightened. More often used to define a particular tightening procedure in which a fastener is first tightened with a preselected torque, and is then tightened further by giving the nut an additional, measured, turn such as "three flats" (180°).

Ultimate strength The maximum tensile strength a bolt or material can support prior to rupture. Always found in the plastic region of the stress-strain or force-elongation curve, and so is not a design strength. Also called *Tensile strength* and ultimate tensile strength.

Ultrasonic extensometer An electronic instrument which measures the change in length of a fastener ultrasonically as, or before and after, the fastener is tightened. (See also *Extensometer.*)

Washer, tension indicating See *DTI.*

Width across flats A principal dimension of nuts, or of bolt heads. (See Fig. B.5.)

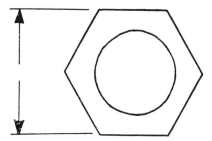

Figure B.5 Width across flats.

Work hardening   The slight increase in hardness and strength produced when a body is loaded past its yield point. Also called strain hardening.

Working load   The tension in a bolt in use; tension produced by a combination of *Residual preload* and a portion (usually) of any *External load*. The *Joint diagram* is usually used to predict the approximate working load a fastener will see in service.

Yield strength   That stress level which will create a permanent deformation of 0.2% or 0.5% or some other small, preselected, amount in a body. Approximately equal to the elastic and proportional limits of the material; a little higher than the proof strength of a bolt. (See *Proof load*.)

C
Sources of Bolting Specifications

AISC   American Institute of Steel Construction, Inc.
400 North Michigan Avenue
Chicago, Illinois 60611.

AISI   American Iron and Steel Institute
1000 16th Street N.W.
Washington, D.C. 20036

AMS   Obtain Aeronautical Material Specifications from:
Society of Automotive Engineers
400 Commonwealth Drive
Warrendale, Pennsylvania 15096

ANSI   American National Standards Institute
1430 Broadway
New York, New York 10018 (or from the ASME)

ASME   American Society of Mechanical Engineers
United Engineering Center
345 East 47th Street
New York, New York 10017

ASTM   American Society for Testing and Materials
1916 Race Street
Philadelphia, Pennsylvania 19103

BS   British Standards Institution
Sales Branch
Newton House
101/113 Pentonville Rd.
London N1, England

BTC   Bolting Technology Council
        c/o Materials Properties Council
        345 East 47th Street
        New York, New York 10017

FF   Obtain Federal Specifications from:
QQ   Commanding Officer
GGG   Naval Publications and Forms Center
        5801 Tabor Avenue
        Philadelphia, Pennsylvania 19120

IFI   Industrial Fastener Institute
        1505 East Ohio Building
        Cleveland, Ohio 44114

ISO   American National Standards Institute
        1430 Broadway
        New York, New York 10018

MIL   Obtain Military Specifications from:
        Commanding Officer
        Naval Publications and Forms Center
        5801 Tabor Avenue
        Philadelphia, Pennsylvania 19120

NAS   National Standards Association
        1321 Fourteenth Street N.W.
        Washington, D.C. 20005

PVRC   Pressure Vessel Research Committee
        c/o Welding Research Council
        345 East 47th Street
        New York, New York 10017

SAE   Society of Automotive Engineers
        400 Commonwealth Drive
        Warrendale, Pennsylvania 15096

WRC   Welding Research Council
        345 East 47th Street
        New York, New York 10017

D
Sources of Bolting Tools and Equipment

The following is not a complete list, nor is it intended to favor these sources over others.

Company	Products
Advance Hydraulics, Inc. Turnpike Industrial Park Westfield, Massachusetts 01085	Hydraulic wrenches
AKO, Inc. 110 Broad Brook Road Enfield, Connecticut 06082 P.O. Box 998 Hartford, Connecticut 06143	Calibration equipment for torque wrenches, multipliers and other bolting tools.
Allen Bradley 624 Alpha Drive Cleveland, Ohio 44143	Single and multispindle air tools. Rockwell's LRM fastening systems.
Atlas-Copco Tools AB S-104 60 Stockholm, Sweden	Air-powered tools Tighten-to-yield systems Torque and/or turn control Single or multispindle air tools
Biach Industries P.O. Box 337 Cranford, New Jersey 07016	Hydraulic tensioners

Company	Products
Bidwell Industrial Group Industrial Tool Division 2055 S. Main St. Middletown, Connecticut 06457	Geared, mechanical wrenches Hydraulic wrenches Hydraulic tensioners Ultrasonic bolt gage
Chicago-Pneumatic Tool Division Utica, New York 13503	Impact wrenches Nut runners Multispindle air tools
Flexitallic Gasket Co. P.O. Box 699 Deer Park, Texas 77536	Hydraulic wrenches Hydraulic tensioners Bolting services
GSE, Inc. 23640 Research Drive Farmington Hills, Michigan 48024	Torque and tension monitors Electronic readout, manual torque wrenches
Ingersoll-Rand Automatic Production Systems Division 23400 Halstead Road Farmington Hills, Michigan 48024	Impact wrenches Nut runners Torque controls Torque-turn controls Torque-time controls Fastener-tensioning systems (tighten- to-yield) Multispindle air tools
Mountz, Inc. 1080 North 11th Street San Jose, California 95112	Manual torque tools Power tools Torque multipliers
Norbar Torque Tools, Ltd. Swan Close Banbury, Oxfordshire, England	Multipliers Torque readout systems Torque wrenches Air-powered multipliers
Skidmore-Wilhelm Manufacturing Co., Inc. 442 South Green Rd. Cleveland, Ohio 44121	Torque wrench calibrators
Snap-On Tools Corp. Kenosha, Wisconsin 53140	Manual torque tools Sockets Power tools Torque multipliers
SPS Technologies Highland Avenue	Single or multispindle air tools Torque wrenches

Company	Products
Sweeney Division Sargent Industries 6300 Stapleton South Drive Denver, Colorado 80216	Hydraulic wrenches Torque multipliers
Thor Power Tool Co. 175 North State St. Aurora, Illinois 60505	Nut runners Torque, control and monitoring Torque-time controls Tension monitors Pulsed tighten to yield Multispindle air tools Portable and fixed-station control systems
X-4 Corporation 59 Hayward Road West Acton, Massachusetts 01720	Multipliers

E
English and Metric Conversion Factors

Conversions for	To obtain	Multiply number of	By
Length, diameter, etc.	mm	in.	25.4
	in.	mm	0.0394
	m	ft	0.305
	ft	m	3.28
Force	N	kgf	9.81
	kgf	N	0.102
	N	lb	4.448
	lb	N	0.225
	kgf	lb	0.454
	lb	kgf	2.203
	g	lb	454
	lb	g	0.0022
	oz	g	0.0352
	g	oz	28.3
Torque	g-cm	lb-in.	0.000868
	lb-in.	g-cm	1150
	g-cm	oz-in.	0.0139
	oz-in.	g-cm	71.88
	N-m	lb-ft	1.36
	lb-ft	N-m	0.738
	kgf-m	lb-ft	0.138
	lb-ft	kgf-m	7.23

Conversions for	To obtain	Multiply number of	By
Pressure or stress	MPa	kgf/mm^2	9.804
	kgf/mm^2	MPa	0.102
	N/mm^2	MPa	1
	KSI	MPa	0.145
	MPa	KSI	6.895
	KSI	kgf/mm^2	1.42
	kgf/mm^2	KSI	0.704
	MPa	psi	6.895×10^{-3}
	N/m^2	Pa	1
	KSI	N/mm^2	0.145
	N/mm^2	KSI	6.895

F

Tensile Stress Areas for English and Metric Threads with Estimated "Typical" Preloads and Torques for As-Received Steel Fasteners

I. GENERAL INSTRUCTIONS

As suggested in Chap. 7, selecting a torque value for a given fastener can be a complex problem. Tables of recommended torque values, however, are common and popular. They're safe to use in noncritical applications and/or such tasks as selecting a tool of the appropriate size. The following table is based on the assumptions that:

1. The fasteners are commercial grade and made of steel.
2. The nut factor (K) is 0.2; i.e., the fasteners are used in as-received condition and are neither cleaned nor lubricated.
3. The fasteners will be tightened by applying torque to the nut, not to the head.
4. The fasteners are tightened to an average stress (in the threaded section) of 25,000 psi (or its metric equivalent of 172.4 MPa).

II. TORQUES FOR DIFFERENT LUBRICANTS OR STRESS LEVELS

If you're using a lubricant, and/or want an average stress different than 25 ksi, you can compute the new torque value from the equation

$$T = \frac{K}{0.2} \times \frac{\sigma}{25,000} \times T_T$$

where T = corrected torque (in.-lb)
        K = nut factor for your application
        σ = average stress in thread region desired in your application
            (psi only; see below for metric)
        T_T = torque value given in this table (in.-lb)

Example: I want to lubricate a $\frac{1}{4}$–20 screw with moly, and tighten it to 60 KSI stress in the threads. What torque should I use?

The nut factor, K, for the moly I'm using is, typically, 0.137 (see Table 7.1, p. 231 ff.). So my new torque T will be

$$T = \frac{0.137}{0.2} \times \frac{60,000}{25,000} \times 39.75 = 65.3 \text{ in.-lb}$$

A similar equation is used for metric units:

$$T = \frac{K}{0.2} \times \frac{\sigma}{172.4} \times T_T$$

where T = corrected torque (N-m)
        K = nut factor (Table 7.1, p. 231 ff.)
        σ = average thread stress desired (MPa)
        T_T = torque from this table (N-m)

III. PRELOADS AND STRESSES FOR DIFFERENT LUBRICANTS OR TORQUES

In a similar fashion, you can compute the preload achieved with a different lubricant and/or torque as follows:
English or metric:

$$F_P = \frac{0.2}{K} \times \frac{T}{T_T} \times F_{PT}$$

where F_P = corrected preload (kip, kN)
        K = nut factor for your application
        T = torque for your application (in.-lb, N-m)
        T_T = torque from this table (in.-lb, N-m)
       F_{PT} = preload from this table (in.-lb, N-m)

If you wish to compute the new stress produced by this new preload, divide the new preload by the tensile stress area given in the table.

Example: I want to compute the preload produced in a $\frac{1}{4}$–20 screw by a torque of 65.3 in.-lb, if I use moly lube. Then I want to compute the resulting stress level in the fastener.

$$F_P = \frac{0.2}{0.137} \times \frac{65.3}{39.75} \times 0.795 = 1.907 \text{ kip}$$

The new stress will be

$$\sigma = \frac{1.907}{0.0318} = 60 \text{ KSI}$$

which agrees with our first example.

IV. TORQUE UNITS

Note that the torques listed in the table are in inch-pounds, not foot-pounds because the short-term torque equation,

$$T = KDF$$

used to derive the table computes torque in inch-pounds unless the nominal diameter D is measured in feet. I've seen many errors in the use of this equation where people dealing with large bolts (and used to working in foot-pounds) automatically expressed D in inches and then took the answer as foot-pounds. So, to avoid this I've used inch-pounds in the table.

If you wish to convert the torque values I've given to foot-pounds, I think it's reasonable to divide the value in the table by 10, even though division by 12 would be theoretically proper. But tables of "recommended torques" can never do more than *approximate* the right torque for your application, so you'll probably get as "good" an answer dividing by 10 as you would be dividing by 12.

In any event, use tables such as this with caution!

English Tensile Stress Areas

Size	Series	Tensile stress area A_S (in.^{2})	Preload (F_P) at 25 KSI (kip)	Torque to achieve 25 KSI (in.-lb)
0 – 80	UNF	0.00180	0.045	0.54
1 – 64	UNC	0.00263	0.0658	0.96
1 – 72	UNF	0.00278	0.0695	1.01

Size	Series	Tensile stress area A_s (in.^{2})	Preload (Fp) at 25 KSI (kip)	Torque to achieve 25 KSI (in.-lb)
2 - 56	UNC	0.00370	0.0925	1.59
2 - 64	UNF	0.00394	0.0985	1.69
3 - 48	UNC	0.00487	0.122	2.42
3 - 56	UNF	0.00523	0.131	2.59
4 - 40	UNC	0.00604	0.151	3.38
4 - 48	UNF	0.00661	0.165	3.70
5 - 40	UNC	0.00796	0.199	4.98
5 - 44	UNF	0.00830	0.208	5.2
6 - 32	UNC	0.00909	0.227	6.27
6 - 40	UNF	0.01015	0.254	7.01
8 - 32	UNC	0.0140	0.35	11.5
8 - 36	UNF	0.01474	0.369	12.1
10 - 24	UNC	0.0175	0.438	16.6
10 - 32	UNF	0.0200	0.5	19.0
12 - 24	UNC	0.0242	0.605	26.1
12 - 28	UNF	0.0258	0.645	27.9
12 - 32	UNEF	0.0270	0.675	29.2
1/4 - 20	UNC	0.0318	0.795	39.75
1/4 - 28	UNF	0.0364	0.91	45.5
1/4 - 32	UNEF	0.0379	0.948	47.4
5/16 - 18	UNC	0.0524	1.310	82.0
5/16 - 20	UN	0.0547	1.367	85.6
5/16 - 24	UNF	0.0580	1.450	90.8
5/16 - 28	UN	0.0606	1.515	94.8
5/16 - 32	UNEF	0.0625	1.563	97.8
3/8 - 16	UNC	0.0775	1.938	145.4
3/8 - 20	UN	0.0836	2.090	156.8
3/8 - 24	UNF	0.0878	2.195	164.6

Size	Series	Tensile stress area A_S (in.^{2})	Preload (F_P) at 25 KSI (kip)	Torque to achieve 25 KSI (in.-lb)
3/8 - 28	UN	0.0909	2.273	170.5
3/8 - 32	UNEF	0.0932	2.330	174.8
7/16 - 14	UNC	0.1063	2.658	232.8
7/16 - 16	UN	0.1114	2.785	244
7/16 - 20	UNF	0.1187	2.968	260
7/16 - 28	UNEF	0.1274	3.185	279
7/16 - 32	UN	0.1301	3.253	285
1/2 - 13	UNC	0.1419	3.548	354.8
1/2 - 16	UN	0.151	3.775	378
1/2 - 20	UNF	0.1599	3.998	399.8
1/2 - 28	UNEF	0.170	4.250	425
1/2 - 32	UN	0.173	4.325	433
9/16 - 12	UNC	0.182	4.550	512
9/16 - 16	UN	0.198	4.950	557
9/16 - 18	UNF	0.203	5.075	571
9/16 - 20	UN	0.207	5.175	583
9/16 - 24	UNEF	0.214	5.350	602
9/16 - 28	UN	0.219	5.475	616
9/16 - 32	UN	0.222	5.550	625
5/8 - 11	UNC	0.226	5.650	706
5/8 - 12	UN	0.232	5.800	725
5/8 - 16	UN	0.250	6.25	781
5/8 - 18	UNF	0.256	6.400	800
5/8 - 20	UN	0.261	6.525	816
5/8 - 24	UNEF	0.268	6.700	838
5/8 - 28	UN	0.274	6.850	856
5/8 - 32	UN	0.278	6.950	869
13/16 - 12	UN	0.289	7.225	994

Size	Series	Tensile stress area A_s (in.^{2})	Preload (F_P) at 25 KSI (kip)	Torque to achieve 25 KSI (in.-lb)
11/16 - 16	UN	0.308	7.700	1,060
11/16 - 20	UN	0.320	8.000	1,101
11/16 - 24	UNEF	0.329	8.225	1,132
11/16 - 28	UN	0.335	8.375	1,152
11/16 - 32	UN	0.339	8.475	1,166
3/4 - 10	UNC	0.334	8.350	1,253
3/4 - 12	UN	0.351	8.775	1,316
3/4 - 16	UNF	0.373	9.325	1,399
3/4 - 20	UNEF	0.386	9.650	1,448
3/4 - 28	UN	0.402	10.05	1,508
3/4 - 32	UN	0.407	10.18	1,527
13/16 - 12	UN	0.420	10.50	1,707
13/16 - 16	UN	0.444	11.10	1,805
13/16 - 20	UNEF	0.458	11.45	1,862
13/16 - 28	UN	0.475	11.88	1,932
13/16 - 32	UN	0.480	12.00	1,951
7/8 - 9	UNC	0.462	11.55	2,021
7/8 - 12	UN	0.495	12.38	2,167
7/8 - 14	UNF	0.509	12.73	2,228
7/8 - 16	UN	0.521	13.03	2,280
7/8 - 20	UNEF	0.536	13.40	2,345
7/8 - 28	UN	0.554	13.85	2,424
7/8 - 32	UN	0.560	14.00	2,450
15/16 - 12	UN	0.576	14.40	2,701
15/16 - 16	UN	0.604	15.10	2,833
15/16 - 20	UNEF	0.620	15.50	2,908
15/16 - 28	UN	0.640	16.00	3,002
15/16 - 32	UN	0.646	16.15	3,030

Size	Series	Tensile stress area A_S (in.^{2})	Preload (F_p) at 25 KSI (kip)	Torque to achieve 25 KSI (in.-lb)
1 - 8	UNC	0.606	15.15	3,030
1 - 12	UNF	0.663	16.58	3,316
1 - 16	UN	0.693	17.33	3,466
1 - 20	UNEF	0.711	17.78	3,556
1 - 28	UN	0.732	18.30	3,660
1 - 32	UN	0.738	18.45	3,690
1 1/16 - 8	UN	0.695	17.38	3,693
1 1/16 - 12	UN	0.756	18.90	4,016
1 1/16 - 16	UN	0.788	19.70	4,186
1 1/16 - 18	UNEF	0.799	19.98	4,246
1 1/16 - 20	UN	0.807	20.18	4,288
1 1/16 - 28	UN	0.830	20.75	4,409
1 1/8 - 7	UNC	0.763	19.08	4,293
1 1/8 - 8	UN	0.790	19.75	4,444
1 1/8 - 12	UNF	0.856	21.40	4,815
1 1/8 - 16	UN	0.889	22.23	5,002
1 1/8 - 18	UNEF	0.901	22.53	5,069
1 1/8 - 20	UN	0.910	22.75	5,119
1 1/8 - 28	UN	0.933	23.33	5,249
1 3/16 - 8	UN	0.892	22.30	5,296
1 3/16 - 12	UN	0.961	24.03	5,707
1 3/16 - 16	UN	0.997	24.93	5,921
1 3/16 - 18	UNEF	1.009	25.23	5,992
1 3/16 - 20	UN	1.018	25.45	6,044
1 3/16 - 28	UN	1.044	26.10	6,199
1 1/4 - 7	UNC	0.969	24.23	6,058

Size	Series	Tensile stress area A_S (in.^{2})	Preload (Fp) at 25 KSI (kip)	Torque to achieve 25 KSI (in.-lb)
1 1/4 - 8	UN	1.000	25.00	6,250
1 1/4 - 12	UNF	1.073	26.83	6,708
1 1/4 - 16	UN	1.111	27.78	6,945
1 1/4 - 18	UNEF	1.123	28.08	7,020
1 1/4 - 20	UN	1.133	28.33	7,083
1 1/4 - 28	UN	1.160	29.00	7,250
1 5/16 - 8	UN	1.114	27.85	7,311
1 5/16 - 12	UN	1.191	29.78	7,817
1 5/16 - 16	UN	1.230	30.75	8,072
1 5/16 - 18	UNEF	1.244	31.10	8,164
1 5/16 - 20	UN	1.254	31.35	8,229
1 5/16 - 28	UN	1.282	32.05	8,413
1 3/8 - 6	UNC	1.155	28.88	7,942
1 3/8 - 8	UN	1.233	30.83	8,478
1 3/8 - 12	UNF	1.315	32.88	9,042
1 3/8 - 16	UN	1.356	33.90	9,323
1 3/8 - 18	UNEF	1.370	34.25	9,419
1 3/8 - 20	UN	1.382	34.55	9,501
1 3/8 - 28	UN	1.411	35.28	9,702
1 7/16 - 6	UN	1.277	31.93	9,180
1 7/16 - 8	UN	1.360	34.00	9,775
1 7/16 - 12	UN	1.445	36.13	10,387
1 7/16 - 16	UN	1.488	37.20	10,695
1 7/16 - 18	UNEF	1.503	37.58	10,804
1 7/16 - 20	UN	1.51	37.75	10,853
1 7/16 - 28	UN	1.55	38.75	11,141
1 1/2 - 6	UNC	1.405	35.13	10,539

Size	Series	Tensile stress area A_s (in.^{2})	Preload (F_P) at 25 KSI (kip)	Torque to achieve 25 KSI (in.-lb)
1 1/2 - 8	UN	1.492	37.30	11,190
1 1/2 - 12	UNF	1.581	39.53	11,859
1 1/2 - 16	UN	1.63	40.75	12,225
1 1/2 - 18	UNEF	1.64	41.00	12,300
1 1/2 - 20	UN	1.65	41.25	12,375
1 1/2 - 28	UN	1.69	42.25	12,675
1 9/16 - 6	UN	1.54	38.50	12,031
1 9/16 - 8	UN	1.63	40.75	12,734
1 9/16 - 12	UN	1.72	43.00	13,438
1 9/16 - 16	UN	1.77	44.25	13,828
1 9/16 - 18	UNEF	1.79	44.75	13,984
1 9/16 - 20	UN	1.80	45.00	14,063
1 5/8 - 6	UN	1.68	42.00	13,650
1 5/8 - 8	UN	1.78	44.50	14,463
1 5/8 - 12	UN	1.87	46.75	15,194
1 5/8 - 16	UN	1.92	48.00	15,600
1 5/8 - 18	UNEF	1.94	48.50	15,763
1 5/8 - 20	UN	1.95	48.75	15,844
1 11/16 - 6	UN	1.83	45.75	15,441
1 11/16 - 8	UN	1.93	48.25	16,284
1 11/16 - 12	UN	2.03	50.75	17,128
1 11/16 - 16	UN	2.08	52.00	17,550
1 11/16 - 18	UNEF	2.10	52.50	17,719
1 11/16 - 20	UN	2.11	52.75	17,803
1 3/4 - 5	UNC	1.90	47.50	16,625
1 3/4 - 6	UN	1.98	49.50	17,325

Size	Series	Tensile stress area A_g (in.^{2})	Preload (Fp) at 25 KSI (kip)	Torque to achieve 25 KSI (in.-lb)
1 3/4 - 8	UN	2.08	52.00	18,200
1 3/4 - 12	UN	2.19	54.75	19,163
1 3/4 - 16	UN	2.24	56.00	19,600
1 3/4 - 20	UN	2.27	56.75	19,863
1 13/16 - 6	UN	2.14	53.50	19,394
1 13/16 - 8	UN	2.25	56.25	20,391
1 13/16 - 12	UN	2.35	58.75	21,297
1 13/16 - 16	UN	2.41	60.25	21,841
1 13/16 - 20	UN	2.44	61.00	22,113
1 7/8 - 6	UN	2.30	57.50	21,563
1 7/8 - 8	UN	2.41	60.25	22,594
1 7/8 - 12	UN	2.53	63.25	23,719
1 7/8 - 16	UN	2.58	64.50	24,188
1 7/8 - 20	UN	2.62	65.50	24,563
1 15/16 - 6	UN	2.47	61.75	23,928
1 15/16 - 8	UN	2.59	64.75	25,091
1 15/16 - 12	UN	2.71	67.75	26,253
1 15/16 - 16	UN	2.77	69.25	26,834
1 15/16 - 20	UN	2.80	70.00	27,125
2 - 4 1/2	UNC	2.50	62.50	25,000
2 - 6	UN	2.65	66.25	26,500
2 - 8	UN	2.77	69.25	27,700
2 - 12	UN	2.89	72.25	28,900
2 - 16	UN	2.95	73.75	29,500
2 - 20	UN	2.99	74.75	29,900
2 1/8 - 6	UN	3.03	75.75	32,194

Size	Series	Tensile stress area A_S (in.^{2})	Preload (F_p) at 25 KSI (kip)	Torque to achieve 25 KSI (in.-lb)
2 1/8 - 8	UN	3.15	78.75	33,469
2 1/8 - 12	UN	3.28	82.00	34,850
2 1/8 - 16	UN	3.35	83.75	35,594
2 1/8 - 20	UN	3.39	84.75	36,019
2 1/4 - 4 1/2	UNC	3.25	81.25	36,563
2 1/4 - 6	UN	3.42	85.50	38,475
2 1/4 - 8	UN	3.56	89.00	40,050
2 1/4 - 12	UN	3.69	92.25	41,513
2 1/4 - 16	UN	3.76	94.00	42,300
2 1/4 - 20	UN	3.81	95.25	42,863
2 3/8 - 6	UN	3.85	96.25	45,719
2 3/8 - 8	UN	3.99	99.75	47,381
2 3/8 - 12	UN	4.13	103.3	49,068
2 3/8 - 16	UN	4.21	105.3	50,018
2 3/8 - 20	UN	4.25	106.3	50,493
2 1/2 - 4	UNC	4.00	100.0	50,000
2 1/2 - 6	UN	4.29	107.3	53,650
2 1/2 - 8	UN	4.44	111.0	55,500
2 1/2 - 12	UN	4.60	115.0	57,500
2 1/2 - 16	UN	4.67	116.8	58,400
2 1/2 - 20	UN	4.72	118.0	59,000
2 5/8 - 4	UN	4.45	111.3	58,433
2 5/8 - 6	UN	4.76	119.0	62,475
2 5/8 - 8	UN	4.92	123.0	64,575
2 5/8 - 12	UN	5.08	127.0	66,675
2 5/8 - 16	UN	5.16	129.0	67,725

Size	Series	Tensile stress area A_s (in.^{2})	Preload (F_P) at 25 KSI (kip)	Torque at achieve 25 KSI (in.-lb)
2 5/8 - 20	UN	5.21	130.3	68,408
2 3/4 - 4	UNC	4.93	123.3	67,815
2 3/4 - 6	UN	5.26	131.5	72,325
2 3/4 - 8	UN	5.43	135.8	74,690
2 3/4 - 12	UN	5.59	139.8	76,890
2 3/4 - 16	UN	5.68	142.0	78,100
2 3/4 - 20	UN	5.73	143.25	78,788
2 7/8 - 4	UN	5.44	136	78,200
2 7/8 - 6	UN	5.78	144.5	83,088
2 7/8 - 8	UN	5.95	148.8	85,560
2 7/8 - 12	UN	6.13	153.3	88,148
2 7/8 - 16	UN	6.22	155.5	89,413
2 7/8 - 20	UN	6.27	156.8	90,160
3 - 4	UNC	5.97	149.3	89,580
3 - 6	UN	6.33	158.3	94,980
3 - 8	UN	6.51	162.8	97,680
3 - 12	UN	6.69	167.3	100,380
3 - 16	UN	6.78	169.5	101,700
3 - 20	UN	6.84	171.0	102,600
3 1/8 - 4	UN	6.52	163	101,875
3 1/8 - 6	UN	6.89	172.3	107,688
3 1/8 - 8	UN	7.08	177.0	110,625
3 1/8 - 12	UN	7.28	182.0	113,750
3 1/8 - 16	UN	7.37	184.3	115,188
3 1/4 - 4	UNC	7.10	177.5	115,375
3 1/4 - 6	UN	7.49	187.3	121,745

Size	Series	Tensile stress area A_S (in.^{2})	Preload (F_P) at 25 KSI (kip)	Torque to achieve 25 KSI (in.-lb)
3 1/4 - 8	UN	7.69	192.3	124,995
3 1/4 - 12	UN	7.89	197.3	128,245
3 1/4 - 16	UN	7.99	199.8	129,870
3 3/8 - 4	UN	7.70	193	130,275
3 3/8 - 6	UN	8.11	202.8	136,890
3 3/8 - 8	UN	8.31	207.8	140,265
3 3/8 - 12	UN	8.52	213.0	143,780
3 3/8 - 16	UN	8.63	215.8	145,665
3 1/2 - 4	UNC	8.33	208.3	145,800
3 1/2 - 6	UN	8.75	218.8	153,200
3 1/2 - 8	UN	8.96	224.0	156,800
3 1/2 - 12	UN	9.18	229.5	160,650
3 1/2 - 16	UN	9.29	232.3	162,610
3 5/8 - 4	UN	9.00	225	163,125
3 5/8 - 6	UN	9.42	235.5	170,738
3 5/8 - 8	UN	9.64	241.0	174,725
3 5/8 - 12	UN	9.86	246.5	178,713
3 5/8 - 16	UN	9.98	249.5	180,888
3 3/4 - 4	UNC	9.66	241.5	181,125
3 3/4 - 6	UN	10.11	252.8	189,600
3 3/4 - 8	UN	10.34	258.5	193,875
3 3/4 - 12	UN	10.57	264.3	198,225
3 3/4 - 16	UN	10.69	267.3	200,475
3 7/8 - 4	UN	10.36	259	200,725
3 7/8 - 6	UN	10.83	270.8	209,870
3 7/8 - 8	UN	11.06	276.5	214,288

Size	Series	Tensile stress area A_S (in.^{2})	Preload (Fp) at 25 KSI (kip)	Torque to achieve 25 KSI (in.-lb)
3 7/8 - 12	UN	11.30	282.5	218,938
3 7/8 - 16	UN	11.43	285.8	221,495
4 - 4	UNC	11.08	277.0	221,600
4 - 6	UN	11.57	289.3	231,440
4 - 8	UN	11.81	295.3	236,240
4 - 12	UN	12.06	301.5	241,200
4 - 16	UN	12.19	304.8	243,840
4 1/8 - 4	UN	11.83	296	244,200
4 1/8 - 6	UN	12.33	308.3	254,348
4 1/8 - 8	UN	12.59	315	259,875
4 1/8 - 12	UN	12.84	321.0	264,825
4 1/8 - 16	UN	12.97	324.3	267,547
4 1/4 - 4	UN	12.61	315.3	268,005
4 1/4 - 6	UN	13.12	328.0	278,800
4 1/4 - 8	UN	13.38	335	284,750
4 1/4 - 12	UN	13.65	341.3	290,105
4 1/4 - 16	UN	13.78	344.5	292,825
4 3/8 - 4	UN	13.41	335	293,125
4 3/8 - 6	UN	13.94	348.5	304,938
4 3/8 - 8	UN	14.21	355	310,625
4 3/8 - 12	UN	14.48	362	316,750
4 3/8 - 16	UN	14.62	365.5	319,813
4 1/2 - 4	UN	14.23	355.8	320,220
4 1/2 - 6	UN	14.78	369.5	332,550
4 1/2 - 8	UN	15.1	378	340,200
4 1/2 - 12	UN	15.3	382.5	344,250

Size	Series	Tensile stress area A_S (in.^{2})	Preload (F_P) at 25 KSI (kip)	Torque to achieve 25 KSI (in.-lb)
4 1/2 - 16	UN	15.5	387.5	349,200
4 5/8 - 4	UN	15.1	378	349,650
4 5/8 - 6	UN	15.6	390	360,750
4 5/8 - 8	UN	15.9	398	368,150
4 5/8 - 12	UN	16.2	405	374,625
4 5/8 - 16	UN	16.4	410	379,250
4 3/4 - 4	UN	15.9	397.5	377,625
4 3/4 - 6	UN	16.5	412.5	391,875
4 3/4 - 8	UN	16.8	420	399,000
4 3/4 - 12	UN	17.1	427.5	406,125
4 3/4 - 16	UN	17.3	432.5	410,875
4 7/8 - 4	UN	16.8	420	409,500
4 7/8 - 6	UN	17.5	437.5	426,563
4 7/8 - 8	UN	17.7	443	431,925
4 7/8 - 12	UN	18.0	450	438,750
4 7/8 - 16	UN	18.2	455	443,625
5 - 4	UN	17.8	445	445,000
5 - 6	UN	18.4	460	460,000
5 - 8	UN	18.7	468	468,000
5 - 12	UN	19.0	475	475,000
5 - 16	UN	19.2	480	480,000
5 1/8 - 4	UN	18.7	468	479,700
5 1/8 - 6	UN	19.3	482.5	494,563
5 1/8 - 8	UN	19.7	493	505,325
5 1/8 - 12	UN	20.0	500	512,500
5 1/8 - 16	UN	20.1	502.5	515,063

Size	Series	Tensile stress area A_S (in.^{2})	Preload (Fp) at 25 KSI (kip)	Torque to achieve 25 KSI (in.-lb)
5 1/4 - 4	UN	19.7	492.5	517,125
5 1/4 - 6	UN	20.3	507.5	532,875
5 1/4 - 8	UN	20.7	518	543,900
5 1/4 - 12	UN	21.0	525	551,250
5 1/4 - 16	UN	21.1	527.5	553,875
5 3/8 - 4	UN	20.7	518	556,850
5 3/8 - 6	UN	21.3	532.5	572,438
5 3/8 - 8	UN	21.7	543	583,725
5 3/8 - 12	UN	22.0	550	591,250
5 3/8 - 16	UN	22.2	555	596,625
5 1/2 - 4	UN	21.7	542.5	596,750
5 1/2 - 6	UN	22.4	560	616,000
5 1/2 - 8	UN	22.7	568	624,800
5 1/2 - 12	UN	23.1	577.5	635,250
5 1/2 - 16	UN	23.2	580	638,000
5 5/8 - 4	UN	22.7	568	639,000
5 5/8 - 6	UN	23.4	585	658,125
5 5/8 - 8	UN	23.8	595	669,375
5 5/8 - 12	UN	24.1	602.5	677,813
5 5/8 - 16	UN	24.3	607.5	683,438
5 3/4 - 4	UN	23.8	595	684,250
5 3/4 - 6	UN	24.5	612.5	704,375
5 3/4 - 8	UN	24.9	623	716,450
5 3/4 - 12	UN	25.2	630	724,500
5 3/4 - 16	UN	25.4	635	730,250
5 7/8 - 4	UN	24.9	623	732,025

Size	Series	Tensile stress area A_S (in.^{2})	Preload (Fp) at 25 KSI (kip)	Torque to achieve 25 KSI (in.-lb)
5 7/8 - 6	UN	25.6	640	752,000
5 7/8 - 8	UN	26.0	650	763,750
5 7/8 - 12	UN	26.4	660	775,500
5 7/8 - 16	UN	26.5	662.5	778,438
6 - 4	UN	26	650	780,000
6 - 6	UN	26.8	670	804,000
6 - 8	UN	27.1	678	813,600
6 - 12	UN	27.5	687.5	825,600
6 - 16	UN	27.7	692.5	831,000

Metric Tensile Stress Areas

Size	Series	Tensile stress area A_S (mm^2)	Preload F_p at 172.4 MPa (kN)	Torque to achieve 172.4 MPa (N-m)
M1.6 × 0.35		1.27	0.219	0.070
M2 × 0.4		2.07	0.357	0.1428
M2.5 × 0.45		3.39	0.584	0.292
M3 × 0.5		5.03	0.867	0.5202
M3.5 × 0.6		6.78	1.169	0.8184
M4 × 0.7		8.78	1.514	1.211
M5 × 0.8		14.2	2.448	2.448
M6 × 1		20.1	3.465	4.158
M6.3 × 1		22.6	3.896	4.908
M8 × 1.25		36.6	6.310	10.1
M10 × 1.5		58.0	9.999	20.0

Size	Series	Tensile stress area A_S (mm^2)	Preload F_P at 172.4 MPa (kN)	Torque to achieve 172.4 MPa (N-m)
M12 × 1.75		84.3	14.533	34.9
M14 × 2		115	19.826	55.5
M16 × 2		157	27.067	86.6
M20 × 2.5		245	42.24	168.96
M24 × 3		353	60.96	292.6
M30 × 3.5		561	96.72	580.4
M36 × 4		817	140.85	1,014
M42 × 4.5		1,120	191.71	1,613
M48 × 5		1,470	253.4	2,433
M56 × 5.5		2,030	349.97	3,920
M64 × 6		2,680	462.03	5,914
M72 × 6		3,460	596.5	8,597
M80 × 6		4,340	748.22	11,968
M90 × 6		5,590	963.72	17,352
M100 × 6		6,990	1,205.08	24,100

G

Basic Head, Thread, and Nut Lengths

Most of the bolts described in these tables are dimensioned as indicated in Fig. G.1. An additional dimension, thread run-out length (L_{ro}), is given for some types of bolt (see Fig. G.2). Some portion of the run-out length, perhaps one-half for the shortest bolts, 90% or so for the longest, should be added to nominal thread length in stiffness or stretch calculations.

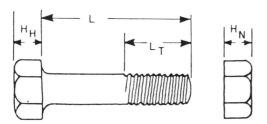

Figure G.1 Bolt dimensions. Note that body length L_B equals basic length L minus thread length L_T.

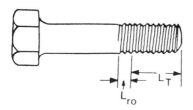

Figure G.2 Thread run-out length L_{ro}.

Type: Square Bolts
Reference: ANSI B18.2.1-1972
Dimensions are in: Inches

Nominal diameter D	Head height H_H	Thread length L_T	
		Bolts 6 in. and shorter	Bolts over 6 in. in length
1/4	11/64	0.750	1.000
5/16	13/64	0.875	1.125
3/8	1/4	1.000	1.250
7/16	19/64	1.125	1.375
1/2	21/64	1.250	1.500
5/8	27/64	1.500	1.750
3/4	1/2	1.750	2.000
7/8	19/32	2.000	2.250
1	21/32	2.250	2.500
1 1/8	3/4	2.500	2.750
1 1/4	27/32	2.750	3.000
1 3/8	29/32	3.000	3.250
1 1/2	1	3.250	3.500

Type: Hex Bolts
Reference: ANSI B18.2.1-1972
Dimensions are in: Inches

Nominal diameter D	Head height H_H	Thread length L_T	
		Bolts 6 in. and shorter	Bolts over 6 in. in length
1/4	11/ 64	0.750	1.000
5/16	7/32	0.875	1.125

Nominal diameter D	Head height H_H	Thread length L_T	
		Bolts 6 in. and shorter	Bolts over 6 in. in length
3/8	1/4	1.000	1.250
7/16	19/64	1.125	1.375
1/2	11/32	1.250	1.500
5/8	27/64	1.500	1.750
3/4	1/2	1.750	2.000
7/8	37/64	2.000	2.250
1	43/64	2.250	2.500
1 1/8	3/4	2.500	2.750
1 1/4	27/32	2.750	3.000
1 3/8	29/32	3.000	3.250
1 1/2	1	3.250	3.500
1 3/4	1 5/32	3.750	4.000
2	1 11/32	4.250	4.500
2 1/4	1 1/2	4.750	5.000
2 1/2	1 21/32	5.250	5.500
2 3/4	1 13/16	5.750	6.000
3	2	6.250	6.500
3 1/4	2 3/16	6.750	7.000
3 1/2	2 5/16	7.250	7.500
3 3/4	2 1/2	7.750	8.000
4	2 11/16	8.250	8.500

Type: Heavy Hex Bolts
Reference: ANSI B18.2.1-1972
Dimensions are in: Inches

Nominal diameter D	Head height H_H	Thread length L_T	
		Bolts 6 in. and shorter	Bolts over 6 in. in length
1/2	11/32	1.250	1.500
5/8	27/64	1.500	1.750
3/4	1/2	1.750	2.000
7/8	37/64	2.000	2.250
1	43/64	2.250	2.500
1 1/8	3/4	2.500	2.750
1 1/4	27/32	2.750	3.000
1 3/8	29/32	3.000	3.250
1 1/2	1	3.250	3.500
1 3/4	1 5/32	3.750	4.000
2	1 11/32	4.250	4.500
2 1/4	1 1/2	4.750	5.000
2 1/2	1 21/32	5.250	5.500
2 3/4	1 13/16	5.750	6.000
3	2	6.250	6.500

Type: Hex Cap Screws (Finished Hex Bolts)
Reference: ANSI B18.2.1-1972
Dimensions are in: Inches

Nominal diameter D	Head height H_H	Thread length L_T			
		L_T for bolts $\leq$ 6 in.	L_T for bolts > 6 in.	L_{ro} for bolts $\leq$ 6 in.	L_{ro} for bolts > 6 in.
1/4	5/32	0.750	1.000	0.400	0.650
5/16	13/64	0.875	1.125	0.417	0.667

(continued)

Nominal diameter D	Head height H_H	Thread length L_T			
		L_T for bolts $\leq$ 6 in.	L_T for bolts $>$ 6 in.	L_{ro} for bolts $\leq$ 6 in.	L_{ro} for bolts $>$ 6 in.
3/8	15/64	1.000	1.250	0.438	0.688
7/16	9/32	1.125	1.375	0.464	0.714
1/2	5/16	1.250	1.500	0.481	0.731
9/16	23/64	1.375	1.625	0.750	0.750
5/8	25/64	1.500	1.750	0.773	0.773
3/4	15/32	1.750	2.000	0.800	0.800
7/8	35/64	2.000	2.250	0.833	0.833
1	39/64	2.250	2.500	0.875	0.875
1 1/8	11/16	2.500	2.750	0.929	0.929
1 1/4	25/32	2.750	3.000	0.929	0.929
1 3/8	27/32	3.000	3.250	1.000	1.000
1 1/2	15/16	3.250	3.500	1.000	1.000
1 3/4	1 3/32	3.750	4.000	1.100	1.100
2	1 7/32	4.250	4.500	1.167	1.167
2 1/4	1 3/8	4.750	5.000	1.167	1.167
2 1/2	1 17/32	5.250	5.500	1.250	1.250
2 3/4	1 11/16	5.750	6.000	1.250	1.250
3	1 7/8	6.250	6.500	1.250	1.250

Type: Heavy Hex Screws
Reference: ANSI B18.2.1-1972
Dimensions are in: Inches

Nominal diameter D	Head height H_H	Thread length L_T			
		L_T for bolts $\leq$ 6 in.	L_T for bolts $>$ 6 in.	L_{ro} for bolts $\leq$ 6 in.	L_{ro} for bolts $>$ 6 in.
1/2	5/16	1.250	1.500	0.481	0.731
5/8	25/64	1.500	1.750	0.773	0.773

Nominal diameter D	Head height H_H	Thread length			
		L_T for bolts $\leq$ 6 in.	L_T for bolts > 6 in.	L_{ro} for bolts $\leq$ 6 in.	L_{ro} for bolts > 6 in.
3/4	15/32	1.750	2.000	0.800	0.800
7/8	35.64	2.000	2.250	0.833	0.833
1	39/64	2.250	2.500	0.875	0.875
1 1/8	11/16	2.500	2.750	0.929	0.929
1 1/4	25/32	2.750	3.000	0.929	0.929
1 3/8	27/32	3.000	3.250	1.000	1.000
1 1/2	15/16	3.250	3.500	1.000	1.000
1 3/4	1 3/32	3.750	4.000	1.100	1.100
2	1 7/32	4.250	4.500	1.167	1.167
2 1/4	1 3/8	4.750	5.000	1.167	1.167
2 1/2	1 17/32	5.250	5.500	1.250	1.250
2 3/4	1 11/16	5.750	6.000	1.250	1.250
3	1 7/8	6.250	6.500	1.250	1.250

Type: Heavy Hex Structural Bolts
Reference: ANSI B18.2.1-1972
Dimensions are in: Inches

Nominal diameter D	Head height H_H	Thread length L_T	
		L_T for any L	L_{ro} for any L
1/2	5/16	1.00	0.19
5/8	25/64	1.25	0.22
3/4	15/32	1.38	0.25
7/8	35/64	1.50	0.28
1	39/64	1.75	0.31
1 1/8	11/16	2.00	0.34

Nominal diameter D	Head height H_H	Thread length L_T	
		L_T for any L	L_{ro} for any L
1 1/4	25/32	2.000	0.38
1 3/8	27/32	2.25	0.44
1 1/2	15/16	2.25	0.44

Nuts
Reference: ANSI B18.2.2-1972
Dimensions are in: Inches

Nominal diameter D	Height of nut (H_N)				
	Square	Heavy square	Hex	Thick hex	Heavy hex
1/4	7/32	1/4	7/32	9/32	15/64
5/16	17/64	5/16	17/64	21/64	19/64
3/8	21/64	3/8	21/64	13/32	23/64
7/16	3/8	7/16	3/8	29/64	27/64
1/2	7/16	1/2	7/16	9/16	31/64
9/16			31/64	39/64	35/64
5/8	35/64	5/8	35/64	23/32	39/64
3/4	21/32	3/4	41/64	13/16	47/64
7/8	49/64	7/8	3/4	29/32	55/64
1	7/8	1	55/64	1	63/64
1 1/8	1	1 1/8	31/32	1 5/32	1 7/64
1 1/4	1 3/32	1 1/4	1 1/16	1 1/4	1 7/32
1 3/8	1 13/64	1 3/8	1 11/64	1 3/8	1 11/32
1 1/2	1 5/16	1 1/2	1 9/32	1 1/2	1 15/32
1 5/8					1 19/32
1 3/4					1 23/32
1 7/8					1 27/32
2					1 31/32

Nominal diameter D	Height of nut (H_N)				
	Square	Heavy square	Hex	Thick hex	Heavy hex
2 1/4					2 13/64
2 1/2					2 29/64
2 3/4					2 45/64
3					2 61/64
3 1/4					3 3/16
3 1/2					3 7/16
3 3/4					3 11/16
4					3 15/16

Bolts
Type: Metric Hex Bolts
Reference: ANSI B18.2.3.5M-1979
Dimensions are in: mm

Nominal diameter D	Head height H_H	Thread length L_T		
		≤ 125	> 125 ≤ 200	> 200
5	3.35-3.58	16	22	35
6	3.55-4.38	18	24	37
8	5.10-5.68	22	28	41
10	6.17-6.85	26	32	45
12	7.24-7.95	30	36	49
14	8.51-9.25	34	40	53
16	9.68-10.75	38	44	57
20	12.12-13.4	46	52	65
24	14.56-15.9	54	60	73
30	17.92-19.75	66	72	85

Nominal diameter D	Head height H_H	Thread length L_T		
		≤ 125	> 125 ≤ 200	> 200
36	21.72–23.55	78	84	97
42	25.03–27.05	90	96	109
48	28.93–31.07	102	108	121
56	33.8–36.2		124	137
64	38.68–41.32		140	153
72	43.55–46.45		156	169
80	48.42–51.58		172	185
90	54.26–57.74		192	205
100	60.1–63.9		212	225

Type: Metric Hex Cap Screws
Reference: ANSI B18.2.3.1M-1979
Dimensions are in: mm

Nominal diameter D	height H_H	Thread length L_T			Thread runout
		≤ 125	> 125 ≤ 200	> 200	
5	3.35–3.65	16	22	35	4.0
6	3.85–4.15	18	24	37	5.0
8	5.10–5.50	22	28	41	6.2
10	6.17–6.63	26	32	45	7.5
12	7.24–7.76	30	36	49	8.8
14	8.51–9.09	34	40	53	10.0
16	9.68–10.32	38	44	57	10.0
20	12.12–12.88	46	52	65	12.5
24	14.56–15.44	54	60	73	15.0
30	17.92–19.48	66	72	85	17.5

Nominal diameter D	Head height H_H	Thread length L_T			Thread runout
		≤ 125	> 125 ≤ 200	> 200	
36	21.62-23.38	78	84	97	20.0
42	25.03-26.97	90	96	109	22.5
48	28.93-31.07	102	108	121	25.0
56	33.80-36.20		124	137	27.5
64	38.68-41.32		140	153	30.0
72	43.55-46.45		156	169	30.0
80	48.42-51.58		172	185	30.0
90	54.62-57.74		192	205	30.0
100	60.10-63.90		212	225	30.0

Type: Metric Hex Flange Screws
Reference: ANSI B18.2.3.4M-1979
Dimensions are in: mm

Nominal diameter D	Head height H_H	Thread length L_T			Thread runout
		≤ 125	> 125 ≤ 200	> 200	
5	5.4	16	22	35	4.0
6	6.6	18	24	37	5.0
8	8.1	22	28	41	6.2
10	9.2	26	32	45	7.5
12	11.5	30	36	49	8.8
14	12.8	34	40	53	10.0
16	14.4	38	44	57	10.0
20	17.1	46	52	65	12.5

Type: Metric Heavy Hex Structural Bolts
Reference: ANSI B18.2.3.7M-1979
Dimensions are in: mm

Nominal diameter D	Head height H_H	Thread length L_T		Thread runout
		≤ 100	> 100	
16	9.25-10.75	31	38	6.0
20	11.60-13.40	36	43	7.5
22	13.10-14.90	38	45	7.5
24	14.10-15.90	41	48	9.0
27	16.10-17.90	44	51	9.0
30	17.65-19.75	49	56	10.5
36	21.45-23.55	56	63	12.0

Type: Metric Heavy Hex Bolts

Reference: ANSI B18.2.3, 6M-1979

Size Range: M12-M36

Head Heights, Thread Lengths: Same as for Metric Hex Bolts (ANSI B18.2.3.5M-1979)

Type: Metric Heavy Hex Screws

Reference: ANSI B18.2.3.3M-1979

Size Range: M12-M36

Head Heights, Thread Lengths, Thread Run-Out: Same as for Metric Hex Cap Screws (ANSI B18.2.3.1M-1979)

Type: Metric Formed Hex Screws

Reference: ANSI B18.2.3.2M-1979

Size Range: M5-M24

Head Heights, Thread Lengths, Thread Run-Out: Same as for Metric Hex Cap Screws (ANSI B18.2.3.1M-1979)

Type: Socket Head Cap Screws (Metric Series)

Reference: ANSI B18.3.1-1978

Size Range: M1.6- M48

Table 3 on p. 8 plus Table 3A on p. 9 of the Reference Gives Data from Which Thread Length Can Be Computed (Many Different, Standard, Grip, and Body Lengths)

Type: Metric Nuts—Various
Reference: British Standards and ISO R272
Dimensions are in: mm

Nominal diameter D	ISO recommendation R272 (1968)	Black hex (typical) BS4190-1967	Precision hex BS3692-1967	Hi-strength hex BS4395-1969
5	4	4	3.70-4	
6	5	5	4.7-5	
8	6.5	6.5	6.14-6.5	
10	8	8	7.64-8	
12	10	10	9.64-10	10.45-11.55
14	11		10.57-11	
16	13	13	12.57-13	14.45-15.55
20	16	16	15.57-16	17.45-18.55
24	19	19	18.48-19	21.35-22.65
30	24	24	23.48-24	25.35-26.65
36	29	29	28.48-29	30.20-31.80
42	34	34	33.38-34	
48	38	38	37.38-38	
56	45	45	44.38-45	
64	51	51	50.26-51	
72	58			
80	64			
90	72			
100	80			

H
Key Equations in Calculator/ Computer Format

I believe that equations in a textbook are easiest to "read" if they are in old-fashioned algebraic format. I have, therefore, used such a format throughout this text. I also believe, however, that, in this day and age, such equations are easiest to USE if they're in a single-line or keyboard format, for input to a computer or pocket calculator. As a result, I have included this last appendix. Here you'll find the key equations of the text repeated in a form appropriate for a BASIC program or a calculator. Terms and units are the same as those used in the text, of course (and may also be found in Appendix A).

Note that in the computer/calculator format used here:

*   signifies multiplication
** signifies exponentiation
/   signifies division

Equation no.	Used to calculate	Algebraic form	Computer/calculator form
2.1	Tensile stress area	$A_S = 0.7854\left(D - \dfrac{0.9743}{n}\right)^2$	$0.7854*(D - 0.9743/n)**2 = A_S$
2.2	Tensile stress area (high strength)	$A_S = \pi\left(\dfrac{D_{min}}{2} - \dfrac{0.16238}{n}\right)^2$	$\pi*(D_{min}/2 - 0.16238/n)**2 = A_S$
2.3	Root area	$A_r = 0.7854\left(D - \dfrac{1.3}{n}\right)^2$	$0.7854*(D - 1.3/n)**2 = A_r$
2.4	Tensile stress	$A_S = 0.7854(D - 0.938p)^2$	$0.7854*(D - 0.938*p)**2 = A_S$
2.5	Tensile stress area (metric-high strength)	$A_S = 0.7854(E_{Smin} - 0.268867p)^2$	$0.7854*(E_{Smin} - 0.268867*p)**2 = A_S$
2.6	Root area (metric)	$A_r = 0.7854(D - 1.22687p)^2$	$0.7854*(D - 1.22687*p)**2 = A_r$
3.2	Shear area of bolt threads (at root)	$A_{TS} = \pi n L_e K_{nmax}$ $\times \left[\dfrac{1}{2n} + 0.57735 \times (E_{Smin} - K_{nmax})\right]$	$\pi*n*L_e*K_{nmax}*(1/(2*n) + 0.57735*$ $(E_{Smin} - K_{nmax})) = A_{TS}$
3.3	Thread engagement length (nut material stronger)	$L_e = \dfrac{2A_S}{\pi n K_{nmax}[1/2n + 0.57735(E_{Smin} - K_{nmax})]}$	$2*A_S/(\pi*n*K_{nmax}*(1/(2*n) + 0.57735*$ $(E_{Smin} - K_{nmax}))) = L_e$
3.5	Shear area of nut threads (at root)	$A_{TS} = \pi n L_e D_{smin}$ $\times \left[\dfrac{1}{2n} + 0.57735(D_{smin} - E_{nmax})\right]$	$\pi*n*L_e*D_{smin}*(1/(2*n) + 0.57735*$ $(D_{smin} - E_{nmax})) = A_{TS}$

Equation no.	Used to calculate	Algebraic form	Computer/calculator form
3.6	Thread engagement length (bolt material stronger)	$L_e = \dfrac{S_{st}(2A_S)}{S_{nt}\pi n D_{smin}[1/2n + 0.57735(D_{smin} - E_{nmax})]}$	$(S_{st}*2*A_S)/(S_{nt}*\pi*n*D_{smin}*(1/(2*n)) + 0.57735*(D_{smin} - E_{nmax}))) = L_e$
3.8	Shear area of threads (pitch line)	$A_{TS} = \pi E_p \dfrac{L_e}{2}$	$\pi*E_p*L_e/2 = A_{TS}$
3.9	Thread engagement length (nut and bolt materials equal strength)	$L_e = \dfrac{4A_S}{\pi E_p}$	$4*A_S/(\pi*E_p) = L_e$
5.3 5.8 5.9 9.2	Change in length of a cylindrical body under a tensile load (several cross sections)	$\Delta L_c = F\left(\dfrac{L_1}{EA_1} + \dfrac{L_2}{EA_2} + \dfrac{L_3}{EA_3}\right)$	$F*(L_1/(E*A_1) + L_2/(E*A_2) + L_3/(E*A_3)) = \Delta L_c$
5.7	Compliance of a group of springs in series	$\dfrac{1}{K_T} = \dfrac{L_1}{EA_1} + \dfrac{L_2}{EA_2} + \dfrac{L_3}{EA_3}$	$L_1/(E*A_1) + L_2/(E*A_2) + L_3/(E*A_3) = 1/K_T$

5.11 4.26	Compliance of bolt-nut-washer system	$\dfrac{1}{K_T} = \dfrac{1}{K_B} + \dfrac{1}{K_N} + \dfrac{1}{K_W}$	$1/K_B + 1/K_N + 1/K_W = 1/K_T$
5.17	Stiffness of concentric joints	$K_{JC} = \dfrac{EA_C}{T}$	$E^*A_C/T = K_{JC}$
	Cross-sectional area of equivalent cylinders	$A_C = \dfrac{\pi}{4}(D_J^2 - D_H^2)$	$0.7854^*(D_J^{**}2 - D_H^{**}2) = A_C$
		$A_C = \dfrac{\pi}{4}(D_B^2 - D_H^2)$ $\quad + \dfrac{\pi}{8}\left(\dfrac{D_J}{D_B} - 1\right)\left(\dfrac{D_B T}{5} + \dfrac{T^2}{100}\right)$	$0.7854^*(D_B^{**}2 - D_H^{**}2) + 0.3927^*$ $(D_J/D_B - 1)^*(D_B^*T/5 + T^{**}2/100) = A_C$
		$A_C = \dfrac{\pi}{4}\left[\left(D_B + \dfrac{T}{10}\right)^2 - D_H^2\right]$	$0.7854^*((D_B + T/10)^{**}2 - D_H^{**}2) = A_C$
5.20	Stiffness of eccentric joint (axial load)	$r_J' = \dfrac{1}{K_{JC}}\left(1 + \dfrac{S^2 A_C}{R_G^2 A_J}\right)$	$(1/K_{JC})^*(1 + (S^{**}2^*A_C)/(R_G^{**}2^*A_J)) = r'$
5.21	Stiffness of eccentric joint (off axis load)	$r'' = \dfrac{1}{K_{JC}}\left(1 + \dfrac{SaA_C}{R_G^2 A_J}\right)$	$(1/K_{JC})^*(1 + (S^*a^*A_C)/(R_G^{**}2^*A_J)) = r''$
7.2	Torque-preload relationship (long form)	$T_{in} = F_P\left(\dfrac{P}{2\pi} + \dfrac{\mu_t r_t}{\cos\beta} + \mu_n r_n\right)$	$F_P^*(P/(2^*\pi) + (\mu_t^*r_t/\cos\beta) +$ $(\mu_n^*r_n)) = T_{in}$
7.4	Torque-preload relationship (short form)	$T_{in} = F_P(KD)$	$F_P^*K^*D = T_{in}$

Equation no.	Used to calculate	Algebraic form	Computer/calculator form
8.3	Slope of pre-load-turn curve	$\dfrac{F_P}{\theta_R} = \left(\dfrac{K_B K_J}{K_B + K_J}\right)\dfrac{P}{360}$	$(K_B{}^*K_J/(K_B + K_J))^*P/360 = F_P/\theta_R$
13.16	Tension created in bolt by differential expansion (conservative)	$F_T = K_B(\Delta L_J - \Delta L_B)$	$K_B{}^*(\Delta L_J - \Delta L_B) = F_T$
13.19	Tension created in bolt by differential expansion (taking bolt-joint stiffness ratio into account)	$F_T = \dfrac{K_B K_J}{K_B + K_J}(\Delta L_J - \Delta L_B)$	$(K_B{}^*K_J/(K_B + K_J))^*(\Delta L_J - \Delta L_B) = F_T$

22.2	Bolt loads for ASME Code rules	$W_{m2} = \pi b G y$	$\pi*b*G*y = W_{m2}$
22.3		$W_{m1} = \dfrac{\pi G^2}{4} P + 2\pi G b m P$	$(\pi/4)*G**2*P + 2*\pi*G*b*m*P = W_{m1}$
19.2	Tightness parameter	$T_P = \dfrac{P}{P^*}\left[\dfrac{L^*_{RM}}{L_{RM}}\right]^a$	$(P/P^*)*((L^*_{RM}/L_{RM})**a) = T_P$
18.1	Threshold stress intensity factor for stress corrosion cracking	$K_{ISCC} = C\sigma\sqrt{\pi a}$	$C*\sigma*((\pi*a)**0.5) = K_{ISCC}$

Note: L^*_{RM} and P^* are symbols used to express reference mass leak rate and pressure. The asterisks are not operators.

Index

[Ultrasonics]
  crack detection, 632
  effects of temperature on, 382,
    393, 400
  general discussion of, 369–418
  instruments used, 374–391
  optimizing the use of, 415–417
  preload control by, 301, 332, 348,
    369–418, 724–725
  problems reduced by, 415
  residual preload measured by,
    305–307, 359, 771
  stretch measured by, 332, 379–384
  where used, 373–374
Units in text, 857–887

Vibration loosening, *see* Self-
  loosening

Washers:
  Belleville, 195
  strain gaged, 341–342

[Washers]
  tension indicating, 342–344, 364
  thick, 36–37
Weight of fasteners, 125, 130
Window control, 287–289
Work done on bolt, 178–180,
  208–209
Working loads on bolts, factors
  affecting, 216–217 (*see also*
  Loads on bolts)
Wrenches:
  hydraulic, 249–250
  impact, 250–251, 254
  torque, 245–247

Yield control, 293–296
Yield strength of bolt, 96–102
Yield strength of joint, 131–134
Yield strength vs temperature,
  107–110
Young's modulus, *see* Modulus of
  elasticity